Climate Change 1995

ECONOMIC AND SOCIAL DIMENSIONS OF CLIMATE CHANGE

Climate Change 1995

Economic and Social Dimensions of Climate Change

Edited by

James P. Bruce
Canadian Climate Program Board

Hoesung Lee
Korea Energy Economics Institute

Erik F. Haites
Margaree Consultants Inc.

Contribution of Working Group III to the Second Assessment Report
of the Intergovernmental Panel on Climate Change

Published for the Intergovernmental Panel on Climate Change

CAMBRIDGE
UNIVERSITY PRESS

Published by the Press Syndicate of the University of Cambridge
The Pitt Building, Trumpington Street, Cambridge CB2 1RP
40 West 20th Street, New York, NY 10011-4211, USA
10 Stamford Road, Oakleigh, Melbourne 3166, Australia

First published 1996

Printed in the United States of America

Library of Congress cataloging-in-publication data are available.

A catalog record for this book is available from the British Library.

ISBN 0-521-56051-9 Hardback
ISBN 0-521-56854-4 Paperback

Also available from Cambridge University Press:

Climate Change 1995 – The Science of Climate Change
Contribution of Working Group I to the Second Assessment Report of the
Intergovernmental Panel on Climate Change. Editors J.J. Houghton, L.G. Meiro Filho,
B.A. Callander, N. Harris, A. Kattenberg and K. Maskell. (ISBN 0-521-56433-6 Hardback;
0-521-56436-0 Paperback)

*Climate Change 1995 – Impacts, Adaptations and Mitigation of Climate Change:
Scientific-Technical Analyses*
Contribution of Working Group II to the Second Assessment Report of the
Intergovernmental Panel on Climate Change. Editors R.T. Watson,
M.C. Zinyowera and R.H. Moss. (ISBN 0-521-56431-X Hardback;
0-521-56437-9 Paperback)

Contents

Foreword

The Intergovernmental Panel on Climate Change (IPCC) was jointly established by the World Meteorological Organization and the United Nations Environment Programme in 1988, in order to: (i) assess available scientific information on climate change, (ii) assess the environmental and socioeconomic impacts of climate change, and (iii) formulate response strategies. The IPCC First Assessment Report was completed in August 1990 and served as the basis for negotiating the UN Framework Convention on Climate Change. The IPCC also completed its 1992 Supplement and *Climate Change 1994: Radiative Forcing of Climate Change and an Evaluation of the IPCC IS92 Emission Scenarios* to assist the Convention process further.

In 1992, the Panel reorganized its Working Groups II and III and committed itself to complete a Second Assessment in 1995, not only updating the information on the same range of topics as in the First Assessment but also including the new subject area of technical issues related to the economic aspects of climate change. We applaud the IPCC for producing its Second Assessment Report (SAR) as scheduled. We are convinced that the SAR, like the earlier IPCC reports, will become a standard work of reference, widely used by policy-makers, scientists, and other experts.

This volume, which forms part of the SAR, has been produced by Working Group III of the IPCC and focuses on the socioeconomic aspects of climate change. It consists of eleven chapters covering the scope of the analysis, decision making under uncertainty, equity issues, intertemporal equity and discounting, applicability of cost and benefit assessments to climate change, social costs of climate change, response options, conceptual issues related to estimating mitigation costs, review of mitigation cost studies, integrated assessment of climate change, and an economic assessment of policy options to address climate change.

As usual in the IPCC, success in producing this report has depended on the enthusiasm and cooperation of numerous busy economists and other experts worldwide. We are exceedingly pleased to note here the very special efforts implemented by the IPCC in ensuring the participation of experts from the developing and transitional economy countries in its activities, in particular in the writing, reviewing, and revising of its reports. The experts from the developed, developing, and transitional-economy countries have given of their time very generously, and governments have supported them in the enormous intellectual and physical effort required, often going substantially beyond reasonable demands of duty. Without such conscientious and professional involvement, the IPCC would be greatly impoverished. We express to all these experts, and the governments that supported them, our grateful and sincere appreciation for their commitment.

We take this opportunity to express our gratitude to the following individuals for nurturing another IPCC report through to a successful completion:

Prof. Bolin, the Chairman of the IPCC, for his able leadership and skilful guidance of the IPCC;

the Co-Chairs of Working Group III, Dr. James P. Bruce (Canada) and Dr. Hoesung Lee (Korea);

the Vice Chairs of Working Group III, Dr. Richard Odingo (Kenya), Dr. Lorents Lorentsen (Norway), and his predecessor Dr. Theodore Hanisch (Norway);

Dr. Erik F. Haites, the Head of the Technical Support Unit of the Working Group and his staff, including Ms. Lori Lawson and Ms. Vanda Dreja as well as Mr. David Francis of Lanark House Communications, the technical editor, and Ms. Kim Massicotte of Carriage Hill Design, the graphic artist for the report;

Mr. W.G.B. (Bill) Smith, who developed and, with the help of his colleagues from Environment Canada – Mr. Mike Malone and Mr. Ralph Horne – implemented the system for collating review comments;

and Dr. N. Sundararaman, the Secretary of the IPCC and his staff, including Mr. S. Tewunga, Mrs. R. Bourgeois, Ms. C. Ettori, and Ms. C. Tanikie.

G.O.P. Obasi
Secretary-General
World Meteorological Organization

Ms. E. Dowdeswell
Executive Director
United Nations Environment Programme

Preface

Responsibilities of Working Group III

Working Group III of the Intergovernmental Panel on Climate Change (IPCC) was restructured in November 1992 to assess "cross-cutting economic and other issues related to climate change." The first Plenary Session of the Working Group was held in Montreal on 4–7 May, 1993. At this session, a proposed plan for the Work Programme was developed and was subsequently approved with a few changes at the IPCC Plenary Session in Geneva, 29–30 June, 1993.

The Work Programme consisted of two parts: (1) an evaluation of emission scenarios to be completed in time for inclusion in the IPCC's 1994 Special Report to the Conference of the Parties (COP) to the Framework Convention on Climate Change (FCCC), and (2) an assessment of the socioeconomic literature related to climate change for the Second Assessment Report.

Working Group III was charged with taking into account a number of considerations, the first of which was:

It will place the socio-economic perspectives of climate change in the context of sustainable development. In particular, and in accordance with the Framework Convention on Climate Change, the work of the Working Group will be comprehensive, cover all relevant sources, sinks and reservoirs of greenhouse gases and adaptation and comprise all economic sectors.

The Working Group was also enjoined to assess available literature in these fields, to avoid policy judgements, and to recognize in its work the adopted Rio Declaration, Agenda '21 and, in particular, the Framework Convention on Climate Change.

The Working Group proceeded, after the May 1993 Plenary, to seek government nominations of experts for writing teams to cover the scope of the two reports (1994 and 1995). IPCC has been fortunate in that countries nominated a number of prominent economists, distinguished social scientists, and other experts. This permitted the Working Group III bureau to form an outstanding set of writing teams. All writing teams included at least one expert from a developing country and some teams had as many as three.

To reach out and learn from an even broader community of experts in economics and the social sciences and to help create awareness and participation in various regions, the Working Group sponsored four workshops, each with a topical and a regional component. These were:

(1) Policy Instruments and their Implications/Asia and Pacific – Tsukuba, Japan, 17–20 January, 1994.
(2) Greenhouse Gas Emissions Scenarios/Latin America and Caribbean – Fortaleza, Brazil, 7–8 April, 1994.
(3) Top-Down and Bottom-Up Modelling: What Can We Learn from Each Approach/Central and Eastern Europe – Milan, Italy, 27–29 April, 1994.
(4) Equity and Social Considerations/Africa – Nairobi, Kenya, 18–22 July, 1994.

Proceedings of Workshops 1 and 4 have been published as additional contributions from Working Group III.

Working Group III also contributed to the IPCC-wide workshop on Article 2 of the FCCC held in Fortaleza, Brazil, 10–15 October, 1994.

Evaluation of Emissions Scenarios

A chapter entitled Evaluation of IS92 Emission Scenarios together with a Summary for Policymakers based on the chapter were published in IPCC's 1994 Special Report entitled *Climate Change 1994*. The Summary for Policymakers was approved and the underlying chapter was accepted by Plenary Sessions of the Working Group in Geneva 6–7 September, 1994, and Nairobi, 7–9 November 1994.

The evaluation of emission scenarios built "heavily on the work of the IPCC 1992 Update report, which contained six greenhouse gas emission scenarios." The peer and government reviews were undertaken concurrently because of the short time frame to meet deadlines for the 1994 Special Report.

The Working Group III Bureau wishes to thank the lead authors of the chapter on evaluation of emissions scenarios – Joseph Alcamo, Alex Bouwman, James Edmonds, Arnulf Grübler, Tsuneyuki Morita, and Aca Sugandhy – for their hard work under severe time constraints.

Assessment of the Socioeconomic Literature

The Working Group III contribution to the IPCC's Second Assessment Report, which follows, contains a Summary for Policymakers and eleven chapters. The Summary for Policymakers draws on the chapters, and its sections follow the sequence of chapters for easy reference. The chapters are an assessment of the available literature in economics, and to a lesser extent in other social sciences, covering the full range of topics identified in Working Group III's approved Work Programme.

For its contribution to the Second Assessment Report, the Working Group followed the IPCC procedures for, first, a peer review and, subsequently, a government and organizations review. Each review resulted in many valuable comments and suggestions and thus in significant revisions to both the chapters and the Summary for Policymakers.

The Summary for Policymakers was revised and approved by Plenary Sessions of the Working Group in Geneva, 25–28 July and Montreal, 11–13 October, 1995. The latter session also accepted the underlying technical report.

According to IPCC procedures the Summary for Policymakers is approved in detail by country representatives. The resulting Summary for Policymakers is thus an intergovernmentally negotiated text. In the course of these negotiations some of the draft text recommended by the Working Group III Bureau was deleted and in a few places, where agreement could not be reached, differing views of the findings are presented. Although the country representatives of the Working Group accept the underlying technical report, it is not reviewed in detail and its contents remain the responsibility of the lead authors. The reader wishing to have a short résumé of the findings of the writing teams is referred to the summaries at the beginning of each chapter.

The Working Group, lead authors, Bureau, and Technical Support Unit of Working Group III hope that this assessment of the socioeconomic literature will provide information of value to all concerned with climate change. In particular, it is our hope that it will assist countries individually and collectively within the FCCC to develop appropriate responses to climate change.

Summary for Policymakers

CONTENTS

SUMMARY FOR POLICYMAKERS

1 Introduction

Working Group III of the Intergovernmental Panel on Climate Change (IPCC) was restructured in November 1992 and charged with conducting "technical assessments of the socio-economics of impacts, adaptation, and mitigation of climate change over both the short and long term and at the regional and global levels." Working Group III responded to this charge by further stipulating in its work plan that it would place the socioeconomic perspectives in the context of sustainable development, and, in accordance with the Framework Convention on Climate Change (FCCC), provide comprehensive treatment of both mitigation and adaptation options while covering all economic sectors and all relevant sources of greenhouse gases and sinks.

This report assesses a large part of the existing literature on the socioeconomics of climate change and identifies areas in which a consensus has emerged on key issues and areas where differences exist.[1] The chapters have been arranged to cover several key issues. First, frameworks for socioeconomic assessment of costs and benefits of action and inaction are described. Particular attention is given to the applicability of cost-benefit analysis, the incorporation of equity and social considerations, and consideration of intergenerational equity issues. Second, the economic and social benefits of limiting greenhouse gas emissions and enhancing sinks are reviewed. Third, the economic, social, and environmental costs of mitigating greenhouse gas emissions are assessed. Next, generic mitigation and adaptation response options are reviewed, methods for assessing the costs and effectiveness of different response options are summarized, and integrated assessment techniques are discussed. Finally, the report provides an economic assessment of policy instruments to combat climate change.

In accordance with the approved work plan, this assessment of the socioeconomic literature related to climate change focusses on economic studies; material from other social sciences is found mostly in the chapter on equity and social considerations. The report is an assessment of the state of knowledge – what we know and do not know – and not a prescription for policy implementation.

Countries can use the information in this report to help take decisions they believe are most appropriate for their specific circumstances.

2 Scope of the Assessment

Climate change presents the decision maker with a set of formidable complications: a considerable number of re-maining uncertainties (which are inherent in the complexity of the problem), the potential for irreversible damages or costs, a very long planning horizon, long time lags between emissions and effects, wide regional variation in causes and effects, an irreducibly global scope of the problem, and the need to consider multiple greenhouse gases and aerosols. Yet another complication arises from the fact that effective protection of the climate system requires global cooperation.

Nevertheless, a number of insights that may be useful to policymakers can be drawn from the literature:

- Analyses indicate that a prudent way to deal with climate change is through a portfolio of actions aimed at mitigation, adaptation, and improvement of knowledge. The appropriate portfolio will differ for each country. The challenge is not to find the best policy today for the next 100 years, but to select a prudent strategy and to adjust it over time in the light of new information.

- Earlier mitigation action may increase flexibility in moving toward stabilization of atmospheric concentrations of greenhouse gases (U.N. Framework Convention on Climate Change, Article 2). The choice of abatement paths involves balancing the economic risks of rapid abatement now (that premature capital stock retirement will later be proved unnecessary) against the corresponding risk of delay (that more rapid reduction will then be required, necessitating premature retirement of future capital stock).

- The literature indicates that significant "no regrets"[2] opportunities are available in most countries and that the risk of aggregate net damage due to climate change, consideration of risk aversion, and application of the precautionary principle provide rationales for action beyond no regrets.

- The value of better information about climate change processes and impacts and society's responses to them is likely to be great. In particular, the literature accords high value to information about climate sensitivity to greenhouse gases and aerosols, climate change damage functions, and variables such as determinants of economic growth and rates of energy efficiency improvements. Better information about the costs and benefits of mitigation and adaptation measures and how they might change in coming decades also has a high value.

- Analysis of economic and social issues related to climate change, especially in developing countries where little work of this nature has been carried out, is a high priority for research. More generally, research is needed on integrated assessment and analysis of decision making related to climate change. Further, research advancing the economic understanding of nonlinearities and new theories of economic growth is also needed. Research and development related to energy efficiency technologies and nonfossil energy options also offer high potential value. In addition, there is also a need for research on the development of sustainable consumption patterns.

A portfolio of possible actions that policymakers could consider, in accordance with applicable international agreements, to implement low cost and/or cost-effective measures to reduce emissions of greenhouse gases and adapt to climate change can include:

- implementing energy efficiency measures including the removal of institutional barriers to energy efficiency improvements;
- phasing out existing distortionary policies and practices that increase greenhouse gas emissions, such as some subsidies and regulations, noninternalization of environmental costs, and distortions in transport pricing;
- implementing cost-effective fuel switching measures from more to less carbon-intensive fuels and to carbon-free fuels such as renewables;
- implementing measures to enhance sinks or reservoirs of greenhouse gases such as improving forest management and land use practices;
- implementing measures and developing new techniques for reducing methane, nitrous oxide, and other greenhouse gas emissions;
- encouraging forms of international cooperation to limit greenhouse gas emissions, such as implementing coordinated carbon/energy taxes, activities implemented jointly, and tradeable quotas;
- promoting the development and implementation of national and international energy efficiency standards;
- promoting voluntary actions to reduce greenhouse gas emissions;
- promoting education and training, implementing information and advisory measures for sustainable development and consumption patterns that will facilitate climate change mitigation and adaptation;
- planning and implementing measures to adapt to the consequences of climate change;
- undertaking research aimed at better understanding the causes and impacts of climate change and facilitating more effective adaptation to it;
- conducting technological research aimed at minimizing emissions of greenhouse gases from continued use of

fossil fuels and developing commercial nonfossil energy sources;
- developing improved institutional mechanisms, such as improved insurance arrangements, to share the risks of damages due to climate change.

Contribution of economics

- Estimates of the costs and benefits of stabilizing greenhouse gas *concentrations* are sensitive to, inter alia, the ultimate target concentration, the emissions path toward this level, the discount rate, and assumptions concerning the costs and availability of technologies and practices.
- Despite its widespread use in economic policy evaluation, Gross Domestic Product is widely recognized to be an imperfect measure of a society's well-being, largely because it fails to account for degradation of the environment and natural systems. Other methodologies exist that try to take these nonmarket values and social and ecological sustainability into account. Such methodologies would provide a more complete indication of how climate change might affect society's well-being.
- Given the interrelated nature of the global economic system, attempts to mitigate climate change through actions in one region or sector may have offsetting economic effects that risk increasing the emissions of other regions and sectors (so-called leakages). These emission leakages can be lessened through coordinated actions of groups of countries.
- The literature suggests that flexible, cost-effective policies relying on economic incentives and instruments, as well as coordinated instruments, can considerably reduce mitigation or adaptation costs, or increase the cost-effectiveness of emission reduction measures.

Equity considerations

In considering equity principles and issues related to greenhouse gas emissions, it is important for policy consideration to take into account in particular Articles 3, 4.2a, and 11.2 of the Framework Convention on Climate Change, Principle 2 of the Rio Declaration, and general principles of international law.

Scientific analyses cannot prescribe how equity should be applied in implementing the Framework Convention on Climate Change, but analysis can clarify the implications of alternative choices and their ethical basis.

- Developing countries require support for institutional and endogenous capacity building, so that they may effectively participate in climate change decision making.
- It is important that both efficiency and equity concerns be considered during the analysis of mitigation and adaptation measures. For the purposes of analysis, it is possible to separate efficiency from equity. This analyti-

cal separation presupposes that (and is valid, for policy purposes, only if) effective institutions exist or can be created for appropriate redistribution of climate change costs. It may be worthwhile to conduct analyses of the equity implications of particular measures for achieving efficiency, including social considerations and impacts.

3 Decision Making Frameworks for Addressing Climate Change

Since climate change is a global issue, comprehensive analyses of mitigation, adaptation, and research measures are needed to identify the most efficient and appropriate strategy to address climate change. International decision making related to climate change as established by the FCCC is a collective process in which a variety of concerns, such as equity, ecological protection, economics, ethics, and poverty-related issues, are of special significance for present and future generations. Treatments of decision making under uncertainty, risk aversion, technology development and diffusion processes, and distributional considerations are at present relatively poorly developed in international environmental economics, and especially in the climate change literature.

Decision making related to climate change must take into account the *unique* characteristics of the "problem": large uncertainties (scientific and economic), possible nonlinearities and irreversibilities, asymmetric distribution of impacts geographically and temporally, the very long time horizon, and the global nature of climate change with the associated potential for free riding. Beyond scientific uncertainties (discussed in Volume 1) and impact uncertainties (Volume 2), *socioeconomic uncertainties* relate to estimates of how these changes will affect human society (including direct economic and broader welfare impacts) and to the socioeconomic implications of emission abatement.

The other dimension that magnifies uncertainties and complicates decision making is *geographical*: climate change is a global problem encompassing an incredibly diverse mix of human societies, with differing histories, circumstances, and capabilities. Many developing countries are in relatively hot climates, depend more heavily on agriculture, and have less well developed infrastructures and social structures; thus, they may suffer more than average, perhaps much more. In developed countries, there may also be large climate change impacts.

The literature also emphasizes that delaying responses is itself a decision involving costs. Some studies suggest that the cost of delay is small, others emphasize that the costs could include imposition of risks on all parties (particularly the most vulnerable), greater utilization of limited atmospheric capacity, and potential deferral of desirable technical development. No consensus is reflected in the literature.

The global nature of the problem – necessitating collective action by sovereign states – and the large differences in the circumstances of different parties raise consequential as well as procedural issues. Consequential issues relate to outcomes whereas procedural issues relate to how decisions are made. In relation to climate change, the existence of an agreed legal framework involves a collective process within a negotiated framework (the FCCC). Accordingly decision making can be considered within three different categories of frameworks, each with different implications and with distinct foci: global optimization (trying to find the globally optimal result), procedural decision making (establishing and refining rules of procedure), and collective decision making (dealing with distributional issues and processes involving the interaction of numerous independent decision makers).

Application of the literature on decision making to climate change provides elements that can be used in building collective and/or market-oriented strategies for sharing risks and realizing mutual benefits. The literature suggests that actions be sequential (temporally distributed), that countries implement a portfolio of mitigation, adaptation, and research measures, and that they adjust this portfolio continuously in response to new knowledge. The potential for transfers of financial resources and technology to developing countries may be considered as a part of any comprehensive analytical framework.

Elements of a market-related strategy concern *insurance and markets for risk*. Pooling risk does not change the risk, but it can improve economic efficiency and welfare. Although insurance capable of sharing climate change risks on a global basis currently does not exist, one of the important potential gains from cooperating in a collective framework, such as the Framework Convention on Climate Change, is that of risk sharing. Creating an insurance system to cover the risks of climate change is difficult,[3] and the international community has not yet established such sophisticated instruments. This does not preclude, however, future international action to establish insurance markets sufficient for some international needs.

4 Equity and Social Considerations

Equity considerations are an important aspect of climate change policy and of the Convention. In common language equity means "the quality of being impartial" or "something that is fair and just." The FCCC, including the references to equity and equitable in Articles 3.1, 4.2.a, and 11.2, provides the context for efforts to apply equity in meeting the purposes and the objective of the Convention. International law, including relevant decisions of the International Court of Justice, may also provide guidance.

A variety of ethical principles, including the importance of meeting people's basic needs, may be relevant to addressing climate change, but the application to relations among states of principles originally developed to guide individual behaviour is complex and not straightforward. Climate change policies should not aggravate existing disparities between one region and another or attempt to redress all equity issues.

Equity involves procedural as well as consequential issues. Procedural issues relate to how decisions are made whereas consequential issues relate to outcomes. To be effective and to promote cooperation, agreements must be regarded as legitimate, and equity is an important element in gaining legitimacy.

Procedural equity encompasses process and participation issues. It requires that all parties be able to participate effectively in international negotiations related to climate change. Appropriate measures to enable developing country parties to participate effectively in negotiations increase the prospects for achieving effective, lasting, and equitable agreements on how best to address the threat of climate change. Concern about equity and social impacts indicates the need to build endogenous capabilities and strengthen institutional capacities, particularly in developing countries, to make and implement collective decisions in a legitimate and equitable manner.

Consequential equity has two components: the distribution of the costs of damages or adaptation and of measures to mitigate climate change. Because countries differ substantially in vulnerability, wealth, capacity, resource endowments, and other factors listed below, the costs of the damages, adaptation, and mitigation may be borne inequitably, unless the distribution of these costs is addressed explicitly.

Climate change is likely to impose costs on future generations and on regions where damages occur, including regions with low greenhouse gas emissions. Climate change impacts will be distributed unevenly.

The Convention recognizes in Article 3.1 the principle of common but differentiated responsibilities and respective capabilities. Actions beyond "no regrets" measures impose costs on the present generation. Mitigation policies unavoidably raise issues about how to share the costs. The initial emission limitation intentions of Annex I parties represent an agreed collective first step of those parties in addressing climate change.

Equity arguments can support a variety of proposals to distribute mitigation costs. Most of them seem to cluster around two main approaches: equal per capita emission allocations and allocations based on incremental departures from national baseline emissions (current or projected). Some proposals combine these approaches in an effort to incorporate equity concerns not addressed by relying exclusively on one or the other approach. The IPCC can clarify scientifically the implications of different approaches and proposals, but the choice of particular proposals is a policy judgment.

There are substantial variations among both developed and developing countries that are relevant to the application of equity principles to mitigation. These include variations in historical and cumulative emissions, current total and per capita emissions, emission intensities and economic output, and factors such as wealth, energy structures, and resource endowments. The literature is weak on the equity implications of these variations among both developed and developing countries.

In addition, the implications of climate change for developing countries are different from those for developed countries. The former often have different urgent priorities, weaker institutions, and are generally more vulnerable to climate change. It is likely, however, that developing countries' share of emissions will grow further to meet their social and developmental needs. Greenhouse gas emissions are likely to become increasingly global, while substantial per capita disparities are likely to remain.

It is important that both efficiency and equity concerns should be considered during the analysis of mitigation and adaptation measures. It may be worthwhile to conduct analyses of the equity implications of particular measures for achieving efficiency, including social considerations and impacts.

5 Intertemporal Equity and Discounting

Climate policy, like many other policy issues, raises particular questions of equity among generations, because future generations are not able to influence directly the policies being chosen today that could affect their well-being, and because it might not be possible to compensate future generations for consequent reductions in their well-being.

Sustainable development is one approach to intergenerational equity. Sustainable development meets "the needs of the present without compromising the ability of future generations to meet their own needs."[4] A consensus exists among economists that this does not imply that future generations should inherit a world with at least as much of every resource. Nevertheless, sustainable development would require that use of exhaustible natural resources and environmental degradation be appropriately offset – for example, by an increase in productive assets sufficient to enable future generations to obtain at least the same standard of living as those alive today. There are different views in the literature on the extent to which infrastructure and knowledge, on the one hand, and natural resources, such as a healthy environment, on the other hand, are substitutes. This is crucial to applying these concepts. Some analysts stress that there are exhaustible resources that are unique and cannot be substituted for. Others believe that current generations can compensate future generations for decreases in the quality or quantity of environmental resources by increases in other resources.

Discounting is the principal analytical tool economists use to compare economic effects that occur at different points in time. The choice of discount rate is of crucial technical importance for analyses of climate change policy, because the time horizon is extremely long, and mitigation costs tend to come much earlier than the benefits of avoided damages. The higher the discount rate, the less future benefits and the more current costs matter in the analysis.

Selection of a social discount rate is also a question of values since it inherently relates the costs of present measures to possible damages suffered by future generations if no action is taken.[5] How best to choose a discount rate is, and will likely remain, an unresolved question in economics. Partly as a consequence, different discount rates are used in different countries. Analysts typically conduct sensitivity studies using various discount rates. It should also be recognized that the social discount rate presupposes that all effects are transformed to their equivalent in consumption. This makes it difficult to apply to those nonmarket impacts of climate change which for ethical reasons might not be, or for practical reasons cannot be, converted into consumption units.

The literature on the appropriate social discount rate for climate change analysis can be grouped into two broad categories. One approach discounts consumption by different generations using the "social rate of time preference," which is the sum of the rate of "pure time preference" (impatience) and the rate of increase of welfare derived from higher per capita incomes in the future. Depending on the values taken for the different parameters the discount rate tends to fall between 0.5% and 3.0% per year on a global basis – using the above approach. Although wide variations in regional discount rates exist, they may still be consistent with a particular global average.

The second approach to the discount rate considers market returns to investment, which range between 3% and 6% in real terms for long-term, risk-free public investments. Conceptually, funds could be invested in projects that earn such returns, with the proceeds being used to increase the consumption for future generations.

The choice of the social discount rate for public investment projects is a matter of policy preference but has a major impact on the economic evaluation of climate change actions.[6] For example, in today's dollars, $1,000 of damage 100 years from now would be valued at $370 using a 1% discount rate (near the low end of the range for the first approach) but would be valued at $7.60 using a 5% discount rate (near the upper end of the range for the second approach). However, in cost-effectiveness analyses of policies over short time horizons, the impact of using different discount rates is much smaller. In all areas analysts should specify the discount rate(s) they use to facilitate comparison and aggregation of results.

6 Applicability of Cost and Benefit Assessments

Many factors need to be taken into account in the evaluation of projects and public policy issues related to climate change, including the analysis of possible costs and benefits. Although costs and benefits cannot all be measured in monetary terms, various techniques exist which offer a useful framework for organizing information about the consequences of alternative actions for addressing climate change.

The family of analytical techniques for examining economic environmental policies and decisions includes traditional project level cost-benefit analysis, cost-effectiveness analysis, multicriteria analysis, and decision analysis. Traditional cost-benefit analysis attempts to compare all costs and benefits expressed in terms of a common monetary unit. Cost-effectiveness analysis seeks to find the lowest cost option to achieve an objective specified using other criteria. Multicriteria analysis is designed to deal with problems where some benefits and/or costs are measured in nonmonetary units. Decision analysis focusses specifically on making decisions under uncertainty.

In principle, this group of techniques can contribute to improving public policy decisions concerning the desirable extent of actions to mitigate global climate change, the timing of such actions, and the methods to be employed.

Traditional cost-benefit analysis is based on the concept that the level of emission control at each point in time is determined such that marginal costs equal marginal benefits. However, both costs and benefits may be hard, sometimes impossible, to assess. This may be due to large uncertainties, possible catastrophes with very small probabilities, or simply lack of consistent methodology for monetizing the effects. In some of these cases, it may be possible to apply multicriteria analysis. This provides policymakers with a broader set of information, including evaluation of relevant costs and benefits, estimated within a common framework.

Practical application of traditional cost-benefit analysis to the problem of climate change is therefore difficult because of the global, regional, and intergenerational nature of the problem. Estimates of the costs of mitigation options also vary widely. Furthermore, estimates of potential physical damages due to climate change also vary widely. In addition, confidence in monetary estimates for important consequences (especially nonmarket consequences) is low. These uncertainties and the resolution of uncertainty over time may be decisive for the choice of strategies to combat climate change. The objective of decision analysis is to deal with such problems. Furthermore, for some categories of ecological, cultural and human health impacts, widely accepted economic concepts of value are not available. To the extent that some impacts and measures cannot be valued in monetary terms, economists augment the traditional cost-benefit analysis approach with such techniques as multicriteria analysis, permitting some quantitative expression of the trade-offs to be made. These techniques do not resolve questions involving equity – for example, determining who should bear the costs. However, they provide important information on the incidence of damage, mitigation, and adaptation costs, and where cost-effective action might be taken.

Despite their many imperfections, these techniques provide a valuable framework for identifying essential questions that policymakers must face when dealing with climate change, namely:

- By how much should the emissions of greenhouse gases be reduced?
- When should emissions be reduced?
- How should emissions be reduced?

These analytical techniques assist decision makers in comparing the consequences of alternative actions, including that of no action, on a quantitative basis – and can certainly make a contribution to resolution of these questions.

7 The Social Costs of Anthropogenic Climate Change: Damages of Increased Greenhouse Gas Emissions

The literature on the subject of this section is controversial and mainly based on research done on developed countries, often extrapolated to developing countries. There is no consensus about how to value statistical lives or how to aggregate statistical lives across countries.[7] Monetary valuation should not obscure the human consequences of anthropogenic climate change damages, because the value of life has meaning beyond monetary considerations. It should be noted that the Rio Declaration and Agenda 21 call for human beings to remain at the centre of sustainable development. The approach taken to this valuation might affect the scale of damage reduction strategies. It may be noted that in virtually all the literature discussed in this section the developing country statistical lives have not been equally valued at the developed country value, nor are other damages in developing countries equally valued at the developed country value. Because national circumstances, including opportunity costs, differ, economists sometimes evaluate certain kinds of impacts differently amongst countries.

The benefits of limiting greenhouse gas emissions and enhancing sinks are (a) the climate change damages avoided and (b) the secondary benefits associated with the relevant policies. Secondary benefits include reductions in other pollutants jointly produced with greenhouse gases and the conservation of biological diversity. Net climate change damages include both market and nonmarket impacts as far as they can be quantified at present and, in some cases, adaptation costs. Damages are expressed in net terms to account for the fact that there are some beneficial impacts of climate change as well, which are, however, dominated by the damage costs. Nonmarket impacts, such as human health, risk of human mortality, and damage to ecosystems, form an important component of available estimates of the social costs of climate change. The literature on monetary valuation of such nonmarket effects reflects a number of divergent views and approaches. The estimates of nonmarket damages, however, are highly speculative and not comprehensive.

Nonmarket damage estimates are a source of major uncertainty in assessing the implications of global climate change for human welfare. Some regard monetary valuation of such impacts as essential to sound decision making, but others reject monetary valuation of some impacts, such as risk of human mortality, on ethical grounds. Additionally, there is a danger that entire unique cultures may be obliterated. This is not something that can be considered in monetary terms, but becomes a question of loss of human diversity, for which we have no indicators to measure economic value.

The assessed literature contains only a few estimates of the monetized damages associated with doubled CO_2 equivalent concentration scenarios. These estimates are aggregated to a global scale and illustrate the potential impacts of climate change under selected scenarios. Aggregating individual monetized damages to obtain total social welfare impacts involves difficult decisions about equity amongst countries. Global estimates are based on an aggregation of monetary damages across countries (damages which are themselves implicit aggregations across individuals) that reflects intercountry differences in wealth and income. This fundamentally influences the monetary valuation of damages. Taking income differences as given implies that an equivalent impact in two countries (such as an equal increase in human mortality) would receive very different weights in the calculation of global damages.

To enable choices between different ways of promoting human welfare to be made on a consistent basis, economists have for many years sought to express a wide range of human and environmental impacts in terms of monetary equivalents, using various techniques. The most commonly used of those techniques is an approach based on the observed willingness to pay for various nonmarket benefits.[8] This is the approach that has been taken in most of the assessed literature.

Human life is an element outside the market, and societies may want to preserve it in an equal way. An approach that includes equal valuation of impacts on human life wherever they occur may yield different global aggregate estimates than those reported below. For example, equalizing the value of a statistical life at a global average could leave total global damage unchanged but would increase markedly the share of these damages borne by the developing world. Equalizing the value at the level typical in developed countries would increase monetized damages several times, and would further increase the share of the developing countries in the total damage estimate.

Other aggregation methods can be used to adjust for differences in the wealth or incomes of countries in calculations of monetary damages. Because estimates of monetary damage tend to be a higher percentage of national GDP for low-income countries than for high-income countries, aggregation schemes that adjust for wealth or income effects are expected to yield higher estimates of global damages than those presented in this report.

The assessed literature quantifying total damages from 2–3°C warming provides a wide range of point estimates for damages, given the presumed change in atmospheric greenhouse gas concentrations. The aggregate estimates tend to be a few percent of world GDP, with, in general, considerably higher estimates of damage to developing countries as a share of their GDP. The aggregate estimates are subject to considerable uncertainty, but the range of uncertainty cannot be gauged from the literature. The range of estimates cannot be interpreted as a confidence interval, given the widely differing assumptions and methodologies in the studies. As noted above, aggregation is likely to mask even greater uncertainties about damage components.

Regional or sectoral approaches to estimating the consequences of climate change include a much wider range of estimates of the net economic effects. For some areas, damages are estimated to be significantly greater and could negatively affect economic development. For others, climate change is estimated to increase economic production and present opportunities for economic development. For countries generally having a diversified, industrial economy and an educated and flexible labour force, the limited set of published estimates of damages are of the order one to a few percent of GDP. For countries generally having a specialized and natural resource-based economy (e.g., heavily emphasizing agriculture or forestry), and a poorly developed and land-tied labour force, estimates of damages from the few studies available are several times larger. Small islands and low-lying coastal areas are particularly vulnerable. Damages from possible large-scale catastrophes, such as major changes in ocean circulation, are not reflected in these estimates. There is little agreement across studies about the exact magnitude of each category of damages or relative ranking of the damage categories.[9] Climate changes of this magnitude are not expected to be realized for several decades, and damages in the interim could be smaller. Damages over a longer period of time might be greater.[10]

IPCC does not endorse any particular range of values for the marginal damage of CO_2 emissions, but published estimates range between $5 and $125 (1990 U.S.) per tonne of carbon emitted now. This range of estimates does not represent the full range of uncertainty. The estimates are also based on models that remain simplistic and are limited representations of the actual climate processes and are based on earlier IPCC scientific reports. The wide range of damage estimates reflects variations in model scenarios, discount rates, and other assumptions. It must be emphasized that the social cost estimates have a wide range of uncertainty because of limited knowledge of impacts, uncertain future technological and socioeconomic developments, and the possibility of catastrophic events or surprises.

8 Generic Assessment of Response Strategies

A wide range of technologies and practices is available for mitigating emissions of carbon dioxide, methane, nitrous oxide, and other greenhouse gases. There are also many adaptation measures available for responding to the impacts of climate change. All these technologies, practices, and measures have financial and environmental costs as well as benefits. This section surveys the range of options currently available or discussed in the literature. The optimal mix of response options will vary by country and over time as local conditions and costs change.

A review of CO_2 mitigation options suggests that:

- A large potential for cost-effective *energy conservation and efficiency improvements* in energy supply and energy use exists in many sectors. These options offer economic and environmental benefits in addition to reducing emissions of greenhouse gases. Various of these options can be deployed rapidly due to small unit size, modular design characteristics, and low lifetime costs.

- The options for CO_2 *mitigation in energy use* include alternative methods and efficiency improvements among others in the construction, residential, commercial, agriculture, and industry sectors. Not all cost-effective strategies are based on new technology; some may rely on improved information dissemination and public education, managerial strategies, pricing policies, and institutional reforms.

- Estimates of the technical potential for *switching to less carbon-intensive fuels* vary regionally and with the type of measure and economic availability of reserves of fossil and alternative fuels. These estimates must also take into account potential methane emissions from leakage of natural gas during production and distribution.

- *Renewable energy technologies* (e.g., solar, hydroelectric, wind, traditional and modern biomass, and ocean thermal energy conversion) have achieved different levels of technical development, economic maturity, and commercial readiness. The potential of these energy sources is not fully realized. Cost estimates for these technologies are sensitive to site-specific characteristics, resource variability, and the form of final energy delivered. These cost estimates vary widely.

- *Nuclear energy*[11] is a technology that has been deployed for several decades in many countries. However, a number of factors have slowed the expansion of nuclear power, including: (a) wary public perceptions resulting from nuclear accidents, (b) not yet fully resolved issues concerning reactor safety, proliferation of fissile material, power plant decommissioning, and long-term disposal of nuclear waste, as well as, in some instances, lower-than-anticipated levels of demand for electricity. Regulatory and siting difficulties have increased construction lead times, leading to higher capital costs for this option in some countries. If these issues, including *inter alia* the social, political, and environmental aspects mentioned above, can be resolved, nuclear energy has the potential to increase its present share in worldwide energy production.

- CO_2 *capture and disposal* may be ultimately limited for technical and environmental reasons, because not all forms of disposal ensure prevention of carbon reentering the atmosphere.

- *Forestry* options, in some circumstances, offer large potential, modest costs, low risk, and other benefits. Further, the potential modern use of biomass as a source of fuels and electricity could become attractive. Halting or slowing deforestation and increasing reforestation through increased silvicultural productivity and sustainable management programmes that increase agricultural productivity, the expansion of forest reserves, and promotion of ecotourism are among the cost-effective options for slowing the atmospheric build-up of CO_2. Forestry programmes raise important equity considerations.[12]

There is also a wide range of available technologies and practices for reducing emissions of *methane* from such sources as natural gas systems, coal mines, waste dumps, and farms. However, the issue of reduction of emissions related to the food supply may imply trade-offs with rates of food production. These trade-offs must be carefully assessed as they may affect the provision of basic needs in some countries, particularly in developing countries.

Most *nitrous oxide* emissions come from diffuse sources related to agriculture and forestry. These emissions are difficult to reduce rapidly. Industrial emissions of *nitrous oxide and halogenated compounds* tend to be concentrated in a few key sectors and tend to be easier to control. Measures to limit such emissions may be attractive for many countries.

The slow implementation of many of the technologically attractive and cost-effective options listed above has many possible explanations, with both actual and perceived costs being a major factor. Among other factors, capital availability, information gaps, institutional obstacles, and market imperfections affect the rate of diffusion for these technologies. Identifying the reasons specific to a particular country is a precondition to devising sound and efficient policies to encourage their broader adoption.

Education and training as well as information and advisory measures are important aspects of various response options.

Many of the emission-reducing technologies and practices described above also provide other benefits to society. These additional benefits include improved air quality, better protection of surface and underground waters, enhanced animal productivity, reduced risk of explosions and fire, and improved use of energy resources.

Many options are also available for *adapting* to the impacts of climate change and thus reducing the damages to national economies and natural ecosystems. Adaptive options are available in many sectors, ranging from agriculture and energy to health, coastal zone management, offshore fisheries, and recreation. Some of these provide enhanced ability to cope with the current impacts of climate variability. However, possible trade-offs between implementation of mitigation and adaptation measures are important to consider in future research. A summary of sectoral options for adaptation is presented in Volume 2.

The optimal response strategy for each country will depend on the special circumstances and conditions which that country must face. Nonetheless, many recent studies and empirical observations suggest that some of the most cost-effective options can be most successfully implemented on a joint or cooperative basis among nations.

9 Costs of Response Options

It must be emphasized that the text in this section is an assessment of the technical literature and does not make recommendations on policy matters. The available literature is primarily from developed countries.

Cost concepts

From the perspective of this section on assessing mitigation or adaptation costs, what matters is the net cost (total cost less secondary benefits and costs). These net costs exclude the social costs of climate change, which are discussed in Section 7. The assessed literature yields a very wide range of estimates of the costs of response options. The wide range largely reflects significant differences in assumptions about the efficiency of energy and other markets, and about the ability of government institutions to address perceived market failures or imperfections.

Measures to reduce greenhouse gas emissions may yield additional economic impacts (for example, through technological externalities associated with fostering research and development programmes) and/or environmental impacts (such as reduced emissions of acid rain and urban smog precursors).

Studies suggest that the secondary environmental benefits may be substantial but are likely to differ from country to country.

Specific results

Estimates of the cost of greenhouse gas emission reduction depend critically on assumptions about the levels of energy efficiency improvements in the baseline scenario (that is, in the absence of climate policy) and on a wide range of factors such as consumption patterns, resource and technology availability, the desired level and timing of abatement, and the choice of policy instruments. Policymakers should not place too much confidence in the specific numerical results from any one analysis. For example, mitigation cost analyses reveal the costs of mitigation relative to a given baseline, but neither the baseline nor the intervention scenarios should be interpreted as representing likely future conditions. The focus should be on the general insights regarding the underlying determinants of costs.

The costs of stabilizing atmospheric concentrations of greenhouse gases at levels and within a time frame that will prevent dangerous anthropogenic interference with the climate system (the ultimate objective of the FCCC) will be critically dependent on the choice of emission timepath. The cost of the abatement programme will be influenced by the rate of capital replacement, the discount rate, and the effect of research and development.

Failure to adopt policies as early as possible to encourage efficient replacement investments at the end of the economic life of a plant and equipment (i.e., at the point of capital stock turnover) imposes an economic cost to society. Implementing emission reductions at rates that can be absorbed in the course of normal stock turnover is likely to be cheaper than enforcing premature retirement now.

The choice of abatement paths thus involves balancing the economic risks of rapid abatement now (that premature capital stock retirement will later be proved unnecessary) against the corresponding risk of delay (that more rapid reduction will then be required, necessitating premature retirement of future capital stock).

Appropriate long-run signals are required to allow producers and consumers to adapt cost-effectively to constraints on greenhouse gas emissions and to encourage research and development. Benefits associated with the implementation of any "no regret" policies will offset, at least in part, the costs of a full portfolio of mitigation measures. This will also increase the time available to learn about climate risks and to bring new technologies into the marketplace.

Despite significant differences in views, there is agreement that energy efficiency gains of perhaps 10 to 30% above baseline trends over the next two to three decades can be realized at negative to zero net cost (negative net cost means an economic benefit). With longer time horizons, which allow a more complete turnover of capital stocks, and which give research and development and market transformation policies a chance to impact multiple replacement cycles, this potential is much higher. The magnitude of such "no regret" potentials depends on the existence of substantial market or institutional imperfections that prevent cost-effective emission reduction measures from occurring. The key question is then the extent to which such imperfections and barriers can be removed cost-effectively by policy initiatives such as efficiency standards, incentives, removal of subsidies, information programmes, and funding of technology transfer.

Progress has been made in a number of countries in cost-effectively reducing imperfections and institutional barriers in markets through policy instruments based on voluntary agreements, energy efficiency incentives, product efficiency standards, and energy efficiency procurement programmes involving manufacturers, as well as utility regulatory reforms. Where empirical evaluations have been made, many have found the benefit-cost ratio of increasing energy efficiency to be favourable, suggesting the practical feasibility of realizing "no regret" potentials at negative net cost. More information is needed on similar and improved programmes in a wider range of countries.

Infrastructure decisions are critical in determining long-term emissions and abatement costs because they can enhance or restrict the number and type of future options. Infrastructure decisions determine development patterns in transportation, urban settlement, and land use and influence energy system development and deforestation patterns. This issue is of particular importance to developing countries and many economies in transition where major infrastructure decisions will be made in the near term.

If a carbon or carbon-energy tax is used as a policy instrument for reducing emissions, the taxes could raise substantial revenues, and how the revenues are distributed could dramatically affect the cost of mitigation. If the revenues are distributed by reducing distortionary taxes in the existing system, they will help reduce the excess burden of the existing tax system, potentially yielding an additional economic benefit (double dividend). For example, those European studies which are more optimistic regarding the potential for tax recycling show lower and, in some instances, slightly negative costs. Conversely, inefficient recycling of the tax revenues could increase costs. For example, if the tax revenues are used to finance government programmes that yield a lower return than the private sector investments forgone because of the tax, then overall costs will increase.

There are large differences in the costs of reducing greenhouse gas emissions among countries because of their state of economic development, infrastructure choices, and natural resource base. This indicates that international cooperation could significantly reduce the global cost of reducing emissions. Research suggests that, in principle, substantial savings would be possible if emissions are reduced where it is cheapest to do so. In practice, this requires international mechanisms ensuring appropriate capital flows and technology transfers between countries. Conversely, a failure to achieve international cooperation could compromise unilateral attempts by a country or a group of countries to limit greenhouse gas emissions. However, estimates of so called leakage effects vary so widely that they provide little guidance to policymakers.

BOX S.1: TOP-DOWN AND BOTTOM-UP MODELS

Top-down models are aggregate models of the entire macroeconomy that draw on analysis of historical trends and relationships to predict the large-scale interactions between the sectors of the economy, especially the interactions between the energy sector and the rest of the economy. Top-down models typically incorporate relatively little detail on energy consumption and technological change, compared with bottom-up models.

In contrast, bottom-up models incorporate detailed studies of engineering costs of a wide range of available and forecast technologies, and describe energy consumption in great detail. However, compared with top-down models, they typically incorporate relatively little detail on nonenergy consumer behaviour and interactions with other sectors of the economy.

This simple characterization of top-down and bottom-up models is increasingly misleading as more recent versions of each approach have tended to provide greater detail in the aspects that were less developed in the past. As a result of this convergence in model structure, model results are tending to converge, and the remaining differences reflect differences in assumptions about how rapidly and effectively market institutions adopt cost-effective new technologies or can be induced to adopt them by policy interventions.

Many existing models are not well suited to study economies in transition or those of developing countries. More work is needed to develop the appropriate methodologies, data, and models and to build the local institutional capacity to undertake analyses.

There has been more analysis to date of emission reduction potentials and costs for developed countries than for other parts of the world. Moreover, many existing models are not well suited to study economies in transition or economies of developing countries. Much work is needed to develop and apply models for use outside developed countries (for example, to represent more explicitly market imperfections, institutional barriers, and traditional and informal economic sectors). In addition, the discussion below and the bulk of the underlying report deal with costs of response options at the national or regional level in terms of effect on GDP. Further analysis is required concerning effects of response options on employment, inflation, trade competitiveness, and other public issues.

A large number of studies using both top-down and bottom-up approaches (see box for definitions) were reviewed. Estimates of the costs of limiting fossil fuel carbon dioxide emissions (expressed as carbon) vary widely and depend on choice of methodologies, underlying assumptions, emission scenarios, policy instruments, reporting year, and other criteria. For specific results of individual studies, see Chapter 9.

OECD Countries. Although it is difficult to generalize, top-down analyses suggest that the costs of substantial reductions below 1990 levels could be as high as several percent of GDP. In the specific case of stabilizing emissions at 1990 levels, most studies estimate that annual costs in the range of -0.5% of GDP (equivalent to a gain of about $60 billion in total for OECD countries at today's GDP levels) to 2% of GDP (equivalent to a loss of about $240 billion) could be reached over the next several decades. However, studies also show that appropriate timing of abatement measures and the availability of low-cost alternatives may substantially reduce the size of the overall bill.

Bottom-up studies are more optimistic about the potential for low- or negative-cost emission reductions, and the capacity to implement that potential. Such studies show that the costs of reducing emissions by 20% in developed countries within two to three decades are negligible to negative. Other bottom-up studies suggest that there exists a potential for absolute reductions in excess of 50% in the longer term, without increasing, and perhaps even reducing, total energy system costs.

The results of top-down and bottom-up analyses differ because of such factors as higher estimates of no-regrets potential and technological progress, and earlier saturation in energy services per unit GDP. In the most favourable assessments, savings of 10–20% in the total cost of energy services can be achieved.

Economies in transition. The potential for cost-effective reductions in energy use is apt to be considerable, but the realizable potential will depend on what economic and technological development path is chosen, as well as the availability of capital to pursue different paths. A critical issue is the future of structural changes in these countries that are apt to change dramatically the level of baseline emissions and the emission reduction costs.

Developing countries. Analyses suggest that there may be substantial low-cost fossil fuel carbon dioxide emission reduction opportunities for developing countries. Development pathways that increase energy efficiency, promote alternative energy technologies, reduce deforestation, and enhance agricultural productivity and biomass energy production can be economically beneficial. To embark upon this pathway may require significant international cooperation and financial and technology transfers. However, these are likely to be insufficient to offset rapidly increasing emissions baselines, associated with increased economic growth and overall welfare. Stabilization of carbon dioxide emissions is likely to be costly.

It should be noted that analyses of costs to economies in transition and developing countries typically neglect the general equilibrium effects of unilateral actions taken by developed countries. These effects may be either positive or negative and their magnitude is difficult to quantify.

It should also be noted that estimates of costs or benefits of the order of a few percent of GDP may represent small differences in GDP growth rates, but are nevertheless substantial in absolute terms.

Preservation and augmentation of carbon sinks offer a substantial and often cost-effective component of a greenhouse gas mitigation strategy. Studies suggest that as much as 15-30% of 1990 global energy-related emissions could be offset by carbon sequestration in forests for a period of 50 to 100 years. The costs of carbon sequestration, which are competitive with source control options, may differ among regions of the world.

Control of emissions of other greenhouse gases, especially methane and nitrous oxide, can provide significant cost-effective opportunities in some countries. About 10% of anthropogenic methane emissions could be reduced at negative or low cost using available mitigation options for such methane sources as natural gas systems, waste management, and agriculture.

10 Integrated Assessment

Integrated assessment models combine knowledge from a wide range of disciplines to provide insights that would not be observed through traditional disciplinary research. They are used to explore possible states of human and natural systems, analyze key questions related to policy formulation, and help set research priorities. Integration helps coordinate assumptions from different disciplines and allows feedbacks and interactions absent from individual disciplines to be analyzed. However, the results of such analyses are no better than the information drawn from the underlying economic, atmospheric and biological sciences. Integrated assessment models are limited by both the underlying knowledge base on which they draw and the relatively limited experiential base.

Most current integrated assessment models do not reflect the specific social and economic dynamics of the developing and transition economies well; for example, none of the existing models addresses most market imperfections, institutional barriers, or the operation of the informal sector in these countries. This can lead to biases in global assessments when mitigation options and impacts on developing or transition economies are valued as if their economies operate like those in the developed countries.

Although relatively new, integrated assessment models of climate change have evolved rapidly. Integrated assessment models tend to fall into two categories: *policy evaluation* and *policy optimization* models. Policy evaluation models are rich in physical detail and have been used to analyze the potential for deforestation as a consequence of interactions between demographics, agricultural productivity, and economic growth, and the relationship between climate change and the extent of potentially malarial regions. Policy optimization models optimize over key variables (e.g., emissions rates, carbon taxes) to achieve formulated policy goals (e.g., cost minimization or welfare optimization).

Key uncertainties in current integrated assessments include the sensitivity of the climate system to changes in greenhouse gas concentrations, the specification and valuation of impacts where there are no markets, changes in national and regional demographics, the choice of discount rates, and assumptions regarding the cost, availability, and diffusion of technologies.

11 An Economic Assessment of Policy Instruments for Combating Climate Change

Governments may have different sets of criteria for assessing international as well as domestic greenhouse policy instruments. Among these criteria are efficiency and cost-effectiveness, effectiveness in achieving stated environmental targets, distributional (including intergenerational) equity, flexibility in the face of new knowledge, understandability to the general public, and consistency with national priorities, policies, institutions, and traditions. The choice of instruments may also partly reflect a desire on the part of governments to achieve other objectives such as sustainable economic development, meeting social development goals and fiscal targets, or influencing pollution levels that are indirectly related to greenhouse gas emissions. A further concern of governments may lie with the effect of policies on competitiveness.

The world economy and indeed some individual national economies suffer from a number of price distortions which increase greenhouse gas emissions, such as some agricultural and fuel subsidies and distortions in transport pricing. A number of studies of this issue indicate that global emissions reductions of 4–18%, together with increases in real incomes, are possible from phasing out fuel subsidies. For the most part, reducing such distortions could lower emissions and increase economic efficiency. However, subsidies are often introduced and price distortions maintained for social and distributional reasons, and may be difficult to remove.

Policy instruments may be identified at two different levels: those that might be used by a group of countries and those that might be used by individual nations unilaterally or to achieve compliance with a multilateral agreement.

A group[13] of countries may choose from policy measures and instruments including encouragement of voluntary actions and further research, tradable quotas, joint implementation (specifically activities implemented jointly under the pilot phase[14]), harmonized domestic carbon taxes, international carbon taxes, nontradable quotas, and various international standards. If the group did not include all major greenhouse gas emitters, then there might be a tendency for fossil fuel use to increase in countries not participating in this group. This outcome might reduce the international competitiveness of some industries in participating countries as well as the environmental effectiveness of the countries' efforts.

At both the international and national levels, the economic literature indicates that instruments that provide economic incentives, such as taxes and tradable quotas/permits, are likely to be more cost-effective than other approaches. Uniform standards among groups of countries participating in an international agreement are likely to be difficult to achieve. However, for one group of countries there has been agreement on the application of some uniform standards.

At the international level, all the potentially efficient market-based instruments could be examined during the course of future negotiations. A tradable quota system has the disadvantage of making the marginal cost of emissions uncertain, whereas a carbon tax (and related instruments)

has the disadvantage of leaving uncertain the effect on the level at which emissions are controlled. The weight given to the importance of reducing these different types of uncertainty would be one crucial factor in further evaluating these alternative instruments. Because of the lack of appropriate scientific knowledge, there would remain a high degree of uncertainty about the results of limiting emissions at specific levels. The adoption of either a tradable quota scheme or international taxes would have implications for the international distribution of wealth. The distributional consequences would be the subject of negotiation. To ensure the practicability of such instruments, there is a need for additional studies on the possible design of tradable quotas and harmonized taxes and on the institutional framework in which they might operate.

Individual countries that seek to implement mitigation policies can choose from among a large set of potential policies and instruments, including carbon taxes, tradable permits, deposit refund systems (and related instruments), and subsidies, as well as technology standards, performance standards, product bans, direct government investment, and voluntary agreements. Public education on the sustainable use of resources could play an important part in modifying consumption patterns and other human behaviour. The choice of measures at the domestic level may reflect objectives other than cost-effectiveness such as meeting fiscal targets. Revenue from carbon taxes or auctioned tradable permits could be used to replace existing distortionary taxes. The choice of instruments may also reflect other environmental objectives, such as reducing nongreenhouse pollution emissions, or increasing forest cover, or other concerns such as specific impacts on particular regions or communities.

Endnotes

1. The Framework Convention on Climate Change defines "climate change" as a change of climate which is attributed directly or indirectly to human activity that alters the composition of the global atmosphere and which is in addition to natural climate variability observed over comparable time periods. The question as to whether such changes are potential or can already be identified is analyzed in Volume 1 of this IPCC Second Assessment Report (SAR).

2. "No regrets" measures are those whose benefits, such as reduced energy costs and reduced emissions of local/regional pollutants equal or exceed their cost to society, *excluding* the benefits of climate change mitigation. They are sometimes known as "measures worth doing anyway."

3. Without knowing the extent of potential impacts, the ability of private markets to insure against losses associated with climate change is unknown.

4. A related (somewhat stronger) concept is that each generation is entitled to inherit a planet and cultural resource base at least as good as that of previous generations.

5. A social discount rate is a discount rate appropriate for use by governments in the evaluation of public policy.

6. Despite the differences in the value of the discount rate, policies developed on the basis of the two approaches may lead to similar results.

7. The value of a statistical life is defined as the value people assign to a change in the risk of death among a population.

8. The concept of willingness to pay is indicative, based on expressed desires, available resources, and information of a human being's preferences at a certain moment in time. The values may change over time. Also, other concepts (such as willingness to accept compensation for damage) have been advanced, but not yet widely applied, in the literature, and the interpretation and application of willingness to pay and other concepts to the climate problem may evolve.

9. Due to time lags between findings in the natural sciences, their use in determination of potential physical and biological impacts, and subsequent incorporation into economic analyses of climate change, the estimates of climate change damage are based mainly on the scientific results from the 1990 and 1992 IPCC reports.

10. See Volumes 1 and 2, the reports of Working Groups I and II.

11. For more information on the technical aspects of nuclear power, see Volume 2.

12. These are addressed in Section 4 above and in Chapter 3.

13. The group could contain only a few, quite a number, or even all countries.

14. See decision 5/CP.1 of the first Conference of the Parties (COP1) to the FCCC.

1

Introduction:
Scope of the Assessment

J. GOLDEMBERG, R. SQUITIERI, J. STIGLITZ, A. AMANO, X. SHAOXIONG,
R. SAHA

Contributors:
S. Kane, J. Reilly, T. Teisberg

CONTENTS

SUMMARY

Climate change presents the decision maker with a set of formidable complications: large uncertainties, the potential for irreversible damages or costs, a very long planning horizon, long time lags between emissions and effects, a global scope, wide regional variation, and multiple greenhouse gases of concern. Irrespective of the possible consequences of climate change, policies that mitigate or assist adaptation to climate change and have zero or negative net costs (*no-regrets* policies) are clearly justified. If the evidence suggests that damages can be expected from climate change, then the expectation of damages provides a rationale for going beyond no-regrets policies to those that incur positive net costs. The principles of risk aversion and portfolio balancing provide a rationale for further steps.

The atmosphere is an international public good, in that all countries benefit from each country's reduction in greenhouse emissions; greenhouse gases are an international externality, in that emissions by one country affect all other countries to some extent.

Both public goods and externalities require a legal framework within which the problems they pose can be addressed. Mechanisms for control of international public goods may include the definition of property rights, the definition of limits to emissions, and a consensus for distributing the same in a fair and equitable manner. If, on the other hand, each agent acts in its individual interest, the result will be too little of the public good and too much of the externality.

A decision process for climate change should be sequential. It should also be able to respond to new information with midcourse corrections and to include insurance arrangements, hedging strategies, and the option value of alternative courses of action. The challenge today is to identify short-term strategies in the face of long-term uncertainty. The question is not, what is the best course over the next 100 years, but rather, what is the best course for the next few years, knowing that a prudent hedging strategy will allow time to learn and change course.

Policy measures to reduce risks to future generations include (1) immediate reductions in emissions; (2) research and development related to new supply and conservation technologies; (3) continued research on how much change is likely and what its effects will be; and (4) investments to assist in adaptation if significant global warming occurs. A well-chosen portfolio of policies will yield greater benefits for a given cost than any one option undertaken by itself. Striking the appropriate balance requires taking into account costs, benefits, and risks.

In an interrelated global economic system, an attempt to reduce greenhouse gas emissions in one region or one sector of the economy may be offset by increases in other regions or sectors. This may occur through the loss of comparative advantage in the carbon-intensive sectors of the regions that limit emissions, through the relocation of industries, or through changes in world energy prices and the resulting shift in consumption. Any control strategy must account for these global effects.

For the purposes of analysis it is useful to separate efficiency from equity. The Framework Convention on Climate Change (FCCC) requires all parties to formulate and implement programmes to mitigate climate change and facilitate adaptation to climate change on the basis of their common but differentiated responsibilities, and taking into account their specific national and regional development priorities, objectives, and circumstances. Developing countries are more likely to be adversely affected economically than the developed countries; moreover, developing countries often lack the financial and technical resources to respond to these changes.

Efficiency requires that emission reductions occur where their cost is lowest, irrespective of who bears the financial responsibility. Efficiency calls for removing energy subsidies, reforming and clarifying property rights that affect energy use and carbon storage, and reducing nongreenhouse externalities that have the side benefit of reducing greenhouse emissions. Efficiency may also be promoted, and greenhouse emissions reduced, by better information dissemination and by addressing capital market imperfections that inhibit the adoption of energy-efficient technology. Dynamic analysis indicates large potential gains from flexibility in the timing of greenhouse reductions to allow for the economical turnover of capital stock and to allow time for the development of low-cost substitutes. Policies that promote efficiency by requiring nations to face the full costs of their actions will also address equity concerns. International mechanisms, such as joint implementation, coordinated economic instruments, carbon taxes, and tradable permits, if appropriately implemented, would promote efficiency.

1.1 Introduction

In recent decades, atmospheric emissions of greenhouse gases have risen significantly. Concentrations are currently about 25% greater than at the beginning of the Industrial Revolution. If current trends continue, concentrations will double from preindustrial levels before the end of the next century and, if unchecked, continue to rise thereafter (IPCC, 1990a).

The scientific community has noted the potentially serious effects of increased concentrations. These climatic effects could, in turn, have further effects on the biosphere, including an increase in mean global temperature, an increase in sea level, changes in agricultural yields, forest cover, and water resources, and a possible increase in storm damage.

Increased concentrations of greenhouse gases are the result of fossil fuel burning, deforestation, livestock raising, and other human activities. Concerted action on the part of individuals and governments will be required to slow the increase in concentrations. Changes in greenhouse gas concentrations and the analysis of the climatic and other physical consequences of those changes lie within the purview of the physical sciences. The role of human activity in generating greenhouse gases, the consequences of those changes for humans, and possible responses lie within the purview of the social sciences.

Climate change impacts are likely to vary dramatically from country to country. A warmer climate could benefit sectors of the economies of some mid- and high-latitude countries. It is possible that anthropogenic warming might heat the atmosphere enough to prevent or delay another ice age. On the other hand, even modest economic losses averaged over the globe could mask large regional losses: a rising sea level and the possibility of increased storm surges could threaten the survival of some small island states and coastal areas and could increase the risk of midcontinent drought and desertification for inland areas on the periphery of deserts. Such changes could promote human migration and major conflicts as well as famine, disease, and increased mortality.

Within the past decade, a consensus has emerged on some key issues in the economics of climate change. This report describes areas of consensus as well as areas of disagreement, the sources of disagreement, and further research that could narrow the range of disagreement. This chapter frames the issue of climate change largely from the point of view of economics but also from that of other social sciences, introducing the more detailed discussions in the chapters to follow.

At least two arguments have been offered to justify the commitment of resources to mitigate climate change. The first arises from fundamental values, the second from decision analysis. They may be summarized as follows:

(1) We have only one planet. Some changes are largely irreversible and may occur rapidly. Prudence calls for avoiding a large-scale experiment with the planet. Thus, avoiding anthropogenic climate change lies beyond the scope of normal economic calculation.

(2) The potential exists for the occurrence of sudden, largely irreversible, nonlinear changes in the global

ecosystem. These would have major economic effects, which would be particularly severe in some countries or regions.

Even if the first view is adopted, economics has much to contribute to the discussion, for the question of cost-effective emission reductions must still be addressed. If the second view is adopted, economics and cost-benefit analysis will clearly be relevant, both in deciding how much mitigation to undertake and in designing the measures.

This chapter, and others in this Assessment Report, draw on the findings of the IPCC's Working Groups I and II, and follow the guidelines provided by the Framework Convention on Climate Change (FCCC). The Convention leaves open a number of important questions that must be addressed at the political level through future negotiations, including reviewing the adequacy of commitments. It is hoped that the findings of this chapter, and the assessment report more broadly, will contribute to these future negotiations by providing an understanding of the costs and consequences of alternative actions and their scientific basis.

1.2 Features of Climate Change

Climate change could impose a variety of impacts on society. Volume 2 of this report analyzes these impacts in detail. They include effects on agriculture, forests, water resources, the costs of heating and cooling, the impact of sea level rise on small island states and low-lying coastal areas, and a possible increase in extreme events (e.g., storms). Although most attention to date has focussed on negative impacts, some impacts will be positive. Beyond these tangible impacts are a variety of intangible impacts,[1] including damages to existing ecosystems and the threat of species losses.[2]

Climate change presents the analyst with a set of formidable complications: large uncertainties, the potential for irreversible damages or costs, a very long planning horizon, long time lags between emissions and effects, a global scope, wide regional variations, and multiple greenhouse gases of concern.

Large uncertainties. Although natural scientists agree that greenhouse gas concentrations are rising, there remain major uncertainties about the impacts on temperature and climate. These are reflected in a wide range of estimates of future global mean temperature increases and in uncertainties about regional climate changes. Estimates of net economic losses for the most likely range of warming over the next century, and the great uncertainties associated with such estimates, are discussed in Chapter 6. Social scientists do not agree on the size of the behavioural responses or economic effects that would follow or on the effect of these changes on human well-being (Manne and Richels, 1992; Peck and Teisberg, 1993; Nordhaus, 1993).[3]

Nonlinearities and irreversibilities. Nonlinearities occur when changes in one variable cause a more than proportionate impact on another variable. Irreversibilities are changes that, once set in motion, cannot be reversed, at least on human timescales. For example, some have suggested that even a modest increase in atmospheric greenhouse gas concentra-

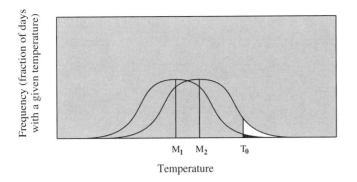

Note: As average temperature increases from M_1 to M_2 the number of days when the temperature exceeds T_0 increases significantly.

Figure 1.1: Effect of an increase in average temperature on the number of days that exceed a specified threshold.

tions could, beyond a certain point, trigger a substantial increase in temperature. Some have hypothesized that "runaway warming" could be triggered by a disruption of the North Atlantic thermohaline circulation or by methane release from thawing of permafrost.[4] Alternatively, even a modest increase in average temperature might significantly increase weather-related agricultural losses because, for many crops, extra days of extreme heat severely limit yields. In Figure 1.1, if the threshold temperature for crop damage is T_0, then even a small increase in the mean temperature, from M_1 to M_2, may greatly increase the number of days above the threshold, represented by the area under the curves to the right of T_0.[5] On the other hand, spending large sums to reduce the risk of climate change may also have largely irreversible consequences, slowing development as it drains resources from other efforts to improve the human condition.

Long planning horizon. Greenhouse gas concentration changes occur over a long period of time; the full consequences of actions taken over the coming decades will be felt increasingly over the next century and in future centuries. The truly long-term nature of the problem is one of the distinctive aspects of greenhouse gas warming. Seldom has the world consciously faced a set of decisions likely to affect our descendants one, two, or three centuries from now.[6] Because the costs of taking action today are borne by the current generation, whereas the benefits that accrue will be felt possibly hundreds of years in the future, the world community is now faced with issues of intergenerational equity on an unprecedented scale. Although society has addressed similar problems concerning trade-offs over periods of fifty or even a hundred years, the long planning horizon for climate change puts the analytic questions at issue in a new light. The length of time involved has one further implication: changes in technology, as well as population and consumption patterns, become of paramount importance.

Long life of capital stock. Every country has made large capital investments in its cities, farms, ports, and other assets. Some of this investment cannot be changed without large costs: low-lying port cities, for example, cannot easily be rebuilt. For other investments, the cost of change will be small.

For diversified agricultural economies, the cost of switching from one annual crop to another will be small if the temperature increase is modest; for less diversified economies, the costs may be larger. In short, both mankind and ecosystems have adjusted to the current climate, and adjustments to accommodate climate change may be costly.

Inertia in the climate system. Atmospheric concentrations, rather than emissions, determine the amount of warming projected by climate models. Concentrations change much more slowly than emissions, meaning that affected nations might not have enough time to prevent impacts from climate change, or to mitigate economic impacts after the effects of climate change become evident. In this respect, the risks of climate change are unlike those of earthquakes or floods. Long time lags and the difficulty of detecting climate change increase the difficulty of reliably determining the magnitude and timing of future effects before they begin to occur.

Global scope. Climate change is a global challenge, which cannot be answered by a single country acting by itself. Mitigation must be coordinated globally. In an interlinked world economy, not only are the actions of a single country, or group of countries, not likely to be sufficient to address the problem, they are likely to be largely offset by actions of other countries. If, for instance, one group of countries reduces timber cutting to increase carbon absorption, the price of lumber will rise, which may induce other countries to increase cutting in their forests.

Moreover, whereas economic analysis generally takes the point of view of a single decision maker or government, the important decisions on climate change will, of necessity, be made by many sovereign governments. Economic and decision sciences are not yet able to predict the outcome of bargaining problems of this type.[7]

Regional variation. Impacts are likely to vary greatly both within and among countries. Some countries and regions will suffer from warming; others will benefit, at least in some sectors. Some cold countries will benefit from a reduction in heating costs and an increase in the length of growing seasons; some warm countries will see a drop in yields from agriculture and forestry; low-lying states are likely to suffer from increased storm surges and flooding.

Aggregation. For the world as a whole, the net effect at any time will be the sum of local effects at many points on the globe, some positive and some negative. Analysts have no way to estimate this sum without detailed local calculations (summarized in Chapter 6).

Multiple gases of interest. The enhancement of the greenhouse effect depends on the concentration of all greenhouse gases, even though most economic modelling to date has limited itself to the implications of changes in CO_2. Because greenhouse gases differ in radiative efficiency and atmospheric lifetime, analysts have devised measures of global warming potential (IPCC, 1990a, 1995) that seek to allow radiative forcing from changes in the concentration of all greenhouse gases to be measured in a commensurable fashion.[8] Studies also demonstrate the important role of sulphur and other aerosols, which cause negative radiative forcing (i.e., cooling) by reflecting incoming solar radiation.[9]

Importance of net emissions. Because greenhouse gas concentrations depend on net rather than gross emissions, changes in forests and other greenhouse gas sinks must be taken into account.[10]

Efficiency vs. equity. From an economist's perspective, how much to reduce emissions is a matter of efficiency (because achieving the proper level of emissions raises net well-being), but who pays is a matter of equity. Economics has much to say about the former, but much less about the latter. Nonetheless, equity considerations will drive many of the policy decisions made under the Framework Convention on Climate Change.

1.3 Contribution of Economics

Economics and the social sciences offer perspectives on climate change not provided by the physical sciences. In the classic definition, economics is the study of the allocation of scarce resources that have alternative uses. Economics emphasizes the importance of trade-offs between different uses of resources, and the forgone value of other uses of a resource, called the opportunity cost. In the context of climate change, this means that (1) costs and benefits matter; (2) resources are not free; and (3) resources used for one purpose are no longer available for other purposes.

This chapter sets out the logic of cost-benefit analysis as applied to climate change. Standard cost-benefit analysis requires (1) a valuing of costs and benefits over time, using *willingness to pay* as a measure of value and (2) a criterion for accepting or rejecting proposals.[11] The standard criterion is the *compensation principle* (Kaldor, 1939; Hicks, 1939), which says that if the project yields positive net benefits, then those made better off could compensate those made worse off with something extra left over. As long as compensation is paid, the result is an unambiguous gain in welfare, without the necessity of weighing effects on different individuals.

Climate change raises difficulties with both requirements. Valuation is difficult because of the difficulty in valuing environmental amenities, which are generally not traded in the market. And the compensation principle will not apply if mechanisms for affecting transfers do not exist, either between countries or regions in one generation, or – especially – between generations. If transfers are not feasible, then the analysis must assign weights to different individuals (for example, the utilitarian welfare function gives equal weight to each person). Only then can conclusions be drawn about net benefits for society as a whole. This issue is addressed in the discussion of equity in Section 1.4.

Beyond these fundamental concepts are ideas, originally from other areas of economics, that may be applied directly to the study of climate change; these include work on risk, dynamics, sequential decision making, public goods and externalities, taxation, and general equilibrium.

1.3.1 Risk

In the past thirty years, much new economic research has focussed on rational responses to risk,[12] including three areas

important to a systematic examination of and rational response to climate change: portfolio theory, insurance, and decision analysis.

1.3.1.1 Portfolio theory
A portfolio manager attempts to get the best return for a given level of risk. One approach is to buy several types of assets whose returns are not correlated or are negatively correlated (that is, whose prices move either independently or in opposite directions). In this respect, climate change policy decisions can be compared to investment portfolio decisions.

When faced with a risk, an individual may (1) act to reduce the chance the unfavourable event will occur; (2) act to reduce the cost if the event does occur; or (3) spread part of the risk to others through insurance. In response to the threat of climate change, nations may (1) reduce the chance that warming will occur by reducing greenhouse emissions (mitigation); (2) adjust to climate change if it does occur (adaptation); or (3) spread part of the risk through insurance. A porfolio approach can be expected to include both mitigation and adaptation actions. These may include government policy reforms, such as reducing fossil fuel subsidies; increased carbon sequestration; reducing emissions of methane and other non-CO_2 greenhouse gases; research and development, which can promote emission reductions or make it easier to adapt to any changes that do occur; and international actions, including joint implementation (one country funding emission reductions in another country) and technology transfer.

A well-chosen portfolio of climate change investments will yield greater benefit for a given cost than any one option undertaken by itself. For an individual country, the issue is how to choose the portfolio of policy measures best suited to its circumstances and to adjust the portfolio over time in response to new developments. Governments will be making climate change decisions for several decades at least. This means that they will have many opportunities to adjust the size (total resources) and mix (choice of measures) of their portfolios of responses. Portfolios may differ from country to country.

1.3.1.2 Risk aversion
Individuals and societies are generally risk-averse when facing large risks; that is, they are willing to pay something to reduce the likelihood of a large risk. The amount they are willing to pay is called the *risk premium* (see Box 1.1).

That individuals and societies are *risk-averse* means that average utility (well-being) is increased by pooling risks, or, equivalently, that people are willing to pay to reduce the risks they face. If society as a whole is risk-averse, then some investments with a negative expected return, for example, a particular investment in climate change mitigation, should be undertaken if they reduce the probability of a loss or the costs of future adaptation.

The magnitude of those expenditures depends on society's degree of risk aversion and the magnitude of the risk. The *risk premium* – the extra amount that society is willing to pay to reduce a risk – is small if the stakes (say, the maximum loss) are small, and large if the stakes are large. An investment of a

---

BOX 1.1: RISK AND UNCERTAINTY

Uncertainty arises when a decision can lead to a range of outcomes.

Expected return or *expected value* of a decision is the mean of the distribution of returns, the amount a person would on average receive as a consequence of the decision.

Risk aversion measures an individual's unwillingness to take risks.

Risk premium is the amount an individual would pay to replace the uncertain distribution of outcomes with the expected value.

Certainty equivalent is the amount that makes an individual indifferent between it and a risky proposition; for a risk-averse person, the certainty equivalent is higher than the expected return; the difference is the risk premium.

---

dollar is justified if it reduces the loss of expected utility by more than a dollar, and not justified if it reduces the loss by less than a dollar. Thus, results reported below focussing on the expected loss of GDP from climate change do not directly address the risk premium. If a possible outcome is a loss of 10%, even though the expected loss is only 3%, then the *certainty equivalent* loss will exceed 3%. A dollar investment that reduces this certainty equivalent loss by more than a dollar should be undertaken. Such an investment could either reduce the average loss, for example, by reducing the probability of the loss occurring (through mitigation actions), or reduce the variance of the loss. For example, some actions that reduce extreme losses will have more than proportionate returns.

Ascertaining the magnitude of the risks, or how they are affected by any particular action, is often difficult in a dynamic setting. One key consideration is how the particular action affects the remaining options – the set of actions available in the future – along with their costs and benefits. Risk-reducing expenditures are referred to as *precautionary investments*. Making precautionary investments has the same effect as buying conventional insurance.

The insurance expenditures associated with mitigation actions and investments are, in a sense, only the differences between the actual expenditures and the *no-regrets* benefits (the benefits other than those associated with greenhouse gas emissions). Thus, investments in fuel-efficient cars may have a direct benefit in reducing the cost of running a car and in reducing its emissions of local air pollutants. The mitigation investment is only the additional investment for climate purposes.

Irrespective of the possible consequences of climate change, policies that mitigate against or assist adaptation to climate change and have zero or negative net costs (no-regrets policies) are clearly justified. If the evidence suggests that damages can be expected from climate change, then the expectation of damages provides a rationale for going beyond no-regrets policies to those that incur positive net costs. The principles of risk aversion and portfolio balancing provide a rationale for further steps. The costs of such policies might be

justified as a risk premium to be paid for the added security of reducing the likelihood of climate change.

Traditional insurance rests on two principles: pooling risks and transferring risks to those more willing or better able to bear them.[13] Because the risks associated with climate change are correlated, pooling risks is less effective than it is in other situations. Nonetheless, differences in predicted regional impacts, implying less than perfect correlation in climate change risks, make possible some degree of risk pooling. This holds irrespective of who pays the cost, because wealthier individuals and countries are better able to bear risk.[14] Many countries likely to be most adversely affected will be developing countries, whereas many of the countries least affected (or positively affected) will be the industrialized countries, which could provide insurance for effects of climate change that might fall harder on less developed economies.[15]

Insurance markets, however, face three problems in addressing climate change. First, they lack a mechanism to transfer some of the risk from those likely to bear it (future generations) to the current generation. Second, losses associated with climate change are likely to be both correlated and large, compared to losses absorbed in a single year by the commercial insurance industry (which itself has been hard pressed in recent years to handle natural disasters). Third, the long-term nature of climate change insurance raises the problem of contract enforcement: Will contracts signed today be enforceable tomorrow? Will the insurers be around to pay claims fifty or one hundred years from now? (Even in the industrialized countries, private markets may be inadequate to insure against losses from a major national disaster today.)

These considerations suggest that private markets will not be able to insure fully against climate change. One possible solution would include international action to establish insurance markets, perhaps with government reinsurance. Should such an insurance market be established, careful attention will have to be given to ensuring that the insured parties engage in appropriate adaptation actions, reducing the losses that might be associated with any greenhouse gas warming.[16]

1.3.1.3 Precautionary investments

A business makes precautionary investments to reduce the total risk of its portfolio. Numerous policy measures are available to reduce risks to future generations from climate change. Four have been most often discussed in recent years: (1) immediate reductions in emissions to slow climate change; (2) research and development focussing on new supply and conservation technology to reduce future abatement costs;[17] (3) continued research to reduce uncertainties about how much change will occur and what effects it will have; and (4) investments in actions to assist human and natural systems to adapt to climate change if it occurs.

Precautionary investments may reduce the risk of climate change itself (mitigation), or they may enhance the ability of future generations to respond, in one of two ways. First, by analogy, an individual may set money aside in a "rainy day fund" when things are going well and allow it to grow over time in order to make it easier to adapt to difficulties that may occur later. Similarly, by increasing investment in productive assets now, countries will have a richer economy to draw on should climate change damages occur later. A policy of precautionary investments means investing more than would otherwise have been invested. Second, investments can be made (including investment in research and development) that would enhance the economy's ability to adapt should climate change damages occur.

Precautionary investments may also enhance the ability of future generations to react. An important reason that people establish savings accounts is to reduce the impact of unfavourable events in the future. Similarly, a society may elect to accumulate capital against the possibility of a large loss from climate change. This is one thread of the debate over discount rates discussed in Chapter 4. Those who argue for a discount rate close to the opportunity cost of capital point out that society may choose between immediate greenhouse gas mitigation, at a cost, and delayed mitigation, with some of the money saved put aside as a savings account for our grandchildren in the event of large climate-induced damages.

1.3.2 Sequential decision making

As a policy question, global climate change is sometimes posed as a choice between (a) doing nothing at all or (b) committing to all-out effort. Given the large current uncertainties about the costs and benefits of greenhouse mitigation, this is the wrong way to frame the issue, as it obscures the choices that should be evaluated. Moreover, in part because option (b) may be perceived as too expensive to get political support, policy paralysis often results.

A more useful formulation is: "Given current knowledge and concerns, what actions should we take over the next one or two decades to position ourselves to act on new information that will become available?" (Lind, 1994). For example, decision makers would like to know if the possibility of irreversible damages, such as might be suffered by low-lying states, justifies undertaking an aggressive abatement programme immediately.[18]

Climate change demands a decision process that is sequential and can incorporate new information. Timing will be a key element, and the date of resolution of uncertainty an important element of the analysis. Figure 1.2 shows schematically the progression from a simple decision to a sequence of linked decisions. In this example, the simple decision might be whether to take aggressive abatement actions now. Let us assume that the uncertainties are resolved in 2005. In the case of sequential decisions, Decision 1 (in 1998) could be whether to take aggressive abatement actions now; Uncertainty 1 (resolved in 2005) might be the cost of mitigation; Decision 2 (in 2010) might be whether to tighten abatement programmes already in place; and Uncertainty 2 (resolved in 2020) might be the relation between greenhouse concentrations and temperature increase. Since both climate change and new knowledge (learning) are continuous processes, actions to address climate change should be adjusted continuously in the light of new information.

A sequential decision-making strategy aims to identify short-term strategies in the face of long-term uncertainty. The next several decades will offer opportunities for learning and

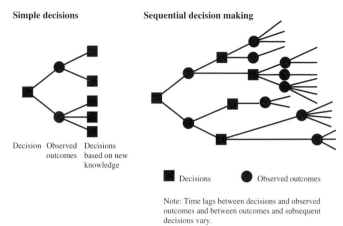

Figure 1.2: Sequential decision making.

making mid-course corrections. The relevant question is not "What is the best course for the next 100 years?" but rather, "What is the best course for the next few years?" because a prudent hedging strategy will allow time to learn and change course.

For example, the choices might be (1) immediate investment in new plant and equipment, (2) aggressive research and development on greenhouse abatement technology, or (3) deferring large investment for ten years, when the nature and size of the threat are better understood, when costs will presumably have dropped owing to the availability of improved technologies, and the job can be done more efficiently.

Inappropriate interim goals increase the total cost of addressing the problem (Richels and Edmonds, 1993). For example, a commitment to certain levels of emissions in certain years fails to take into account the effects of temporary economic disturbances on GDP, and thus on emissions and, hence, on the cost of controls.

Because of the high cost of being wrong in either direction, the *value of information* about climate change is likely to be great. In particular, the value of information about the sensitivity of temperature to CO_2 increases, the temperature damage function, the GDP growth rate, and the rate of energy efficiency improvement is likely to be high (Chao, 1992; Peck and Teisberg, 1992; Manne and Richels, 1992; Nordhaus, 1993).[19]

The presence of uncertainty along a dynamic path creates an *option value,* the value of preserving choices for the future. In climate change, the term has been used in two different ways. One stresses the irreversibilities of climate change: mitigation expenditures now preserve the option of avoiding adaptation expenditure later. The other stresses irreversibilities in investment and the cost of premature turnover of capital. Any action taken today changes the options available later or, more precisely, changes the consequences of any future action. Sequential decision making focusses on how those consequences are affected if one action is taken today rather than another, and on how these consequences are affected if actions are taken in parallel rather than serially.

1.3.3 Dynamics

The problem of greenhouse gas warming involves additions to concentrations resulting from net emissions over extended periods of time. Thus the analysis must focus on dynamics. Dynamic analysis involves three stages: the dynamic processes involved, the trade-offs, and judgments concerning those trade-offs.

Dynamic analyses have led to important insights. For example, atmospheric concentrations of greenhouse gases and, therefore, their effect on temperatures depend on the total amount emitted over a period of years. A given concentration target can be achieved by a variety of emission timepaths. Timepaths that provide for the economical turnover of existing capital stock and time to develop low-cost substitutes are likely to be less costly. This suggests large potential gains from flexibility in the timing of emission reductions.[20]

1.3.3.1 Kaya identity
The driving forces in emissions of any greenhouse gas can be seen in the following identity for carbon dioxide emissions (Kaya, 1989):

$$CO_2 = CO_2/E \times E/Q \times Q/L \times L$$

| CO_2 carbon dioxide emissions | CO_2/E carbon dioxide emissions per unit energy | E/Q energy per unit output | Q/L output per capita | L population |

or, expressed in rates of change:

$$d \ln CO_2/dt = d \ln CO_2/E /dt + d \ln E/Q /dt + d \ln Q/L /dt + d \ln L /dt$$

that is, the percentage rate of change in carbon dioxide emissions is equal to the rate of change in carbon dioxide emissions per unit energy plus the rate of change in energy requirements per unit output plus the rate of change in output per capita plus the rate of change in population.[21]

This identity clarifies different approaches to reducing emissions. For a developed country with a stable or slowly growing population, as long as the ratio of emissions to output declines at least as fast as productivity rises, CO_2 emissions will not increase. Because of the substantial potential for energy efficiency improvements, this seems feasible for most developed countries. Many opportunities exist for increasing end-use efficiency, represented by the second term on the right-hand side – for example, a shift from cars to public transportation and from less to more fuel-efficient cars and homes.[22] (The same term may also reflect structural shifts in the economy, as in the case of a shift away from energy-intensive industries.) Fuel switching (for example, from coal-based electricity to oil, gas, hydroelectric, wind, or geothermal), represented by the first term, also offers the potential to limit CO_2 emissions in many countries.

For many developing countries, emissions will increase unless energy efficiency and greenhouse gas emissions per unit of energy change to offset growth in per capita output and population. For many developing countries with rapidly growing populations, pressures for economic development will make it difficult to direct capital from investments with higher greenhouse gas emissions to those with lower greenhouse gas emissions.[23]

Evidence for two other essential issues is currently limited. First, to what extent will improvements in energy efficiency require net increases in investment beyond the resources saved from reduced energy usage; in other words, how much does aggressive emission reduction depress economic growth? Order-of-magnitude calculations suggest the presence of only limited trade-offs, at least for the near term. Second, do developing countries have the institutional capacity to achieve the desired increases in emission efficiency?

1.3.3.2 Nonrenewable resources, backstop technologies, and emission reduction strategies
In principle the atmosphere is a renewable natural resource. However, the longevity of the greenhouse gases and the relationships between stocks and flows mean that, for practical purposes, it may be better treated as an exhaustible natural resource, although one in which welfare depends not just on the flow out of the stock but on the stock itself.[24]

The central problem with natural resources, whether renewable or not, is timing. Many renewable resources possess a maximum sustainable rate of exploitation. Exploitation cannot long exceed this maximum rate without depleting the stock and ultimately reducing the harvest. Corresponding to the sustainable flow rate is a steady-state stock. If, initially, the actual stock exceeds the steady-state stock, then the flow can exceed the maximum sustainable flow for a while. The question is how to distribute this excess over time. In addition, even when the stock is at the sustainable level, it may be desirable in times of emergency to exceed the maximum sustainable flow. For a renewable resource, this can be done, though only at the expense of decreased flows later.[25]

Timing of reductions in greenhouse gas emissions should reflect differences in costs, discounting (to evaluate those costs), and risk. If technological change will make future emission reductions much less costly, some reductions should be postponed.[26] Conversely, research on learning effects shows that if actions taken today will lower costs faced tomorrow, then these dynamic benefits should be included in the calculus (Arrow, 1962; Atkinson and Stiglitz, 1980). In the context of climate change, if emission constraints stimulate technical or other developments that help to lower the costs of continuing or additional emission abatement, then reductions should be accelerated (Grubb *et al.*, 1993a, 1993b, 1995). If discount rates are high, the costs borne by future generations will carry less weight than if discount rates are low. In the presence of risk, nonlinearities, or irreversibilities, the principle of risk aversion suggests a strategy of early mitigation.

The theory of nonrenewable resources contains a second set of lessons for climate change policy. Suppose primary energy sources are divided into four groups: coal, gas and oil, biomass, and noncarbon sources. As a first approximation, gas and oil may be taken to be exhaustible. This means that total carbon emissions from gas and oil are fixed, or at least confined to narrow bounds; the question is not how much will be consumed, only when.[27]

As an example, Figure 1.3 depicts three phases of fuel use in an economy. In the first phase, the economy relies on nonrenewable resources, such as oil and gas; in the second, on coal and perhaps nuclear power; in the third, on backstop

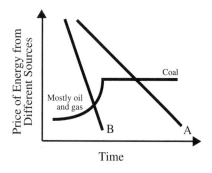

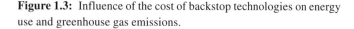

Note: If the price of the nonfossil backstop technology is given by **A,** all oil and gas reserves will be exhausted and coal will be used for some time before being displaced by the backstop technology. If the price of the nonfossil backstop is given by **B,** it will be adopted before the oil and gas reserves are exhausted, resulting in lower greenhouse gas emissions.

Figure 1.3: Influence of the cost of backstop technologies on energy use and greenhouse gas emissions.

technologies,[28] such as biomass combined with noncarbon sources. The switch points depend on rising energy prices and improving technology, which lowers the costs of the backstops. The figure shows alternative long-run scenarios. In alternative A, the price of the backstop technology falls sufficiently slowly that, for a time, the economy relies on coal. In alternative B, the price of the backstop technology falls fast enough to eliminate the intervening stage of primary reliance on coal.

Since, to a first-order approximation, the total carbon load from oil and gas is fixed (and limited), the total carbon load on the atmosphere is ultimately primarily related to coal usage. From this perspective, an important uncertainty is the pace at which the cost of the backstop decreases. If it decreases fast enough, the intermediate stage of coal dependence will be short, and the total carbon load low, whereas if the price decreases slowly, the carbon load could be much larger.

From this perspective, rational use of gas and oil is an important part of a risk strategy, for it provides insurance against the possibility of a delay in the arrival of backstop energy sources.[29] This, in turn, has important implications for the "leakage" debate, which asks whether, in the event that the developing countries impose carbon taxes but the undeveloped countries do not, the latter's response will largely offset emission reductions made by the former. It is also possible that lower prices for gas and oil will induce coal-rich countries to decrease their reliance on coal, thus possibly producing negative leakages. But to the extent that such leakage occurs, the insurance provided by greater conservation of oil and gas is eliminated, and it is this insurance that should be an essential part of a dynamic strategy. If the backstop arrival is delayed, then earlier fuel switching by China and India, to mention the most pertinent examples, will yield no long-run benefits. On the other hand, if the backstop arrival is early, the whole issue is largely moot. Thus, leakage needs to be looked at, not from the static perspective of what occurs in a single year, but from a dynamic perspective that corresponds to the long-run nature of the global climate problem.

1.3.4 *International public goods*

The atmosphere is an *international public good,* in that atmospheric concentrations are the result of combined actions by all countries. A pure public good (Samuelson, 1954) has two properties: *nonrivalry* and *nonexcludability* (see Box 1.2).[30] The atmosphere has both characteristics. That means that if one country's greenhouse gas reductions thwart global warming, all countries benefit.[31]

Since countries will be affected differently by climate change, the benefits of avoiding greenhouse warming will also differ from country to country. Further, some actions will simultaneously affect local atmospheric conditions (providing local public goods) and greenhouse concentrations. For example, actions that reduce urban driving improve local air quality and at the same time reduce greenhouse gas emissions.[32]

1.3.4.1 *Property rights*

An important strand of economic thought associates externalities with a failure to assign property rights (Coase, 1960). Assigning property rights to the atmosphere is particularly difficult, however, since this would require the agreement of many sovereign states. Tradable greenhouse emission permits, discussed below, can be thought of as an attempt to resolve the problem by explicitly assigning property rights to greenhouse emissions.

Establishing property rights for emissions is not the same thing as considering the climate system itself as a public good. The FCCC refers to the climate system (not the atmosphere) as a "common concern of humankind." Atmospheric concentrations are the net result of emissions plus the contributions and effects of other factors such as the oceans, forests,

---

BOX 1.2: PUBLIC GOODS AND EXTERNALITIES

Externality: An externality, or spillover, arises when the private costs or benefits of production differ from the social costs or benefits. Because the social costs or benefits are external to the private costs that firms face, the economy will tend to produce too little of a public good (like education) and too much of a public bad (like pollution).

Public good: A public good has two properties: nonrivalry and nonexcludability. Nonrivalry means that additional consumers do not have to compete with each other to use the good and therefore drive up its cost: the marginal cost of an additional individual using the good is zero. Nonexcludability means that the marginal cost of exclusion – of stopping an individual from enjoying the good – is prohibitive. Public goods thus permit "free riding." Lighthouses, for example, are public goods: when lighthouse services are provided to one person, others may enjoy the same services without cost.

Market failure: Private markets may sometimes fail to provide a good at the most desirable level: private markets alone will likely provide too few lighthouses and too much pollution.

and agricultural activity. Establishing property rights over emissions does not provide guidance for the consideration of these other influencing factors. Thus, establishing property rights for the atmosphere is only one of the mechanisms necessary to regulate climate change.

1.3.4.2 *Paying for an international public good*

Who should pay for a global public good? Every country faces this question internally in determining who should pay for the public goods it provides. Who should, for instance, pay for pollution control within a country? Economists generally agree on the following principles:

First, for the purposes of analysis, *it is useful to separate efficiency from equity*. The implication of this principle is that because pollution is a social cost of production (and consumption), everyone should be made to pay the full social costs of the pollution they generate. Thus, if there is a social cost to a unit of greenhouse gas emissions, that cost is the same no matter who produces the emissions. All should pay the full social costs of their actions, whether rich or poor. In this perspective, corrective (Pigouvian) taxes should be imposed uniformly.

Second, *it is inappropriate to redress all equity issues through climate change initiatives,* although climate change should not aggravate disparities between one region and another.

No scientific consensus exists on the framework for deciding the burden of financing mitigation and adaptation. At least four approaches have been proposed to determine how the burdens of taxation should be shared. One approach looks at *benefits*: Just as those who benefit from private goods must pay for them, those who benefit from a public good should be made to pay for it. The principle has some force when large differences in preferences exist within any income class. Providing a particular public good benefits some of those individuals more than others, creating inequalities in the absence of benefit taxes. A major problem frequently encountered, particularly for pure public goods, is that it may be difficult to determine who benefits. It is, in general, possible to ascertain the *economic benefits* of mitigation, and these are likely to be quite unequally distributed. But this principle, by itself, does not fully determine who should bear the costs. Appropriately designed mitigation strategies will produce a surplus of benefits over costs, a surplus that must somehow be divided.

A second approach looks at ability to pay. It is often held that richer countries (or individuals) should pay more than poorer ones. This approach sometimes rests on the claim that all people are entitled to a certain minimum consumption (Dasgupta, 1982). But this principle does not answer the question of how much extra the richer countries should pay.

A third approach is based on contribution to the problem. Because the industrialized countries have contributed more than two-thirds of the stock of anthropogenic greenhouse gases in the atmosphere today, this approach seems to suggest that they have a larger responsibility for bearing the costs. On the other hand, by the time greenhouse gas concentrations double from preindustrial levels, the developing countries are projected to be contributing more than half of annual emis-

sions, and roughly half of the total stock in the atmosphere (IPCC, 1990a; Cline, 1992). Thus, under this criterion, the developing countries might eventually pay far more of the mitigation costs than under the other principles described earlier.

Economists have turned to a fourth approach – the social welfare function – to answer the question of how much extra different parties should pay, as well as the question of how to distribute the surplus. The discussion of equity below differentiates between the Rawlsian and utilitarian approaches. In this case, both approaches yield similar results: In the absence of incentive problems, both imply that all of the surplus should be allocated to the poorer countries, or that all of the burden of effort should be borne by the richer countries.[33]

Yet a different approach holds that social scientists as such have nothing to say about these ethical issues. Coase (1960), for instance, approaches the problem of externalities by emphasizing that (a) in the absence of bargaining costs, an efficient solution can be obtained by assigning property rights[34] and (b) this solution is independent of how property rights are assigned.[35] Coase also emphasizes the importance of transaction costs, which will often influence the choice of policies.

A simple approach that yields efficiency but does not require redistribution (and is thus consistent with the two principles enunciated above) requires coordinated tax rates so that all countries face the same energy prices. This approach makes the cost of emitting an extra tonne of carbon equal across all countries, with each country retaining the revenues thus generated.[36] The net cost of such a tax (ignoring the benefits from reduced greenhouse gas emissions) will, in general, be smaller for poorer countries, as a percentage of their national output. The burden of the tax is progressive in its distribution across countries, even though the tax is levied at the same rate in all countries.[37]

Accounting for past emissions. Article 3.1 of the Framework Convention on Climate Change directs the Annex I (i.e., developed) countries to take the lead in responding to the threat of climate change "on the basis of equity and in accordance with their common but differentiated responsibilities and respective capabilities." Some have argued in addition that because the industrialized countries have been the major contributors to current levels of greenhouse gases, they should bear most of the costs of mitigation. This view says that costs should be borne, not in proportion to benefits expected, but in proportion to contribution to pollution. This argument, however, is not based on the principle of economic efficiency. Efficiency requires that incentives be prospective (forward looking), not retrospective.[38] No incentive effects result from imposing charges based on *past* actions. Whether to charge nations that contributed CO_2 to the atmosphere is an issue of ethics, not efficiency.[39]

The controversial issues of population growth and consumption patterns, although central to economic development, bear on climate change largely through their effects on emissions. Population growth in developing countries may also exacerbate the ecological and socioeconomic impacts of climate change. At the same time, high per capita consumption in industrialized countries, where populations have nearly stabilized, will also affect mitigation and adaptation costs and strategies. The relation of population to sustainable development, although important, is beyond the scope of this paper.[40]

1.3.4.3 Enforcement

Both externalities and public goods need a legal framework within which the problems they pose can be addressed. Without compulsory taxation, there is an incentive for each individual to be a free rider, though there is some empirical evidence that the free rider effect may not be as significant as economists have previously assumed (Bohm, 1993). In the absence of compulsory taxation, externalities can only be addressed with well-defined property rights (Coase, 1960)[41] and a legal system that enforces compensation for externalities.

Enforcing compliance with international legal agreements presents a number of legal and political problems. Many states resist compulsory use of the judicial process; this provides an incentive for free riding. The FCCC provides several means of settling disputes, including judicial recourse and arbitration. It also requires the parties to consider establishing a "multilateral consultative process" to assist in implementation of the Convention and to anticipate and prevent confrontations concerning compliance and enforcement.

1.3.4.4 Knowledge

A key element in addressing the problem of global warming is knowledge – knowledge about climate science as well as about the economic and social aspects of impacts, mitigation, and adaptation. Much of this knowledge is in the nature of an international public good. Developing ways of increasing energy efficiency will benefit all countries. Although in some cases, those countries engaged in the research will be able to appropriate for themselves a significant fraction of the *private* benefits (mostly in reduced energy costs), they will not be able to appropriate the broader social benefits, except through an energy tax (or permit fee) that is high enough that the price fully internalizes the emissions externality. Even then, the social benefits of innovation tend to far exceed the private benefits.

This suggests the need for an international agreement to fund basic research and subsidize applied research, particularly in energy-related technologies for the developing countries. There need not be a central funding agency, or a central directorate determining which research should be undertaken. This calls for some mechanism, possibly including joint implementation, for sharing research results and for ensuring that the fruits of this research are made available.

1.3.5 Efficiency

With some exceptions noted below, efficiency and equity can be analyzed separately.[42] Analysts agree that any actions responding to climate change should be cost-effective: no matter who bears the cost of emission reductions, reductions should occur where their cost is lowest.[43] Because of the low energy efficiency in many developing countries, many have proposed that more attention be paid to emission reductions there.

Mechanisms for reducing emissions equitably and efficiently, including joint implementation, tradable permits, and coordinated tax policies, are discussed below and in Chapter 11. All these approaches, however, attempt to confront all individuals and producers in all countries with the same cost of emissions. Emission control is an international public good, in that greenhouse emission reductions have the same effect wherever they occur. Just as efficiency in the production of steel or any other commodity requires that all consumers and producers face the same price, so too with emissions. This can be achieved either through coordinated energy taxes or through tradable permit requirements; but unless the rules are applied in a systematic way to both developed and developing countries, emission reductions will be inefficient.

Partial participation in an international emission reduction programme will significantly reduce its effectiveness. The growth of international trade has resulted in important links between the developed and developing countries, and the total effects of any policy undertaken in the former can only be evaluated in terms taking into account the systemic responses, including responses from the developing countries, as discussed under the heading of "general equilibrium" in Section 1.3.6 below.[44]

1.3.5.1 Bankable permits

Efficiency imposes several requirements. One, just described, is that at any moment, the costs of reducing emissions should be minimized. The second is *intertemporal efficiency*: The marginal cost of reducing emissions at two points in time must be the same. If it will cost less to reduce emissions at some future date, adjusting for time discounting, option values (risk), and impacts on atmospheric concentrations, then the reduction schedule should be adjusted accordingly.

Intertemporal efficiency would be promoted by allowing banking of permits (allowing a source to use fewer permits in one year and more in another), and by the development of futures and options markets. A bankable permit system would address some equity issues between developed and developing countries, including the concern among developing countries that delays in mitigation now by the industrialized economies would leave a greater burden for the future.

1.3.5.2 Exchange/risk efficiency

When different parties to an agreement hold widely differing views about risk and the probability of loss, significant efficiency gains can result from transferring risk among them. In Section 1.3.1.2 we discussed the importance of establishing an international insurance market for those facing the threat of losses under global warming and noted the advantages of establishing a market within which countries that are less concerned about these risks can assume more of the insurance burden. Any efficient international system for addressing the problems of global climate change must include both mitigation and insurance obligations. Governments that believe they have a comparative advantage in assuming climate risks can assume a larger share of those risks, trading off other obligations and substantially reducing the overall costs of responding to climate change.

1.3.5.3 *Comprehensiveness*

Efficiency also requires that the cost of reducing all greenhouse emissions be minimized. This principle implies that any mitigation programme must include not only all greenhouse gases (taking into consideration their heat-trapping potentials and atmospheric lifetimes[45]) but also carbon sinks.

Finally, and perhaps most controversially, it implies that mitigation strategies should focus on all elements of the Kaya identity, ensuring that the marginal cost of reductions is the same for each of the possible strategies. Thus, population control may be an element in a long-term mitigation strategy, no less than a shift in the composition of production or an increase in the energy efficiency of the economy.

1.3.6 *General equilibrium*

General equilibrium theory, an important element of economic research over the last century, demonstrates the advantage of looking beyond first-stage effects. It offers two important insights for climate change analysis. First, the various parts of an economic system are interrelated; perturbations to one part have ramifications for other parts, which may be quite distant. Second, when all the reverberations are taken into account, the net effect of an action may be markedly different from the initial (and intended) effect.

One implication of general equilibrium theory has already been noted: Taxes imposed on one part of the global economy may have little if any effect on global emissions; they may simply result in a relocation of economic activity. Increasingly, the world's economic system must be viewed from a global perspective. Location of economic activities is determined primarily by relative factor prices, taking into account certain specific locational advantages and specialized competencies. If, for example, the OECD countries impose carbon taxes on energy-intensive industries, those industries may relocate outside the OECD. Further, if greenhouse mitigation puts an economic drag on the developed countries, developing countries would be affected through trade.

If different countries have different obligations to reduce greenhouse emissions, different implicit tax rates will result. This will interfere with world economic efficiency – decreasing world real output – possibly with little effect on total greenhouse gas emissions. For example, the most energy-intensive activities – such as aluminum production – may well relocate to developing countries.[46]

Whereas many of the energy-economy-carbon models described in subsequent chapters attempt to estimate the magnitude of such carbon "leaks" (where carbon-intensive production moves to areas in which it is least regulated), most are based on standard international trade models, in which the location of production of various goods and machines is fixed (Whalley and Wigle, 1991).[47] Thus, estimates of carbon leakages are based only on commodity substitution.[48] But in the very long run, which is the time span of interest for an analysis of global warming, leakages may well be higher, owing to the relocation of industries. No consensus now exists on the magnitude of long-run leakages.

1.3.6.1 *Intertemporal substitution*

General equilibrium issues also arise when production can be shifted from one period to another. For example, in a partial equilibrium analysis, a tax on gas or oil raises the price of the fuel taxed, thereby reducing its consumption, and thus associated emissions. But over the long run an exhaustible natural resource like gas or oil has, by definition, an inelastic supply (ignoring for the moment extraction costs, which, in the case of gas and oil, are small relative to the price). The general theory of incidence argues that when a commodity is in inelastic supply, a tax affects the price, but not the level of consumption. That is, *when all countries impose a tax on gas or oil, the producer price of oil falls, by an amount just equal to the tax.* The full incidence falls on producers; as a first approximation, *the level of consumption – and thus the level of emissions – remains unchanged.* Obviously, if the tax is large enough, the price will fall below the cost of extraction for some output, and supply will be reduced. Moreover, in the case of exhaustible natural resources, taxes may affect the timing of consumption of the oil and gas (Stiglitz, Dasgupta, and Heal, 1980).[49]

1.4 Equity

Who will be allowed to increase their greenhouse emissions and who will pay for greenhouse gas abatement and adaptation are among the most contentious issues in climate change. Equity issues such as these have immediate implications for policy as well, because the initial allocation of emission rights and emission constraints will largely determine the distribution of costs. The Framework Convention on Climate Change explicitly directs the parties to consider the problem of equity:

> The Parties should protect the climate system for the benefit of present and future generations of humankind, on the basis of equity and in accordance with their common but differentiated responsibilities and respective capabilities. Accordingly, the developed country Parties should take the lead in combating climate change and the adverse effects thereof. . . .
>
> [Account must be taken of] the differences in their starting points and approaches, economic structures and resource bases, the need to maintain strong and sustainable economic growth, available technologies and other individual circumstances, as well as the need for equitable and appropriate contributions by each of these Parties to the global effort regarding [the Convention's] objective.[50]

1.4.1 *General issues*

Within the limits of cost-benefit analysis, equity arises because of the principle of compensation, discussed in Section 1.3. For example, suppose it could be shown that a business-as-usual path produced higher total benefits than a path with lower greenhouse emissions. We could not conclude from this that the world as a whole would be better off. Indeed, if the losers are not compensated, and their loss is counted more heavily than the winners' gain, the world as a whole would be worse off. Or suppose the costs of warming fall predominantly on one group or one generation, while the benefits accrue to another. (For example, some have speculated that the costs of damages from warming would fall largely on developing countries, without a compensating increase in benefits (Parikh, 1994).) Unless the gainers actually compensate the losers for their losses, cost-benefit analysis cannot conclude that the change has, on balance, been good for society. Compensation is particularly difficult if future generations bear most of the costs, because no "fund for future greenhouse victims" exists (Birdsall and Steer, 1993). Thus, some have argued that, in the absence of mechanisms to make these transfers, we should not rely on possible future transfers from gainers to losers but should instead insist that the gainers pay the costs up front. An alternative explanation addresses these equity concerns by assigning different weights, perhaps based on economic status, to changes in consumption of different individuals (Atkinson, 1970).

No consensus exists among either economists or philosophers about the appropriate ethical responses to the changes that would come with global warming. Should, for instance, owners of resources be compensated for the losses they incur as a result of mitigation actions and the consequent change in prices or values? Economists have often argued no, for several reasons. First, some wealth is not a reward for productive activity, but merely an accident. It is not because country X did something that $200 billion worth of minerals or oil was discovered to lie beneath its territory. One position is that these random allocations of wealth have actually contributed to world inequality, and that eliminating a part of these windfall gains would, from the perspective of an egalitarian social welfare function, be welfare-increasing. Another view denies that government policies would be taking value from assets that by right should be there; until now, according to this view, these resources have simply been underpriced, not fully reflecting the social costs imposed by their use. In this view, actions to discourage overuse would simply rectify a previous mistake.[51]

Whether investors or countries should be compensated for the adverse effects on their market values remains controversial. There is, however, reasonable consensus on three general principles: First, workers in adversely affected sectors may need assistance to switch occupations (the market failure here is that workers cannot purchase insurance against these kinds of adverse shocks). Second, gradual transitions may significantly lower the absolute cost of the transition. For instance, workers leave jobs through natural attrition; if those leaving are not replaced, the industry will be scaled down, with no transition cost to any individual worker. Third, the magnitudes of the uncompensated redistributions associated with any change in policy are often correlated with the magnitude of the political opposition.

Most policy changes produce winners as well as losers. If the relative price of natural gas increases, owners of natural gas deposits may actually be better off. In economies with progressive taxation of either capital gains or consumption, some part of those gains is implicitly shared more broadly.

Virtually all policies discussed below also have different effects on different groups. Residents of very hot and very cold climates consume more energy for heating and cooling, and thus would be worse off, relative to those in more moderate climates. In some countries, city dwellers can choose less energy-intensive modes of transport than those in the countryside.

Many of these impacts will be reflected in capital costs. Thus, the value of land is likely to rise in temperate climates and to fall in extreme climates. (Similar points can be raised, of course, about the costs imposed by climate change itself.) This capitalization effect has both favourable and unfavourable implications. In the long run, residents of colder climates are likely to consume less heating fuel, but perhaps at a higher price, which would leave them on balance about where they started. Energy prices are likely to rise and land rents to fall, in an almost offsetting way. On the other hand, current owners of land would bear the full brunt of the present discounted value of all future increases in taxes (or the tax-equivalent cost of regulations designed to reduce energy utilization). As a result, unless policy changes are introduced gradually, dramatic changes in land values may occur, with possibly large effects on financial institutions and the economy as a whole. Anticipation of these policies would partially offset these effects.[52]

National security. Climate change itself may affect the national security of many countries. At the same time, policies to reduce greenhouse emissions may affect the export earnings and therefore the national security of the energy exporting countries.

Although countries may be willing (or forced) to accept changes in national wealth as a result of changes in world prices induced by the response to the threat of global warming, countries are less likely to be willing or able to accept what they may perceive as implied threats to national security over which they have some control. Thus, a country with a large endowment of coal becomes more vulnerable if it comes to rely on imported oil or gas – supplies of which could be cut off in time of war. These countries may feel it imprudent to switch, even if private economic gains were to be had; and they are particularly unlikely to switch if the benefits take the form of an international public good.

Increasing world political stability would clearly address these concerns. But even were that successful, it would not suffice. To increase national security, policies will need to focus on increasing energy efficiency and reducing energy demand. This is an example of the necessity of allowing sufficient flexibility in the design of an international structure for greenhouse gas emissions that the particular circumstances of each country can be appropriately taken into account.

Benchmarks. The earlier discussion of equitable distribution of the burdens of responding to greenhouse gases, employing generally accepted principles of public finance, avoided the concept of benchmarks, that is, setting target emission reductions in relation to past emissions. These have played an important role in international negotiations. Indeed, the only quantitative target in the FCCC (Article 4.2a and b) requires developed country parties to aim to return their emissions of CO_2 and other greenhouse gases not controlled by the Montreal Protocol to 1990 levels by the year 2000. Although this target suggests equitable treatment, were it to be accepted as legally binding on developed countries without further qualification, it would in fact result in unequal burdens, as it fails to take into account relative incomes and therefore imposes unequal tax rates.

But this criterion can also be criticized as inequitable in a more fundamental sense, since it pays no attention to past efforts at achieving energy efficiency, or to other circumstances that might affect the implied tax rate. For instance, a country that during the preceding ten years had made every effort to increase energy efficiency and switch consumption to less energy-intensive commodities would face the burden of reducing its emissions still further. Because the marginal cost curve for emission reductions rises steeply beyond a certain point, the implied tax rate would be considerably higher than for a country that had previously encouraged overconsumption of energy, for example, by energy subsidies. For the second country, achieving the emission targets might only require elimination of the energy subsidies, a policy with an implied negative tax rate. For a similar reason, countries with large endowments of hydroelectric power may find it relatively difficult to meet an emission target of this form.

Public finance theory has focused extensively on "second-best" policies, recognizing the difficulty of achieving first-best objectives of either economic efficiency or distributive justice. In this context, the central issue is whether alternatives to benchmarking exist. When the U.S. government recently issued tradable emission permits for sulphur dioxide, it took account of emission reductions already achieved. Benchmarking reflects information that would not be reflected in a simple criterion such as a particular emission level per unit population or GDP. Thus, a strong case can be made for including benchmarking, if not in the final allocation of permits (obligations), at least in the transition rules.

1.4.2 Intergenerational equity

Efforts to control greenhouse emissions will provide benefits primarily for our grandchildren and their descendants. We face a difficult task in estimating and judging what aspects of climate and environment they will value and how best to preserve those aspects for them. If we take aggressive action to limit climate change, they may regret that we did not use the funds instead to push ahead development in Africa, to better protect the species against the next retrovirus, or to dispose of nuclear materials safely. Chapter 4 addresses directly the most important issue in intergenerational equity: choice of an appropriate discount rate.

A similar argument applies to actions with differential impacts on different regions. If greenhouse warming turns out to be a major threat to developing countries and if the developed countries fail to reduce emissions aggressively now, the developing countries may suffer additional damage later. Alter-

natively, if the developed countries choose to embark on an aggressive control regime now, and if this cuts into their growth rates, the result will shrink export markets for developing countries and thus reduce growth there. In addition, if developed countries view their greenhouse efforts as, in effect, aid to developing economies, they may cut back on other programmes (sanitation, water, education for women, etc.) that have a more immediate impact on life expectancy, health, and well-being.

1.4.3 Within-country equity

Most discussions of equity and climate change have so far focussed on developed and developing country issues or on issues between one country and another, but issues of equity within a country are also important, and indeed play a central role in the political debates about appropriate responses to climate change. Most policy recommendations involve large within-country losses for certain groups. For instance, any policy leading to less use of coal and lower producer prices for it will lead to large losses for coal mine owners and workers.[53]

The net efficiency gains (in reduced emissions) relative to the distributive effects may differ markedly across resources. Thus, if the price elasticity of world oil supply is small, a tax on oil will be reflected in the prices received by producers and have little effect on the cumulative consumption of oil, though it may result in some short-run substitution against coal. Policies aimed at bringing closer the date of arrival of substitutes for fossil fuels could lead to an increase in current emissions, though long-run effects on atmospheric concentrations would be positive.

1.5 Economics of Policy Actions

Earlier sections set forth a basic framework for analyzing policies related to global climate change, including a combination of mitigation, adaptation, and possibly climate engineering. Striking the appropriate balance requires taking into account the costs, benefits, and risks associated with each strategy. For instance, setting aside risk, governments should reduce emissions to the point at which a dollar of extra spending would yield a dollar of expected savings from preventing damages imposed by climate change or would save an extra dollar of expected costs of adaptation. Adding risk and sequential decision making complicates the analysis but leaves the basic principles unchanged. Because of the lasting impact of climate change and the magnitude of the resulting economic uncertainties, most policy analysis has focussed on a narrower set of questions:

(1) What actions would improve economic efficiency (including the social costs of implementing the policy) *and* reduce net greenhouse gas emissions? How much could emissions be reduced by these means?

(2) Beyond these zero-cost options, what are the least-cost methods of reducing greenhouse gas emissions? What

do the cost curves look like?[54] What are the alternative policy measures, and how do they compare?

(3) What are the essential ingredients of an adaptation strategy, and to what extent will market forces, on their own, provide the appropriate adaptive responses?

1.5.1 Zero-cost options

A variety of inefficiencies in the energy sector – many of them government-induced – would, if eliminated, increase economic efficiency and reduce greenhouse gas emissions at the same time. How large is the reservoir of conservation opportunities? Proponents of the two major approaches to the question have debated this point for more than a decade. *Top-down models* extrapolate observed behaviour into the future. *Bottom-up models* combine cost estimates derived from engineering analyses with economic models of individual choice. Top-down models generally show significant costs to reducing greenhouse emissions in the future.[55] Bottom-up, or technology-specified models, have been used to show the existence of significant reductions in the cost of energy as new low-emission technologies are adopted. Some proponents of bottom-up models argue that emission reductions can be achieved at essentially no cost.[56]

Much of the disagreement turns on empirical estimates. Economists have catalogued the unintended consequences of government regulation. Many have also identified important market failures that could give rise to inefficiencies within the private sector itself. The next two sections will examine each of these effects.

1.5.2 Policy reform

A variety of government reforms could enhance energy efficiency, including removing energy subsidies, reforming or clarifying property rights, reducing nongreenhouse gas externalities, and administrative reforms.

1.5.2.1 Removing energy subsidies

Energy subsidies induce inefficient energy use, reducing the total output of the economy as well as increasing CO_2 emissions.[57] Shah and Larsen (1991) estimated world energy subsidies in 1990 to have been $230 billion. They calculated that their elimination would reduce global carbon emissions by $9\frac{1}{2}\%$ in addition to improving allocative efficiency and thereby generating a welfare gain in subsidizing countries. Burniaux *et al.* (1992) obtained similar results using the GREEN model, concluding that the elimination of all existing distortions on energy markets would yield an increase in world real income of 0.7% per year in addition to cutting world emissions by 18% in 2050 (Dean, 1994). Agricultural subsidies also distort the outcome, especially by affecting the size of forests.

1.5.2.2 Property rights reform

One responsibility of governments is to define property rights and enforce contracts. Ill-defined property rights encourage overconsumption of resources. A clearer definition of property rights could be particularly important in helping to decrease deforestation, for example, while improving economic efficiency. Uncertainties about *future* property rights may also contribute to economic inefficiency. Thus, for example, in those developing countries where large forests are owned by a few large landowners, excessive deforestation may result from the landowners' fear that their tenure will be limited.

1.5.2.3 Administrative reforms

Defining property rights and eliminating energy subsidies are two important actions governments can take to reduce greenhouse emissions. At the same time, many less sweeping reforms can improve economic efficiency and simultaneously reduce greenhouse gas emissions. For example:

Pricing of government-produced electricity. Many governments price electricity not at the market price but at the cost of production. Economists generally recommend that electricity, like any good, be priced not at its cost of production but at the competitive price. In countries with a mix of plants, this means that electricity from all sources should be priced the same – at the highest marginal cost of production.[58]

Land use and other regulation. Changes in land use policy can also reduce energy consumption (especially for transportation, space heating, and air conditioning) and thus greenhouse gas emissions.

Full utilization of nonfossil fuel energy sources (taking account of other environmental impacts). When hydroelectric power generation, which does not increase greenhouse emissions, can be cost-effectively expanded without other environmental effects, it should be done.[59]

1.5.2.4 Regulating nongreenhouse externalities

Many activities producing greenhouse emissions also generate pollution of other types. For example, fossil fuel combustion releases conventional air pollutants; rush hour auto use contributes to road congestion. In the presence of these spillover effects or externalities, market solutions will not properly reflect the externalities generated, leading to the overconsumption of environmental resources. Energy taxes, congestion pricing, or tradable permits can correct these market signals, resulting in lower emissions of both greenhouse gases and other pollutants. Some reforms, such as congestion pricing, also reduce the need for roads and other physical capacity.

1.5.2.5 Special problems of economies in transition

The economies in transition provide special opportunities for mitigating greenhouse emissions. In the former Soviet bloc, high energy subsidies and other price distortions affected energy usage directly, as well as indirectly through the composition of output (i.e., a bias towards heavy industrial production). Spotty environmental regulation meant that Eastern Bloc nations lacked the environmental controls common in the OECD. The capital shortage of the past decade has contributed to the problem through a general deterioration of physical capital stock.

Although these problems are largely of governments' making, the remedy is likely to rely on a combination of public and private actions: effective environmental regulation, elimination of government-caused price distortions, and an economic environment in which foreign and domestic investment can enhance the efficiency (including energy efficiency) of the economy. For example, many analysts believe that cutting methane leakage from gas pipelines will yield both high economic benefits and cost-effective reductions in greenhouse emissions (IPCC, 1990a).

1.5.2.6 Examples of policies that affect efficiency

Policies that cause individuals not to take into account the full social costs of their actions often result in greater energy use and greenhouse emissions. The National Action Plans of many countries have revealed examples of such policies and have also suggested remedies,[60] including:

Unit pricing of waste disposal to encourage recycling. The *life-cycle* social cost of consuming a good includes its costs of production plus disposal. Most consumers, and many businesses, pay a flat fee for trash disposal; with a flat fee, the marginal cost of throwing away an extra pound of trash is zero. By moving from flat fees to unit pricing, the actual price consumers pay to buy and dispose of a good will more closely match its full life-cycle social cost.

Pay-at-the-pump automobile insurance. In most countries, drivers pay automobile insurance yearly or monthly. Once the premium is paid, the marginal insurance cost of driving an extra mile is zero, even though driving more does increase the chance of being in an accident. As an alternative, drivers could be required to pay a portion of their insurance bill at refuelling. With pay-at-the-pump insurance, a tax would be levied on gasoline and earmarked to pay for insurance premiums. This would raise the cost of gasoline at the pump, but lower auto insurance premiums.[61]

Eliminating subsidies for auto travel. Many countries subsidize auto travel in various ways. In some industrialized countries, employers may provide parking to employees at no cost or lower-than-market cost, thus lowering the relative price of commuting by car relative to public transportation. A distortion arises when governments tax income spent on public transportation but not income implicit in the parking subsidy.

Eliminating subsidies that increase housing size. In some industrialized countries, home mortgage interest is tax-deductible. In all but a few countries, the implicit income on owner-occupied housing is not taxed. These tax provisions encourage individuals to consume more housing space than they otherwise would. In cold or hot climates, where more housing space requires more energy for heating and cooling, this tax treatment increases CO_2 emissions.

Eliminating subsidies for trucking. Studies suggest that virtually all road damage is caused by heavy trucks, which pay only a portion of the expense of building and maintaining the road system. Many countries thus subsidize trucking compared with rail or barge transport, probably increasing greenhouse emissions.[62]

1.5.3 Market failures and government responses

Policies exist that would increase economic efficiency at the same time that they reduce greenhouse emissions. For exam-

ple, in some countries, fuel prices do not reflect the full social cost of fuel burning. Taxes can correct this market failure. There is less agreement about whether, *given market prices,* firms fail to take advantage of all the energy efficiency opportunities available to them. This controversy underlies the bottom-up versus top-down controversy treated at greater length in Chapter 8. Engineers have identified a host of seemingly profitable actions that would also save energy. Many economists, however, view this as evidence that the engineering analysis has omitted characteristics important to consumers.

The substantial differences in practices both within and between countries suggest scope for significantly increasing energy efficiency. Moreover, even best practices within a country may not put it at the technological frontier. In deciding whether to adopt a new production process, businesses look only at the private costs and benefits. Many technologists, however, conclude that, even considering private costs only, firms should be undertaking many energy efficiency improvements. This section attempts to reconcile the different schools of thought by reference to information-based market imperfections as well as the criteria by which businesses make decisions.

Information dissemination. Acquiring information is costly. Moreover, providing and disseminating information has many features of a public good (Stiglitz, 1988). In the absence of government intervention, there will be too little production and dissemination of information. This is particularly true for information with widely dispersed impacts, as opposed to information about, for example, the production of certain chemicals, which is primarily of value to a few companies.[63]

Moreover, both theory and evidence support the view that markets, on their own, do not provide an efficient level of disclosure of information (see, *inter alia,* Stiglitz, 1975a; Grossman and Stiglitz, 1981). Indeed, some evidence indicates that markets may try to obfuscate relevant information.[64] This provides the rationale for government provision of information, or laws that in many countries require disclosure of interest rates and other consumer-relevant information, including appliance energy consumption.

Bureaucratic structure and limited scope of attention. In recent years, economic and organizational theory[65] has emphasized that large organizations are not, in general, run by owners; that the managers, even with the best-designed incentives, do not in general maximize the firm's market value; and that among the principal scarce factors within an organization are time and attention. How managers direct their attention has much to do with what the firm does.[66] The information services and disclosure requirements noted above, as well as a number of other government programmes focusing on energy efficiency in consumer products, electric lights, and motors, help focus management attention on energy efficiency. The marginal managerial time required to make efficient energy decisions may be small, and focussing attention on this issue – when information is being freely provided through government and other sources – may thus yield *private* returns well beyond these slight marginal costs.

Returns to scale and system effects (network externalities). Some technologies might be economically attractive at a large scale of production but not on the much smaller scale on which they might initially be adopted. Other technologies exhibit dynamic scale economies: Unit cost falls over time as a function of the cumulative output of firms or industries. Technology "networks" may also affect diffusion rates. For example, cars and trucks powered by electricity, natural gas, methanol, or other alternative energy sources, require a re-fuelling infrastructure, which itself competes for resources with the conventional fuel infrastructure already in place.[67]

Building codes can be justified both in terms of these effects on network externalities and in terms of information failures. Consumers often have limited information concerning the construction of their houses, and obtaining the information after the house is completed is often difficult. Even were they to be provided with construction details, they would have difficulty interpreting the implications.

Capital market imperfections. A major explanation of the difference between best-practice and actual-practice technology is that bottom-up models often compute cost-effectiveness using a discount rate substantially lower than the cost of capital calculated by firms.[68] Studies of implicit discount rates consistently show that households and firms use discount rates substantially above the market rate for long-term government bonds. Two explanations have been offered:

(1) *Risk:* Interest rates facing firms and households reflect the risk premium that lenders require to compensate them for the probability of default. Firms often use discount rates that include a risk premium to reflect the riskiness of projects.

(2) *Capital constraints*: Individuals and firms often face rationing in capital markets, both for credit and equity. Recent research has provided a rationale for this rationing based on the fact that information is imperfect and costly.[69]

These capital market problems have one important implication: Models analyzing best-practice, cost-effective technologies using discount rates lower than those typically employed by firms will overestimate the rate of dissemination of these technologies and underestimate the perceived costs (to the firms and households adopting these technologies) of mitigation strategies.

But these capital market problems raise three other questions: (1) Are firms rational in using such high discount rates? (2) Does the use of such high discount rates imply a market failure? (3) If so, will government intervention improve on the market outcome?

Economists emphasize that an analysis of the costs and benefits of a project must separate four issues: timing, risk, capital constraints, and information. Discount rates are only to be used for timing. Risk should be treated by converting costs and benefits into certainty equivalents, then discounting costs and benefits for each year at the relevant discount rate.[70] Higher risks should not result in higher discount rates.[71] Similarly, capital constraints should be reflected in the shadow price of capital, not in the discount rate.[72] Because of limited

information (and a version of the "winners' curse"[73]) firms often require threshold rates of return significantly greater than the market rate of interest. In doing so, they may confuse time and information risk; that is, the rules of thumb firms use to evaluate investments may sometimes lead to market inefficiencies, including some perhaps in the area of energy-efficient technologies.

Even were firms to follow the economists' guidelines, in the presence of capital constraints, market outcomes would not, in general, be socially efficient (i.e., they are not constrained to be Pareto optimal). There may be significant discrepancies between social and private returns on investment (even apart from the externalities associated with greenhouse gases or technological diffusion). This provides part of the rationale for possible government interventions in capital markets.[74] Though these capital market imperfections imply that there is no presumption that market allocations are efficient, there is no consensus that they lead to significant underinvestment in energy-efficient technologies in particular.[75]

1.5.3.1 Revising national accounts

Some have suggested revising the conventional systems of national accounts to incorporate full social pricing of resources. An early contribution suggested a new measure of economic welfare based on consumption that increases quality of life (Nordhaus and Tobin, 1973). These authors and others recognized that national income accounting, widely adopted after World War II, measures aggregate income and expenditure flows but does not incorporate environmental costs and benefits.

Many researchers have noted deficiencies in standard national income accounts. First, national income accounts do not, in general, provide an adequate measure of welfare; second, they do not provide the correct information for making policies relevant to sustainable development. Sustainable development is concerned with society's resources; an economy is growing when its resource base (capital stock combined with natural resources) is growing. GDP does not, and is not intended to be, a measure of resource availability. Firms have two sets of accounts – cash flow (income) statements and balance sheet statements. GDP is a statement of the former type.

Standard accounting procedures require that firms, in an attempt to present an accurate account of "true income," take account of depreciation. GDP measures gross output; it does not take into account depreciation, either of natural or physical capital stocks. The reason is simply that it is hard to get accurate measures of depreciation. Net national product, however, does consider depreciation, the change in capital stock. And it is this account that should be most subject to criticism, since it accounts for changes in the physical capital stock but not in other capital assets, in particular, environmental assets and natural resources.

Conventional national income accounting does not fully report three categories of resource expenditures: (a) defensive expenditures, either for pollution prevention before the fact or for cleanup after the fact (although these expenditures are not separately reported, they are counted in GDP); (b) consumption of environmental goods (such as exhaustible resources);[76]

and (c) conflicting uses of environmental services (such as the atmosphere, used by producers as an input into production and by households as a consumption good).

One proposal would include in GDP the effect of changes in quality of the environment. In Eastern Europe and the former Soviet Union, steady increases in reported postwar GDP masked the effects of decades of environmental degradation; for part of that period, environment-adjusted GDP almost certainly declined.[77]

However, important conceptual problems in defining levels and changes of environmental assets, complicate the task of modifying national accounts. First, the stock of natural resources has no obvious definition. Although most geologists would agree on the size of coal stocks – their location is known and their *in situ* value can be estimated – this cannot be said for oil or minerals.[78] Second, environmental assets, such as air quality, present another set of problems, because no market prices exist to value the asset.

Four approaches are commonly used to calculate changes in the natural environment (Peskin and Lutz, 1990):

(1) The environmental expenditure approach, used recently in the United States, which subtracts pollution abatement expenditures from GDP;

(2) The physical accounting approach, used in Norway and France, which establishes satellite accounts using physical units of measurement to account for flows and stocks of resources;

(3) The depreciation approach, which adjusts gross and net product by subtracting the value of natural resource depletion (Repetto, 1989; El Serafy, 1989); and

(4) The comprehensive approach, which uses both physical measures and value (United Nations Statistical Office 1992).

Another measure of broad-based welfare, although it does not include environmental amenities, is the UN's human development index or HDI (UNDP, 1992). The HDI gives a composite measure of human development by combining three key indicators: longevity (measured by life expectancy at birth), education (measured by adult literacy and mean years of schooling), and income (real GDP per capita adjusted for purchasing power). Although the HDI is not directly related to global environmental issues, both global warming and abatement policies may affect it.[79]

1.5.4 Innovation

Standard competitive analysis argues that, *given all required information and technology,* market economies produce efficient outcomes. But recent economic analyses have shown that, in general, market economies need not result in the efficient allocation of resources either to information production and dissemination or to innovation. The first of these issues was discussed earlier. The second is more complex.

In the absence of intellectual property rights, firms would have less incentive to innovate. With standard patent terms, firms are not able to appropriate all the returns from their innovative activity. Setting the optimal patent life involves bal-

ancing off the inefficiencies resulting from the exercise of monopoly power during the duration of the patent (static inefficiencies) with the increased incentives for innovation.[80] Largely because innovators seldom appropriate all the returns from their innovations, there is a general consensus that markets provide insufficient incentives for research and development, and the greater the unrewarded spillovers, the greater the undersupply of innovation.[81] The fact that spillovers are likely to be greater at more basic levels of research suggests a role for government in subsidizing basic and near-basic research. In the same way, the high cost of establishing intellectual property rights impedes the transfer of technology to developing countries.

Still, there is a general consensus among economists that the patent system provides a better basis for financing applied research than do government grants, largely because of the difficulties government has in picking those innovations most likely to produce high returns. Consequently, it should be asked if market failure because of insufficient innovation is more likely than market failure for other reasons. In other words, are there any special grounds for arguing for government research and development subsidies, provided the government has corrected energy prices to reflect the externalities generated? Obviously, in the absence of such corrections, market incentives to provide energy-saving innovations will be distorted, just as market incentives to adopt energy-saving technologies are reduced.[82] (Tradable permits have effects similar to those of corrective taxes: They encourage firms to place a higher value on new technologies that reduce emissions because reductions will require them to purchase fewer permits.)

Innovation is important, because it provides perhaps the best opportunity for low-cost methods of reducing emissions. Several studies have confirmed the impact of accelerated deployment of advanced energy technologies on the future rate and timing of anthropogenic climate change.[83]

1.5.5 Carbon taxes and tradable permits

Economic efficiency requires all agents in the economy to pay the full marginal social costs of their actions. But firms and households are not charged for the additional warming potential they add to the atmosphere, and so do not pay the full social costs they impose. Two economic instruments can correct this market failure: carbon taxes and tradable permits. (Note on terminology: More detailed treatments, as in Chapter 11, often reserve the term "emission permits" for domestic instruments and "emission quotas" for international instruments. This section treats the issue of tradable emission rights in a general way, most often using the term "permits"; whether the meaning is domestic or international should be clear from the context.)

Implementing either one of these instruments could raise issues of national sovereignty, because to be effective either carbon taxes or permits would require the creation of institutions with the authority to allocate, administer, and enforce agreements. Although these issues lie outside the immediate concern of economics, they are relevant to any discussion of practical implementation.

A tradable permit scheme involves a determination of the total level of permits and a distribution of the initial allocation, with emission levels for any firm limited to the number of permits held. The initial distribution may be made by an auction or allocation according to benchmarks (e.g., per capita emissions as of a given date), or by historical emission levels ("grandfathering"). Alternatively, emission rights could be grandfathered in at current levels and gradually shifted over to a per capita allocation as of a given date.

Once permits are distributed among the regulated entities, a market is set up, allowing companies to buy and sell permits according to their plants' planned emissions. The cost of production then includes not only the costs of conventional inputs, but also the costs of additional permits to offset additional emissions. Plants whose cost of mitigation is low will find it relatively easier to abate pollution rather than to buy permits. Plants with higher costs of mitigation will have a greater preference for buying permits than for abating pollution. The price of the permits, which are artificially created scarce resources, is determined by the market. With the use of tradable permits, companies have an incentive to improve the efficiency of their production and thereby reduce their emission levels, as they can sell excess permits on the market and generate revenue.[84]

Although permits thus create a marginal cost of production related to the marginal emissions, carbon taxes impose a tax directly on the marginal emissions. Both systems thus force producers and households to face the true social costs of their actions.[85] In principle, either could be adjusted to achieve the level of emissions desired, although adjustments of this sort may be difficult in practice.

The initial allocation of permits will largely determine the distribution of costs of abatement (Chapter 3 discusses these issues of equity at greater length) and at the same time influence the growth path of participants' economies. For example, an allocation based on population at a given date would provide an incentive for population control.[86]

Imposing carbon taxes can have large distributive consequences. A system of grants can largely offset these distributive consequences, but such offsetting grants might well not be made. Providing tradable permits equal to existing levels of emissions seemingly makes no firm or household a loser. But granting permits in that way effectively represents a grant of money (such permits have monetary value) in a way that may not accord well with standard ethical principles. For instance, by embarking on an ambitious programme to reduce emissions, a firm may qualify for *fewer* permits than it would otherwise. Not only does this violate ordinary notions of fairness, but anticipation of granting permits in this way would, accordingly, have strong adverse effects on emissions.[87]

Although presenting a political impediment to its introduction, the fact that a tax has large distributive consequences is not necessarily an argument against it. Some argue that those who failed to pay the full social costs of their actions earlier are not therefore entitled to special allotments now.

Once it is recognized that the distribution of permits across countries will inevitably be decided by some principle other than current levels of emissions, then it becomes clear that both taxes and tradable permits will have distributive consequences. An agreement among countries to impose uniform corrective taxes, with each country retaining its own revenue, would have few consequences for redistribution between countries, and the burden of the tax would, as noted earlier, likely be progressive.

Governments in the developed countries might decide to use some of the revenues so generated to encourage activities that benefit less developed countries (such as research and development directed at technology appropriate for developing economies) or to provide other forms of assistance. Decisions about such uses could be made bilaterally or collectively. In contrast, decisions about how tradable permits would be allocated across countries would have to be made multilaterally. Arriving at a formula for distributing these property rights may be far more difficult than arriving at a tax rate and a procedure for its revision, as any such formula may entail substantial redistribution.

1.5.5.1 *A double dividend?*
Many measures that reduce emissions of greenhouse gases also yield other environmental or economic benefits, such as reductions in emissions of other pollutants and energy savings. These are called secondary benefits or double dividends. Some analysts argue that using carbon tax revenues to reduce existing distortionary taxes is an economic double dividend.

Revenues from carbon taxes may allow a reduction in distortionary taxes elsewhere in the economy. If the (compensated) elasticity of demand of labour is relatively high, and the revenues from the carbon tax are used to reduce taxes on labour income, then there would be a double dividend from the carbon tax in the reduced deadweight loss from the labour tax, which would otherwise be significant.

At least three objections have been raised to this idea. First, conceptually, rationalizing the tax system by reducing the most distortionary taxes is certainly a worthy goal but is not equivalent to imposing a carbon tax. Distortions can be reduced without a carbon tax, and a carbon tax could be imposed without reducing the existing distortions. Second, empirically, if the (compensated) labour supply elasticity is relatively low, then the deadweight loss from the labour tax is low, and the commensurate welfare gain is reduced. Third, politically, if carbon tax revenues are used to offset the existing deficit rather than to reduce taxes on labour and capital, then the carbon tax acts more like an ordinary tax increase, increasing distortions from taxation to pay for budget items with a lower return than the extra burden imposed. The gains to total welfare (reductions in deadweight loss) depend on the welfare losses associated with these other distortionary taxes, as well as the cross-elasticities of demand between carbon and other taxed commodities.[88]

Even though carbon taxes may have a positive effect on economic welfare, they can at the same time have a negative effect on *measured* economic growth, since those measures typically do not include the value of environmental degradation. Researchers differ on the size of the loss. The wide spread in the numerical results, however, should not obscure agreement among researchers on a number of important points. All models used in the major comparison studies to date have projected, first, that intervention would be required to achieve the emission targets; second, that the size of the required tax increases with the stringency of the carbon limit; and third, that the size of the appropriate carbon tax varies over time, even for the same emission or concentration target.[89]

1.5.5.2 Energy taxes

Energy taxes as a means of controlling greenhouse emissions must be viewed as "second-best" taxes, in that they do not directly tax the externality, greenhouse emissions. Whereas carbon taxes directly penalize the externality-generating activity, less targeted alternatives, such as energy taxes, may be politically more acceptable. Carbon taxes reduce emissions, first directly, by moving up the demand curve; second indirectly, by encouraging consumers to switch to less carbon-intensive energy sources. On the other hand, energy taxes work through the first path by reducing total energy consumption. But to the extent that certain kinds of energy, like hydroelectric, have, at least in the short run, a relatively inelastic supply, there will be a major impact on oil, gas, and coal; and to the extent that oil and gas supplies are best described by a model of an exhaustible natural resource, with relatively low extraction costs, most of the supply reduction will occur in coal. Thus, indirectly, there will be a considerable amount of switching, through the indirect effects.

1.5.5.3 Tradable permit markets

In order for systems of emission permits to achieve reductions in emissions efficiently, there needs to be a market for emissions *across international boundaries*. There is some debate about the role of government or international organizations in establishing a market for such emissions. Some believe that there are private incentives for the establishment of markets; others contend that government can play a key market facilitation role through establishing centralized clearinghouses for information or even providing for permit banking (storage) or brokerage (trading) to facilitate trades between private parties. These services would prove especially useful in the more complex international context.

1.5.5.4 Combining taxes with tradable permits

Although carbon taxes and tradable permits are typically presented as alternatives, policymakers may prefer to combine them. The major disadvantage is the additional administrative cost. The advantage is more subtle: The market value of tradable permits is reduced as taxes increase. With an optimal carbon tax, and with a tradable permit supply set equal to the optimally chosen level of emissions, the price of a permit should be zero. More generally, the greater the tax, the less the value of a permit (for a fixed supply of permits), and thus, the less the distributive consequences of alternative rules for allocating the initial endowments of permits. Another possible combination would utilize permits for large sources and a tax (set to equal the permit price) on small sources.

1.5.5.5 Intertemporal patterns of taxation

If the target is the long-run atmospheric concentrations of greenhouse gases, then climate change damages will be approximately the same for emissions in any particular year, although the optimal carbon tax must be adjusted for differences in costs, discounting, and risk.[90] The focus on concentrations also implies that early reductions are more valuable than later reductions.

For exhaustible natural resources such as oil, economic efficiency requires that those deposits with the lowest cost of extraction be extracted first. Hotelling (1931) argued that competitive equilibrium implies that rents (price minus costs of extraction) must rise at the rate of interest. The price of the backstop technology (an energy source assumed to be available in unlimited quantities at a certain price after a certain date, such as electricity from solar photovoltaic cells) determines the set of resources to be ultimately exploited, namely, all resources for which the cost of extraction is less than the price implied by the backstop technology's price. Thus, it is the tax on oil or gas at the date of switching to the backstop technology that determines the ultimate amount of oil and gas that will be extracted, and thus the total burden of CO_2 placed on the atmosphere by oil and gas. If that were the only matter of concern, one could simply announce a commitment to impose such a tax sometime in the future when relevant backstop technologies become available and competitive. That announcement would, if believed, have an immediate effect on current prices.

1.5.6 Regulatory approaches

Regulation of greenhouse emissions may take many forms, including fuel restrictions, technology standards, and various economic incentives. Chapter 11 discusses these options in detail. Economists have long argued for the use of economic incentives for environmental management, although governments have so far relied on traditional regulations almost exclusively, as traditional approaches have been more acceptable to the public and industry.

Proponents of the traditional approaches often claim that these approaches "force" technology (i.e., stimulate technological innovation or refinement), with less redistribution than forcing technology through taxes. Thus, if automobile makers are required to attain a certain mileage standard, they will meet the standard; on the other hand, gasoline taxes might have to rise significantly to reduce fuel consumption by the same amount. Evidence for the claim of technology forcing, however, is equivocal. In several cases in which industry failed to meet the applicable standards, regulators withdrew the standard in the face of unacceptably high economic costs. The apparent advantage of technology forcing – one large instrument of coercion instead of the subtle and continuous incentives provided by market forces – is often in fact a disadvantage.

There are other disadvantages to the traditional approach as well. First, traditional regulations do not in general result in economic efficiency, since those in one sector face implicit or explicit incentives at the margin that differ from those in other sectors. Second, traditional regulations fail to account for off-setting private responses that may neutralize the regulation's intended effects and even cause environmental harm. Third, traditional regulations provide no incentives for exceeding the given target, even when doing so might result in little additional cost.[91]

Traditional regulations that focus on inputs and technology rather than outputs have the further disadvantage of not directing research toward meeting performance objectives at least cost. For instance, when stack gas scrubbers are required, research will be directed at producing scrubbers at least cost, rather than reducing emissions at least cost. Hence, a dynamic inefficiency is added to the obvious static inefficiencies. Finally, because of the nature of the regulatory process, traditional regulatory designs are more likely to be captured by special interest groups.[92]

It should be kept in mind that government policies to address external effects (market imperfections) are likely to generate their own external effects, in part because of the difficulty in identifying the gainers and losers from changes in policy. Thus, the likely cost of "market failure" must be compared to the likely costs imposed by governments' attempts to remedy the problem.

1.6 Sustainable Development

The concept of sustainable development was formulated about 1980 as a response to the apparent conflict between environmental concerns and the need for economic growth, especially in developing countries. At the time, preserving biodiversity and maintaining environmental quality seemed incompatible with a five- or tenfold increase in world output, as would be necessary if per capita incomes of the developing countries were eventually to approach those enjoyed by the developed countries now. The sustainable development debate rekindled interest in the question of resource scarcity, originally addressed in the economics literature by Malthus (1798) and revived in the policy arena with the publication of *The Limits to Growth* (Meadows and Meadows, 1972). Recently, the field of "ecological economics" has extended this approach (Howarth and Norgaard, 1992).

A variety of definitions of sustainable development have been proposed. The Brundtland Commission offered this interpretation (World Commission on Environment and Development, 1987):

> Sustainable development is development that meets the needs of the present without compromising the ability of future generations to meet their own needs.

Although the Commission clearly had in mind environmental considerations, its report did not spell out exactly what sustainable development included.

1.6.1 *The economic concept of sustainable development*

Although sustainable development began as an ethical principle, it is at the same time an economic concept, focussing on two issues: (1) intertemporal equity and (2) capital accumulation and substitutability.

Intertemporal equity. Robert Solow's definition of sustainable development (Solow, 1992), which focusses on intertemporal equity, has enjoyed wide currency among economists. Sustainable development, he argues, requires that future generations be able to be at least as well off as current generations. The central implication is that any environmental degradation should be offset by increases in capital stock sufficient to ensure future generations at least the same standard of living. Sustainable development does not preclude the use of exhaustible natural resources but requires that any use be appropriately offset.

In practice, sustainability as defined by Solow provides few constraints on growth paths for the developed countries, so long as steady increases in productivity continue. Technical change alone, without further capital accumulation, may well sustain future living standards and offset any effects of environmental degradation. To see this with a numerical example, note that even if estimates of adaptation costs are taken to be 1-3% of GDP should significant warming occur, and if even moderate rates of technical progress of 1-1.5% per annum continue to occur, then future generations 45 to 70 years from now will have twice the income of the current generation. Even with no discounting, it would be hard on this account alone to justify the sacrifice of further consumption by this generation in order to enhance the standard of living of the future generation.

Capital accumulation and substitutability. To what extent can technology, skills, and capital equipment substitute for a decline in exhaustible resource stocks or a decline in per capita environmental amenities? Solow's definition, in common with much economic theory to date, implicitly assumes that substitutes exist or could be found for all resources. Pearce (1989, 1991) argued that if substitution possibilities are high, as most evidence from economic history indicates, then no single resource is indispensable, and intertemporal equity stands as the only crucial issue. If, on the other hand, human and natural capital are complements or only partial substitutes for each other (e.g., if, because of the irreversibility of extinction,[93] capital accumulation is only a partial substitute for biodiversity), then different classes of assets must be treated differently, and some assets are to be preserved at all costs.

Pearce *et al.* (1994) distinguished between strong and weak sustainability. Weak sustainability requires that any depletion of natural capital be offset by increases in human-produced capital – the Solow criterion – or by the substitution of other forms of natural capital, such as renewable assets in place of nonrenewable assets. Strong sustainability requires that some natural capital, being irreplaceable, must be preserved.[94] It has been argued that there are no close substitutes for the atmosphere and the climate it produces, implying no substitution possibilities and hence the need to preserve the atmosphere.

1.6.2 Implications of sustainable development for developing countries

In many developing countries, Solow's definition would not be viewed as acceptable, since it seems to place no weight on their aspirations for growth and development. Developing countries have also implicitly criticized the debate over substitutability for the same reason: If some natural assets must be preserved at any cost, then there may be no trade-off with development. Tariq Osman Hyder of Pakistan, a leading spokesman for the G-77 group of developing nations, has emphasized the importance of economic growth in achieving sustainable development:

> None of these linked [development] issues can be resolved unless and until there is broad-based development in the South. Only such broad-based development can provide the foundation of international security. The Northern approach is to attack the symptoms, with a residual emphasis on poverty eradication. But the international community must insist on addressing the underlying causes for concern. Development, environmental protection, peace, and security are indivisible. (Hyder, 1992)

Similarly, the G-77 and China emphasized the need for economic growth in the following statement on sustainable development and the environment introduced during the INC-2 negotiations in 1991:

> Protection of the global climate against human-induced change should proceed in an integrated manner with economic development in light of the specific conditions of each country, without prejudice to the socioeconomic development of developing countries. Measures to guard against climate change should be integrated into national development programmes, taking into account that environmental standards valid for developed countries may have inappropriate and unwarranted social and economic costs in developing countries. (Hyder, 1992)

Endnotes

1. Some analysts believe that the justification for costly and more restrictive actions rests on intangible costs (Nordhaus, 1993). Intangible refers to the difficulty of measuring; intangible costs are related to such factors as migration, comfort, health, leisure activities, urban infrastructure, and air pollution (Fankhauser, 1994; Cline, 1992). A warmer climate would improve human comfort in cold areas, and in the winter generally, while decreasing comfort in warm areas. Mearns *et al.* (1984) calculate a threefold increase in heat waves for a 1.7°C. rise in U.S. mean temperature. It is not yet clear whether net comfort averaged over the globe will rise or fall for a given rise in temperature. Chapter 6 covers these issues in more detail.

2. Both the natural rate of species loss and the human contribution to the process are difficult to estimate (U.S. EPA, 1989). Predicting the effect of climate change on species distribution is more difficult still. The magnitude and even the sign of these intangibles remains in dispute, for uncertainty about the duration and types of environmental changes that would be caused by climate change makes the long-term projection of species change highly complex. Population pressures have added to pressure on ecosystems, particularly in the Third World. Climate change may exacerbate these damages, particularly in Africa, where environmental degradation has been particularly pronounced during the last fifteen years (UN, 1989).

3. For greater warming – such as might be experienced over the next two centuries, or which could occur sooner if modest near-term warming triggers the release of large amounts of carbon from the biosphere and oceans to the atmosphere – the probability of large losses increases. These losses refer to global losses, which represent an aggregation of individual losses. Impacts will differ across individuals, groups, and regions. For some, the probability of large losses will be high over the next century, despite the expectation that aggregate global losses will be small.

4. Recent ice core data from Greenland point to the occurrence of earlier temperature rises of several degrees within a few decades (IPCC, 1995). Reasons for these sudden changes are still not understood, but might have come from changes in deep ocean currents. The other side of this debate holds that on the whole the biosphere is homeostatic or self-correcting. This "Gaia hypothesis" compares the biosphere to a living being: Once moved away from equilibrium, self-correcting forces naturally move it back to equilibrium (Lovelock, 1979). The two hypotheses are not necessarily inconsistent. Within a range of variation, homeostatic properties could dominate, even if stability were not guaranteed outside that range.

5. A warmer climate might also increase climate variability, though climatologists cannot say with assurance whether climate will become more or less variable daily and seasonally. The normal variability of existing climate also makes it difficult to detect any warming that might be occurring. The "signal-to-noise" problem makes it possible for observers to mistakenly consider a normal extreme event as evidence of a trend or to fail to see a trend in a noisy data series. Because of the signal-to-noise problem, the scientific community is unable to indicate confidently, for example, whether the extremely warm years of the 1980s are evidence of climate change or not (IPCC, 1992; Solow, 1990).

6. Countries make century-long choices implicitly, for example, when they choose population policies, policies affecting long-term capital formation and productivity growth, or policies to protect environmental assets.

7. Bargaining theory has contributed some basic principles, however, such as the importance of threat points (i.e., the outcome in the absence of an agreement).

8. Although the global warming potential (GWP) measure, endorsed by the IPCC, is useful in formulating comprehensive approaches to greenhouse mitigation policies (Stewart and Wiener, 1990), some analysts have recently criticized it on the grounds that GWP implicitly makes the opportunity costs of an increment in radiative forcing equal for all periods in the future (Schmalensee, 1993). If all greenhouse gases had the same rate of decay, then this problem would not arise.

9. Aerosols have very short lifetimes, so their effects are far more regional than those from longer-lived greenhouse gases. It may not be possible to develop an analogue of global warming potential for aerosols.

10. Forests cannot be expanded indefinitely, however. Thus, increased carbon sequestration is not a permanent solution to increasing greenhouse gas emissions.

11. Advances in cost-benefit analysis have allowed the introduction of risk and equity issues in a systematic way.

12. Knight's often-quoted distinction (Knight, 1921) separates risk, for which the probabilities of different outcomes are known, from uncertainty, in which either the probabilities are unknown or some potential outcomes are not specified.

13. This transfer of risk is made easier by dividing the risk into small parts, so that any individual faces only a small risk and thus requires only a small risk premium in compensation.

14. This conclusion assumes decreasing absolute risk aversion (about which there is a general consensus). Even if developing countries were adversely affected, so long as the adverse economic effects were proportionately smaller, then, assuming decreasing relative risk aversion (about which there is less consensus), the developed countries would be in a position to insure the undeveloped.

Insurance contracts may also be created because of differences in judgments concerning the probability of the insured event occurring. This suggests the potential of an important principle to be invoked in future international negotiations or agreements: Countries that believe that the risks of climate change are low, and are therefore seemingly unwilling to take strong actions to mitigate these risks, ought to be willing to provide insurance against climate change at low cost, since it has, from their perspective, an actuarially low value (Chichilnisky and Heal, 1993), although enforcement issues may complicate the problem.

Insurance markets, if appropriately designed, have one further advantage: They encourage actions that diminish the potential size of any loss, because insurance firms have an interest in minimizing the cost of losses. Some of the losses associated with climate change can be easily avoided or reduced (e.g., by making ocean-front houses more durable to reduce vulnerability to storm damage). In designing insurance for climate change risks, either the insurance should be based on exogenous events (e.g., not on the dollar losses incurred, but on the rise in sea level), or the insurance companies should be given broad discretion to require the insured to undertake actions to mitigate losses.

15. Chapter 6 provides estimates of expected regional damage.

16. In the absence of such requirements, moral hazard problems arise. It may be desirable to focus government intervention not on the primary insurance market but on the reinsurance market. See recent U.S. government analyses of failures in insurance markets for natural disasters and the design of appropriate responses to these.

17. Computer modellers participating in the Energy Modeling Forum (EMF) examined an "accelerated R&D" scenario in which the cost of nonelectric backstop falls from $100 to $50 per barrel of oil equivalent, and the cost of the electric backstop falls from 75 to 50 mills per kWh. The four models used were remarkably consistent in their estimates of economy-wide costs, reporting GDP losses falling by 65% for the 20% emission reduction scenario (EMF, 1993).

18. Weitzman *et al.* (1981), cited in Lind (1994), make these points in formulating a sequential decision strategy for developing synthetic fuels.

19. Manne and Richels (1992), Nordhaus (1993), and Peck and Teisberg (1993) all report a high value for better scientific information on climate change, including the cost and timing of new supply and conservation strategies.

20. Richels and Edmonds (1993) provide a demonstration of this proposition; they calculate relatively low costs for stabilizing CO_2 concentrations if flexibility in timing is allowed, compared to capping and stabilizing emissions to achieve the same atmospheric concentration.

21. Alternatively, the Kaya identity may be written

$$
\begin{array}{cccc}
\text{Growth rate of} & = \text{growth rate} & - \text{ decline in energy} & - \text{ emissions} \\
CO_2 \text{ emissions} & \text{of output} & \text{per unit output} & \text{per unit of} \\
 & & & \text{energy use}
\end{array}
$$

That is, CO_2 emissions will not rise as long as output grows no faster than the combined decline in energy intensity per unit of production and CO_2 emissions per unit of energy use. This formulation applies most usefully to the developed countries.

22. Chapter 8, Estimating the Costs of Mitigating Greenhouse Gases, treats the important issue of inertia and technology.

23. Note that energy-efficient development paths for developing countries have been proposed (Goldemberg *et al.,* 1988).

24. That is, for conventional exhaustible resources, there is a stock, S. Welfare depends on flows out of the stock each year:

$$ U(S_1 - S_o, S_2 - S_1, \ldots\ldots S_{t+1} - S_t, \ldots\ldots) $$

where U is utility or welfare, and S_t is the stock at the end of period t.

In the case of climate change, welfare depends *directly* only on the stock of carbon in the atmosphere, though indirectly also on emissions, through the consumption of goods.

$$ U(S_o, C_1(S_1 - S_o), S_1, C2(S_2 - S_1), \ldots\ldots S_t, C_t(S_{t+1} - S_t), \ldots\ldots) $$

25. Even when the flow exceeds the long-run sustainable level, it will not be optimal to reduce the flow instantaneously, unless there are zero costs of adjustment. For the atmosphere, a sustainable stock of greenhouse gases means stable concentrations. Current emissions are estimated to be at about twice the level consistent with stable concentrations.

26. Postponing action may lead to some irreversible damages, such as the flooding of low-lying states.

27. As noted, this is a simplification. Recent literature in resource economics often treats fossil fuels as depletable rather than exhaustible; that is, price affects incentives for exploration, and some marginal wells would not be drilled if oil prices fell too low. It is also a simplification to treat coal reserves as inexhaustible. With the more realistic assumptions of a steeply rising supply curve for oil and gas reserves (over the next century or two) and a rather flat supply curve for coal, the results quoted still hold.

28. Energy economic models often use the assumption that alternative sources of energy will become available at some future date from nonfossil (noncarbon-emitting) sources. The means to supply this energy is often called *backstop technology*. The modeller must specify the quantity, cost, and date of availability of the backstop technology.

29. For analyses of market and optimal responses to uncertainty about the arrival of backstop technologies, see Dasgupta, Gilbert, and Stiglitz (1979) and Dasgupta and Stiglitz (1981).

30. For counterexamples to the received wisdom, see Coase (1960).

31. Formally, if A measures the quality of the atmosphere, then each individual's or country's welfare, U^j, is a function of its own consumption, C^j, and the shared public good, A: $U^j(C^j, A)$. This does not mean that value changes in A are the same for all individuals and countries; that is $(\partial U^j/\partial C^j)/(\partial U^j/\partial A)$ may differ in magnitude, and even in sign.

32. Although there may also be trade-offs. Reductions in gases that contribute most to local pollution may sometimes be achieved at the expense of increased emissions of greenhouse gases.

33. This kind of optimization problem was first studied by Edgeworth (1881). The importance of incentive effects for the analysis of distributional issues was first emphasized by Mirrlees (1971). There are, obviously, important incentive effects: If the less developed countries were able to classify any expenditure that had some effect on mitigation as a mitigation expenditure, with the cost borne by the developed countries, they would have an incentive to undertake excess expenditures of this type. The GEF (Global Environmental Facility) directly addresses this issue by providing funds only for incremental costs, that is, those costs that go beyond what would have been the efficient level of expenditures if the public good benefits of greenhouse gas mitigation were ignored.

34. The importance of the assumption that bargaining or transaction costs are absent (and perfect information is present) has only gradually come to be recognized. See Stiglitz (1988) for an elementary textbook treatment.

35. For instance, it makes no difference whether smokers or nonsmokers are given the property rights to air. Rather, it is whether smokers value smoking more or less than nonsmokers value clean air that will determine whether smoking occurs. How property rights are assigned *does* make an important difference for the distribution of welfare. Coase's conclusion that outcomes are unrelated to the initial assignment of property rights obviously ignores potentially important income effects.

A slight extension of this perspective says that social scientists should simply *describe* the outcome of the bargaining process by which property rights are assigned. Beginning with the important work of Nash (1955), a variety of bargaining theories has been developed, most of which emphasize the importance of "threat points" – the outcomes which arise in the absence of a bargaining agreement – to the determination of the eventual outcome. In this case, the fact that the net losses of many developed countries may be limited relative to those of many of the less developed countries suggests a bargaining solution in which much more of the costs of mitigation are borne by the less developed countries than under the "social welfare function" allocations described earlier.

36. In the case of small taxes, these are "compensated" taxes and have no welfare effect, though they have a substitution effect, and therefore do reduce pollution.

37. The loss in welfare (ignoring the benefits from reduced greenhouse gas warming) are the Harberger triangles, and can thus be shown to be proportional to the product of the elasticity of demand for energy and the share of energy in national output. Since poorer countries are likely to have less access to alternatives that increase the elasticity of demand, and since the share of energy is larger in richer countries, the burden of the tax is progressive.

38. The *polluter-pays principle* endorsed by the OECD is exclusively prospective.

39. Many have challenged the ethical basis for assigning responsibility based on past damages. Using either an egalitarian social welfare function approach or a Rawlsian "behind the veil of ignorance" analysis (Rawls 1971) leads to the rejection of the polluter-pays principle. Since at the time the relevant actions are taken, the polluter is not cognizant of the effects, such fees have no incentive effects, but rather appear as random taxes, lowering each person's expected utility, and in particular the expected utility of the worst-off individual.

There is a further ethical issue: It is generally difficult to ascertain who actually benefits from escaping the obligation of paying for the pollution. It need not be the individual, firm, or country actually engaging in the pollution-generating activity. In competitive markets, when firms are not charged the full social costs of production, product prices will fall, giving consumers a substantial fraction of the benefits.

40. Manne and Richels (1992) show that, under the IPCC emission scenarios, even the most drastic controls on emissions from developed countries would be insufficient to stabilize greenhouse gas concentrations without some means of controlling emissions from developing countries.

41. Public goods exist when property rights are not or cannot be clearly assigned. The atmosphere is an international public good because assigning property rights to the atmosphere is difficult for one nation acting alone, and particularly difficult when many sovereign states must agree among themselves. Tradable greenhouse emission permits, discussed below, attempt to resolve the problem by explicitly assigning property rights to greenhouse emissions.

42. Coase's discussion of externalities (1960), emphasized the separability of efficiency and equity issues. Though there have been several important qualifications to Coase's conjecture, emphasizing the importance of public goods, imperfect information, and transaction costs, the basic insight still remains applicable here.

43. Chapter 7 discusses this issue at greater length.

44. These concerns are not just theoretical possibilities, as the following two examples illustrate. Assume that the developed countries impose high energy taxes, but the developing countries fail to do so. Energy-intensive industries, such as aluminum, migrate from the developed to the developing countries. But energy efficiency in the developing countries is much less than in the industrialized countries, so the total energy used to produce a tonne of aluminum could increase substantially. Although economic efficiency would call for locating energy intensive industries where energy efficiency is greatest, a system of partial controls would result in energy-intensive industries being located where energy efficiency is lowest. Similarly, the reduced energy consumption by the developed countries will result in lower *producer* prices of oil and gas, leading to increased consumption of energy in the developing world, partially offsetting any energy conservation induced in the industrialized countries.

45. The precise manner in which this should be done is a technical matter, treated in the literature on Global Warming Potential (see, e.g., IPCC, 1990a, 1995). To the extent that there are large differences in atmospheric lifetimes, then the relative weighting of different greenhouse gases should change over time, since the "shadow price" associated with effects on relative concentrations at different dates will differ.

46. Similarly, if the developed countries restrict forest cutting, the price of lumber may rise, inducing the developing countries to cut down more of their own trees. Thus, total global carbon sequestration may not increase. Furthermore, if hardwood forests in the less developed countries are the least desirable ones to cut down from an ecological or economic perspective, as some researchers have concluded, then environmental and economic efficiency will decrease if these forests are exploited more intensively (Edmonds and Reilly, 1983).

47. Only a few models take into account international capital flows. Thus, most models do not address issues of industry relocation (McKibben and Wilcoxen, 1992). It is also doubtful whether computed general equilibrium models can realistically represent capital relocation from one country to another. Chapter 11 provides a more complete discussion of leakages.

48. In the case of production of highly substitutable commodities, carbon leakage will be much greater.

49. Whether taxes, in fact, have this effect depends in part on the shape of the demand curves. With intertemporal separability in demand curves, constant elasticity, and no backstop technology, a constant ad valorem tax has no effect on the pattern of consumption.

Coal presents markedly different issues, not so much because of its greater emissions per unit energy, but because of its higher cost of extraction-to-price ratio. Lowering producer prices may result in less coal being consumed, provided alternative energy sources become available. Thus, taxes on coal are likely to have significant general equilibrium as well as partial equilibrium effects; the increase in the price of coal will lead to a substitution of gas and oil. If alternative energy sources are not available, such policies will only affect the intertemporal timing of coal consumption (given the much more limited resources of gas and oil). But even that might be of some value in reducing long-run greenhouse gas emissions, as the ability to extract energy from coal may increase significantly over time. Analyzing the optimal intertemporal structure of taxes to minimize long-run ambient levels of greenhouse gases, taking into account both intertemporal substitution and substitution across energy sources, is a complicated technical issue that to date has not been adequately analyzed.

50. United Nations Conference on Environment and Development, Framework Convention on Climate Change, May 9, 1992.

51. The most difficult problem is posed by investors who invested in these resources under a previous regime (where these resources were not taxed). Do they have any special claim to compensation for a "change in regime." Changes in demands and supplies occur for virtually all resources and are an inevitable part of the risks in investing. Most economists would argue that arbitrary and capricious changes in policies contribute to business uncertainty, and therefore have an adverse effect on economic growth, but reasoned changes in policies in response to changes in information are an inevitable part of business risk.

52. For example, some electric utilities in the U.S. are already making decisions in anticipation of some future policies to limit greenhouse emissions.

53. These issues also arise among countries. Countries with large coal deposits will find the value of their natural wealth eroded and, quite naturally, will be less enthusiastic about international agreements that reduce coal use or lower producer prices.

54. Studies show variation in GDP losses across models. For example, it is estimated that stabilizing emissions at their 1990 levels would reduce U.S. GDP by 0.2% to 0.8% in the year 2010 – roughly

a $20 billion to $80 billion loss for that year. Estimates of the costs of reducing emissions by 20% below 1990 levels in the year 2010 range from 0.9% to 1.7% of GDP. Aggregated models (top-down) have generally reported higher costs, whereas disaggregated models (bottom-up) have shown lower costs. Chapter 9 contains a more complete discussion.

These GDP losses occur when carbon taxes lead to investments that are more expensive than those that would take place in the absence of the taxes. The higher the carbon taxes, the greater the investment in price-induced conservation and the greater the extent of fuel switching toward less carbon-intensive substitutes.

The overall impact of a carbon tax will depend not only on the size of the tax but also on the uses to which the revenues are put. In the standard Energy Modeling Forum (EMF) scenarios, it was assumed that tax revenues would be redistributed in a neutral manner (i.e., without affecting the marginal tax rates). There are, of course, numerous ways in which tax revenues can be used. These include reducing budget deficits; reducing marginal rates of income, payroll, corporate, or other taxes; granting tax incentives to preferred activities; or increasing the level of government expenditures. The costs of the tax will vary widely depending on how the revenues are recycled.

55. Top-down models estimate that for developing countries there exist low-cost options for reducing emissions in the near term, but eventually costs would exceed 1% to 2% of GDP (EMF, 1993). For economies in transition, because of historical inefficiencies and energy subsidies, there exist large opportunities to reduce emissions at little or no cost. For developing countries, problems of informal economies make hard estimates difficult, but the cost of stabilizing emissions would likely be large enough to cut into economic growth.

56. Recent comparisons indicate that the most important differences between top-down and bottom-up models arise from differences in input parameters rather than from differences in model structure.

57. Government institutions and regulations often hinder the efficient use of energy. Developing countries are least able to absorb the costs of these inefficiencies. Thus, although some developing countries argue that they cannot afford to reduce greenhouse emissions, or cannot eliminate the subsidies to poor people implicit in below-market energy prices, the same countries often have the most to gain from reforming government-caused inefficiencies. At least in the short run, international agreements committing countries to eliminate at least the most egregious of these practices might go a long way to addressing the problem of emission reductions.

58. This may also be a problem with electricity generated by the private sector, as regulation has historically set price equal to average cost, rather than allowing it to match the competitive price. In many countries, the increase of competitive pressures has moved electricity prices closer to the marginal cost of production.

59. Further examples include:

Eliminating regulations impeding efficient energy utilization. Many, perhaps most, countries have a host of regulations that increase energy use as they impede economic efficiency. For instance, the U.S. has had a policy of restricting oil exports from Alaska. Whatever the merits of that policy, it has forced Japan to import oil from Indonesia and Saudi Arabia. World oil transportation costs have thus been greatly increased at the expense of the American economy. Another example of government reform, included in the U.S. Action Plan (Clinton and Gore, 1993), encourages efforts to expand and improve natural gas markets through continued regulatory reform. These reform efforts include guidelines to allow greater natural gas use in the summer in coal- and oil-fired power plants.

Other regulation. Unintended effects of many tax, expenditure, and other policies have contributed further to inefficiencies in land use. Among the unfortunate effects of the U.S. Superfund programme for the management of hazardous wastes, for example, has been the creation of large unoccupied holes in the centres of major cities.

60. In actual practice, some of these proposals would likely present practical obstacles, including the cost of implementation.

61. Pay-at-the-pump insurance also has several potential drawbacks. For example, most such proposals fail to link the factors that most influence insurance rates (driver history, vehicle location, vehicle repair costs) with actual insurance payments.

62. Consider the following thought experiment: Compare an optimally designed road system that only carries cars with an optimally designed road system that also carries trucks. The incremental cost of carrying trucks is, in most countries, much larger than the proportionate share of the cost they bear in fuel taxes and other fees.

63. For example, in many countries, governments have taken an active role in the dissemination of information to the agriculture sector. These programmes are in some measure responsible for the large increase in agricultural productivity in countries with agricultural extension services.

64. This is because those who would be at a competitive disadvantage under "true" disclosure have an incentive to add "noise" and because there are strong market forces for product differentiation. In markets with homogeneous commodities, profits will be driven to zero (in Bertrand competition), even with a limited number of suppliers. For a discussion of these and related issues, see Salop (1977), Salop and Stiglitz (1977, 1982), and Stiglitz (1988).

65. The standard reference in the organizational literature is March and Simon (1958). Economic theories emphasizing the nonvalue maximizing behaviour of managers include those of Baumol (1959) and Marris (1964). The principal agent literature (Ross, 1973; Stiglitz, 1974) provided the informational microfoundations for understanding the divergences of interests. See Stiglitz (1988). A more recent overview is provided by Stiglitz (1991) and the symposium in the *Journal of Economic Perspectives,* Spring 1991.

66. The facts that time is a scarce commodity and that decision making in large organizations is decentralized do not in themselves constitute a market failure; they do not prove that resources are not efficiently allocated, given the real constraints facing society, which include time. However, Greenwald and Stiglitz (1984, 1988) have established a very general theorem showing that when information is imperfect and costly, market equilibrium is, in general, not Pareto efficient. Thus, there is no presumption concerning the efficiency of the market economy, even in the absence of the kinds of externality and public goods problems that are associated with greenhouse gases. For a more extended discussion, see Stiglitz (1994).

Recent advances in the economics of information have provided sounder microfoundations for these theories of the firm. And indeed, the importance of the limitations on the availability of information, and the consequent importance of attention-directing efforts, applies to individuals as well as to organizations. Some studies have suggested that the limited success of the special tax provisions in the U.S. designed to encourage savings (IRA accounts) was primarily

due to the competitive efforts of banks to recruit these accounts and the attention that savings got as a result.

67. Network externalities are manifested in other ways. Builders fail to install energy-efficient light bulbs, because customers dislike them and stores do not carry replacements; and stores do not carry them because the demand for them is too low.

When there are important network externalities, market equilibria are frequently inefficient. The economy might, for instance, get "stuck" in the wrong equilibrium. Government action can, in these instances, "force" the economy to move from one equilibrium to another.

68. This is not the only explanation of differences between bottom-up and top-down models. There are several other features of market behaviour that bottom-up models often ignore.

(a) *Hidden costs:* Consumers value a range of attributes difficult to include in an engineering model. For example, auto buyers value not only initial costs and fuel economy (which computer models can easily calculate) but also performance, safety, and durability, which they typically do not.

(b) *Divergence between laboratory and in-use performance:* Especially for new technologies, actual energy use often differs significantly from energy use calculated in the laboratory. It is the latter on which purchasers focus.

(c) *Variation across individual consumers:* Engineering models generally assume an average consumer, but actual consumers may display a wide range of characteristics and usage patterns. Except when demand functions are linear in the relevant variables, the consumption of the "average" individual is not equal to the average consumption, and what is optimal for the average person may not be optimal for a significant fraction of the population.

69. For a survey, see Jaffee and Stiglitz (1990). The basic theory of credit rationing was developed in Stiglitz and Weiss (1981), and the theory of credit rationing is further developed in Greenwald *et al.* (1984) and Myers and Maljuf (1984).

70. This generally accepted methodology is, for instance, reflected in the guidelines issued by the Office of Management and Budget in the U.S. for the evaluation of projects and regulations. The applied literature does not address the question of whether this procedure is appropriate in the presence of certain types of time and risk nonseparabilities.

71. Though if the variance of the net benefits is increasing over time in a particular manner, the differences in the two methodologies may not be large.

72. Again, under certain restrictive conditions, where the shadow value of a capital constraint is changing systematically over time, the differences in the two methodologies may not be great.

73. See Wilson (1977). The "winner's curse" describes the tendency for the winner of an auction to fail to realize a competitive return; that is, in retrospect, to have paid too much. Though the original discussion of winner's curse focussed on bidding in auctions, it has subsequently come to be applied to a range of other market phenomena.

74. For a discussion of the role of the state in capital markets, see Stiglitz (1994).

75. In some industrialized countries, energy-efficient home mortgage lending may help correct the problem. Lenders generally set criteria for the maximum loan amount based on the borrowers' ability to repay, which, in turn, depends on income and wealth. The fact that a particular expenditure would enhance efficiency and reduce utility bills is not given special attention. Energy-efficient mortgages provide funds to households to make energy efficiency–enhancing investments that are intended to pay for themselves by reducing utility bills by an amount equal to or greater than the interest payments. With capital constraints, builders may have an incentive to trade off initial capital costs for higher maintenance costs (lower energy efficiency). Building codes specifying minimal levels of energy efficiency and full disclosure of expected life-cycle energy costs may help address these market distortions.

76. A country that rapidly depletes its natural resources may show a high rate of growth under conventional income accounting, but a lower rate of growth when resource depletion is taken into account. Repetto (1989, 1991) calculated resource-adjusted GDP for several countries that were rapidly harvesting their stocks of hardwoods and other resources, arguing that conventional measures sharply overstated GDP.

77. Daley and Cobb (1989) have even claimed that U.S. per capita GDP, when adjusted for environmental damage, was stagnant between 1950 and 1986. This assertion is hard to reconcile with the steady improvement in most measures of environmental quality since 1970, when measurement standards were established.

78. Analysts now use two methods to estimate stocks. The first assumes a fixed stock of a natural resource such as oil. Consumption of oil then depletes the stock by the amount of consumption. The second begins by treating discovered reserves as the asset. Thus, additions to reserves increase the asset, while consumption reduces it. If in any given year, new discoveries match resource utilization, then according to this method no net depletion has occurred.

79. A number of difficult conceptual problems face the analyst defining levels and changes in levels of these assets.

First, how should the "stock" of natural resources be defined? Coal poses perhaps the easiest situation. The location of coal reserves is known. Costs of extraction are high, so the rents (the value of coal *in situ*) are low. The depletion can be measured not by the coal used times the market price, but the coal used times the *in situ* value. But for oil and other minerals, information about where reserves are located is vital. Two models have been proposed. One sees the world as having a fixed stock of natural resources (say oil). When one uses oil, one is depleting this stock. Thus, to calculate the value of depletion, one does not need to know the entire stock; the flow (the amount of oil consumed) provides an accurate measure of the change in stock.

The alternative model looks at the size of discovered reserves. Reserves are treated as the asset. Additions to reserves thus are viewed as increasing the resource base. If, in any given year, new discoveries match resource utilization, then there is no net depletion. This is the approach being taken by the U.S. Department of Commerce. This accounting framework would be correct if there were an infinite supply of the resource (reflected in zero rents). The essential "capital" good is information about where the resource is located.

Environmental assets – such as air quality – present another set of problems, because there are no market prices to value the asset. Dynamic optimization problems of the kind described earlier can be used to calculate shadow prices. How sensitive these shadow prices are to specific assumptions remains to be investigated.

Accounting systems do not, however, have to aggregate all information together. Just as information about longevity and other indicators of well-being (see below) serve to complement information from national income accounts concerning standards of living, so too can information about *physical* environmental measures be used to complement information from the extended national income accounts.

80. There is some concern that excessively broad and long patents may actually impede innovation. When technological progress occurs by building on previous innovations, later innovators require the permission of earlier innovators to realize the returns on their innovation. Although advocates of broad patent coverage argue that the parties always arrive at efficient bargaining solutions, critics point out that the outcomes of bargaining models with incomplete information often entail large inefficiencies.

81. Matters are more complicated, since the patent does not reward the innovator with his marginal contribution – the increase in the present discounted value of benefits as a result of the innovation occurring earlier than it otherwise would have occurred. For a fuller discussion, see Stiglitz (1994) and Dasgupta and Stiglitz (1980).

82. If less developed countries fail to implement fully a set of corrective taxes or tradable permits, or if less developed countries fail to adopt and effectively enforce intellectual property rights, there will be insufficient incentives to produce energy- and emission-saving innovations, particularly those appropriate for the level of technological knowledge, human capital, and factor prices in those countries. If less developed countries do take these actions, there is concern that they will result in higher prices for innovations, and thus the pace of adoption will be retarded. An effective form of aid, targeted to reducing greenhouse gas emissions, might take the form of subsidies directed at producing appropriate energy-saving and emission-reducing technologies for LDCs.

83. Edmonds *et al.* (1994) have studied the importance of available advanced energy technologies such as those proposed by Johannson, *et al.* (1993). Edmonds *et al.* use the Edmonds-Reilly-Barnes model for energy-related greenhouse gas emissions; the MAGICC model for atmospheric composition, climate response, and sea level rise; the IPCC scenario IS92a (IPCC, 1992) as the reference base case, and five alternative energy scenarios that are far more advanced than today's energy supply and transformation technologies. The five energy scenarios are:

- advanced fossil fuel technologies
- advanced liquefied hydrogen fuel cells
- advanced hydrogen fuel cells without liquefied hydrogen
- low-cost biomass
- accelerated rate of exogenous end-use energy intensity improvement

Combined, the energy technologies reduce annual emissions from fossil fuel use to levels that stabilize atmospheric concentrations below 550 ppmv (i.e., double the concentration prior to the Industrial Revolution). The tax rate used, which was assumed to apply globally, was the marginal cost of stabilizing fossil fuel carbon emissions in the reference case. With values reflected for only carbon dioxide emission reductions, the estimated cost of global emission reductions grew from approximately $35 (U.S.) billion in 2005 to $230 (U.S.) billion per year in the year 2095. With advanced fossil fuels, low-cost solar electric power, low-cost fuel cell vehicles, the present discounted value of adding low-cost biomass fuel to the energy technology bundle is almost half a trillion dollars (U.S.). The present discounted value of the advanced energy technologies taken together is $1.8 trillion (U.S.).

The introduction of advanced biomass energy production technology was found to play a key role in reducing emissions. Biomass energy at $2.00/GJ, growing to become the core energy supply technology by 2050, could significantly reduce emissions. This possibility highlights the potential role of technology development and deployment relative to that of fiscal and regulatory intervention.

These results should be viewed as illustrative rather than predictive. In this analysis, the gains from introduction and deployment of advanced energy technologies depend on the sequence of technologies evaluated in the study.

84. The literature has identified three types of permit systems. The ambient permit system (APS) works on the basis of permits defined according to exposure at the receptor points. Each polluter, then, may face quite complex markets – different permit markets according to different receptor points, and hence different prices. The simpler emission permit system (EPS) issues permits on the basis of source emissions and ignores what effects those emissions have on the receptor points. Within a given region or zone, the polluter would have only one market to deal with and one price. Finally, there is the pollution offset (PO) system, wherein the permits are defined in terms of emissions and trade takes place within a defined zone. However, the standard has to be met at all receptor points. The exchange value of the permits is then determined by the effects of the pollutants at the receptor points. The PO system thus combines characteristics of the EPS and the APS (Pearce and Turner, 1990). These distinctions are of limited relevance for greenhouse gases, where what is of concern is global emission levels. The specific location of the emissions is of no concern.

85. The choice between taxes and tradable permits depends on the objectives of the policymaker and the nature of the uncertainty about the marginal cost and marginal benefit curves for carbon emission reductions (Weitzman, 1974). Theory tells us that if the nature of the curves is known with very little certainty, but the marginal cost curve is known to be relatively steeper (i.e., a change in the level of pollution allowed brings about a greater change in the marginal costs of mitigation than in the marginal benefits) then taxes should be the policy of choice. This is because, in this case, an erroneous estimation of the optimal tax rate will lead to a relatively small deviation from the optimal pollution level. On the other hand, an erroneous estimation of the optimal level of total emissions in a permit scheme will lead to a relatively large deviation from the optimal cost of the permits.

If the marginal benefit curve is known to be relatively steeper than the marginal cost curve, however, tradable permits are the better option. Here, an erroneous estimation of the optimal tax rate will lead to a relatively large deviation from the optimal level of emissions, whereas an erroneous estimation of the optimal level of emissions in a permit strategy will lead to a relatively small deviation from the optimal cost of the permits.

In the case of greenhouse gas emissions, the time horizon for adjustment is sufficiently long that many of these uncertainties become less important. If the tax rate initially chosen yields too high a level of emissions in one year, it can be increased, and the net impact of the erroneous initial estimate on global warming (or the total cost of achieving a given level of atmospheric concentration) will be negli-

gible. In any case, as the earlier discussion of sequential decision making has emphasized, there is likely to be a need for continued revisions in either tax rates or permit levels.

Still, there is some argument that the required adjustments under a permit scheme may be less burdensome (Tietenberg, 1992). For instance, if the authority feels that the old standard needs some tightening it may enter the market itself and buy some of the permits, holding them out of the market.

There must be effective, competitive markets in tradable permits if such schemes are to achieve efficient outcomes. There are transaction costs of running such schemes, just as there are transaction costs associated with collecting tax revenues. Whether transaction costs give one system a decided advantage over the other is not clear. There seems to be no compelling reason to believe that good markets in tradable permits would not develop.

86. It is possible to design allocations of trading permits that (a) on average, impose no net burden on developing countries (thus conforming to the ability to pay principle); (b) provide those economies that are growing faster per capita with commensurately greater permits, thus imposing no net drag on economic growth, provided the economy exhibits an increase in fuel efficiency at least equal to the average of fast-growing, less-developed countries; and (c) rewards those economies that are able to reduce greenhouse gas emissions faster than benchmark rates, either through greater control of population growth, larger increases in energy efficiency, or switching from higher- to lower-carbon fuels.

The extent to which individual circumstances of countries should be taken into account in setting benchmarks remains a question for international negotiations. To the extent that high emissions are due to natural endowments (e.g., the availability of coal rather than natural gas as a source of energy), a persuasive case can be made for benchmarks to reflect initial emission levels. To the extent that high emissions are due to inappropriate energy pricing policies, the case that benchmarks should reflect initial emission levels is far more tenuous.

87. Similarly, some developing countries have asked, should the developed countries be given higher levels of permits, simply because they have, in the past, been the chief source of greenhouse gases?

88. Standard tradable permit schemes essentially take the revenue from a carbon tax and distribute it to current user emitters, rather than using the revenue to reduce other taxes. An alternative to these standard schemes is for the government to auction off the tradable permits.

If taxing carbon leads to reduced labour supply or reduced savings, then government revenues from wage or capital taxes may be reduced, more than offsetting the direct revenue gain from the carbon tax. Cross-elasticities of this magnitude are unlikely, though any such cross-elasticity will reduce the net gain from the carbon tax. The magnitude of the double dividend has been the subject of some dispute, with Goulder (1994) and Repetto *et al.* (1992) taking opposite views.

89. Two main studies provide insights into the root of the variance in estimates of the economic effects of carbon taxes: the Energy Modeling Forum Study-12 and the OECD comparison project. In each case, sophisticated sensitivity analyses were run by standardizing key economic assumptions and using these in conjunction with common reference-case scenarios of reductions. The magnitude of the effect on economic growth in these studies depends both on assumptions concerning the effect of carbon taxes on savings and labour supply, and the induced investment to offset the higher energy prices. If higher energy prices do not lead to much capital substitution, and if the cross-elasticity with savings and labour is low, then the likely effect on economic growth will be small.

The OECD model comparison project was conducted to compare economy-wide estimates of the effects of carbon taxes. Time horizons as well as the key economic assumptions on growth, population, and resource prices, and the reduction scenarios for six global models were standardized. The global models compared were the GREEN model, the IEA model, and four North American models (the Edmonds-Reilly Model (ERM), the Global 2100 model of Manne and Richels (MR), Rutherford's Carbon Rights Trade Model (CRTM), and the Whalley-Wigle model) (Dean, 1994).

Because of differing assumptions about several key considerations, the models showed significant variation in tax rates and costs for the same amount of emission reduction. Several factors explain the differences between model results. The most important factors are:

- the degree of substitution between fuels – the ease with which producers and consumers can switch from high-carbon-content fuels to low-carbon-content fuels

- expectations about future energy prices and taxes

- the speed of emission reduction

- the way in which revenue is recycled

- the treatment of the removal of energy subsidies

- assumptions regarding backstop technology and a host of other technical and economic factors

Because of the varying approaches to these questions, the range of estimated tax rates and costs is quite wide. For a 45% reduction in baseline emissions by 2020, the required tax would be in the range of $150–325 per tonne of carbon and the cost might be in the range of 1.5–2.9% of world GDP. A 70% reduction in baseline emissions by 2050 could require a tax of between $230 and $880 per tonne and a loss in world GDP of 2.4–3.8% (Dean, 1994).

The required carbon taxes and associated costs vary significantly across regions in all of the models. This indicates that the same proportional reductions in emissions across all regions would give rise to very different costs in different regions and would thus be globally inefficient – with great potential for savings in the global cost of reducing emissions through the use of emission trading between countries or regions or a global carbon tax.

Three insights emerging from the OECD study (Dean, 1994) are:

- Small amounts of emission reduction can probably be achieved with low taxes;

- Large reductions can only be achieved at high tax rates (i.e., marginal reduction costs rise with emission reductions);

- Carbon-free backstop technologies are likely to slow the rise of the carbon tax, or halt it altogether, if they are available at constant marginal cost.

The Energy Modeling Forum-12 (Impact of Carbon Emission Control Strategies) examined the cost of reducing CO_2 emissions (EMF, 1993). A diverse group of economic models, employing common assumptions for selected numerical inputs, were used to analyze a standardized set of emission reduction scenarios. In all, fourteen top-down models participated in the study.

The EMF model comparison provides the most comprehensive application of top-down methodologies to date. The study addresses a wide range of policy questions. How large are emissions likely to grow in the absence of controls? How much market intervention will be required to meet alternative targets? What will be the price tag? In exploring economic costs, the modellers were asked to examine the impacts of timing, research and development, and revenue recycling.

The EMF exercise provides a wealth of useful information for policymaking. Although the focus was primarily on the U.S., many of the insights are applicable to developed countries in general.

In selecting parameters for standardization, the EMF study focussed on what were felt to be the most influential determinants of mitigation costs. These included GDP, population, the fossil fuel resource base, and the cost and availability of long-term supply options. In addition, although the EMF models differed considerably in their technology representation, the study attempted to impose uniformity with regard to world oil prices, the oil and gas resource base, and the cost of backstop technologies. For its reference case, EMF adopted the average of the 1990 IPCC high and low economic growth cases (IPCC, 1990c). To be consistent with the IPCC scenarios, the study also adopted the population growth projections of Zachariah and Vu (1988).

The modellers generally used taxes based on the carbon content of the fossil fuels to achieve a prescribed emission reduction. The magnitude of the tax provided a rough estimate of the degree of market intervention that would be required to achieve the carbon emission target. Estimates ranged from $20 to $140 per tonne for the carbon taxes required to hold emissions at 1990 levels in 2010. Estimates of the carbon taxes required to reduce emissions by 20% below 1990 levels in 2010 ranged from $50 to $330 per tonne.

Two parameters are particularly important in explaining the differences in tax projections: the price elasticity of energy demand and the speed with which the capital stock adjusts to higher energy prices. Neither was controlled in the EMF experiments. Those models using lower price elasticities required higher taxes to achieve the same emission goal. Those models that assumed greater malleability of capital required lower taxes.

90. Rapid capital stock retirement may add to the cost of immediate CO_2 reductions. Also, reductions may be cheaper later because of technical changes in the intervening years.

91. Indeed, the "ratchet effect," a commonly observed phenomenon in command and control economies (Stiglitz, 1975b; Weitzman, 1978), leads firms to do no more than just satisfy the target.

92. Two examples from the experience of the United States Clean Air Act: (1) the expensive and evidently counterproductive requirement in the 1977 amendments that coal-burning power plants install stack-gas scrubbers even if they burn low-sulphur Western coal has been attributed to an unusual alliance between environmental activists and Eastern U.S. high-sulphur coal interests; (2) the mandates for ethanol in the motor fuels market are generally attributed to the political power of the corn and ethanol interests.

93. It has been argued that new species and varieties produced by genetic engineering may be able to more than offset the loss in genetic variability from climate impacts, particularly because of the increasing capacity to direct those mutations toward socially desirable objectives.

94. Other views represented in the debate on sustainable development but not generally accepted in the economics profession include those of Daly (1991) and Daly and Cobb (1989), and those of the new field of ecological economics (Costanza, 1991).

References

Arrow, K.J., 1962: The economic implications of learning by doing, *Review of Economic Studies* **29,** 155–173.

Atkinson, A.B., 1970: On the measurement of inequality, *Journal of Economic Theory,* **2,** 244–263.

Atkinson, A.B., and J.E. Stiglitz, 1980: *Lectures on public economics,* McGraw-Hill, New York.

Baumol, W.J., 1959: *Business behavior, value, and growth,* chapters 6–8. Macmillan, New York.

Birdsall, N., and A. Steer, 1993. Act now on global warming – but don't cook the books, *Finance and Development* **30**(1), 6–8.

Bohm, P., 1993: Making carbon-emission quota agreements more efficient: Joint implementation vs. quota-tradability. Paper presented at Institute for Applied Systems Analysis (IIASA), Laxenburg, Austria, October.

Burniaux, J.M., J.P. Martin, and J. Oliviera Martins, 1992: The effects of existing distortions in energy markets on the costs of policies to reduce CO_2 emissions: Evidence from GREEN. OECD Economic Studies no. 19, OECD, Paris.

Chao, Hung-po, 1992: Managing the risk of global climate catastrophe. Electric Power Research Institute, Palo Alto, CA.

Chichilnisky, G., and G. Heal, 1993: Who should abate carbon emissions? An international viewpoint. National Bureau of Economic Research (NBER) working paper no. 4425, Cambridge, MA. August.

Cline, W.R. 1992: *The economics of global warming.* Institute for International Economics, Washington, DC.

Clinton, W., and A. Gore, 1993: *The climate change action plan.* Executive Office of the President, Washington, DC.

Coase, R., 1960: The problem of social cost, *Journal of Law and Economics* **1,** October, 1–44.

Costanza, R., 1991: *Ecological economics: The science and management of sustainability,* Columbia University Press, New York.

Daly, H., 1991: *Steady-state economics,* 2d ed., Island Press, Washington, DC.

Daly, H., and J. Cobb, 1989: *For the common good: Redirecting the economy toward community, the environment, and a sustainable future,* Beacon Press, Boston.

Dasgupta, P., 1982: *The control of resources,* Basil Blackwell, Oxford.

Dasgupta, P., and J.E. Stiglitz, 1980: Industrial structure and the nature of economic activity, *Economic Journal* **90**(358), June, 266–293.

Dasgupta, P., and J.E. Stiglitz, 1981: Resource depletion under technological uncertainty, *Econometrica* **49**(1). January, 85–104.

Dasgupta, P., R. Gilbert, and J.E. Stiglitz, 1979: Energy resources and research and development, in *Erschöpfbare Ressourcen,* S. Brand, ed., Duncker and Humboldt, Berlin.

Dean, A., 1994: The effectiveness of carbon taxes at the international level. In *Climate change: Policy instruments and their implications:* Proceedings of the Tsukuba Workshop of IPCC Working Group III, 17–20 January, 1994, pp. 46–59, Environmental Agency of Japan, Tokyo.

Edgeworth, F.Y., 1881: *Mathematical psychics,* C. Kegan Paul & Co., London.

Edmonds, J., and J. Reilly, 1983: Global energy and CO_2 to the year 2050, *The Energy Journal* **4,** 21–47.

Edmonds, J., M. Wise, and C. MacCracken, 1994: Advanced energy technologies and climate change: An analysis using the global change assessment model (GCAM), Global Environmental Change Program, Pacific Northwest Laboratory. Washington DC. Draft.

El Serafy, S., 1989: The proper calculation of income from depletable natural resources. In Ysuf J. Ahmad, S. El Serafy, and E. Lutz, *Environmental accounting for sustainable development,* World Bank, Washington, DC.

EMF (Energy Modeling Forum), 1993: *EMF-12 global climate change: Impacts of greenhouse gas control strategies,* Stanford University, Palo Alto, CA.

Fankhauser, S., 1994: The social costs of greenhouse gas emissions: An expected value approach. In N. Nakicenovic *et al.* (1994).

Goldemberg, J., T. Johannson, A. Reddy, and R. Williams, 1988: *Energy for a sustainable world.* Wiley Eastern Limited, New York.

Goulder, L., 1994: Environmental taxation and the "double dividend": A reader's guide, Stanford University Department of Economics, Palo Alto, CA. Draft.

Greenwald, B., and J.E. Stiglitz, 1984: Pecuniary and markets externalities: Toward a general theory of the welfare economics of economies with imperfect information and incomplete markets. NBER Working Paper No. 1304. National Bureau of Economic Research, Palo Alto, CA.

Greenwald, B., and J.E. Stiglitz, 1988: Pareto inefficiencies of market economies: Search and efficiency wage models. *American Economic Review* **78**(2), May, 351–355.

Greenwald, B., and J.E. Stiglitz, 1989: Impact of the changing tax environment on investments and productivity, *The Journal of Accounting, Auditing, and Finance* **4**(3), Summer, 281–301.

Greenwald, B., J.E. Stiglitz, and A. Weiss, 1984: Informational imperfections in the capital market and macroeconomic fluctuations, *American Economic Review* **74**(2), May, 194–199.

Grossman, S., and J.E. Stiglitz, 1981: Stockholder unanimity in the making of production and financial decisions, *Quarterly Journal of Economics* **94,** May, 543–566.

Grubb, M., Thierry Chapuis, and Minh Ha Duong, 1995: The economics of climate change: Implications of adaptability and inertia for optimal climate policy. Forthcoming.

Grubb, M., J. Edmonds, P. ten Brink, and M. Morrison, 1993a: The costs of limiting fossil-fuel CO_2 emissions: A survey and analysis, *Annual Review of Energy and the Environment* **18,** 397– 478.

Grubb, M., Minh Ha Duong, and Thierry Chapuis, 1993b: Optimising climate change abatement responses: On inertia and induced technology development, IIASA Workshop on Integrative Assessment of Mitigation, Impacts, and Adaptation to Climate Change, Laxenburg, Austria, October.

Hicks, J.R., 1939: The foundations of welfare economics, *Economic Journal* **49,** December, 696–712.

Hotelling, H., 1931: The economics of exhaustible resources, *Journal of Political Economy,* April, 137–175.

Howarth, R., and R. Norgaard, 1992: Environmental valuation under sustainability. *American Economic Review* **82**(2), May, 473–477.

Hyder, T.O., 1992: Climate negotiations, the North/South perspective. In *Confronting climate change: Risks, implications, and responses,* I. Mintzer, ed., Cambridge University Press, Cambridge.

IPCC (Intergovernmental Panel on Climate Change), 1990a: *Climate change: The IPCC scientific assessment,* J.T. Houghton, G.J. Jenkins, and J.J. Ephraums, eds., Cambridge University Press, Cambridge.

IPCC (Intergovernmental Panel on Climate Change), 1990b: *Climate change: The IPCC impacts assessment.* Report Prepared for IPCC by Working Group I, World Meteorological Organization and United Nations Environment Programme, New York.

IPCC (Intergovernmental Panel on Climate Change), 1990c: *Emissions scenarios prepared by the response strategies working group of the IPCC. Report of the expert group on emissions scenarios.* World Meteorological Organization and United Nations Environment Programme, New York.

IPCC (Intergovernmental Panel on Climate Change), 1992: *The supplemental report to the IPCC scientific assessment,* J.T. Houghton, B.A. Callander, and S.K. Varney, eds., Cambridge University Press, Cambridge.

IPCC (Intergovernmental Panel on Climate Change), 1995: *Climate change 1994: Radiative forcing of climate change and an evaluation of the IPCC IS92 emission scenarios,* J.T. Houghton, L.G. Meira Filho, J. Bruce, Hoesung Lee, B.A. Callander, E. Haites, N. Harris, and K. Maskell, eds., Cambridge University Press, Cambridge.

Jaffee, D., and J.E. Stiglitz, 1990: Credit rationing. In B. Friedman and F. Hahn, eds., *Handbook of monetary economics,* pp. 837–888, Elsevier Science Publishers, Amsterdam.

Johannson, J., Goldemberg, R. Williams, and H. Kelley, 1993: *Renewable energy: Sources for fuels and electricity,* Island Press, Washington.

Kaldor, N., 1939: Welfare propositions of economics and interpersonal comparisons of utility, *Economic Journal* **49,** December, 549–552.

Kaya, Y., 1989: Impact of carbon dioxide emission control on GNP growth: Interpretation of proposed scenarios, Intergovernmental Panel on Climate Change/Response Strategies Working Group, May.

Knight, F., 1921: *Risk, uncertainty, and profit.* Houghton Mifflin, Boston.

Lind, R.C., 1994: Intergenerational equity, discounting, and the role of cost-benefit analysis in evaluating global climate policy. In N. Nakicenovic *et al.* (1994).

Lind, R.C., K.J. Arrow, G.R. Corey, P. Dasgupta, A.K. Sen, T. Stauffer, J.E. Stiglitz, J.A. Stockfisch, and R. Wilson, 1982: *Discounting for Time and Risk in Energy Policy.* Washington, DC: Resources for the Future.

Lovelock, J., 1979: *Gaia: A new look at life on earth,* Oxford University Press, London.

Malthus, T., 1798: *Essay on the principle of population* (reprinted Macmillan, London, 1926).

Manne, A.S., and R.G. Richels, 1992: *Buying greenhouse insurance: The economic costs of CO_2 emission limits,* MIT Press, Cambridge, MA.

March, J., and H. Simon, 1958: *Organizations,* Wiley, New York.

Marris, R., 1964: *The economic theory of managerial capitalism,* Macmillan, New York.

McKibben, W., and P. Wilcoxen, 1992: The global costs of policies to reduce greenhouse gas emissions. Brookings Discussion Papers in International Economics no. 97, Brookings Institution, Washington, DC.

Meadows, D.H., and D.L. Meadows, 1972: *The limits to growth: A report to the club of Rome's project on the predicament of mankind,* Universe Books, New York.

Mearns, L.O., P.W. Katz, and S.H. Schneider, 1984: Extreme-high temperature events: Changes in their probabilites with changes in mean temperature, *Journal of Climate and Applied Meteorology,* **23,** December, 1601–1613.

Mirrlees, J., 1971: An exploration in the theory of optimum income taxation, *Review of Economic Studies* **38,** April, 175–208.

Myers, S., and N. Maljuf, 1984: Corporate financial and investment decisions when firms have information that investors do not have, *Journal of Financial Economics* **13**(2), June, 187–221.

Nakicenovic, N., W.D. Nordhaus, R. Richels, and F.L. Toth, eds., 1994: *Interactive assessment of mitigation, impacts, and adaptation to climate change.* Proceedings of a workshop held on 13–15 October 1993 at the Institute for Applied Systems Analysis, Laxenburg, Austria.

Nash, J.F., 1955: The bargaining problem, *Econometrica* **18,** 155–162.

Nordhaus, W.D., 1993: *Managing the global commons: The economics of climate change,* MIT Press, Cambridge, MA.

Nordhaus, W.D., and J. Tobin, 1973: Is growth obsolete? *Income and Wealth,* vol. 38, National Bureau of Economic Research, New York.

Parikh, J.K., 1994: Vulnerability vs. responsibility in climate change outlook. Proceedings, Australian Bureau of Agricultural and Resource Economics (ABARE), 1–4 Feb., 1994, Canberra Australia.

Pearce, D.W., 1989: Sustainable development: Towards an operational definition and its practical implications. OECD Economics and Statistics Department/Environmental Directorate Joint Seminar on the Economics of Environmental Issues. 2–3 Oct., Paris.

Pearce, D.W., 1991: Evaluating the socio-economic impacts of climate change: An introduction. In OECD, *Climate change: Evaluating the socio-economic impacts,* chapter 1, Paris.

Pearce, D.W., G. Atkinson, and W. DuBourg, 1994: The economics of sustainable development, *Annual Review of Energy and the Environment* **19,** 457–474.

Pearce, D.W., and R.K. Turner, 1990: *Economics of natural resources and the environment,* Johns Hopkins University Press, Baltimore.

Peck, S., and T. Teisberg, 1992: CETA: A model for carbon emissions trajectory assessment, *The Energy Journal,* **13**(1), 55–77.

Peck, S., and T. Teisberg 1993: Global warming uncertainties and the value of information: An analysis using CETA, *Resource and Energy Economics,* **15**(1), 87–102.

Peskin, H., and E. Lutz, 1990: A survey of resource and environmental accounting in industrialized countries. The World Bank Environment Working Paper no. 37, August.

Polasky, S., A.R. Solow, and J.M. Broadus, 1993: Searching for uncertain benefits and the conservation of biological diversity, *Environmental and Resource Economics* **3,** 171–181.

Rawls, J., 1971: *Theory of justice,* Belknap Press, Cambridge, MA.

Reddy A., and J. Goldemberg, 1990: Energy for the developing world, *Scientific American,* **263**(3), 111–118.

Repetto, R., 1989: *Wasting assets: Natural resources in the national income accounts,* World Resources Institute, Washington, DC.

Repetto, R., 1991: *Accounts overdue: Natural resource depletion in Costa Rica,* World Resources Insitute, Washington, DC.

Repetto, R., R. Dower, R. Jenkins, and T. Geoghegan, 1992: *Green fees: How a tax shift can work for the environment and the economy,* World Resources Institute, Washington, DC.

Richels, R, and J. Edmonds, 1993: The economics of stabilizing atmospheric CO_2 concentrations. Draft manuscript.

Ross, R., 1973: The economic theory of agency: The principal's problem, *American Economic Review* **63,** 134–139.

Rothenberg, J., 1993: Time comparisons in public policy analysis of global climate change: An economic analysis, MIT Press, Cambridge, MA.

Salop, S., 1977: The noisy monopolist: Imperfect information, price dispersion, and price discrimination, *Review of Economic Studies* **44**(3), October, 393–406.

Salop, S., and J.E. Stiglitz, 1977: Bargains and ripoffs: A model of monopolistically competitive price dispersion, *Review of Economic Studies* **44**, October, 493–510.

Salop, S., and J.E. Stiglitz, 1982: The theory of sales: A simple model of equilibrium price dispersion with identical agents, *American Economic Review,* Dec., 1121–1130.

Samuelson, P., 1954: The pure theory of public expenditure, *Review of Economics and Statistics* **36**, Nov., 387–389.

Schmalensee, R., 1993: Comparing greenhouse gases for policy purposes, *The Energy Journal* **14**(1), 245–255.

Shah, A., and B. Larsen, 1991: *Carbon taxes, the greenhouse effect and developing countries,* World Bank, Washington, DC.

Solow, A.R., 1993: The response of sea level to global warming. In *The world at risk: Natural hazards and climate change,* R. Bras, ed. AIP Conference Proceedings 277, pp. 38–42, American Institute of Physics, New York.

Solow, A.R., 1990: Is there a global warming problem? In *Global warming: Economic policy responses,* R. Dornbusch and J. M. Poterba, eds., MIT Press, Cambridge, MA.

Solow, R.M., 1974: Intergenerational equity and exhaustible resources, *Review of Economic Studies* (symposium volume), pp. 29–45.

Solow, R.M., 1992: An almost practical step toward sustainability, *Resources for the Future,* Washington, DC.

Steer, A., and E. Lutz, 1993: Measuring environmentally sustainable development, *Finance and Development* **30**(4), 20–23.

Stewart, R., and J.B. Wiener, 1990: A comprehensive approach to climate change: Using the market to protect the environment. In *The American Enterprise,* Vol.1.

Stewart, R.B., and J.B. Wiener, 1992: The comprehensive approach to global climate policy: Issues of design and practicality, *Arizona Journal of International and Comparative Law,* 9(1), 83–113.

Stiglitz, J.E., 1974: Incentives and risk sharing in sharecropping, *Review of Economic Studies* **41,** April, 219–255.

Stiglitz, J.E., 1975a: Information and economic analysis. In *Current economic problems,* Parkin and Nobay, eds., pp. 27–52, Cambridge University Press.

Stiglitz, J.E., 1975b: Incentives, risk, and information: Notes toward a theory of hierarchy. *Bell Journal of Economics* **6,** Autumn, 552–579.

Stiglitz, J.E., 1977: Monopoly and the rate of extraction of exhaustible resources, *American Economic Review* **66**(4), Sept., 655–661.

Stiglitz, J.E., 1982: The efficiency of the stock market equilibrium, *Review of Economic Studies* **49,** 241–261.

Stiglitz, J.E., 1985: Information and economic analysis: A perspective, *Economic Journal* **95,** Supplement, 21–45.

Stiglitz, J.E., 1988: Economic organization, information, and devel-opment. In *Handbook of Development Economics,* H. Chenery and T.N. Srinivasan, eds., pp. 94–160. Elsevier Science Publishers, Amsterdam.

Stiglitz, J.E., 1991: Symposium on organizations in economics, *Journal of Economic Perspectives* **5**(2), 15–24.

Stiglitz, J.E., 1994: *Whither socialism?* Wicksell Lectures, MIT Press, Cambridge, MA.

Stiglitz, J.E., and P. Dasgupta, 1971: Differential taxation, public goods, and economic efficiency, *The Review of Economic Studies* **37,** 151–174.

Stiglitz, J.E., P. Dasgupta, and G. Heal, 1980: The taxation of exhaustible resources. In *Public policy and the tax system,* G.A. Hughes and G.M. Heal, eds., George Allen & Unwin, London.

Stiglitz, J.E., and A. Weiss, 1981: Credit rationing in markets with imperfect information, *American Economic Review* **71**(3), June, 393–410.

Tietenberg, T., 1992: Introduction and overview. In *Innovation in environmental policy: Economic and legal aspects of recent developments in environmental enforcement and liability,* by T. Tietenberg, pp. 1–17, Elgar, Aldershot, UK.

UNDP (United Nations Development Programme), 1992: *Human development report 1992,* Oxford University Press, Oxford.

United Nations, 1989: African alternative framework to structural adjustment programmes for socio-economic recovery and transformation, UNECA/OAU, E/ECA/cm.15/6/Rev 3. United Nations, New York.

United Nations Statistical Office, 1992: *SNA draft handbook on integrated environmental and economic accounting,* provisional version. New York.

U.S. EPA (Environmental Protection Agency), 1989: Sea level rise. In *Report to Congress on the potential effects of global climate change on the United States,* Washington, DC.

Weitzman, M.L., 1974: Prices vs quantities, *Review of Economic Studies,* **41**(4), October, 477–491.

Weitzman, M.L., 1978: Optimal rewards for economic regulation, *American Economic Review* **68**(4), Sept., 683–691.

Weitzman, M.L., W. Newey, and M. Rabin, 1981: Sequential R&D strategy for synfuels, *The Bell Journal of Economics* **12**(2), 574–590.

Whalley, J., and R. Wigle, 1991: Cutting CO_2 emissions: The effects of alternative policy approaches, *Energy Journal* **12**(1), 109–124.

Wilson, C.A., 1977: A model of insurance markets with incomplete information. *Journal of Economic Theory* **16**(2), Dec., 167–207.

World Commission on Environment and Development, 1987: *Our common future,* Oxford University Press, London.

Zachariah, K.C., and M.T. Vu, 1988: *World population projections 1987–88.* Johns Hopkins University Press, Baltimore.

2

Decision-Making Frameworks for Addressing Climate Change

Principal Lead Authors:
K.J. ARROW, J. PARIKH, G. PILLET

Lead Authors:
M. GRUBB, E. HAITES, J.C. HOURCADE, K. PARIKH, F. YAMIN

Contributors:
P.G. Babu, G. Chichilnisky, S. Faucheux, G. Froger, W. Hediger, F. Gassmann, S. Kavi Kumar, S.C. Peck, R. Richels, C. Suarez, R. Tol

CONTENTS

SUMMARY

This chapter discusses possible decision-making frameworks related to climate change. It begins with a review of some of the unique features of the climate change problem and their implications for decision making. Then two interrelated approaches to decision making – optimizing quantitative models (decision analysis) and negotiation – are described. The chapter concludes by considering the specific implications of the preceding discussion for climate change decisions under the Framework Convention on Climate Change (FCCC).

Decision making related to climate change must take into account the unique characteristics of climate change: large scientific and economic uncertainties; long time horizons; nonlinearities and irreversibility of effects; the global nature of the problem; social, economic, and geographic differences among the affected parties; and an agreed framework to address the issue.

Decision analysis uses quantitative techniques to identify the "best" choice from among a range of alternatives. A review of the real world limitations of quantitative decision models and the consistency of their theoretical assumptions with climate change decision making highlights the following points:

- There is no single decision maker in climate change. Because of differences in values and objectives, parties participating in a collective decision-making process do not apply the same criteria to the choice of alternatives. Consequently, decision analysis cannot yield a universally preferred solution.

- Decision analysis requires a complete and consistent utility valuation of decision outcomes. In climate change, many decision outcomes are difficult to value and a global welfare function does not exist, so quantitative comparisons of decision options are not meaningful.

- Decision analysis may help keep the information content of the climate change problem within the cognitive limits of decision makers. Without the structure of decision analysis, climate change information becomes cognitively unmanageable, limiting the ability of decision makers to analyze the outcomes of alternative actions rationally.

- The treatment of uncertainty in decision analysis is quite powerful, but the probabilities of uncertain decision outcomes must be quantifiable. In climate change, objective probabilities have not been established for many outcomes, and subjective probabilities would be controversial, so climate change decisions cannot fully satisfy this requirement.

- Because of the large uncertainties and differences between parties, there may be no "globally" optimal climate change strategy; nevertheless, the factors that affect optimal single-decision-maker strategies still have relevance to individual parties.

The lack of an individual decision maker, utility problems, and incomplete information suggest that decision analysis cannot serve as the primary basis for international climate change decision making. Although elements of the technique have considerable value in framing the decision problem and identifying its critical features, decision analysis cannot identify globally optimal choices for climate change abatement. Decision analysis suffers fewer problems when used by individual countries to identify optimal national policies.

The Framework Convention on Climate Change establishes a collective decision-making process within which the parties will negotiate future actions. Although some features of the decision-making process are set out in the Convention, many are still undecided. It becomes important, then, to examine negotiation and compromise as the primary basis for climate change decisions under the Convention. Important factors affecting negotiated decisions include the following:

- Excessive knowledge requirements in negotiated environmental decisions may stand in the way of collective rational choice. This difficulty could be reduced by making the negotiation process itself more manageable through the use of tools like stakeholder analysis or by splitting accords into more easily managed clusters of agreements.

- Since society has no consistent probability threshold for ignoring particular risks, it may be vulnerable to surprise when risks are uncertain. In climate decisions, this vulnerability could be reduced by relating event scenarios to explicit probabilities of surprise.

- In the face of long-term uncertainties, sequential decision making allows actions to be better matched to outcomes by incorporating additional information over time. Sequential decision making also minimizes harmful strategic behaviour among multiple decision makers.

- Improved information about uncertain outcomes may have very high economic value, especially if that information can create future decision options.

- There are currently no effective mechanisms for the sharing of risks related to climate change and their associated economic burdens. International risk sharing could yield substantial benefits for global economic and social welfare.

The Convention is, first and foremost, a framework for collective decision making by sovereign states. Given this collective decision mechanism and the uncertainties inherent in the climate problem, several recommendations emerge. Climate actions under the FCCC should be sequential; countries should implement a portfolio of mitigation, adaptation, and research measures; and they should adjust this portfolio continuously in response to new knowledge. The value of better information is potentially very large. To distribute the risks of losses related to climate change efficiently, new insurance mechanisms may be warranted.

2.1 Introduction

Decision makers face unavoidable choices in addressing the threat of climate change – even doing nothing is a choice. This chapter attempts to shed some light on the nature and context of climate change decisions and effective means of making those decisions. It explores quantitative analysis and negotiation in climate change decision making and suggests ways to enhance the decision process within the Framework Convention on Climate Change (FCCC).

The first section revisits some of the unique features of the climate change problem as set out in Chapter 1. These include:

- uncertainty about the impacts of climate change, abatement costs, and other factors

- long time horizons and a substantial time lag between emissions and impacts

- irreversibility of effects

- the global nature of the problem, which necessitates collective action by sovereign states

- social, economic, and geographic differences among the affected parties

- an agreed on legal framework for climate change responses

The next section describes quantitative analytic models of decision making, including their underlying assumptions and their applicability to climate change decisions. It discusses the real world limitations of quantitative decision models and highlights those aspects of climate change decision making that are likely to fall outside the theoretical assumptions required by such models. The discussion explores two interrelated approaches to decision making:

- *Optimizing quantitative models,* where clear decision structures and well-understood decision parameters can serve as the primary basis for decision making

- *Negotiation,* where complex decision structures and poorly understood or conflicting decision parameters prevent a purely analytic resolution of decision problems

Quantitative optimization and negotiation can be complementary. This section reviews the academic literature on the economic, behavioural, and organizational aspects of the decision process, highlighting those aspects of a decision scenario best suited to quantification or negotiation. The context of the climate problem and the literature on decision making suggest that quantitative approaches are unlikely to produce universally acceptable solutions, so negotiation will be important in reaching collective climate change decisions.

The chapter concludes by considering the specific implications of the previous discussion for climate change decisions under the Convention. First and foremost, the FCCC is a framework for collective decision making by sovereign states. Given this collective decision mechanism and the uncertainties inherent in the climate problem, several recommendations emerge. Climate actions under the FCCC should be sequential; countries should implement a *portfolio* of mitigation, adaptation, and research measures; and they should adjust this portfolio continuously in response to new knowledge. To efficiently distribute the risks of losses related to climate change, new insurance mechanisms may also be warranted.

2.2 The Context for Climate Change Decision Making

All international decisions are complex, but those related to climate change are particularly complicated because of society's limited understanding of the problem, the long time frames involved, and the global nature of climate effects and climate treaties.

2.2.1 Impact uncertainties

Making decisions about climate change requires an understanding of the impact of uncertainties. Although they do not affect the essential need for a decision, these uncertainties complicate the decision process and limit the ability of decision makers to identify superior options.

There are three principal areas of uncertainty related to climate change impacts:

- *Scientific uncertainties* obscure relationships between emissions and atmospheric concentrations of greenhouse gases, the dynamics of climate feedback, and the effects of climate change on global temperature, ecological cycles, sea level, and the occurrence of weather events. These uncertainties are explored in Volume 1 of this report.

- *Socioecologic uncertainties* obscure how climate change will affect the relationship between human societies and the biosphere, particularly where human welfare is strongly affected by nature. Such relationships include agricultural production, fishing, and the spread of disease. These issues are treated in Volume 2.

- *Socioeconomic uncertainties* obscure the economic and social welfare effects of climate change and abatement. These uncertainties, which affect the economic valuation of resources, international trade, technological change, and other socioeconomic interactions, are the focus of the present volume.

2.2.1.1 *An illustration of uncertainty in climate decisions*
Figure 2.1 illustrates a hypothetical relationship between climate parameters and impact uncertainty. Using mean temperature change (ΔT) as a proxy for a set of critical climate variables, Figure 2.1 outlines probability distributions for different levels of temperature change.

As the figure shows, for lesser temperature changes global adaptation to climate impacts may be relatively easy, with a very small risk of serious social or ecologic damage. The likelihood of severe consequences increases with rising temperature, becoming extremely likely at $\Delta T = 6°C$. As the average

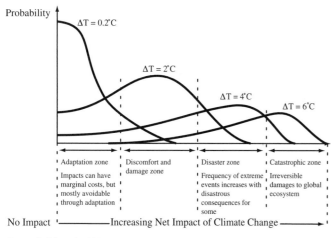

Source: Parikh *et al.* (1995)

Figure 2.1: Schematic linkage between scientific uncertainties and net impacts.

temperature change increases, the probability that the impacts will remain within the adaptation zone becomes smaller. In reality the shape or magnitude of these probability curves is not known, but the illustration is valuable in framing the role of uncertainty within the decision problem.

2.2.2 Time horizons

Much of the uncertainty associated with climate change arises from the extraordinarily long time horizons involved. Few other decision problems so obviously require a decision perspective spanning many decades or even centuries.

Long time scales affect both the scientific and the technological aspects of climate change. Time scales may be similar for the development and diffusion of climate change abatement technology. Historically it has taken about 50 years for a new energy carrier to move from 1% to 50% penetration of its potential market (Häfele *et al.*, 1981), and that is probably after several prior decades of research and development. On the other hand, long time scales present considerable opportunity for innovation – time for the accumulation of incremental improvements in abatement techniques or for the emergence of revolutionary, environmentally benign technologies.

Long time scales also increase uncertainties about the social impacts and proper economic valuation of climate effects. Climate impacts will be imposed on future generations and in different countries with different value systems from those we have now. The values and requirements of future society are not known. It is possible that adverse climate impacts 50 years from now may be considered incommensurate with some level of monetary compensation established today. Schelling (1994) has argued that such intergenerational equity is a fundamentally political concern, albeit with ethical dimensions. But intergenerational issues have practical implications for the identification and implementation of climate change strategies. For example, Chichilnisky (1993a) has called for abatement philosophies that include specification of a "safe

minimum standard" to avoid "dictatorship of the present" over future generations. Decision making may also need to address the distinction between passing on future *benefits* and imposing future *damages* – a distinction found in most ethical and legal systems but absent from the conventional economic evaluation of future impacts.

2.2.3 International diversity and climate-related vulnerability

Climate change is a global problem encompassing a diverse mix of human societies – each with a distinct geography, culture, political system, and economic status. These differences affect both the exposure and the vulnerability of individual countries to climate impacts:

- Some regions will be subject to above-average temperature, sea-level, storm, or other changes; others will experience below-average impacts or even impacts differing in sign from the global average;

- The impact on human societies will differ according to myriad factors, such as the amount of low-lying or arid land they occupy and their degree of dependence on agriculture or aquatic resources;

- The ability of societies to absorb climate impacts will differ according to the strength of their infrastructure, social structure, and other local factors.

Many developing countries are in relatively hot climates. They are more dependent on farming, and have less well-developed infrastructures and social structures than developed countries. As a consequence, developing nations may realize much more severe welfare loss due to climate change than wealthier nations. Figure 2.2 illustrates hypothetically how the same degree of climate change may impose greater risks and welfare loss on poor countries. The figure outlines two probability distributions for the welfare impacts of a given climate change: one for a developed nation, the other for an underdeveloped nation. Because of its relative poverty, the underdeveloped country has fewer adaptatation options and is much more susceptible to economic disaster than the developed country.

Economic dependence on agriculture could be one good indicator of a nation's relative exposure to severe climate-related welfare losses. As a share of domestic product, agriculture ranges between 16% and 64% in low-income developing countries compared to between 12% and 37% in middle-income developing countries, and 3% in the U.S. (World Bank, 1994). A study by Parry and Rosenzweig (1994) illustrates how climate change is likely to alter the distribution of global food production, even if there is little impact on the global total. In their scenarios cereal production in developing countries could fall by 6–12% by 2050 whereas in industrialized countries it could rise by 2–14%. The recent African famines illustrate how critical such changes in distribution can be.

Agricultural dependence is only one dimension of the differences between countries that are likely to be important in

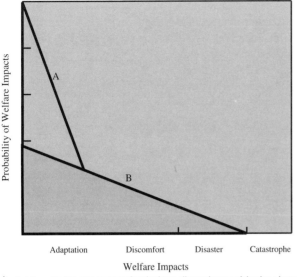

A   A rich country has more resources for adapting to climate change and thus has a better chance of avoiding extreme impacts.

B   The same level of climate damage can bring economic disaster to a poor country.

Source: Parikh *et al.* (1995)

Figure 2.2: Different welfare impacts for the same climatic damage: an illustration.

climate change. Chapter 3 explores these differences among countries further, including:

- wealth and consumption
- emissions – past, present, and future
- the distribution of and vulnerability to climate change
- endowment with resources that may be affected by responses to climate change

Such differences will be reflected in the attitudes that countries bring to international climate change negotiations. For example, European countries may focus most on the possible costs of abatement, whereas developing countries in Africa and South America may be most concerned with the burden of adaptation and vulnerability. Island states may be most threatened by a major loss of coastal land mass. Oil exporters may be most concerned about their potential loss of revenue from abatement strategies that reduce international fossil fuel consumption. An understanding of such differences in national perceptions, capabilities, and objectives must inform the decision process, particularly where those decisions must be reached collectively.

2.2.3.1 Ecocentrism vs. anthropocentrism
Most international discussions and IPCC reports take a strictly *anthropocentric* view of climate effects: Resources are valued only in terms of their value for human recreation, medicine, and other aspects of human welfare. *Ecocentrism,* espoused in some religious and environmental philosophies, views *homo sapiens* as just one of the species on Earth, expected to share the biosphere in balance with others. Climate change is likely to affect the habitat of all flora and fauna in

the integrated global biosphere. An ecocentric viewpoint questions society's moral authority to make decisions affecting Nature as a whole. Irrespective of whether one shares such a viewpoint – which is a fundamental value judgment – ecocentrism highlights the complexity of biospheric interactions with which climate change may interfere.

2.2.4 Existing decision framework

The Framework Convention on Climate Change went into effect in 1994. The Convention establishes many general commitments for the parties to coordinate environmental activities and to review their own policies to see if they encourage emissions. It also allows parties' obligations to be modified in response to new information. Most important, the Convention establishes a general framework for ensuring that implementation is meeting the Convention's objectives. To this end it specifies a set of institutional arrangements for climate change decision making: a decision-making body, advisory groups, administrative support, and a financial mechanism. But the Convention's specification of a climate decision process is not complete, leaving open many critical decision issues for resolution in the future. (The provisions of the Convention related to decision making are summarized in Section 2.4.)

2.3 Quantitative Models of Decision Making

The integration of utility theory, probability, and mathematical optimization in the study of decision processes has yielded quantitative models of decision making. These models seek to explain the organization, valuation, and selection that occurs when one possible course of action is chosen over others. Quantitative decision models attempt to improve decision making by clarifying complex decisions and making the best use of available information. Although they suffer from practical limitations, such models can offer valuable insight into many aspects of environmental decision problems like the climate change problem. Apart from their ability to generate numerical results, such models "provide a conceptual framework (or several) for relating means to ends . . . for identifying the existing technical alternatives and for inventing new ones" (Raven as quoted in Morgan and Henrion, 1990). They can serve as an idealized reference case by which to evaluate actual decision structures and decision choices.

This section reviews the fundamentals of decision analysis and its underlying theoretical assumptions. It examines the validity of these assumptions and discusses the implications of violations of these assumptions in the climate change decision context. The discussion shows that, in collective decision-making situations with varying decision criteria, quantitative analytic models may not be able to identify universally optimal decisions. In these cases negotiation may have the dominant influence on decision making. The section concludes with an examination of negotiation in collective decision-making situations, such as climate change, which have no clear analytic solution.

2.3.1 An overview of decision analysis

Decision analysis is a formal quantitative technique for identifying "best" choices from a range of alternatives. Decision analysis requires the development of explicit influence structures (trees) specifying a complete set of decision choices (nodes), possible outcomes, and outcome values. Uncertainty concerning the outcomes of possible choices is incorporated into decision structures explicitly by assigning probabilities to individual outcomes.

Figure 2.3 is an example of a simple decision tree for the comparison of three climate change abatement alternatives. In Figure 2.3, the choice is represented by a square node and uncertain outcomes by round nodes, with costs and probabilities as indicated on the "branches." For example, Option 2 will cost $30 to implement. It has an 80% chance of yielding a zero value outcome and a 20% chance of yielding an outcome of −$300, and so forth.

Analysis of the decision depicted in Figure 2.3 would proceed as follows: Option 2 will cost $30 to implement, has an 80% chance of yielding a zero value outcome and a 20% chance of yielding a -$300 outcome, so its "expected value" is:

$$-\$30 + .8(\$0) + .2(-\$300) = -\$90.$$

Similarly, the expected value of Option 1 is -$110. The expected value of doing nothing is -$340. If the goal here is to minimize cost, then the best choice is to implement Option 2 since it is expected to cost less than the other options.

The decision analysis example in Figure 2.3 is trivial. It is easy to imagine more elaborate decision trees with many more branches, sequential decisions, continuous probability distributions, and other complications. Nonetheless, the figure illustrates all the essential features of decision analysis relevant to this discussion.

2.3.1.1 Assumptions of decision analysis

It may be apparent from the preceding example that decision analysis operates under a restrictive set of assumptions. In particular, decision analysis assumes the following:

- *There is a single decision maker.* Decision analysis does not make the actual decisions, it only lays out the options in a consistent manner so that they can be compared. In decision analysis an individual (or a perfectly cooperating group of individuals) must choose an option based on some selection criterion.

- *Decision alternatives are limited.* To construct a complete decision tree, the number of choices and their possible outcomes must be finite (and known).

- *Valuation of alternatives is consistent.* Decision analysis yields expected values for different options. For these values to be strictly comparable, they must be expressible in the same units (such as dollars, injuries, ΔT, etc.).

- *Choices are rational.* Decision analysis assumes that decision makers will choose rationally among consistently valued alternatives.

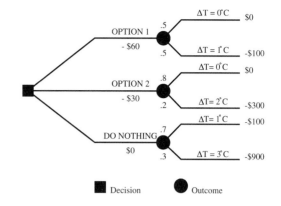

Figure 2.3: Simple decision tree.

- *Uncertainties are quantifiable.* Uncertainties must be quantified in decision analysis models, either discretely or continuously, in order to calculate expected values.

It is generally accepted that these assumptions will not be perfectly satisfied in actual decision analyses, but often will be satisfied approximately. If violations of these assumptions are minor, decision analysis may still yield superior solutions compared to other decision-making approaches. If any of the violations is serious, the effectiveness of decision analysis as a stand-alone decision tool is compromised – decision problems become intractable due to complexity, ambiguity, or incompleteness. In these cases, decision analysis loses its ability to generate a specific result (although it retains much of its conceptual power).

2.3.2 Decision analysis and climate change

As noted at the start of this section, decision analysis can be a powerful tool for defining the structure of decision scenarios and for understanding the barriers to making optimal choices. Thinking about climate choices as decision analysis problems may highlight those aspects most likely to affect the decision process. This section explores the assumptions of decision analysis in climate change and the implications when aspects of those decisions fall outside the decision analysis model.

2.3.2.1 Selection criteria and risk attitudes

One of the strengths of decision analysis is that it employs a uniformly valued set of outcomes, so options can be compared directly. Given a single decision maker and consistent comparisons, identifying a superior choice is straightforward. "Consistency" in this case means "behaviour in accordance with some ordering of alternatives in terms of relative desirability" (Arrow, 1951a; Blaug, 1992). In decision analysis, consistency of choice is implied by adhering to an explicit selection criterion when identifying preferred alternatives.

For analytical purposes it is often (but not always) assumed that all important choice attributes can be expressed in monetary terms. Then the choice of action can be based on the expected monetary values of alternatives using a particular selection criterion. Different selection criteria point to different optimal choices. Some of the most common criteria include (after Morgan and Henrion, 1990):

- *Maximizing multiattribute utility,* which specifies a function that evaluates outcomes in terms of all their important utility attributes (see 2.3.2.3 below), including uncertainties and risks. The alternative with maximum utility is selected (Savage, 1954; Arrow, 1987).

- *Cost-benefit,* which estimates the costs and benefits of alternatives in economic terms and chooses the one with the highest net benefit. (Chapter 5 examines this criterion in greater detail.)

- *Cost effectiveness,* which selects a desired performance level, perhaps on noneconomic grounds, and chooses the option achieving this level at lowest cost. Cost-effectiveness analysis can aid choice between options but, unlike the cost-benefit criterion, cannot indicate whether or not any of the options are worth doing (MIT, 1986).[1]

- *Minimax loss,* which finds the loss associated with the extreme event for each alternative, then selects the strategy that minimizes the worst loss. This criterion amounts to expecting the worst (very severe impacts or very expensive mitigation measures) and avoiding those strategies that could generate the worst outcomes (Wald 1950).

- *Maximin gain,* which finds the gains associated with each action and tries to maximize the minimum of these gains (Wald, 1950).

- *Minimax regret,* which chooses measures of "regret" and minimizes the maximum regret (Savage, 1951). This criterion emphasizes the cost of making the wrong decision – for example, the cost of relying heavily on new technologies that do not materialize to mitigate climate change.

- *Bounded cost,* which sets the maximum budget to devote to risk management activity.

The choice of a selection criterion itself is subjective and exogenous to the decision analytic model. One of the strongest influences on this choice is a decision maker's attitude toward "risk" – his or her exposure to a chance of loss (Random House, 1966). Evidence indicates that most (but not all) individuals prefer a certain outcome to an uncertain outcome, even though both outcomes may have the same expected value. Such an attitude reflects "risk aversion." Risk-averse individuals would prefer decision criteria that minimize the chance of the worst possible outcome, whereas "risk-seeking" individuals would prefer criteria that maximize the chance of the best possible outcome.

Risk attitudes do not directly affect decision analysis, but they can affect the valuation of decision options. When risks are known, the expected values in a decision analysis can still be used to identify best choices (Ramsey, 1928; von Neuman and Morgenstern, 1947). When risks are not known, however, expected values may no longer serve as a reliable indicator of relative preferability (Savage, 1951, 1954). This problem is exacerbated when changes in the decision environment are irreversible, because irreversible changes may eliminate future options. Arrow and Fisher (1974) and Henry (1974) have

shown that, under such circumstances, selection based on expected values is inappropriate.

2.3.2.2 *Multiple decision makers*

Decision analysis assumes that there is one decision maker. The problem introduced by multiple decision makers is the potential lack of a consistent valuation of decision alternatives, with the result that a *universally* "best" solution cannot be identified. Even with a common assessment of the expected values of possible outcomes, differences in risk attitudes and other factors may cause multiple decision makers to prefer different selection criteria. Since any set of decisions preferred by one party may not be acceptable to another, the parties will have to negotiate an agreed course of action – one that is likely to differ from either of the analytically derived solutions they originally preferred. This negotiation introduces a host of subjective factors (negotiating skill, for example) into the decision process, which decision analysis is ill equipped to handle.

In the case of climate change, there is no single decision maker responsible for choosing preferred actions. In fact, there are two sets of collective decision makers: one at the international level and one at the national level. According to the Framework Convention on Climate Change (see Section 2.4), international parties will collectively negotiate climate decisions. There must also be a set of decision makers within each country collectively responsible for defining that country's positions in the international negotiations. As noted in Section 2.2, the international parties may be uniquely affected by numerous climate change-related factors, so they may not share a consistent set of goals and will probably not cooperate perfectly. But even within a country different government agencies, individuals, and firms may have conflicting views about that country's optimal policy for addressing climate change.

2.3.2.3 *Identification and valuation of decision alternatives*

The economic theory of choice holds that when an individual can take a limited range of actions, he or she "has in mind an ordering of all possible consequences" and chooses "that action whose consequences are preferred to those of any other" (Arrow, 1951a). These individual preferences are often expressed in terms of multiattribute utility functions, where "utility" is defined as an "amount of satisfaction" (Keeney and Raiffa, 1976; Eeckhoudt and Gollier, 1995). The "attributes" may include economic value, risk, aesthetics, peace of mind, and other criteria affecting an individual's overall satisfaction. Choice theory assumes that, in evaluating uncertain situations, the decision maker uses the monetary value of final outcomes as a substitute measure of their utility. (Section 2.3.2.1 assumed the opposite conversion – of utility to economic value. In either case decision options should have the same relative values and, given the same selection criterion, should lead to the same choices.) The valuation of outcomes in terms of utility enables individuals to rank their preferences. If individuals are perfectly knowledgeable and consistent in their choices, the most valuable option – in terms of utility – is always chosen (Price, 1993). Utility does not define the decision criteria; it requires only that the criteria be consistently applied. Thus, an entrepreneur wishing to maximize profits will always choose the action that yields the highest profit after accounting for risk and other factors.

The incorporation of choice theory in decision analysis is theoretically appealing since a well-defined set of outcome utilities leads to a straightforward analytic solution – a "best" choice in terms of expected utility. Unfortunately, expression of utility as a function in international decision making is problematic. Arrow's general impossibility theorem (Arrow, 1951a) suggests that a global social welfare function cannot exist. Consequently, decisions cannot be based on an analysis requiring such a function. Even if such utility functions did exist, some decision outcomes, like energy security and biological diversity, can be difficult to value in economic terms. When multiple decision makers are involved, this valuation problem may become so severe that quantified decision outcomes lose their meaning.

In the context of climate change, the assumption that the collective decision makers will be completely informed about all possible decision options and consequences is questionable. As noted earlier in this chapter, climate change may result in a multitude of subtle and interrelated ecologic, economic, and sociologic consequences. The effects of these consequences on individual countries will vary widely according to their individual vulnerabilities. It appears unlikely, therefore, that the international parties will be able to identify a complete set of consequences for all affected parties around the globe. Furthermore, there is no agreement in the literature as to how a utility function (e.g., an index of social welfare) would be constructed or applied to such a set of consequences.

2.3.2.4 Rational choice
Decision analysis assumes that decision makers will make rational choices. Dawes (1988) defines rational choice as meeting three criteria:

- It is based on the decision maker's current state of being.
- It is based on the possible outcomes of the available choices.
- It treats uncertainty in a manner consistent with probability theory.

If any of these criteria are violated, it is possible for a decision maker to reach contradictory conclusions about what to choose, even though the conclusions are based on the same preferences and evidence. For example, an *irrational* decision maker might prefer one course of action if choices were considered from first to last, but prefer a different course of action if the same choices were considered from last to first. Thus, an irrational decision maker can decide that a particular choice is simultaneously desirable and undesirable (Dawes 1988).

Simon (1957) and Arrow (1987) have shown that, since decision makers have finite cognitive resources, their ability to

gather and process information is limited – that is, their rationality must be *bounded*. When information content is within human ability to process it, decisions may be rational. But when information content exceeds cognitive limits, decisions must necessarily be irrational, according to the definition above. For example, in situations where it is not possible to specify all options and possible outcomes in advance, one strategy would be to collect information in a predetermined manner for a limited time, then select the best alternative (Dawes, 1988). This is a bounded rational approach, since it makes the best choice given the limited information. It is not a *strictly* rational approach because it imposes an (arbitrary) time constraint to limit the information available to decision makers. In this example, it is conceivable that setting a different deadline would have resulted in different decisions, even though the true set of options and outcomes would be unchanged. Bounded rationality, then, addresses the ways in which the decision *process,* rather than the decision input, influences decision making (see also Vercelli, 1991; Faucheux and Froger, 1994).

Simon suggests that cognitive limitations usually stem from the limited adequacy of scientific theories to predict relevant phenomena (Simon, 1987). Since climate change faces precisely these kinds of prediction problems, it is likely that cognitive limitations will have to be addressed in any comprehensive climate decision process.

2.3.2.5 Uncertainty
Arrow (1971) defines uncertainty as incomplete knowledge of the state of the world. This incomplete knowledge implies that individuals cannot predict with precision all the consequences of their actions. Uncertainty in decision processes is inescapable. It "affects all fundamental variables that determine behaviour, explain choices and bring about decisions" (Kessler, in Eeckhoudt and Gollier, 1995). Morgan and Henrion (1990) argue that "policies that ignore uncertainty about technology, and about the physical world, often lead in the long run to unsatisfactory technical, social, and political outcomes." They view the consideration of uncertainty in decisions as a valuable means of:

- identifying important decision factors and sources of disagreement
- hedging or planning for contingencies
- understanding information provided by experts
- reconsidering repeated decision problems

One of the most powerful aspects of decision analysis is its explicit inclusion of uncertainty in the decision process. If the outcomes associated with possible actions are uncertain, probabilities can be assigned to those outcomes, based either on scientific knowledge (objective) or personal judgment (subjective). When it is not possible to arrive at unique probability distributions for future decision outcomes, they may sometimes be constructed from the known probabilities of past events. Once outcome probabilities are established for possible actions, they are used to calculate the expected values of given decisions. These expected values serve as the basis for identifying preferred choices using a particular selection criterion.

Decision analysis requires that uncertainty be quantified. Without probabilities for decision outcomes, expected value calculations cannot be made and the technique loses its analytic foundation. Unfortunately, in the context of climate change, objective data on the probabilities of all decision outcomes are not available. Subjective probabilities could conceivably be used, but it is unlikely given the range of views related to climate change held by different interest groups that collective agreement could be reached on subjective probabilities for different outcomes. With no historical evidence available about the occurrence of particular climate events, equal outcome probabilities could initially be assigned, then updated when relevant new scientific information becomes available (Caselton and Luo, 1994), but Heap *et al.* (1992) have questioned this approach.

Climate change is not a situation of total ignorance – we have some idea of plausible outcomes and their relative probability. It is inappropriate, therefore, to sacrifice the power of decision analysis to address uncertainty simply because some uncertainties are not quantifiable. For example, analyses of alternative climate change scenarios may reveal that some actions are preferred to others regardless of uncertainty. It is probably more reasonable to use decision analysis, given the best objective or subjective probabilities available, as an element in a broader decision process that accounts in some other way for those phenomena the decision analysis excludes.

2.3.2.6 *Conclusions about decision analysis*
The preceding sections have touched on a number of important considerations in evaluating the climate change problem and the application of decision analysis to that problem. (These aspects of decision analysis have also been discussed in detail in Parikh *et al.*, 1994, 1995.) The main points can be summarized as follows:

- There is no single decision maker in climate change. Differences in values and objectives prevent collective decision makers from preferring the same selection criterion for decision alternatives – so decision analysis cannot yield a universally preferred solution.

- Decision analysis requires a complete and consistent utility valuation of decision outcomes. In climate change, many decision outcomes are difficult to value and a global welfare function does not exist, so quantitative comparisons of decision options are not meaningful.

- Decision analysis may help keep the information content of the climate change problem within the cognitive limits of decision makers. Without the structure of decision analysis, climate change information becomes cognitively unmanageable, limiting the ability of decision makers to rationally analyze the outcomes of alternative actions.

- The treatment of uncertainty in decision analysis is quite powerful, but it requires that the probabilities of uncertain decision outcomes be quantifiable. In climate change, objective probabilities have not been established for many outcomes, and subjective probabilities would be controversial, so climate change decisions cannot fully satisfy this requirement.

- There may be no "globally" optimum climate change strategy because of the large uncertainties and differences between parties, but the factors that affect optimal single decision maker strategies still have relevance to individual parties.

The lack of an individual decision maker, utility problems, and incomplete information suggest that decision analysis cannot serve as the primary basis for international climate change decision making. Although elements of the technique, such as the construction of influence diagrams, have considerable value in framing the decision problem and identifying its critical features, decision analysis cannot identify globally optimal choices for climate change abatement. Decision analysis suffers fewer problems when used by individual countries to identify optimal national policies, since decision scenarios are probably less complex and differences among collective decision makers may be less extreme. Without an effective quantitative approach to decision optimization, climate change decision makers will have to rely on negotiation to choose their responses to the problem.

2.3.3 *Negotiating climate change decisions*
The Framework Convention on Climate Change establishes a collective decision-making process within which the parties will negotiate future actions. Although some features of the decision-making process are set out in the Convention, many are still undecided. The preceding discussion has shown that, since decisions related to climate change involve many decision makers with different objectives, it may not be possible to find a decision-analytic outcome that is "best" from every perspective. It becomes important, then, to examine negotiation and compromise as the primary basis for climate change decisions under the Convention.

2.3.3.1 *Collective decisions in environmental accords*
Environmental accords tend to be complex integrated packages addressing a multiplicity of interrelated issues – like possible actions, implementation procedures, cost sharing and enforcement – that are difficult to negotiate internationally. Much of the protracted nature of environmental negotiation is due to scientific uncertainty and the (*a priori*) need for continued learning about causes and effects to establish good collective policy. These agreements tend to be incremental because of the political uncertainty in devising equitable approaches to addressing environmental problems at the global level. They typically involve a succession of negotiations: before, during, and after the accord is reached.

Although there have been a number of international environmental accords in recent years, the negotiation of these accords has not gotten any easier. Reaching superior decisions about environmental problems in a collective decision process poses the following theoretical challenges:

- *Collective irrationality* – Where nonmarket resources and cooperation are involved, a fully rational collective decision cannot be defined (Pillet, 1996).

- *Uncertainty* – The literature does not provide general procedures for incorporating differing assessments of uncertainty into a collective decision-making process (Price, 1993).

- *Economic transfers* – At this time, there are no effective mechanisms for the distribution of economic burdens associated with environmental change.

In addition to these issues, a recent study of environmental treaties by Sjöstedt (1993) found that the complexity of environmental negotiations has been growing over time as a result of the following factors:

- Environmental issues have become increasingly linked to other policy concerns, such as trade and economic development. The implications of these linkages are often poorly understood.

- Many new nongovernment parties have become important participants in the negotiations.

- Bridging the differences between industrialized and developing countries has been a continuing problem, given their views on the equity of particular solutions.

- International environmental accords face excessive national ratification delays and limited compliance (IIASA, 1993).

Negotiations related to climate change share all these theoretical and practical characteristics. Inasmuch as the effectiveness of the FCCC is hampered by these factors, decision makers may need to consider ways that these problems can be addressed in the collective decision process.

2.3.3.2 Collective rationality and the negotiation process

The concept of individual rationality discussed in Section 2.3.2.4 can be extended to collective rationality in multiparty decision scenarios. Economists sometimes accept some form of "collective preference" as a means of allowing collective decision makers to reach rational decisions (even if the individual decision makers are not strictly rational). Other economic models, such as game theory models and principal agent models, assume that individual agents behave and interact rationally, so the decisions they reach jointly are also rational. But when individuals are dealing with nonmarket resources (like clean air), or when cooperation is required, knowledge requirements in related decision problems exceed those usually required for rational collective choice. If limitations on knowledge and computational capacity do not apply equally to all parties in the collective decision process, it makes the problem even worse, since asymmetric information

introduces gaming and other strategic behaviour which may stand in the way of efficient outcomes.

One way to address these collective cognitive barriers is to simplify the process of negotiation itself. A number of tools, like cognitive mapping, simulation modelling, rule-based systems, and stakeholder analysis, have been developed to diagnose and facilitate negotiation (IIASA, 1993). Stakeholder analysis, for instance, uses information on the position, interest, and priorities of stakeholders to identify the range of differences among them and to increase the potential for the formation of coalitions. As noted in Section 2.3.3, decision analysis techniques can also be used to support negotiation. For example, multiattribute utility techniques have been reoriented by researchers at IIASA to evaluate coalition building and preference adjustment in the UN Conference on Environment and Development. Sjöstedt (1993) has suggested even more innovative approaches to facilitate negotiation, such as using third parties or developing reasoning heuristics.

For the most part, studies have focussed primarily on negotiation prior to and during the development of an international accord. But postagreement negotiation is often required to sustain dialogue on issues that cannot, by their nature, be resolved by a single agreement. Postagreement negotiations present policymakers with somewhat different concerns, such as progressive reframing of problems, adjusting strategies and perceptions, and refining solutions. Although it has received little attention from researchers, postagreement negotiation can also be improved, for instance, by involving domestic stakeholders in negotiations from the beginning (and not just at the postagreement stage), or by modifying the structure of the accords themselves to create more manageable clusters of single-issue agreements (Sjöstedt 1993). Regardless of the stage of negotiation, by improving the decision process these techniques are intended to help collective decision makers reach an acceptable agreement (but not necessarily the *best* agreement).

2.3.3.3 Uncertainty and surprise

Decision makers must rely on measures of uncertainty as a basis for addressing potential outcomes. But in negotiated agreements, the treatment of uncertainty is often very specific to the nature of the uncertainty. Without the quantitative structure of a decision-analytic model, communicating uncertainty to decision makers in a manner that facilitates more objective judgments becomes difficult. The literature offers no general procedures for incorporating risk in a consensus-building decision process. Shlyakhter *et al.* (1995) have explored a central question along these lines: Are decision makers better served by probability distributions for potential impacts or by a set of "best-guess" deterministic impact scenarios?

Probability distributions for different climate change outcomes (developed subjectively or objectively) provide decision makers with the maximum amount of information regarding impact uncertainty. Deterministic scenarios filter and discretize that information by assuming that all uncertainties are resolved before decisions are made. For example, a "most likely" impact scenario could be generated by projecting the state of the world when all stochastic variables are set at their mean levels. A set of scenarios spanning the range of anticipated outcomes could be developed in a similar manner. Parties in a collective negotiation process can attach different weights to the scenarios, reflecting their particular vulnerabilities in the given scenarios, their preferred decision criteria, or other factors.

Deterministic scenarios and probability distributions are distinct forms of addressing uncertainty; they can lead to different decision outcomes. According to Shlyakhter *et al.* (1995), these techniques can be reconciled by developing two types of deterministic scenarios – those that are *probable,* and those that are *possible* but improbable:

> Those risks that fall below a particular threshold of probability – and are thereby ignored by a particular group or society – can be called *de minimis* risks. . . . Although there is no clear definition of a *de minimis* risk, it can generally be seen to be closely akin to a related concept, the probability of surprise.

When an event is perceived as extremely improbable, society may take no action to avert or otherwise prepare for it. In these circumstances, the society is vulnerable to surprise should the event occur. If the event is truly improbable, then surprise must also be improbable. But if the assumption of *de minimis* risk is based on erroneous estimates of improbability, the chance of surprise may be unacceptably high. This leads to the question: "At what probability of a serious effect should society take action?"

Shlyakhter *et al.* (1995) note that "how societies and governments decide what constitutes *de minimis* risk in particular situations . . . is largely a matter of political judgement." In an effort to identify a more objective threshold for *de minimis* risks, the authors denote surprise scenarios as those where the true value of a particular parameter appears at least 2.6 standard deviations away from its current "best guess" value. For a normally distributed random variable, the probability that the "true" value is more than 2.6 standard deviations from the best guess is just 1%. According to this formula, then, *de minimis* risks would be those with less than a 1% chance of occurring. For comparison, Shlyakhter *et al.* (1995) report that public opinion polls show many people are unconcerned about a 5% chance of a climate-related catastrophe within their lifetime, but are concerned about a 1% chance of a nuclear accident. Shlyakhter, Valverde, and Wilson (1995) further report that an airliner with a calculated chance of failure around 5% in its 30-year life would not be allowed to fly in commercial service.

2.3.3.4 *Uncertainty and strategic behaviour*
Real world decision makers are faced not only with uncertainties but with difficult scientific, technical, ethical, and political controversies. Allais (1953) pointed out long ago that, apart from increasing the complexity of decision making, uncertainty creates manoeuvering room for strategic behaviour in controversial collective negotiations. For instance, cautious parties may feel they can do better in the long run by playing it safe for the present and thus increasing the probability that

they will be able to take advantage of more favourable deals in the future (Pratt, 1995).

Historical lessons from environmental disputes point to three pitfalls arising from strategic behaviour:

- Preemption by the short term of the long term because of unchanged current behaviour
- Dictatorship of the long term over the short term due to premature (arbitrary) decisions
- Paralysis of collective action by endless disputes

Strategic behaviour in negotiation is possible because the valuation of decision outcomes is dynamic – there is a distinction between long-term and short-term interests, and decisions themselves are made over time. One means of avoiding the pitfalls of strategic behaviour is to acknowledge and incorporate the dynamic nature of collective decision making by means of a sequential decision process.

2.3.3.5 *Sequential decision making under uncertainty*
Sequential decision making is called for when choosing short-term strategies under long-term uncertainty. Such decision scenarios, as pointed out by Grubb, Lave, Dowlatabadi, Hourcade, and other authors in IIASA (1994), are often characterized by:

- inertia in technological trends
- endogenous uncertainties
- the possibility of surprise
- diversity of beliefs and expectations

The specification of the relationships among choices makes decision analysis a potentially powerful technique for evaluating sequential decisions. Although the intractability of complex decision trees has limited the application of this technique in environmental problems, several studies have analyzed sequential climate change decisions using simple models, typically involving two or three decision points, choices, and possible outcomes. According to these analyses, in the face of uncertain and irreversible outcomes, the best decision lies somewhere between the most extreme options (i.e., doing nothing and aggressive action), given the outcome probabilities available at the time (Singer *et al.,* 1991; Manne and Richels, 1992, 1993; Hammitt *et al.,* 1992; IUCC, 1993; Richels, 1994; Peck, 1994). Such a strategy hedges against being surprised by those events that fall below a particular threshold of probability.

This sequential decision strategy can be described as an "act–learn–act" approach. Initial actions are taken without knowing exactly how the world will respond. As time passes, new information is incorporated into the decision process. Additional actions are taken based on this updated information, and so on. Since global environmental processes, scientific research, and developments related to mitigation and adaptation are all continuous, decisions on environmental policies should be adjusted continuously in the light of new information – and not only at prespecified intervals. Note that the "act–learn–act" approach is distinct from a simple hedging strategy, since the latter does not necessarily incorporate learning between decisions.

In climate change, complete information to support long-term decision making is unavailable, and the costs of delay are potentially high (due to forgone abatement opportunities and damages due to increased emissions). Given the collective nature of climate change decisions, the flexibility of the international community to react to new information is essential. The best decision strategy, then, may be the "act–learn–act" strategy, with initial decisions based on the best currently available information. In fact, prefatory statements in the FCCC include citations of existing and ongoing analysis as the basis of the Convention. So the actions undertaken in the Convention can already be viewed as the basis of an "act–learn–act" approach, with the first action being the initiation of studies to inform the decision process.

2.3.3.6 *The value of improved information*

In sequential decision making, additional information on the nature or likelihood of potential outcomes may allow actions to be better matched to those outcomes. Assuming that subjective outcome probabilities may be altered by learning, research is valuable to the extent that it refines subjective probabilities in a way that improves future decisions.

The value of new information (perfect and imperfect) can be computed. (When parties are risk-neutral, this value is equivalent to expected utility as discussed in Section 2.3.2.3.) Suppose that expected values for a decision option are as follows:

Without information:	$150
With imperfect information:	$325
With perfect information:	$500.

This hypothetical example illustrates a perfectly general result: The value of perfect information ($500 – $150 = $350) clearly exceeds the value of imperfect information ($325 – $150 = $175). It also indicates that some information – albeit imperfect – is better than no information at all (Eeckhoudt and Gollier, 1995). Under "act–learn–act" decision making, the value of new climate study information depends on changes in the probabilities assigned to alternative decision options before and after the study. "If the probabilities of (alternative) scenarios remain equal, then the value of the study is zero; if, on the other hand, only one scenario can be selected, a study might be worth as much as 100 billion dollars" (Shlyakhter *et al.*, 1995).

In "act–learn–act" decision making, uncertainty is not fully resolved before a decision is taken, so the resolution of the uncertainty must be viewed as either unacceptably costly or not helpful to the decision. But reducing uncertainty in a way that can create future options (without obligating parties to take a particular action) should be highly valued. Accordingly, uncertainty should prompt decision makers to focus on the timing of crucial investment decisions and to accelerate those near-term activities that create options (Dixit and Pindyck, 1995). Hourcade and Chapuis (1994) have shown that, after accounting for climate-related surprises in a sequential decision model, a resolution of uncertainty had high value relative

to no-regret potentials and technical innovation. Peck and Teisberg (1993) and Manne and Richels (1992) show similar results.

In collective decision making, the choice of actions may be strongly affected by friction between the long-term nature of the problem (including the required time for resolving uncertainties) and the relatively short tenure of decision makers. For example, to avoid negative political repercussions, decision makers may choose to postpone a costly or otherwise unpopular abatement activity until after their official terms expire. Uncertainty can be an excellent justification for such a postponement, even though the delay may be costly in the long run. In the case of climate change, for example, it is assumed that the damage potential will remain unknown until 2010–2020; if that is so, then appropriate emission levels can be determined only at the end of this period. To avoid costly postponement, emission decision makers should be explicit about the decision uncertainties as they are understood at the time, as well as the potential long-term implications of their short-term decisions.

2.3.3.7 *Economic transfers*

Collective assessments must address international economic transfers. The need to compare costs and benefits in collective decisions, for example, requires careful consideration of both the local and global ramifications of an action. Abatement "costs" may include not only domestic expenditures but also economic losses in the import/export market, a slowing of industrial development, and other effects. Similarly, benefits may include not only reductions in local impacts but also the overall survival of vulnerable states and the stimulation of abatement technology development useful to other countries. The global aspects of environmental problems require that domestic costs and benefits be defined relative to other abatement opportunities in other regions. Cost-benefit assessments, therefore, must be informed by an understanding of different perceptions and priorities among countries (Chapter 3), the importance of time horizons in different regions (Chapter 4), economic transfers, and, perhaps most important, the effects of an action on global cooperation itself.

Although the need to consider the interests of all affected parties greatly complicates the collective decision process, collective action yields benefits that could not be achieved if the problem were addressed in a more fragmented manner. One of the important potential gains from cooperating in a collective accord is the distribution of risk when outcomes are uncertain. A group has much greater possibilities for coping with risk than an individual, which is why organizations have always been best suited to undertake risky activities. As Eeckhoudt and Gollier (1995) point out:

> This is partly due to the opportunities of diversification within the group and partly due to the transfer of risk towards the least risk-averse members (or the richest members if absolute risk aversion is decreasing). . . . Without this diminution of risk aversion in an economy, thanks to the creation of risk pools, many risky projects would not have been undertaken and we would undoubtedly not have

known the economic expansion that we have observed over the last two centuries.

Since individuals are generally willing to pay for a reduction in risk, risk itself is costly (apart from the potential losses being risked). Transferring risk from one country to another or sharing risk collectively does not change the risk itself, but it can improve the overall welfare of the parties exposed to that risk. The literature indicates that "the transfer of risk is a potential source of large improvements in economic efficiency and social welfare" rather than a zero sum game (Eeckhoudt and Gollier, 1995).

Unfortunately, effective mechanisms for the appropriate redistribution of these international economic burdens are not currently available. Economic theory demonstrates that, where damages are borne collectively, individual responsibility for action leads to suboptimal outcomes. There is a strong incentive for each party to rely on the others to act – in other words, to be a free rider. A conventional approach to reducing the cost of risk is insurance, where relatively small premiums paid by all parties are able to compensate those who suffer losses. But there is no global market for national insurance against environmental losses. Even if the institutions were available to distribute such risk, expecting agreement on the appropriate sharing formula would be unrealistic.

Given the political and economic ramifications associated with international transfers (due to the imposition of payments on electorates, for example), reaching any collective agreement to address climate change will be difficult. Completely satisfying all parties is probably impossible. Nevertheless, there is a clear need to consider fair and feasible transfers in any comprehensive decision process in order to attract widespread participation and insure international support. Sharing the costs of actions in a sequential path of acting and learning should lead to collective support for the globally optimal climate strategy. (Chapter 3 discusses at some length conditions needed for equity among the parties in collective decision making.)

2.3.3.8 *Conclusions about negotiated decision making*

The limitations of decision analysis and the collective decision structure of the FCCC make negotiation the dominant element in climate change decison making. The preceding discussion has identified several important factors affecting these decisons:

- Excessive knowledge requirements in negotiated environmental decisions may stand in the way of collective rational choice. This cognitive burden could be reduced by facilitating the negotiation process itself through the use of tools like stakeholder analysis or by splitting accords into more manageable clusters of agreements.

- Since society has no consistent probability threshold for ignoring particular risks, it may be vulnerable to surprise when risks are uncertain. In climate decisions, this vulnerability could be reduced by considering event scenarios relative to an explicit probability of surprise.

- When faced by long-term uncertainties, sequential decision making allows actions to be better matched to outcomes by incorporating additional information over time. Sequential decision making also minimizes harmful strategic behaviour among multiple decision makers.

- Improved information about uncertain outcomes may have very high economic value, especially if that information can create future decision options.

- There are currently no effective mechanisms for sharing the risks related to climate change and their associated economic burdens. International risk sharing could yield substantial benefits for global economic and social welfare.

A consideration of the climate change problem from the perspectives of quantitative optimization and collective negotiation has yielded many general insights into the nature of climate decision making, the obstacles it faces, and potential means of addressing those obstacles. The chapter will now conclude with a more specific treatment of key issues currently faced by the Convention.

2.4 Implications for National Decision Making Under the FCCC

Because of decision uncertainties and the differing interests and values of international parties, there is no unique globally optimum response to climate change. Action will be taken as a result of collective negotiation under the FCCC. In this process, each party will judge appropriate responses according to sometimes different perceptions of what might constitute an optimum strategy both for itself and for the world as a whole. This is not a pursuit of unfettered self-interest. The purpose of negotiations is for countries to agree to act differently than they would in the absence of an agreement in order to realize the common benefits from collective action. Given the decision context set out in Section 2.2, this section elaborates on the implications of the decision-making issues introduced in Section 2.3 in the context of global negotiations on climate change.

2.4.1 *The Framework Convention on Climate Change*

The Framework Convention on Climate Change went into effect on 21 March 1994. As of April 1995, it had been ratified by some 128 parties. The Convention provides the legal, institutional, procedural, and normative framework for the international community to consider responses to the threat of climate change and its impacts.

2.4.1.1 *Objectives and commitments*

The FCCC's objective provides a fundamental reference point for decisions by the parties. The objective of the Convention (Article 2) is

to achieve stabilization of greenhouse gas concentrations in the atmosphere at a level that would prevent dangerous anthropogenic interference with the climate system. Such a level should be achieved within a time frame sufficient to allow ecosystems to adapt naturally to climate change, to ensure that food production is not threatened and to enable economic development to proceed in a sustainable manner.

The objective, and other provisions of the Convention, make clear that the parties must consider both mitigation and adaptation to climate change.

The Convention established general commitments which bind all parties and specific commitments that apply to developed country parties. All parties must develop emission inventories, implement mitigation/adaptation programmes, support technology transfer, promote sustainable management of greenhouse gas sinks, and account for climate change in social, economic, and environmental policies, where feasible. In addition to these commitments, developed countries must "aim" to return to 1990 levels of greenhouse gas emissions and take the lead in modifying longer-term emission trends. Unfortunately, the language of the Convention leaves unclear the precise nature and extent of some of its specific commitments.

The Convention also requires developed parties to coordinate relevant economic and administrative instruments to achieve the Convention's objectives and to review their own policies to see if they encourage increased emissions. Each country's contribution to emission reduction depends on a number of factors, including its economic structure, resource base, starting point, individual circumstances, and equity. The Convention allows the Conference of the Parties (COP) to weaken or strengthen the parties' obligations under the treaty in response to scientific information on climate change as well as relevant technical, social, and economic information. Parties agree to pay particular attention to supporting international and intergovernmental efforts to strengthen systematic observation and national scientific and technical research capacities and capabilities.

2.4.1.2 Institutional arrangements for decision making

The Conference of the Parties is the supreme decision-making body of the Convention, responsible for keeping the implementation of the Convention under regular review and for ensuring that implementation is meeting the Convention's objective. The COP is supported by the Subsidiary Body for Scientific and Technological Advice (SBSTA) and the Subsidiary Body for Implementation (SBI). The SBSTA is intended to link the scientific, technical, and technological assessments and information provided by competent international bodies with the policy-oriented needs of the COP. The SBI is intended to develop recommendations to assist the COP in its review and assessment of the implementation of the Convention and in the preparation and implementation of its decisions. The institutional arrangements for decision making also include a financial mechanism which functions under the guidance of the COP.

The COP was intended to adopt rules of procedure for itself and its subsidiary bodies by consensus at the first Conference of the Parties (COP 1). These rules were intended to provide decision-making procedures, including procedures

for matters not covered by the Convention, and could have included specified majorities required for the adoption of particular decisions. Consensus could not be reached on the whole set of rules at COP 1 (although decisions were still made). Negotiations between parties on the rules of procedure continue.

The Convention's overall decision-making machinery includes a Secretariat to support the Convention. It also includes a multilateral consultative process, which has not been finalized but is intended to assist resolution of questions regarding the implementation of the Convention. COP 1 decided to establish an ad hoc open-ended working group of technical and legal experts to study all issues relating to the establishment of this process for COP 2. Finally, in the event of a formal dispute between parties, the Convention provides for the possibility of conciliation, arbitration, or recourse to the International Court of Justice.

The need for sequential decision making is reflected in the Convention's provisions for review, assessment, elaboration of commitments, and other such procedures. In this context, COP 1 established an open-ended ad hoc group of parties "to begin a process to enable it [the COP] to take appropriate action for the period beyond 2000 . . . through the adoption of a protocol or another legal instrument." The FCCC also says much about decision-making procedures, including the need for transparency, publication of reports, and wide participation. But the lack of formal, agreed rules of procedure leaves open important aspects of how the parties will make decisions.

2.4.2 International transfers in climate change

Section 2.3.3.7 raised the issue of international transfer as a potential obstacle to reaching collective decisions. Such transfers have already emerged as one of the most difficult issues facing the FCCC community. The possible nature, degree and role of international transfers related to climate change must be considered in any comprehensive climate decision. This requirement arises for a number of reasons, including the following (see also Chapter 3):

- The FCCC already mandates that developing countries will receive the "full agreed incremental costs" of measures taken under the Convention.

- Efficiency arguments suggest that some of the cheaper abatement opportunities may lie in developing countries. This has resulted in calls to allow joint implementation between developed and developing countries as a way of meeting emission commitments.

- Equity arguments based on the concept of "environmental space" hold that the industrialized world has emitted the great majority of persistent greenhouse gases. Consequently, those who have occupied an "unfair" share of this space should in some way compensate the others. This is one interpretation of the "polluter pays" principle.

- A more generalized efficiency and equity case for transfers arises from the long-term need to define fair "emission rights" (tradable quotas) and to allow countries to exchange these on mutually beneficial terms (Parikh, 1994b; see also Chapter 3).

- Efficiency considerations suggest that risks associated with climate impacts should be shared through an international insurance mechanism (see below).
- Ethical arguments and some principles of international law suggest that countries should be liable for environmental damage they impose on others.

International transfers, in one form or another, are likely to serve as both the building blocks of globally optimal action and the cement of global cooperation. Nevertheless the political and managerial difficulties surrounding such transfers need to be understood and respected by all parties if the process is not to collapse into an unproductive struggle over resource transfers (Parikh and Painuly, 1994; Parikh, 1995).

Table 2.1. *Natural disasters and associated insured losses*

Event	Year	Insured Losses (billion U.S.$)
Hurricane Gilbert	1988	0.05
Hurricane Hugo	1989	5.8
Winter storms (Europe)	1990	10.0
Summer storms (Colorado)	1990	1.0
Hurricane Bob	1991	0.62
Hurricane Andrew	1992	15.5
Hurricane Iniki	1992	1.6
East coast storms (U.S.)	1993	1.6
Midwest floods (U.S.)	1993	0.76

Source: Weilenmann (1994).

2.4.3 Sequential climate decisions

The most important benefit of applying sequential decision making to the climate problem is that the FCCC "need not be overly concerned with . . . inability to predict the economic and technical system several decades into the future: uncertainty is important only to the extent that it confounds near-term decision making. Today's decisions appear to be relatively insensitive to some of the more controversial longer-term uncertainties in the greenhouse debates" (Manne and Richels, 1993). This insensitivity of some short-term climate decisions to long-term uncertainties is fortunate, since diverging expectations about the long run, like appropriate economic development time horizons and expert disagreement about the large-scale use of competing energy sources (such as biofuels or nuclear energy), could impede the formation of a consensus for action.

The FCCC negotiations do not have to resolve controversies on long-term issues like sea level rise or force premature agreements on difficult disputes about burden sharing. The objective of the first step of the decision sequence is to put society as far as possible in the position to postpone technological or institutional "lock-in" and to use the extra negotiation time to increase options and reach wider consensus on how to approach the more difficult longer-term decisions.

The literature identifies five types of short-term decisions apt to improve society's decision-making capabilities in the future:

- Investing in climate research simply because of the high economic value of scientific information
- Financing technology research and development through government-led programmes (Chapter 9)
- Inducing technical change through market incentives (Arrow, 1962; Grubb *et al.*, 1995)
- Making low-cost abatement decisions to increase learning time where risks are controversial or where the potential for surprise raises the value of information and new options (Hourcade and Chapuis, 1994; Chapter 8)
- Preventing bifurcation toward high carbon-intensive development paths (Chapter 8)

The goal of the near-term effort should be to make these decisions with an eye toward reconciling three long-term needs: stable greenhouse gas concentrations, scientific knowledge, and technical and consumption patterns that enhance flexibility in managing transition.

2.4.4 Instruments for international insurance

Losses associated with individual natural disasters have been rising (even in wealthy countries). Of these losses, those due to weather have been rising more quickly than those due to earthquakes (Yokohama World Conference on Natural Disaster Reduction, 1994). A list of major storms and associated insured losses over the period 1988–1993 is shown in Table 2.1. Storms with insured losses greater than $1 billion were unknown before 1989, but six of the nine events listed in Table 2.1 had insured losses in excess of this amount. Hurricane Andrew alone involved losses of $15.5 billion.

The reason why losses due to weather-related events are higher than those due to earthquakes is not clear and cannot be conclusively related to climate change. An increase in the frequency and severity of extreme events as atmospheric concentrations of greenhouse gases rise is not clear from climate models.

Insurance as a means of sharing risks is well suited to situations where the likelihood of a damaging event for any individual party is relatively small but the potential damages are large. Since the relationship between weather-related events and climate change is not known, insurance to cover the risks of climate change, *per se,* is probably not feasible or necessary. However, insurance to cover the damages associated with weather-related events is desirable. To the extent that the frequency or severity of weather events turns out to be affected by climate change, such insurance would be a form of climate change insurance. But since a meaningful premium for private insurance against climate change losses cannot be calculated, insurers have responded to weather disasters by withdrawing or restricting coverage in regions that are particularly prone to such events. If climate change does contribute to those events, the affected regions bear the costs of actions by the rest of the world.

Section 2.3.3.7 noted reasons why international insurance against climate impacts could enhance welfare. However, Wilford (1993), Chichilnisky and Heal (1993), and others point out several reasons why commercial insurance markets cannot adequately cover risks associated with climate change. First, there is no international market in which individuals or countries can insure themselves against losses from climate change or related abatement policies. Even if such a market existed, insurance on a country-by-country basis would miss many potential benefits from collectively sharing risks. To the limited extent that insurance could cover climate risks, the insurance premiums would probably be borne inequitably by the parties exposed to those risks. Establishing an appropriate form of global insurance could thus increase both efficiency and equity by reducing exposure to risk and the cost to individual countries of bearing that risk.

2.4.4.1 Financial markets for risk

Climate change risks impose particular requirements for insurance. One option would be a mutual insurance contract – an agreement between parties subject to similar risks that those who suffer losses will be compensated by others. Such insurance is used, for example, in agricultural cooperatives. In the context of climate change, this type of insurance contract would be a binding agreement in which countries that suffer greater-than-average (or expected) climate-related losses would be assisted by those suffering less-than-average losses.

A second type of insurance contract acknowledges that the overall nature and distribution of some climate-related risks are uncertain. In such circumstances, the formal treatment of an appropriate insurance structure is complex (Arrow, 1953). It requires defining "risk securities" for each possible outcome that pay out only if that outcome is realized. In climate change, such insurance would require each country to make compensation commitments as insurance against a particular climate outcome. To distribute the risks efficiently, countries would then be allowed to trade these securities. To the extent that the perception of risks varies, such an approach would amount to betting on particular climate outcomes (Pillet, 1994).

By allowing for different beliefs about risks, risk securities and mutual insurance would permit a more efficient distribution of those risks. For example, a country genuinely believing that climate change is unlikely to have serious global impacts would be more prone to hold those securities that pay out under these conditions (Heal, 1993). A formal two-country treatment is summarized in Chichilnisky and Heal (1993) and Chichilnisky (1994). These studies also note that creating risk securities may provide an objective test of the honesty of national positions on the risks of climate change. If, as part of a negotiating ploy to avoid onerous abatement commitments, a country were to argue that climate change does not involve substantial risks, that country would have to be prepared to hold associated high payoff securities. Hence, there would be an economic penalty for misrepresentation of true beliefs. These penalties could offset some of the incentive to free ride on other countries' efforts to reduce greenhouse emissions. Risk securities also have potential for improving

equity in both decision processes and outcomes. The feasibility, credibility, and equity of such securities has yet to be established and may be an important research topic.

The international community is a long way from having sophisticated instruments like risk securities. Nevertheless, there have been proposals for specific international insurance funds, particularly to help the most vulnerable countries cope with climate impacts. Although it is closer to a liability scheme than to an insurance contract, an AOSIS proposal submitted to the First Conference of the Parties calls for insurance pool contributions to be collected in 2004 – provided the rate of global mean sea level rise has by then reached an agreed figure. If sea levels have not risen substantially by that time, a review of conditions would be undertaken. This AOSIS proposal was presented to the negotiators of the Climate Convention but was excluded from the final treaty. The AOSIS and related proposals are outlined in Chapter 3.

2.4.5 Portfolios of climate actions

Actions to limit the impacts of climate change may be required for the next century or longer. Numerous measures are potentially available to address these impacts, but their effects are uncertain and no single action appears to be clearly superior to the others. Under uncertainty, a portfolio of measures will, on average, yield a better outcome than any individual action. The decision problem, then, is to choose a portfolio of measures to achieve climate change goals at minimum cost, accounting for the risks associated with different measures. The portfolio may include some relatively high cost measures to diversify the risks.

Many investment decisions have implications for climate change because they may lock in commitments to energy use or other greenhouse gas sources for several decades. The potential climate implications of these investments do not justify delaying such decisions. However, the potential climate change implications of transportation infrastructure, land use decisions, energy-using equipment, and similar investments should be considered when the alternatives are being evaluated. The decisions should be based on the best information available at the time and should properly reflect the value of future flexibility.

Some of the measures in the climate change portfolio will also have relatively long lifetimes. The extent to which resources are locked in by these measures should be considered when their climate change benefits are evaluated. Although some individual measures may be inflexible, other measures can be chosen to ensure that the overall portfolio has sufficient flexibility. Thus, decision makers will be able to adjust a well chosen portfolio frequently in response to new information, even though some of the measures are locked in for relatively long periods.

2.4.5.1 Climate portfolio options

The key to selecting a climate change action portfolio is to understand how measures interact over time. Mitigation measures provide future flexibility, technology research can lower the cost of future action, and climate research can provide better information on the nature of the actions required. In principle, the measures available to countries (individually or jointly) to limit climate change and its impacts include:

- Implementing low cost measures, such as energy efficiency, to reduce emissions of greenhouse gases
- Phasing out existing distortionary policies, such as some fossil fuel subsidies, that reduce welfare and increase greenhouse gas emissions directly or indirectly
- Switching from more to less carbon-intensive fuels or to carbon-free fuels to reduce emissions of greenhouse gases
- Enhancing or expanding greenhouse gas sinks or reservoirs, such as forests
- Implementing existing techniques (and developing new ones) for reducing methane and nitrous oxide emissions from industrial processes, landfills, agriculture, fossil fuel extraction, and transportation
- Instituting forms of international cooperation, such as joint implementation, technology transfer, and tradable quotas to reduce the cost of limiting greenhouse gas emissions
- Planning and implementing measures to adapt to the consequences of climate change
- Undertaking additional research on climate change causes, effects, and adaptation (Economic studies suggest that such research can yield high returns by reducing uncertainty about actions to address climate change)
- Conducting technological research to enhance energy efficiency, minimize emissions of greenhouse gases from fossil fuel use, and develop commercial nonfossil energy sources (In the long run, the cost and timing of availability of nonfossil energy technologies is one of the major determinants of the cost of addressing climate change)
- Developing institutional mechanisms, such as insurance, to share the risks of damages due to climate change

The specific policy measures available vary from country to country. Countries will select a portfolio of climate change measures that reflect, implicitly or explicitly, their individual objectives and constraints. A country may look for the economically optimal portfolio of climate change measures, but the impacts of the portfolio on different economic groups, international competitiveness, international equity, and intergenerational equity are likely to come into play.

2.4.5.2 *Choosing the best climate action portfolio*
There is no operational model with which to identify the optimal portfolio of climate change policies for a country. Nonetheless, the limited literature offers some insights concerning the costs and benefits of possible measures. For example:

- Analyses of the costs of reducing greenhouse gas emissions have been undertaken for many countries and sec-

tors (see Chapter 9). Once other environmental benefits (such as lower emissions of other pollutants) and economic benefits (such as reduction of existing distortionary taxes) are accounted for, nearly every study finds some measures to reduce greenhouse gas emissions at very low or negative cost.

- Peck and Teisberg (1993) and Manne and Richels (1993) have estimated the value of spending on climate research. They use highly stylized models for the analysis, but find that the expected return is several times the current level of climate research spending. Peck (1994) has also shown that current spending in the U.S. is not allocated in an optimal manner.
- Results from Energy Modeling Forum 12 (Energy Modeling Forum, 1993) suggest that reducing the costs of future nonfossil energy technologies could reduce the costs of achieving emission reduction targets by as much as two-thirds. This suggests a potentially large economic return from technology research. Funding for technology research related to climate change should be considered as a risk premium, not as subsidies to be randomly allocated.
- Several researchers have compared the costs of unilateral action and international cooperation to address climate change. The analyses consistently show large economic returns from international cooperation. Such cooperation, however, requires mechanisms for transferring resources among countries. Otherwise, countries with high marginal costs of emission reduction might find it advantageous to be free riders until a complete international climate agreement is concluded (Pillet *et al.,* 1993).
- Analyses of phasing out existing inefficient emission policies suggest that emissions can be reduced with net economic benefits. Larsen and Shah (1992) estimate that global CO_2 emissions would be reduced by between 4 and 5% if all energy subsidies were removed. An OECD study estimates that removal of energy subsidies would reduce global emissions 18% from projected levels in 2050 while increasing global real incomes by 0.7% annually (Burniaux *et al.,* 1992).

Decision makers need to decide on the level of climate change spending and the allocation of that total among the available measures. The level of climate change spending is likely to reflect both international commitments and domestic considerations such as the need for adaptation measures and economic development strategies based on mitigation technologies.

Conceptually the mix of climate change measures should be adjusted so that the risk-cost ratio is equal at the margin. However, only sketchy information is available on the costs, benefits, and risks of alternative measures. Governments can make subjective judgments about the merits of different measures and adjust the portfolio incrementally as new information becomes available.

Endnote

1. The ultimate objective of the Framework Convention on Climate Change is "stabilization of greenhouse gas concentrations in the atmosphere at a level that would prevent dangerous anthropogenic interference with the climate system. Such a level should be achieved within a time-frame sufficient to allow ecosystems to adapt naturally to climate change, to ensure that food production is not threatened and to enable economic development to proceed in a sustainable manner" (Article 2). Conceptually, cost-benefit analysis could be used to identify the stabilization level and date that yield the largest net benefit, although in practice the information needed to perform such calculations is not available. Having chosen a stabilization level and date, regardless of how that decision is made, cost-effectiveness analysis could be used to choose among strategies for achieving the objective.

References

Allais, M., 1953: Le comportement de l'homme rationnel devant le risque, critique des postulats et axiomes de l'école américaine, *Econometrica* **21**: 503–546.

Arrow, K.J., 1951a: *Social choice and individual values,* Wiley, New York.

Arrow, K.J., 1951b: Alternative approaches to the theory of choice in risk-taking situations, *Econometrica* **19**: 404–437. Reprinted in K.J. Arrow, 1984: *Collected papers of Kenneth J. Arrow, 3, Individual choice under certainty and uncertainty, pp.* 5–41, The Belknap Press of Harvard University Press, Cambridge, MA.

Arrow, K.J., 1953: The role of securities in the optimal allocation of risk bearing, *Économétrie,* Colloques internationaux du Centre National de la Recherche Scientifique, **11**, 41–47 (original in French). Reprinted in K.J. Arrow, 1983: *Collected Papers of Kenneth J. Arrow,* Volume 2, *General equilibrium,* pp. 46–57.

Arrow, K.J., 1962: The economic implications of learning by doing, *Review of Economic Studies, 29,* 155–173. Reprinted in K.J. Arrow, 1985: *Collected papers of Kenneth J. Arrow,* Volume 5, *Production and capital,* pp. 157–180.

Arrow, K.J., 1963: *Social choice and individual values,* 2d ed., Wiley, New York. Reprinted by Yale University Press, New Haven.

Arrow, K.J., 1971: Exposition of the theory of choice under uncertainty, In *Decision and organization,* C.B. McGuire and R. Radner, eds., pp. 19–55, North-Holland, Amsterdam. Reprinted in K.J. Arrow, 1984: *Collected papers of Kenneth J. Arrow,* Volume 3, *Individual choice under certainty and uncertainty,* pp. 172–208.

Arrow, K.J., 1983–1985: *Collected papers of Kenneth J. Arrow.* The Belknap Press of Harvard University Press, Cambridge, MA.

Arrow, K.J., 1987, 1990: Economic theory and the hypothesis of rationality, *The new Palgrave,* pp. 25–37, Norton, New York and London.

Arrow, K.J., 1994: Discounting and climate change prospects, Communication at the IPCC Working Group III Writing Team II Montreux Meeting, March 3-6. Montreux, Switzerland.

Arrow, K.J., and A.C. Fisher, 1974: Environmental preservation, uncertainty and irreversibility, *Quarterly Journal of Economics,* **88,** 312–319. Reprinted in K.J. Arrow, 1985, *Collected papers of Kenneth J. Arrow,* Volume 6, *Applied economics,* pp. 165–173.

Ayres, R.U., and P.M. Weaver, 1994: The case for proactive risk management. In *Environmental Hazard: Liability or Opportunity?* Report of the SANDOZ/INSEAD Colloquium held 29 June 1993, Paris: INSEAD, Centre for the Management of Environmental Resources, pp. 3–9.

Beltratti, A., 1994: Environmental problems and attitudes towards risk and uncertainty. In *Steps Towards a Decision Making Framework to Address Climate Change. Report of the IPCC Working Group III Writing Team II Montreux Meeting, March 3-6,* Pillet and Gassmann, eds., pp. 67–74. Würenlingen & Villigen (Switzerland): PSI-Bericht 94-10, Paul Scherrer Institute.

Björkman, M., 1984: Decision making, risk taking and psychological time: Review of empirical findings and psychological theory, *Scandinavian Journal of Psychology,* **25,** 31–49.

Blaug, M., 1992: *The methodology of economics – Or how economists explain,* Cambridge surveys of economic literature, 2d ed., Cambridge University Press, Cambridge, MA.

Bray, M., 1987: Perfect foresight, *The new Palgrave,* pp. 144–148, W.W. Norton & Company, New York and London.

Broome, J., 1992: *Counting the cost of global warming,* Cambridge University Press, Cambridge.

Burniaux, J.M., J.P. Martin, G. Nicoletti, and J.O. Martins, 1992: GREEN, Working Paper 115, Organization for Economic Cooperation and Development (OECD), Paris.

BUWAL/OFEFP, 1994: *Le réchauffement planétaire et la Suisse: bases d'une stratégie nationale / Global warming and Switzerland: Foundations for a national strategy,* Federal Office of Environment, Forests and Landscape (BUWAL/OFEFP), Berne (Switzerland).

Caselton, B. and W. Luo, 1994: Dempster-Shafer theory and decision making under near-ignorance, Working Paper, University of British Columbia and British Columbia Hydro, Vancouver.

Chichilnisky, G., 1993a: What is sustainable development? Research paper, Columbia University, New York.

Chichilnisky, G., 1993b: Markets with endogenous uncertainty: Theory and policy, Working Paper, Stanford University, Department of Economics, Stanford, CA.

Chichilnisky, G., 1994: Global environmental risks and financial instruments. In *Steps towards a decision making framework to address climate change. Report of the IPCC Working Group III Writing Team II Montreux Meeting, March 3-6,* Pillet and Gassmann, eds., pp. 99–109, Würenlingen & Villigen (Switzerland): PSI-Bericht 94-10, Paul Scherrer Institute.

Chichilnisky, G., and G. Heal, 1992: Financial markets with unknown risks, First Boston Working Paper Series, Columbia Business School, New York.

Chichilnisky, G., and G. Heal, 1993: Global environmental risks, *Journal of Economic Perspectives* **7**(4), 65–86.

Choucri, N. (ed.), 1993: *Global accord – Environmental challenges and international responses,* MIT Press, Cambridge, MA.

Cline, W.R., 1992: *The economics of global warming,* Institute for International Economics, Washington, DC.

Cline, W.R., 1993: Greenhouse policy after Rio: Economics, science, and politics. In *Costs, impacts, and benefits of CO_2 mitigation,* Y. Kaya, N. Nakicenovic, W.D. Nordhaus, and F.L. Thoth, eds., Proceedings of a Workshop held on 28-30 September 1992 at IIASA, Laxenburg, Austria, pp. 41–56, IIASA, Laxenburg, Austria.

Dawes, Robyn M., 1988: *Rational choice in an uncertain world.* Harcourt Brace Jovanovich, Orlando, FL.

Dempster, A.P., 1967: Upper lower probabilities induced by a multivalued mapping, *Ann. Math. Statist.* **37**: 355–374.

Dixit, A.K., and R.S. Pindyck, 1995: The options approach to capital investment, *Harvard Business Review* **73** (3): 105–115.

Dowlatabadi, H., and M.G. Morgan, 1993a: A model framework for integrated studies of the climate problem, *Energy Policy,* **21,** 209–221.

Dowlatabadi, H., and M.G. Morgan, 1993b: Integrated assessment of climate change, *Science,* **259,** 1813 & 1932.

Edmonds, J.A., and J.M. Reilly, 1986: *Uncertainty in future global energy use and fossil fuel CO_2 emissions 1975 to 2075,* Prepared for the U.S. Department of Energy, Washington, DC.

Edmonds, J.A., and J.M. Reilly, 1993: Global energy and CO_2 to the year 2050, *The Energy Journal,* **4:** 21–47.

Eeckhoudt, L., and C. Gollier, 1995: *Risk – Evaluation, management and sharing.* Translated by V. Lambson, Harvester Wheatsheaf, New York.

Energy Modeling Forum, 1993: *Reducing global carbon emissions – Costs and policy options.* EMF-12. Stanford University, Stanford, CA.

Fankhauser, S., 1994: Global warming damage and the risk of a climate catastrophe. In *Steps towards a decision making framework to address climate change. Report of the IPCC Working Group III Writing Team II Montreux Meeting, March 3-6.* Pillet and Gassmann, eds., pp. 19–29, Würenlingen & Villigen (Switzerland): PSI-Bericht 94-10, Paul Scherrer Institute.

Faucheux, S., and G. Froger, 1994: Decision-making under environmental uncertainty. In *Steps towards a decision making framework to address climate change. Report of the IPCC Working Group III Writing Team II Montreux Meeting, March 3-6.* Pillet and Gassmann, eds., pp. 51–65, Würenlingen & Villigen (Switzerland): PSI-Bericht 94-10, Paul Scherrer Institute.

Fishburn, P.C., 1987, 1990: Utility theory and decision theory, *The new Palgrave,* pp. 303–312, Norton, New York and London.

Fisher, I., 1930: *The theory of interest,* Macmillan, New York.

Froger, G., and E. Zyla, 1994: Decision making for sustainable development: Orthodox or system dynamics models? *Proceedings of the International Symposium on Models of Sustainable Development,* Université Sorbonne-Panthéon C3E and afcet, Paris, March 16–18, 1994, Volume II, pp. 1061–1074.

Gassmann, F., 1993: Non-linear dynamical systems, an introduction. In *Some physico-chemical and mathematical tools for understanding of living systems.* Greppin *et al.,* eds., University of Geneva, Geneva.

Gassmann, F., 1994: Abrupt change scenarios. In *Steps towards a decision making framework to address climate change. Report of the IPCC Working Group III Writing Team II Montreux Meeting, March 3-6.* Pillet and Gassmann, eds., pp. 3–9, Würenlingen & Villigen (Switzerland): PSI-Bericht 94-10, Paul Scherrer Institute.

Grinols, E.L., 1985: Public investment and social risk-sharing, *European Economic Review* **29,** 303–321.

Grubb, M., T. Chapuis, and M. Ha Duong, 1995: The economics of changing course, *Energy Policy,* **23**(4/5), 417–424.

Häfele, W., J. Anderer, A. McDonald, and N. Nakicenovic, 1981: *Energy in a finite world,* Ballinger, Cambridge, MA.

Haites, E.F., 1994: Portfolio analysis model of climate change decision making. In *Steps towards a decision making framework to address climate change. Report of the IPCC Working Group III Writing Team II Montreux Meeting, March 3-6.* Pillet and Gassmann, eds., pp. 111–114, Würenlingen & Villigen (Switzerland): PSI-Bericht 94-10, Paul Scherrer Institute.

Hall, C.A.S. (ed.), 1995: *Maximum power: The ideas and applications of H.T. Odum,* University Press of Colorado, Boulder, CO.

Hammitt, J.K., R.J. Lempert, and M.E. Schlesinger, 1992: A sequential-decision strategy for abating climate change, *Nature* **357:** 315–318.

Heal, G., 1993: Valuing the very long run: Discounting and the environment. Discussion paper, Business School, Columbia University, New York.

Heap, S.H., M. Hollis, B. Lyons, R. Sugolen, and A. Weale, 1992: *The theory of choice—A critical guide.* Blackwell, Oxford, UK, and Cambridge, MA.

Hediger, W., 1994: On the opportunity cost of climate change and policy. In *Steps towards a decision making framework to address climate change. Report of the IPCC Working Group III Writing Team II Montreux Meeting, March 3-6.* Pillet and Gassmann, eds., pp. 41–49, Würenlingen & Villigen (Switzerland): PSI-Bericht 94-10, Paul Scherrer Institute.

Henry, C., 1974: Investment decisions under uncertainty: The irreversibility effect, *American Economic Review,* **64,** 1006–1012.

Hope, C.W., J. Anderson, and P. Weinman, 1993: Policy analysis of the greenhouse effect – An application of the PAGE model, *Energy Policy* **21,** 327–338.

Houghton, J.T., G.J. Jenkins, and J.J. Ephraums (eds.), 1990: *Climate change. The IPCC scientific assessment.* Cambridge University Press, Cambridge.

Hourcade, J.C., 1993: Economic issues and negotiation on global environment. Some lessons from the recent experience on greenhouse effect. In *Trade, innovation, environment,* C. Carraro, ed., Kluwer Academic Publ., Dordrecht.

Hourcade, J.C., 1994: Analyse économique et gestion des risques climatiques, *Natures, Sciences, Sociétés* **2**(3), 202–211.

Hourcade, J.C., and T. Chapuis, 1994: No-regret potentials and technical innovation: A viability approach to integrative assessment of climate policies. In *Integrative assessment of mitigation, impacts and adaptation to climate change,* N. Nakicenovic, W.D. Nordhaus, R. Richels, and F.L. Toth, eds., IIASA Workshop Proceedings, Laxenburg, Austria.

IIASA (International Institute of Applied Systems Analysis), 1993: IIASA's options, *Processes of International Negotiation,* June 1993, 4–12.

IIASA (International Institute of Applied Systems Analysis), 1994: *Integrative assessment of mitigation, impacts and adaptation to climate change,* N. Nakicenovic, W.D. Nordhaus, R. Richels, and F.L. Toth, eds., IIASA Workshop Proceedings, IIASA, Laxenburg, Austria.

INSEAD/SANDOZ, 1994: *Environmental hazard: Liability or opportunity?* Report of the SANDOZ/INSEAD Colloquium, 29 June 1993. INSEAD, Paris.

IPCC, 1994: *IPCC technical guidelines for assessing climate change impacts and adaptations,* University College London and Center for Global Environmental Research, National Institute for Environmental Studies, London and Tsukuba, Japan.

IUCC, 1993: *Insurance against climate change? The AOSIS proposal,* Climate Change Fact Sheet 238, UNEP/WMO, Geneva.

Johansson, P.-O., 1993: *An introduction to modern welfare economics,* Cambridge University Press, Cambridge.

Joint Climate Project, 1992: *Joint climate project to address decision makers' uncertainties.* Science and Policy Associates, Inc., Washington, DC.

Jungermann, H., and F. Fleischer, 1988: As time goes by: Psychological determinants of time preferences. In *The Formulation of Time Preferences in a Multidisciplinary Perspective,* G. Kirsch, P. Nijkamp, and P. Zimmermann, eds., Edward Elgar, Aldershot, UK.

Kates, R.W., C. Hohenemser, and J.X. Kasperson (eds.), 1985: *Perilous progress: Managing the hazards of technology.* Westview Press, Boulder, CO.

Kaya, Y., N. Nakicenovic, W.D. Nordhaus, F.L. Toth (eds.), 1993: *Costs, impacts, and benefits of CO_2 mitigation,* Proceedings of a workshop held on 28-30 September 1992 at IIASA, Laxenburg, Austria. IIASA, Laxenburg, Austria.

Keeney, R.L., and H. Raiffa, 1976: *Decisions with multiple objectives: Preferences and value tradeoffs.* Wiley, New York.

Kneese, A.V., 1968: *Economics and the quality of the environment – Some empirical experiences.* Reprint Number 71, Resources for the Future, Washington, DC.

Kohlas, J., L. Cardona, R. Haenni, U. Hänni, B. Anrig, and N. Lehmann, 1994: Defensible reasoning and uncertainty management systems. In *Research Projects 1993-1994,* pp. 11–12, University of Fribourg (Switzerland), Institute of Informatics.

Krutilla, J.V., 1967: Conservation reconsidered, *American Economic Review* **57,** 777–786. Reprinted in Smith (ed.), *Environmental resources and applied welfare economics – Essays in honor of John V. Krutilla,* V. Smith, ed., pp. 263–273, Resources for the Future, 1988, Washington, DC.

Larsen, B. and A. Shah, 1992: *World fossil fuel subsidies and global carbon emissions,* Policy Research Working Paper Series, No. 1002, World Bank, Washington, DC.

Lind, R.C., K.J. Arrow, G.R. Corey, P. Dasgupta, A.K. Sen, T. Stauffer, J.E. Stiglitz, J.A. Stockfisch, and R. Wilson, 1982: *Discounting for time and risk in energy policy.* Resources for the Future, Washington, DC.

Luce, R.D., and H. Raiffa, 1957: *Games and decisions.* Wiley, New York.

Manne, A.S., and R.G. Richels, 1992: *Buying greenhouse insurance.* MIT Press, Cambridge, MA.

Manne, A.S., and R.G. Richels, 1993: CO_2 *hedging strategies – The impact of uncertainty upon emissions,* Prepared for the OECD/IEA Conference on the Economics of Climate Change, Paris, 14–16 June.

Markandya, A., and D.W. Pierce, 1991: Development, the environment, and the social rate of discount, *The World Bank Research Observer* **6** (2), 137–152.

Meyer, A., and A. Sharan [undated]: *Equity and survival – Who provides global benefit; Who causes global disbenefit? A basis for the reform of global institutional arrangements,* Global Commons Institute, London.

MIT, 1986: *The MIT dictionary of modern economics,* D.W. Pearce, ed., 3d ed., MIT Press, Cambridge, MA.

Molina, M.J., 1994: Science and policy interface, *Business & The Contemporary World* **6**(2), 126–128.

Morgan, M.G., and M. Henrion, 1990: *Uncertainty. A guide to dealing with uncertainty in quantitative risk and policy analysis.* Cambridge University Press, Cambridge.

National Research Council, 1983: *Risk assessment in the federal government: Managing the process.* National Academy Press, Washington, DC.

Nordhaus, W.D., 1982: How fast should we graze the global commons? *American Economic Review, AEA Papers and Proceedings* **72** (2), 242–246.

Nordhaus, W.D., 1991: To slow or not to slow: The economics of the greenhouse effect, *The Economic Journal* **101,** 920–937.

Nordhaus, W.D., 1992: *The DICE model: Background and structure of a dynamic integrated climate-economy model of the economics of global warming.* Foundation Discussion papers 1009 & 1019, Yale University, New Haven.

Nordhaus, W.D., and G. Yohe, 1983: Future carbon dioxide emissions from fossil fuels. In *Changing climate,* National Research Council, National Academy Press, Washington, DC.

Odum, H.T., 1983: *Systems ecology.* Wiley, New York.

Parikh, J., 1994a: Joint implementation and sharing commitments: A Southern perspective. In *Integrative assessment of mitigation, impacts and adaptation to climate change,* N. Nakicenovic, W.D.

Nordhaus, R. Richels, and F.L. Toth, eds., IIASA Workshop Proceedings, IIASA, Laxenburg, Austria.

Parikh, J., 1994b: North-South issues for climate change, *Economic and Political Weekly,* 5-12 November, 2940–2943.

Parikh, J., 1995: Joint implementation and North-South cooperation for climate change, *International Environmental Affairs,* **7**(1), 22–41.

Parikh, J., and J. Painuly, 1994: Population, consumption pattern, and climate change: A socio-economic perspective from the South, *Ambio,* **23**(7), 434–437.

Parikh, J., *et al.,* 1993: *Natural resource accounting,* IGIDR mimeograph, Indira Gandhi Institute of Development Research, Bombay.

Parikh, J., P.G. Babu, and K. Parikh, 1994: Decision making framework to address climate change. In *Steps towards a decision making framework to address climate change. Report of the IPCC Working Group III Writing Team II Montreux Meeting, March 3-6.* Pillet and Gassmann, eds., pp. 77–97, Würenlingen & Villigen (Switzerland): PSI-Bericht 94-10, Paul Scherrer Institute.

Parikh, J., K.J. Arrow, G. Pillet, *et al.,* 1995: Decision making framework to address climate change. In *Poverty, environment, and economic development,* N.S.S. Narayana and A. Sen, eds., Interline Publications, Bangalore, India.

Parikh, K., *et al.,* 1988: *Towards free trade in agriculture.* Kluwer Academic Press, Dordrecht.

Parry, M.L., and C. Rosenzweig, 1994: Potential impact of climate change on world food supply, *Nature,* **367,** 133–138.

Paté-Cornell, M.E., 1992: Risk analysis and relevance of uncertainties in nuclear safety decisions. In *Public regulation: New perspectives on institutions and policies.* E. Bailey, ed., pp. 227–253, MIT Press, Cambridge, MA.

Peck, S.C., 1993: The implications of non-linearities in global warming damage costs. In *Costs, impacts, and benefits of CO_2 mitigation,* Y. Kaya, N. Nakicenovic, W.D. Nordhaus, and F.L. Toth, eds., pp. 209–210, Proceedings of a Workshop held on 28-30 September 1992 at IIASA, Laxenburg (Austria). IIASA, Laxenburg.

Peck, S.C., 1994: Candidate framework for analysis of climate change decisions and uncertainties, Communication at the IPCC Working Group III Writing Team II Montreux Meeting, 3–6 March. Montreux (Switzerland).

Peck, S.C., and T.J. Teisberg, 1993: Global warming uncertainties and the value of information: An analysis using CETA, *Resource and Energy Economics,* **15,** 71–97.

Pethig, R. (ed.), 1994: *Valuing the environment: Methodological and measurement issues.* Kluwer Academic Publishers, Amsterdam.

Pigou, A.C., 1932: *The economics of welfare,* Macmillan, London.

Pillet, G., 1994: Betting on climate states – The impact of climate change prospects on decision making under "strong" uncertainty and collective risks. In *Steps towards a decision making framework to address climate change. Report of the IPCC Working Group III Writing Team II Montreux Meeting, March 3-6.* Pillet and Gassmann, eds., pp. 115–137, Würenlingen & Villigen (Switzerland): PSI-Bericht 94-10, Paul Scherrer Institute.

Pillet, G., 1996: *Welfare economics and social choice within an ecological framework* (in French), University of Fribourg, Fribourg (Switzerland).

Pillet, G., and F. Gassmann (eds.), 1994: *Steps towards a decision making framework to address climate change. Report of the IPCC Working Group III Writing Team II Montreux Meeting, March 3-6.* Würenlingen & Villigen (Switzerland): PSI-Bericht 94-10, Paul Scherrer Institute.

Pillet, G., and H.T. Odum, 1987: *E³ – Energie, ecologie, economie.* Georg, Geneva.

Pillet, G., W. Hediger, S. Kypreos, and C. Corbaz, 1993: *The economics of global warming – National and international climate policy: The requisites for Switzerland.* Würenlingen & Villigen (Switzerland): Paul Scherrer Institut, PSI Bericht Nr. 93-02, 2d ed., 1994.

Pratt, J.W., 1995: "Foreword" to L. Eeckhoudt and C. Gollier, 1995: *Risk – Evaluation, management, and sharing,* trans. V. Lambson, Harvester Wheatsheaf, New York.

Price, C., 1993: *Time, discounting and value,* Blackwell, Cambridge, MA and Oxford, UK.

Querner, I., 1993: *An economic analysis of severe industrial hazards.* Physica-Verlag, Berlin.

Ramsey, F.P., 1928: A mathematical theory of savings, *The Economic Journal* **38,** 543–565.

Random House, 1966: *The Random House dictionary of the English language,* Jess Stein, ed., Random House, New York.

Richels, R., 1994: Decision making under uncertainty and the global climate debate, Contribution to the IPCC Working Group III Writing Team II Montreux Meeting, 3-6 March.

Ríos, S. (ed.), 1994: *Decision theory and decision analysis: Trends and challenges,* Kluwer Academic Publishers, Dordrecht.

Safra, Z., 1987: Contingent commodities, *The new Palgrave.* pp. 22–24, Norton, New York and London.

Sandmo, A, 1972: Discount rates for public investment under uncertainty, *International Economic Review* **13,** 287–302.

Savage, H.A., 1951: The theory of statistical decisions, *American Statistical Association Journal* **46,** 55–67.

Savage, H.A., 1954: *The foundations of statistics,* Wiley, New York.

Schelling, T.C., 1994: Intergenerational discounting. In *Integrative assessment of mitigation, impacts, and adaptation to climate change,* N. Nakicenovic, W.D. Nordhaus, R. Richels, and F.L. Toth, eds. IIASA, Laxenburg (Austria).

Schmalensee, R., 1993: Symposium on global climate change, *Journal of Economic Perspectives* **7**(4), 3–10.

Schmalensee, R., 1993: Symposium on global climate change, *The Economic Journal* **38,** 543–565.

Schmeidler, D. and P. Wakker, 1987: Expected utility and mathematical expectation, *The new Palgrave,* pp. 70–78, Norton, New York and London.

Schubert, R., 1994: Climate change and discount rates. In *Steps towards a decision making framework to address climate change. Report of the IPCC Working Group III Writing Team II Montreux Meeting, March 3-6,* Pillet and Gassmann, eds., pp. 33–40, Würenlingen & Villigen (Switzerland): PSI-Bericht 94-10, Paul Scherrer Institute.

Shackle, G.L.S., 1967: *Décision, déterminisme et temps,* Centre d'Econométrie de la Faculté de Droit et des Sciences Economiques de Paris, Dunod, Paris.

Shafer, G., 1976: *A mathematical theory of evidence.* Princeton University Press, Princeton, NJ.

Shlyakhter, A., L.J. Valverde, Jr., and R. Wilson, 1995: Integrated risk analysis of global climate change, *Chemosphere,* **30,** no. 8, 1585–1618.

Simon, H.A., 1957: *Models of man.* Wiley, New York.

Simon, H.A., 1987: Bounded rationality, *The new Palgrave,* pp. 15–18, Norton, New York and London.

Singer, F., R. Revelle, and C. Starr, 1991: What to do about greenhouse warming, *Cosmos,* pp. 28–33.

Sjöstedt, G. (ed.), 1993: *International environmental negotiation.* Sage Publications, Newbury Park, CA.

Smith, K., 1991: Allocating responsibility for global warming: The natural debt index, *Ambio* **20**(2), 95–96.

Solow, R.M., 1986: On the intergenerational allocation of natural resources, *Scandinavian Journal of Economics,* **88,** 141–149.

Solow, R.M., 1992: *An almost practical step toward sustainability,* Resources for the Future, Washington, DC.

Sugden, R., and A. Williams, 1990: *The principles of practical cost-benefit analysis,* Oxford University Press (1st publication, 1978), Oxford.

Svenson, O., 1991: The time dimension in perception and communication of risk. In *Communicating risk to the public,* R.E. Kasperson and P.J. Stallen, eds., Kluwer, Dordrecht.

Tol, R.S.J, 1994: The damage costs of climate change: Towards more comprehensive calculations, *Environmental and Resource Economics,* **5,** 353–374.

Tooley, M.J., 1994: Sea-level changes and impacts. In *Environmental hazard: Liability or opportunity?* Report of the SANDOZ/INSEAD Colloquium held 29 June 1993, pp. 10–16, INSEAD Centre for the Management of Environmental Resources, Paris.

Ulph A., and D. Ulph, 1994: Global warming: Why irreversibility may not require lower current emissions of greenhouse gases. Discussion paper, University of Southampton, UK.

UN Framework Convention on Climate Change, 1992: Published by IUCC/UNEP (Information Unit on Climate Change), Geneva.

U.S. EPA (U.S. Environmental Protection Agency), 1990: *Reducing risk,* Relative Risk Reduction Strategies Committee of the US EPA Science Advisory Board, Environmental Protection Agency, Washington, DC.

Vercelli, A., 1991: *Methodological foundations of macroeconomics: Keynes and Lucas,* Cambridge University Press, Cambridge.

Von Neuman, J., and O. Morgenstern, 1947: *Theory of games and economic behavior,* Princeton University Press, Princeton, NJ.

Wald, A., 1950: *Statistical decision functions.* Wiley, New York.

Walley, P., 1991: *Statistical reasoning with imprecise probabilities,* Chapman and Hall, London.

Weilenmann, U., 1994: Insurable risk associated to climate change. In *Steps towards a decision making framework to address climate change. Report of the IPCC Working Group III Writing Team II Montreux Meeting, March 3-6.* Pillet and Gassmann, eds., pp. 11–17, Würenlingen & Villigen (Switzerland): PSI-Bericht 94-10, Paul Scherrer Institute.

Wilford, M., 1993: Insurance against sea level rise. In *The global greenhouse regime: Who pays?,* P. Hayes and K. Smith, eds., UNU Press/Earthscan, London.

World Bank, 1994: *World Development Report 1994,* The World Bank, Washington, DC.

Yaari, M., 1987: The dual theory of choice under risk, *Econometrica* **55,** 95–116.

Yokohama, 1994: *Yokohama World Conference on Natural Disaster Reduction,* Information document, International Directorate for Natural Disaster Reduction, United Nations, Geneva.

3

Equity and Social Considerations

T. BANURI, K. GÖRAN-MÄLER, M. GRUBB, H.K. JACOBSON, F. YAMIN

CONTENTS

SUMMARY

Equity and social considerations are central to discussions of steps to be taken to implement the Framework Convention on Climate Change, both intrinsically and because widespread participation is essential if the objectives of the Convention are to be gained. Countries are unlikely to participate fully unless they perceive the arrangements to be equitable. This applies particularly to equity among regions and countries, but equity within countries, and associated social considerations, are also important influences on what is possible and desirable. Mitigating and adapting to climate change will require actions on the part of individuals. Governments will find it easier to comply with international obligations if their citizens feel that the obligations and benefits of compliance are distributed equitably. And richer countries are unlikely to burden their poorer citizens to benefit relatively rich citizens in poor countries.

Issues relating to equity among regions and countries stem from the substantial differences that exist among countries. Countries differ not only in terms of size, resources, population, and wealth, but also in terms of emissions of greenhouse gases, vulnerability to climate change, and institutional capabilities to respond effectively to climate change. In general, the implications of climate change for developing countries differ from those for developed countries because the former are generally poorer, emit much less per capita, have contributed less to past emissions, and have shorter policy time horizons. Moreover, their institutions are often weaker, they face other urgent priorities, and they are generally more vulnerable to climate change. But there are substantial variations within both the developed and the developing countries, and a rigid delineation of equity issues along developed and developing country lines is inappropriate and may be highly damaging in the long run.

The framework convention

The concept of equity is prominent in the Framework Convention on Climate Change because of the need to gain widespread adherence. The Convention itself provides considerable guidance for applying the concept to take account of the many differences among countries, particularly those between developed and developing countries. Such equity considerations are reflected in the requirement for developed country parties to take the lead and to assist developing country parties in coping with both the costs of abatement and the costs of adaptation to the adverse effects of climate change and, correspondingly, in the recognition that developing coun-

tries' emissions are relatively low and will need to grow to meet their legitimate social and developmental needs. Issues of procedural equity are reflected in the need for "equitable and balanced representation" and transparent governance in the financial and other mechanisms. However, the application of equity to specific circumstances will require further elaboration of the Convention's principles and obligations, many of which were designed to be ambiguous and remain so.

The role of analysis

Scientific analyses cannot prescribe how equity and social considerations should be applied, but analysis can clarify the implications of alternative choices and their ethical basis. There are a variety of meanings of equity and different philosophical and policy approaches to it. On some issues many different equity principles point to similar implications and offer clear guidance, whereas on others they may conflict. In either case, there is a need for judgment, drawing on concepts of equity.

Equity concerns both "process" issues and "outcomes" in terms of the distribution of costs and benefits internationally. Indices such as Gross World Product aggregate wealth independently of distribution. However, assessing aggregate welfare (utility) requires valuing and aggregating differential impacts among countries. This is an issue of ethics and politics, not economics. In global assessments, therefore, separation of international economic analysis from explicit equity considerations is only possible if effective institutions exist for appropriate (compensating) international redistribution.

Impacts and the costs of coping

In general, climate change seems likely to impose greater risks and damage on poorer regions. Thus, it may exacerbate inequalities in the absence of compensating measures. This would violate a number of ethical principles, including potentially those drawing on basic needs and Rawlsian approaches, particularly with respect to transboundary impacts of some actors upon others. There are few, if any, ethical systems in which it is acceptable for one individual knowingly to inflict potentially serious harm on another and not accept any responsibility for helping or compensating the victim. Given this, the monetary evaluation of global impacts has an ethical dimension in which the willingness of countries to accept compensation for imposed climate-related damages is a relevant consideration. This consideration has not been reflected in damage estimates so far.

An effective international insurance mechanism could be one way of reducing both international and intergenerational inequities arising from climate change. There is some economic as well as ethically motivated literature on this, though many practical, institutional, and political issues remain to be resolved.

Distribution of emissions and abatement costs

The need for emission reductions raises equity issues distinct from those of distributing the costs of coping. The Convention lays out bases of common but differentiated responsibilities, and the initial aggregate implications of this with respect to developed and developing countries. Much of the broader debate and literature also focusses on issues of equity between developed and developing countries, leaving serious lacunae concerning the application of equity within these groups, which, are themselves very diverse. This is of immediate concern for developed countries, given their obligations under the Convention and the Berlin Mandate to take the lead.

The literature on possible emission obligations examines many different approaches. Many sources highlight the past "overuse" of the atmosphere by industrialized nations, but others dispute this and/or its relevance to current decisions. For future entitlements, the dominant contrast is that between approaches that focus on burdens related to changes from current emission levels and various interpretations of per capita emission entitlements. Debates over potential payments have mostly concerned different interpretations of the polluter pays principle and indices of ability to pay. In reality, feasible and fair criteria are likely to involve a negotiated and evolving mix of these approaches.

Institutional and procedural fairness

Institutional weaknesses inhibit the ability of developing countries to participate effectively in international negotiations. Assistance to help these countries develop a greater capacity to assimilate and analyze information and proposals, and to participate effectively in international discussions, would increase the prospects for achieving effective, lasting, and equitable agreements on how best to address the threat of climate change.

Social considerations

Social considerations, and the experience of implementing structural adjustment policies, point to the need to consider and target specific groups for special consideration. Countries (such as island and other low-lying states or dryland regions) and special groups within society that are especially vulnerable to climate change (such as the poor, and sometimes women or children, or specific occupations or regions) – in other words, those on whom the costs of abatement and coping would be especially burdensome – merit special attention.

Overall, concern about equity and social impacts points to the need to strengthen institutional capacities, particularly in developing countries, to make and implement collective decisions in a legitimate and equitable manner. These institutional capacities surely include developing resources to analyze equity and social issues more thoroughly, and to integrate these perspectives better with the insights of other disciplines.

3.1 Introduction

In common language equity means "the quality of being fair or impartial," or "something that is fair and just" (Flexner, 1987). It has been a central preoccupation of social and political thought through the ages, and it is a consideration of some considerable importance in the Framework Convention on Climate Change (FCCC).

Although science cannot prescribe or decide what actions would be equitable under the Climate Convention, science, religion, and philosophy can illuminate the meaning of equity and clarify the choices that the parties to the Convention face. This is the task of this chapter. It considers concepts of equity and issues that must be addressed in efforts to apply these concepts. It then views these broad concepts within the tradition of international law and the specific context of the Framework Convention, and considers the several ways in which this text assigns specific meaning to equity. It next analyzes in detail several specific aspects of equity: international equity in coping with the impacts of climate change and associated risks, international equity in efforts to limit climate change, equity and social considerations within countries, and equity in international processes. Equity among generations (or intergenerational equity) is the subject of the next chapter.

3.1.1 The role of equity

This analysis focusses on equity issues relating to climate change responses that might be considered by the international community. Equity issues exist at both the national and international levels, but there is an important difference between these contexts. Within countries institutions exist to address a wide range of issues of common interest to members of that society. The institutions have developed in part to provide a way of taking decisions about what constitutes acceptable behaviour and about the distribution of wealth and resources, which in most countries are redistributed through regulation of market structures, backed by legal codes, and by taxation with the intent of promoting social good. Equity – in the form of views about what constitutes justice – has an important influence on these institutions and their decisions and is a measure of their legitimacy. The actual strength and perceived legitimacy of these institutions vary widely, but they nevertheless provide an existing framework within which policies to address climate change at the national level can be developed and implemented.

By contrast, institutions at the international level are relatively weak. In responding to climate change, the international community faces unavoidable decisions about the distribution of effort and burdens and what constitutes acceptable behaviour in circumstances where the internal behaviour of one state may directly affect all others. Although, on a modest scale, there are precedents, climate change is unique in the scale and scope of its potential implications and the coordinated international responses it requires. Yet international institutions dedicated to coordinating such responses are relatively new or in the process of being established as a result of the FCCC. Moreover, because the examination of international equity issues is still in its infancy, these may need greater explicit analysis and consideration by international institutions. To the extent that they are implemented at the national level, international response strategies will also have implications concerning equity within countries that should be considered. Acceptance of burdens decided at the international level will depend in substantial measure on their perceived legitimacy at the national level.

No international agreement has ever been founded purely on a logical consideration of equity issues. A host of other factors, ranging from basic economic and political power structures to accidents of timing and personalities, influences the outcome. But the converse – the cynical view that equity considerations play no role at all in the real world of international politics – is not true either. Many authors have argued that the long-term, cross-cutting strategic and global nature of the climate problem makes equity issues central to any solution. Indeed, a broad view of self-interest can also often point towards explicit consideration of equitable outcomes because of the longer-term risks that grossly inequitable behaviour may pose to stability and cooperation in the international system.

3.1.2 Concepts of equity

Since Confucius, Plato, Aristotle, and the Vedantic and Biblical texts, theorists who have dealt with economic, political, and social issues have developed and explored concepts of equity. Several broad points emerge from this extensive literature. The first is that equity applies to two separate categories of issues. It applies to both procedural issues – how decisions are made – and consequentialist issues – the outcomes of decisions.

3.1.2.1 Procedural equity
Procedural equity has two components. The first relates to participation, the idea that those who are affected by decisions should have some say in the making of those decisions through either direct participation or representation.

The second relates to process, most notably the principle of equal treatment before the law: Similar cases must be dealt with in a similar manner, and exceptions must be made on a principled basis. In this sense, the principle of equity not only requires that law should govern decisions but also provides guidance in how laws should be applied. However, all the circumstances in which a law would be applied cannot be foreseen at the time of its formulation. Thus, starting with Aristotle, theorists have argued that laws be applied in an equitable manner to achieve what the legislators would have intended in the specific circumstances of a particular case (Shapiro, 1990). The concept of equity also embodies a higher notion of justice that goes beyond the rules, no matter how fairly they were devised. The Anglo-American common law tradition, for example, often introduces equity into judicial decisions to correct a potential injustice by too rigid an application of the law.

The principle of equal treatment before the law is closely allied to notions of basic, minimum rights for individuals. For instance, John Rawls (1971) has argued that all individuals have equal rights to as extensive a system of political liberties as is possible without diminishing the liberties of others. A similar but stronger view has been put forward by Robert Nozick (1974), who has argued that all individuals have a sphere of moral rights in which no one, including the state, is allowed to interfere, irrespective of the consequences that might arise.

It is important to recognize that these particular theories were originally developed for dealing with questions of justice within a state. Rawls, for instance, draws a sharp distinction between the principles of justice that prevail among persons within a society and "justice between states" to which his theory was not intended to apply (Stone, 1993). Consequently, the application of these theories to the subject of international justice – justice between states – presents problems (Van Dyke, 1975; Stone, 1993). Not least is the fact that as holders of the rights of sovereignty, it is nation-states, rather than individual human beings, that negotiate the nature of international commitments. This is so despite the fact that there is a huge amount of cross-border interaction between individuals, corporations, and international nongovernmental groups, and such nonstate actors can play important roles.

Nevertheless, by extrapolation and analogy, these theories offer insights about the application of procedural equity between states. For example, the notion that procedural equity demands that basic rights (of individuals or states), however they are defined, must be respected in decision making is commonly accepted in domestic and international law. Article 2, paragraph 7, of the UN Charter, for instance, states that nothing contained in the Charter shall authorize the United Nations to intervene in matters which are essentially within the domestic jurisdiction of any state, but that the principle should not prejudice the application of enforcement measures. Other elements of procedural equity between states are discussed in more detail in Section 3.2, on international law.

3.1.2.2 Consequentialist equity

Consequentialist equity has to do with the outcome of decisions, particularly the distribution of burdens and the allocation of benefits. There are several broad traditions defining the meaning of equity in this sense (Young, 1994). They may be summarized in the following categories: parity, proportionality, priority, classical utilitarianism, and Rawlsian distributive justice.

Parity is a formula for equal distribution of burdens or benefits. Parity demands that all claimants receive equal shares; it is closely associated with egalitarianism.

Proportionality is a principle that dates back at least to Aristotle; it asserts that burdens or benefits should be distributed in proportion to the contributions of claimants.

Priority argues that those with the greatest need should be advantaged. This forms the basis of the "basic needs" approach, which puts the emphasis on the absolute right of individuals to goods and services necessary to sustain their lives at some minimum standard of well-being. This would include potable water, minimum nutrition, and health care and general environmental resources.

Classical utilitarianism proposes that burdens and benefits should be distributed to achieve the greatest good for the greatest number. This Benthamite formula can be expressed mathematically in terms of maximizing total utility, which requires the measurement and comparison of utilities, an issue which will be considered below.

Rawlsian distributive justice (Rawls, 1971) carried the concept of utilitarianism a step further, arguing for an equal distribution unless an unequal distribution operates to the benefit of the least advantaged.

No society has ever had complete consensus on any one of these approaches alone as an adequate criterion for defining consequentialist equity. Some (such as basic needs) are incomplete, prohibiting certain outcomes but not helping choices between other options. Single principles may also not be appropriate or practical as the only standard because, among other reasons, cases and individuals are rarely identical and burdens and benefits may not be divisible; or, if they are, they may not be divisible into shares that are susceptible to precise, cardinal measurement, thus making it impossible to apply the principles directly.

In practice, when societies try to achieve equity, they do so in nuanced and subtle ways, applying several criteria and seeking to achieve a balance among them. The balance is also affected by self-interest: In the real world, people tend to seek and to emphasize principles that may advantage them. Nevertheless, the principles are useful guides to what might constitute equitable decisions.

These consequentialist principles were developed in the context of specific societies, not internationally. The literature on theories of international consequentialist equity is more recent and is largely derived by extension of the above principles (Beitz, 1979; for a review in the context of climate change, see Paterson, 1994/1996). Illustrating the likely consequences of their application may clarify the choices that have to be made in seeking equitable solutions in the implementation of the Framework Convention on Climate Change.

Both procedural and consequentialist equity issues are complicated by a wide variety of cultural and societal assumptions about ethics, the environment, and development. The existence of these different and sometimes conflicting principles, and the need for compromise between them, is considered by Rayner (1993, 1994), who argues that a number of fundamentally different "world views" are adopted towards climate change. These views not only span different ideas of equity but also differ concerning basic assumptions about the urgency of abatement action and appropriate management strategies, and can be correlated with different institutional types identified by cultural theory.[1] The process of developing a response is seen as a process of compromise between these different world views, each of which tries to influence policy to correspond more closely to its own perceptions – perceptions which again tend to be influenced by interests.

3.1.3 Utility, equity, and economic efficiency

Welfare economics is based on utilitarianism. It requires some measure of individual welfare (utility), as a function of various factors such as the amount of goods and services that the individual can access, different aspects of the individual's physical and spiritual environment, and rights and liberties. Such a "utility function" aggregates a rather long list of factors affecting individual well-being into one single measure of welfare. However, constructing an aggregate measure of these utilities for many different individuals is a much more difficult task. The definition and aggregation of utilities present a complex ethical issue in connection with evaluating the global welfare loss associated with climate change impacts and measures to limit them.

There is no inherent conflict between economics and most conceptions of equity. But a conflict can arise because of differing ideas or assumptions, sometimes hidden, about how individual utilities should be defined, compared, and aggregated. It is important to understand that economics itself cannot resolve these differences. Explicit discussions of equity are essential because they reflect differing ideas about how – and indeed whether – individual utilities should be measured and aggregated.

Arrow (1951) addressed the fundamental question of whether individual preferences can be aggregated in a reasonable way into overall societal preferences. He concluded that, in general, it is impossible to add individual preferences together to produce a social welfare function if we require the resulting aggregation to satisfy some very natural and reasonable conditions, such as preventing individuals from holding dictatorial powers. Thus, it is generally not possible to deduce "objectively" a socially preferred distribution of well-being from individual preferences. However, if it is known that these preferences are restricted to certain types, then it may still be possible to combine them in a consistent and reasonable way to form a social ordering (see Sen, 1984).

With respect to certain environmental considerations, there does indeed seem to be a rough consensus as to what constitutes an equitable distribution of welfare over time. This is suggested by the apparent agreement on the concept of sustainable development, as introduced by the World Conservation Strategy (IUCN 1980) and popularized in the report of the World Commission on Environment and Development – the Brundtland Commission – in 1987.

The central idea behind sustainable development is that the present generation should not make changes that reduce the possibilities for future generations to achieve comparable well-being. The concept has received widespread support internationally, as evidenced by its inclusion in the Rio Declaration and Agenda 21. It is specifically mentioned in paragraph 5 of Article 3 of the Framework Convention on Climate Change. Nevertheless, despite its widespread acceptance, there is no universal agreement as to the precise meaning of the concept, and, as a result, its application is not straightforward.

The comparison and aggregation of utilities across different countries and across different individuals is also contentious. The Gross National Product (GNP) indicator avoids this by focussing simply on the total measured consumption in a country; in principle, it lays no claim to represent welfare directly, nor does it claim that aggregating GNP across different countries is a valid measure of global welfare. In practice,

however, maximizing GNP does often become a primary focus of policy and economic analysis. This implicitly embodies an assumption either that a given amount of additional wealth is equally valuable to everyone or that the additional welfare can and will be redistributed to fulfill some more explicit measure of aggregate social welfare. The latter goal is achieved by a balance between maximizing GNP and the establishment of institutions and processes charged with redistribution, social protection, and provision of various social goods.

Because such processes and institutions are weak or nonexistent internationally, the debate about whether and how to compare national utilities internationally is of central importance. Views range from asserting that countries should act as if they value all countries equally (i.e., assume equivalent utility functions and aggregate all with the same weight) to asserting that utilities can and should not be estimated and aggregated at all across countries, that countries bear no responsibility for the welfare of others. International negotiations are to an extent about trying to reach a compromise between these two extremes, especially concerning policy on issues like climate change, where the activities of one country may directly affect another.

It is in this issue of whether and how to aggregate separate utilities that an apparent conflict between equity and economic efficiency can arise. Whether it does or not depends on how efficiency is defined. "Pareto efficiency," for example, describes situations in which no one can be made better off without making anyone else worse off. Pareto efficiency is thus generally neutral with respect to equity because it allows a wide variety of possible distributions. More often, however, the term "economic efficiency" describes the maximization of "something" with the resources available. Maximizing GNP – or perhaps "World Product" (WP) – could involve highly inequitable outcomes, which might well imply lower global welfare, depending on how welfare is defined in relation to the distribution of wealth. With appropriate international transfers, however, it could also allow a much fairer and ethically benign world with a real gain in global welfare.

Equity is thus essential to climate change discussions, because there is no consensus about whether and how to measure and aggregate welfare within and, still less, between countries. On the contrary, there are fiercely competing views grounded in differing interests and beliefs. The optimal policy is thus inherently a matter for debate, negotiation, and compromise between conflicting interests and ethical philosophies.

3.2 Equity in International Law and in the Framework Convention

This section examines how the basic rights and obligations of states established under international law through treaties, custom, general legal principles, and judgments and awards of courts and international tribunals provide a framework for consideration of issues concerning procedural and distributive equity (Cheng, 1990; Franck and Sughrue, 1993; Sands, 1995; Schachter, 1977; Tarlock, 1992; Weiss, 1993). In addition, the Framework Convention on Climate Change contains principles and specific provisions concerning equity. These provisions have important implications for the implementation of the Convention, including the elaboration of further commitments and mechanisms for burden sharing on an equitable basis.

3.2.1 *International legal framework*

One of the basic tenets of international law is the sovereign equality of all states. Each state has jurisdiction over its territory and has the right freely to choose and develop its political, social, economic, and cultural systems, including the right to develop its own policies and laws regulating the exploitation of its natural resources. As a corollary to these principles, each state has a duty to refrain from threatening the territorial integrity of another and the obligation not to intervene in matters within the domestic jurisdiction of any other state.

According to Principle 2 of the Rio Declaration on Environment and Development:

> States have, in accordance with the Charter of the United Nations and the principles of international law, the sovereign right to exploit their own resources pursuant to their environmental and developmental policies, and the responsibility to ensure that activities within their own jurisdiction or control do not cause damage to the environment of other States or of areas beyond the limits of national jurisdiction. (United Nations, 1993)

Each state may devise its own climate change policies and programmes. Accordingly, the extent to which national measures are equitable in allocating costs and benefits among various regions, economic sectors, social groups, or individuals within its territory is primarily a domestic matter. The allocation of responsibilities between states for mitigation and adaptation and mechanisms to implement these is, however, an international matter and subject to the general rules and principles of international law. For states that are parties to the Framework Convention on Climate Change, the allocation of these responsibilities is also subject to the specific equity provisions of the Convention. It is important, therefore, to understand the meaning of equity in international law, in particular, the factors that have been included in the concept of equity and their practical procedural and distributive consequences.

3.2.2 *The ICJ and the concept of equity*

The International Court of Justice (ICJ) explained the legal nature of equity in its judgment on the 1982 Continental Shelf Case involving Tunisia and Libya. It stated:

> Equity as a legal concept is a direct emanation of the idea of justice. The Court whose task is by definition to administer justice is bound to apply it. (ICJ, 1982)

The Court explained that "the legal concept of equity is a general principal [sic] directly applicable as law." This means that equity can be a source of law as well as a consideration for the sensible application of the law (Cheng, 1987). In its 1982

judgment, the Court recognized that equity was relevant when it was called on to choose among several possible interpretations of the law. The Court interpreted this as meaning, not that it could fashion new law, but that when it could choose among several interpretations of the law it was bound to choose the interpretation "which appears, in the light of the circumstances of the case, to be the closest to requirements of justice" (ICJ, 1982, p. 60, para. 71). The Court also recognized that in international law the application of equity must take into account all the legal and factual circumstances relevant to the case in hand.

Because individual cases may involve unique procedural and distributive elements, the application of equity cannot be generalized as a set of principles, and factors relevant to the application of equity to one context cannot necessarily be transposed and applied in another. The following examination of equity in different international environmental contexts is intended to provide background information about the role of equity in benefit- and burden-sharing arrangements between states concerning access to and use of natural resources. The legal insights and practical experience gained by states and the ICJ in dealing with equity in these contexts may be relevant for climate change. It is important to note that these insights and experiences do not bind the parties to the FCCC, who may negotiate further agreements about the appropriate role to be given equity in the context of the Convention.

3.2.3 *Equity in the context of the continental shelf*

The ICJ has examined the role of equity in a series of cases between states concerning use of the continental shelf and rights of access to it. In a dispute between Malta and Libya (ICJ, 1985), the Court said that the concept of equity included the principle of good faith negotiations to resolve disputes between parties. This interpretation highlights the important procedural components of the concept, which, the Court stated, also included "the principle of respect due to all such relevant circumstances [and] the principle that although all States are equal before the law and are entitled to equal treatment, equity does not necessarily imply equality."

Concerning distributive equity, it went on to declare that the application of equitable principles cannot be used for "refashioning geography or compensating for the inequalities of nature." The Court pointed out that, so far as the law concerning continental shelf delimitation is concerned, "equity does not necessarily imply equality . . . nor does it seek to make equal what nature has made unequal." The Court also went on to state that equity includes "the principle that there can be no question of distributive justice." Accordingly, in that case the Court rejected Malta's claim to a greater share of continental shelf based on its argument that it was resource-poor and had greater socioeconomic and developmental needs than oil-rich Libya.

In a recent ICJ case between Norway and Denmark (ICJ, 1993), the Court again considered the relevance of socioeconomic factors, including population, and rejected these as irrelevant in determining a state's entitlement to continental shelf resources. In a case between the U.S. and Canada (ICJ, 1984), however, the Court indicated that where the overall result might entail "disastrous repercussions on the subsistence and economic development of the populations concerned," it may be inequitable not to take such factors into account.

It is not possible to derive general conclusions about equity from the foregoing, as the Court has stressed that each case must be examined in the light of its legal and factual circumstances. What this body of law does make clear, however, is that the Court has not yet had to deal with interpreting obligations that are related to or conditional on the consideration of complex factors such as socioeconomic development or the needs of present and future populations. Moreover, a closer reading of these cases illustrates the Court's reluctance to use equity as a basis to achieve distributive justice on a wider scale. This suggests that, in the absence of clear legal rules requiring the consideration of factors such as socioeconomic development and population, the ICJ may not necessarily regard them as relevant or of paramount importance in other contexts where disputes concerning access to and use of natural resources raise wide-ranging distributive justice concerns.

This approach may have significant legal and practical consequences in the climate change context, where implementation of a range of parties' obligations under the Convention is conditional on taking into account their socioeconomic development, national needs, and a wide-ranging list of geographical factors. It is also particularly important in the context of negotiating future greenhouse gas emission reductions or devising joint implementation systems, as both raise fundamental distributive questions about the basis on which countries are entitled to continue their emissions.

3.2.4 The Law of the Sea

The role of equity in the legal regime established for the deep seabed in Part XI of the 1982 United Nations Conference on the Law of the Sea (UNCLOS) differs markedly from the principles and rules concerning the continental shelf. UNCLOS aims at distributing the benefits of exploitation as widely as possible because it incorporates a notion of equity that includes a substantial element of distributive justice. The equitable principle that inspires this regime is that the "Area" (the deep seabed) and its resources are the "common heritage of mankind" (Birnie and Boyle, 1992). Article 140 provides, for example, that activities in the Area "shall be carried out for the benefit of mankind as a whole, irrespective of the geographical location of States, whether coastal or land-locked, and taking into particular consideration the interests and needs of developing States and of people who have not attained full independence."

Unlike the law on continental shelf delimitation, these provisions expressly call for the consideration of socioeconomic factors and economic needs, particularly those of developing countries or other states disadvantaged by geography, in apportioning benefits. However, no definition of equity was included in UNCLOS. To give effect to these provisions, parties

to UNCLOS may have to elaborate equitable criteria for sharing any financial and other benefits arising from exploitation of the deep seabed. These criteria may have implications for climate change issues.

3.2.5 The ozone regime

Perhaps the clearest international example for the application of equity is the 1987 Montreal Protocol on Substances that Deplete the Ozone Layer, which arose as a result of the 1985 Vienna Convention for the Protection of the Ozone Layer. Many developing countries had argued that, in view of their marginal contribution to the ozone problem, limited financial resources, and more pressing developmental concerns, they should not be expected to take on the same commitments as richer developed countries whose emissions had caused the ozone problem (Franck and Sughrue, 1993; Tarlock, 1992). Participation in the Montreal Protocol of a large number of developing country parties is widely viewed as a measure of its success in addressing the equity concerns of these countries and in balancing environmental needs with economic imperatives and flexibility for industrial producers of ozone-depleting substances.

The use of innovative legal techniques to implement these equity concerns distinguishes the Montreal Protocol from other conventions. The following list provides a brief outline of the way in which these concerns are given practical expression in the Protocol's substantive provisions. These "equitable" techniques include:

- differentiated standards for developed and developing country parties, including the provision of grace periods for compliance for the latter, allowing increased production by developed country parties to enable developing country parties to meet their "basic domestic needs," and allowing developing country parties to determine their emission entitlements on a per capita basis

- financial assistance to developing country parties, over and above overseas development assistance, to cover "all agreed incremental costs" and enable compliance

- transfer of technology, especially of best available, environmentally safer substitutes under fair and most favourable conditions, facilitated by the Protocol's financial resources if necessary

- limited operation of a tradable permit or joint implementation scheme to achieve "industrial rationalization" to minimize economic disruption and provide flexibility to producers of ozone-depleting substances

- an acknowledgment that the ability of developing country parties to comply is conditional on the "effective implementation" of financial cooperation and technology transfer obligations by the developed countries

As a result of its success in attracting the participation of developing countries, the Montreal Protocol approach was extensively discussed as a "model" for the Framework Convention on Climate Change (Benedick, 1991; Handl, 1990).

3.2.6 *The Framework Convention on Climate Change*

What is the role of equity in the Framework Convention on Climate Change so far as rights and responsibilities to protect the climate system are concerned? Unlike the Montreal Protocol, which mentions equity only once in the Preamble, the Climate Convention includes references to equity three times in its substantive provisions. The first of these, in Article 3.1, states:

> The Parties should protect the climate system for the benefit of present and future generations of humankind, on the basis of equity and in accordance with their common but differentiated responsibilities and respective capabilities. Accordingly, the developed country Parties should take the lead in combating climate change and the adverse effects thereof.

This principle, which is intended to provide guidance in implementing all the provisions of the Convention, mentions equity in the context of burden sharing between all parties, and, in particular, between developed and developing country parties. It also suggests that equity requires consideration of the responsibilities of present generations to future ones as part of burden-sharing arrangements.

Equity also appears in Article 4.2(a) which requires developed country parties to commit themselves to

> adopt national policies and take corresponding measures on the mitigation of climate change. . . . These policies and measures will demonstrate that developed countries are taking the lead in modifying longer-term trends in anthropogenic emissions consistent with the objective of the Convention . . . taking into account the differences in these Parties' starting points and approaches, economic structures, available technologies and other individual circumstances, as well as the need for equitable and appropriate contributions by each of these Parties to the global effort regarding that objective.

The use of the term "equitable" in reference to the specific commitments of developed countries reflects the intention of the parties that equity should be applied not only between developed and developing countries but among developed countries as well. Finally, Article 11.2 requires the Convention's financial mechanism to "have an equitable and balanced representation of all Parties within a transparent system of governance."

Equity in this context appears to reflect developing country concerns. These concerns are of an essentially procedural nature, reflecting the fact that the implementation of procedural elements may be essential for guaranteeing distributive outcomes that are perceived to be equitable.

What then is the significance of equity in the Convention and what practical consequences flow from its mention? It is clear that the application of equity in these contexts is intended to respond to quite different concerns. Equity cannot, therefore, have one meaning, as its meaning will depend on the legal and factual circumstances of particular situations. Particular disputes will themselves depend on the interpretation of the nature and extent of the parties' obligations under the Convention.

The terms "equity" and "equitable" in the Framework Convention on Climate Change are closely related to virtually all its other substantive provisions. According to rules set down in the Vienna Convention on the Law of Treaties (1969), these terms must be interpreted in the light of the Convention's overall objective, approach, and context. The Framework Convention's objective, as stated in Article 2, is to stabilize greenhouse gas concentrations at a level that "would prevent dangerous interference with the climate system." It then goes on to specify that "such a level should be achieved within a time frame sufficient to allow ecosystems to adapt naturally to climate change, to ensure that food production is not threatened and to enable economic development to proceed in a sustainable manner." The objective of the Convention thus bounds the way in which its provisions should be implemented and equity should be achieved.

In addition, the Convention itself requires the parties to use the principles contained in Article 3 to achieve its objectives and guide implementation of its provisions. Equity is mentioned in the context of Article 3.1, which concerns the principle of common but differentiated responsibilities for the climate system, which the preamble states is a "common concern of humankind." However, this is only one of five principles found in Article 3. The others include the right to promote sustainable development, the precautionary principle, the need to take into account the specific needs and special circumstances of developing country and vulnerable parties, and the commitment to promote a supportive and open international economic system.

Much of the meaning of these complex principles, including concepts such as "common concern of humankind," remains open to interpretation. It is clear, however, that these interlocking concepts and principles bound the way in which the parties' obligations can be interpreted and in which equity can be applied in a particular case.

The application of equity is also bounded by the structure of differentiated commitments set out in Articles 4.1–4.5 of the Convention, which distinguish between developed and developing country parties and those with "an economy in transition." All developed country parties, including those with economies in transition (listed in Annex 1 of the FCCC),[2] are required to take the lead in mitigating climate change (Article 4.2 (a)). These parties

> may implement such policies and measures jointly with other Parties and may assist other Parties in contributing to the achievement of the objective of the Convention and, in particular, that of this subparagraph.

Developed country parties and other developed parties listed in Annex II of the FCCC (the European Union and the member countries of the Organization of Economic Cooperation and Development) must transfer technology and financial resources to developing country parties to enable the latter to implement their more limited commitments to combat climate change (Articles 4.3 and 4.5). They are also obligated to assist developing country parties that are particularly vulnerable to the adverse effects of climate change in meeting the costs of adaptation (Article 4.4). In view of their limited financial resources, however, parties with economies in transition are not obligated to provide such assistance.

Articles 4.6–4.10 provide a range of factual or other circumstances that must be given consideration with respect to the implementation of the parties' differentiated commitments. These articles are, therefore, of particular relevance in considering what factors might or might not count as relevant "equitable factors" in a particular case where implementation of commitments is in question.

For example, Article 4.6 provides that a certain amount of "flexibility" must be given to parties with economies in transition in the implementation of their Article 4.2(a) obligations. Article 4.7 makes the implementation of the developing country parties' commitments conditional on the implementation of the developed country parties' financial and technology transfer commitments and recognizes that "economic and social development and poverty eradication are the first and overriding priorities of the developing country Parties."

Article 4.8 requires the parties to give full consideration to the specific needs and concerns of the developing country parties with respect to a broad list of geographical, biological, and economic factors, such as whether a country is a small island, is prone to natural disasters or desertification, or is highly dependent on income from fossil fuel consumption or production. Article 4.9 provides that full account must be taken of the needs of the least developed countries for funding and technology transfer. Finally, Article 4.10 provides that consideration should be given to all parties whose economies are highly dependent on income generated from the production or consumption of fossil fuels or energy-intensive products for which there are serious difficulties in switching to alternatives.

By differentiating obligations and by including the foregoing factors, the Framework Convention on Climate Change appears to have dealt comprehensively with the equity concerns. On a practical level, however, the implementation of the Convention on an equitable basis will require further agreement between parties about the significance of these factors, the relative weight to be given to each in particular situations, and the precise meaning of commitments undertaken by each of them. This, in turn, will require agreement about the meaning of principles, such as the principle of common but differentiated responsibilities, the right to sustainable development, and concepts such as "common concern of humankind," which is a new concept in international law. Reaching agreement about these matters will have a critical bearing on how the benefits and burdens of combatting climate change are allocated between states.

The continental shelf cases and the general rules of international law concerning procedural equity suggest that states enter into good faith negotiations to resolve differences about access to natural resources and their use. Further negotiations between parties to the FCCC would certainly assist the implementation of the Convention on an equitable basis. To the extent that the parties do not enter into such negotiations or where disagreements persist, the general rules and principles of international law will remain relevant to resolving disputes about equity.

3.3 Principal Differences Among Regions and Countries

3.3.1 Introduction

Equity is not the same thing as equality, but issues of international equity are clearly related to a variety of differences between countries. This section summarizes ways in which countries differ and that are relevant to the question of equity between countries in responding to climate change. Subsequent sections review some of the issues and conclusions that writers have drawn concerning implications for climate change policy. Five main dimensions of difference are cited in the literature:

- wealth and consumption
- emissions – past, present, and future
- impacts – the distribution of and vulnerability to climate change
- social considerations and institutional capabilities
- endowment with resources that may be affected by responses to climate change

This section of the chapter considers each in turn.

3.3.2 Wealth and consumption

Wealth is one of the most obvious and pervasive differences between countries. Much of the literature on international equity starts from this issue, and the statistics need little elaboration. In terms of annual average income, measured by gross national product (GNP) at market exchange rates, more than half of the world's population (58.7%) live in the forty-two countries that are classified as "low-income" in the World Bank's *World Development Report 1994* (World Bank, 1994). These countries have an average per capita gross national product (GNP) of $390. The 15.2% of the world's population that live in the twenty-three countries that the *World Development Report* classifies as "high-income economies" have an average per capita GNP of $22,160, almost sixty times that of the low-income economies. The remaining sixty-seven "middle-income economies" have an average per capita GNP of $2,490, just slightly more than one-tenth that of the "high-income" countries.

Such comparisons give an exaggerated impression because they do not reflect wide variations in purchasing power between countries that are not reflected in exchange rates. Attempts to correct for this, giving income estimates based on "purchasing power parity" as a more accurate measure, still highlight very wide disparities in average real per capita income.

These differences have a direct bearing on the issue of climate change in various ways. Activities of the poor that result in emissions of greenhouse gases are those that relate most closely to "basic needs," often at little more than subsistence levels – energy for cooking or keeping tolerably warm, emissions from agricultural activities, perhaps energy for light to enable reading, and occasionally for travel by public transport. Emissions from the rich tend to be dominated by activities such as driving private cars, home central heating, and energy embodied in a wide variety of manufactured goods and the use of such goods. The welfare impacts of cutbacks of greenhouse gas emissions may thus differ greatly according to the level of personal wealth (WCED, 1987; CDCGC, 1992). Cutbacks of greenhouse gas emissions could also have significant impacts on countries where the production of hydrocarbon fuels accounts for a substantial portion of their national income.

Obviously there are great variations within countries, but in sum the same broad issues apply concerning wealth disparities between countries. The aggregate relationship between wealth, consumption of a variety of natural resources, and emissions of various pollutants has been explored methodically by a number of authors. Using cross-sectional data, Parikh *et al.* (1992) shows that consumption of a wide variety of resources – many of which involve emissions of greenhouse gases in their extraction, processing, and application – is closely related to the level of wealth. Williams *et al.* (1987), Drucker (1986), and others have argued that, in many developed economies, a decoupling of wealth from the volume of material consumption has occurred since the 1970s and that this tendency will accelerate. Others have disputed this interpretation, however (Herman *et al.,* 1987). Nevertheless, the aggregate relationship at middle and lower income levels seems undisputed. The *World Development Report 1992* (World Bank, 1992) notes that emissions of some pollutants decline beyond a certain stage of economic development. However, the report does not claim that such a point has yet been reached in the case of CO_2.

Not only is wealth one of the most important correlates of greenhouse gas emissions (at present), but it also has a very important bearing on vulnerability to climate change. Richer countries, because they are richer, will tend to find it easier to deal with the costs of coping with climate change and measures to abate climate change. Poorer countries will tend to be more vulnerable to climate change for a variety of reasons considered below.

Poverty also has an important bearing on national priorities and the time scales considered in policy. Economists have long noted that personal discount rates decline with rising income. Richer people can afford to look further ahead, have greater security, and can afford to invest more for the future (though, in fact, the pattern of investment as a fraction of wealth is very variable). Poor people tend to be focussed more on short-term concerns, including striving to ensure they can meet basic needs. This has important implications for the potential equity impact of policies to address climate change within countries, as noted earlier in this chapter, and the same applies at the level of national economic and political systems.

Thus, in poorer countries interest rates tend to be higher, capital is scarcer, and the whole focus of policy and politics tends to be on meeting pressing short-term needs, ranging from poverty alleviation and employment generation to the management of fiscal crises, which are often partly driven by the needs of debt repayment. At the industrial level, the focus may, for example, be on the scramble to construct infrastructure and capacity fast enough to meet rapidly rising demand, rather than the more considered examination of optimal investments over longer periods that may be possible in richer countries. Thus, the context for both actual investment behaviour and broader public policy is strongly affected by national wealth in ways that are directly relevant to the climate problem (Mathur, 1991; Ewah, 1994).

Some of these issues were addressed by the Special Report on Developing Countries of the Intergovernmental Panel on Climate Change (IPCC, 1990), which stated that "the priority for the alleviation of poverty continues to be an overriding concern of the developing countries; they would rather conserve their financial and technical resources for tackling their immediate economic problems than make investments to avert a global problem which may manifest itself after two generations." Article 4.7 of the Framework Convention on Climate Change also recognizes that "economic and social development and poverty eradication are the first and overriding priorities of the developing country Parties" and these will influence implementation of their commitments.

3.3.3 Patterns of greenhouse gas emissions

A second important dimension of difference is that countries vary widely in the nature and degree of their contribution to climate change. Contributions to climate change span many different gases and sources.[3] Countries have different capacities as sinks for absorbing carbon emissions. Although the range of sources and sinks included in analyses or agreements may not in itself be regarded as an equity issue, it is clear the conclusions that may be drawn from different ways of aggregating and presenting the data can be. This section summarizes the main approaches that have been taken and the results obtained.

During the negotiation of the FCCC, many of these considerations came to be discussed in terms of whether the Convention should adopt a "comprehensive" approach, and, if so, whether it should focus on "gross" greenhouse gas emissions or "net" emissions derived by subtracting the removal of greenhouse gases by sinks from total emissions by sources (Bodansky, 1993). Adoption of the comprehensive approach means that all sources and sinks of different greenhouse gases have to be considered in formulating policy. Accordingly, global warming potentials are calculated for each gas to permit emissions of different gases to be compared according to a single metric.

In the course of the negotiations, many developing countries viewed the comprehensive approach as inequitable, arguing that methane emissions from subsistence agriculture should not be compared with carbon dioxide emissions, because the former are "survival emissions" which cannot be controlled without irreparable social and economic damage whereas the latter are due in large part to profligate lifestyles in the developed countries (Bodansky, 1993). Subak (1993)

BOX 3.1: THE WRI DATA CONTROVERSY AND ALLOCATION OF SINK CAPACITY

In 1990 the World Resources Institute (WRI, 1990) published an extensive set of data on national greenhouse gas emissions from which they concluded that responsibility for climate change was shared widely between countries in the industrialized and developing world. Their methodology and conclusions were vigorously attacked by Agarwal and Narain (1991), who claimed that the analysis was inherently biased against poor countries. They presented an alternative approach, concluding that developing countries bore no responsibility for the problem but were, in fact, contributing to cleaning up emissions from the industrialized world. The dispute attracted widespread academic, public, and political attention.

There were important disagreements over the accuracy of data and the range of gases included (Ahuja, 1992; Thery, 1992), and it has now been recognized that emissions from deforestation, and probably methane, in many developing countries are much lower than in the original WRI estimates. But the heart of the dispute was over the assignment of "natural sink capacity" – the natural processes that remove greenhouse gases.

Because of all the difficulties associated with estimating, comparing, and projecting greenhouse gas removal over time, WRI took the measured increase in concentration of each gas – which gives a direct measure of the radiative change – and assigned this to each country in proportion to estimated emissions in that year (or an average of recent years).

Agarwal and Narain raised two central objections to this measure. First, it took no account of the distribution of historical emissions, which are in fact largely responsible for current concentration levels. Second, by distributing the net concentration in proportion to gross emissions, the natural reabsorption – the "sink capacity" – was implicitly being allocated in proportion to emissions. This, they contended, was a grossly inequitable approach, because the sink capacity was a natural, global common resource that should be allocated equally to all people. By dividing the total annual absorption of each greenhouse gas (the difference between gross emissions and measured atmospheric increase) in proportion to national population, they showed that most developing countries' shares of the global reabsorption of greenhouse gases on this basis exceeded their actual gross emissions. Far from bearing any responsibility for the problem, they concluded, developing countries were in fact helping to clean up the excessive emissions from industrialized countries.

A difficulty with Agarwal and Narain's approach is that the rate at which greenhouse gases are removed from the atmosphere depends on the concentration gradient and therefore the level of gross emissions: The more that is emitted, the more is removed. Thus, it does not follow from Agarwal and Narain's approach that concentrations would decline if all countries did emit at the per capita rate of the poorer countries. After a few years it is likely that concentrations would still rise, but at a much slower rate. It is a complex process, however, and it is widely believed that the proportion absorbed declines with rising global emissions and could fall sharply if human emissions started to saturate natural sink processes. The real relationship between "gross" and "net" emissions, and how the latter should be assigned, is thus very complex.

To avoid this complication, and in keeping with much of the literature, the data in this chapter are given in terms of gross emissions in considering current and projected emission rates. Most analysis recognizes that from a policy perspective it will also be necessary to consider anthropogenic stimulation of sinks (primarily tree planting) and probably human impacts on the ability of natural ecosystems to act as sinks for greenhouse gases. For the present at least, these impacts are not thought to be major contributions to the overall flow of greenhouse gases, though the spread of land management for agriculture and other purposes, and desertification, have probably reduced the sink capacity in many diverse countries.

---

gives more general examples of how different countries might prefer different selections of cases to be included according to interest. Smith (1994) argues that climate change agreements should not incorporate any biotic sources or sinks (even if affected by human actions), both on grounds of practicality and equity – the latter argument being that the estimates require a baseline date against which to measure carbon accumulation or loss, which is likely to favour those regions (i.e., Europe and North America) that deforested in earlier ages.

Developing countries also argued that the Convention should include a principle that all states have an equal right to ocean sinks, as these are part of the global commons (Bodansky, 1993). This principle did not find its way into the Convention, and the question of entitlements to global sinks therefore remains open. The Convention does, however, mention the comprehensive approach favourably in both the preamble and Article 3.3. Notwithstanding this mention, the final language of the Convention leaves open the possibility that policies and targets for individual gases can be adopted.

Many of the equity concerns about how sources and sinks should be dealt with, and which gases should be controlled on what timescale, continue to manifest themselves in fierce methodological disputes about the calculation, aggregation, and presentation of data. This was vividly illustrated in an important debate about data relating to emissions and sinks presented initially by the World Resources Institute (See Box 3.1).

In terms of controlling emissions, it is relevant to note that CO_2 is the biggest anthropogenic contributor to radiative change to date and is projected to continue to be so. Methane is also significant, though it decays much more rapidly, emissions are much more uncertain, and in recent years the growth rate in the atmosphere has slowed dramatically. CFC emissions, almost entirely from industrialized countries, grew very rapidly until the late 1980s, but the overall impact on radiative forcing, taking into account indirect effects, is still uncertain (IPCC, 1992, 1995). Also, since these emissions are now being rapidly phased out under the Montreal Protocol and amendments, with limited interim growth in developing countries, they are less relevant to future policy. The replacement of CFCs by HCFCs, which have a high radiative impact and are not controlled by the Montreal Protocol, however, is a matter of concern, and their increased use deserves close scrutiny. Similar remarks apply to sulphur dioxide, which, though not a greenhouse gas, is thought to have a considerable indirect impact on the radiative balance. The role of other anthropogenic gases is not big enough to affect the equity issues discussed here.

3.3.3.1 Historical and cumulative emissions

Cumulative past emissions account for the buildup of gases in the atmosphere. Smith (1993) provides the most thorough account of how indices of past emissions can relate to current concentrations and gives data for cumulative carbon emissions in the period 1950–86. From these he has derived estimates of the "natural debt" that each country has drawn from (over)use of the atmospheric capacity; he justifies the 1950 cutoff on both ethical and practical grounds. The most extensive databases for cumulative emissions of CO_2 and methane are those developed by the International Institute of Applied Systems Analysis (IIASA). These include estimates of industrial and biotic carbon emissions back to 1800, as well as estimates of methane emissions, presented to a useful but manageable degree of regional breakdown. Their results are shown in Table 3.1.

These results suggest that North America accounts for 33% of the contribution from fossil fuels, Europe 26%, and the former USSR 14%. The industrialized countries together account for 84% of the total. When estimates of biotic (mostly deforestation) emissions are included, the North American figure is reduced to about 30%, and the contribution from a number of developing countries becomes significant. In total, the industrialized countries account for about two-thirds of cumulative carbon emissions.

Including methane makes very little difference to these results, partly because of the shorter residence time in the atmosphere. Based on current populations, the ratio of cumulative emissions per capita between the industrialized and developing nations is more than 10 to 1.

These estimates can be subjected to the criticisms that Agarwal and Narain (1991) raised in connection with an index of current net emissions (see Box 3.1). By "discounting" emissions to reflect the removal of gases, there is an implicit assignment of the earth's past sink capacity in proportion to emissions. Fujii (1990) and Meyer (1995) furnish calculations to show that if the total CO_2 absorption were assigned on an equal historic per capita basis, most developing countries are in fact "in credit" – their cumulative emissions are smaller than the global average per capita absorption, and so on this basis their past contribution is not merely small but actually negative.

Some logical problems with this latter approach are noted in the accompanying box. If a more directly physical definition of "relative contribution to the build-up of greenhouse gases is used,"[4] the contribution of developing countries is positive, but probably less than that indicated in Table 3.1. On any mea-

Table 3.1. *Historic CO_2 and methane contributions by region, 1800–1988 (in percentages)*

Region	Industrial CO_2	Total CO_2	CO_2 + CH_4
1. OECD North America	33.2	29.7	29.2
2. OECD Europe	26.1	16.6	16.4
3. Eastern Europe	5.5	4.8	4.7
4. Former USSR	14.1	12.5	12.4
5. Japan	3.7	2.3	2.3
6. Oceana	1.1	1.9	1.9
7. China	5.5	6.0	6.3
8. India	1.6	4.5	4.8
9. Other Asia	1.5	5.0	5.2
10. N. Africa & Mid-East	2.2	1.7	1.8
11. Other Africa	1.6	5.2	5.2
12. Brazil	0.7	3.3	3.3
13. Other Latin America	3.2	6.5	6.5
Developed Countries (1-6)	83.8	67.8	66.9
Developing Countries (7-13)	16.2	32.2	33.1
World	100.0	100.0	100.0

Source: Grubler and Nakicenovic, 1991.

sure, the contribution of the world's poorer regions to the total buildup of greenhouse gases over the past century is modest, and even more so when considered in relation to population.

3.3.3.2 Current emissions

Because of the somewhat more rapid growth of developing country emissions in the last two to three decades, responsibility for the problem is much more sensitive to the index chosen for current emissions than it is for cumulative contributions. The most important issue is whether one focusses on total or per capita emission rates. Developing countries tend to focus on the latter, as an index that highlights the extent of disparities between industrialized and developing societies in emissions associated with individual lifestyles, and consequently as a way of emphasizing their argument that profligate and unsustainable lifestyles in the industrialized countries are the root of the problem. Commentators from industrialized countries more often point out that, in terms of climate change, it is total emissions that matter, and that per capita indices ignore the important element of gross population itself as a causal factor.[5]

In reality, both aspects matter, and the most illuminating way of presenting the data is in a form that displays both simultaneously. This is done in Figure 3.1, in which 1993 emissions of CO_2 from fossil fuels are illustrated on a graph of per capita emissions against population. The area of each block, as the product of the two, represents total fossil CO_2 emissions from the country or region.

Projection against the per capita axis illustrates the scale of these disparities, not only between developed and developing countries but also within groups. North America and Australia emit between 4.5 and 6.0 tC (tonnes of carbon) per person on average. The figure for the industrialized regions of the "Old

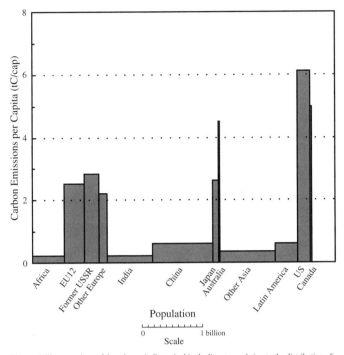

The graph illustrates three of the primary indices cited in the literature relating to the distribution of current emissions among the major countries or regions indicated beneath each block. The height of each block shows per capita CO_2 emissions from fossil fuel consumption (all uses) in 1993; the width of each block is proportional to the estimated population in 1993; and the area of each block, as the product of per capita emissions and population, is proportional to total fossil CO_2 emissions. The qualitative impact of other sources and gases (for which data are less well developed) is outlined in the text.

Source: Grubb (1990) updated by the author using data from *BP Statistical Review of World Energy, 1994* and *World Population Prospects*.

Figure 3.1: Carbon emissions per capita and population, 1993.

World" – including Russia and Central and Eastern Europe after the contractions of recent years – is between 2 and 3 tC per capita. Average per capita emissions in many parts of East Asia and Latin America are in the range 0.5-1.0 tC per capita, though some of the Asian "tiger" economies have already risen well above this. Per capita emissions on the Indo-Pakistan subcontinent and in much of Africa rarely exceed 0.3tC per capita – a tenth or less of the per capita emissions in the industrialized world. Others – such as the Middle East and various island states – vary widely within this range.

Considering the population axis gives a counterpoint to this picture. Many of the blocks that are much lower in Figure 3.1 (low per capita emissions) are also much wider (high population). The developing economies of Asia alone account for more than half the global population. Despite their lower per capita levels, this makes them significant contributors to the total (the area of the individual blocks shown in Figure 3.1). Altogether in 1993 the OECD countries accounted for about 50.5% of global fossil carbon emissions, with about half of this being from the U.S. The former USSR and Eastern European countries accounted for 17% (with half of this being from Russia), and the developing countries contributed just under a third of gross fossil carbon emissions (with over 40% of this from China).

Another basis for comparing emission profiles is to consider the "productivity" of emissions – emissions per unit of economic output as conventionally defined. This tends to receive more emphasis in developed countries, which are generally more efficient. It again has its counterpart in consider-

ing total economic output, indicating what some developing countries characterize as overconsumption. Economists in the industrialized countries tend to put the spotlight on the lower efficiency levels in developing countries, whereas others attack the overall high level of consumption in the developed countries as the root of the problem. Again, it is possible to demonstrate both dimensions on a single graph of emission intensity versus wealth. In this, a complicating factor arises from the uncertainties over appropriate exchange rates: The use of purchasing power parities gives a very different impression from comparing GNP at market exchange rates (see Box 3.2).

Data for sources other than fossil CO_2 (see, for example, SEI, 1992) are, as noted, much more uncertain, but in general the disparities between industrialized and developing countries are greater for CFCs, and less for all other sources, than those noted here for fossil CO_2. Most notably, almost all emissions from deforestation – thought to be in the range of 15-25% of total CO_2 releases in the early 1990s – are from a relatively small group of developing countries; anthropogenic methane emissions are probably predominantly from China (fossil and agricultural methane) and South and East Asia (agricultural methane). However, emissions of this gas are fairly broadly distributed, with most continents each contributing about 10% of the anthropogenic total (SEI, 1992; Subak, 1993). How the warming contribution of these emissions compares with fossil CO_2 data depend on the data and the basis of comparison chosen, as discussed in the IPCC report, *Climate Change 1994*. As noted, these regional comparisons obscure both large disparities in per capita emissions and important national and subnational variations within each group (IPCC, 1995; Subak, 1993).

3.3.3.3 Future emissions

The pattern of emissions is changing – it has indeed changed markedly in the last five years – and is expected to continue changing. A very large number of emission scenarios have also been prepared by various authorities, with a variety of assumptions, timescales, and degrees of regional breakdown.

From an equity perspective it is important to distinguish "business as usual" projections from assumptions about the location of abatement, and to have a sufficient regional breakdown to relate emission patterns to economic and other regional differences. The scenarios of the 1990 IPCC Scientific Assessment (unlike the 1992 update) had sufficient clarity on both counts, but were criticized for embodying highly inequitable assumptions, both in the reference scenario and in the apportioning of emission reductions in abatement scenarios. Parikh (1992) noted that the reference scenario "permitted" a substantial increase in North American emissions but considered that "cuts for the South are already built into the reference models," compared with the rates of growth that might otherwise be expected. "The stabilization scenarios," she argued, "stabilize the lifestyles of the rich and adversely affect the development of the poor."

The IPCC 1992 scenarios span a very wide range, with little geographical detail, and thus do not embody a clear view of the distribution of emission trends. A more recent and thorough development of emission scenarios for fossil fuels is that

BOX 3.2: INTERNATIONAL COMPARISONS AND EMISSION INTENSITIES

The national emission intensity can be defined as the ratio of carbon emissions to GNP. International comparisons then depend on how GNP in different currencies is compared. If market exchange rates are used, output from many of the developing countries appears very small. The first graph (Figure 3.2) gives examples of emission intensity for key regions and countries, employing a technique similar to that used in Figure 3.1 for population and per capita emissions. On this basis, emission intensity in China is about five times that of the U.S., and in general the emission intensity in the developing world is much higher than in OECD countries.

It is well known, however, that exchange rates tend to undervalue the real purchasing power of currencies in poor economies, particularly where there are strong exchange controls; it is, for example, possible to live quite well in some poor countries on the exchange rate equivalent of a few U.S. dollars a day. Estimates of purchasing power parity (PPP) attempt to compare currencies in a way that equilibrates purchasing power. For OECD countries the difference in GNP calculated at PPP and market exchange rates is modest, usually 10-30%; for some developing countries the difference in GNP is extremely large.

The second graph (Figure 3.3) illustrates what happens when emission intensities are calculated on a PPP basis, using estimates from the World Bank. On this basis, emission intensities in China and Africa are close to (and straddle) that of the U.S., whereas that of India is comparable with the European Union.

For several reasons, however, neither index can be taken as giving an authoritative comparison of energy efficiency. PPP estimates themselves are not only approximate (they can vary considerably between sources) but are not always more appropriate for comparing energy intensities. That is because fossil carbon emissions arise largely from industrial production, often involving goods traded at market exchange rates, whereas PPP is driven by consumption comparisons.

Also, carbon intensity can be very different from energy intensity, particularly in developing countries, due to the high use of biomass fuels. Few would dispute that, overall, energy does tend to be used much less efficiently in most developing countries and that much could be done over time to improve efficiencies in these countries, and indeed everywhere. The point is more that glib and aggregated statistical comparisons, whether on a GNP or PPP basis, are unlikely to help the international debate.

Note: The graphs show selected regions and countries because data on PPP are incomplete. Data for formerly centrally planned European countries are not included because rapid economic (and emission) changes, incompatible definitions, and statistical inadequacies make comparisons of available data of dubious value and rapidly outdated.

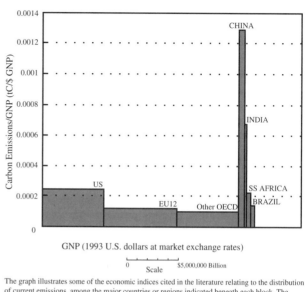

The graph illustrates some of the economic indices cited in the literature relating to the distribution of current emissions, among the major countries or regions indicated beneath each block. The height of each block shows CO_2 emissions intensity (emissions per unit GNP) from fossil fuel consumption (all uses) in 1993; the width of each block is proportional to the GNP in 1993; and the area of each block, as the product of emissions intensity and GNP, is proportional to total fossil CO_2 emissions. GNP is estimated using market exchange rates. Figure 3.3 shows the data for GNP on a purchasing power parity basis (see text). The qualitative impact of other sources and gases (for which data are less well developed) are outlined in the text.
Source: Authors, derived from data in *BP Statistical Review of World Energy 1994* and *World Development Report 1994.*

Figure 3.2: Carbon emissions per unit GNP and total GNP (market exchange rates)

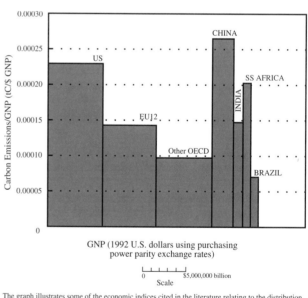

The graph illustrates some of the economic indices cited in the literature relating to the distribution of current emissions, among the major countries or regions indicated beneath each block. The height of each block shows CO_2 emissions intensity (emissions per unit GNP) from fossil fuel consumption (all uses) in 1993; the width of each block is proportional to the GNP in 1993; and the area of each block, as the product of emissions intensity and GNP, is proportional to total fossil CO_2 emissions. All GNP data are on purchasing power parity basis. The qualitative impact of other sources and gases (for which data are less well developed) is outlined in the text. The more limited regions, as compared with Figure 3.1, are due to the limited data and the difficulty of aggregating PPP-adjusted GNP from the source data.
Source: Authors, derived from data in *BP Statistical Review of World Energy 1994* and *World Development Report 1994.*

Figure 3.3: Carbon emissions per unit GNP and total GNP (purchasing power parity exchange rates)

by the World Energy Council (WEC, 1993), which developed regional scenarios both from a global perspective and with regional teams. Interestingly, their reference scenario was also criticized by participants from the developing world and former USSR for being too optimistic about the prospects for improved energy efficiency in these regions. Taking account of these concerns led to a variant of the reference scenario with much higher emissions, particularly in South Asia, Central and Eastern Europe, and Africa.

Whichever variant of future scenarios is considered – and especially for the higher growth cases favoured by many in the developing world – the vast majority contain a clear message. The ambiguities about whether the developing world holds significant responsibility for contributing to climate change will not persist for long. Even in relatively restrained scenarios of emission growth from the developing world, such as the original WEC reference case, fossil CO_2 emissions alone from these countries will equal those of the OECD by 2020. And in many scenarios, fossil CO_2 emissions from the current developing nations will exceed today's total global emissions before the year 2050, despite still being considerably lower in per capita terms. Furthermore, in terms of the distribution of abatement efforts, there are significant spillover effects both positive and negative.[6] Edmonds (1993) has explored the impact of various possible abatement "coalitions," and Bradley *et al.* (1994) note on the basis of a range of abatement scenarios that "anthropogenic climate change cannot be controlled by the OECD nations alone."

Emissions from deforestation do not – and probably cannot – keep pace with projections of fossil fuel emissions. The projections reviewed in the IPCC 1992 assessment almost all stay below 2 GtC/year, and decline after 2030. In comparison, projections of global fossil fuel emissions rise above 10 GtC/year by that date in many "reference" scenarios. Methane projections are highly uncertain but also suggest slower growth. Clearly, neither deforestation nor methane releases is likely to alter substantially the overall trend in the distribution of emission sources. Whatever the past and current responsibilities and priorities, it is not possible for the rich countries to control climate change through the next century by their own actions alone, however drastic. It is this fact that necessitates global participation in controlling climate change, and hence, the question of how equitably to distribute efforts to address climate change on a global basis.

3.3.4 Differential impacts, extreme vulnerability, and the valuation of impacts

3.3.4.1 General features

Another aspect – distinct but again partly correlated with wealth – is the likely differential impact of climate change. Both Volume 2 of this report and Chapter 6 of the present volume agree that impacts may vary considerably between countries.

These impacts are a product of the degree and nature of the physical change, the degree to which the society depends on the natural resources affected, and its institutional and social

capabilities for handling change. Some physical aspects of climate change could be beneficial, particularly in cooler climates and in a long-run equilibrium, but the transition may still be difficult, especially if it must contend with increased climatic variability.

In hotter climates and developing countries with economies that depend more heavily on natural resources, the impacts are widely expected to be adverse (albeit to varying degrees) for three main reasons:

(1) Most developing countries are in tropical regions, and projected climatic changes seem unlikely to improve either the quality of the physical environment or agricultural productivity; indeed, modelling studies indicate that climate change will reduce relative agricultural output in the developing world (Chapters 2, 6; see also Volume 2).

(2) Developing economies are much more dependent on agriculture and other aspects of natural resource flows.

(3) Institutional and social structures within developing countries tend to be weaker and hence less able to cope with change. Developing countries also have fewer financial resources for investing in more robust infrastructure. Meyer-Abich (1994) concludes that this on its own is sufficient reason "to prevent greenhouse gas-induced climate change, because if it happens, present inequalities would be irresponsibly increased."

3.3.4.2 Extreme vulnerabilities

In addition, some regions may be exceptionally vulnerable to climate change. Small island states and countries with low-lying coastal areas are obviously vulnerable, not only because of their greater susceptibility to sea level rise but also their heavy dependence on natural resources for domestic use and for trade that may be affected by climate change (Saha, 1994). In addition to these adverse economic impacts, all or part of these countries may be destroyed or rendered uninhabitable by sea level rise (which is projected to continue well beyond the end of the next century, even if action is taken to curb emissions) and attendant salt water intrusion. Storm surges and other changes (such as subsidence) may exacerbate these impacts. The potentially devastating impact of tropical storms on low-lying regions was graphically illustrated by the 1991 Bangladesh cyclone, which claimed more than 200,000 lives (del Mundo, 1992) and perhaps twice that number. Such impacts clearly could have profound consequences for both human communities and entire ecosystems (see, for example, Broadhus, 1993).

Potentially extreme impacts may also occur in very dry regions that may already be on the margins of existence. The central fear relates to the adequacy of food and water supply. Despite a global food surplus in the 1980s, millions still starved because of a lack of local resources, infrastructure, and social stability. Although models suggest that global food production can be maintained under climate change, agricultural production in the developing world may decline. Indeed, most models project a drying of continental interiors, and the African region may be particularly vulnerable. Sidibe (1995) gives a review of possible implications for Mali; see also the assessments in Volume 2 of this report.

If such projections prove true, climatic change will create "environmental refugees." Even without the worst projected impacts, problems of both domestic and international migration are likely to be exacerbated. Myers (1993, 1994) cites estimates that there are about 10 million environmental refugees at present, and on the basis of a survey of projected impacts in vulnerable regions, estimates that this figure could rise to 150 million by the middle of the next century as a result of climate change. He sketches the immense social, economic, and political costs implicit in such movements, "pushing the overall cost far beyond what we can realistically envisage in the light of our experience to date . . . it requires a leap of imagination to envisage 150 million destitutes abandoning their homelands, many of them crossing international borders." Again, the poor seem most likely to suffer, though clearly such movements might also trigger broader ethnic or even international conflicts that could envelop whole societies.

These extreme vulnerabilities raise important equity issues, not only because they represent potentially great suffering but also because those most vulnerable are to a large extent those contributing very little to its causes. For small islands, but also for many other developing countries, adaptation to climate change may be a more important, and more urgent, task than planning and undertaking national mitigation policies. The language of the FCCC does not give mitigation absolute priority, as it accepts that some countries and communities will have to place more emphasis on adaptation. However, the acceptance that the developed country parties must assist particularly vulnerable developing country parties in meeting the costs of adaptation indicates agreement that it would be inequitable for members of the international community that were highly vulnerable to climate change to bear the costs of that vulnerability alone. Possible implications of this exceptional vulnerability for international responses that aimed at coping with climate change and minimizing its impacts are considered further in Section 4.

The common thread in all the scenarios of extreme vulnerability and impacts is the emphasis on the social dimension. A society with strong social institutions and some financial resources should, with rare exceptions, be able to develop and implement strategic plans for reducing vulnerability, albeit at a cost. But these conditions do not exist in some of the most vulnerable regions and are not easy to import. Furthermore, there is an internal distributional dimension to physical impacts such as storm surges, hurricanes, sea level rise, and changing climate patterns. The burden of adapting to these events will fall more heavily on the parts of the communities that may already be disadvantaged and unable to adapt without further hardship.

There is also a potentially vicious interaction between social instability and climate change. Suliman (1994) argues powerfully that this has been an important part of the Sudanese tragedy: What was originally an ethnic and religious conflict has evolved over a period of three decades into a conflict about resources, driven by a combination of climatic

change and sometimes inappropriate policies advanced by international lending agencies.

According to Suliman, climatic change (the occurrence of which in the area is uncontested; only the cause is disputed) altered the distribution of resources around which fragile social institutions were beginning to develop, thus undermining their stability and helping to plunge the Sudan back into civil war.

It may be much more cost-effective for the rich world to help protect such regions directly than by additional mitigation of emissions. However, the scale of existing inequalities, the resistance to additional foreign aid, and the mixed efficacy of international development assistance suggest that it may be facile to rely on this alone.

3.3.4.3 *Global valuation of impacts*

The possibility of extreme physical impacts and complex social reactions poses dilemmas for evaluating aggregate impacts. As discussed in the introduction to this chapter, there is no single right way to do this. It can only be done by making ethical judgments about what to measure, and how to measure and compare the welfare of different people in different countries across time. Considering only impacts on World Product is a poor enough indicator because (as outlined in Chapter 1 and discussed in Chapter 7) World Product makes no pretence to be a real measure of welfare. But if it is predominantly poorer people who are affected, it is even worse as an indicator, because poor people add very little to aggregate wealth, and impacts on them are similarly discounted (Ayres and Walter, 1991).

Attempts to allow for "nonmarket" impacts, such as those on human health, social stability, and environmental quality, require assumptions to be made about how to value and aggregate such impacts on different people in different countries. Economics seeks classically to establish a monetary equivalent for these impacts (e.g., Fankhauser, 1993), as discussed further in Chapters 5 and 6. Scaling impact costs according to relative GNP implies valuation of human impacts based on average earning power. In welfare economics this approach is contentious, as it is widely acknowledged to relate poorly even to the more established "willingness to pay" criterion, and that itself is acknowledged to have major limitations. Particularly for poorer people, "willingness to pay" to avoid environmental damage may give a very different (and much lower) valuation than "willingness to accept" compensation for such damage.

The implicit valuation of human welfare, and indeed of risk to life itself, differs greatly between countries. This is partly because it is constrained by aggregate wealth, and partly by the blunt fact that societies clearly do not value people in other countries equally. But one of the most significant aspects of climate change is that activities of people in one country can have destructive consequences for people in others. Any aggregation that evaluates and aggregates impacts in relation to national wealth (such as impacts on World Product, plus nonmarket impacts related to national GNPs) in effect yields the result that the impact is less significant if it is poor people, or people in poorer countries, who suffer. If poorer countries are more vulnerable and are likely to suffer more, the dominant impacts are likely to be precisely in those regions. The question of how these impacts are valued thus becomes central and cannot be passed off simply as inevitable economic logic (Grubb, 1993, 1995).

The question is even more difficult in the case of those small islands and low-lying coastal areas where climate change may lead to loss of territory and profound social impacts, including destruction of indigenous cultures and whole communities, with the loss of their attendant heritage and spiritual, cultural, and social values. This loss in human diversity does not appear to command the attention it deserves. No amount of monetary compensation may be sufficient for communities that have had no choice but to suffer such impacts.

This issue in itself could have important implications for the balance between mitigation and adaptation. Although significant populations may be involved, particularly if migratory pressures spread the human impacts, their wealth and hence "willingness to pay" (constrained by their ability to pay) are negligible on a global scale. On most utilitarian formulations, their potential suffering thus registers little. By the same token, they have scarcely contributed to the problem that threatens them. Some commentators dispute the appropriateness of even attempting economic valuation of such extreme impacts, particularly when they are inflicted by the actions of others. Evaluation in terms of "willingness to accept" compensation for such damages can preserve the ethical basis of monetary evaluation of impacts, but impact assessments have not yet been developed on this basis. Also, a claim for infinite compensation would return the issue to one of the ethics of imposing damages that the victim claims are not compensable by money. Adams (1995) argues that seeking to monetize impacts in these circumstances risks entrenching disputes rather than resolving them, and should be abandoned in favour of negotiations that seek to forge greater consensus about the appropriate values to employ.

Evaluating damages from climate change therefore becomes a highly political question of how much the world cares about imposing such risks on some of the poorest and most vulnerable people. If the success of international responses is measured in terms of aggregate impacts on global wealth, their position will not feature. If it is judged by how well it protects the weakest, and minimizes the most severe suffering, an entirely different approach may be called for (Meyer-Abich, 1994). This obviously fundamentally reflects the ethical debates surrounding utilitarianism and the Rawlsian response outlined above. Thus, it needs to be recognized that attempts to quantify the costs associated with climate change involve inherently difficult and contentious value judgments, and different assumptions may greatly alter the resulting conclusions.

3.3.5 Social and institutional dimensions

Countries are not equal in their capabilities to deal with the challenges posed by climate change. Some are strong nation-

states, with a large degree of societal consensus and strong institutions that can formulate and implement policy effectively and that can act to protect weak and vulnerable groups. Others may be riven by internal differences and have fragile governing institutions that may be unable, or unwilling, to formulate and implement effective policy or to protect the most vulnerable groups – indeed, these institutions may themselves be vulnerable to the stresses that climate change may bring.

These differences affect both the vulnerability to climate impacts, as discussed, and the ability to formalize and implement efficient abatement policies. Juma *et al.* (1994), for example, emphasize how these factors affect efforts to transfer technologies. Some of these issues are discussed more fully in Section 3.6.

More broadly, the social dimensions of both climate change impacts and responses deserve more consideration. For example, there is relatively little literature on public perceptions and social responses to climate change and how different social groups view global environmental problems. Such research would appear necessary as a first step to increasing public participation in ameliorating these problems (Lofstedt, 1994).

3.3.6 Differing resource endowments

Finally, countries differ widely in their endowment of resources that may be affected by efforts to mitigate climate change and in their current reliance on production or conversion of these resources. Coal resources and production are relatively widespread globally, though with major variations within regions. Remaining oil reserves are heavily concentrated in the Middle East, North and Central America, and the former Soviet Union. For some Middle Eastern countries especially, oil is the only significant export and source of foreign exchange. Proved gas reserves are also quite concentrated in the Middle East and Russia (though their value would probably be enhanced by mitigation efforts). Forests and timber exports are likewise concentrated, and can economically, as well as environmentally and socially, be very important to some countries in both the developed and developing world. The value of such resources can be affected by action to limit climate change, though the actual equity implications may be complex, as discussed later in this chapter.

3.3.7 Conclusions

This section has illustrated several ways in which countries differ and that are relevant to formulating climate change responses. A number of these group along a developed/developing country axis. The implications for developing countries are different because they are generally poorer, have contributed much less to past emissions, and still emit much less per capita. They may also have shorter policy time horizons, weaker institutions, and be more vulnerable to climate change. Yet, in other respects, the divisions are much more complex. There are substantial variations among both developed and developing countries in terms of absolute and per capita emissions, the likely impacts of climate change, institutional strengths and preferences, and endowment with natural resources that may be affected by mitigation. Similarly, in the next century some developing countries will continue to make marginal contributions to the problem, whereas emissions from others may well start to dominate it. A rigid developed/developing country delineation of equity issues in the debate is thus inappropriate and may indeed be highly damaging in the long run, though it inevitably permeates many components of the debate at present.

The rest of this chapter concerns the implications of the various differences for what might constitute equitable responses. In structuring it, we draw on the observation by Shue (1992) that the equity issues in responding to climate change comprise at least four distinct questions:

(1)   What is the distribution of the costs of coping with climate change?
(2)   What is the distribution of future emissions?
(3)   What is the distribution of the costs of measures to limit emissions?
(4)   What is the background allocation of resources and capabilities and hence the consequent basis for fair bargaining and fair processes?

All these questions are interrelated. For example, a fair distribution of future emissions may depend somewhat on how much countries help others with the costs of coping. A fair distribution of abatement costs depends on various background circumstances, and debates about allocating the "burden of abatement" in general, and joint implementation and tradable emission quotas in particular, highlight a particularly strong link between the second and and third questions. Few authors have distinguished these different components as distinct equity issues (though several note that the distribution of paying for abatement could be very different from the distribution of where abatement efforts – and hence emissions – are located). But they are distinct, and it is useful to address the equity issues each raises in turn, even if two or more of them may in practice be tied or even addressed through the same instruments.

This next section (Section 3.4) considers the question of the costs and risks of coping. Section 3.5 considers the common issues surrounding distributing emissions and the costs of abatement before summarizing specific proposals concerning each in turn. The final sections then examine internal equity and procedural equity issues.

3.4 Distributing the Costs of Coping: Impacts, Risks, and International Insurance

The previous section has noted the fact, analyzed more fully in the 1990 IPCC Scientific Assessment, that some countries are likely to be much more affected by climate change than others. To an important degree, poor countries may be much more vulnerable than richer countries. But all projections of impacts are fraught with uncertainty, and in some scenarios parts of the developed world could also be badly affected.

Thus, the issue is also about sharing and minimizing risks. This section reviews these issues.

3.4.1 Nature of impacts and uncertainties

Predictions of climate change remain highly uncertain. The uncertainty about the future average temperature is high enough, but it is local effects that really matter, and predictions of regional changes are even more uncertain and can vary substantially among different climate models. Fluctuations in the frequency of extreme events will not, for the most part, be predictable (except in a probabilistic sense), and will therefore add to the uncertainty.

It is not likely that such uncertainties can be resolved quickly, and it may always remain impossible to make reliable local predictions or to associate particular local impacts with human-induced climate change, because to an extent climate is inherently chaotic. Variations may be extremely sensitive to initial conditions (see Gleick, 1987, for a popular discussion of dynamic systems and their inherent unpredictability).

Thus, in discussing climate impacts it is better to refer to "expectations" of costs and benefits, in the mathematical sense of an average of possible outcomes, and to recognize simultaneously the importance of uncertainties. This implies there is another cost item – the cost of bearing risk. Thus, we can categorize the distributive issues arising from climate change impacts into four groups:

(1)   aggregate welfare impacts over time (intertemporal distribution of expected global utility)
(2)   welfare impacts between countries (international distribution of expected utility)
(3)   welfare impacts within countries (interpersonal distribution of expected utility within countries)
(4)   the distribution of the cost of risk bearing associated with each of the above

The question of aggregate intertemporal distribution is the subject of the next chapter. With respect to impacts within countries, over and above those arising from policies promoting redistribution, governments generally accept responsibility for trying to manage, respond to, and mitigate the worst impacts of extreme events, as in the case of government relief programmes for victims of major storms. In addition, big disasters in the developing world will also often mobilize international relief efforts, though these tend to be quite short-lived. After the initial effort, the local people, with some assistance from the government, tend to be left to rebuild their lives.

The new aspects introduced by human-induced climate change thus concern the broader impacts on international welfare distribution and the cost of associated risks. We consider each in turn.

3.4.2 Paying for the costs of coping

If there are no international transfers or other assistance connected directly with the measures to cope with actual or potential impacts of climate change, each nation in effect is expected to cope using its own resources. The available literature appears to be unanimous in considering this inequitable, for two main reasons.

The first arises directly from the observation above that poor countries are likely to be much more vulnerable to climate change. Climatic change is thus likely to exacerbate existing inequalities. Most ethical systems would in any case expect richer parties to contribute more to addressing a common problem. The rationale is much stronger still when the cause is shared but the impacts are disproportionately large on the least well-off parties.

The second factor, which applies irrespective of this inequitable distribution of impacts (but is much reinforced by it), is the direct causal link: the fact that the actions of some people (emitting greenhouse gases) may directly harm others. There are few, if any, ethical systems in which it is acceptable for one individual knowingly to inflict potentially serious harm on another and not accept any responsibility in helping or compensating the victim or to pay in some other way. An analogous principle in international law provides that a state may exploit its resources but has the responsibility to ensure that activities in its jurisdiction do not cause harm to the environment of other states or to areas beyond its national jurisdiction. This principle is generally accepted as a rule of customary international law. Its application to the climate change context is problematic, however, as it is extremely difficult for one state to prove before an international court or tribunal a direct causal link between another state's emissions and its own environmental damage. Although, upon signing the Framework Convention on Climate Change, a number of small island countries indicated that they may rely on traditional rules of international law to address damage to their environment from climate change, it is widely recognized that it would be factually and legally difficult to use these rules successfully.

This difficulty points to the need to develop general rules of international law regarding liability and compensation. However, in the absence of new rules of liability, the traditional "fault-based" rules of responsibility of international law could provide a basis for vulnerable, generally poorer, states adversely affected by climate change to receive compensation from richer states whose past and present emissions of greenhouse gases have caused environmental harm. The issue of historical emissions could be of direct relevance here, as climate impacts are a function of atmospheric concentrations, which depend strongly on cumulative emissions. Although the Convention's preamble recognizes that the largest share of historical and current global emissions of greenhouse gases originated in developed countries, and that per capita emissions in developing countries are relatively low and will need to grow to meet their social and development needs, its legal ramifications remain unclear.

3.4.3 *Sharing risks*

The question of who should pay is not solely about equity; it also concerns more classical economic issues about sharing risks. Chichilnisky and Heal (1993), drawing on Cass, Chichilnisky, and Cass *et al.* (1991), note that the risks involved have four important characteristics. Such risks are:

(1) poorly understood – we cannot assign clear probabilities to different impacts
(2) endogenous – they are created or affected by our own actions
(3) collective – given climate impacts may affect large numbers of people; consequently, the risks are not statistically independent and, hence, may not be greatly reduced by "pooling"
(4) irreversible – impacts cannot in general be reversed and, indeed, we may be committed to these long in advance

This results in a complex combination of issues of equity and decision making under uncertainty. Decision making, when the probabilities of different outcomes are known or reasonably estimated ("risk"), is a field of extensive economic study. Rational decision making when the probabilities of different outcomes are not known is more difficult but is still a subject of considerable analysis. These issues are discussed further in Chapter 2.

One criterion for decision making under extreme uncertainty is the "minimax" criterion (see Chapter 2), which involves selecting the strategy that minimizes the worst outcome. Its basis is similar to that of Rawls's criterion for justice. It recommends avoiding those strategies that could generate the worst outcomes. The minimax principle bears some resemblance to the precautionary principle expressly included in the Convention in Article 3.3, which provides that

> The parties should take precautionary measures to anticipate, prevent, or minimize the causes of climate change and mitigate its adverse effects. Where there are threats of serious or irreversible damage, lack of full scientific certainty should not be used as a reason for postponing such measures, taking into account that policies and measures to deal with climate change should be cost-effective so as to ensure global benefits at the lowest possible cost.

A philosophic and economic examination of risk bearing, including the maximin principle, may provide insights relevant to the application of the precautionary principle. The maximin criterion is, however, rather conservative because it takes into account only the worst outcome and not the nature of the other outcomes from a strategy. In practice, climate change is not a situation of total ignorance – we have some idea of plausible outcomes and their relative probability. This increases the appropriateness of approaches based on subjective probabilities, combined with an appropriate measure of risk aversion to reflect the cost of bearing risk (as outlined in Chapter 2).

Climate risk is an equity issue for several reasons. There are good reasons to believe that the distribution of the cost of risk bearing is very uneven. First of all, industrialized countries have, in general, better insurance markets. This means that individuals in those countries can reduce their cost of risk bearing substantially (but not completely, partly because global warming is a collective risk and partly because of the incompleteness of markets). These possibilities are not open to most individuals in developing countries, and they therefore have a higher cost of risk bearing. Second, the cost of risk bearing may be much higher in some developing countries than in industrialized countries because of the nature of the impacts. For example, when considering sea level rise, a range from 20 cm to 60 cm may imply a quite low cost of risk bearing in a country like the U.S., and the expected damage would be a good proxy for the total cost of sea level rise. For a country like Bangladesh, however, the difference between 20 cm and 60 cm may mean a substantial difference in utility, which is not accounted for in the expected damage. Thus the cost of risk bearing in Bangladesh is much higher than in the U.S.

The same holds true for countries that may be threatened by increases in both the frequency and the severity of droughts. The variation in utility will probably be much greater than the variation in actual damages among different outcomes, and therefore the cost of risk bearing will be high. In the absence of empirical studies on the cost of risk bearing the values involved are highly uncertain, but the nature of the conclusions seems reasonable.

3.4.4 Policy implications

Most of the discussion of the economic and equity aspects of climate change impacts has focussed on the "expected" changes over time and between and within countries. But uncertainties combined with irreversibilities – both in abatement and in impacts from climate change – make the issue of risk bearing also of great importance. Analysis is technically difficult but not impossible, and more study is needed. For example, the combination of the idea of basic needs and risk bearing could have far-reaching implications. If we accept that individuals have the right to some basic goods – food, shelter, access to environmental resources – and if emitting greenhouse gases threatens the availability of these basic goods, then those responsible for the emissions should be prepared to protect potential victims against losing the right to basic goods.

The Framework Convention on Climate Change mentions the concept of insurance in Article 4.8, which provides that

> Parties shall give full consideration to what actions are necessary under the Convention, including actions related to funding, insurance and the transfer of technology to meet the specific needs and concerns of developing country Parties arising from the adverse effects of climate change and/or the impact of the implementation of response measures.

In addition, Article 4.4 provides that the developed country parties must "assist developing country Parties that are particularly vulnerable to the adverse effects of climate change in meeting the cost of adaptation to those adverse effects."

This is a specific example, but at least two general conclusions can be drawn from the above discussion:

(1)   It is necessary to include the cost of risk bearing in any equity discussion related to global warming. In particu-

lar, estimates of the costs of impacts should also include this cost component in addition to expected cost.

(2)   Policy should seek to establish appropriate global insurance against impacts from climate change.

The appropriate form of insurance is not a simple matter, however. The characteristics of climate change discussed above lead to particular requirements for approaches to insurance, some economic principles of which are summarized in Chapter 2.

Also relevant is the fact that there are three broad types of costs that may be incurred in coping with climate change:[7]

(1)   costs related to preparation (or protection) to minimize the potential impacts of climate change, for example, through the construction of sea defences, improved water storage or water transfer capabilities, or changing the kinds or diversity of crops

(2)   costs arising directly as a result of changes in the resource base (e.g., loss of land to the sea, other loss of agricultural land) and the social costs that may follow from this (Suliman, 1994)

(3)   costs caused directly by extreme events, such as freak storms and droughts – which can be considered a component of both 1 and 2

Although the underlying equity issue of who should pay is similar in each case, some of the practical issues differ. Even if there is agreement on the underlying equity principles, major practical differences and disputes may arise because of all the uncertainties surrounding climate change. General protection measures (type 1 costs) have multiple benefits, and views will differ on the extent to which they should be funded primarily on the basis of concerns about climate change. It is likely that victims of changes in the resource base (type 2) and extreme events (type 3) will attribute their suffering to human-induced climate change more readily than those responsible for the emissions. This raises important issues about fair processes for deciding about the probability of events being tied to human-induced climate change, which none of the literature addresses explicitly.

The different nature of these costs has also led to different proposals about how to tackle them. Notably, there have been suggestions (for example, in the Toronto Conference, 1988, and the Beijing Declaration, 1992) for a "climate fund" to help developing countries pay for protection measures of type 1. International funding of such measures can be considered both as a matter of equity and as a way of reducing overall risks.

Also, the Alliance of Small Island States (AOSIS) has proposed creating an insurance pool to provide some cover against the impacts of sea level rise and associated salt intrusion and storm surge damage. This was elaborated as a proposed annex to the draft Climate Convention but removed at a late stage under pressure of the Rio deadline, and is discussed by Wilford (1993). AOSIS proposed that the insurance pool be created with international funding from governments. Two precedents are noted:

- The 1971 International Convention on the Establishment of an International Fund for Compensation for Oil Pollution Damage, which is funded by agreed levies on oil importers.
- The 1963 Brussels Supplementary Convention (an OECD agreement, amended by a 1964 protocol) to the 1960 Paris Convention on Third Party Liability in the field of Nuclear Energy. This is funded half on the basis of GNP and half on the basis of nuclear capacity.

The draft Insurance Annex to the Framework Convention on Climate Change proposed, by analogy with the nuclear precedent, that funding be based on a combination of GNP and CO_2 emissions. The latter component was suggested both on grounds of responsibility and to give an incentive to limit emissions.

The main criterion for entitlement to claim put forward in the draft annex was proved loss attributable to sea level rise, including storm surges. There is no reason why such an insurance pool should not, in principle, be available for countries other than small islands and for damage other than sea level rise. It would appear more equitable to extend the scope of such a fund to insurance against all kinds of climate impacts, though determining cause and effect may be more difficult in cases other than sea level rise. The literature does not appear to explore this issue, nor does the question of compensation or other support for other kinds of climate coping appear to have been explored more fully, though some elements are implicit in the more general economic treatment of insurance (see Chapter 2).

3.5 Distributing Future Emissions and Abatement Costs

3.5.1 Introduction

Issues of how future emissions (or abatement efforts) should be allocated, and who should pay for abatement, form some of the most contentious equity issues in climate policy. As noted above, the two questions are closely intertwined. In the absence of separate international transfers associated with abatement, the initial distribution of emission constraints will largely determine the distribution of costs. Much of the literature has focussed on the question of future emission allocation, but some of it has addressed more directly the distribution of costs. This section considers, first, issues which are common to both topics, and then, in turn, some of the specific proposals for fair allocation of emissions and abatement costs.

The applied research literature on these questions – relating specifically to climate change policy – is limited compared to that on climate science, impacts, or the economics of responses. There are many short articles and statements of positions made in newspapers, newsletters, "nonpapers" at negotiating sessions, and similar documents, and also some important official declarations developed through extensive discussions to reflect consensus positions among groups of countries (e.g., the Beijing Declaration and various regional statements). Also, the UNCED agreements themselves contain innumerable references related to equity. Most of this discussion is cast in developed/developing country terms, with important additions concerning the interests of particular groups.

This section attempts to reflect this material, although recognizing the impossibility of ensuring comprehensive coverage and drawing heavily on the limited deeper analysis of international equity issues in the more conventional academic literature of books, substantive reports, and articles in peer-reviewed journals. Interestingly, of those who have contributed such analyses, few are professionally employed as academics in the discipline which might seem to be of greatest direct relevance – moral philosophy.

3.5.2 Principles for shorter-term action

In approaching the topic it is useful to distinguish short-term from long-term allocation issues. In the long term, decisions will have to be taken – by design or default – about the overall allocation of greenhouse gas emissions. The appropriate allocation depends on the conception of equity, and different ethical approaches may have very different implications. But in the short term, the position is simpler. All ethical systems, and all the applied literature, appear to point in the same direction. As Shue (1992) has put it:

> Even in an emergency one pawns the jewellery before selling the blankets. . . . Whatever justice may positively require, it does not permit that poor nations be told to sell their blankets [compromise their development strategies] in order that the rich nations keep their jewellery [continue their unsustainable lifestyles].

The Framework Convention on Climate Change certainly requires that developed country parties take the lead in combatting climate change and its adverse effects. It also recognizes that the extent to which developing countries implement their commitments under the Convention depends on the effective implementation by the developed country parties of their commitments to transfer financial and technological resources and must fully take into account that, for the developing countries, economic and social development and poverty eradication are the first and overriding priorities. These provisions flow from the Convention's principle of common but differentiated responsibilities and the parties' respective capabilities, rather than being predicated on the rich countries' unsustainable lifestyles.

The observations about what is not fair in the initial distribution of emission obligations extends to other measures of abatement effort. In particular, some earlier and overly simplified economic studies assumed that a uniform carbon tax, as an indicator of abatement effort, would distribute the "burden" equally, fairly, and efficiently. None of these three suppositions has stood up to closer scrutiny in the literature:

- A uniform carbon tax would not impose equal burdens on different countries because of differences in existing tax structures (Shah and Larsen, 1992), resource endowments (Rutherford, 1993), and stages of development.

- Even in a scheme involving "equal burdens," most observers argue that a uniform carbon tax would not be fair because of the many differences outlined in Section 3.3, notably differences in historical and current emissions and in current wealth and consequent priorities.

- A uniform tax would also be Pareto-inefficient in the absence of optimal international lump-sum transfers, because of the differing marginal utilities between countries at different stages of development combined with the "global commons" characteristics of climate change (Chichilnisky and Heal, 1994). In similar arguments, Uzawa (1994) suggests differentiated carbon taxes as an equitable and efficient approach.

The literature thus demonstrates a degree of consensus about relative responsibilities and equitable directions of response in the short term and some elements of what would not be equitable in longer-term allocations.

A more difficult debate about equity in short-term responses concerns how to share efforts within the industrialized world. Clearly, the cost of reducing emissions by a given amount varies between countries. Factors that can increase the costs of reducing energy-related emissions from a given baseline date include population growth, development based heavily on domestic low-cost resources, or a starting point of low per capita emissions due to relatively lower economic development or an already carbon-free electricity system. One recent study, for example, claims that these factors would make emission reductions in Australia and New Zealand more costly than for many other OECD countries (Chisholm and Moran, 1994). Although the results of any specific modelling studies need to be treated with caution because of the inherent uncertainties and scope for biasing assumptions towards national interests, there is no doubt that there are complex issues of fair burden sharing among OECD countries that the next stage in negotiation will have to start addressing.

This debate could be usefully illuminated by some of the broader literature about longer-term principles for burden sharing. For beyond the short term, and in the attempt to develop more precise allocation principles, there is a far wider range of perspectives, which derive from a range of ethical approaches. We consider first some of these underlying ethical assumptions, and then consider more specific proposals for allocation of future emission entitlements and abatement costs.

3.5.3 *Longer-term allocation: Underlying principles*

General ethical principles underlying discussions about fair allocation fall into several categories. Several authors have sought to classify and in some cases critique equity principles applied to this question. Those attempting a fairly broad coverage include d'Arge (1989), Rose (1990), Young (1990), Grubb *et al.* (1992), Rose and Stevens (1993), Ghosh and Jaitly (1993), and Bhaskar (1993). Most focus upon direct national emissions, though Fermann (1993) also raises questions about assignment of emissions in traded goods. Rose (1990) and Rose and Stevens (1993) provide a very different list of allocation categories. The latter translate these to operational rules and model their impacts, though the criteria themselves comprise a mix of outcome, process, and nondistributional elements that are not always clearly defined, and some of which could overlap. Simonis (1994), amongst others, relates some simple rules to previous agreements. Some of the principles sketched here are in part an attempt to formulate more specific interpretations of utilitarianism, as well as to apply alternative or complementary ethical frameworks.

3.5.3.1 *Egalitarianism*
This reflects the underlying parity principle that each human being should have equal rights, in this case with respect to access to common global resources. It underlies several specific proposals considered in the next section. The Convention notes that the per capita emissions of developing countries are relatively low, but its substantive provisions do not propose equal per capita emissions as a prescriptive criterion for emission entitlements.

3.5.3.2 *Basic needs and Rawlsian criteria*
A basic needs approach in the present context involves allowing countries the right to emit the minimum levels of greenhouse gases needed to meet the basic needs of their citizens, defined as the minimum consumption levels needed to support full participation in society, and then requiring countries to buy (or pay taxes on) the rights to emission levels above these. It would perhaps be close to the allocation of emission permits according to population, although basic needs could vary from country to country depending on climate and other matters. Chichilnisky (1977) introduced, formalized, and developed empirically the idea of development aimed at satisfaction of basic needs while creating a society intrinsically compatible with the environment.

This approach can be related to a Rawlsian philosophy. At an absolute minimum it suggests that developing countries be left at least as well off under an emissions control regime as they would be in its absence; a stronger interpretation is that the regime should be used specifically to improve their position (Benestad, 1994).

The only quantitative target in the Convention requires developed country parties to aim to return their emissions of carbon dioxide and other greenhouse gases not controlled by the Montreal Protocol to 1990 levels by the end of the decade. This objective, which is difficult to regard as a legally binding commitment, applies a uniform target based on historical or current emissions, and is often referred to as the "grandfathered emissions" approach (Bodansky, 1993). Developed country parties with economies in transition are allowed "a certain degree of flexibility" with implementation of this obligation. Developing countries are not required to limit their current emissions, and the Convention's preamble recognizes that developing countries' emissions will grow to meet their social and economic development needs. These provisions could implicitly support some kind of concept of basic needs emission entitlements, varied according to national circumstances. However, it is important to bear in mind that the Convention's negotiators could not agree on the bases of present and future entitlements to emissions (Bodansky, 1993). The Convention does not, therefore, preclude the relevance of the basic needs approach to future discussions about emissions entitlements.

3.5.3.3 *Proportionality and polluter-pays principle*
Proportionality reflects the ancient Aristotelian principle that people should receive in proportion to what they put in and pay in proportion to their contribution to damage caused. This idea has a clear potential relationship with the polluter-pays principle, which is formulated as a principle of economic efficiency. In the context of international pollution, this principle can be interpreted in a number of ways, as discussed below.

3.5.3.4 *Historical responsibilities*
Many authors consider that allocation should be strongly influenced by the patterns of past emissions. This has been a particularly persistent theme in debates between developed and developing countries. Hyder (1992) draws together several of the underlying rationales in expressing the perception of many developing countries:

> It is difficult for most of the developing countries to accept the proposition that they should enter into commitments which would adversely bind them, either now or later on, for the sake of a problem caused by the developed countries – who neither wish to equitably share the remaining emission reserves in the atmosphere, nor to share (even in a small way) the benefits and resources that they have built up by plundering the world's greenhouse gas reservoir capacity.

This statement illustrates that several issues are involved in the general assertions about historical responsibility: ownership of the emissions potentially responsible for transboundary impacts, prior use of a finite stock (the atmosphere), and current monopolization of the benefits derived from these activities.

3.5.3.5 *Comparable burdens and ability to pay*
Another general approach to emission allocation is based on the sentiment, frequently stated but not subjected to more detailed analysis, that allocation should affect all countries similarly or involve "comparable burdens" or "sharing the effort equally." This is not to imply that countries should incur the same monetary costs (e.g., per capita). Some conceive of it as a fixed proportion of GNP (Burtraw and Toman, 1992), others in terms of a more general measure of ability to pay, including some measures that are more complex (e.g., the UNDP's Human Development Index or the Physical Quality of Life Index noted below in Section 3.5.5.6).

The Convention's principle of common but differentiated responsibilities and respective capabilities embodies elements of proportionality and historical responsibility as well as the comparable burdens/ability-to-pay approach.

3.5.3.6 *Willingness to pay*
Another principle is simply that countries should bear costs based on what they are – or objectively "should" be – willing to pay, given the potential damages they face from climate change. According to this principle, contributions should be determined by a combination of ability to pay (reflecting current wealth) and national benefits gained in terms of reduced climatic stresses and level of general concern about climate change.

This approach reflects the principle of welfare economics that subjectively perceived benefits ought to affect the distribution of burdens (and hence allocations). Dorfman (1991) draws the important distinction that for a common property resource (degradation of the atmosphere), willingness to pay should be based not on how a country values a given reduction in its own emissions, but instead, how much it would value a reduction in world-wide aggregate emissions. Presumably those valuing such reductions to a greater extent should be more willing to abate. This basically accords with Kant's categorical imperative, which states that one should act as if the maxim of one's act were to become a universal law of nature. Barrett (1992) analyzes an idealized application of this Kantian rule for abatement.

One concern is that this could lead to a situation of "victim pays," though in the greenhouse case the wealthiest countries may be both the greatest polluters and the most willing to pay. At present an intractable difficulty is the lack of knowledge concerning the regional impacts of climate change, and in practice such a criterion would imply piecemeal negotiations in which each country had an incentive to underestimate the potential damages from climate change, and overestimate the costs of abatement, to obtain a lax allocation for itself.

3.5.3.7 *Applications of distribution theory in welfare economics*
A very large welfare economics literature exists on the distribution of income within countries, and clearly there is limitless practical experience on how countries handle distribution issues. It is an interesting question whether and how any of this analysis and experience should be brought to bear on the issue of climate change. None of the climate change literature at present appears to attempt this.

3.5.3.8 *Conclusions*
The various approaches outlined offer general criteria, which have been most widely raised in the context of wealth and other differences between developed and developing countries, but which can in principle be extended to reflect other differences. Although complex, the provisions of the Convention are consistent with or reflect a number of these approaches. Neither the Convention nor these approaches, however, give specific rules to determine a unique, maximally equitable allocation of emission entitlements or abatement costs. We turn now to consider specific, equity-based proposals for allocating future emissions and/or abatement costs.

3.5.4 Specific emission allocation proposals

This section considers proposals for fair distribution of emissions – and by implication abatement effort – separated from any substantive financial transfers related to climate change. The subsequent section then considers other approaches which consider directly the question of "who should pay" abstracted from the question of distribution of emissions.

There are three reasons for dividing the issues in this way. First, it provides a useful way of classifying the literature. Second, it provides an analytically clear delineation that may help to remove some of the confusion in understanding the literature. Much of the economics literature assumes that an international market – mediated by large international transfers to produce a global least-cost response – exists or will naturally come into being. This debatable assumption explains much of the division between economic and other perspectives. Third, by the same token, it provides a politically useful clarification. Many are skeptical of the prospect that an international fund, with payments through it on some equitable basis in support of a globally optimal response, will in reality ever be created, and this explains much of the fear about inequitable global responses. Focussing initially on allocation in the absence of separated big international transfers helps to clarify these issues.

3.5.4.1 Ad hoc proposals

Various ad hoc allocation schemes have been proposed. Krause *et al.* (1990) suggest that an overall cumulative limit for emissions of 300 GtC over the period 1985–2100 be established and divided equally among the current industrialized and developing countries. The general characteristics of this allocation are argued to be ethically appealing – it requires substantial cutbacks from the industrialized world and allows for some interim growth for developing countries – but no more detailed division is offered and no more formal justifications for a 50:50 aggregated division of future emissions are presented.[8] Westing (1989) suggests land area as a basis for allocation, but without any particular justification and with some perverse consequences (Grubb *et al.*, 1992).

More widely suggested is some proportion of GNP; Wirth and Lashof (1990) and Cline (1992), for example, suggest that this could be a component of an allocation on the grounds that carbon emissions are partially tied to economic activity. One difficulty is that the relationship is quite variable between countries for a variety of reasons, and pure GNP allocation faces the objection that on its own it would directly reward the richer countries, thus exacerbating inequalities (especially since poorer countries generally have lower GNP/carbon ratios). There could also be technical difficulties associated with definitions and manipulation of GNP comparisons (Grubb, 1989).

3.5.4.2 Equal per capita entitlements

The most widely cited specific proposal for allocation in the literature is that derived directly from egalitarianism, suggesting that all human beings should be entitled to an equal share of the atmospheric resource. This takes two general forms, with minor variations.

Equal contemporary entitlements – allocation in proportion to national population – is proposed by Grubb (1989), Bertram *et al.* (1989), Bertram (1992), Epstein and Gupta (1990), and Agarwal and Narain (1991), among others. The net effect in all cases would be to give developing countries, with much lower per capita emissions, a substantial excess entitlement, whereas the industrialized countries, with emissions well above the global average, would have a deficit.

The major objections are raised partly on ethical and practical "comparable-burden" arguments (since this approach would imply a huge adjustment burden on developed countries, to which they are unlikely to agree) and partly on grounds of concerns that such allocation might "reward" population and population growth. Proponents tend to argue that any such effect would be negligible compared to other factors influencing population; but to avoid any inducement to population growth, Grubb (1989) suggests that allocations should be restricted to population above a certain age. This has been criticized for "discriminating against children" (though, like all international allocation proposals, it says nothing about distribution within countries). Grubb *et al.* (1992) note a wider range of possibilities for avoiding any incentive to population growth, including "lagged" allocations (where the allocation for a given year corresponds to the country's population a specified number of years before that date) or allocations apportioned to a fixed historical date or including an explicit term related inversely to population growth rate.

The second general form of equal per capita entitlements involves several proposals for equal historical or stock entitlements. Fujii (1990), Ghosh (1993) and Meyer (1995) propose that everyone should have an equal right to identical emissions, regardless of country and generation. Ghosh (1993) argues that this can be derived from a range of diverse ethical principles including Rawlsianism, traditional utilitarianism, and even libertarianism.[9] The general justifications for historical responsibilities in various forms are discussed further below. Grubler and Fujii (1991) and Meyer (1995) have developed detailed systems of accounts involving intergenerational transfer of the responsibility for past excess per capita emissions by the industrialized countries, with the former allowing for progressive absorption of past emissions; den Elzen *et al.* (1993) apply the idea with an energy model to a total carbon budget defined over 1800–2100. All these studies indicate that the industrialized countries have "overused" the atmosphere in the past, and in most such approaches have built up a large "debt" whereas the developing countries are in "credit."

Solomon and Ahuja (1991) propose that allocations should be based on "national historical per capita emissions" – the "natural debt" index as introduced by Smith and discussed in Section 3.3.3.1 above – but do not elaborate quite what this means in terms of emission allocation. (Smith's own application of the "natural debt" concept is in terms of obligation to pay and is discussed in Section 3.5.5.5 below.)

Egalitarian allocation, in either the contemporary or historical form, would be compatible with some of the more general principles noted above (such as Rawlsian and basic needs) but not others (such as comparable burdens and willingness to pay).

3.5.4.3 *Status quo*

At the opposite end of the spectrum, Young and Wolf (1992), for example, have suggested, but not necessarily advocated, an alternative principle for emission allocation based on the "status quo." According to this view, past emitters should not only be held harmless but their current rate of emissions would constitute a status quo right established by past usage and custom. Analogies are drawn with the common law principles of adverse possession, such as "squatter's rights." Ghosh (1993) rejects this argument emphatically, stating that "pollution rights have no common law sanction."

International law has allowed a state to establish preferential rights to fisheries, a natural resource that is otherwise a common, limited resource, where another state has recognized that state's long-established use and its particular economic dependence on that resource. Although the law concerning the allocation of fisheries cannot be transposed to the climate change context, it is conceivable that the general principles of equity could be invoked to justify developed countries' use of the climate system at present day rates, particularly in the absence of challenges to such use from developing countries (Yamin, 1994). Yamin notes that an important general equitable principle in this regard is the principle of "estoppel," which, according to Franck and Sughrue (1993), "imposes a duty on States to refrain from engaging in inconsistent conduct vis-à-vis other States" and suggests that a failure by developing countries to contest present levels of emissions of developed countries could later come to be considered as an implicit acceptance by them of the right of developed countries to such emission levels (Yamin, 1994).

A strict status quo allocation – proportionate to current emissions – would violate all the more general principles outlined in the previous section. In fact, no one in the literature appears to advocate strict status quo as an equity principle in its own right. It has, however, received widespread reference as a basis or starting point position from analysts taking a pragmatic or game-theoretic approach (e.g., Barrett, 1992), in that it is the default allocation that would arise in the absence of agreement. It is also the only specific allocation basis which does not automatically impose big burdens on the industrialized countries. In such treatments the ethical basis, where considered, is usually referred back to concepts of equal burdens. It should be emphasized that in this context status quo is not a static concept, but rather can refer to starting from a baseline of what emissions would have been in the absence of abatement. It could thus allow for more rapid growth from developing countries.

3.5.4.4 *Mixed systems and conclusions*

The number of specific, single-criterion proposals for emission allocation is thus rather limited. A comparative study by Burtraw and Toman (1992) also considers just the two main dimensions of equal per capita entitlements and equal percentage cuts. There are many other proposals that approach the question directly in terms of "who should pay," and these are considered below. Before turning to them, we shall touch on an emerging theme in the literature, the possibility of combining different criteria.

Wirth and Lashof (1990) propose allocations based on an equal (50:50) mix of population and GDP, and Cline (1992) proposes an alternative allocation which consists of a weighted combination of population, GNP, and current emissions.

A form in which historical contributions are considered as a determinant of emissions allocation, other than historical egalitarianism, is a scheme examined by Grubler and Nakicenovic (1991) in which countries have to cut back from current levels in direct proportion to their responsibility for past increases. In a sense, therefore, this is a mixed system. In requiring cutbacks from all countries – albeit more from the richer countries – it is probably inconsistent with the other principles sketched in the previous section (though modifications could be considered to address this).

Grubb and Sebenius (1992), supported by Shue (1993), examine a mixed system in more detail, and an almost identical formulation was proposed independently by Welsch (1993). These authors suggest an allocation which links egalitarian and status quo/comparable burden principles by combining population and current emission factors. They do not, however, specify an equal weighting of these components. Rather, they argue that the weighting accorded to population should increase over time towards a purer per capita allocation. Grubb and Sebenius suggest that the weighting would have to be determined through negotiation, the outcome of which would partly reflect the strength of the competing equity arguments employed. Welsch (1993) proposes a 50-year transition from current emissions in the year 2000 to per capita allocation in the year 2050 and presents sample calculations of the distribution of costs this could involve.

The various allocation proposals present an apparently confusing array. Paterson (1994/1996) seeks to trace each of the main approaches back to different theories of international ethics. Grubb (1995) suggests that the single-criterion approaches can be grouped in terms of two main "focal" approaches: equal per capita allocation, which subsumes the more general Rawls/Beitzian and basic needs principles, and approaches related to the status quo, which subsume the principles of willingness to pay and comparable burdens. These focal approaches could be identified broadly with developing and developed country preferences respectively, but in reality all the longer-term proposals defy simple delineation along such lines.

3.5.5 *Proposals for distributing abatement costs including international transfers*

The proposals in the previous section focussed on "fair" approaches to distributing emission "entitlements." Financial transfers were considered only insofar as they might arise from exchanging these entitlements. Most analysts, however, suggest that both equity and efficiency considerations create a case for large international resource transfers as part of any regime for substantial reductions in greenhouse gas emissions.

Three main approaches have been taken to consider how such transfers might occur and how large they should be.

3.5.5.1 *Transfers on the basis of the Climate Convention*

One is based upon the principles outlined in the Framework Convention on Climate Change. With respect to paying for abatement, the Convention effectively divides the parties into three groups. Annex 1 countries (all the industrialized countries) are expected to bear their own abatement costs. Those also listed in Annex II (industrialized countries minus those with economies in transition) agree, in addition, to fund the efforts of the third group, the developing countries. The Convention states that Annex II countries "shall provide new and additional financial resources to meet agreed full costs incurred by developing country Parties" in preparing reports under the Convention (Article 12), and "shall also provide such financial resources, including for the transfer of technology, needed by the developing country Parties to meet the full incremental costs of implementing measures that are covered by paragraph 1 of Article 4 and that are agreed [between that country and the funding agency]."

This wording is very delicate. Exactly how the term "agreed full incremental costs" is to be interpreted, and just who has to agree and how, are questions that are exercising the minds of experts in many different countries and institutions and that raise both process and outcome equity issues in their own right. Some of these issues will need to be addressed by the Convention's Conference of the Parties.

In addition to using the Convention's financial mechanism to channel the financial and other resource flows, a number of countries have proposed using its provisions on joint implementation to permit an additional avenue of funding and technology to assist abatement efforts in developed and developing countries. The concept of joint implementation via offset investments in third countries and the concept of tradable permits or "entitlements," which are the subject of strenuous debate at present, are considered further below.

3.5.5.2 *Transfers determined by emission allocation and tradable entitlements*

A second approach to international transfers is to let them be generated directly by determining a fair initial allocation, and allowing these "entitlements" to be traded (UNCTAD, 1992).

The initial allocation would be the principal means of determining resource transfers. Since most of the allocation proposals noted above would give poorer countries entitlements greater than their needs, and leave richer countries with a deficit, this would generate a transfer of resources to the poorer countries. The "outcome" equity issues involved thus parallel those noted above in discussing emission allocation. Indeed, many of the specific allocation proposals noted were only advanced in the context of their being initial endowments for such a tradable entitlement system, on the grounds that this would not only be more efficient but would also be the only feasible way of introducing allocations radically different from current emission patterns.

A number of concerns have been raised about the practical and legal aspects of such systems, including their possible fairness in operation without institutional safeguards. Several analysts, including both those supporting and those opposing tradable entitlement schemes, have raised concerns about the legal and equity implications of allowing countries to sell forever a "right" to emit (Yamin, 1993a). For this reason, it has been proposed that entitlements should be either leasable (Grubb, 1989; Agarwal and Narain, 1991) or issued with a finite lifetime (probably overlapping) and reissued periodically according to the agreed allocation scheme (Bertram, 1992; Grubb and Sebenius, 1992). Negotiating and overseeing any such system would also involve a range of other process equity issues parallel to those considered below.

3.5.5.3 *Direct payment accountability approaches*

The third approach is to seek a basis to determine directly "who should pay" for abatement, rather than treat this as a consequence of future emission allocations. To a large extent the general equity issues involved parallel those involved in considering fair emission allocations, with ideas based, for example, on interpretations of Rawlsian criteria, equal burdens, and willingness to pay. But the limited literature addressing the question in this form places relatively more emphasis on two other components: responsibility for emissions (often including past emissions) and ability to pay for abatement. Smith *et al.* (1993) draw this distinction directly in terms of indices of historical responsibility and ability to pay. Grubb (1990) draws a more general distinction between responsibility-based and burden-based criteria.

3.5.5.4 *Polluter-pays principle*

A general basis for responsibility may be considered in terms of the "polluter-pays" principle. The OECD formally adopted a form of polluter-pays principle in 1974 as a guide to environmental policy. It noted, in particular, that the costs of reducing pollution

> should be reflected in the cost of goods and services which cause pollution in production and/or consumption. Uniform application of this principle . . . would encourage the rational use and the better allocation of scarce environmental resources and prevent the appearance of distortions in international trade and investment. (OECD Council, 14 November 1974 C(1974) 223, Paris, 1974)

The essential concern of this principle is that polluters should bear the costs of abatement without subsidy. The principle does not explicitly state that all emitters of a common pollutant should pay in proportion to their emissions, and, in fact, the literature seems remarkably opaque about how the principle should be applied in a context like climate change. The principle clearly points towards responsibility-based rather than burden-based criteria, proportional in some way to emissions: The polluter should pay, but on what basis, who should receive how much, and for what purposes?

A critical distinction that is rarely clarified is whether the principle applies to gross payment or net payment. Burtraw and Toman (1992) assume that it applies to net payments, so that each country should pay for abatement in proportion to its contemporary emissions. This, they note, would be "regressive against national income, a characteristic that is bound to spark developing country opposition," and Chichilnisky and Heal (1994) and others have pointed out that such allocation of payment may be neither efficient nor fair. Other authors assume that the principle means that gross payments should be proportional to emissions, leaving how the resulting revenues should be distributed as a separate question of efficiency and equity. They also consider different bases for payment, which we discuss next.

3.5.5.5 Historical responsibility and natural debt

Smith *et al.* (1993) propose that responsibility for paying should be determined on the basis of the "natural debt" index described in Section 3.3.3.1, namely, in proportion to total cumulative emissions since a specified date. Because this principle, in itself, would result in all countries bearing some responsibility for paying (though very much less for developing countries), they suggest a lower threshold for "basic needs" emissions. This suggestion is consistent with the arguments of Agarwal and Narain (1991) and others.

The principle of using cumulative historical emissions directly as a component in determining payment (or future emission allocations) is considered by these authors as a natural and important matter of equity, and many others also argue the central equity importance of historical emissions in more general terms. Some of these authors recognize a variety of potential practical difficulties, such as how far back the emission estimates should go; whether the natural decay (reabsorption) of emissions should be taken into account and, if so, how; which gases should be included, given the highly variable quality and in some cases complete absence of data; and how to relate the emissions to scale (e.g., cumulative population – and, if so, how defined – or current population, etc.). In part this reflects different potential definitions of the "natural debt" concept. However, at the same time as espousing the Fujii formulation of intergenerational egalitarianism, Ghosh (1993) criticizes the "natural debt" concept as a basis of historical responsibility on the grounds that it is an abstract, environment-centred focus which "does not acknowledge responsibility to others for one's actions [and] does not relate to fairness across human beings."

Historical responsibility as an equity principle has strong support in the literature and politically in developing countries (see, for example, the quotation from Hyder in Section 3.5.3.4). However, Young (1990) and Grubb *et al.* (1992) note three counterarguments about the equitability of what Grubb *et al.* characterize as "making present generations pay, by virtue of their geographical location, for the activities of previous generations" in that country. According to this view:

- Past generations were largely unaware of the consequences of their actions and had no incentive to limit emissions (this issue has some analogies with the issue of strict versus fault-based liability in legal regimes).

- It is not always clear who has benefited from historical emissions, given that the patterns of production, trade, consumption, and migration shift over time and are intricately interwoven. In some cases boundary changes could also create major difficulties for allocating past emissions to current states (as with recent changes in Eastern Europe).

- Although development has generally bequeathed the greatest benefits to descendants in the same country, important benefits have spread much more widely. Positive externalities associated with development (e.g., accumulated knowledge) as well as negative ones (such as pollution) can be transmitted both internationally and intertemporally.[10]

In response to some of these criticisms, Smith *et al.* (1993) defend historical responsibility on the grounds that "we are asking the present generation to take responsibility for the future. . . . If we dismiss historical responsibility, what is to keep the next generation from doing so?" Or, as Hayes (1993) puts it in the same volume, "the current generation of leaders cannot disavow its obligation to pay off its natural debt . . . at the same time as it claims to be adopting the principle of intergenerational equity." Ghosh also rejects objections to historical responsibility as partly inaccurate[11] but mostly irrelevant to the Fujii principle of equal historical per capita entitlements, since the Fujii principle is based not on fault, blame, or compensation but on an egalitarian principle of access.

These viewpoints, and the different elements noted in the quotation from Hyder above, illustrate several different dimensions to the debate about historical responsibility. A clear statement by Bhaskar (1993) links several of these themes and lays the prime emphasis on the fact that

> the current generation [in developed countries] are the prime beneficiaries of resource transfers from previous generations. . . . Developing countries have a claim to part of the transfers, simply because they were made possible by the excessive use of global environmental resources in previous generations in the developed countries. . . . If the current generation accepts assets from their parents, then it is incumbent upon them to also accept the corresponding liabilities.

Grubb *et al.* (1992) in turn note the curious feature that in such reasoning, a "primary justification for considering historical emissions is based upon current wealth. And it is intended to compensate for future increased costs to developing countries, if they are constrained. . . . Tentatively, we suggest that the ethical divide between historical criteria and those focused upon current emissions and current relative ability to pay may not be as large as it appears."

3.5.5.6 Incorporating ability to pay

In addressing the question of "ability to pay," several authors have made reference to GDP or the UN Scale of Assessments. Smith *et al.* (1993) suggest an ability-to-pay element which is proportional to GNP (on a purchasing power parity basis) for all countries subject to a threshold value, determined by Morris's Physical Quality of Life Index (PQLI). Countries with a PQLI below the threshold would be exempt.

On the basis of this, the authors argue that countries should have an "obligation to pay," based on the two elements of historical responsibility and ability to pay (both with lower thresholds). They suggest that equal weight be given to these two components and observe that the resulting obligation to pay could be either the product or the sum, although they favour the latter.

3.5.5.7 *Utilitarian formulation*

A wholly different quantified approach to the question of who should pay for abatement is provided by Chichilnisky and Heal (1994), who apply the concepts of classical economics to construct a strictly utilitarian formulation that seeks to maximize the sum of utilities across all countries, in other words, to aggregate world utility. They argue that the independence of distribution and efficiency so central to the welfare economics of markets for private goods is not a characteristic of markets for environmental public goods such as the atmosphere. As mentioned previously, these authors noted that, in a situation without international transfers, efficient marginal abatement costs would vary between countries. If unrestricted lump-sum transfers are allowed, however, in order to recover the classical solution in which marginal abatement costs are equalized to achieve efficiency, only certain distributions of payment can maximize the aggregate utility. Specifically, they show that for the case of two countries in which both have the same standardized utility function (i.e., both value a given consumption level – and atmospheric quality – equally), then:

> At a Pareto-efficient allocation, the fraction of income which each country allocates to carbon emission abatement must be proportional to that country's income level, and the constant of proportionality increases with the efficiency of that country's abatement technology.

This is the only distribution-of-abatement effort, for a given global total abatement and set of welfare weights, in which no country can be made better off without making another worse off. (Note, however, that it can only define relative contributions between different countries when the weights assigned to welfare in each country are specified.)[12] The result, in turn, implies that the total resources each country should put towards abatement would increase as the square of the national income. The paper implies (but does not prove) that a similar conclusion would hold for the more general case of other utility functions (provided these are still similar between countries) and more countries.

3.5.5.8 *Application of funds and transfers*

These efforts to address questions of who should pay still do not address the questions of who should receive and for what purposes. Smith *et al.* (1993) imply that the payments they consider would go to an international fund, which would then be used to finance abatement at the lowest marginal cost. In applying this index to evaluating transfers, however, Hayes (1993) develops crude aggregate cost curves which combine abatement costs with protection against sea level rise (but not other components of adaptation), aggregating developed and developing country blocs. Combining this index and cost function leads him to conclude that "the incremental cost of abatement and coastal protection in the South that is justifiably the responsibility of the North" is a few tens of billions of dollars annually. (In doing so, he draws on the transboundary impacts as part of the justification for using historical emissions. This, as noted, is an important ethical issue but one that is distinct from that of paying for abatement.) Chichilnisky and Heal (1993) also imply that expenditures would be directed towards the global least-cost options, whatever they may be.

3.5.6 *International trade and the incidence of abatement costs*

The above analyses all focus on highly simplified pictures of the abatement problem, in which the only interrelationship between countries is the impact of greenhouse gas emissions and deliberate financial transfers associated with paying for abatement (or, in the context of Section 3.3, adaptation). In reality, abatement involves actions that affect the consumption of many goods that are traded internationally, and this in itself has other indirect consequences.

One concern has been that abatement action by the industrialized world will itself indirectly harm developing countries by slowing economic growth and thereby lessening demand for a variety of traded goods from the developing countries. Given the modest estimates of the impact of abatement on economic growth rates – at least for abatement to the degree currently being considered (see Chapter 8) – it is likely (but remains to be demonstrated) that this second-order effect would be negligible compared with three other international effects:

(1)   the ultimate benefits from reducing the rate of climate change itself; for the reasons discussed in this chapter, those benefits are likely to be greatest for the developing countries

(2)   the impact on the location of specific industries if countries take differential actions; for example, if some regions introduce carbon/energy taxes and offset them against capital or labour tax reductions, energy-intensive industries will tend to locate elsewhere whilst the number of lower-than-average energy consuming industries may increase

(3)   the impact on the traded volume and price of the commodities that give rise to greenhouse gases – particularly fossil fuels

This last factor may introduce substantial variations between countries, depending on how restrictions are applied. If – as is generally assumed – abatement policies are applied to consumption, the likely effect is to reduce the international trade volume and price of coal and (to a lesser extent) oil. This effect will tend to benefit net importers of these fuels and reduce the revenues to fossil fuel exporters (see Chapter 8). More generally, "the relative value of non-fossil energy resources would be increased; that of high carbon resources decreased" (Grubb, 1990). Extensive abatement efforts could thus have serious implications for those countries that depend heavily on coal exports or (to a lesser extent) oil exports – an equity issue already raised in international discussions.

The effects are complex, however, particularly concerning oil. First, one initial response to reduce CO_2 emissions could be to burn more fuel oil in power stations in place of coal. Furthermore, gas – which in power stations emits about half as much CO_2 as coal – is very likely to be favoured in CO_2 abatement strategies, and gas resources are often associated with oil deposits and also with many coal mines. The overall impact also depends very much on the time horizon considered. Heal (1984) noted that, in general, aversion to climatic risks would tend to flatten the use of carbon resources over time: Peak and initial usage would be lower, but usage would also fall more slowly, partly because the resource base would be extended over time. Manne and Rutherford (1994) provide a more detailed energy analysis, concluding that "carbon restrictions tend to depress oil prices in the near term, but to increase them in the long term . . . oil is less carbon-intensive than coal-based synthetic fuels, and hence oil enjoys a premium." They also find that carbon restrictions improve the prospects and price for gas exports.

Such indirect impacts of abatement on fossil fuel exporters (and importers) are thus potentially important but are frequently not simple.

3.5.7 *Joint implementation by industrialized and developing countries*

Joint implementation (JI) has emerged as one of the major debates in the post-Rio negotiations. The term is generally applied to the idea that industrialized countries might invest in projects in other countries, particularly developing countries, to reduce or sequester greenhouse gas emissions. Industrialized countries consider that the main incentives arise if some or all of the emission savings can be credited towards the donor country's emission target. Many economists tend to see JI as a natural way of improving economic efficiency, by allowing industrialized countries to invest in abatement wherever it is cheapest.

Industrialized/Developing Country Joint Implementation (IDCJI) – also referred to in the literature as North-South Joint Implementation (NSJI) – might involve project-level investment, appraisal, and crediting of emission savings by a "donor" country on the territory of a "host" country. This involves a number of equity issues that underpin the political debate, many of which illustrate equity issues discussed in this chapter. (Note that IDCJI avoids some of the equity issues concerning proposals for international tradable permits, in which the issues would arise primarily, though not exclusively, from the allocation of initial emission rights between legally equivalent parties. Dubash (1994) explores the many differences and argues that IDCJI is an attempt to lower the political costs associated with allocating tradable permits, "at the price of higher social and environmental costs.")

A number of technical and economic concerns about IDCJI have been raised, particularly with respect to the need to mea-sure savings against a "counterfactual" baseline (i.e., an estimate of what would have happened in the absence of the JI project), thus raising questions about what is to prevent inflated baseline estimates or other distortions. Another concern involves uncertainties about project success: Who has the responsibility for project success, and what happens if it fails? Administrative costs arising from both these factors are a further issue. Reviews of these technical issues are given by Loske and Oberthur (1993) and Selrod and Torvanger (1994), the latter being much more optimistic than the former, and in a number of books and conference proceedings (e.g., Kuik *et al.,* 1994; Ramakrishna, 1994). Here, however, the focus is on the underlying equity issues, which can be classified along the lines set out in this chapter.

3.5.7.1 *Allocation issues*
Analysts have raised the following concerns about the implications of IDCJI for emission abatement responsibilities and associated costs:

- IDCJI could impose hidden administrative and/or opportunity costs on host countries which are only partly recovered through the investment itself (Maya, 1994).

- IDCJI could "skim off" the cheapest projects, so that, if and when developing countries are required to adopt emission constraints in the future, they will be faced with higher marginal abatement costs (e.g., Parikh, 1994, and many others).

- If IDCJI allows industrialized countries to continue increasing their own emissions, it will perpetuate – and perhaps even exacerbate – global inequalities in per capita emissions rather than tending towards any long-term convergence (Parikh and Gokarn, 1993; Loske and Oberthur, 1993).

- By reducing the pressure on the industrialized countries to take domestic action, IDCJI may reduce incentives for technical advance and innovation that could lower costs for all in future abatement (Loske, 1991). A slower pace of technical innovation may itself differentially increase the future costs to developing countries (Ghosh and Puri, 1994).

To each of these concerns there are responses that results will depend (in part, at least) on how the framework for IDCJI is defined and on how future commitments are structured to take account of any previous JI activity (e.g., Vellinga and Heintz, 1993; Metz, 1994; Ghosh and Puri, 1994). This does illustrate, however, that JI does not really overcome the basic political issue of allocating future emission rights, though it delays (at the expense of further complication) the need to address this core equity issue concerning the role of developing countries.

3.5.7.2 *Procedural issues*
Efforts to address the technical and allocation issues also highlight concerns about more procedural equity issues. Since a JI project cannot (or is unlikely to) proceed without the con-

sent of the government of the host country, inadequate state participation *per se* is not a primary matter of concern, but issues of information and power asymmetries are raised in the literature.

Concerning information, it has been pointed out that, because donor countries tend to have a far greater research and analysis capability, they have a comparative advantage in any negotiations about framing JI projects. Pachauri (1994) comments that "the biggest apprehension that exists on the issue of JI relates to small countries that lack the capacity to evaluate the implications of specific projects." Such countries could accept projects that later prove to be against their interests.

Power asymmetries could also result in developing countries being pushed into projects on disadvantageous terms. Short-term financial pressures, for example, could lead to host countries giving up land to JI projects that could have far greater productive use in later years. JI projects may also be driven by the interests of technology suppliers in the donor countries, perhaps at the expense of equipment manufacturers in the host country (Maya, 1994).

Again, protection against at least some of these potential inequities can be developed, with the emphasis in the literature being on the need for an open, transparent, and multilateral framework for JI agreements between parties to the Convention (e.g., Ghosh and Puri, 1994; Markandya, 1994), with clear, legally binding rules (Yamin, 1993b), and also on projects that meet a range of development needs (Imbree, 1994).

Others, however, express concern that such an extensive multilateral framework may involve such high administrative costs as to render the idea impotent. And, conversely, it could be considered inequitable to expect industrialized countries to engage in stronger commitments whilst barring them from exploring alternative ways of achieving the same global environmental benefit. Therefore, there is also an onus on developing countries to clarify their concerns to allow a basis for a regime that is considered fair and practical by all parties. The purpose here, however, is not to argue for or against IDCJI, but to illustrate the central place, and range, of relevant equity issues in any practical moves towards transnational implementation of measures to address climate change.

3.6 Equity Within Countries

3.6.1 Introduction

This section examines issues within countries. Not only is this important in its own right; it can also shed a new light on the issues discussed in the other sections of this chapter. As noted earlier (Section 3.1.1 above), institutions to address a wide range of common interests exist primarily within countries, and equity is an important influence and measure of the strength and perceived legitimacy of these institutions. Although the nature and strength of these national institutions is far from uniform, they do provide the main working model for incorporating equity into international relations.

A national focus also allows lessons to be drawn from the experience of economic reform programmes, especially the structural adjustment programmes in developing countries, but also programmes of economic restructuring in industrialized as well as industrializing countries. Furthermore, unless intranational equity issues are addressed explicitly, it may be impossible to mobilize public opinion for amelioratory action in both developed and developing countries, as it is only at the intracountry level that we can begin to explore nontechnocratic approaches to collective action. Since policies and actions to combat climate change would in any event begin at the national level, the examination of intracountry considerations could help individual countries by clarifying the range of issues involved.

The provisions on education, training, and public awareness in the Convention call upon the parties to develop educational programmes, provide public information, and promote public participation in addressing climate change and its effects and developing adequate responses. However, it must be acknowledged that intracountry equity does not figure prominently in the Framework Convention on Climate Change. It is being introduced here to facilitate the search for optimal national and international policies. It is not the purpose of this section to introduce an entirely new element into international negotiation. Nor is its purpose to sidestep the central concern of the Convention and also of this chapter – the issue of international equity and differentiated responsibility. Nor, indeed, is the intention to legitimize a new period of colonialism in which the scope for sovereign national action is restricted unduly by international agreements on the nature and form of intracountry equity.

As mentioned earlier, equity considerations have become a part of climate change discussions primarily because of the awareness of the diverse situations of both countries and individuals – not only in terms of their contribution to climate change but also of the impact of the change on their well-being, their individual and collective capacities to cope with this impact, and on their abilities to undertake mitigating action. Furthermore, there is also a growing awareness that the combination of large-scale changes and iniquitous arrangements will necessarily lead to conflicts and even wars, which will further aggravate the adverse effect on social well-being. Finally, it is also acknowledged that the burden of mitigation must not be placed disproportionately on the poor and vulnerable groups.

The above considerations have influenced decisions in a number of areas. In addition to the agreement to limit emissions to minimize the magnitude of the change, there is also an emerging consensus on the need to compensate the worst-affected groups and countries, especially the small island states. There is support as well, through Agenda 21, for the strengthening of coping systems in vulnerable countries. The agreement to distribute the burden of mitigation in an equitable manner also falls into this category. Finally, the awareness of the need for adjustment has also led some to recommend that countries should begin addressing major sources of inequity.

To state all of this somewhat differently, equity considerations are germane to the climate change issue in two different contexts – equitable distribution of the costs (or benefits) of the change, and equitable contribution to amelioratory and preventive actions. Neither experience is without precedent in national contexts. There is an extensive literature on the distributional impacts of the business cycles, supply shocks (e.g., the oil price increases in the 1970s), economic development, and episodes of structural adjustment and liberalization. In all such cases, the questions are similar to the ones being raised in the present situation, namely, how is the burden of the change to be shared, and who should pay for amelioratory or preventive actions. This literature can be divided into two groups, that which looks at governmental policies within a given institutional context, and that which examines the institutional context itself. In addition to this, there is also a theoretical literature on conditions that will give rise to equitable outcomes within countries.

3.6.2 Policy-oriented literature

The classic example of policy-oriented literature is the UNICEF publication on structural adjustment in developing countries, *Adjustment with a Human Face* (Cornia et al., 1987). To quote the subtitle of this text, the object of the exercise is both "protecting the vulnerable and promoting growth." This book is a part of a critique that emerged in the 1980s against the current conventional approaches. Since then, the critique has gradually come to represent the new mainstream position.

Structural adjustment or "stabilization" becomes necessary when the aggregate expenditure (or consumption) of a country begins systematically to exceed its aggregate income. The result is the accumulation of debt at an unsustainable and unserviceable rate. In the 1970s, these difficulties were seen as part of the short-term problems of cyclical fluctuations and were responded to with the short-term measures of stabilization, on the pattern of similar problems and policy responses in developed countries. Subsequently, it was recognized that the problem was of a longer-term duration and structural rather than cyclical; in other words, the problems emerged not because of the normal functioning of a business cycle, but because the economic and political structure of some countries was designed to produce these results on a systematic basis. In either case, the symptoms were a growing public debt, deficits in government budgets and international payments, and accelerating inflation.

In the absence of a policy response, these symptoms were likely to be cured through the "classical adjustment" mechanism, in which the debt squeeze brings about an economic contraction and thus a reduction in expenditures. Since the publication of Keynes's *General Theory of Employment, Interest, and Money* (1936), this has been recognized as a suboptimal path because of its high social and economic cost. The main policy response in the 1970s and 1980s was that of a managed economic contraction through budget cuts (especially on such "nonessential" items as social services, consumer subsidies, and general administration), reduction in domestic credit, devaluation, and the liberalization of foreign trade and payments regimes (see Banuri, 1992).

The performance of these policies has come under considerable scrutiny (see e.g., Taylor, 1993; Banuri, 1992; Cornia and Jolly, 1984; Cornia et al., 1987). Earlier critics argued that the orthodox adjustment programmes were both ineffective and unnecessarily painful. They proposed heterodox, or supply-side approaches, which aimed to increase production directly through investment promotion, labour training, direct export promotion, and income policies. Many of these recommendations are extremely sound, but experience shows that they have to be combined in varying proportions with orthodox demand contraction policies. Also, the hope for a win-win solution, in which the costs of adjustment would be avoided altogether has not been vindicated by experience. The economics profession appears to have come to the conclusion that some degree of economic contraction is inevitable during structural adjustment, if only to dampen the inflationary expectations built up during periods of prosperity.

As a result, the attention of many of the critics shifted to the protection of important areas during such contractions. These critics argue that the conventional approach is suboptimal, in that it leads to the marginalization and immiserization of vulnerable groups – women, children, the poor, the rural population, labour, and the aged population – who are forced to carry a disproportionate burden of the adjustment. As part of their argument, Cornia and Jolly (1984) and Cornia et al. (1987) documented successful cases in which government policies managed to protect vulnerable groups while bringing structural adjustment. The result of these studies has been a remarkable turnaround in the conventional wisdom on structural adjustment. From a situation of almost total neglect of the social sector and vulnerable groups in the 1980s, current policies include the protection of these sectors and groups as integral elements of structural adjustment programmes.

However, even the modified approach, which seeks to incorporate social and equity concerns, has not been without its critics. A number of writers have suggested that government-directed and targeted programmes are not successful in alleviating poverty or delivering social services in many contexts. This criticism has been launched in particular by the Public Choice Theory school (Buchanan and Tollison, 1984; Krueger and Bates, 1993) and has been accompanied by calls for privatization and deregulation. Other critics have used similar arguments to recommend institutional development programmes. We shall take up some of these criticisms in the next subsection.

The analogy of structural adjustment to climate change is quite striking. The climate change threat has also emerged because the planetary consumption of certain materials (mainly fossil fuels) has begun systematically to exceed sustainable levels. In this case, there is the accumulation of global warming gases in the atmosphere, again at an unsustainable rate. The analogy of the classical adjustment mechanism is ecological homeostasis, which may bring about a new planetary equilibrium, albeit with far fewer humans and after the payment of a very high human and social cost. The Framework Convention on Climate Change can be seen as analogous to an agreement to undertake a structural adjustment programme.

In other words, the experience of structural adjustment in developing countries has many lessons to offer for the foreseeable adjustment, both at a global level and in industrialized countries.

- The first lesson is that "blind" or neutral policies have not performed well. As a result, the need to protect the social sector and some targeted groups has become widely accepted. In the climate change context, this means that such "blind" policies as a carbon tax, or even tradable permits, will have to be supplemented with policies to protect vulnerable groups and vulnerable sectors.

- In general, the experience of structural adjustment shows that policy responses need to be sensitive to the economic, political, and institutional contexts of various countries. Reliance on a formulaic approach has often been ineffective if not counterproductive.

- The experience with heterodox approaches to stabilization suggests that these need to be combined in varying proportions with the orthodox, demand-contraction policies. In the climate change context this means that pure win-win solutions based on technological optimism – namely the expectation that technological improvements alone will suffice to reduce emissions to sustainable levels – are unlikely to produce the desired results unless combined with steps to reduce aggregate consumption.

- Another lesson is that, in many cases, even the modified approach may be deficient unless accompanied by institutional strengthening programmes.

Structural adjustment and stabilization are, however, the most recent of the many arenas for the debate over equity concerns. As is apparent from the above remarks, the experience of stabilization was not restricted to developing countries. In industrialized countries, similar questions have been discussed in the context of business cycles and stabilization. More generally, the discussion of trade-offs between equity and growth also falls in this category.

In this context, equity has been taken to mean income distribution. There is a literature on the appropriate index to be used to measure income distribution. Starting with the Lorenz Curve and the Gini coefficient, a number of improvements have been proposed. The Gini coefficient is an index of the inequality of relative income shares, with zero representing complete equality and one representing complete inequality. Subsequently, attention shifted to other indicators, such as absolute poverty – measured in terms of the number of people below a nutrition-based poverty line and quality of life (Morris, 1979) – and basic human needs, namely education, food, health, water supply, sanitation, and housing (Streeten, 1981).

The conventional view of the relationship between equity and growth was derived from Kuznet's (1955) observation of an inverted-U relationship (the Kuznet's curve) between GNP per capita and the Gini coefficient. This observation was based on both time series and cross-section data. However, subsequent studies have added substantial qualifications to this result (Ahluwalia and Chenery, 1974).

The thrust of these arguments is that economic development and structural transformation affect income distribution within countries in specific ways. The earlier view was that income distribution would worsen with economic growth before getting better. However, the more recent evidence suggests that the pattern of change is sensitive to policy choices and institutional conditions in individual countries. Countries that followed proactive policies, and favoured investment in social sectors (education, health, social welfare) were able to pursue economic growth and social development simultaneously. In other words, at least insofar as the experience of development is concerned, a win-win solution is possible. There are, however, considerable differences over the factors contributing to win-win solutions. Neoliberal writers (e.g., Balassa, 1993; Krueger, 1993) attribute the success of newly industrializing countries to market-based, outward-oriented policies. Others look to institutions of governance as key elements (see, e.g., Wade, 1992; Alam, 1989; Banuri, 1992).

3.6.3 Institution-oriented literature

The common element of the above analyses is their rootedness in a technocratic approach to problem solving. As a result, they are focussed on what the government can or cannot deliver, either in terms of policies or of outcomes. Dirigistic arguments favour direct interventions by centralized states in the form of government investment, protection of domestic economic activity, social cost-benefit analyses, and others. Even neoliberal arguments that favour market-based approaches view the government as virtually omnipotent in bringing about the structural transformation towards a free market economy.

An alternative to this approach is the literature that takes governance as an entry point to the analysis. It covers a broad range of ideas, including capacity building and technical assistance (UNDP, 1993), community participation and empowerment, people-centred development (Korten, 1987, 1990), and sustainable development (Banuri et al., 1994). Its unifying theme is the emphasis on collective action and development of institutional and decision-making capacities. At smaller units of aggregation, this focus leads to the analysis of participation, empowerment, and community development. At the middle level, it translates into the argument for the strengthening of various organs of civil society, including institutions of education, research, and monitoring. At the governmental level, the focus is on administrative reform and capacity development.

The relationship of these approaches to the climate change debate on equity issues derives from the analysis that lies behind them. The main idea is that inequitable outcomes are produced by institutions that fail to protect the rights of vulnerable or marginalized groups in society. Similarly, countries cannot take advantage of global opportunities, or suffer disproportionately from adverse global processes, if they lack the institutional capacity to protect their interests. Finally, countries that exhibit superior performance on one indicator of social welfare also perform better on virtually every other indicator. This means that rather than trade-offs between different goals, there is a clustering of performance around various goals. This can only be explained by institutional factors that favour some countries over others. The solution for countries that perform poorly is to enhance their national institutional capacity to protect their rights and, more generally, to pursue collective goals. Analogously, the solution for weak or marginalized groups is also to enable them to make collective decisions and protect their rights.

This provides a somewhat different slant on the literature cited earlier. It could be argued that the need for structural adjustment arises when the sum total of property rights exceeds the available aggregate endowment of assets or resources. The solution then is to reduce the command over goods and services by reallocating or eliminating some property rights. These might include the rights of the government (generated through money creation), or those of groups favoured by distributional policies or subsidies, or even of investors who are allowed to borrow at highly subsidized rates. The argument of the critics is that in this reallocation of rights, it would be difficult for the vulnerable groups to protect themselves, partly because of their very weakness, and partly because many of their rights are likely to be informal in character.

One strand of this argument has been made by Sen (1976) and various others in terms of "entitlements." In Sen's approach, crises, such as famines, occur not because of a shortage of food but because of a failure of entitlements, namely rights to commodities. However, as Gore (1993) has argued, in much of his work Sen focuses on entitlements conferred by the formal legal system. Although in the examination of intrafamily distribution (Sen, 1984) he extends the analysis to include informal entitlements as well, this does not seem to have affected the exclusive focus on the formal legal system in the remaining corpus of his work.

Be that as it may, two points are significant in this debate. First, that it may be more germane to look at property rights, whether formal or informal, rather than consumption or production patterns alone. Second, that there is an interrelationship between economic and ethical considerations. As Gore points out, "for Sen, policies to counter hunger and famine must ultimately be justified through foundational arguments about what is valuable. Equally, ethical assessment of institutions, such as property rights, requires empirical analysis of causes and effects" (Gore, 1993).

From this perspective, the development process appears as a trend towards the formalization of property rights and in many cases the expropriation of the customary rights of local communities in favour of centralized states. This was justified on the grounds that it would facilitate the pursuit of national goals, such as development or equity. The institutional perspective would argue for the reconstitution of local property rights, primarily to enable the protection of the rights of vulnerable communities.

These considerations are relevant for the examination of climate change issues. Insofar as the climate change issue is also one of property rights (or entitlements) exceeding the total endowment of planetary sinks, the response will necessarily include a curtailment of aggregate entitlements. The result is that blind policies are likely to reallocate the property rights in an inequitable manner by curtailing the entitlements of the poor disproportionately. This is true for countries as well as for groups within countries. Given the fact that the developing world will be affected more by climate change, that it has lesser flexibility, is closer to subsistence, and lacks the institutional infrastructure to protect lives and livelihoods, its rights will be curtailed disproportionately in any adjustment episode. The much greater vulnerability of the developing countries is starkly apparent when one considers that the 1991 cyclone in Bangladesh killed more than 200,000 people, whereas in 1992 Hurricane Andrew, of the same intensity, killed only 34 in the U.S. The same disproportionality affects the vulnerability of groups within countries, for example:

- At an individual level, the choice of protecting the rights of various people will be influenced by cultural arrangements, which favour men over women.

- In most countries, declines in food availability will fall disproportionately on the politically weaker and relatively passive groups, on the poor, the children, women, older people, the unemployed, and the rural people.

- Not only is this likely to produce conflict, but the emergence of conflict will further discriminate against the weaker and more peaceable groups (Suliman, 1992).

- Major technological or political changes in agriculture result typically in expropriation of tenants and landless poor. Attempts to offset these consequences through land reforms have had only limited impact (Moyo, 1994; Sobhan, 1993).

Equally, although there is rhetorical support for amelioratory actions, though often only at elite levels, this has translated neither into a broad legitimacy of concerted state action in industrialized countries nor into change in behaviour towards more conservationist patterns. Similarly, the awareness of the primary responsibility of the industrialized countries for the ecological threat seems to have absolved the affluent groups in the developing countries of their responsibility to respond to it.

In other words, it may be idealistic to expect that the nation-state as currently constituted will respond equitably in the face of disasters and large-scale dislocations. The same considerations apply to the international situation, where the degree of institutional weakness, the inability of the vulnerable to fight back, and absence of a collective morality are even more pronounced. Writers concerned with this weakness have argued that the only solution is to develop the decision making capacity of vulnerable groups. This has led to programmes of community development, rural development, and capacity building. At national levels, it might ask for a broad-based programme, involving the government as well as nongovernment and private sector institutions (see Banuri *et al.,* 1994).

This still leaves open the question whether there is an ethical framework that can enable national societies, and indeed the global human society, to respond to the emerging problems in an equitable manner.

3.6.4 The question of ethics

Equity, justice, fair play are all concepts with strong moral and ethical overtones. The literature that has been discussed exhibits these overtones either explicitly or implicitly. It is fair to say, however, that the policy-oriented literature seeks an ethically neutral stance. It is based on a technocratic approach, which focusses on the means or instruments to given or agreed goals. In this approach, the only ethical considerations pertain to the goals, not to the means through which the goals are to be achieved. The goal of this literature is to determine a universal formula for decision making. Differences between philosophers pertain not to this search for a universal formula but to the contents of the formula.

This distinction is not simply that between outcomes and processes. For example, Nozick (1974) argues for liberty as the primary principle, not only in its own right but as the definitive feature of equity. This is generally taken to be an example of process-oriented ethics. However, it too is consequentialist, in that it assumes a transcendental goal, which can be achieved through the appropriate institutions – in this case free markets. Similarly, Rawls (1971) analyzes the concept of justice from an individualistic viewpoint to argue for a Solomonic veil of ignorance that would induce each individual to seek a maximin solution. If this could be agreed, presumably we could also agree to establish institutions (a government, a judiciary) that would put our values into action.

Both these perspectives, which are two of the most salient recent analyses of the issue of justice, derive from a "consequentialist" position, in which values have to be justified by outcomes – even if they are defended by processes. Hannah Arendt (1961) has made a scathing critique of such approaches, by arguing that they are based on a confusion between meanings and values as well as between values and goals. The result is that everything is reduced in the end to means. In this approach, culture and values become mere instruments for the achievement of particular goals (e.g., development). Indeed, even equity is but an instrument for the pursuit of such goals as economic growth or political stability.

This is not the only possibility however. An alternative approach, which would be sensitive to Arendt's criticism, sees culture and values (including equity) as foundational and as developed through experience. This is implicit in Rawls's argument, when he says that "the primary subject of justice is the basic structure of society" (Rawls, 1971). Following this, Michael Sandel (1982) has commented that "For a society to be just in this strong [Rawlsian] sense, justice must be constitutive of its framework and not simply an attribute of certain of the participants' plans of life." This is the definition of a moral community and not a society of strangers.

Arendt herself has argued against transcendental principles by invoking judgment as an alternative entry point – contextual, subjective, interactive, and constructive of community. Judgment requires being able to judge particulars without subsuming them under general rules. Such concepts as justice, courage, truth, goodness, are not based on general rules but on concrete experiences. If one takes this point of view, the search for universal rules is dangerous as well as futile. Justice lies in experience, not in transcendental rules. Equity and efficiency do not collide as contrasting goals. We cannot have efficiency in an inequitable society.

However, this perspective presupposes an alternative institutional structure, in which the experience of value creation can become possible again. This structure must build on small, decentralized communities, in which the scope for judgment is reestablished. This would go against the broad trends of twentieth-century development, which have been in the direction of greater centralization and concentration and have also been accompanied by the transfer of all collective rights to the state. In the realm of economics, therefore, they have left an amoral universe, in which equity can only be an instrument for the achievement of some other goal, not a value in its own right.

3.7 Procedural Fairness in International Climate Change Processes

3.7.1 Introduction

The various differences between countries outlined in this chapter have implications for the way in which international responses to climate change are developed. In the context of the different equity issues discussed in the introduction, issues of procedural equity are important as well as the consequences of decisions taken. Indeed, the two ultimately are likely to be inseparable. Oran Young (1990) notes that

> One of the most robust findings of the social sciences is that . . . [countries in international negotiations] . . . can and often do fail to realize feasible joint gains, thus ending up with outcomes that are suboptimal (and sometimes highly destructive) for all concerned.

Part of the reason is the lack of trust and adequate information that can arise from negotiations that are, or are perceived as, operating in an unfair or inequitable way. The result can be – and often is – that countries do not even agree to things that would benefit each and every one of them. Hence, the importance of procedural equity. An overview of post-Rio issues by Jaitly (1993) illustrates a balance of developing country concerns regarding procedural as well as outcome issues.

One important element of procedural equity has been mentioned in Section 3.2, namely, the principle of sovereignty in international affairs, coupled with a responsibility towards the impact of internal actions on others. The difficulty with climate change is that it is an issue in which the activities of each country have impacts on all others; hence, the need for processes through which states can seek agreement on mutual changes to reduce such impacts, and/or to compensate accordingly.

These principles apply to the negotiations, associated processes like the IPCC, and any institutional structures and procedures created through negotiations. The principle of seeking consensus among all participating states for international agreements makes it still more important that participants feel that negotiations are conducted fairly and that they feel able to participate effectively, and that states take part in good faith. Countries that do not perceive a process to be fair have great power to obstruct it, ensuring that negotiations make little progress.

Procedural equity is an aspect of institutions, and the construction of international institutions is partly about establishing structures that command widespread adherence because they offer adequately fair representation. The painstaking negotiations over the Global Environmental Facility, for example, can be seen largely as negotiation about fair representation.

The overall literature on institutions is very large, and there is an extensive literature (especially in the U.S.) about institutions for managing global environmental affairs (e.g., Millennium, 1990; WRI, 1991; Choucri, 1994). The academic literature specifically on practical aspects of equity in international institutions for managing global environmental affairs is, however, much smaller. Among the most important studies in this corpus is the report of the IPCC Special Committee on the Participation of Developing Countries (IPCC, 1990). This identified five areas of particular concern that might inhibit adequate involvement: "insufficient information, insufficient communication, limited human resources, institutional difficulties, and limited financial resources." Here we group these more broadly under "participation" and "information," each including the question of adequate resources. "Resources" is used in the broad sense of the word; it encompasses human resources, financial resources, institutions, and infrastructure that can underlie adequate participation in and information for effective negotiations.

These issues are important. The cost of inadequate participation is likely to be low adherence to any resulting agreement. The cost of inadequate information is likely to be a suspicious and obstructive attitude that reduces negotiations to a crawl. We consider each in turn.

3.7.1.1 Participation

Participation in international processes – at the simple level of having nationals or other representatives present – is restricted chiefly by limitations on human and financial resources, though delays in obtaining visas and the sheer difficulty of travel from some developing countries can also be important constraints. For some processes (like the IPCC), financial constraints on travel from poorer countries are addressed through special funds supported by developed countries, though inevitably there are recurring concerns about their adequacy. The matter of human resources is more difficult to deal with, since it depends partly on long-run education and reflects the contrary priorities in many developing countries outlined in Section 3.3.2.

A related problem can be the appropriateness of participants and the load on them. Many of the smaller and/or poorer developing countries had just a few delegates covering the whole range of negotiations surrounding the Rio Earth Summit, compared with several hundred, for example, from the U.S. They could not possibly have had the specialized knowledge required to participate effectively.

That said, it is obviously unrealistic to expect universal participation in everything. Indeed it would be a huge waste of scarce human and financial resources and would in a sense be unfair to participants with more urgent priorities. In practice, the smaller countries – in both the industrialized and developing world – are selective about their participation through choice as well as need, without necessarily adverse consequences. Smaller countries may be able to rely on larger neighbours that share common interests or may otherwise be able to ensure that their concerns are reflected through delegated or grouped responsibilities. Luxembourg, for example, generally has little difficulty signing on to agreements in whose formation it played no direct part. And, as noted earlier, addressing climate change requires widespread but not necessarily universal participation. Where universal participation is impractical or even undesirable, what matters is ensuring that a fair balance of interests and perceptions is represented.

A related problem is that of institutional coordination within countries. Climate change involves an extraordinary diversity of issues and interests, ranging from meteorology to international trade. Frequently, governments have trouble coordinating policy between different internal groups – and in the international arena it is indeed not uncommon for meteorologists from developing countries to find themselves discussing issues that have more to do with international trade policy.

Concerns about the ability to participate effectively underlie the frequent insistence of many developing countries on rules of procedure, such as the development of official negotiating timetables without informal meetings outside this agenda. To bigger and more advanced countries this can be an intense irritation and may seem an unnecessary brake on progress (which such procedures may indeed turn out to be). To weaker and poorer participants they are seen as a protective device to give them a chance to keep up with discussions. Again, it is an area where a fair compromise between adequate progress and comprehensive participation is called for.

Another aspect of participation is that of nongovernmental organizations. The Rio processes were marked by unprecedented involvement of NGOs, which are widely considered to have contributed positively to the process. Brenton (1994) lists their involvement as one of the four major forces for international cohesion in environmental policy. Their legitimacy and impact derive from a widespread membership, strong international links, and extensive information, when these attributes are brought to bear positively on the process. Environmental NGOs tend to be relatively concerned about international ethical responsibilities. In international negotiations, they frequently help to counter financial and political pressures that exist in some of the richer countries against change or acceptance of international responsibilities. They also frequently support the positions of some of the poorer countries by providing information and other kinds of logistic support.

3.7.1.2 *Information*

Perhaps a bigger problem than adequate participation is the need for adequate information for those who participate. Climate change issues are complex. The work of the IPCC, for example, spans issues ranging from atmospheric science to ecology to economics. OECD representatives, in particular, can draw on a huge infrastructure of specialist knowledge and analysis in developing policy positions and arguing for them (even if researchers still feel that insufficient attention is paid to their work, and negotiators feel deluged by research much of which is peripheral or naive). Many developing countries feel seriously disadvantaged by not having adequate research capacity to draw on. It is striking, for example, that the Indian Minister of Finance expressed such a concern (on a different issue), despite India's having one of the most advanced policy research capabilities in the developing world (Singh, 1993). Communication is a related issue. Heidenreich (1995) illustrates how the absence of "mass distribution" communication facilities places African researchers at a continuing disadvantage.

Again this reflects a mix of problems: a lack of human and financial resources, limited education infrastructure, and weak communications. The IPCC Special Committee recommended a number of steps that could be taken, including support for regional centres and seminars and the development of regional specialists to help improve information and analytic capabilities in developing countries, particularly as they affect the ability of these countries to get to grips with the climate change issue (IPCC, 1990).

The importance of adequate information cannot be overstressed. Many bad policy decisions can be traced to inadequate information, and so can many of the difficulties in international negotiations. The feeling that some participants are much better informed than others can, and frequently does, lead to a suspicious and obstructive attitude in negotiations that ultimately may be to the detriment of all. It was noted above that in international affairs, countries often do fail even to reach agreements that would be of mutual benefit. Lack of adequate information – compounded by the distrust this can help to generate – is one root cause of this. Negotiations are difficult enough when each country is seeking only to pursue a narrowly defined self-interest. They can be even more difficult when states do not have a clear idea of where their self-interest lies.

3.7.2 *Links between internal and international processes*

The focus in this discussion has been on state actors and government representatives. But fair representation in international discussions also requires that those attending fairly represent interests within their own countries, and their effective participation also requires them to draw on the human and other resources of the nongovernmental sector. There is

thus increasing acceptance that nongovernmental organizations have an important role to play in the process: in harnessing and analyzing information; in developing international links; in assessing the implications of proposed policies; in changing cultural attitudes, including communication across different sectors of society; and in observing and monitoring the implementation of decisions agreed internationally (e.g., Haas *et al.,* 1993; Brenton, 1994; Choucri 1994).

A number of these elements have been discussed in Section 3.6 above. Given the complexity of the climate change issue, this points to the ultimate inseparability of issues such as fair and informed participation between the national and international levels. Addressing climate change – both adaptation and abatement – may require changes at many different levels of society, and the implementation and acceptance of policy at many different levels.

Thus, we come full circle to the observation at the beginning of this chapter. Perceived justice, in terms of representation and consideration of different perspectives in processes and in big decisions, is one of the basic measures of legitimacy for governmental institutions. Equivalent structures internationally are weak or nonexistent. But the climate issue forces recognition of global interdependence. For negotiations to be effective, they need to command widespread acceptance as fair processes, reaching decisions that are fair compromises between widely divergent views of what constitutes an equitable outcome, and thus reflecting the range of issues addressed in this chapter.

3.8 Conclusions

Equity and social considerations are central to discussions of steps to be taken to implement the Framework Convention on Climate Change because widespread participation is essential if the objectives of the Convention are to be gained. This is why the concept of equity is so prominent in the Convention. Countries are unlikely to participate fully unless they perceive the arrangements to be equitable. This applies particularly to equity among regions and countries, but equity within countries, and associated social considerations, are also important influences on what is possible and desirable. Mitigating and adapting to climate change will require actions on the part of individuals. Governments will find it easier to comply with international obligations if their citizens feel that the obligations and benefits of compliance are distributed among them equitably. And richer countries are unlikely to burden their poorer citizens to benefit relatively rich citizens in poor countries.

Scientific analyses cannot prescribe how social issues should be taken into consideration and how equity should be applied in implementing the Convention, but analysis can clarify the implications of alternative choices and their ethical basis. There are a variety of meanings of equity, and there are various principles that have been designed to achieve equity. On some issues different equity principles point to similar responses, suggesting clear guidance, whereas on others they may conflict. In either case, there is a need for judgment, drawing on concepts of equity.

Issues relating to equity among regions and countries stem from the substantial differences that exist among countries. Countries differ not only in terms of size, resources, population, and wealth, but also in terms of emissions of greenhouse gases, vulnerability to climate change, and institutional capabilities to respond effectively to climate change. There is no single method of aggregating these differences and no decision rule for dealing with them. The Framework Convention on Climate Change provides considerable guidance for applying the concept of equity to take account of the many differences among countries. However, the application of equity to specific circumstances will require further elaboration of the Convention's principles and obligations, many of which were designed to be ambiguous and remain so.

Equity issues involved in responding to climate change may be divided into four distinct categories:

- distributing the costs of adaptation
- distributing future emission rights
- distributing the costs of abatement
- ensuring institutional and procedural fairness

The Framework Convention on Climate Change offers some guidance on all these issues. It requires developed country parties to take the lead in limiting their emissions whilst recognizing that developing countries' emissions are relatively low and will need to grow to meet their legitimate social and developmental needs. It also requires developed country parties to assist developing country parties in coping with both the costs of abatement and the costs of adaptation to the adverse effects of climate change. Such assistance must be provided by the Convention's financial mechanism, which must have "an equitable and balanced representation" and transparent governance to ensure procedural equity between donors and recipients.

This focus on developed/developing country interaction is also apparent in much of the literature, which seeks to clarify considerations of equity between developed and developing countries much more than among developed countries. Since developed countries have obligations under the treaty to take immediate action, this is a serious lacuna. There are, however, a variety of specific analyses that propose schemes for distributing the costs of coping and abatement as well as emission rights and that analyze the distributional effects of these schemes across the range of countries.

Social considerations and the experience of implementing structural adjustment policies point to the need to consider and target specific groups for special consideration. Countries (such as island and other low-lying states or dryland regions) and special groups within society that are especially vulnerable to climate change (such as the poor, and sometimes women or children, or specific occupations or regions) – in other words, those on whom the costs of abatement and coping would be especially burdensome – merit special attention.

Concern about equity and social impacts points to the need to strengthen institutional capacities, particularly in developing countries, to make and implement collective decisions in a legitimate and equitable manner. These institutional capacities surely include developing resources to analyze equity and social issues.

Institutional weaknesses also inhibit the ability of developing countries to participate effectively in international negotiations. Assistance to help these countries develop a greater capacity to assimilate and analyze information and proposals, and to participate effectively in international discussions, would increase the prospects for achieving effective, lasting, and equitable agreements on how best to address the threat of climate change.

Endnotes

1. Four main views are identified. A "hierarchical" view believes that a problem like climate change needs to be managed rationally and that this can be achieved by suitable management institutions backed by good science and judgment. A "market" view believes that the problem should be characterized primarily as an economic externality, and that the solution lies in creating property rights and a market structure which enable market actors to internalize these costs in their transactions. An "egalitarian collective" view places the emphasis on equal rights in a limited and perhaps fragile biosphere, emphasizing the need for radical abatement and abstinence from excessive consumption. A "fatalist" view considers that humans can adapt to whatever changes are caused and probably cannot manage abatement effectively anyway; it thus argues for no action.

2. Annex I lists the European Community and 36 states: the 24 members of the OECD in 1992 (Australia, Austria, Belgium, Canada, Denmark, Finland, France, Germany, Greece, Iceland, Ireland, Italy, Japan, Luxembourg, the Netherlands, New Zealand, Norway, Portugal, Spain, Sweden, Switzerland, Turkey, the UK, and the U.S.) and 12 European states that are undergoing the process of transition to market economies (Belarus, Bulgaria, the Czech Republic, Estonia, Hungary, Latvia, Lithuania, Poland, Romania, the Russian Federation, Slovakia, and Ukraine).

3. Smith and Ahuja (1990) present succint and strong reasons for considering the range of gases in assessing individual projects, together with a general framework indicating a basis on which such comparisons might be developed. At the aggregate level required for international comparisons and considerations of equity, the issues are somewhat different. The best data, and most analysis, have focussed on the major contribution from CO_2 – especially from fossil fuels, for which national emissions are generally known to within ± 5% (uncertainty range of 10%) or better. CFC emissions – measured as defined in the Montreal Protocol – are known to similar accuracy. Other sources are more uncertain. National estimates of CO_2 and other emissions from deforestation and other land use changes can vary by at least ± 30%, both for individual countries and in aggregate. Most methane emissions are similarly uncertain. Contributions from other gases (N_2O and various other gases involved in different aspects of atmospheric chemistry) are still more uncertain, but the global impact of these other emissions probably totals less than 10% of the major greenhouse gases. Further difficulties are introduced in attempting to compare the radiative impact of different gases, as discussed in earlier IPCC reports.

4. This would apply, for example, if one considered the impact of a single country's emissions in the absence of any other emissions or evaluated the marginal impact of adding each country's emissions in turn, with all the others fixed. None of the literature appears to attempt such an assessment explicitly, and scientific understanding of the carbon cycle may not be sufficient to perform such calculations with confidence.

5. To a degree there are parallels here with the debate over the allocation of sinks (see box "WRI Data"). Looking only at total emissions is similar to the WRI index; focussing on per capita emissions gives a very different view more closely related to that of Agarwal and Narain. Considering national per capita emissions relative to the global per capita mean (so they can be negative as well as positive), as advanced by Mukherjee (1992), is directly equivalent to the approach advocated by Agarwal and Narain.

6. Action by one group of countries to limit fossil fuel emissions, for example, will tend to depress international fuel prices, which may accelerate emission growth elsewhere if other regions do not participate. Conversely, it is likely to accelerate technology developments, which may then diffuse, acting in the opposite direction. The net effect is difficult to predict and depends on timescales and models of elasticities, technology development, and diffusion.

7. The Coastal Zone Management subgroup of the IPCC Response Strategies Working Group suggests slightly different division of "retreat," "accommodation," and "protection" in response to sea level rise. The classification given here reflects the different kinds of costs incurred. A variety of instruments may be appropriate for seeking to share, insure against, or redistribute these different kinds of impact costs.

8. If the remaining global fossil carbon budget were shared according to strict person-year equity, including historic emissions, industrialized countries would not have any emission rights left. A reasonable compromise between international equity and practical feasibility would be to allocate 150 GtC each to industrialized and developed countries (Krause *et al.,* 1990).

9. Note that this does not imply equal benefits over time, as technology development and other factors enable more value to be extracted from a given level of emissions. The issue has some elements in common with issues of natural resource depletion and the intergenerational equity issues discussed in Chapter 4, but the literature has not explored this in any depth.

10. As one relevant indication, the peak energy intensities achieved by countries in the process of industrialization have steadily declined over the century, as countries developing later can draw on better (and more recent) technologies developed elsewhere (Häfele *et al.,* 1981).

11. Ghosh (1993) agrees in theory that positive as well as negative externalities are associated with development transfer from one country to another, but points out that any such application of the principle "must be comprehensive. . . . All external contributions of all persons over all time must be accounted for . . . allowing for those developing countries which were the cradle of civilisation to claim resource transfers. . . . It is difficult to devise practical ways of implementing this principle."

12. It specifies directly how a country's effort should vary as a function of its own income level and abatement efficiency. If the welfare weights are equal across countries, then it defines directly the optimal relative contribution from each country.

References

Adams, J., 1995: *Cost-benefit analysis: part of the problem, not the solution,* Green College Centre for Environmental Policy and Understanding, Oxford.

Agarwal, A., and S. Narain, 1991: *Global warming in an unequal world,* Centre for Science and Environment, New Delhi.

Ahluwalia, M.S., and H.B. Chenery, 1974: *The economic framework.* In World Bank, *Redistribution with growth,* Oxford University Press, New York.

Ahuja, D.R. 1992: Estimating national contributions of greenhouse gas emissions: The CSE-WRI controversy, *Global Environmental Change* **2**, 3.

Alam, M.S. 1989: *Governments and markets in economic development strategies: Lessons from Korea, Taiwan and Japan,* Praeger, New York.

Arendt, H., 1961: *Between past and future,* Viking Press, New York.

Arendt, H., 1968: *Imperialism.* Harcourt Brace Jovanovich, New York.

Arrow, K.J., 1951: Alternative approaches to the theory of choice in risk-taking situations, *Econometrica* **19**, 404–437.

Arrow, K.J., 1953: *Social choice and individual values,* Coles Foundation Monographs, Wiley, New York.

Arrow, K.J., and M. Kurz, 1970: *Public investment, the rate of return and optimal fiscal policy,* Johns Hopkins University Press, Baltimore, MD.

Ayres, R.U., and J. Walter, 1991: The greenhouse effect: Damages, costs and abatement, *Environmental and Resource Economics* **1** (3), 237–270.

Balassa, B.A., 1993: *Policy choices for the 1990s,* New York University Press, New York.

Banuri, T. (ed.), 1992: *Economic liberalization: No panacea.* Clarendon Press, Oxford.

Banuri, T., G. Hyden, C. Juma, and M. Rivera, 1994: *Operationizing sustainable human development: A guide for the practitioner,* UNDP, Bureau of Policy and Programme Evaluation (BPPE), New York.

Barrett, S., 1992: "Acceptable" allocations of tradable carbon emission entitlements in a global warming treaty. In UNCTAD, *Combating Global Warming,* UNCTAD, Geneva.

Beitz, C., 1979: *Political theory and international relations,* Princeton University Press, Princeton.

Benedick, R.E., 1991: *Ozone diplomacy: New directions in safeguarding the planet,* Harvard University Press, Cambridge, MA.

Benestad, O. 1994: Energy needs and CO_2 emissions: Constructing a formula for just distributions, *Energy Policy* **22**(9), 725–734.

Bertram, G., 1992: Tradeable emission permits and the control of greenhouse gases, *Journal of Development Studies* **28** (3), 423–446.

Bertram, G., R.J. Stephens, and C.C. Wallace, 1989: The relevance of economic instruments for tackling the greenhouse effect, Report to New Zealand Ministry of the Environment, August.

Bhaskar, V., 1993: Controlling global warming: Distributional issues, Delhi School of Economics, New Delhi.

Birnie, P., and A. Boyle, 1992: International law and the environment, Clarendon Press, Oxford.

Bodansky, D., 1993: The UN Framework Convention on Climate Change: A commentary, *Yale Journal of International Law* (Summer) **18**, 451–558.

Bradley, R.A., J. Edmonds, M. Wise, H. Pitcher, and C. MacCracken, 1994: Controlling carbon: Equity, efficiency and participation in possible future agreements to control fossil fuel carbon emissions, presented at IPCC Workshop on Equity and Social Considerations, Nairobi, July.

Brenton, T., 1994: *The greening of Machiavelli,* Royal Institute of International Affairs/Earthscan, London.

Brierly, J.L., 1963: *The law of nations: An introduction to the international law of peace,* H. Waldock, ed., Oxford University Press, Oxford.

Broadhus, J.M., 1993: Possible impacts of, and adjustments to, sea-level rise: The cases of Bangladesh and Egypt. In *Climate and sea level change: Observations, projections and implications,* R.A. Warrick *et al.,* eds., Cambridge University Press, New York.

Brown, P.G., 1992: Climate Change and the Planetary Trust, *Energy Policy* **20**, 208–222.

Buchanan, J.M., and R.D. Tollison (eds.), 1984: *The theory of public choice,* Vol. 2. University of Michigan Press, Ann Arbor.

Burtraw, D., and M.A. Toman, 1992: Equity and international agreements for CO_2 constraint, *Journal of Energy Engineering* **118**(2), 122–135.

Cass, D.G., G. Chichilnisky, and H. M. Wu, 1991: Individual risks and mutual insurance, Working Paper CARESS, University of Pennsylvania, Department of Economics.

CDCGC (Commission on Developing Countries and Global Change), 1992: *For Earth's sake,* International Development Research Centre, Ottawa, Canada.

Chayes, A., E.B. Skolnikoff, and D.G. Victor, 1992: *A prompt start: Implementing the Framework Convention on Climate Change,* MIT Center for International Studies, Cambridge, MA.

Cheng, B., 1987: *General principles of law as applied by international courts and tribunals,* Grotius Publications Ltd., Cambridge.

Cheng, B., 1990: Equity, special considerations and the Third World, *CoJielp* **1**, 57.

Chichilnisky, G., 1977: Economic development and efficiency criteria in the satisfaction of basic needs, *Applied Mathematical Modeling,* **1**, (6) 275–304.

Chichilnisky, G., and G. Heal, 1993: Global environmental risks, *Journal of Economic Perspectives* **7**(4), 54–86.

Chichilnisky, G., and G. Heal, 1994: Who should abate carbon emissions? An international viewpoint, *Economic Letters* **44**, 443–449.

Chisholm, A., and A. Moran, 1994: *Carbon dioxide emissions abatement and burden sharing among the OECD countries,* Occasional Paper B26, Tasman Institute, Melbourne, Australia.

Choucri, N. (ed.), 1994: *Global accord,* MIT Press, Cambridge, MA.

Cline, W.R., 1992: *The economics of global warming,* Institute for International Economics, Washington, DC.

Cornia, G.A., and R. Jolly (eds.), 1984: *The impact of world recession on childhood,* Oxford University Press, Oxford.

Cornia, G.A., R. Jolly, and F. Stewart, 1987: *Adjustment with a human face: Protecting the vulnerable and promoting growth,* Oxford University Press, Oxford.

Craig, P., H. Glasser, and W. Kempton, 1993: Ethics and values in environmental policy: The said and the UNCED, *Environmental Values,* **2**(2), 137–158.

d'Arge, R., 1989: Ethical and economic systems for managing the global commons. In *Changing the world environment,* D. Botkin *et al.,* eds., Academic Press, New York.

Dasgupta, P., and G. Heal, 1979: *Economic theory and exhaustible resources,* Cambridge University Press, Cambridge.

del Mundo, F., 1992: Bangladesh in the eye of a storm, *Refugees,* **89**, 29–31.

den Elzen, M., M. Janssen, J. Rotmans, R. Swart and B. de Vries, 1993: Allocating constrained global carbon budgets, *International Journal of Global Energy Issues* **4**(4), 287–301.

Dommen, E. (ed.), 1993: *Fair principles for sustainable development,* Edward Elgar, Aldershot, UK, and Brookfield VT, U.S.

Dorfman, R., 1991: Protecting the transnational commons, mimeo, Harvard University, June.

Drucker, F., 1986: The changed world economy, *Foreign Affairs,* **64**(1), 768–791.

Dubash, N., 1994: Commoditizing carbon: Social and environmental implications of trading carbon emissions entitlements, Master's Thesis, University of California, Berkeley, CA.

Edmonds, J., 1993: *Carbon coalitions,* Battelle Pacific Northwest Laboratory, Washington DC.

Engel, J.R., and J.G. Engel, 1990: *Ethics of environment and development: Global challenge and international response,* Belhaven Press, London.

Epstein, J., and R. Gupta, 1990: *Controlling the greenhouse effect: Five global regimes compared,* Brookings Institution, Washington, DC.

Ewah, O.E., 1994: Africa's decline and greenhouse politics, *International Environmental Affairs* **6**(2), 133–148.

Fankhauser, S., 1993: The economic costs of global warming: Some monetary estimates. In Kaya *et al.,* eds., proceedings of IIASA workshop on the costs, impacts and benefits of CO_2 mitigation, International Institute of Applied Systems Analysis, Laxenburg, Austria, 1992.

Fermann, G., 1993: Climate change, burden-sharing criteria and competing conceptions of responsibility, *International Challenges* **12**(4), Norway.

Flexner, S.B., ed., 1987: *The Random House Dictionary of the English Language,* 2d ed. (unabridged), Random House, New York.

Franck, T., and D. Sughrue, 1993: Equity as fairness, *Georgetown Law Journal,* **81**, 563.

Fujii, Y., 1990: An assessment of the responsibility for the increase in the CO_2 concentrations and intergenerationl carbon accounts, IIASA Working Paper WP-05-55, International Institute of Applied Systems Analysis, Laxenburg, Austria.

Ghosh, P., 1993: Structuring the equity issue in climate change. In *The climate change agenda: An Indian perspective,* A.N. Achanta, ed., Tata Energy Research Institute, New Delhi.

Ghosh, P., and A. Jaitly, 1993: *The road from Rio: Environmental and development policy issues in Asia,* Tata Energy Research Institute, New Delhi.

Ghosh, P., and J. Puri (eds.), 1994: *Joint implementation of climate change commitments: Opportunities and apprehensions,* Tata Energy Research Institute, New Delhi.

Gleick, J., 1987: *Chaos: Making a new science,* Penguin Books, New York.

Gore, C., 1993: Entitlement relations and "unruly" social practices: A comment on the work of Amartya Sen, *Journal of Development Studies* **29** (3): 429–460.

Grubb, M.J., 1989: *The greenhouse effect: Negotiating targets,* Royal Institute of International Affairs, London.

Grubb, M.J., 1990: *Energy policies and the greenhouse effect,* Vol. 1: *Policy appraisal,* Royal Institute of International Affairs, London.

Grubb, M.J., 1993: The costs of climate change: Critical elements. In *Proceedings of IIASA Workshop on the Costs, Impacts and Benefits of CO_2 Mitigation,* Kaya *et al.,* eds. IIASA, Laxenburg, 1992.

Grubb, M.J., 1995: Seeking fair weather: Ethics and the international debate on climate change, *International Affairs* **71**(3), 463–496.

Grubb, M.J., and J. Sebenius, 1992: Participation, allocation, and adaptability in international tradeable emission permit systems for greenhouse gas control. In *Climate change: Designing a tradeable permit system,* OECD, Paris.

Grubb, M., J. Sebenius, A. Magalhaes, and S. Subak, 1992: Sharing the burden. In *Confronting climate change: Risks, implications, and responses,* I.M. Mintzer, ed., pp. 305–322, Cambridge University Press, Cambridge.

Grubler, A., and Y. Fujii, 1991: Intergenerational and equity issues of carbon account, *Energy* **16**(11/12), 1397–1416.

Grubler, A., and N. Nakicenovic, 1991: *International burden-sharing in greenhouse gas reduction,* Environmental Policy Division, World Bank.

Gyawali, D., 1993: Risk in poverty and fairness in development, IIASA Workshop on Risk and Fairness, International Institute of Applied Systems Analysis, Laxenburg, Austria, June.

Haas, P.M., R.O. Keohane, and M.A. Levy (eds.), 1993: *Institutions for the Earth: Sources of effective international environmental protection,* MIT Press, Cambridge, MA.

Häfele, W., J. Anderer, A. McDonald, and N. Nakicenovic, 1981: *Energy in a finite world,* Ballinger, Cambridge, MA.

Hammond, A., E. Rodenberg, and W. Moomaw, 1991: Calculating rational accountability for climate change, *Environment* **33**(1), 11–35.

Handl, G., 1990: International efforts to protect the global atmosphere: A case of too little, too late? *European Journal of International Law* **1**(1/2), 250–257.

Hayes, P., 1993: North-South transfer. In Hayes and Smith, 1993.

Hayes, P., and K. Smith (eds.), 1993: *The global greenhouse regime: Who pays?* UNU Press/Earthscan, London.

Heal, G., 1984: Interaction between policy and climate: A framework for policy design under uncertainty. In *Advances in applied microeconomics,* V. Smith and A.D. White, eds., pp. 151–158, J.A.I. Press, Greenwich, CT.

Heidenreich, A., 1995: Equity, communication and climate change: Access to information and communications networks in Africa, presented at IPCC Workshop on Equity and Social Considerations, Nairobi, July.

Heintz, R., and H. Markus, 1993: Joint implementation: Exploiting resources or sacrifice, *Change* **17**, 1–3.

Herman, R., S.A. Ardekani, and J.H. Asubel, 1987: Dematerialization. In *Technology and Environment,* J.H. Asubel and H.E. Sladovich, eds., National Academy Press, Washington DC.

Hyder, T.O., 1992: Climate negotiations: The North/South perspective. In I. Mintzer, ed., *Confronting climate change: Risks, implications and responses,* Cambridge University Press, Cambridge, UK.

ICJ (International Court of Justice), 1982: *ICJ Reports: Case concerning the Continental Shelf* (Tunisia/Libya), ICJ, The Hague.

ICJ (International Court of Justice), 1984: *ICJ Reports: Gulf of Maine case,* ICJ, The Hague.

ICJ (International Court of Justice), 1985: *ICJ Reports: Case concerning the Continental Shelf (Malta/Libya),* ICJ, The Hague.

ICJ (International Court of Justice), 1993: *ICJ Reports: Jan Mayen case, ICJ,* The Hague.

Imbree, S., 1994: Investing in less greenhouse gas intensive development: What joint implementation could be. In Ghosh and Puri (1994).

IPCC (Intergovernmental Panel on Climate Change), 1990: Report of the Special Committee on the Participation of Developing Countries. IPCC, Geneva.

IPCC (Intergovernmental Panel on Climate Change), 1992: *Climate change 1992: The supplementary report to the IPCC scientific assessment,* J.T. Houghton, B.A. Callander, and S.K. Varney, eds., Cambridge University Press, Cambridge.

IPCC (Intergovernmental Panel on Climate Change), 1995: *Climate change 1994: Radiative forcing of climate change,* J.T. Houghton, L.G. Meira Filho, J. Bruce, Hoesung Lee, B.A. Callander, E. Haites, N. Harris, and K. Maskell, eds., Cambridge University Press, Cambridge.

IUCN (International Union for the Conservation of Nature), 1980: *World conservation strategy,* International Union for the Conservation of Nature, UNEP, Geneva.

Jaitly, A., 1993: An overview of post-Rio political economy issues. In Ghosh and Jaitly (1993).

Juma, C., J.B. Ojwang, and P. Karani, 1994: Equity considerations in the climate debate: Technology transfer, presented at IPCC Workshop on Equity and Social Considerations, Nairobi, July.

Kaplan, G.T., and C.S. Kessler, 1989: *Hannah Arendt: Thinking, judging, freedom,* Allen and Unwin, Sydney, Australia.

Keynes, J.M. 1936: *The general theory of employment, interest and money,* Macmillan, London.

Korten, D.C., 1987: *Community management: Asian experience and perspectives,* Kumarian Press, West Hartford, CT.

Korten, D.C. 1990: *Getting to the 21st century: Voluntary action and the global agenda,* Kumarian Press, West Hartford, CT.

Krause, F., W. Bach, and Jon Kooney, 1990: *Energy policy in the greenhouse: From warming fate to warming limit,* vol. 1, Earthscan, London and IPSET, El Cerrito, CA.

Krueger, A.O., and R.H. Bates (eds.), 1993: *Political and economic interactions in economic policy reform: Evidence from eight countries,* Blackwell,, Oxford, UK, and Cambridge, MA.

Kuik, O., P. Peters, and N. Schrijver, 1994: *Joint implementation to curb climate change,* Kluwer, Dordrecht.

Kuznet, S., 1955: Economic growth and income inequality, *American Economic Review* **45** (1) 1–28.

Lane, C., and R. Moorehead, 1994: *Who should own the range: New thinking on pastoral resource tenure in dryland Africa,* International Institute for Environment and Development (IIED), London.

Lofstedt, R.E., 1994: Human dimensions of global warming: They should not be ignored, Mimeo, Centre for Environmental Study, University of Surrey, UK.

Loske, R., 1991: Ecological taxes, energy policy and greenhouse gas reductions: A German perspective, *Ecologist* **21** (July/August), 173–176.

Loske, R., and S. Oberthur, 1993: Joint implementation under the Climate Change Convention, *International Environmental Affairs,* **6**(1) 45–58.

Manne, A.S., and T.F. Rutherford, 1994: International trade in oil, gas and carbon emission rights: An intertemporal general equilibrium model, *The Energy Journal* **15**(1), 31–56.

Markandya, A., 1994: JI: The way forward or a negotiator's nightmare? In *Joint implementation of climate change commitments: Opportunities and apprehensions,* P. Ghosh and J. Puri, eds., Tata Energy Research Institute, New Delhi.

Mathur, A., 1991: India: Vast opportunities and constraints. In *Energy policies and the greenhouse effect,* vol. 2, *Country studies and technical options,* M. Grubb *et al.,* eds., pp. 395–430, Royal Institute of International Affairs, London.

Maya, S., 1994: Joint implementation: Cautions and options for the South, paper presented at the International Conference on Joint Implementation, Groningen, The Netherlands, June.

Meier, G.M., 1984: *Leading issues in economic development,* Oxford University Press, New York.

Metz, B., 1994: Joint implementation as a financing instrument for global reductions in greenhouse gas emissions. In Ramakrishna (1994).

Meyer, A., 1995: The unequal use of the global commons, presented at IPCC Workshop on Equity and Social Considerations, Nairobi, July.

Meyer-Abich, K., 1994: Winners and losers in climate change. In *Global ecology: A new arena of political conflict,* W. Sachs, ed., Zed Books, London and Atlantic Highlands, NJ.

Millennium, 1990: Global environmental change and international relations, **19** (Winter), 337–476.

Morris, M.D. 1979: *Measuring the condition of the world's poor: The physical quality of life index,* Overseas Development Council, Washington, DC; Pergamon, New York.

Moyo, S., 1994: *Land entitlements, tenure and climate change: The Zimbabwe case,* Institute of Development Studies, Harare.

Mukherjee, N., 1992: Greenhouse gas emissions and the allocation of responsibility, *Environment and Urbanisation* **4**(1), April.

Myers, N., 1993: Environmental refugees in a globally warmed world, *Bioscience* **43,** (11).

Myers, N., 1994: Environmental refugees and climate change: Estimating the scope of what could well become a prominent international phenomenon, presented at IPCC Workshop on Equity and Social Considerations, Nairobi, July.

Nozick, R., 1974: *Anarchy, state, and Utopia,* Basic Books, New York.

Pachauri, R., 1994: The economics of climate change: A developing country perspective. In *OECD: The economics of climate change,* proceedings of an OECD Conference, Paris, pp. 171–179.

Parikh, J., 1992: IPCC strategies unfair to the South, *Nature* **360,** 507–508.

Parikh, J., 1994: Joint implementation and sharing commitments: A Southern perspective. In *Integrative assessment of mitigation, impacts, and adaptation,* Nakicenovic *et al.,* eds., IIASA, Laxenburg, Austria.

Parikh, J., 1995: North-South cooperation in climate change through joint implementation, *International Environmental Affairs,* **7**(1), 22–43.

Parikh, J., and S. Gokarn, 1993: Climate change and India's energy policy options, *Global Environmental Change* **3,** 276–292.

Parikh, J., K. Parikh, and S. Gokarn, 1991: Consumption patterns: The driving force of environmental stress, IGIDR report prepared for United Nations Conference on Environment and Development (UNCED), Indira Gandhi Institute of Development Research, Bombay, India.

Paterson, M., 1994/1996: International justice and global warming. Proceedings of 1994 workshop published in *The ethical dimensions of global change,* B. Holden, ed., Macmillan, London, forthcoming 1996.

Pieterse, J.N., 1994: Globalization as hybridization, Working Paper Series No. 152, Institute of Social Studies, The Hague.

Putnam, R., R. Leonardi, and R.Y. Nanetti, 1992: *Making democracy work: Civic traditions in modern Italy,* Princeton University Press, Princeton.

Ramakrishna, K. (ed.), 1994: *Criteria for joint implementation under the Framework Convention on Climate Change,* Woods Hole Research Center, Woods Hole, MA.

Rawls, J., 1971: *A theory of justice,* Harvard University Press, Cambridge, MA.

Rayner, S., 1993: Governance and the global commons, discussion paper no. 8, Centre for the Study of Global Governance, LSE, London.

Rayner, S., 1994: A conceptual map of values for climate change decision making, presented at IPCC Workshop on Equity and Social Considerations, Nairobi, July.

Rose, A., 1990: Reducing conflict in global warming policy – the potential of equity as a unifying principle, *Energy Policy* **18,** 927–935.

Rose, A., and B. Stevens, 1993: Equity aspects of marketable permits for greenhouse gases: Global dimensions of the Coase Theorem, mimeo, Pennsylvania State University, University Park, PA.

Rutherford, T., 1992: Welfare effects of fossil carbon restrictions: Results from a recursively dynamic trade model. OECD Working Paper 112, April.

Saha, V.L., 1994: Dilemma of small island and coastal states, presented at IPCC Workshop on Equity and Social Considerations, Nairobi, July.

Sandel, M.J., 1982: *Liberalism and the limits of justice,* Cambridge University Press, Cambridge.

Sands, P., 1995: *Principles of international environmental law,* Manchester University Press, Manchester.

Schachter, O., 1977: *Sharing the world's resources,* Columbia University Press, New York.

SEI (Stockholm Environment Institute), 1992: *National greenhouse gas accounts: Current anthropogenic sources and sinks,* S. Subak, P. Raskin and D. von Hippel, eds., SEI, Boston, MA.

Selrod, R., and A. Torvanger, 1994: What might be minimum requirements for making the mechanism of joint implementation under the Climate Change Convention credible and operational? Draft discussion paper, CICERO (Centre for International Climate and Environment Research), Oslo.

Sen, A., 1976: Famines as failures of exchange entitlements, *Economic and Political Weekly* **11,** 31–33.

Sen, A., 1984: *Resources, values and development,* Basil Blackwell, Oxford.

Shah, A., and B. Larsen, 1992: Carbon taxes, the greenhouse effect and developing countries, World Bank Working Paper series no. 957, World Bank, Washington DC.

Shapiro, I., 1990: *Political criticism.* University of California Press, Berkeley, CA.

Shue, H., 1992: The unavoidability of justice. In *The international politics of the environment: Actors, interests and institutions,* A. Hurrel and B. Kingsbury, eds., pp. 373–397, Clarendon Press, Oxford.

Shue, H., 1993: Subsistence emissions and luxury emissions, *Law and Policy,* **15**(1), 39–59.

Sidibe, H., 1995: Impact des variations climatiques sur l'évolution des conditions de vie dans les communautés pastorales touareg et fulbe de la région lacustre de l'Issa-Ber au Mali, presented at IPCC Workshop on Equity and Social Considerations, Nairobi, July.

Simonis, U., 1994: Towards a Houston Protocol, or how to allocate CO_2 emissions reductions between North and South, Wissenschaftszentrum Berlin (WZB), Science Centre Berlin, Germany.

Singh, Hon. S.M., 1993: Valedictory address. In Ghosh and Jaitly (1993).

Smith, K.R., 1993: The basics of greenhouse gas indices. In Hayes and Smith (1993).

Smith, K.R., 1994: Preindustrial missing carbon and current greenhouse responsibilities. In *Preindustrial human environmental impacts,* D.K. Kammen, K.R. Smith, A.T. Rambo, and M.A.K. Khalil, eds., pp. 1135–1143, Special Issue of *Chemosphere* **29**(5), 827–1143.

Smith, K.R., 1995: The natural debt: North and South. In *Climate change: Developing Southern Hemisphere perspectives,* T. Giambelluca and A. Henderson-Sellers, eds., Wiley, Chichester, UK.

Smith, K.R., and D. Ahuja, 1990: Toward a greenhouse equivalence index: The total exposure analogy, *Climatic Change* **17** (1), 1–7.

Smith, K.R., J. Swisher, and D. Ahuja, 1993: Who pays to solve the problem and how much? In Hayes and Smith (1993); also working paper no. 1991–22, World Bank Environment Department.

Sobhan, R., 1993: *Agrarian reform and social transformation: Preconditions for development,* Zed Books, Atlantic Highlands, NJ.

Solomon, B.D., and D.R. Ahuja, 1991: International reductions of greenhouse-gas emissions: An equitable and efficient approach, *Global Environmental Change,* **1**(5), 343–350.

Stephan, G., R. van Nieuwkoop, and T. Wiedner, 1992: Social incidence and economic costs of carbon limits, *Environmental and Resource Economics,* **2,** 569–591.

Stone, C., 1993: *The gnat is older than man, global environment and human agenda,* Princeton University Press, Princeton, NJ.

Streeten, P., 1981: *First things first,* Oxford University Press, New York.

Subak, S., 1993: Assessing emissions: Five regimes compared. In Hayes and Smith (1993).

Suliman, M., 1992: *Civil war in Sudan: The impact of ecological degradation,* Institute for Africa Alternatives, London.

Suliman, M., 1994: Climate change and conflict, presented at IPCC Workshop on Equity and Social Considerations, Nairobi, July.

Tarlock, D., 1992: Environmental protection: The potential misfit between equity and efficiency, *University of Colorado Law Review,* **63,** 871–900.

Taylor, L., 1988: *Varieties of stabilization experience,* Clarendon Press, Oxford.

Taylor, L., 1993: *The rocky road to reform: Adjustment, income distribution, and growth in the developing world,* MIT Press, Cambridge, MA.

Thery, D., 1992: Should we drop or replace the WRI global index? *Global Environmental Change* **2**(3), 88–89.

UNCTAD (United Nations Conference on Trade and Development), 1992: *Combating global warming,* UNCTAD/RDP/DFP/1, Geneva and New York.

UNDP (United Nations Development Programme), 1993: *Rethinking technical cooperation: Reforms for capacity building in Africa Annex. Statistics on technical cooperation,* UNDP, New York.

United Nations, 1993: *Earth summit, Agenda 21, The United Nations programme of action from Rio: The final text of agreements negotiated by governments at the United Nations Conference on Environment and Development (UNCED),* 3-14 June 1992, Rio de Janeiro, Brazil. United Nations, New York.

Uzawa, H., 1994: Equity in economic theory: Implications for climate change, presented at IPCC Workshop on Equity and Social Considerations, Nairobi, July.

Van Dyke, V., 1975: Justice as fairness: For groups? *The American Political Science Review* **69**(2) (June), pp. 607–614.

Vellinga, P., and R. Heintz, 1994: Joint implementation: A phased approach. In Ramakrishna (1994).

Vienna Convention on the Law of Treaties, 1969: *ILM* **8,** 679.

Wade, R., 1992: *Governing the market: Economic theory and the role of government in East Asian indust.rialization,* Princeton University Press, Princeton.

Walzer, M., 1983: *Spheres of justice: A defense of pluralism and equality,* Basic Books, New York.

WCED (World Commission on Environment and Development), 1987: *Our common future,* Oxford University Press, Oxford.

WEC (World Energy Council), 1993: *Energy for tomorrow's world,* Kogan Page, London.

Weiss, E.B., 1993: International environmental law: Contemporary issues and the emergence of a new world order, *Georgetown Law Journal* **81,** 675–710.

Welsch, H., 1993: A CO_2 agreement proposal with flexible quotas, *Energy Policy* **21**(7), 748–756.

Westing, A.H., 1989: A law of the air, *Environment* **31,** (3) (April).

Wilford, M., 1993: Insuring against sea level rise. In Hayes and Smith (1993).

Williams, R.H., E.D. Larson, and M.H. Ross, 1987: Materials, affluence, and industrial energy use, *Annual Review of Energy 12,* Annual Reviews Inc., Palo Alto, CA.

Wirth, D.A., and D.A. Lashof, 1990: *Ambio* **19** (October), 305–310.

World Bank 1992: *World Development Report, 1992.* World Bank, Washington, DC.

World Bank, 1994: *World Development Report, 1994,* World Bank, Washington, DC.

World Council of Churches, 1994: *Accelerated climate change: Sign of peril, test of faith,* WCC, Geneva, Switzerland.

WRI (World Resources Institute), 1990: *World resources 1990-91,* Oxford University Press, New York.

WRI (World Resources Institute), 1991: *Greenhouse warming: Negotiating a global regime,* WRI, Washington, DC.

Yamin, F., 1993a: Joint implementation and the Framework Convention on Climate Change: Legal, institutional and procedural issues, FIELD Working Paper, Foundation for International Environmental Law and Development, London.

Yamin, F., 1993b: The Framework Convention on Climate Change: The development of criteria for joint implementation, FIELD Working Paper, Foundation for International Environmental Law and Development, London.

Yamin, F., 1994: Principles of equity in international environmental agreements with special reference to the Climate Change Convention, Presented at IPCC Workshop on Equity and Social Considerations, Nairobi, July.

Young, H.P., and A. Wolf, 1992: Global warming negotiations: Does fairness count? *The Brookings Review* **10**(2).

Young, O.R., 1990: Global environmental change and international governance, *Millennium* **19.**

Young, P., 1994: *Equity in theory and practice,* Princeton University Press, Princeton: NJ.

4

Intertemporal Equity, Discounting, and Economic Efficiency

K.J. ARROW, W.R. CLINE, K-G. MALER, M. MUNASINGHE,
R. SQUITIERI, J.E. STIGLITZ

CONTENTS

SUMMARY

The discount rate allows economic effects occurring at different times to be compared. It plays a vital role in public policy analysis of actions with varying time paths of costs and benefits. It is particularly important in climate change: Because of the very long times involved in climate change decisions, the choice of a discount rate powerfully affects the net present value of alternative policies, and thus the policy recommendations that emerge from climate change analysis.

Two major approaches are used to determine the appropriate discount rate for climate change analysis. The normative or ethical perspective (called the *prescriptive approach* in this chapter) begins with the question, "How (ethically) should impacts on future generations be valued?" The positive perspective, called here the *descriptive approach,* begins by asking, "What choices involving trade-offs across time do people actually make?" and, "To what extent will investments made to reduce greenhouse gas emissions displace investments elsewhere?"

The prescriptive approach tends to generate relatively low discount rates and thus favours relatively more spending on climate change mitigation. The descriptive approach tends to generate somewhat higher discount rates and thus favours relatively less spending on climate change mitigation.

Although economists support the concept of discounting for climate change analysis, they continue to debate which of the two approaches is correct, and the parameters to be used in calculating the rate. These choices in turn significantly affect the conclusions of the analysis.

4.1 Introduction

The discount rate allows economic effects occurring at different times to be compared. Discounting converts each future dollar amount associated with a project into the equivalent present dollar amount. The discount rate is generally positive because resources invested today can be transformed into, on average, more resources later; this holds for investments in both physical capital (e.g., machines) and human capital (e.g., education).

Greenhouse gas control programmes may be viewed as an investment: Money is spent today to reduce the costs of climate change tomorrow. If the real rate of return on investment in greenhouse gas controls exceeds the rate of return on investment in machines or education, then future generations would be better off if less were invested today in machines and education and more in controlling greenhouse gas emissions. The converse also holds, provided that the money is spent on investment rather than consumption.

Because of the very long times involved in climate change decisions, the choice of a discount rate powerfully affects the net present value of alternative policies, and thus the policy recommendations that emerge from climate change analysis.[1]

The benefits of greenhouse gas abatement accrue decades or even centuries in the future. For this reason, use of a high discount rate results in a low present value of actions that slow climate change. For example, at a discount rate of 7% annually (as is commonly used in short-horizon project analysis), damages of $1 billion 50 years hence have a present value of only [$1 x 10^9]/[1.07^{50}] = $33.9 million; the same damages 200 years hence have a present value of only $1300. Thus the use of too high a discount rate will result in too little value placed on avoiding climate change and too little investment in climate change programmes. Conversely, applying too low a discount rate to climate change programmes will result in too much investment in them and crowd out better uses of the resources.

Determining the appropriate discount rate involves issues in normative as well as positive economics. These two perspectives raise very different questions. From a normative or ethical perspective, the key question might be: "How (ethically) should impacts on future generations be valued?" From a positive perspective, the appropriate question might be: "To what extent will investments made to reduce greenhouse gas emissions displace investments elsewhere?"

The debate is often confusing, in part because three separate issues are being addressed: how to discount the *welfare* or *utility* of future generations, how to discount future dollars, and how to discount future pollution. Further, the argument often combines questions of efficiency and questions of ethics. Although economists can make no special claim to professional expertise in questions of ethics, they have developed rigorous methods for analyzing the implications of ethical judgments.

Climate policy raises particular questions of equity among generations, as future generations are not able to influence directly the policies being chosen today that will affect their well-being (Mishan, 1975; Broome, 1992). Moreover, it may

not be possible to compensate future generations for reductions in well-being caused by current policies, and, even if feasible, such compensation may not actually occur.[2]

4.1.1 Areas of agreement and disagreement

Economists are in general agreement that cost-benefit analysis, including discounting, is useful in examining policies with long or complex time paths, or policies whose effects extend across generations (see, for example, Layard, 1976; Cline, 1992; Lyon, 1994). At the same time, cost-benefit analysis, and the techniques that go with it, including discounting, focus on economic efficiency, and therefore have limitations as a guide to policy.[3]

The trade-off between consumption today and consumption in the future raises two central questions: first, how to think about this trade-off; second, what numerical value to attach to it. Many economists subscribe to a general framework that focusses on the social marginal utility of consumption today compared with consumption in the future. In this framework, the discount rate can be expressed as:

$$d = \rho + \theta g, \tag{4.1}$$

where d is the discount rate, ρ is the rate of pure time preference (also called the utility discount rate, a measure of the difference in importance attached to utility today versus utility in the future),[4] θ is the absolute value of the elasticity of marginal utility (a measure of the relative effect of a change in income on welfare), and g is the growth rate of per capita consumption. Equation 4.1 provides a way to think about discounting that subsumes many related subtopics, including treatment of risk, valuing of nonmarket goods, and treatment of intergenerational equity.

This equation sets out explicitly the two reasons for discounting future consumption: either (1) one cares less about tomorrow's consumer than today's, or about one's own welfare tomorrow than today (reflected in the first term, ρ); or else (2) one believes tomorrow's consumer will be better off than today's (reflected in the second term, θg). For a discussion of the derivation of equation 4.1, see Annex 4A.

Economists are in general agreement about several empirical issues that affect the discount rate, including the range of returns to investment, and the average interest rates earned and paid by consumers.

There is also a general consensus about certain basic principles of discount rate analysis. Most economists believe that considerations of risk can be treated by converting outcomes into *certainty equivalents,* amounts that reflect the degree of risk in an investment,[5] and discounting these certainty equivalents. There is general agreement that in evaluating competing projects, all spending, including investment, is to be converted into consumption equivalents first, then discounted (Arrow and Kurz, 1970; Lind *et al.,* 1982). Environmental impacts may be incorporated by converting them to consumption equivalents, then discounting. Many people expect the relative price of environmental goods to increase over time, which would have consequences equivalent to adopting a lower discount rate for such goods at unchanged prices.

However, given appropriate estimates of relative prices, there is no reason to explicitly modify the discount rate for environmental goods (see Annex 4A). In addition, economists generally believe that future generations could be compensated for some loss of environmental amenities by offsetting accumulations of capital.[6]

Economists disagree, however, about several other issues that affect the choice of a discount rate, including key parameters such as the likely rate of future per capita economic growth and the changing relative scarcity of environmental goods. These calculations require economic judgments about the degree of economic efficiency reflected in market outcomes, the extent of constraints on policy, and the proper approach to distributional concerns. Disagreements on these points drive the differences in conclusions about the discount rate.

The next section presents the two most prominent approaches to discounting for climate change analysis, together with the reasons for their differing conclusions. The two approaches start from very different places. What is called below the *prescriptive approach* begins with ethical considerations. What is called the *descriptive approach,* on the other hand, begins with evidence from decisions that people and governments actually make.[7]

4.2 Prescriptive Approach

The prescriptive approach, which is usually associated with a relatively low discount rate, begins with a *social welfare function* (an algebraic formulation that "adds up" the consumption of different individuals, yielding a measure of the well-being of society as a whole) constructed from ethical principles. Those who hold the prescriptive view typically argue that market interest rates often provide a poor indicator of the marginal trade-offs to society, because of market imperfections and suboptimal tax (and sometimes expenditure) policy, and because of constraints on policy, especially the difficulty in making transfers to future generations.

In the absence of such limitations, the social marginal utility of consumption would be the same at each point in time, and the social marginal rate of substitution between consumption today and consumption tomorrow would equal the market rate of interest. In the presence of such limitations, the social marginal rate of substitution will in general differ from market rates of interest.

Some advocates of the prescriptive approach use the term *social rate of time preference* (SRTP) to refer to the discount rate they derive. In this chapter, the term SRTP will be reserved for the discount rate derived from the prescriptive approach. Using this new expression, equation (4.1) is

$$SRTP = \rho + \theta g$$

The first term, ρ, reflects discounting of the utility of future generations. This term is sometimes said to represent discounting for impatience or myopia; alternatively, it may represent discounting for empathetic distance (because we may feel greater affinity for generations closer to us). See Annex 4A.3 for more discussion on the pure rate of time preference.

The second term, θg, reflects discounting for rising consumption (or consumption equivalents). If per capita consumption is growing at rate g, then an extra unit of consumption in the next period should be discounted by the term θg to take account of the lower marginal utility of consumption at higher consumption levels. Even if present and future generations are given equal weight, so that pure time preference is zero ($\rho = 0$), future consumption would still be discounted if later generations are expected to be better off; in this case, an extra unit of consumption would not be worth as much in the future as it is today. For example, if technical change continues at the pace of the last century, with productivity and living standards doubling about every thirty years, then this high value of g would push up the SRTP. This means that an additional unit of consumption by future generations, who would be much richer than we are, would count much less than an additional unit of consumption today.[8]

The SRTP approach values the total change in consumption at each date, not just the direct outputs of the project. Where mitigation projects displace other investment, future consumption must be reduced by the consumption that the displaced investment projects would have generated. (This requires an explicit analysis of the project's effects on consumption and investment.) The SRTP is then applied to net consumption. In effect, all results are converted to their consumption equivalents, then discounted at the SRTP.

The prescriptive approach arrives at the following conclusions:

(1) It is appropriate to apply a discount rate to public and private investment, including regulatory decisions. This discount rate should be derived from ethical considerations, reflecting society's views concerning trade-offs of consumption across generations.

(2) Because of practical limits on the feasibility of intergenerational transfers, and in the absence of optimal tax policy,[9] the SRTP will in general fall below the producer rate of interest.

(3) The cost of a greenhouse mitigation project must include the forgone benefits of other competing investments not undertaken. This means that costs should be adjusted for the shadow price of capital, the present value of the future consumption yielded by a unit of capital. If a mitigation project would displace private investment, and returns to both projects accrue to the same generations, then it is appropriate to use the opportunity cost of capital – the return that the private investor would have received from the forgone capital investment. Only after doing this will it be appropriate to use the social rate of time preference to discount consumption.

4.2.1 Discount rate estimates – Prescriptive approach

If the pure rate of time preference (ρ) is zero, then high rates of productivity increase (and thus high g), of the order of 1.5%, plus high (absolute) values of the elasticity of marginal utility (θ)[10] imply a social discount rate of about 3%. With low rates of productivity increase, of the order of 0.5%, and low (absolute) values of the elasticity of marginal utility, the social discount rate is of the order of 0.5%. In a gloomy scenario, in which future output and consumption decline, then g and thus the SRTP may be negative (Munasinghe, 1993). Also, the discount rate need not be constant over time even if ρ and θ are constant, since g need not be constant.

The economic literature on global warming has used a range of discount rates. To follow the approach suggested by Cline (1992), with a zero rate of pure time preference (ρ), and using the central case consumption growth rate of 1.6% per capita from the IPCC scenarios (IPCC, 1992), multiplied by an elasticity of marginal utility (θ) of 1.5, gives an SRTP of 2.4%. If, instead, it is assumed that per capita growth is only 1% (perhaps because of slower growth after 100 years), or if $\theta = 1$, then the SRTP becomes 1.5%. After taking account of the share of resources coming out of capital (20% economy-wide versus 80% out of consumption) and taking into account the opportunity cost of displaced capital and depreciation, the effective discount rate becomes 2 to 3%.

A higher SRTP may apply to developing countries with higher rates of productivity growth. If labour productivity increases by 5-8% per year, as experienced by the high-growth countries of Asia, and with an elasticity of marginal utility of 2, discount rates of the order of 10 to 16% could be justified. Similarly, low-income countries close to subsistence levels could have high elasticities of marginal utility (this assumes a rapid fall-off of marginal utility from the extremely high initial levels associated with privation), so that their SRTPs could be high even if they were experiencing slow growth over long periods. These distinctions imply that developing countries may be less willing than industrialized countries to assume abatement costs now in anticipation of climate change benefits later.[11]

These discount rates apply to consumption only. They can be used only after the forgone benefits of other investments not made (i.e., the opportunity costs) have been included in the costs of the programme. If the forgone investments would have produced a high return, then calculated output and future consumption will suffer, making the mitigation programme relatively less attractive.

Critics of the prescriptive approach note that the opportunity cost of capital (the market rate of return) usually exceeds the SRTP. This suggests that society should not make decisions on the basis of a 2% discount rate, because in doing so we would be forgoing better alternative investments. Prescriptionists argue that the SRTP does not equal the market rate of interest because important alternatives are not feasible – in particular, because society cannot set aside investments over the next three centuries, earmarking the proceeds for the eventual compensation of those adversely affected by global warming. Accordingly, if the SRTP is 1 to 2%, a climate change investment returning 2% is better than no investment at all. Critics of the prescriptive approach also point out that a discount rate of 2% is glaringly inconsistent with observed behaviour (e.g., government spending on education or research, or development assistance by donor countries). To

this, prescriptionists reply that just because the government fails to allocate resources in one area on the basis of ethical considerations, that is no reason to insist that decisions in other areas be consistent with that initial decision.

4.3 Descriptive Approach

The other widely employed approach focusses on the (risk-adjusted) opportunity cost of capital. Most global warming optimization models (e.g., Nordhaus, 1994; Peck and Teisberg, 1992; Manne *et al.*, 1993) rely on the descriptive approach, which rests on three arguments:

(1)   Mitigation expenditures displace other forms of investment. Advocates of the descriptive approach advise decision makers to choose the action that leads to the greatest total consumption (Nordhaus, 1994).[12]

(2)   If the return on mitigation investments lies below that of other investments, then choosing other investments would make current and future generations better off. Transfers to future generations, if necessary, are to be considered separately.

(3)   The appropriate social welfare function to use for intertemporal choices is revealed by society's actual choices (hence the name, descriptive approach). Believing that no justification exists for choosing an SWF different from what decision makers actually use, advocates of the descriptive approach generally call for inferring the social discount rate from current rates of return and growth rates (Manne, 1994).

Critics have questioned all three arguments.[13]

4.3.1   Formulation of the descriptive approach

The descriptive approach looks at investments in the real world, and sets the discount rate accordingly. The descriptive approach implicitly aims to maximize the economic resources available to future generations, allowing them to decide how to use these resources. Nordhaus (1994), Lind (1994), Birdsall and Steer (1993), Lyon (1994), and Manne (1994), among others, have all stressed the importance of the opportunity cost of capital, noting that even apparently small differences in rate of return result in large differences in long-run results. Over 100 years, an investment at 5% returns 18 times more than one at 2%. Thus, where some redistribution of future returns is possible, society would be foolish to forgo a 5% return for a 2% return.

Birdsall and Steer of the World Bank (1993) explain the need to direct investment to the most productive uses, warning against use of too low a discount rate:

We feel that meeting the needs of future generations will only be possible if investable resources are channelled to projects and programmes with the highest environmental, social, and economic rates of return. This is much less likely to happen if the discount rate is set significantly lower than the cost of capital.[14]

Wildavsky (1988) explains the point in the context of health and safety regulations:

> Insofar as we today should consider the welfare of future generations, our duty lies not in leaving them exactly the social and environmental life we think they ought to have, but rather in making it possible for them to inherit a climate of open choices – that is, in leaving behind a larger level of general fluid resources to be redirected as they, not we, see fit.

To advocates of the prescriptionist approach who claim that on ethical grounds, it is difficult to support a rate of pure time preference much above zero (Cline, 1992), advocates of the descriptive approach point to actual behaviour of individuals and nations. For example, development assistance budgets for the OECD countries average about 0.25% of GDP – certainly inconsistent with the ethical arguments used to justify the assumption that $\rho = 0$.[15]

Further, as Manne (1994) demonstrates, a low SRTP implies a high rate of investment: A discount rate of 2% implies far more investment than actually occurs in any country now, and thus would require a big jump in savings rates to finance.[16] But tax policy in most OECD countries significantly depresses investment, which raises the return to investment at the margin, and is therefore inconsistent with a low SRTP. What conclusion to draw from this evidence depends on whether tax policy is viewed as a constraint or as the result of optimizing an SWF. Most advocates of the descriptive approach hold the latter view. Descriptionists also emphasize that in the presence of multiple departures from perfect competition, the piecemeal fix proposed in the prescriptive approach may make matters worse rather than better.

Advocates of the descriptive approach have debated whether to use the producer interest rate i (the private rate of transformation between investment today and investment in the future), the consumer interest rate r (equal to the producer rate after taxes), or something in between. The choice depends in large part on the degree of distortion introduced in the tax system.

The rate of return on corporate capital, equities, and even bonds can be thought of as including a risk premium for various uncertainties, including the risk of inflation. The very low return on short-term government bonds has the lowest risk component and, some would argue, is closer to the risk-adjusted rate we are seeking.

4.3.2 Returns to investment and discount rate estimates – Descriptive approach

A review of World Bank projects estimated a real rate of return of 16% at project completion; one study found returns of 26% for primary education in developing countries. Even in the OECD countries, equities have yielded over 5% (after corporate and other taxes) for many decades, which is comparable to a pretax rate of at least 7% (see Table 4.1).[17] Note that although average rates of return are observed, decisions are based on marginal rates of return.

Table 4.1. *Estimated returns on financial assets and direct investment*

Asset	Period	Real return (%)
High-income industrial countries		
Equities	1960–84 (a)	5.4
Bonds	1960–84 (a)	1.6
Nonresidential capital	1975–90 (b)	15.1
Govt. short-term bonds	1960–90 (c)	0.3
U.S.		
Equities	1925–92 (a)	6.5
All private capital, pretax	1963–85 (d)	5.7
Corporate capital, posttax	1963–85 (e)	5.7
Real estate	1960–84 (a)	5.5
Farmland	1947–84 (a)	5.5
Treasury bills	1926–86 (c)	0.3
Developing countries		
Primary education	various (f)	26
Higher education	various (f)	13

Sources: Quoted in Nordhaus, 1994: (a) Ibbotson and Brinson, 1987, updated by Nordhaus, 1994; (b) UNDP, 1992, Table 4, results for G-7 countries; (c) Cline, 1992; (d) Stockfisch, 1982; (e) Brainard *et al.,* 1991; (f) Psacharopoulos, 1985.

4.4 Conflicts Between the Two Approaches

Much of the disagreement between the prescriptionist and descriptionist views turns on the question of compensation among generations. The descriptive approach assumes compensation from one generation to another for any loss of environmental amenities, implicitly leaving unanswered whether compensation is likely to occur.[18] The prescriptionist view implies not only that transfers to future generations are constrained, but that climate change policies are the only way to make these transfers (Manne, 1994). The descriptionist view argues for choosing the path that maximizes consumption, making transfers among generations separately out of the larger present value of consumption. The alternative – overriding market prices on ethical grounds – opens the door to irreconcilable inconsistencies. If ethical arguments, rather than the revealed preferences of citizens, form the rationale for a low discount rate, cannot ethical arguments be applied to other questions? If it is argued, on ethical grounds, that it is unethical to pay rents (royalties) to oil companies, does that mean that cost-benefit calculations should use $2 for the price of oil (Nordhaus, 1994)?

4.5 Conclusion: What Can Discounting Contribute to Climate Change Analysis?

The prescriptive approach can be interpreted as doing as much as is economically justified to reduce the risk of climate change; the descriptive approach can be interpreted as maxi-

BOX 4.1: EXAMPLE OF PROJECT EVALUATION USING PRESCRIPTIVE AND DESCRIPTIVE APPROACHES

Suppose a greenhouse mitigation project is under consideration. If undertaken now, it will cost $1 million. If not undertaken, a new sea wall might be required in year 50, costing $10 million. If it is necessary, building a sea wall would avoid damages of $1 million per year.

Capital cost	$1 million
Time until damages begin	50 years
Cost of sea wall, year 50	$10 m
Avoided damages, years 50, 51, 52, 53, . . .	$1 m/yr
Opportunity cost of capital	5%

The decision maker has four options:

(a) Do nothing (year 0), do nothing (year 50)
(b) Do nothing (0), build sea wall if necessary (50)
(c) Mitigation project (0), do nothing (50)
(d) Other investment (0), build sea wall if necessary (50)

The stream of benefits is as follows:

Option (year)	0	...	50	51	52	...
(a)	0	...	0	0	0	
(b)	0		-10	1.0	1.0	...
(c)	-1		1.0	1.0	1.0	...
(d)	-1		11.5	1.0	1.0	...
			-10			
			=1.5			

At discount rates below 10%, option (b) dominates option (a) – if the sea level rises, it is better to build the sea wall than do nothing. Option (d) dominates option (c), as investing the $1 million in year 0 at 5% yields $11.5 million in year 50, enough to build the sea wall with $1.5 million left over. Thus, the descriptive approach would point to option (d) or (b). But option (d) may be institutionally infeasible, as there may be no way to put aside $1 million today and leave it untouched for 50 years as a Fund for Future Greenhouse Victims. If (d) is infeasible, as advocates of the prescriptive approach might suggest, then the decision maker must choose between (b) and (c). In summary, then, descriptionists would choose between (b) and (d), whereas prescriptionists would choose between (b) and (c). In either case, the choice will depend on the value attached to the consumption between years 1 and 49, which depends on the consumption rate of discount.

average rates of return to capital. Refinements to the descriptive approach would take into account limitations on intergenerational transfers, including the absence of lump sum redistributive taxes.

The discount rate is particularly important in climate change analysis. Because of the very long times involved in climate change decisions, the choice of a discount rate powerfully affects the net present value of alternative policies, and thus the policy recommendations that emerge from climate change analysis.

The prescriptive approach tends to generate relatively low discount rates and thus favours relatively more spending on climate change mitigation. The descriptive approach tends to generate somewhat higher discount rates and thus favours relatively less spending on climate change mitigation.

Although economists support the use of discounting for climate change analysis, they continue to debate which of the two approaches is correct, and the parameters to be used in calculating the rate. These choices in turn significantly affect the conclusions of the analysis.

Annex 4A: Methodological Notes on Discounting

4A.1 Intertemporal maximization of well-being

In an influential series of articles, Koopmans (1960) conducted a series of thought experiments on intertemporal choice to see the implications of alternative sets of ethical assumptions in plausible worlds. He suggested that we can have no direct intuition about the validity of discounting future well-being, unless we know something concrete about feasible development paths.

Koopmans considered the set of feasible consumption paths (from the present to the indefinite future) and the corresponding set of welfare or "well-being" paths. These paths could then be ordered to select the optimum path of well-being, according to the criterion:

$$z = \int_{t=0}^{\infty} W(c_t) e^{-\rho t} dt \qquad (4A.1)$$

with $\rho > 0$, where W is welfare, and c_t is consumption at time t. Correspondingly, the discount rate for the time path of consumption is:

$$i_t = i(c_t) = \rho + \theta(c_t) [dc_t/dt]/c_t \qquad (4A.2)$$

where $\theta(c_t)$ is the elasticity of marginal well-being, or marginal utility, at time t (Arrow and Kurz, 1970). (Note that whereas the main text treats this term as a constant, it is explicitly considered to vary with the level of consumption in the treatment here.) Along a full optimum path, the consumption rate of discount equals the productivity of capital (i.e., the social rate of return on investment; in this case, i_t equals the producer rate of interest). This is the Ramsey Rule (Ramsey, 1928).

A convenient form of W is one giving a constant elasticity of marginal utility, such as:

$$W(c) = c^{-\theta} \qquad (4A.3)$$

mizing the economic resources available for future generations and allowing them to decide how to use the resources. Both include the opportunity cost of capital – directly in the case of the descriptive approach, indirectly in the case of the prescriptive approach, which takes account of the full impact on consumption and, thus, of the cost of any displaced investment (see example of project evaluation in Box 4.1). The prescriptive approach looks at the risk-adjusted marginal return to capital, which may be considerably lower than observed

As discussed in the text, the greater the rate of pure time preference (ρ), the lower the weight accorded to future generations' well-being relative to that of the present generation. Mirrlees' (1967) computations introduced this possibility ($\rho > 0$) as a way of countering the advantages to be enjoyed by future generations should the productivity of capital and technological progress prove to be powerful engines of growth.

A higher value of θ means greater emphasis on intergenerational equity. As $\theta \to \infty$, the well-being function in (4A.1) resembles more and more the Rawlsian max-min principle; in the limit, optimal growth is zero.

In (4A.3), $W(c)$ has no minimum value. If $\rho = 0$, this ensures that very low future consumption rates would significantly affect aggregate intertemporal welfare. On the other hand, if ρ were positive, low rates of consumption by generations sufficiently far in the future would not be penalized by the optimal path criterion in (4A.1). This means that unless the economy is sufficiently productive, optimal consumption will tend to zero in the very long run. Dasgupta and Heal (1974) and Solow (1974a) showed in a model economy with exhaustible resources that optimal consumption declines to zero in the very long run if $\rho > 0$ and in the absence of technical change, but that it increases to infinity if $\rho = 0$.

It is in such examples that notions of sustainable development can offer some analytical guidance. If by sustainable development we mean that the chosen consumption path should never fall short of some stipulated positive level, then it follows that the value of ρ would need to be adjusted downward in a suitable manner to ensure that the optimal consumption path meets the requirement. This was the substance of Solow's remark (see Solow, 1974b) that in the economies of exhaustible resources the choice of ρ can be a matter of considerable moment.

So far an assumption underlying this discussion has been that well-being or utility has not been bounded. If we impose bounds on well-being, other results obtain, because of the mathematical properties of the space of bounded sequences. For such sequences present value calculations are not rich enough to capture all the subtleties of evaluation of a utility stream. Instead, one must add another term to the present value. This second term will in general have the form of a long-term average. It could be approximated by minimum requirements for the long-run stocks of environmental resources. This formulation attempts to account for both basic levels of human needs and limitations on total resources.

4A.2 *Consumption versus investment discount rate*

Sandmo and Dreze (1971) address the choice of the correct rate of discount to use in the public sector when there are distortions in the economy, for example, in the form of taxes, which prevent the equalization of marginal rates of substitution and transformation in the private sector. Under certain assumptions, the corporate tax drives a wedge between the marginal rate of time preference of consumers and the marginal rate of transformation in private firms.

They find that for a closed economy:

$$1+r < 1+i < 1+[r/(1-t)] \qquad (4A.4)$$

where r is the consumer interest rate, t is the tax rate, and i is the public sector's discount rate. This rate should thus be a weighted average of the rate facing consumers and the tax-distorted rate used by firms. Since $1 + r$ measures the marginal opportunity cost of transferring a unit of resources from private consumption, and since $1+[r/(1-t)]$ is the measure for transfers from private investment, a unit of resources transferred from the private to the public sector should be valued according to how much of it comes out of consumption and how much out of investment.[19]

The general idea of the prescriptive approach is to calculate impacts on consumption and to find the appropriate discount factor for discounting those changes. We are, in effect, taking consumption as our standard of measure. This is convenient and natural, but there are other ways of performing the calculations, using other measures. If these other measures are used, relative prices over time (discount factors) will differ from those associated with the consumption measure.

By the same token, if, for example, systematic relationships exist between the outputs and inputs of a project and the total changes in consumption they induce, and if consumption changes over time, then instead of discounting total consumption impacts at the SRTP, one could calculate the direct impacts using another discount factor. The discussion above of the Sandmo-Dreze formulation is a case in point. These alternatives do not provide prescriptions, only alternative formulas for arriving at the same point.

The discrepancy between public evaluation of a marginal dollar to future generations and individuals' own intertemporal evaluations can arise even in the case of very simple social welfare functions. Thus, assume that there is a utilitarian social welfare function, which simply adds up the utility of successive generations, and for simplicity, assume each generation lives for only two periods. The tth generation's utility is represented by a utility function of the form:

$$U^t(c^t_t, c^t_{t+1}) \qquad (4A.5)$$

where the first argument refers to consumption during the first period of the individual's life, the second to consumption during the second period. Then observed market rates of interest refer to how individuals are willing to trade off consumption over their own life. These may or may not bear a close correspondence to how society is willing to trade off consumption across generations. The former (the investment discount rate) corresponds to U^t_2/U^t_1, whereas the latter (the consumption discount rate) corresponds to U^{t+1}_1/U^t_1.

If the government has engaged in optimal intertemporal redistribution and does not face constraints in imposing lump sum (i.e., nondistorting) taxes on each generation, then the two discount rates will be the same and equal to the marginal rate of transformation (in production, i.e., the return to investment). But whenever either of these conditions is not satisfied, then market rates of interest facing consumers (measuring their own marginal rates of substitution) need bear no close relationship to society's marginal rate of substitution across generations. Diamond and Mirrlees (1971) show that if the only reason for the discrepancy between producer and consumer interest rates is optimally determined commodity taxes, and there are no after-tax profits, possibly because there is a 100% pure profits tax, then the government should use the producer rate of interest. Stiglitz and Dasgupta (1971) have shown that this result does not hold if either of these assumptions is dropped.

Under certain circumstances (in particular, the existence of optimal intergenerational lump sum transfers), asymptotically the producer rate of interest will equal the pure rate of time preference of society. More generally, when the government must resort to distortionary taxes, not only is this not true, but the rates of discount employed may reflect distributional considerations (see Stiglitz, 1985).

4A.3 The social rate of time preference

As stated in the main text, the social rate of time preference (SRTP) is composed of pure time preference (ρ) and a discount rate that takes into account falling marginal utility as the level of consumption rises (θg); or $SRTP = \rho + \theta g$.

Pure time preference. The earliest economics literature, in addressing these issues, argued that the appropriate value of ρ was zero (Ramsey, 1928). Ramsey based his argument on the ethical presumption that all individuals, including those living in different generations, should be valued the same. The argument since then has advanced only slightly. Some have argued that the discount rate should be adjusted for the probability of extinction. Plausible estimates of this effect would add very little to the discount rate. Others have pointed out that a positive discount rate is needed for acceptable optimization results: In the absence of a discount factor, the sum of future utilities may be infinite, so that the mathematics of maximizing a social welfare function are ill-defined. Because even a very small positive discount rate, however, would resolve the mathematical issue, this objection has little practical moment.

In a society in which income levels are not expected to rise, impatience may still cause a household (or the present generation) to discount the future (generation), that is, to prefer consumption today to consumption tomorrow; in discounting terms, this means equating a smaller amount of consumption today with a larger amount in the future. In his classic paper on optimal saving, Ramsey (1928) judged that any allowance for pure time preference ($\rho > 0$) "is ethically indefensible and arises merely from the weakness of the imagination." Correspondingly, he argued that future generations should have equal standing with the current generation; there was no moral or ethical basis for giving less weight to the welfare of future generations than to that of the current generation.

For an individual, some nonzero value of pure time preference can make sense, because he or she has a finite life and thus uncertainty about being alive to enjoy future consumption. Nonetheless, for a life span of 70 years, pure time preference at even 1% per annum implies that consumption at the end of life is worth only half that at the beginning. Evidence also suggests that individuals' discount rates may change over

time, with lower discount rates being used for longer time horizons.

Considerations for society as a whole are different. The social welfare function approach asks: If society values different generations in a particular way (reflected in the social welfare function), how should changes in consumption in different generations be compared? Ramsey's analysis focussed on the ethical presumption that consumption by all generations should have equal value. But this does not exclude the possibility that, as a matter of *description,* the current generation gives less value to consumption of future generations.

Diminishing marginal utility. The second term on the right side of equation (4.1) (θg) raises two questions. First, what are reasonable expectations concerning increases in per capita income (growth rate g in the equation)? Second, how should intertemporal differences in expected consumption per capita be translated into social weights, that is, marginal valuations of dollars of future income? This second question refers to the parameter θ, the elasticity of marginal utility, which tells how rapidly the additional utility from an extra unit of consumption drops off as consumption rises.

No consensus on the first question has emerged. Although no consensus has emerged on the answer to the second question, there is a generally accepted method for approaching the issue. The evaluation of any individual's consumption can be summarized by a utility function of the form $U = U(c)$ where the parentheses indicate that U, utility, is a function of c, per capita consumption. Marginal utility is positive ($U'(c) > 0$), but declines as consumption rises ($U''(c) < 0$). A new shirt, for example, benefits a pauper more than a prince. That is why if consumption of some future generation is higher, the marginal valuation of its consumption will be lower. The question is, how much lower? Formally, the answer is given by the elasticity of marginal utility (θ) or:

$$[dU'/U']/[dc/c].$$

Individuals in their day-to-day decision making reveal information about their perceptions concerning their own utility functions in at least two different contexts: behaviour towards risk and behaviour towards intertemporal allocation of consumption. In both contexts, there seems to be a consensus that elasticities of marginal utilities lie in the range of 1 to 2, even though the empirical studies require strong assumptions about the specific form of the utility function (symmetric and time separable). Thus, one of the most commonly used utility functions, the logarithmic, implies $\theta = 1$, meaning that if income rises by 1% the marginal utility of consumption falls by 1%. Attempts by Fellner (1967) and Scott (1989) to estimate this elasticity place it somewhat higher, at 1.5, whereas recent estimates reviewed by Pearce and Ulph (1994) place it in the vicinity of 0.8.

Just as the choice of the rate of pure time preference (ρ) has important implications for intergenerational equity, as discussed above, so does the choice of the elasticity of marginal utility. The more weight the society gives to equity between generations, the higher the value of θ. Thus, a value of, say, 3, would mean that it would require a 30% rise in the next generation's per capita consumption to warrant a 10% reduction in that of the present generation; or, under a bleaker outlook, that if the future generation is expected to be poorer than the present, the present would be prepared to accept a 30% reduction in consumption to secure a 10% increase in that of the future generation (so long as the two relative consumption levels did not reverse). Even $\theta = 1$ gives some emphasis to equity, however. When $\theta = 1$, a 10% reduction in the richer generation's income will be an acceptable trade-off for a 10% increase in that of the poorer generation, even though the absolute reduction of the one exceeds the absolute increase of the other (because the absolute consumption base of the one is larger than that of the other).

Risk. Utility may also be discounted for risk. The standard treatment of risk in models involving impacts over a single individual's life is not to raise the discount rate for riskier projects, but instead to convert probabilistic consumption patterns into their certainty equivalents and then discount the results at the standard rate. The same should be true for the pure time preference component of the SRTP when discounting across generations. This component should remain unchanged with respect to risk, and the influence of risk should instead be incorporated in the stream of expected consumption effects.

There would seem to be an argument for varying the growth-based component of the SRTP with respect to risk, however. If there is uncertainty about the rate of per capita income growth, g, then consider the effect on the component θg in the SRTP. Suppose two scenarios each have 50% probability: per capita income growth of 1% and per capita growth of 2%. There will be two resulting possible streams of marginal utility over time. The stream of expected value of marginal utility will be the average of these two streams. But if marginal utility is a convex function of consumption, this average will be greater than the stream of marginal utility generated by considering the simple average growth rate over time, 1.5%. That is, with diminishing marginal utility, at any point in time marginal utility along the path for 1.5% growth will be closer to that of the 2% growth path than to that of the 1% growth path. Correspondingly, the expected marginal utility path lying halfway between the two scenarios will coincide with the marginal utility stream for a growth rate closer to 1% than to 2%. Essentially, the expected value of marginal utility is greater than the marginal utility of expected income. On this basis, there would be grounds for reducing the growth-based component of the SRTP under circumstances of risk. Because the risk in predicting per capita growth on centuries-scale horizons is high, this consideration is particularly relevant for the problem of global warming.

Other arguments. Empathetic distance provides another rationale for discounting. Rothenberg (1993) and Schelling (1993) have suggested that although nonzero pure time preference might make sense for an initial two or three decades, beyond a certain future point it makes no sense to apply further discounting of consumption for pure time preference. Thus, "as the future recedes . . . single generations come to be perceived more and more as homogeneous entities" (Rothen-

berg). Similarly, "time may serve as a kind of measure of distance. . . . Beyond certain distances there may be no further depreciation for time, culture, geography, race, or kinship" (Schelling). A graph of the fraction of face value accorded to each successive generation (for constant real consumption) would thus be a series of declining, successively shallower steps that eventually reach a horizontal plateau. A deep plateau signifies major discounting for empathetic distance; a horizontal line beginning and remaining at unity is the zero pure time preference rate across generations recommended by Ramsey. Policy based on empathetic distance (a shelf lower than unity) may be more defensible in a normative sense when the action is refraining from conferring a windfall gain (as in penurious aid budgets) than when it involves the imposition of windfall damage (as in global warming's effects on future generations).

Another argument for nonzero pure time preference is that setting the rate at zero could imply that the present generation should accept near-starvation consumption levels and correspondingly low utility because, with even very small returns on investment, an endless stream of future generations could enjoy increased consumption and (to a lesser degree) utility as a result. To some extent, however, this concern is already addressed in the overall discount rate equation (4.1). As noted, the first term in that equation discounts utility (pure time preference), but the second term additionally discounts consumption to take account of falling marginal utility. The present generation is protected against an optimizing programme setting its consumption near zero if the elasticity of marginal utility (θ) is large enough and marginal utility drops off fast enough to rule out impoverishment of the present generation for gains to future generations.

Koopmans (1966) and Mirrlees (1967) have expressed the concern that zero time preference would imply unacceptably low levels of current consumption, and even no consumption under some circumstances. Even positive but very low discount rates might, in the absence of technological progress, lead to unreasonably high savings rates. (This illustrates a general problem with models founded on utilitarianism: They may imply very large sacrifices from one generation or group.) These models might well be seen as providing arguments that the rate of time preference is greater than zero, though they do not go far in specifying its proper magnitude.[20]

4A.4 The social welfare function

Economists have long debated the equity of discounting distant future benefits (Ramsey, 1928; Mishan, 1975; Rawls, 1971; Sen, 1982). The usual approach to issues of equity since Bergson (1938) has involved the choice of a social welfare function, and arguments about the choice among alternative social welfare functions have turned on the ability to derive a particular function from sound theoretical principles (seemingly plausible axioms) and on the resulting reasonableness of its derived implications.

Although all social welfare functions have been criticized for assuming interpersonal comparability of utility, there seems to be no way of addressing the ethical issues involved in making decisions affecting different generations without making some assumptions implicitly or explicitly about interpersonal comparability. Two polar views are represented by the utilitarian approach, in which social welfare is the sum of utilities, and the Rawlsian approach, in which social welfare reflects the welfare of the worst-off individuals. Whereas the utilitarian approach can be derived from what many view as a persuasive axiomatic (theoretical) structure (Harsanyi, 1955), the Rawlsian approach is derived from a "max-min" strategy (maximize the minimum outcome for any given party). Although this strategy is popular in game theory, it does not rest on widely accepted axiomatic principles.

The Rawlsian max-min principle is the strongest in assuring (the least fortunate groups of) future generations levels of consumption at least as great as that of (the least fortunate groups of) the current generation. It is consistent with the Brown-Weiss (1989) approach noted below. The max-min criterion permits inequality in consumption between individuals (or in this case, between generations) only if it improves the position of the poorest. In the absence of technical change this would imply that consumption per head should be the same for all generations. By contrast, the utilitarian criterion allows future consumption, in principle, to fall below current consumption, provided the current generation is made sufficiently better off as a result. Correspondingly, it also allows for decreases in present consumption, provided the future generation is made sufficiently better off as a result.

The Rawls and utilitarian social welfare functions can be viewed as limiting cases of more general social welfare functions embracing social values of equality (Atkinson, 1970; Rothschild and Stiglitz, 1973). In practice, so long as there is sufficient scope for technological change, optimizing any egalitarian social welfare function over time yields increases in consumption per capita. Moreover, with any of the approaches, earlier generations are entitled to draw down the pool of exhaustible resources so long as they add to the stock of reproducible capital.

Within the individualistic utilitarian social welfare approach, there is still the question of the appropriate value of ρ, the pure rate of discounting future utility relative to current utility. Ramsey and others have argued that there exists no ethical basis for treating different generations differently; thus ρ should be zero. The individualistic social welfare function, accepted by most economists as the basis of ethical judgments, accepts individuals' own relative valuations of different goods. It does not place separate valuations on unequal access to particular goods, other than through their effects on the affected individuals. Although this probably represents the consensus view, some economists have insisted that for particular goods, individuals' valuations need not be the basis of societal valuations. For instance, Tobin (1970), in what he called specific egalitarianism, argued that society might argue for greater equality in distribution of health care than would be reflected in individuals' own evaluations. Most economists, however, reject this view.

Sen (1982) similarly suggests a basis for not discounting when environmental effects are in question. He argues that a fundamental right of the future generation may be violated when the environment is degraded by the present generation, and that the resulting "oppression" of the future generation is inappropriate even if that generation is richer than the present and has a lower marginal utility of consumption. In this framework, intertemporal equity for environmental questions requires "a rejection of 'welfarism,' which judges social states exclusively by their personal welfare characteristics." It should be noted that this recommendation leads to paradoxes and inconsistencies.

4A.5 *Departures from "first-best" assumptions*

Analysis of optimal tax and expenditure policy occurs in a hypothetical "first-best" world, with complete markets and optimal redistribution policy (i.e., in which redistribution occurs only through lump sum taxes that do not change relative prices). In such a world, the discount rate will equal the marginal product of capital (i.e., the value of the additional output resulting from an additional unit of investment), which will equal the interest rate faced by both producers and consumers (Lind *et al.*, 1982).

Because the real world economy may differ in important respects from the first-best world, the literature also addresses departures from the first-best assumptions. Taxes drive a wedge between i, what producers pay to borrow, and r, what consumers receive on their savings. If money for public investment comes entirely from other investment, then the discount rate should be the producer interest rate i. If the money comes entirely from consumption, then the discount rate should be the consumer interest rate r. If the money comes partly from investment and partly from consumption, then the appropriate discount rate will fall somewhere between r and i; the exact answer requires an explicit analysis of how climate policy affects investment and consumption.

In the general case in which costless intergenerational transfers are not possible, no single discount rate can be applied. Rather, project-specific discount rates are required. Market rates are no longer a reliable indicator of the appropriate discount rate, which may be greater than or less than the before-tax return on investment (Stiglitz, 1982). In this general case, no theoretical rule connects the discount rate to any observed market rate, although market rates still contain valuable information that should be used in arriving at a discount rate.

Economists have long recognized that a competitive market equilibrium yields a Pareto-efficient outcome under appropriate conditions (perfect competition, no externalities, etc.). The distribution of income that it yields, however, does not in general maximize any particular social welfare function. It is a well-recognized function of government, therefore, to intervene in the distribution of income, for example, by establishing programmes for the very poor. Prescriptionists note that the *intertemporal* distribution of welfare that emerges from the market will not, in general, maximize any particular social welfare function either. Although it is a legitimate function of government to intervene to change the intergenerational distribution of welfare, there is no presumption that the government has in fact intervened to make the observed resource allocations maximize intertemporal social welfare. Moreover, in the case of climate change, no one government exists to make these decisions.

Prescriptionists emphasize that the market rate of interest – the relative price of consumption of one generation in one year of its life to its consumption in another year – will not in general equal the SRTP. In standard life-cycle models, with no technological progress and an economy in steady state, there would be no discounting for society's purposes: Each generation is identical, so the marginal utility of consumption of each is the same. Nonetheless, the market rate of interest will be positive in any efficient equilibrium under certain reasonable assumptions about utility functions (such as individual impatience and zero bequest motive; Diamond, 1965). In such models the market rate of interest would thus always overestimate the SRTP. Under some special conditions, with governments intervening with nondistorting taxation to optimally redistribute income across generations, observed market rates of interest will accord with the SRTP. But these are highly specialized conditions (see Stiglitz, 1985; Pestieau, 1974). The market rate of interest remains relevant because it reflects the opportunity cost of capital, which strongly affects the changes in consumption generated by any change in policy.

4A.6 Special considerations for discounting in government projects

A large literature has debated whether, for small changes in consumption levels, observed rates of interest provide the appropriate basis of trading off government expenditures and changes in consumption of individuals of different generations at different dates. In the simplest case, in which there is no taxation, there are no market distortions, and a single individual living forever (or else "dynastic" utility functions in which individuals take full account of their descendants' welfare), society's intertemporal discount rate will correspond to that of the representative individual, and his or her trade-offs across time would be given by the market rate of interest. But these assumptions are not generally satisfied, as evidenced by the marked discrepancy between the lower interest rates on savings typically facing consumers and the higher rates earned on investments by producers.

Some of the disagreement arises from confusion about what is being discounted. The social discount rate approach discounts changes in consumption at different dates. The producer interest rate approach discounts direct cash flows from the project. The two need not be inconsistent.

If a government is comparing two projects of equal cost, producing a result in the same year, then a comparison of the rates of return would provide an appropriate basis for choosing among projects. Cline (1992) proposes using a shadow price of capital set equal to the present discounted value of an annuity paying equal annual installments over a lifetime of N years (set at 15 years for the lifetime of typical capital equipment), with a return of r equal to the rate of return on capital, and discounted at the SRTP. With plausible ranges for N, r, and SRTP, the shadow price of capital can range from 2 to over 10 units of consumption equivalent per unit of capital (Lyon, 1994).

If a public project were to displace a private project of equal cost, the same reasoning would imply that the government should only undertake the public project if the rate of return exceeded the rate of return in the private sector (Stiglitz, 1982). More generally, when the government undertakes a project, complex general equilibrium effects can be expected. The full consumption effects of these changes (or their consumption equivalents) need to be calculated, and then discounted using the SRTP (Arrow and Kurz, 1970; Feldstein, 1970; Bradford, 1975; Stiglitz, 1982). This approach uses a shadow price of capital to convert all investment effects into their consumption equivalents, and then uses the SRTP to discount the resulting stream of consumption equivalents (Lind *et al.*, 1982; Gramlich, 1990).

For some projects, an adjusted discount factor, the public sector discount rate, is appropriate. A large literature addresses how the adjustment is to be made. One approach emphasizes the effects on consumption versus investment, deriving a weighted average of the consumption and investment rates of return, with weights depending on the respective importance of the sources of finance (Sandmo and Dreze, 1971).

4A.7 The environment and discounting

The essence of social discounting is to convert all effects into their consumption equivalents at the proper relative prices and then to discount the resulting stream of consumption equivalents at the social rate of time preference. Incorporating environmental effects thus does not change the discount rate itself, but does require special attention to the proper relative pricing of environmental goods over time. Although there is a generally accepted approach to valuing goods, there is less consensus concerning valuation of environmental impacts, other than those valued solely for their impacts on the production of goods.

The question is addressed within the public finance literature in terms of the valuation of public goods. Assume consumers have utility functions of the form $U = U(c,G)$ where G is some public good (e.g., quality of the environment). Then marginal rates of substitution between different values of c at different dates may bear no correspondence to marginal rates of substitution between different values of G at different dates. This implies that there is no justification for discounting environmental degradation at market rates of interest. The appropriate procedure entails converting the environmental change into equivalent contemporaneous consumption benefits and discounting those.

Technical progress and structural change over the past several decades have resulted in improvements in several measures of environmental quality in the developed countries. Moreover, recorded reserves of many "exhaustible resources" have actually increased over the past century, accompanied by a fall in their real prices. This provides evidence that continued growth in per capita incomes will result in improved environmental quality in at least some dimensions. Some have supposed, however, that environmental degradation will occur as society grows (Weitzman, 1993). If this occurs or if the environment is an income elastic good on which people are willing to spend relatively more as their income rises, then the marginal rate of substitution between environmental quality and private goods will systematically change over time, towards a higher relative marginal value of the environment. The result is equivalent to using a low (or even negative) discount rate for environmental amenities with prices unchanged. However, this process involves properly valuing future environmental benefits in arriving at the future flow of equivalent consumption, and does not change the discount rate to apply to the consumption stream.

Much of the environmental literature critical of cost-benefit analysis, in contrast, argues for a zero discount rate without seeming to recognize the distinction between a zero rate of pure time preference (ρ) and a zero discount rate (see, e.g., Daly and Cobb, 1989). But from equation (4.1), so long as consumption growth is positive there will be a nonzero SRTP. Similarly, some modern philosophers seem to make the same mistake (e.g., Parfit, 1983; Cowen and Parfit, 1992).

Finally, there has been considerable discussion about the proper discounting method for environmental projects of institutions such as the Global Environmental Facility of the World Bank (see, e.g., Munasinghe, 1993). The method that follows from the social cost-benefit approach is to obtain consumption equivalents of the environmental effects over time and then apply the appropriate discount rate. Within a fixed institutional investment budget, it may be that the collection of potential projects that successfully passes a cost-benefit test on this basis more than exhausts available funds. If so, efficient trade-offs within the menu of projects will appropriately involve cutoffs at a higher shadow price in funds drawn from the institutional budget – but always with benefit evaluation based on the consumption equivalence principle just outlined.

4A.8 Discounting and sustainable development

The Brundtland Commission called for "sustainable development," defined as economic activity that "meets the needs of the present without compromising the ability of future generations to meet their own needs" (United Nations, 1987). Similarly, Brown-Weiss (1989) argued from the standpoint of international law that "each generation is entitled to inherit a planet and cultural resource base at least as good as that of previous generations."

A consensus exists among economists that this does not imply that future generations should inherit a world with at least as much of *every* resource. Such a view would preclude consuming any exhaustible natural resource. The common interpretation is that an increase in the stock of capital (physical or human) can compensate for a decline in the stock of a natural resource. Under most calculations, given the savings rates of all but the lowest-saving countries in the world, most countries now pass this test of sustainability.

Economics has recognized the concept of sustainability for some time. Hicks (1938) used the idea in defining net national income. Neoclassical growth theory (Phelps, 1961; Robinson, 1962) advanced the idea of sustainability in its formulation of the "Golden Rule," which is that configuration of the economy giving the highest level of consumption per head that can be maintained indefinitely. A recent extension has proposed the "Green golden rule" (Beltratti *et al.,* 1993). The recent economic debate on sustainable development has focussed on two issues: (1) intertemporal equity and (2) capital accumulation and substitutability. The extent to which natural and cultural resources are substitutable is critical to this analysis and is contentious. Many economists (for example, Pearce and Turner, 1990) stress the need for sustainability limits on the use of resources that future generations will need but cannot create.

Robert Solow's definition of sustainable development (Solow, 1992) focusses on intertemporal equity: Sustainable development requires that future generations be able to be at least as well off as current generations. Sustainable development does not preclude the use of exhaustible natural resources, but requires that any use be appropriately offset. Likewise, any environmental degradation must be offset by an increase in productive capital sufficient to enable future generations to obtain at least the same standard of living as those alive today.

Solow's definition, and much of economic theory to date, implicitly assumes that substitutes exist or could be found for all resources. If substitution possibilities are high, as most evidence from economic history indicates, then no single resource is indispensable, and intertemporal equity stands as the only crucial issue (Pearce and Turner, 1990). If, on the other hand, human and natural capital are complements or only partial substitutes, then different classes of assets must be treated differently, and some assets are to be preserved at all costs.

In many developing countries, Solow's definition would not be viewed as acceptable, since it seems to place no weight on their aspirations for growth and development. Further, formal models analyzing optimal development paths using a max-min (Rawlsian) criterion would focus exclusively on the welfare of the less developed countries. (Note that in Rawls' formulation, $\theta = \infty$.) But the remedy would be simple: immediate massive redistribution from the developed to the developing countries, with no environmental justification required. Even if there were limits on the transfers, this remedy would suggest that all the costs of mitigation – including those occurring within the developing economies – be borne by the developed countries.

Even the utilitarian approach ($\theta < \infty$) would tend to lead to higher general income transfers to poor countries than presently observed. Adherents of the descriptive approach might ask why the utilitarian construct is appropriate when considering intergenerational equity (as in the identification of the SRTP suggested in equation (4.1)) if it is not applied in practice across or even within countries now. In one sense, this question is another application of the principle suggested above, that in the absence of optimal redistribution intervention by government, observed market rates (in this case of transfers from rich to poor nations) will not necessarily or likely equal social rates. Alternatively, the equity norm suggested here may not be widely shared by governments or voters.

Despite the political constraints on present-day transfers from rich to poor countries, the time-discounting concepts of the utilitarian approach (and the SRTP in particular) remain valid. Thus, consider a matrix with two rows, developed nations and developing nations, and two columns, present and future. The SRTP can appropriately be applied between the two columns along each row, even if there is a barrier to its application between the two rows. Leaders and electorates in developing countries have cause for concern about their descendants just as do their counterparts in developed countries. As noted above, however, the value of the SRTP is likely to be higher for the developing nations than for the developed.

Endnotes

1. Identifying the appropriate discount rate has been discussed in the context of general cost-benefit analysis (Chapter 5) for many years (Dasgupta *et al.*, 1972; Harberger, 1973; Little and Mirrlees, 1974; Sen, 1967; Stiglitz, 1982). More recently, social scientists have debated the precise rate to use for global climate analysis (Broome, 1992; Cline, 1992; Nordhaus, 1991).

2. Direct intergenerational transfers could be made through a fund to compensate future greenhouse victims; without some such mechanism, however, there is no guarantee that such transfers will be made. Compensation will occur indirectly, however, if we bequeath a richer economy to our children and grandchildren.

3. In particular, an efficient policy is unequivocally better than an alternative only if those who are made better off under the efficient policy actually compensate those made worse off. More general treatments of cost-benefit analysis do incorporate distributional considerations.

4. When applied to discounting the utility of different generations, ρ is referred to as the social rate of pure time preference.

5. The *certainty equivalent* is the certain result that would make an individual indifferent between it and the uncertain outcome. Issues of equity can be treated analogously through the use of "equity equivalents" (Atkinson, 1970; Rothschild and Stiglitz, 1973).

6. The alternative view, which could be called environment-specific egalitarianism, says that each good must be valued in isolation from all others. This view stresses the need for limits to the use of resources that will be needed or desired, but cannot be created, by future generations (Pearce and Turner, 1990). In the extreme, this belief, known as specific egalitarianism, argues that environmental goods (and in some cases, each environmental good) must be treated separately from all other goods and that each generation should enjoy the same level of environmental benefits as previous generations.

The mainstream view in economics holds that future generations can be compensated for decreases in environmental goods by offsetting accumulations of other goods (though increasing scarcity of some environmental goods will require increasing amounts of capital to offset the loss of an additional unit of the environmental good). Environmentalists may favour restricting the use of nonreproducible environmental resources in a way entirely consistent with the mainstream view, in that risk aversion in the matter of environmental quality will affect the rate at which society trades environmental goods for other goods. Only in the limiting case of infinite risk aversion will no trade-offs be made. Thus, adherents of environment-specific egalitarianism may back the same policies as risk-averse adherents of the mainstream view.

A related issue is whether decision makers should accept the current generation's valuation of the future benefits of environmental goods, as reflected in the market. Even those who believe the answer is no may accept trading off environmental for other goods, though those trade-offs may not be well reflected in current market prices.

7. The economist Thomas Schelling (1993) argues against the way discounting is generally applied to climate change projects. Schelling notes that discussions of discounting within the context of climate change policy often confuse three ideas: (1) discounting for consumption enjoyed in the future; (2) discounting for risk; and (3) discounting for consumption by others.

Schelling points out that one thinks differently about one's own consumption than about the consumption of others, and that a key feature of the climate change problem is that those likely to bear the cost of mitigation (the developed countries) differ from those likely to enjoy the benefits (the currently developing countries). Thus, says Schelling, we should recognize that climate change mitigation is more like foreign aid than it is like the usual public investments to which we apply discounting. Foreign aid budgets are low because the donors do not have strong feelings of concern for the beneficiaries. In the absence of evidence to the contrary, says Schelling, there is no reason to impute much stronger moral sentiments to those who will be paying for climate change mitigation.

8. The empirical problem of uncertainty in forecasting *g,* the growth rate, has yet to be resolved. The post-1973 slowing of productivity increases in many OECD countries suggests the need for a reexamination of historical trends and perhaps a reduction in the recommended discount rate. These considerations have become particularly important with the addition of intergenerational distributional effects. Low-income groups within developed countries have seen a sharp reduction in per capita income growth; this would lead to lower discount rates. On the other hand, some developing countries now enjoy high per capita income growth, suggesting a higher discount rate. At 7% per capita income growth, and with $\theta = 1.5$, the discount rate would exceed 10% even with ρ set to zero.

9. Optimal tax policy is intertemporally and distributionally optimal.

10. Standard estimates put this elasticity between one and two. Such estimates are based on an additive social welfare function using elasticities of marginal utility revealed by behaviour toward risk. Though specialists debate the appropriateness of the assumptions, no generally accepted view supports a different value of θ.

11. Other factors, however, might push the calculations the other way, such as the likelihood of higher relative future damage from global warming for the developing countries (see Chapter 6).

12. This will be the path that satisfies the intertemporal efficiency conditions (Lind *et al.,* 1982):

(1) Production: the marginal rate of transformation in production between one period and the next, and thus the marginal product of capital, equals the producer rate of interest for all goods: $MRT_j (t,t + 1) = i$, that is, the marginal rate of transformation for any good *j* from period *t* to period *t* + 1 equals the producer rate of interest *i*.

(2) Consumption: the ratio of the marginal utility of consumption in period *t* to the marginal utility of consumption in period *t* + 1 equals 1 plus the consumer interest rate *r*, or $MUC_k(t)/MUC_k(t + 1) = 1 + r$.

(3) Overall: the consumer interest rate equals the producer interest rate for all goods, for all consumers, in all time periods; that is, $r = i$.

13. Critics have noted (a) that is is not in general the case that mitigation expenditures displace other forms of investment on a dollar-for-dollar basis; (b) that the second argument can be read as a statement of the compensation principle, which holds that one need not ask if compensation has actually been paid, only whether it could be paid, so that questions of distribution and efficiency can be separated; and (c) that the third argument assumes the presence of lump sum redistributive mechanisms (in the absence of which the social marginal rate of substitution may not equal the opportunity cost of capital) and a degree of rationality in collective decision making that may not be plausible. Society may not engage in optimal intergenerational redistribution; yet in evaluating a policy, it may still wish to consider explicitly intergenerational effects. Taken to an extreme, argument (c) would suggest that the social marginal utility of the rich must equal that of the poor; otherwise, governments would have redistributed income already.

14. It might be argued that resources could still be channelled to the best projects using a lower discount rate, by employing a *shadow price of capital,* reflecting the scarcity of capital. The issues of the intertemporal price and the *current* scarcity price of capital can, in principle, be separated.

15. Technically, indifference to inequality between countries at a given time implies instead that the other key parameter, the elasticity of marginal utility (θ), is zero.

16. That is, if the social welfare function implied a 2% discount rate, and the government employed policies to maximize social welfare, then the savings rate would be very high.

Manne uses a standard growth model to examine the relation between discount rates and savings rates in the context of developed economies. He finds that discount rates of 1 or 2% imply an unrealistically rapid near-term increase in the rate of investment. Manne thus concludes that a discount rate this low is grossly inconsistent with observed or plausibly anticipated behaviour. On the other hand, prescriptionists might interpret Manne's analysis as showing simply that the intertemporal equilibrium established by market economies differs markedly from that corresponding to the solution of an intertemporal maximization problem based on a social welfare function derived from ethical considerations. But even if savings could be increased enough to drive the discount rate to 1 or 2%, climate change investments would still have to compete with many other public and private investments offering higher returns.

17. Some care must be taken in inferring the appropriate opportunity cost of capital from observed market rates of return. First, many standard measures reflect average rates of return rather than the relevant marginal rates. Second, most investments carry some risk. Cline (1992) observes that investors purchase both safe government bonds yielding about 1.5% real, and stocks, yielding 5-7% real; he argues that this suggests a risk premium of 3.5-5.5%. Thus, if the average observed return to capital is 7%, and if the marginal return is less than the average (as one would expect), then the certainty equivalent opportunity cost would be less than 3.5%. On the other hand, it has also been argued that this calculation holds only if it is assumed that households allocate assets efficiently (an assumption that prescriptionists deny in other contexts); that bonds have risks quite different from either stocks or climate mitigation investments; and, thus, that this comparison is invalid (Nordhaus, 1994).

18. In contrast, the predominant view fifty years ago held that a project should be considered desirable if the winners *could possibly* compensate the losers, whether or not this compensation actually occurred (Kaldor, 1939; Hicks, 1939). This "compensation principle" (which is no longer accepted) would support the view that the discount rate should be the producer cost of capital – the rate that investments would have earned elsewhere in the economy. If a dollar invested in education, research and development, or new factories yields a return of 10%, and climate mitigation yields 5%, then converting climate mitigation investment to something more productive would yield higher total returns, implying that everyone could be made better off. The compensation principle would be satisfied. But compensation may not actually be paid, and future generations will probably not benefit from knowing that they *might have been* made better off.

Economists consider two cases: (1) Pareto improvements – changes, including compensation actually paid, that make everyone better off; these are obviously desirable; and (2) changes that produce some winners and some losers. To address the second case, economists generally use a social welfare function, typically showing some preference for greater income equality (that is, increasing equality raises social welfare). A considerable literature, building on the work of Rothschild and Stiglitz (1973) has added precision to this idea. In choosing an SWF, economists also generally assume separability. That is, the SWF can be written $W = W(U_1, \ldots)$. The ethical idea underlying this assumption is that society's willingness to substitute consumption between individuals i and j does not depend on the utility or income of individual k, a form of the assumption of the independence of irrelevant alternatives. Economists also generally assume consumer sovereignty. That is, each individual's utility (entering the SWF) is determined by that person's own judgments, not the judgments of society more generally.

19. For an open economy, the elasticity-adjusted rate on foreign loans also enters the calculus. However, for analysis of a global issue, this extension is probably inappropriate, as globally the economy is closed.

20. Alternatively, these models may suggest that the problem lies in the assumption about technical change. If little or no technical change had been the rule in recent centuries, society might have evolved toward the high savings rates that seem so implausible given actual historical experience.

References

Arrow, K.J., and M. Kurz, 1970: *Public investment, the rate of return and optimal fiscal policy,* Johns Hopkins University Press, Baltimore.

Atkinson, A.B., 1970: On the measurement of inequality, *Journal of Economic Theory,* **2,** 244–263.

Beltratti, A., G. Chichilnisky, and G. Heal, 1993: Sustainable growth and the green golden rule, National Bureau of Economic Research, Working Paper 4430, August.

Bergson, A., 1938: A reformulation of certain aspects of welfare economics, *Quarterly Journal of Economics,* **52,** 310–334.

Birdsall, N., and A. Steer, 1993: Act now on global warming – But don't cook the books, *Finance and Development,* **30**(1), 6–8.

Brainard, W.C., M. Shapiro, and J. Shoven, 1991: Fundamental value and market value. In *Money, macroeconomics, and economic politics: Essays in honor of James Tobin,* W.C. Brainard, W.D. Nordhaus, and H.W. Watts, eds., pp. 277–307, MIT Press, Cambridge MA.

Bradford, D.F., 1975: Constraints on government investment opportunities and the choice of discount rate, *American Economic Review* **65**(50), 887–899.

Broome, J., 1992: *Counting the costs of global warming,* White Horse Press, Cambridge.

Brown-Weiss, E., 1989: In *Fairness to future generations: International law, common patrimony, and intergenerational equity,* United Nations University, Tokyo.

Cline, W.R., 1992: *The economics of global warming,* Institute for International Economics, Washington DC.

Cline, W.R., 1993: Modelling economically efficient abatement of greenhouse gases, paper presented at the United Nations University Conference on Global Environment, Energy, and Economic Development, September 1993, Tokyo.

Cowen, T., and D. Parfit, 1992: Against the social discount rate. In *Philosophy, politics and society: Series 6, Future Generations,* In P. Laslett and J. Fishkin (eds.), Yale University Press, New Haven CT.

Daly, H.E., and J.B. Cobb, 1989: *For the common good,* Beacon Press, Boston MA.

Dasgupta, P., S.A. Marglin, and A.K. Sen, 1972: *Guidelines for project evaluation,* UNIDO, New York.

Dasgupta, P., and G.M. Heal, 1974: The optimal depletion of exhaustible resources, *Review of Economic Studies,* Symposium volume.

Diamond, P.A., 1965: The evaluation of infinite utility streams, *Econometrica* **33,** 170.

Diamond, P., and J. Mirrlees, 1971: Optimal taxation and public production, 1 and 2, *American Economic Review,* **61** (March and June), 8–27, 261–278.

Feldstein, M.S., 1970: Financing in the evaluation of public expenditure, Discussion Paper No. 132, Harvard Institute for Economic Research (August), Cambridge MA.

Fellner, W., 1967: Operational utility: The theoretical background and a measurement. In *Ten economic studies in the tradition of Irving Fisher,* John Wiley and Sons, New York.

Gramlich, E.M., 1990: A guide to cost-benefit analysis, 2nd ed., Prentice Hall, Englewood Cliffs NJ.

Harberger, A.C., 1973: *Project evaluation: Collected essays,* Markham, Chicago.

Harsanyi, J.C., 1955: Cardinal welfare, individualistic ethics, and interpersonal comparisons of utility, *Journal of Political Economy,* **63,** 309–321.

Hicks, J.R., 1938: *Value and capital,* Clarendon Press, Oxford.

Hicks, J.R., 1939: The foundations of welfare economics, *Economic Journal* **49,** (December), 696–712.

Ibbotson, R.G., and G.P. Brinson, 1987: *Investment markets,* McGraw-Hill, New York, updated in Nordhaus (1994).

IPCC (Intergovernmental Panel on Climate Change), 1992: *The supplementary report to the IPCC Scientific Assessment,* J.T. Houghton, B.A. Callander, and S.K. Varney, eds., Cambridge University Press, Cambridge.

Kaldor, N. 1939: Welfare propositions of economics and interpersonal comparisons of utility, *Economic Journal* **49** (December), 549–552.

Koopmans, T.C., 1960: Stationary ordinal utility and impatience, *Econometrica* **28,** 287–309.

Koopmans, T.C., 1966: *On the concept of optimal economic growth, econometric approach to development,* Rand McNally, Chicago.

Layard, R. 1976: Introduction. In *Cost-benefit analysis,* Baltimore, MD: Penguin Books, Baltimore.

Lind, R.C., 1994: Intergenerational equity, discounting, and the role of cost-benefit analysis in evaluating global climate policy. In *Integrative assessment of mitigation, impacts, and adaptation to climate change,* Proceedings of a workshop held on 13-15 October 1993, N. Nakicenovic, W.D. Nordhaus, R. Richels, and F.L. Toth, eds., International Institute of Applied Systems Analysis, Laxenburg, Austria.

Lind, R.C., K.J. Arrow, G.R. Corey, P. Dasgupta, A.K. Sen, T. Stauffer, J.E. Stiglitz, J.A. Stockfisch, and R. Wilson, 1982: *Discounting for time and risk in energy policy,* Resources for the Future, Washington, DC.

Little, I.M.D., and J.A. Mirrlees, 1974: *Project appraisal and planning in developing countries,* Heinemann, London.

Lyon, R.M., 1994: Intergenerational equity and discount rates for climate change analysis, prepared for IPCC WG III meeting, Nairobi, Kenya, 18-23 July, 1994, draft.

Manne, A.S., 1994: The rate of time preference: Implications for the greenhouse debate. In *Integrative assessment of mitigation, impacts, and adaptation to climate change,* Proceedings of a workshop held on 13-15 October 1993, N. Nakicenovic, W.D. Nordhaus, R. Richels, and F.L. Toth, eds., International Institute of Applied Systems Analysis, Laxenburg, Austria.

Manne, A.S., R. Mendelsohn, and R.G. Richels, 1993: MERGE – A model for evaluating regional and global effects of greenhouse gas reduction policies, Electric Power Research Institute, Palo Alto CA.

Mirrlees, J.A., 1967: Optimum growth when technology is changing, *Review of Economic Studies* **34,** 95–124.

Mishan, E.J., 1975: *Cost-benefit analysis: An informal introduction,* Allen & Unwin, London.

Munasinghe, M., 1993: *Environmental economics and sustainable development,* World Bank, Washington DC.

Nakicenovic, N., W.D. Nordhaus, R. Richels, and F.L. Toth (eds.), 1994: *Integrative assessment of mitigation, impacts, and adaptation to climate change,* Proceedings of a workshop held on 13-15 October 1993, International Institute of Applied Systems Analysis, Laxenburg, Austria.

Nordhaus, W.D., 1991: To slow or not to slow: The economics of the greenhouse effect, *Economic Journal* **101**(407), 920–937.

Nordhaus, W.D., 1993: Reflections on the economics of climate change, *Journal of Economic Perspectives,* **7**(4), 11–25.

Nordhaus, W.D., 1994: Managing the global commons: The economics of climate change. Mimeo, MIT Press, Cambridge, MA.

Parfit, D., 1983: Energy policy and the further future: The identity problem. In *Energy and the future,* D. MacLean and P.G. Brown, eds., pp. 38–58, Rowman and Littlefield, Totowa, NJ.

Pearce, D.W., and R.K. Turner, 1990: *Economics of natural resources and the environment,* Harvester Wheatsheaf, London.

Pearce, D.W., and D. Ulph, 1994: Estimating a social discount rate for the United Kingdom, mimeo, Centre for Social and Economic Research on the Global Environment, University College London and University of East Anglia.

Peck, S.C., and T.J. Teisberg, 1992: CETA: A model for carbon emissions trajectory assessment, *Energy Journal* **13**(1), 55–77.

Pestieau, P.M., 1974: Optimal taxation and the discount rate for public investment in a growth setting, *Journal of Public Economics,* **3,** 217–235.

Phelps, E., 1961: The golden rule of accumulation: A fable for growth men, *American Economic Review* **51** (September), 638–643.

Psacharopoulos, G., 1985: Returns to education: A further international update and implications, *Journal of Human Resources* **20** (Fall), 583–604, cited in Nordhaus (1994).

Ramsey, F.P., 1928: A mathematical theory of saving, *Economic Journal* **138**(152), 543–559.

Rawls, J., 1971: *A theory of justice,* Harvard University Press, Cambridge MA.

Robinson, J., 1962: A neo-classical theorem, *Review of Economic Studies* **29** (June), 219–226.

Rothenberg, J., 1993: Economic perspectives on time comparisons. In *Global accord: Environmental challenges and international responses,* C. Nazli, ed., MIT Press, Cambridge MA.

Rothschild, M., and J.E. Stiglitz, 1973: Some further results in the measurement of inequality, *Journal of Economic Theory,* **6,** 188–204.

Sandmo, A., and J.H. Dreze, 1971: Discount rates for public investment under uncertainty, *International Economic Review* **2,** 169–208.

Schelling, T.C., 1993: Intergenerational discounting, mimeo, University of Maryland, College Park (November).

Scott, M.F., 1989: *A new view of economic growth,* Clarendon Press, Oxford.

Sen, A., 1982: Approaches to the choice of discount rates for social benefit-cost analysis. In *Discounting for time and risk in energy policy,* R.C. Lind et al., eds., pp. 325–353, Resources for the Future, Washington, DC.

Sen, A., 1967: Isolation, assurance and the social rate of discount, *Quarterly Journal of Economics* **81.**

Solow, R., 1974a: Intergenerational equity and exhaustible resources, *Review of Economic Studies,* Symposium volume.

Solow, R., 1974b: The economics of resources or the resources of economics, *American Economic Association: Papers and Proceedings* **64**(2), 1–14.

Solow, R., 1992: *An almost practical step toward sustainability,* Resources for the Future, Washington, DC.

Stiglitz, J.E., 1985: Inequality and capital taxation, mimeo, IMSSS Technical Report #457. Stanford University. (July).

Stiglitz, J.E., 1982: The rate of discount for cost-benefit analysis and the theory of the second best, in *Discounting for time and risk in energy policy,* R.C. Lind *et al.,* eds., pp. 151–204, Resources for the Future, Washington DC.

Stiglitz, J.E., and P. Dasgupta, 1971: Differential taxation, public goods, and economic efficiency, *The Review of Economic Studies* **37**(2), 151–174.

Stockfisch, J.A., 1982: Measuring the social rate of return on private investment. In *Discounting for time and risk in energy policy,* R.C. Lind *et al.,* eds., pp. 257–271, Resources for the Future, Washington, DC.

Tobin, J., 1970: On limiting the domain of inequality, *Journal of Law and Economics,* **13** (2) (October), 263–277.

UNDP (United Nations Development Programme), 1992: *Human development report 1992,* Oxford University Press, Oxford.

United Nations, 1987: *Our common future,* The Report of the World Commission on Environment and Development (Brundtland Report), Oxford University Press, Oxford.

Weitzman, M.L., 1993: On the environmental discount rate, *Journal of Environmental Economics and Management,* 26, **(2),** 200–209.

Wildavsky, A., 1988: *Searching for safety: Studies in social philosophy and policy, Series,* No. 10, Bowling Green State University, Bowling Green, OH.

5

Applicability of Techniques of Cost-Benefit Analysis to Climate Change

M. MUNASINGHE, P. MEIER, M. HOEL, S.W. HONG, A. AAHEIM

The authors would like to acknowledge the helpful comments provided by P. Brown, E. Haites, L. Lorentsen, I. Mintzer, J.G. Oh, and R. Raufer.

CONTENTS

SUMMARY

In public policymaking, decision makers routinely compare the perceived costs and perceived benefits of an action. Cost-benefit analysis (CBA) provides an analytical framework that seeks to compare the consequences of alternative policy actions on a quantitative rather than a qualitative basis. The basic principles are well understood and straightforward: For an action to be justified, the costs of the action should be less than the benefits derived therefrom.

Traditional cost-benefit analysis requires that all costs and all benefits be expressed in a common monetary unit to facilitate the comparison. Modern cost-benefit analysis also includes techniques such as cost-effectiveness analysis and multicriteria analysis to analyze trade-offs when some of the benefits and/or costs can be quantified but cannot be expressed in monetary units.

This chapter examines how and under what circumstances CBA can make a contribution to the resolution of the following central questions now facing decision makers about global climate change.

By *how much* should emissions of greenhouse gases be reduced?

Provided costs and benefits can be estimated with adequate accuracy, this question is answerable by CBA: The measures whose marginal costs are less than the marginal benefits should be implemented. The marginal benefits are the marginal damage costs avoided. However, the difficulty of estimating the marginal damage costs of climate change make the practical application of CBA difficult. Although such analyses could be performed from a national perspective, it is a fundamental premise of CBA that the global perspective is the proper one.

When should emissions be reduced?

This question is more complicated because it also involves judgments about uncertainty. If the marginal damage cost (benefit) is known with certainty, and if future technological advances that might significantly change the marginal cost curve are known with certainty, then the timing of abatement is given by that portfolio of implementation options that maximizes the present value of avoided damage costs (benefits) less abatement costs. This is a relatively straightforward calculation; however, neither damages nor costs are known with certainty. Consequently, extensions of CBA – such as decision analysis – are required.

How should emissions be reduced?

This is closely related to the question of how much emissions should be reduced and is directly addressed by CBA. Attempts to determine the extent of reductions must usually consider the specific methods that might be used to reduce emissions. A bottom-up, empirical estimation of the marginal cost curve involves analysis of the broad spectrum of possible abatement options from which marginal cost curves can be derived. Top-down models also require explicit consideration of specific policy or technology options. Clear economic analysis and identification of the most cost-effective abatement options through the use of CBA are critical for practical policymaking.

Who should reduce emissions?

None of the family of techniques considered in this chapter can by itself resolve the question of who should reduce emissions – which involves considerations of equity. However, these techniques do provide a framework for understanding the trade-offs to be made between equity and economic efficiency.

The value of CBA

Practical application of cost-benefit analysis to climate change is difficult because of the global and intergenerational nature of the problem. It is further complicated by the difficulties of valuing some categories of ecological, cultural, and human health impacts. Nevertheless, CBA remains a valuable framework for identifying the essential questions that policymakers must face when dealing with climate change. The CBA approach forces decision makers to compare the consequences of alternative actions, including that of no action, on a quantitative basis. Indeed, the most important benefit of applying CBA may be the *process itself* (which forces a rigorous approach to decision making) rather than the predicted outcome (which always depends on the particular assumptions and techniques used).

5.1 Introduction

In public policymaking, decision makers routinely compare the perceived costs and perceived benefits of an action. Nevertheless, their decisions are frequently made on intuitive and qualitative grounds. Cost-benefit analysis (CBA) provides an analytical framework that seeks to compare the consequences of alternative policy actions on a quantitative rather than a qualitative basis. The basic principles are well understood and straightforward: For an action to be justified, the costs of the action should be less than the benefits derived therefrom.[1] If there are several alternatives, then one ought to pick that option whose benefits most exceed the costs.[2] Traditional cost-benefit analysis requires that all costs and all benefits be expressed in a common monetary unit to facilitate the comparison. Modern cost-benefit analysis also includes techniques such as cost-effectiveness analysis and multicriteria analysis to analyze trade-offs when some of the benefits and/or costs can be quantified but cannot be expressed in monetary units.

The objective of this chapter is to examine how and under what circumstances CBA can make a contribution to the resolution of the central questions now facing decision makers about global climate change:

(1)   By *how much* should emissions of greenhouse gases be reduced?
(2)   *When* should emissions be reduced?
(3)   *How* should emissions be reduced?
(4)   *Who* should reduce emissions?

CBA can at least theoretically and conceptually answer the first three questions. The fourth question is one of equity and is not amenable to resolution by CBA, even in simple, traditional applications not complicated by the complexities of the climate change problem.[3]

Section 5.2 defines more carefully what is meant by CBA. The term has come to encompass a wide variety of specific techniques. We also review the basic concepts. In Section 5.3 we examine the unique features of global warming and climate change as they pertain to decision making. Section 5.4 presents a discussion of the application of CBA to the climate change problem in light of these unique features. In Section 5.5 we discuss the key issues: risk, uncertainty, irreversibility, valuation, discounting, equity, and multiple criteria.

5.2 Cost-Benefit Analysis

Cost-benefit analysis is a generic term that presently subsumes a wide body of specific techniques. The method was developed initially as a means to evaluate projects that were limited in scale, geographic extent, and time span. Such traditional project level CBA (see Box 5.1) is too narrow to be relevant for evaluating climate change issues. However, the original technique has been extended to cover applications of increasing complexity. Modern CBA, more widely defined, includes a family of approaches that are more useful in assessing climate change options. Indeed, one of these approaches, cost-effectiveness analysis, has been widely used in climate change studies.

However, to evaluate cost-effectiveness, it is crucial to clarify how the policy target is defined, because there are several options in the global climate change context. Most recent analyses of mitigation costs have focussed on a target based on future emission levels,[4] such as stabilization of the emission of certain greenhouse gases by a given year. However, it might be more relevant to express the targets in terms of concentrations of atmospheric greenhouse gases at some future time.

To change the target for climate policy from emissions to atmospheric concentrations indicates a radically different cost-effectiveness strategy. A stabilization of CO_2 emissions at present levels is not sufficient to stabilize the atmospheric concentrations. Richels and Edmonds (1995) have compared the costs of reaching a particular concentration by 2100 for a variety of strategies. They show that a given concentration in 2100 could be achieved at a considerably lower cost if emissions were not stabilized immediately. The reason is that a more gradual reduction of emissions would avoid the economic shock that would follow a sudden stabilization, enable future advanced technologies to be utilized to a larger extent, and facilitate the postponement of sizable abatement costs.

Another possible target, also affecting the cost-effectiveness of alternative measures, refers to the physical consequences of climate change. Apart from the fact that predictions of these consequences are far more difficult to make than forecasts of emissions and atmospheric concentrations of greenhouse gases, the effects of regional differences would have to be included if targets were based on the consequences. For instance, whether climate change contributes to sea level rise or to an increase in the frequency of rain storms will be of quite different importance to people in Nepal and the Netherlands. In this context, an additional problem arises as to how to assess benefits from abatement of different consequences for different countries.

A further refinement of modern CBA is multicriteria analysis (MCA), a body of techniques developed to deal with the difficulties of economically valuing certain types of impacts (see box). Indeed, even if one attempted to place economic value on certain impacts – such as human life – not everyone would agree that it is appropriate to do so.[5] Moreover, cost-benefit analysis presupposes that the relevant costs and benefits are those that ultimately affect human welfare.[6]

Such views further support the need for MCA-based approaches to decision making. Similarly, there are concerns that monetized values themselves may be inaccurately estimated, and, in any case, such values might not reflect welfare. However, the question of who is affected, and how they will perceive the impact, is an issue that needs careful definition within an MCA analysis.

As noted earlier, conventional CBA cannot provide answers about the optimum level of equity in the same way that it provides answers about the optimum level of economic efficiency. But MCA *can* identify the *trade-offs* between equity

---

BOX 5.1: TECHNIQUES OF MODERN COST-BENEFIT ANALYSIS (CBA)

Traditional project level CBA

CBA evolved as a technique to evaluate and compare project alternatives. In the early years of its application, there was little concern with externalities, and the analysis took into account just the direct costs of projects and the direct benefits. As it was originally developed for use in industrial market economies, market prices provided appropriate guidance on how to evaluate benefits. When the World Bank began to apply the technique to nonmarket economies, where prices were subject to significant distortions, shadow pricing techniques provided simple corrections. For example, if an oil-importing country kept the domestic price of oil at artificially low levels, CBA would require the use of the border price, not the domestic price, as a basis for valuing oil.

One of the central concepts in CBA is that of discounting, which addresses the fact that costs and benefits may not occur at the same point in time. For example, whereas costly actions to avoid future climate change may need to be taken in the near future, most of the benefits of such actions will occur in the distant future. Discounting enables one to take into account the time value of money. In the case of evaluating simple investment alternatives over shorter time horizons (e.g., 15 years), the use of the opportunity cost of capital as the discount rate to be applied to both costs and benefits is well established and uncontroversial. However, in the case of complex public policy applications, particularly those whose time horizons are very great and involve environmental impacts or the depletion of natural resources – essential characteristics of the global climate change issue – there is sharp disagreement as to what discount rate is appropriate. This issue is dealt with in Section 5.5.2.1.

Cost-effectiveness analysis

As CBA began to be applied to much broader contexts, and particularly to the comparison of alternative *portfolios* of projects and to broad policy choices, the increasing complexity made it desirable to keep the level of benefits constant and to analyze the problem simply in terms of finding the most effective, or "least-cost" option to meet the desired level of benefits. This has the additional advantage that benefits in some cases need not be explicitly valued. For example, in power sector planning, models are applied to identify the capacity expansion plan whose present value of system costs is minimized, given some exogenously specified time path of electricity demand and some exogenously specified level of reliability. As we shall see below, this is the variant of CBA that has seen the most widespread application to the climate change problem, in which one seeks to identify the least-cost option to achieve given levels of greenhouse gas emission reductions, without any explicit attempt to specify what the *benefits* of that level of emission reduction may be.

Multicriteria analysis

The most basic requirement for the application of CBA is that both costs and benefits can be given economic value. This is typically a two-step process: first the costs and benefits must be quantified in terms of the physical measures that apply, and then those physical impacts must be valued in economic terms. Some applications of valuation techniques are likely to be controversial. Putting a value on human health and illness has been a major problem in the practical application of cost-benefit analysis in the past, even in those situations where one can agree on the levels of increased morbidity and mortality that might be caused by some policy or project. Efforts to place economic value on the loss of biodiversity have been equally difficult. Recognizing this problem has led to the development of so-called multicriteria analysis (MCA) techniques. These are expressly designed to deal with multiple objectives, of which economic efficiency may be only one. MCA is a particularly powerful tool for quantifying and displaying the trade-offs that must be made between conflicting objectives.

Decision analysis

MCA addresses certain shortcomings of conventional CBA (like valuation problems), but it does not necessarily deal more effectively with uncertainty. This complication has led to the development of a further extension of CBA known as decision analysis. Here the focus is expressly on how one makes decisions under conditions of uncertainty. These techniques find application in a wide variety of situations, from decision making in the high-risk field of wildcat oil drilling to analysis of financial options. As we shall see below, such techniques provide a rational approach for dealing with irreversibility, one of the more important characteristics of the climate change problem.

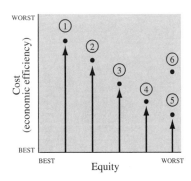

Figure 5.1: Multicriteria analysis.

objectives and economic efficiency, as suggested by Figure 5.1. Thus, the best equity result (indicated by option 1, say an equal per capita sharing of the burden of greenhouse gas emission reduction) may have the highest cost, whereas the worst equity result (indicated by option 5, say based on the present distribution of emissions) may have the lowest economic cost.[7] Nevertheless, even MCA requires a quantification or at least an ordinal ranking of the noneconomic efficiency criteria, as suggested in the figure. However, even such a noncardinal ranking may prove problematic when a global issue like climate change requires comparisons across countries and cultures.[8]

More generally, it is increasingly accepted that the pursuit of sustainable development will require recognition of goals related to economic efficiency, social equity, and environmental protection (Munasinghe, 1993). Economic valuation of the impacts of climate change on certain social and environmental aspects (e.g., biodiversity or cultural assets) will be difficult, and MCA-related approaches will be needed to make the trade-offs among otherwise noncompensable costs and benefits.

5.2.1 Basic concepts

An economically efficient policy for emissions reduction is one that maximizes the net benefits (i.e., the benefits of reduced climate change less the associated costs of emissions reductions).[9] Economic theory indicates that emission reduction efforts should be pursued up to the level where the environmental benefits of an additional unit of reduced warming (the marginal benefit) is equal to the cost of an additional unit of emission reduction (the marginal cost). Figure 5.2 illustrates the concepts of total and marginal costs in simplified form – the marginal cost at any level of emission reduction is equal to the slope of the total cost curve at the same level.

The shape of the total cost and benefit curves reflects the idea of diminishing returns. Each additional unit of emission reduction will have a higher unit cost: The first 10% reduction can be done cheaply, but the next 10% will cost considerably more, and so on.[10] Thus, the abatement cost curve is upward sloping as shown in Figure 5.2. Similarly, the marginal benefit (avoided greenhouse gas damages) falls as emission levels are reduced. The consequence of the foregoing is that the total cost (TC) has a minimum at the point where the positive slope

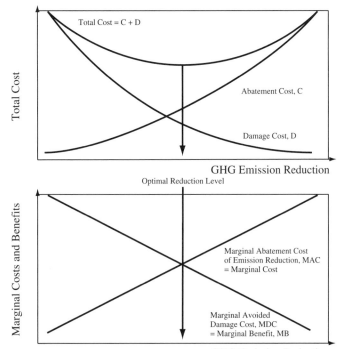

Figure 5.2: Total and marginal costs and emission reductions.

of the abatement cost curve equals the negative slope of the damage cost curve.[11]

The foregoing analysis ignores many complications. For example, the emission of a unit of CO_2 may give rise to a varying stream of environmental costs that must be discounted to yield a present value aggregate. The environmental damage function may be discontinuous and nonconvex. Because of technological progress, abatement costs may change over time, depending on when the technologies are applied. Similarly, abatement costs may exhibit economies of scale (e.g., mass production of solar photovoltaic cells), resulting in a marginal cost curve that actually declines beyond a certain point. Finally, the abatement costs are net costs, to the extent that certain technologies (e.g., renewables) may produce other (nonclimate-related) benefits and costs – the so-called joint products complication (discussed below).

5.3 Unique Features of Climate Change

Several important characteristics define the context in which traditional CBA is applied. The first is that costs and benefits arise within a time span typically no more than 15 to 25 years, corresponding roughly to the physical life of most projects over which benefits are derived. The second is that the elements of uncertainty are relatively tractable and can often be characterized by probability distributions.

These characteristics are very different in the context of climate change. The relevant time spans extend to a century or more. The uncertainties are extremely large, and few elements of uncertainty are amenable to characterization as probability distributions. Moreover, the cascaded uncertainty implied in each link of the chain of causality greatly amplifies the total uncertainty in the final outcome, namely, the extent of damage caused by climate change.

5.3.1 The complex chain of causality

Figure 5.3 shows the chain of causality. It begins with emissions of greenhouse gases. Although the most important of these is CO_2, it should be noted that there are many other gases that contribute to the climate change phenomenon, including methane and CFCs. Estimates of CO_2 emissions from fossil fuel combustion are fairly straightforward, but emissions from other sources are subject to much higher uncertainty. Moreover, the separation of anthropogenic and natural causes of climate change is much more difficult than in the case of other important regional/global pollution issues (such as CFCs or nuclear wastes). Indeed, the calculus of CBA may be significantly affected by natural events such as major volcanic eruptions.

The first link in the chain of causality (1 in the diagram) is between emissions and the resultant ambient concentration of CO_2 in the atmosphere. Unlike other pollutants, which are subject to complex chemical transformations in the atmosphere,[12] the calculation of the ambient concentration increase that follows from a given increment of emissions of CO_2 is relatively simple. However, because of the role of natural sources and sinks (particularly the ocean), even this calculation is subject to a considerable degree of complexity and uncertainty.[13]

The next link (2), between atmospheric concentration and temperature, is subject to much greater scientific uncertainty. The greenhouse effect itself – the trapping of outgoing infrared radiation – is subject to additional factors that are highly complex, particularly feedback effects, such as those from clouds, that complicate the calculation of equilibrium temperatures.

The subsequent link (3), between temperature increase and physical effects such as sea level rise, involves many different components, all of which are somewhat difficult to calculate. For example, calculating the rise in sea level associated with increases in mean sea temperature proves to be far from simple in some cases (as noted in Volume 1 in the case of the West Antarctic ice sheet). There are also large time lags associated with sea level rise, which are likely to continue for several centuries after the concentration of greenhouse gases has been stabilized. Even more complicated are estimates of how precipitation patterns might change, especially the spatial and temporal distribution of rainfall. Perhaps of even greater concern in developing countries will be the changes in the patterns of extreme weather events, to which they are particularly vulnerable.

If these physical effects (climatic and sea level changes) are understood in general terms, quantifying their impacts on flora, fauna, and human beings (link 4) is much more difficult. To illustrate, suppose it were possible to predict the change in precipitation and temperature regime for some given region.

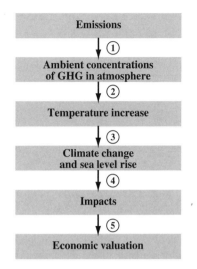

Figure 5.3: The chain of causality.

What can be said about the shifts in vegetation patterns? Although general poleward shifts of vegetation and agricultural production zones can be predicted, quantifying the effect is quite difficult. What is the impact on biodiversity? On wetlands? On the water table? On human communities? Clearly these are very difficult assessments to make.

Finally, to estimate the damage costs, one must be able to value these effects (link 5). Some valuation tasks will be relatively straightforward: For example, the cost of engineering structures to protect against sea level rise is relatively easy to establish, given the existing experience in this field (in such countries as the Netherlands). Other calculations will be more complex, but at least tractable, as, for example, the calculation of increased cooling and decreased heating costs or the impact on electricity demand of increased irrigation pumping requirements associated with drier climates in some regions. But a very large number of potentially important impacts will be very difficult to value – such as the impacts on forests, wetlands, and biodiversity, especially if the physical, biological, and social effects have not been accurately quantified.

5.3.2 Other special features

Beyond the complexity of the causal chain and the high degree of uncertainty surrounding the issue, what makes the analysis of climate change so different from other environmental analysis problems? The main reasons can be summarized as follows.

Greenhouse gases are stock, not flow, pollutants. Many pollutants (e.g., sulphate aerosols) have relatively short atmospheric lifetimes. As a result, the damages they cause are closely related to the current rate of emissions, and reducing emissions at major sources will likely have a relatively rapid impact on atmospheric concentrations and their associated effects. Greenhouse gases, however, have relatively long atmospheric lifetimes, and concentrations therefore respond slowly to changes in emissions. Consequently, the damages they cause are a result of the current atmospheric concentrations that have built up due to all past emissions – in other words, they are related to the stock of the pollutant, not to the current rate of emissions.[14] Similarly, a change in emissions of a greenhouse gas at any time will affect the atmospheric concentrations of this gas, and thus the climate, in all future periods. Therefore, in the context of the climate problem, global climate change is a lagged function of the emissions of the various greenhouse gases. At any time, atmospheric concentrations (or stocks) of these gases depend on the whole history of their emissions up to that time.

To calculate the marginal environmental cost of increased current emissions of a greenhouse gas, one must first calculate the physical impact of such an emission increase on the future atmospheric concentration of the gas. This will depend on the physical characteristics of the gas, which affect how rapidly an increased atmospheric concentration depreciates. There are large differences between different greenhouse gases, with atmospheric lifetimes varying from roughly fifteen years for methane[15] to more than 100 years for N_2O and some CFCs.

Once the impact of current emissions of a greenhouse gas on future atmospheric concentrations has been calculated, one can in principle calculate the effect of the increase in current emissions on future climate development. If one has specified a function which measures the monetary cost of climate change, one may calculate the incremental costs and thus obtain a present value measure of the marginal cost of increased current emissions of a greenhouse gas.

It is clear from the description above that the marginal monetary cost of greenhouse gas emissions is a complex concept. Several assumptions of an economic nature must be made, such as the appropriate discount rate and the monetary costs of climate change for the whole future. In particular, the relative importance of different greenhouse gases is much more complex than is implied by a simple physical conversion index such as the global warming potential (GWP) discussed in Volume 1 of this report. In the context of the climate problem, a reasonable definition of the importance of a greenhouse gas relative to, say, CO_2, is the marginal environmental cost of current emissions of this gas relative to the corresponding marginal cost of CO_2. It follows from the discussion above that the relative importance of greenhouse gases depends on a number of economic assumptions that must be made.[16]

Inertia and irreversibility. Since emissions of greenhouse gases in any one year represent a relatively small fraction of the total global stock, the system has great inertia. This means that even if all emissions went to zero, it would be decades, if not centuries, before the stock of greenhouse gases was reduced significantly. Therefore, decisions about emission levels are effectively irreversible, at least over the 100–200-year timespan of interest. In other words, failure to reduce emissions in the short to medium term may be irreversible in the sense that once the effects of climate change become apparent, it will then be too late to do anything about it.

Global characteristics. Most environmental pollution problems are local or regional in scale.[17] Thus, the benefits of emission reductions generally accrue to the same geographic areas as bear the costs. The damages associated with greenhouse gases, however, depend on the global greenhouse gas concentration, which is largely independent of the regional meteorological patterns that usually define the geographic scope of other transnational environmental problems such as acid rain. Therefore, the distribution of the benefits of emission reduction is global, not local. *Per contra,* even a country that emits no greenhouse gases can incur the damages of emissions by other countries.

Geographical distribution of impacts. Poorer nations are likely to be the most vulnerable to the impacts of climate change, since they lack the resources to protect themselves against sea level rise, extreme weather events, or desertification. In contrast, the impacts of acid rain from emissions in some poorer countries or regions may be concentrated in richer countries[18] (e.g., the effects of emissions from Eastern Europe on western Germany[19]). These richer areas therefore have powerful incentives to promote emission reduction programmes in the source areas, including the provision of financial assistance.[20] However, in the case of greenhouse gas emission reductions, countries such as the U.S. (where there is the *perception* that the *direct* impacts of climate change on the U.S. itself are relatively small) may have fewer obvious economic incentives to reduce emissions.[21] Considerations of humanitarian solidarity and equity alone are unlikely to be sufficient motivators.

Absence of actual impact data. In the case of climate change, unlike almost all other environmental externalities, actual impact data are scarce, and estimates of physical impacts are based entirely on the predictions, judgments, and models of scientists.[22] Only once (or if) climate change does in fact occur, will the impacts be known. The evidence of cause and effect will be difficult to substantiate, because of the likelihood that, at least initially, changes will be incremental. It should be noted, however, that there does exist a significant body of verifiable scientific theory that underlies the estimates and models of scientists.

Nonlinearity. Global climate change is determined by complicated interactions involving a chain of nonlinear linkages (i.e., greenhouse gas emissions, atmospheric concentration, temperature change, climate system feedbacks, and physical impacts). Therefore, climate change phenomena and risks are likely to be much more nonlinear than the relationship between conventional emissions and more local pollution.

Very long time frame. The very long time frames involved in climate change cause some normally external variables to become internal factors of change. For example, the economic impact of sea level rise depends on the size of the population living in low-lying coastal areas, which may decrease once a sea level rise becomes evident. The costs of such adaptation mechanisms may be especially difficult to estimate.

5.4 Cost-Benefit Analysis in the Context of Climate Change

In light of these unique characteristics of the climate change problem, what can we say about the suitability of CBA? How and under what circumstances can CBA make a contribution to the central questions now facing decision makers, in particular:

- By *how much* should emissions be reduced?
- *When* should emissions be reduced?
- *How* should emissions be reduced?

The fundamental problem in applying CBA to the climate change problem follows directly from the chain of causality discussed in Figure 5.3 above: whereas estimating the costs of emission reduction involves the beginning of the chain, estimating the benefits (the avoided damages) involves the very end of the chain. Since there is some level of uncertainty associated with each of the links, estimates at the last stage of the chain are subject to compound uncertainties that may be very large indeed.

5.4.1 Estimates of the marginal cost curve

Marginal abatement cost (MAC) curves for greenhouse gas emission reductions have been derived for many industrialized countries, but for only a few developing countries.[23] Figure 5.4 shows such a curve for Thailand. This curve is derived by a rank ordering of the individual measures by cost per tonne of CO_2 saved (the height of each block), with the width of each block representing the tonnes of CO_2 so saved. The shape of the MAC curve, when smoothed, is indeed of the type indicated in Figure 5.2, a result confirmed by many other examples.[24] Generally, these studies rely on known or near-term technical options and ignore effects due to joint products (see Section 5.4.4), economies of scale, and capacity building that might reduce the upward slope of the cost curve. Such marginal cost curves depend on discount rate and price assumptions and may evolve over time as options are taken up and new technologies develop. A more detailed discussion of estimates of the MAC curves is contained in Chapters 8 and 9.

What is interesting in these (and other studies[25]) is the significance of "below the line" options (i.e., where MAC is negative but still upward sloping). These are measures that appear to have negative costs associated with them – in other words, when these options are implemented, both costs and emissions go down, relative to the reference case.[26] Compact fluorescent lighting, other energy-efficient devices, and demand-side management measures typically fall into this category, and in developing countries, measures such as reducing electricity transmission and distribution losses or instituting vehicle maintenance programmes also appear here. These, then, are measures that should be implemented in any least-cost energy development strategy, even in the absence of any desire to reduce greenhouse gas emissions. The fact that they are not implemented reflects either market failures, the influence of powerful vested interests, or other factors, such as high transaction costs (which might exceed the potential benefits).[27] Indeed, in developing countries, there may be more such "below the line" options because education and information about the availability and benefits of options in this segment of the MAC curve are less available than in developed countries.

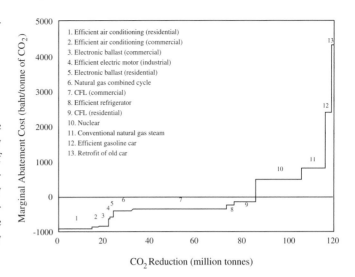

Figure 5.4: The marginal abatement curve for Thailand.

This issue highlights a practical problem for the Global Environment Facility (GEF). The GEF has been set up as the funding mechanism, on an interim basis, to provide financial resources needed by developing countries to meet the "full incremental costs incurred" in complying with their obligations under the Framework Convention on Climate Change.[28] If the GEF is to fund the "incremental costs" of greenhouse gas reduction measures, this presupposes that a baseline can be unambiguously defined, against which such incremental costs are to be measured.[29] But under such a definition, should demand-side management (DSM) programmes that have negative incremental costs – such as energy-efficient lighting – be funded by the GEF?[30]

5.4.2 The marginal benefits curve

It follows from the discussion of the previous section that the level of uncertainty in benefits is much greater than the level of uncertainty in costs. The practical implication of this, depicted in Figure 5.5, is that the optimum point of emission reduction is also subject to significant uncertainty. Whether the marginal benefits curve is MB_1, or MB_2, or MB_3 has a much greater impact on the location of the optimum point of emission reduction (indicated by points A, B, and C) than the much smaller uncertainties in the marginal cost curve. Therefore, with such wide uncertainty about the economic optimum, CBA may not be very helpful. In this situation, an arbitrary level of emission reduction at P has a very high probability of at least meeting the criterion that MB (marginal benefit) > MC (marginal cost). By contrast, C may represent a risk-averse, "precautionary" level of abatement, in which the expected value of MC may be greater than the expected value of MB (see Section 5.5 and Figure 5.10 for further details).

In fact, there exist few if any estimates of the benefits curve, and most of the estimates that do exist are not much more than single point estimates for some presumed level of greenhouse gases in the atmosphere. A complete discussion of benefit estimates is provided in Chapter 6.

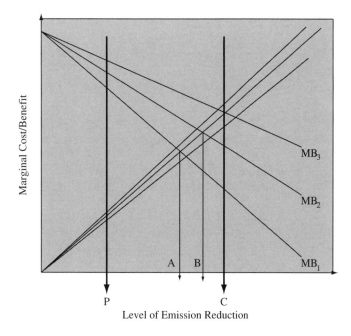

Figure 5.5: Uncertainty in the benefits curve.

A number of estimates (see, for example, Nordhaus, 1994b) suggest that the economic impact of climate change on the industrialized countries is quite small. However, Nordhaus, as well as others, is quick to concede that such results may not apply to developing countries, where typically much larger shares of national income are in agriculture, and much of it in subsistence farming in marginal areas where even small changes in climate may have devastating effects and where the ability of the farming population to adapt may be very small indeed.[31] The types of human settlements most vulnerable to climate change are concentrated in developing countries, including low-income communities, residents of coastal lowlands and tropical islands (such as coastal Bangladesh or the Maldives), populations in areas already significantly affected by desertification, and the urban poor in squatter settlements.[32]

Unfortunately, to date, most of the discussions of impacts of climate change on the developing countries have been qualitative in nature,[33] and there is an urgent need for more quantitative estimates. According to a recent GEF survey of country studies (Fuglestvedt *et al.,* 1994),[34] there are now almost as many studies underway on effects as on mitigation, but it is difficult to ascertain how many of these studies will result in the sort of quantitative information necessary to construct an impact cost curve. Certainly none of the studies and papers published to date for developing countries contains quantitative estimates of the type derived by Nordhaus for the U.S.[35]

5.4.3 *Measuring costs and benefits*

Most of the work on the benefit (avoided cost) side, and much of the work on the cost side, assumes that the relevant measure by which to evaluate options is loss of GDP. This raises a separate set of questions about the extent to which GDP (or

loss of GDP growth) is the appropriate measure.[36] It is well established (Weitzman, 1976; Brekke, 1994) that GDP is not a welfare measure *per se,* but rather a convenient way to aggregate goods and services that clearly contribute to welfare. Nevertheless, Figure 5.6 summarizes some of the estimates that have been made for GDP impacts. Chapter 9 discusses such estimates in more detail.

Since GDP is an imperfect measure of social welfare, it should not be surprising that the cost-effectiveness of policy options can differ significantly when analyzed in terms of their impact on GDP and on social welfare. Recent studies have found this to be the case for industrialized countries such as Norway (Alfsen *et al.,* 1992) as well as for developing countries such as Sri Lanka (Meier *et al.,* 1993; see Table 5.1). The "consumer impact" is an estimate of the impact or social welfare not reflected in traditional GDP measures. At 10-50% of the total, the consumer impact is clearly a significant fraction of social welfare.

Of course, the magnitude of such impacts as those shown in Table 5.1 is a function of how much of the technology is adopted, how soon, and what discount rate is used. The effect of a single 25 MW wind plant will obviously be much less than if a 300 MW wind farm displaces a plant using imported coal in the near term. Generally, what these results do underscore is the need for great caution when interpreting the so-called costs of greenhouse gas emission reductions attributable to specific technologies.

Equally important are the concerns over the extent to which conventional measurements of GDP growth take into account the depletion of natural resources ("green accounting"). For many developing countries this is a particularly important issue, and there is a growing literature suggesting that conventional accounting substantially overstates GDP growth. The corollary is that estimates of GDP impacts of climate change that *fail* to take account of changes in the rate at which natural resources are depleted, or changes in environmental impacts, may be understating the true effect of these impacts.[37]

Some of these problems are evident from the numerical estimates. In Figure 5.6 we display the results of recent studies of the impact of emission reduction strategies on loss of GDP.[38] As is evident, these estimates show considerable variation. Even so, none of these estimates reflects "green accounting" or any margin for no-regrets options, and only one (Glomsrod *et al.,* 1990) takes into account joint products. Most of these studies are national studies. Many researchers argue that unilateral actions by the U.S. or by OECD countries are likely to be less effective than global action, and that unilateral actions are likely to exaggerate the impact on GDP.[39] However, these studies also neglect the benefits of the development of advanced technologies in the market economies, which could then be more quickly adapted in the developing world.

Traditional CBA basically relies on a partial equilibrium analysis, in which only the relevant portion or subsystem of the more complete economic system is studied in depth, and many parameters (and prices) outside this system are taken as exogenously fixed. For questions relating to climate change (whose consequences may have major effects on prices),

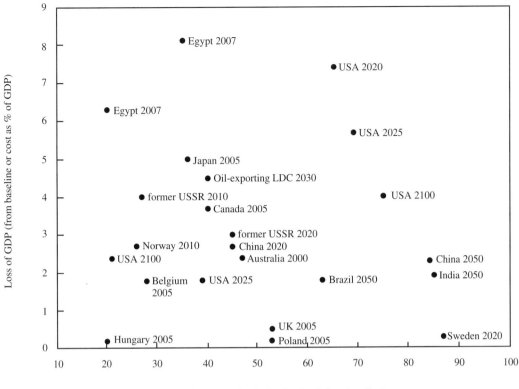

Figure 5.6: Impact of GHG emission reduction on GDP.

Table 5.1. *Welfare losses due to different greenhouse gas abatement options in Sri Lanka (in $/tonne of avoided carbon)*

	No coal	Wind	CFL[a]
Supply cost impact	16	67	-472[b]
Consumer impact	10	66	-66
Total	26	131	-538

[a]Compact fluorescent lighting.
[b]Negative numbers indicate a benefit.
Source: Meier *et al.* (1993)

the more far-reaching and inclusive general equilibrium approach – in which the properties and relationships of the entire economic system are analyzed – is appropriate. For this reason, computable general equilibrium models are widely used in climate change studies.

5.4.4 *The joint products problem*

A related question concerns the joint benefits and costs of emission reduction. Options that reduce greenhouse gas emissions may also provide significant changes in other pollutants whose impacts are of a quite different scale. For example, the substitution of renewable energy technologies for coal will reduce not just CO_2 emissions – a *global* benefit – but also SO_2 and particulate emissions, thus bringing a reduction in *local*

environmental damages. On the other hand, the increased use of renewable energy technologies (such as hydroelectric generation), which also reduce CO_2 emissions, may also impose new and different local environmental *costs* (such as loss of biodiversity associated with reservoir inundation).[40]

Estimates of the importance of joint benefits and costs vary widely (in part, because valuing the costs and benefits of other environmental impacts may be subject to similar difficulties of valuation and scientific uncertainty as greenhouse gas emissions).[41] A recent study (Alfsen *et al.*, 1992) found that the joint products of carbon reduction (reduction in environmental damages to forests and lakes, health damages, reduced traffic congestion, road damage, etc.), offset about 30-50% of the initial abatement costs in the case of Norway.[42] A British study (Barker, 1993) found that the secondary benefits exceeded the cost of the greenhouse gas abatement measures.

Adding the benefits of reductions in conjoint pollutants to the MB measure is one possibility. The other is to subtract these benefits from the MC measure. Since most of such benefits go to the country that bears the costs of abating the CO_2 emissions, while the benefits of lower CO_2 emissions go to all countries, it follows that the preferred approach is to adjust the marginal cost curves.

5.4.5 *The aggregation problem*

We now turn to the matter of implementing policy. Consider first Figure 5.7, which depicts the situation faced by a hypo-

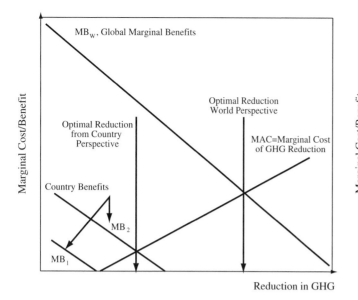

Figure 5.7: The marginal cost and benefits curve for an industrial country.

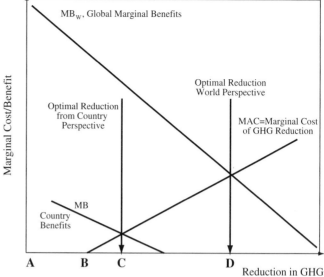

Figure 5.8: The marginal cost and benefits curve for a developing country.

thetical country. Its marginal cost curve is upward sloping, in the manner discussed above. It begins with negative values (the "no-regrets" or "win-win" options), again in the manner indicated earlier.

Suppose Figure 5.7 depicts the situation for an industrialized country, say the U.S. Suppose that indeed Nordhaus is correct in his estimates that the benefits to the U.S. are rather small. If MB_1 were true, then the optimal policy would be to implement only the win-win policies – demand-side management, more efficient end-use devices, economically efficient pricing, and the like. If MB_2 were true from the U.S. perspective, then perhaps only small emission reductions are warranted. We assume, of course, an appropriate adjustment for the evaluation of the benefits of the reduction of joint costs.

However, the benefits of emission reductions are likely to be much higher to the rest of the world, and hence the curve MB_w lies as shown, far above the MB_1 and MB_2 curves. Therefore, when the global benefits of emission cost reductions are taken into account, the optimum level of emission reduction shifts to the right, as shown. How one persuades decision makers in the U.S. (and in other industrialized countries) to take a global perspective is the main question.[43]

Consider now the situation for a developing country (Figure 5.8). Again the global benefits curve lies far above the curve for an individual country. Evidence from empirical studies suggests that a far larger portion of the MC curve is in the no-regrets zone. Several multicountry studies point to this result.[44]

In the case of developing countries, the GEF funding mechanism provides the means for shifting the level of emission reduction undertaken by any specific country to the right. But in what range does the GEF mechanism operate? Conceptually, it should operate in the range CD; that is, it should provide funding only beyond the optimal level from the country

perspective. But since the MB curve is so difficult to determine, an easier definition, operationally, is for the GEF to operate in the range BD; that is, from the conventional "least-cost" point that would normally apply, and in which any benefits, even locally, from greenhouse gas reductions are simply taken as zero. For example, wind plants would not normally lie in the least-cost expansion path for an electric utility (if environmental externalities are valued at zero); GEF provides the mechanism to fund such technologies. Yet, in fact, the GEF has also funded projects in the range AB – such as the energy-efficient lighting project in Mexico.[45]

The difference between the optimal level of reduction from the national perspective and that from the global perspective lies at the heart of the practical problem of implementation. A related but different issue – involving equity – concerns one of the premises of CBA, namely that sunk costs and past actions are not relevant. Yet developing countries argue that in the case of the climate change problem, past emissions are relevant, because it is the industrial countries, not the developing countries, that have accounted for the bulk of the anthropogenic greenhouse gases present in the atmosphere (e.g., see Munasinghe, 1991).

5.4.6 Systemic evaluation

The comparison of abatement cost estimates shown in Table 5.2 also points to the importance of systemic evaluation. The cost per tonne of avoided emissions for both wind and the no-coal option are much lower in the Sri Lankan than in the GEF or Indian estimates for the exact same technology. The reason lies in the fact that the incremental cost of, say, wind power, depends on what technology is being substituted, and this will be very different from case to case. For example, in Sri Lanka, imported coal baseload plants largely determine the future system expansion cost. This is much more expensive than in

Table 5.2. *Comparisons of cost estimates for CO_2 abatement (in \$/tonne of CO_2)*

	Study type	Wind	No coal
Sri Lanka	Systemic, single 300 MW wind farm	67-131	16-26
GEF	Generic	116-223	45-89
India	Supply cost	150-600	

Sources: For Sri Lanka, Meier *et al.* (1993); for India, Hossein and Sinha (1993). The generic estimate is from London Economics (1992).

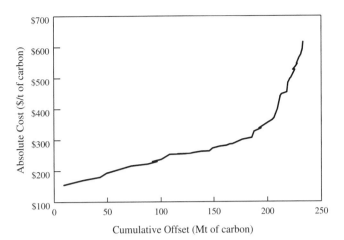

Figure 5.9: Supply cost curve for wind farms in India.

India, which has available relatively large quantities of low-cost domestic coal. Similarly, in the generic cost estimates made by the GEF for no-coal options, the presumed substitute fuel is domestic oil or gas, not imported oil (as is the case for Sri Lanka). The point is that there likely exist very large, country-specific variations in estimates of the cost of greenhouse gas emission abatement through given technologies. Joint costs and benefits that are local in nature will very likely vary even more from place to place.

Another way of making the same point is by noting that single-point cost estimates for many technologies can be quite meaningless, because the supply curve for even an individual technology is not flat. For example, as Hossein and Sinha (1993) have shown in a recent study of wind and hydro for India, the supply curves have the expected classical upward sloping shape as illustrated in Figure 5.9 for the supply cost of wind farms in India. Such static cost curves neglect the counterargument that, in a more dynamic analysis, cost may decline due to economies of scale and technological advances.

5.5 Issues

5.5.1 Risk, uncertainty, and irreversibility

Our knowledge about how anthropogenic emissions of greenhouse gases affect global temperature, what kind of effects a change in global temperature may have, and how efforts to mitigate climate change may work is clearly restricted. How different greenhouse gases react in the atmosphere is not fully understood, and even if exact predictions of the average increase in global temperature could be made, the different regional effects of these increases will be exceedingly difficult to foresee. There is also considerable uncertainty about the economic and social effects of abatement measures, which are decisive for determining their associated costs and benefits.

One cannot, therefore, evaluate climate measures without taking these uncertainties into account. On the other hand, acknowledgment of vast uncertainties should not lead to an inert attitude but rather to the development of rational strategies for acting under uncertainty. Economic analysis under uncertainty aims at developing strategies for decision makers who face uncertainties in future costs and benefits. The uncertainty

in the outcome of a variable is often described by a probability attached to each possible outcome. In some cases, the probability distribution is objectively presented; in such cases one normally talks about risk. More often, subjective probability distributions are assumed, in which case one talks about uncertainty.

In Section 5.4 we noted that the level of uncertainty is greater in the benefits curve than in the cost curve (recall Figure 5.5). Figure 5.10 illustrates the practical consequence of uncertainty in a different way. Figure 5.10a depicts the *cost-benefit analysis* of the optimum level of emission reduction. The optimal reduction, R_{opt}, is given by the point at which MB = MC. If the marginal mitigation cost and marginal damage cost functions were known with certainty or if, in the absence of risk aversion, one were to use expected values of the uncertain mitigation cost and damage functions, then the optimum degree of emission reduction would be as shown.

However, given uncertainty in the marginal mitigation cost and damage functions, the optimal emission reduction cannot be determined precisely. It could lie anywhere within a relatively wide range. Uncertainty in the damage function and risk aversion lead one to a "precautionary approach," which requires more stringent emission reductions, lying to the right of the expected value R_{opt} and roughly determined by the intersection of the cost curve and some notional upper envelope estimate of the damage function, as indicated by C in Figure 5.5.

Figure 5.10b illustrates the case in which the damages are sufficiently uncertain that a marginal damage function cannot be defined. The risks associated with various emission levels are considered, using the best available evidence. This information, together with the associated costs, is used to select an emission reduction, R_{ASM}, that constitutes an *affordable safe minimum standard.* Analytically, such a standard would have to be based on a multicriteria analysis.

Finally, in Figure 5.10c the emission reduction R_{AS} is based solely on a scientific assessment. This corresponds to the first of the two views of the sustainability approach discussed in Chapter 6. Since the obligation to avoid harm is absolute, the cost of avoiding harm is irrelevant. The benefits of

a) Cost-Benefit Analysis

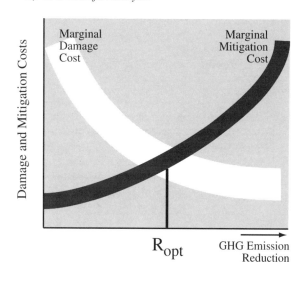

b) Affordable Safe Minimum Standard

c) Absolute Standard

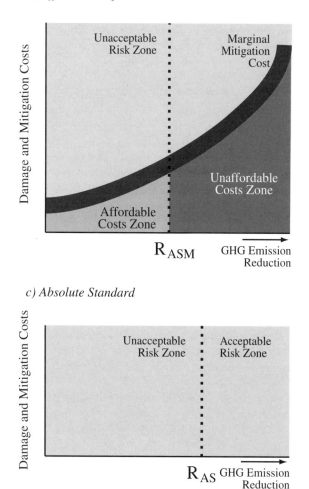

Figure 5.10: Methods and rules for determining mitigation targets.

control are so large that consideration of costs is unhelpful. This can be termed the *absolute standards* approach.

The attitude towards risky outcomes is often expressed in terms of a risk premium, which is the minimum compensation required to accept a lottery with expected return *x* compared to a certain return of *x*. Such a premium may be required because the decision maker prefers certain to uncertain outcomes, but may also occur if the net benefit of his investment is nonproportionate in quantity. For instance, suppose a country commits itself to reduce emissions by 100 tonnes and considers two alternative measures. One alternative reduces emissions by 100 tonnes with certainty. The other measure has a 50% chance of reducing emissions by 200 tonnes. Even if the expected reductions are equal (i.e., 100 tonnes), it is evident that a decision maker will prefer the certain alternative unless the uncertain alternative is considerably less costly, as there is a chance of having to impose additional measures if unlucky. Assessments of the uncertainty involved in the analysis of climate change may therefore be of great importance to decision making.

However, given the uncertainty it is also important to emphasize that some actions will be less attractive than others. The mirror of the above argument is that some actions may be acceptable, even with negative net expected benefit, if they reduce the uncertainty. One way to do so is to spread the risk among several measures. Then, if one fails to meet its expected target, another may satisfy the achievements that were expected. The total risk would thus be lower than if all efforts were concentrated on one single measure.

One is scarcely able to account for all the uncertainties involved when analyzing climate change. In some cases it may be equally difficult to assess a probability distribution of a variable as a single expected estimate. Ignorance about certain effects of a measure indicate that they should be left out of the analysis. However, CBA may be of great help to decision making even when limited to effects on which reasonably well-founded value estimates can be provided.

One issue that has attracted a lot of interest in the literature of environmental economics is the problem of making irreversible decisions under conditions of uncertainty. Because of the long atmospheric lifetimes of greenhouse gases, irreversibility is a central characteristic of the climate change problem. A former, more benign climate cannot be reestablished if decisions lead to worse-than-expected climatic outcomes.

To incorporate the cost of irreversible effects in the CBA, a vast amount of information is required about such factors as how the uncertainty evolves over time. Such information is rarely available, but we may assume that increased knowledge about climate change will narrow the range of uncertainties in the future. This possibility also enhances the importance of keeping future options open. This is an extra cost of irreversibility – the so-called quasi-option value – and it suggests that decision makers should follow flexible strategies when faced with uncertainty.

The strategy that leaves most future options open is difficult to determine, however. Investing in abatement of climate change opens possibilities for increasing emissions at a later stage if the effects turn out to be less serious than expected. On the other hand, it reduces the potential use of adaptation. In addition, it is difficult to predict the effect of increased knowledge, especially whether anticipated outcomes will become more certain. When dealing with many of the effects of climate change, ignorance is perhaps a more appropriate concept than uncertainty. Increased knowledge may change ignorance to uncertainty, but the range of possibilities will not necessarily narrow for that reason.

In applying CBA to the global climate change problem, and in particular to the evaluation of alternative policies to optimize net benefits, several major sources of uncertainty need to be considered:[46]

- *Uncertainty about the actual rates of emission.*[47] Most of the many studies cited earlier make assumptions about current and future rates of emissions, taking some "business-as-usual" case as a starting point. Current CO_2 emissions from fossil fuel use may be reasonably well known, but the level of uncertainty increases as one considers levels of possible fossil fuel use in the more distant future,[48] the impact of deforestation,[49] or the emissions of other greenhouse gases such as methane.[50]

- *Uncertainty about the costs of emission reduction.* Again there are significant differences between estimating such costs in electric utility systems (which are relatively well known for conventional technologies) and those elsewhere, particularly for reforestation.

- *Uncertainty about scientific linkages.* As already noted in Section 5.3, there exists a chain of scientific uncertainty (see Figure 5.3). The extent to which these uncertainties can be resolved by further research is itself subject to uncertainty (especially in light of the previously noted fact that by its very nature, *ex ante* verification of models by actual data is difficult). Thus, it is unclear that similar arguments invoked in the context of other environmental problems have validity for global climate change. In contrast, the argument that further understanding of atmospheric chemistry, or of the chemistry of lake acidification, or of the exact nature of forest damage mechanisms was necessary before very costly efforts were undertaken to control SO_2 and NO_x emissions did at least have the merit that data on the actual damage of acid rain could be found.

- *Uncertainty in valuing the costs and benefits of the physical impacts.* Here there may exist quite large variations in the level of uncertainty: For example, evaluating the cost of protective dikes to protect against sea level rise or estimating the opportunity cost of inundated land is subject to significantly less uncertainty than estimating the impact on agriculture or on biodiversity. However, estimating the costs of more extreme climate conditions (e.g., more intense storms) will be very difficult.

- *Uncertainty about the assumptions underlying policy options.* A number of studies, for example, have esti-mated the impact on greenhouse gas emissions of eliminating subsidies on coal or electricity, using assumed values of price elasticity.[51] General equilibrium models require all kinds of assumptions about the elasticities of substitution: The actual values used in these numerical simulations are either based on historically estimated elasticities or on the judgment of the modeller. In either case, there is uncertainty about the extent to which such values match actual behaviour.

- *Uncertainty about the effectiveness of policies.* For example, the proposition that a certain level of carbon tax will in fact result in a certain hypothesized fuel substitution makes a number of assumptions about the functioning of markets. As noted earlier, the very fact that many apparently "no-regret" options are not being implemented suggests a higher level of market imperfection than economists like to admit, and/or substantially higher transaction costs or discount rates.

- *Uncertainty about joint benefits and costs.* As noted earlier, joint benefits and costs may be a very significant factor in evaluating options for greenhouse gas abatement. However, these joint benefits and costs are also subject to significant uncertainties (and also measurement problems).

The application of valuation techniques is difficult even where the impacts themselves can be quantified with relative confidence. But in the case of global climate change, uncertainties in economic valuation techniques may be significantly smaller than the *scientific* uncertainties that surround the impacts of increasing concentrations of greenhouse gases. Advocates of immediate mitigation action to reduce emissions argue that even if the probabilities of some of the important impacts are unknown and subject to great uncertainty, they are not zero. Low probability/high impact events are especially complex to model, and concern about such events may also provoke extreme risk-averse behaviour. Further, the process of climate change, once underway, will be irreversible (at least during a period measured in centuries), and the damages that may result are so catastrophic that action may be warranted even in the absence of more precise scientific knowledge about the impacts.

A good analogy is a nuclear power plant accident. The Chernobyl incident notwithstanding, the risk of a catastrophic accident is extremely small. But the cost of a major accident is undoubtedly very large. Both the probabilities[52] and the costs of the impacts are very difficult to estimate. But even if one could agree on appropriate values to use, and one were able to calculate an expected value, decision making on the basis of the expected value may still not reflect the preferences either of the public or of decision makers. The consequences of even an extremely unlikely event may be perceived as so undesirable (especially in the case of extreme risk aversion), that "normal" decision rules may simply not be viewed as appropriate. In other words, cost-benefit analysis must deal not just with expected values, but also with the risk preferences of the decision makers and those they represent.

The traditional and simple way of incorporating uncertainty considerations in CBA has been through sensitivity analysis. Using optimistic and pessimistic values for different variables can indicate which variables will have the most pronounced effects on benefits and costs. Although sensitivity analysis need not reflect the probability of occurrence of the upper or lower values, it is useful for determining which variables are most important to the success or failure of a project. Indeed decision makers often assign probabilities (even if only implicitly) to the various outcomes. Admittedly, the sheer magnitude of the costs of catastrophic climate change will make the sensitivity analysis problematic.

One might note that for certain types of uncertainty, something akin to risk insurance is available in the form of futures and options markets. For example, one can hedge against uncertainty in the future price of oil by transactions in the oil futures markets, but their efficient functioning depends on there being some balance between those who are buyers of oil (who are primarily interested in protecting themselves against a rise in the price), and sellers of oil (who are interested in protecting themselves against a fall in the price).[53] (See Box 5.2.)

Finally, extreme uncertainty might also influence the nature of the economic instruments employed for policy purposes. For example, a price-oriented mechanism (such as a carbon tax) which limits economic dislocation might be preferred over a quantity-based mechanism (such as tradable permits) in a situation where there are both uncertain control costs and an uncertain environmental response (see Lave and Gruenspecht, 1991).[54]

5.5.2 *Valuation*

The robustness of a cost-benefit analysis depends critically on how reliable the values attached to each item are. The prices of marketable goods and services express social values as long as the goods in question are not rationed and there are no externalities.[55] For nonmarketed goods and services, such as many environmental services, values have to be estimated in order to aggregate costs and benefits and obtain an overall evaluation of choice of policy. Estimated prices may depend on the methodology chosen to create them, and one should therefore interpret with caution results which include such prices.

One reason why avoiding climate change may have a value is that climate change will cause a change in economic activities. Sea level rise will force people to move, for example, or more turbulent weather conditions may increase the need to rebuild damaged structures or replace damaged materials. A second reason is that people attach subjective values to the climate where they live – values that are to some extent reflected in the notions of "good" and "bad" weather.

However, it is difficult to assess these values. To simplify somewhat, one may base an estimate on the anticipated costs of achieving a certain target at observed market prices: for instance, the minimum abatement cost of attaining the same level of greenhouse gas emissions as a previous year. Alternatively, one may estimate the willingness to pay for reaching such a target. In neither of the cases, however, would one be able to assess the benefits in terms of avoided future damages.

Valuation of environmental effects in CBA may be helpful in attaining cost-effective decisions. However, the valuation should be based on a reasonably well-founded methodology. Speculative assumptions will not contribute to decision making. A measure that yields negative net benefits according to an analysis may be worthwhile if effects that are assumed to be positive but not explicitly valued in the calculation are well documented. Decision makers will normally manage to consider more than one measure simultaneously. CBA usually simplifies the decision by aggregating several effects, but there is no necessity for all effects to be aggregated into one single measure.

There are several fundamentally different types of costs and benefits that must be addressed, each of which requires somewhat different approaches to quantification and valuation:

- *Mitigation actions taken before the actual impacts are observed.* These are primarily a matter of reducing emissions or of removing greenhouse gases through reforestation. The vast majority of the applications of cost-benefit analysis have addressed the question of how best to achieve given levels of emission reduction. Valuation issues generally do not arise in this category.

- *Costs of mitigation actions taken after impacts become apparent.* These will necessarily occur in the future, and would be undertaken only if (1) climate change actually does occur and (2) climate change does indeed result in specific impacts. The cost of dikes to prevent inundation of coastal areas is a typical example in this category. Based on actual experience (such as in the Netherlands), the cost of such mitigation actions are relatively easy to establish.[56] Climatic engineering options, such as painting roads and roofs white or putting particles into the stratosphere also fall into this category. Again, there are few valuation issues here.

- *Costs (and benefits) of adaptation.* Society will adapt with varying degrees of pain to many of the impacts of climate change – indeed, society has already adapted to changes in climate that have occurred in the past. For example, climate change will affect crop yields and may result in poleward shifts in the distribution of cultivated land. Some areas will gain, and some will lose, and the consequences become an equity issue (between regions and countries) as much as a cost issue. Estimates of net losses for U.S. agriculture, for example, suggest a tolerable impact for the U.S. as a whole, but significant regional variations.[57] Some of the costs of adaptation will vary, depending on *ex ante* actions (e.g., the development of drought- or saline-resistant crops).[58]

- *Costs (and benefits) of nonadaptation.* In some cases, adaptation may not be possible or the cost of mitigation may be higher than the loss incurred in its absence. For ex-

BOX 5.2: APPLICATIONS OF DECISION ANALYSIS

Option analysis

In conventional CBA analysis, the usual decision rule is to take some action if the expected benefits exceed the expected costs. Depending on the degree of irreversibility present, a more appropriate rule is to take the action when benefits exceed costs by an amount at least equal to the value of the forgone option. Suppose some investment depends on some assumptions that are subject to great uncertainty – such as future world oil prices. If one makes an investment decision that is largely irreversible – such as building a large hydroelectric power project – then one loses the flexibility associated with waiting to learn more about the factors that affect oil prices. Preserving that flexibility has some economic value, namely the so-called option value. In financial and commodity markets such options to buy (and sell) are traded, with option prices determined by the market itself. But option value theory is now being applied to other fields involving capital-intensive investments, such as power generation.[1]

In applying these concepts to the climate change problem, there are many key differences. First, in one sense the problem is exactly opposite to that faced by, say, the power sector. In climate change, one loses flexibility if one does *not* make short-term investments to reduce greenhouse gas emissions. However, investment in reductions now is not free: Resources have to be diverted from other uses, and better emission technologies may be available in the future. Thus to some extent, committing now to current technology restricts the option to use better technology later. Second, unlike the financial and commodity context, there is no marketplace to set the value of the option.[2]

Decision analysis and hedging strategies

Among the early attempts at applying decision analysis to the climate change problem are those of Manne and Richels,[3] who have developed an approach for determining the optimal hedging strategy. The paradigm they use is that of a portfolio of insurance options:[4] What combination of insurance should be bought, if indeed any at all? What portion goes to R&D to resolve scientific uncertainties? What portion goes to the development of new supply and conservation technologies to reduce abatement costs? And what portion goes to immediate abatement of emissions? In particular, they focus on the value of information[5] and on how much accuracy is needed in climate modelling and impact assessment. Clearly, with *perfect information,* the best course of action can be charted immediately, and there is no need to hedge bets. Manne and Richels conclude that the need for precautionary near-term emission reductions is inversely related to the sustained commitment to R&D to develop better climate information (which reduces the need to hedge against an uncertain and potentially hostile future). However, given the inherent predictive uncertainty of climate change (and in particular the reliability of indicators), the limitations of such approaches need to be recognized.

---

[1]See, for example, Crousillat and Martzoukos (1991). This study reviewed power sector investment decisions in Costa Rica, Hungary, and West Africa. For a general review, see Dixit and Pindyck (1994).

[2]A marketplace might emerge if a tradable emission permit system were to be instituted. Chao and Wilson (1993) outline a means of using option values to quantify the flexibility associated with the purchase of a tradable emission permit instead of fixed capital investment in control technology.

[3]See, for example, Manne and Richels (1992).

[4]Lave (1991) notes that the concern about global climate change is not concentrated just in the rich nations but in the upper income groups in those nations. These are the same groups that voluntarily purchase insurance to protect themselves against other losses, such as those related to health, floods, and earthquakes. Persuading poor people to buy flood or earthquake insurance is exceptionally difficult even in the developed countries.

[5]The value of information under uncertain conditions is a concept much used in the private sector. For example, before embarking on the expensive proposition of drilling a wildcat well, oil drillers must decide how much ought to be spent on much less expensive prior survey work: Will general magnetic surveys suffice or are more expensive seismic surveys needed? Neither yields perfect information. For details of how such decision-theory models are applied in this field, see, for example, Newendorp (1976).

Table 5.3. *Potential impacts to be valued (for the U.S.)†*

Systems	Potential Impacts
Forests and terrestrial vegetation	Migration of vegetation Reduction in inhabited range Altered ecosystem composition
Species diversity	Loss of diversity Migration of species Invasion of new species
Coastal wetlands	Inundation of wetlands Migration of wetlands
Aquatic ecosystems	Loss of habitat Migration to new habitat Invasion of new species
Coastal resources	Inundation of coastal development Increased risk of flooding
Water resources	Changes in supplies Changes in drought and floods Changes in water quality + hydropower production
Agriculture	Changes in crop yields Shifts in relative productivity and production
Human health	Shifts in range of infectious disease Changes in heat-stress and cold-weather afflictions Changes in fertility due to stress
Energy	Increase in cooling demand Decrease in heating demand Changes in hydropower output
Transportation	Fewer disruptions in winter transportation Increased risk for summer inland transportation Risks to coastal roads
Weather-related damages	Damages related to changes in the frequency and severity of extreme weather events like storms

†Systems and potential impacts as listed in OTA (1993), except for weather-related damages.

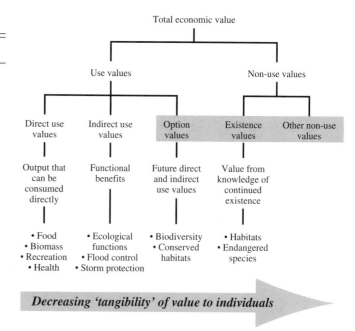

Figure 5.11: Categories of economic values attributed to environmental assets.

$$TEV = UV + NUV$$

or

$$TEV = [DUV + IUV + OV] + [NUV]$$

Figure 5.11 shows this disaggregation of *TEV* in schematic form. Below each valuation concept, a short description of its meaning and a few typical examples (based on a tropical rainforest) of the environmental resources underlying the perceived value are provided. Option values, non-use values, and existence values are shaded as a caution that some ambiguities are associated with defining these concepts. (As shown in the examples, they can spring from similar or identical resources, and their estimation can be interlinked also.) However, these concepts of value are generally quite distinct. Option value is based on how much individuals are willing to pay today for the option of preserving the asset for future (personal) direct and indirect use (see Box 5.3). In the context of uncertainty, quasi-option value is said to define the value of preserving options for future use in the expectation that knowledge – about the potential benefits or costs associated with the option (see Pearce and Turner, 1990; Fisher and Hanemann, 1987) – will grow over time. This approach may be quite relevant, given the great uncertainties associated with climate change. Existence value is the perceived value of the environmental asset unrelated either to current or optional use (i.e., the value it has simply because it exists). A variety of valuation techniques may be used to quantify the above concepts of value.[61]

ample, in some areas, the cost of construction of dikes may be far higher than the value of the land lost to coastal flooding; in such a case, the relevant cost to be estimated for purposes of CBA is the value of the land lost.

It is thus in the third and fourth of these categories that valuation issues arise. Table 5.3 lists the specific categories of impacts that may be encountered as a result of global climate change.

Conceptually, the total economic value (*TEV*) of a resource consists of its use value (*UV*) and non-use value (*NUV*).[59] Use values may be broken down further into the direct use value (*DUV*), the indirect use value (*IUV*), and the option value (*OV*) (potential use value).[60] One needs to be careful not to double-count both the value of indirect supporting functions and the value of the resulting direct use. One major category of non-use value is existence value (*EV*). Thus

The basic concept of economic valuation underlying all these techniques is the willingness to pay (WTP) of individuals for an environmental service or resource.[62] As shown in the box, valuation methods can be categorized according to which type of market they rely on and by considering how they make use of actual or potential behaviour.

BOX 5.3: TAXONOMY OF VALUATION TECHNIQUES

Type of Behaviour	Type of Market		
	Conventional market	Implicit market	Constructed market
Based on actual behaviour	Effect on production Effect on health Defensive or preventive costs	Travel cost Wage differences Property values Proxy marketed goods	Artificial market
Based on intended behaviour	Replacement cost Shadow project		Contingent valuation

Effect on production. An investment decision often has environmental impacts, which in turn affect the quantity, quality, or production costs of a range of productive outputs that may be valued readily in economic terms.

Effect on health. This approach is based on health impacts caused by pollution and environmental degradation. One practical measure related to the effect on production is the value of human output lost due to ill health or premature death. The loss of potential net earnings (called the human capital technique) is one proxy for forgone output, to which the costs of health care or prevention may be added.

Defensive or preventive costs. Often, costs may be incurred to mitigate the damage caused by an adverse environmental impact. For example, if drinking water is polluted, extra purification may be needed. Then, such additional defensive or preventive expenditures after the fact could be taken as a minimum estimate of the benefits of mitigation beforehand.

Replacement cost and shadow project. If an environmental resource that has been impaired is likely to be replaced in the future by another asset that provides equivalent services, then the costs of replacement may be used as a proxy for the environmental damage – assuming that the benefits from the original resource are at least as valuable as the replacement expenses. A shadow project is usually designed specifically to offset the environmental damage caused by another project. For example, if the original project was a dam that inundated some forest land, then the shadow project might involve the replanting of an equivalent area of forest elsewhere.

Travel cost. This method seeks to determine the demand for a recreational site (e.g., number of visits per year to a park) as a function of variables like price, visitor income, and socioeconomic characteristics. The price is usually the sum of entry fees to the site, costs of travel, and opportunity cost of time spent. The consumer surplus associated with the demand curve provides an estimate of the value of the recreational site in question.

Property value. In areas where relatively competitive markets exist for land, it is possible to decompose real estate prices into components attributable to different characteristics like house and lot size, air and water quality. The marginal WTP (willingness to pay) for improved local environmental quality is reflected in the increased price of housing in cleaner neighbourhoods. This method has limited application in developing countries, since it requires a competitive housing market, as well as sophisticated data and tools of statistical analysis.

Wage differences. As in the case of property values, the wage differential method attempts to relate changes in the wage rate to environmental conditions, after accounting for the effects of all factors other than environment (e.g., age, skill level, job responsibility, etc.) that might influence wages.

Proxy marketed goods. This method is useful when an environmental good or service has no readily determined market value, but a close substitute exists which does have a competitively determined price. In such a case, the market price of the substitute may be used as a proxy for the value of the environmental resource.

Artificial market. Such markets are constructed for experimental purposes, to determine consumer WTP for a good or service. For example, a home water purification kit might be marketed at various price levels, or access to a game reserve may be offered on the basis of different admission fees, thereby facilitating the estimation of values.

Contingent valuation. This method puts direct questions to individuals to determine how much they might be willing to pay for an environmental resource, or how much compensation they would be willing to accept (WTA) if they were deprived of the same resource. The contingent valuation method (CVM) is more effective when the respondents are familiar with the environmental good or service and have adequate information on which to base their preferences. Recent studies indicate that CVM, cautiously and rigorously applied, could provide rough estimates of value that would be helpful in economic decision making, especially when other valuation methods are unavailable.

Source: Munasinghe (1993).

Valuation techniques obviously need to be selected with some care, and in particular one must recognize that a given valuation technique may not necessarily capture the entire value. For example, if the replacement cost approach is being used to value the loss of forest area being inundated by a dam, it would likely capture only the *use* value. The value of biodiversity loss involved in the loss of primary forest, or a developed ecosystem, may not be included.[63]

We note that these valuation techniques have been developed for more conventional environmental impact analysis and would require significant modification and/or careful interpretation when applied to issues connected with global climate change (e.g., long-term intergenerational impacts, biodiversity loss, welfare comparisons across cultures or where there are wide gaps between gainers and losers, etc.). Nevertheless, whatever the difficulties, the importance of valuation remains, and the development of better techniques should be viewed as an important item in the overall climate change research agenda. Certainly, ignoring an impact because it cannot be satisfactorily valued carries high risk and is one of the reasons for the use of MCA (see below).

5.5.2.1 Discount rate
We noted in the introduction that CBA requires a very specific and explicit way of dealing with time. The first principle is that past (or "sunk") costs are ignored, based on the premise that, since past decisions cannot be changed, they have no bearing on decisions regarding the efficiency of resource use that are to be made in the present or in the future.[64]

The second principle is that a discount rate is applied to future costs and benefits to yield their present values. The issue of choosing an appropriate discount rate has been discussed in the context of general CBA for many years (Dasgupta *et al.*, 1972; Harberger, 1976; Little and Mirrlees, 1974; Sen, 1967). The long-term perspective required for sustainable development suggests that the discount rate might play a critical role in intertemporal decisions concerning the use of environmental resources (Arrow, 1982). We briefly discuss below several key issues relating to discount rates. The topic is dealt with more fully in Chapter 4.

Compared with most other economic investment decisions, the time perspective of measures aimed at mitigation of climate change is considerably longer. Cline (1992) suggests a 200-300-year time horizon for climate policy decisions, whereas investments in economic activities seldom need more than a 25-year horizon. This longer time horizon makes assumptions about how the economic and the environmental systems will develop and the discounting of future values critical to the evaluation of measures.

The discount rate denotes the social opportunity cost of capital. It reflects the net impact on total social benefits if one unit of present output is withdrawn from consumption and instead is invested elsewhere (for instance, in production or abatement). The criterion for optimal social and economic development is that the marginal total benefits from the different investments should be equal regardless of what the investments are aiming at. In other words, the social discount rate should be equal for all investments. If not, it would be possible to reallocate resources and attain a higher social benefit without any cost. Thus, the discount rate expresses a condition for dynamic (or intertemporal) efficiency.

The discount rate also provides a signal to decision makers who evaluate single projects or measures to take decisions in accordance with dynamic efficiency (over time). Even if one accepts the requirement to apply the same discount rate for marginal projects within a given time period, there are many potential optimal levels of the discount rate. This level depends, *inter alia,* on the social preferences about present versus future consumption that may be reflected in an intertemporal welfare function. The formulation of this function has been the subject of an extended debate in which questions about intertemporal comparisons as well as the current distribution of welfare have been raised.

It is worth emphasizing that, although externalities related to climate change will affect the social rate of discount, it is not sufficient merely to adjust this discount rate to take full account of climate change in a CBA. One must include also the "price" of the environment, which may increase substantially over time. As a consequence, future impacts from climate change may be quite important to present day decisions, even in "discounted terms": A 5% increase in the price of the environment will fully counteract the effect of a 5% discount rate.

To conclude, discounting is necessary in order to compare costs and benefits at different time periods. Attempts to avoid discounting or to apply a different discount rate for climate measures than for other investments will inevitably result in an inefficient policy. However, it is difficult to pick out the correct social discount rate, as there are no practical observations of such a rate. Furthermore, discount rates may depend on the future scenario that is assumed, and could vary over time – in particular, very long-term discount rates may be lower as economic growth rates saturate and decline (see Munasinghe, 1993).

5.5.2.2 CBA and equity
The benefits and costs of climate change mitigation strategies may accrue to different countries (and to different regions within larger countries) in different ways and at different times. How one reconciles these differences is therefore one of the central dilemmas facing policymakers, and it involves some crucial equity issues (see Chapter 3).

Thus, although CBA can provide answers on who should engage in how much abatement based solely on the criterion of maximizing economic efficiency, it must be recognized that some deviation from the global least-cost solution as obtained by CBA may have to be accepted to get international agreement. As indicated earlier in Figure 5.1, there will likely be a trade-off between equity objectives and economic efficiency. CBA can help define the trade-off curve, but it cannot provide an answer to what combination of economic efficiency and equity is necessary to get international agreement. However, whether there is a trade-off between equity and efficiency, and what the properties of this trade-off are, depend on what policy instruments are available. For example, if one permits side payments (in lump sum form) between countries, the efficient allocation of emissions across countries could be achieved independently of the equity issue.

Several concerns shape equity perceptions and the ability to obtain international agreements. Effective action to control climate change depends on a degree of international agreement. Therefore, a first obstacle to whatever mechanism might be agreed on is national sovereignty – to what extent will sovereign nations subject themselves to enforcement actions by others? Even simple agreements for joint implementation of projects involving two countries have run into difficulties (as discussed later in this section).

The second obstacle is the heterogeneous nature of the effects of greenhouse warming. Although the most widely cited measure of climate change is the average increase in global temperature, climate change affects different countries differently. In addition, the costs or response measures and their economic implications vary greatly among countries, particularly as a function of the level of development. Therefore, perceptions of the benefits of global cooperation will differ greatly.

A third obstacle is posed by strategic incentives. If some countries take the lead and set up a greenhouse gas control agreement, others have an incentive to free-ride and abstain from joining, as they cannot be excluded from the benefits such an agreement creates. If countries act selfishly in this way, few will become party to an agreement. Instead, most countries will not cooperate, and no general agreement will be reached even if all countries were to benefit from it. (This is the well-known prisoner's dilemma from game theory.[65]) Some argue that the overwhelming historical contribution to the build-up of greenhouse gases from developed countries constitutes an "environmental debt," that cannot be conveniently ignored using the traditional "sunk cost" approach. If past contributions to greenhouse gas emissions are considered from an equity viewpoint, establishment of appropriate side payment mechanisms from developed to developing countries, including financial assistance and technology transfer, could facilitate more enthusiastic cooperation by developing countries in efforts to mitigate climate change.

The question of joint implementation illustrates the limits of CBA in this regard. The motivation for joint implementation is a straightforward result of CBA. If a country – say the U.S. – decides to make an effort to reduce greenhouse gas emissions, and if reductions can be obtained at lower costs abroad, then the U.S. should initiate projects in other countries to minimize overall costs (Aaheim, 1993). The receiving country has nothing to lose if the additional cost of such a joint implementation project is covered by the investing nation. Yet many countries and organizations have reacted with skepticism to such an approach, for reasons that are political and based on equity concerns rather than on economic efficiency. The reasons include mistrust of the true willingness of the industrial countries to mitigate climate change, the belief that the ultimate reduction in greenhouse gases under such a regime would be negligible, and the suspicion that joint implementation gives industrial countries an opportunity to "buy themselves out of their problems" at the expense of the developing countries.[66] Indeed, some developing countries fear that

Table 5.4. *Criteria for choosing a strategy*

1.	Flexibility
2.	Urgency
3.	Low cost
4.	Irreversibility
5.	Consistency
6.	Economic efficiency
7.	Profitability
8.	Political feasibility
9.	Health and safety
10.	Legal and administrative feasibility
11.	Equity
12.	Environmental quality
13.	Private vs. public sector
14.	Unique or critical resources

Source: EPA (1989).

joint implementation investment might partly substitute for traditional forms of financial and donor assistance, and that such agreements might preclude the right of their own future generations to emit greenhouse gases.

Brazil and other countries have advanced a further reason for host country skepticism of joint implementation projects, namely that Annex I countries to the Framework Convention on Climate Change would invest in all the low-cost/high-return projects, and thus when non-Annex I countries were eventually required to curb emissions they would find the cheapest and best options already taken up.

5.5.3 Multicriteria analysis

Even the staunchest advocates of cost-benefit analysis would concede that economic efficiency (or economic value) is not the sole criterion in setting public policy, and that policymakers rightfully need to consider a broader set of objectives. Unfortunately, there is much confusion about what constitutes a coherent set of objectives. Table 5.4, taken from a major United States Environmental Protection Agency (EPA) study, lists the criteria suggested as constituting the basis for selecting public policy. The authors point out that the first four criteria listed – flexibility, urgency, irreversibility, and low cost – "would generally be given the highest priority." Note that many of these criteria overlap each other, and economic efficiency is among them!

Simple applications of CBA tend to focus only on economic efficiency. However, in more recent extensions, traditional CBA concepts are embedded in MCA, which expressly allows more than one objective and expressly addresses risk and uncertainty, thereby providing an integrating mechanism for most of the criteria listed. Multicriteria analysis techniques first gained prominence in the 1970s, when the intangible environmental externalities lying outside conventional cost-benefit analysis (CBA) methodologies were increasingly recognized. It also met one objective of modern decision makers, who preferred to be presented with a range of feasible alternatives, as opposed to one "best" solution. MCA also allows for the appraisal of alternatives with differing objectives and varied costs and benefits, which are often assessed in differing units of measurement.

Of the criteria listed in Table 5.4, criteria 1, 2, 3, 4, 5, and 6 can all be treated by modern decision analysis. Indeed, questions of timing (urgency), flexibility (or robustness), capital constraints ("low cost") are all central elements of the approach. Also, criterion 13 is really part of 6. (In the text, the authors of the EPA report amplify the criterion as follows: "Does the strategy minimize governmental interference with decisions best made by the private sector?") Furthermore, modern valuation techniques permit substantial parts of criteria 9 and 12 to be included in the economic analysis as well. As conceded by the EPA report (p. 393), "if the principal costs and benefits can be quantified in monetary terms, economic theory provides a rigorous procedure for making trade-offs between present and future costs, and for considering uncertainty, profitability, and most of the other criteria."

There is also a need to separate the basic goals of public policy – such as economic efficiency and equity – which surely have primacy, from implementation issues such as legal and administrative feasibility, which are generally secondary. The premise of CBA analysis is that one looks first at the primary objectives and then asks how many of the primary objectives one may have to sacrifice to achieve practical implementation. This principle has become accepted in many areas of policymaking. For example, the starting point for setting electric utility rates is to calculate the economically efficient tariff (based on marginal costs) and then make adjustments to protect low income groups (through lifeline rates, special provisions for disconnection in the event of nonpayment, etc.). The essence of the approach is not that noneconomic issues are ignored, but that the trade-offs between economic efficiency and equity (or indeed other objectives not readily monetized) are explicitly quantified and displayed in such a manner that decision makers are made aware of how much of one objective is traded off in the interests of the other.

Indeed, one of the advantages of MCA is that it forces political decision makers to look at the trade-offs between their major objectives rather than attempting to boil down everything into a single number, particularly where valuation techniques may be controversial. Nowhere is this more important than in the valuation of risk to human life.[67]

The application of MCA methods involves the following steps:

(1) Selection and definition of attributes, say A_i, $i = 1, \ldots n$, selected to reflect important planning objectives. Although the two major relevant attributes in the context of the global climate change problem are cost and greenhouse gas emission reductions, we have already noted that strategies to control emissions may have other side effects, some positive and some negative, that may also be difficult to value and that might therefore require consideration of additional attributes (such as biodiversity and equity).

(2) Quantification of the levels A_{ij} of the i attributes estimated for each of the j alternatives. In this quantification,

Table 5.5. *Technology interventions for greenhouse gas emission reductions*

Option	Comments	Symbol
Wind energy	305 MW total	**wind**
Minihydro		**miniHy**
DSM: energy efficient refrigerators		**EEF**
DSM: compact fluorescents		**CFL**
Transmission & distribution loss reduction	10% T&D loss goal (in place of 12%) by 2000	**TD+**
	12% goal delayed to 2003	**TD–**
Max hydro	Builds both reservoirs in the Upper Kotmale project; 144 MW high dam version of Kukule	**maxHy**
Clean coal technology	Pressurized fluidized bed combustion-combined cycle units; assumed for all coal units after 2000	**PFBC**
	With pessimistic capital cost assumptions	**PFBC–**
Clean fuels	Use imported low-sulphur residual oil for diesels (0.5% S by weight rather than 2.5% S)	**low S oil**
	Use low sulphur (0.5%) coal (rather than 1% S coal)	**low S coal**
FGD systems	Model free to choose optimal generation mix; coal plants must have FGD systems	**FGD**
	FGD systems forced onto basecase solution	***FGD**
No coal	Model free to choose least-cost combination of diesels + hydro	**noCoal**

full consideration must be given to discounting issues, for noneconomic and economic attributes alike. At this stage of the analysis, trade-off curves are powerful tools for communicating with decision makers. They are particularly relevant in a situation, such as the climate change problem, where the quantification of benefits may be difficult and where decision makers must act largely on the basis of trading-off short-term costs against certain levels of greenhouse gas emission reduction.

(3) Determination and application of a decision rule, which amalgamates the information into a single overall value or ranking of the available options or which reduces the number of options for further consideration to a smaller number of candidate plans. Where amalgamation is contemplated, attribute levels are first translated into a measure of value, $v_i(A_{ij})$ (also known as the attribute value function).[68] This is sometimes combined with a normalization procedure, usually on a scale of 0 to 1 (in which the lowest value of the attribute value function is assigned 0, the highest 1). Subsequently, weights w_i for each attribute must be determined to arrive at the overall amalgamation.

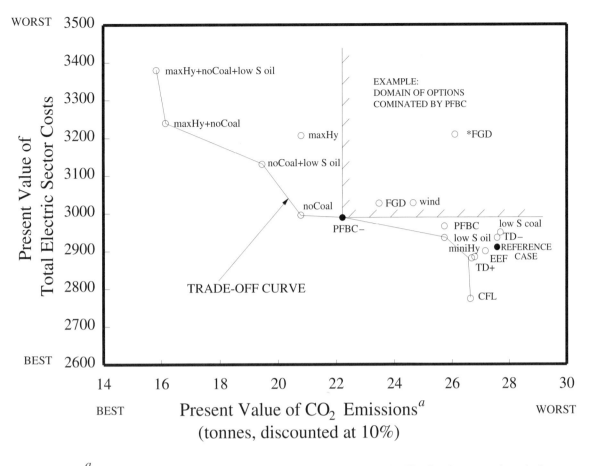

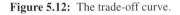

^{a}Calculating the present value of tonnes of CO_2 emissions over the life of each measure is equivalent to assuming that the damage due to a tonne of CO_2 remains constant over this period. It is not necessary to know the monetary value of the damage due to CO_2 emissions since the analysis only involves a comparison of options.

Figure 5.12: The trade-off curve.

Trade-off curves are a particularly useful tool for the analysis of energy-environmental policy options. Figure 5.12, taken from a recent study of options for greenhouse gas emission reductions in Sri Lanka (Meier *et al.,* 1993), illustrates the essential concepts. The figure is a plot of two attributes – greenhouse gas emissions and total system costs – for the technology options identified in Table 5.5. Each point represents a perturbation of the reference case, defined as the official 1993 basecase capacity expansion plan of the national power company, the Ceylon Electricity Board (CEB).

The *trade-off curve* is the set of options that are not dominated by others (sometimes referred to as the "noninferior set"). These are the options that are "closest" to the origin, and therefore represent the "best" set of options that merit further attention.[69]

Several useful concepts arise here. First is the concept of *dominance*.[70] Pressurized fluidized bed combustion (PFBC – a clean coal technology) is said to dominate the options in the sector shown, namely flue gas desulphurization (FGD) and wind. PFBC has better costs and better (i.e., lower) greenhouse gas emissions, and is thus preferred over the other op-

tions under both criteria. If only these two attributes mattered, there would be no reason to select any of the dominated options in place of PFBC.

Another perspective is gained by dividing the solution space into quadrants with respect to the reference case (Figure 5.13). The options that fall into quadrant III are the "win-win" options, which are better than the reference case in *both* attributes. In this case, minihydro, energy-efficient refrigerators, transmission and distribution system loss reduction, and compact fluorescents all fall into this quadrant, providing both cost and emission gains. Such win-win solutions were mentioned earlier, in Section 4, in connection with the empirical estimates of the MAC curves (e.g., Figure 5.4). These "below-the-line" options in the MAC curves are equivalent to the options in quadrant III of a multiattribute analysis.[71]

Finally one should note that MCA leads to *implicit* valuations whenever two options are compared. For example, in the case of Figure 5.13, a decision maker who prefers option Y (maximum hydro + no coal) to option X (no coal + low sulphur oil) makes an implicit valuation of the concomitant re-

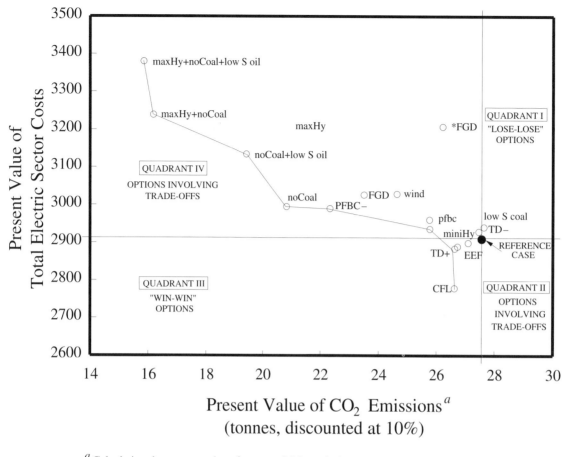

Source: Meier *et al.* (1993).

Figure 5.13: "Win-win" options.

duction of greenhouse gas emissions in terms of the increased costs (i.e., equal to the slope of the trade-off curve between X and Y or about 200$/ton of CO$_2$).[72]

It must also be noted that the choice of criteria in an MCA will depend on each country's short- and long-term development plans. Despite a common global objective of stabilizing atmospheric concentrations of greenhouse gases, developing countries may use different criteria because of immediate or urgent needs to ensure food supplies and service debt requirements. Consequently different countries may place different weights on the attributes.

5.6 Conclusion

Cost-benefit analysis has many advocates but also many detractors. Certainly the rather narrowly defined traditional approaches to CBA, developed originally for project-level decision making with planning horizons typically no more than 20 years, clearly have difficulty in dealing with the very long time frames and high levels of uncertainty encountered in the climate change context. This chapter has interpreted modern CBA more broadly to encompass a family of decision-analysis techniques that includes cost-effectiveness analysis, multicriteria analysis, and decision analysis, in addition to traditional cost-benefit analysis.

Despite the current limitations of these various techniques, modern CBA (broadly defined) remains the best framework for identifying the essential questions that policymakers must face when dealing with climate change. The CBA approach forces decision makers to compare the consequences of alternative actions, including that of no action, on a quantitative basis. To the extent that some impacts and measures cannot be valued monetarily (e.g., biodiversity), extensions of the traditional CBA approach, such as multicriteria analysis, permit some quantitative expression of the trade-offs to be made.

Decision analysis also provides many insights for dealing with uncertainty. Flexible policies are essential when faced with large uncertainties. Increased knowledge may narrow uncertainty, but the range of options may not necessarily increase.

Finally, the most important benefit of applying CBA is not necessarily the predicted outcome (which always depends on assumptions and the particular technique used) but the *process itself* (which establishes a framework for gathering information and forces an approach to decision making that is based on rigorous and quantitative reasoning).

Endnotes

1. However, it is not universally accepted that cost-benefit analysis is appropriate to the analysis of policy options to address global climate change. A major report recently issued by the U.S. Office of Technology Assessment (OTA, 1993) contains an extensive discussion of how adaptation strategies should be chosen, yet manages to avoid all mention of cost-benefit analysis *per se*. It talks about how one might minimize vulnerability to climate change and about insurance strategies, but avoids the central question of how one might determine the amount of insurance one wishes to buy. Similarly, priorities may be set on noneconomic grounds, and CBA could be used in a secondary role (see e.g., Turner, 1991).
2. Although this needs a bit of modification in the presence of capital constraints, which may limit selection of the "best" single project.
3. However, as we shall see later, extensions of CBA *can* help in identifying the *trade-offs* between economic efficiency and equity.
4. The degree of emission abatement is reported in such studies in two rather different ways. The first is as a reduction from some baseline – itself defined as the trajectory of greenhouse gas emissions for some postulated business-as-usual scenario. The second is in terms of reductions from some reference year (e.g., "reduce greenhouse emissions to 80% of their 1990 levels by 2010").
5. The ethical and epistemological aspects of the climate change problem are not addressed here. For further discussion, see, for example, Brown (1992).
6. A recent survey of economists and scientists knowledgeable about the climate change problem elicited typical views at the extremes of this spectrum (Nordhaus, 1994a). One respondent argued, "the existence value of species is irrelevant – I don't care about ants except for drugs," while another cautioned that "loss of genetic potential might lower the income of the tropical regions substantially." In Section 5.5 we address the different types of values – use, option, existence – in more detail.
7. But see below for a discussion of the difficulties of making cross-country comparisons of costs.
8. There may also be some outcomes that are inefficient, namely those that lie inside the frontier of efficient points shown in Figure 5.1. Such an inefficient point is represented by option 6. This is discussed further in the presentation of multicriteria analysis, below.
9. It should be noted that the algebra of cost-benefit analysis can be expressed in many different ways: the cost-benefit ratio, net present value, or the internal rate of return are all different ways of doing the arithmetic. However, particularly in situations involving portfolios of potential actions, and where shortages of capital may constrain the choice, great care must be paid to rigorous application of the principles; otherwise, different methods may yield different decisions. Maximizing net present value subject to applicable re-

source constraints is the most useful approach for climate change analysis.
10. Perhaps the simplest and most intuitive example of why the marginal cost of emission reduction increases with increasing levels of reduction is the removal of pollutants from wastewater. The first 60% can be easily removed by a settling basin: Large particles simply settle by gravity, and all that is required is a structure in which the process can take place. The next 30%, however, requires biological treatment. This involves not just a physical structure, but pumps to aerate the water to promote the growth of aerobic bacteria. The next 5-6% requires chemical treatment, with high operating costs arising from the use of chemical agents. Although 100% removal is theoretically possible, it would require complete distillation, which is extremely expensive.
11. Although we show the marginal cost and benefits of emission reduction as linear in Figure 5.2, this need not be so. For example, where abatement costs are subject to economies of scale, there might be sections of the MAC curve that have a form other than that shown in Figure 5.2. But as noted elsewhere (e.g., Figure 5.4), empirical studies of the marginal cost curves frequently do exhibit the stylized shapes shown in Figure 5.2.
12. The acidity of precipitation is influenced by complex interactions between sulphur oxides, related oxidation products, and NO_x.
13. For example, the presence of CFCs could affect climate change not only directly through their global warming potentials, but also indirectly through the impact of stratospheric ozone depletion on biota-like nanoplankton – which in turn influence oceanic CO_2 uptake. Similarly, the degree of reliance on fossil fuels would affect CO_2 emissions directly and CO_2 absorption indirectly – via the effects of acid rain on forests and biomass.
14. There are some exceptions, notably for radioactive wastes, which also have extremely long lifetimes. Thus the total environmental risk, and the scale of the disposal problem at any one time, is not so much dependent on current rates of production of nuclear wastes as on the total stock.
15. IPCC (1995) estimates the lifetime of methane at 14.5 ± 2.5 years.
16. See, for example, Hoel and Isaksen (1993, 1994) for a further discussion and numerical calculations.
17. There are exceptions here as well, most notably the phenomenon of acid rain, which is largely a long-range phenomenon often involving emissions in one country and acid rain damage in another. However, to the extent that lake acidification completely destroys aquatic ecosystems, one could argue that at least some of the impacts are irreversible, although even here the impacts are generally of a fairly local nature. Long time periods may also elapse between the onset of acid rainfall and actual visible damage.
18. For new data on emissions and acid deposition rates in Asia, see World Bank (1993).
19. To be sure, there are exceptions. Even the richer countries of Europe are affected by mutual pollution problems (e.g., acid rain in Scandinavia from the UK, or the severe water pollution problems in the Rhine Basin involving Switzerland, Germany, France, and the Netherlands, or the dumping of wastes in the North Sea). However, in most cases where international pollution issues involve richer countries, much better institutional mechanisms exist for addressing these problems (e.g., the EU in Europe) than are available for resolving environmental disputes between rich and poor.
20. For a discussion of the relationship between economic assistance for restructuring in Eastern Europe and assistance to guarantee desired environmental standards, see, for example, Amman *et al.* (1992).
21. However, the indirect impacts (for example, large-scale immigra-

tion from Mexico that might follow from agricultural devastation in that country) may ultimately prove to be much more serious for the U.S. than the direct consequences of sea level rise or higher energy bills for air conditioning, but such impacts are also very difficult to quantify and many regard them as speculative. Proper CBA analysis would correct for such distorted perspectives.

22. In some cases laboratory experiments – such as growing plants in CO_2-enriched atmospheres – do provide some actual data for predictions.

23. A recent UNEP review of greenhouse gas abatement costing studies concluded that "the state of abatement costing studies in developing countries is wholly inadequate even to draw preliminary conclusions concerning possible costs and the impact of different abatement options. It is a body of analysis which is only just beginning, and which may take many years to mature towards consensus even on very rough estimates and understanding of the key issues" (UNEP, 1992).

24. See, for example, the review of eleven studies by London Economics (1992).

25. See, for example, Moreira *et al.* (1992) for Brazil; Sitnicki *et al.* (1991) for Poland; or Meier *et al.* (1993) for Sri Lanka.

26. Unfortunately, there is some confusion in terminology with respect to this point. Some (e.g., London Economics, 1992) use the term "no regrets" to describe policies for which MB > MAC, that is, for which the marginal benefits exceed the marginal costs. Others use the term only where MAC < 0, that is, for those options that are "below the line" in the empirical cost curves of the type shown in Figure 5.5. However, since on both the cost and the benefit side there will be some netting out (e.g., to account for joint costs and benefits), the criterion MAC< 0 is arbitrary, whereas MAC< MB is well defined.

27. The fact that such "no-regrets options" are in fact observed is much debated. What may be calculated as monetary benefits need not necessarily be regarded as benefits by decision makers – there may be other, nonmonetary costs involved. The huge subsidies given to European agriculture illustrate the point that more than monetary benefits might affect decisions. In the case of developing countries, the unavailability of finance may constrain the ability to implement some of these options. For example, until recently, obtaining finance through export credits for power generation expansion has been much easier than financing energy efficiency measures.

28. These methodological problems have been recognized by the GEF, which has initiated a research programme to find an operational approach for measuring and agreeing on full incremental costs within the context of the Framework Convention on Climate Change. This is the so-called PRINCE study (Programme for Measuring Incremental Costs for the Environment); see King (1993).

29. Another problem here is that the concept of "least cost," as well as the integrated planning process to achieve it, may be quite complex (see, for example, Munasinghe, 1990; Meier, 1990; Crousillat, 1989). Such a solution, typically obtained by fairly sophisticated optimization models, may be "least-cost" only for a very narrow band of input assumptions; and if these assumptions prove to be different, then an investment programme predicated on the "least-cost" plan may ultimately be distinctly nonoptimal.

30. In the words of King (1993),

> How can the adoption of apparent win-win solutions be stimulated? Such solutions are sometimes referred to as negative incremental cost projects because they are economically viable in their own right. The dilemma arises because these projects are often not being funded. On the one hand, if the GEF restricts itself to those projects that have positive incremental cost while the bulk

of negative incremental cost options remains unfunded, it risks becoming irrelevant to the main solution to the global environmental problem. On the other hand, providing grant finance for economically viable projects effectively makes a net transfer to the country, which is not the purpose of new and additional funding; worse, it provides a perverse incentive to potential recipient countries to delay economic reform.

31. This is likely to be true even though higher CO_2 concentration itself may promote plant growth.

32. For further discussion, see, for example, IPCC (1990) or Lave and Vickland (1989). However, as also noted by Nordhaus (1994b), the fertilizing effect of atmospheric CO_2 is a particularly strong mitigating factor for agricultural nations, particularly where water is a limiting factor. Although the extent and quantitative importance of CO_2 as a fertilization agent are controversial, the balance of the evidence is positive.

33. See, for example, Glieck (1989) or Pachauri (1991).

34. This report identifies thirty-nine studies (or elements of studies) focussed on inventories of greenhouse gases, forty-eight studies of effects, and forty-six studies of mitigation.

35. For example, in a report prepared by the Tata Energy Research Institute for the India Ministry of Environment and Forests (TERI, 1991), the chapter "Adaptive Strategies for India in the Perspective of Climate Change," which elaborated on the impacts associated with given levels of sea level rise and mean temperature, was entirely qualitative in nature, and no cost estimates were attempted. Similarly, neither the volume on climate change published by the Asian Energy Institute (Pachauri and Behi, 1991) nor the country studies conducted by UNEP (reported in UNEP, 1992) contains any country-specific estimate of costs associated with specific levels of greenhouse gas accumulation in the atmosphere. The Asian Development Bank is currently sponsoring a multicountry project in Asia that is attempting to establish costs of potential effects.

36. There are several related issues. For example, should one use nominal exchange rates, trade-weighted rates, or purchasing power parity rates in making cross-country comparisons of costs or impacts?

37. The question of "green accounting" is one of valuing and aggregating marketable and nonmarketable goods (see, for example, United Nations, 1993, or Munasinghe *et al.*, 1995). The many different proposals for "green accounting" reflect the difficulties involved. Rather than speculate about valuation or poorly founded proxies for nonmarketable goods, MCA provides an alternative approach.

38. These results are for studies that estimate both emission reductions and GDP impacts relative to some baseline, i.e., relative to the trajectory of emissions in the absence of the policies followed. For many countries, particularly the developing nations, even significant reductions from such baselines may still imply increases in the absolute quantity of greenhouse gases emitted. Other studies report results in terms of greenhouse gas emissions in some future target year, relative to emissions in some prior year (e.g., 2010 emissions 10% less than 1990 emissions).

39. For further discussion see UNEP (1992), p.72. Studies that have examined the question of unilateral vs. multilateral approaches include Proost and van Regemorter (1990), Edmonds and Reilly (1985), and Rutherford (1992).

40. Additionally one might note that, although the environmental risks of current technologies are fairly well established, those that apply to new technologies – including some (such as fusion or new nuclear reactor cycles) that are sometimes proposed as solutions to the climate change problem – are less certain.

41. See Chapter 6 (Section 6.7) for a more extensive discussion of the secondary benefits of reducing greenhouse gas emissions. Results reported there range from $2/tonne of carbon abated to over $500/tC.

42. For example, in one alternative (the "Treaty Alternative"), Alfsen *et al.* (1992) estimated the loss in GDP at 3.1 billion Kroner and the loss of private consumption at 2.6 billion. However, joint product benefits were estimated at 2.4 billion (with a range of 1.0-3.8 billion).

43. For purposes of clarity, we have drawn only a single marginal cost curve (MAC) in Figure 5.7. It is quite possible, however, that different countries will also have different marginal cost curves (although the differences in costs are likely to be smaller than the differences in benefits).

44. For example, studies by Burgess (1990) and Larsen (1993) address the impact of eliminating price subsidies. They estimate the level of reduction in greenhouse gas emissions by applying assumed price elasticities to the difference between the subsidized and unsubsidized prices. Burgess, using the difference between the actual average cost of electricity and the estimated long-run marginal cost (LRMC) and applying an assumed long-run price elasticity of -1, estimates the reduction in greenhouse gas emissions for eleven countries, including the U.S., China, India, and some small developing countries such as Tanzania and Peru. Not surprisingly, the bulk of the total carbon emission savings of 124 million tons/year (mtpy) comes from coal fuel savings, of which India accounts for 11.9 mtpy, China 26.6 mtpy, and the U.S. 85.4 mtpy. Larsen, performing the same analysis but from the perspective of *fuel* prices, applies estimated own- and cross-price elasticities for the different fossil fuels to the difference between an appropriately adjusted border price and the domestic subsidized fuel price, and, more significantly, includes the countries of the former USSR and Eastern Europe. In this analysis, the former USSR (917 mtpy) and Poland (105.2 mtpy) dominate the results; indeed, the combined estimated impact of India (54 mtpy) and China (45.4 mtpy) together is less than that for Poland.

45. The GEF has recently introduced the term "type I project" for those which operate in the range AC (i.e., for which national economic benefits are greater than national costs) and "type II project" for those in the range CD (i.e., for which national economic benefits are less than national economic costs, but global benefits are such that the project is justified under GEF criteria). For further discussion, and arguments for why the GEF should give priority to type II projects, see, for example, Anderson and Williams (1993).

46. This is a partial list. Other problems include the assumption that noncarbon-based materials are benign with respect to the greenhouse gas problem, uncertainty concerning natural sources, and uncertainty over the spatial distribution of physical impacts.

47. See IPCC (1995), Chapter 6, for further discussion of uncertainty in future emissions.

48. One need only remember how dramatically growth rates and energy/GDP ratios fluctuated in the decade following the 1973 oil crisis to recognize the hazards of such forecasts.

49. This issue is made more complicated by the fact that much deforestation is presently driven by the need to expand pasture land, which in turn implies higher methane emissions from livestock.

50. For a full discussion of uncertainty in emissions, see, for example, Ebert and Karmali (1992).

51. See note 45.

52. The U.S. Nuclear Regulatory Commission has made repeated attempts to establish probabilities for specific kinds of accidents through a technique known as fault tree analysis. Despite the appearance of scientific rigour, the resultant probability estimates remain highly controversial. For examples of the use of cost-benefit analysis by the U.S. NRC, see, for example, Mubayi *et al.* (1991) or Abrahamson *et al.* (1989).

53. Precautionary expenditures may also be influenced by aversion to catastrophic risks. A recent example is the hundreds of billions of dollars per year spent in the U.S. to avert nuclear attack.

54. Perhaps there is a possible role for financial markets in insuring (and pricing) environmental risks (just as utilities can insure or self-insure against environmental damages caused by accidents at plant sites), but since the damage estimates related to climate change are very difficult to assess, the opportunities for such an approach seem limited, even if occurrence probabilities for various damages could be estimated more accurately.

55. If prices of marketed goods are distorted (e.g., due to arbitrary taxes and subsidies), it will be necessary to use shadow prices – usually the set of economic opportunity costs or efficiency prices – to determine their correct economic value (for details, see Dasgupta *et al.,* 1972; Little and Mirrlees, 1974).

56. It should be noted that the distinction between a mitigation measure and an adaptation measure may not always be clear cut.

57. "Preliminary results suggest that although U.S. crop production could decline, supplies would be adequate to meet domestic needs" (EPA, 1989). It might well be pointed out, however, that this reflects the very narrow perspective of the study: U.S. grain exports represent a significant supply of food for developing countries, and were the U.S. surplus to decline, developing countries might well be concerned about the use of food exports as a political weapon.

58. Particularly difficult to value is the cost of forced adaptation and population movements – a problem already encountered in cost-benefit analyses of the impact resulting from the creation of large reservoirs where significant numbers of individuals must be forcibly relocated (World Bank, 1994).

59. For further details, see Munasinghe (1993) and Pearce and Warford (1993).

60. The issue of option values and irreversibility in CBA has received increasing attention in the literature, starting with Arrow and Fischer (1974) and continuing more recently with Chichilnisky and Heal (1993).

61. For a recent overview of techniques suitable for valuing environmental costs and benefits, especially in developing countries, see Munasinghe (1993).

62. The theoretically correct measure of WTP is the area under the Hicksian demand curve, which describes the relationship between the price and quantity demanded of the environmental resource, keeping the level of consumer utility intact. Problems of measurement may arise because the commonly estimated demand function is the Marshallian one, which indicates how demand varies with the price of an environmental good, while keeping the user's income level constant. In practice, it has been shown that the Marshallian and Hicksian estimates of WTP are in good agreement for a variety of conditions, and in a few cases the Hicksian function may be derived once the Marshallian demand function has been determined (Willig, 1976; Braden and Kolstad, 1991). What people are willing to accept (WTA) in the way of compensation for environmental damage is another measure of economic value that is related to WTP. WTA and WTP could diverge (Cropper and Oates, 1992). In practice, either or both measures are used for valuation.

63. Contingent valuation methods in particular are somewhat controversial and need great care in their application to produce credible results.

64. One also needs to take note of the fact that this principle is also not universally accepted, particularly by political leaders. The principle that "past sacrifices ought not to be in vain" is frequently invoked as relevant to present decisions. Indeed, in the climate change debate, developing countries correctly note that it is the developed countries that are largely responsible for the present levels of greenhouse gases in the atmosphere and that this past behaviour *is* relevant in the search for equitable solutions in the future. However, these observations point to the *equity* dimension of the problem and do not affect how one ought to seek the economically efficient options.

65. However, for a criticism of this argument, see Brown (1992), who points out that some European countries have already taken unilateral action to reduce greenhouse gases in the hope that others will follow. Moreover, other game-theory paradigms have been proposed for modelling international environmental negotiations. For example, Carraro and Siniscalo (1992, 1993) propose a "chicken game" framework belonging to the class of coordination games.

66. For a discussion of these issues, and a discussion of how CBA can make a contribution to the evaluation of such projects, see Aaheim (1993). Concerns about "market justice" and related considerations are outlined in Rose (1990).

67. Differential valuation of human lives is strenuously opposed by some: if a U.S. life is worth $1.5 million, then so is everyone else's. On the other hand, when making comparisons of per capita GDP, economists are increasingly turning to purchasing power parity adjustments in an attempt to make more valid comparisons of economic level. MCA avoids these difficulties by moving judgments about the value of human life from the domain of technical assessment into the domain of political decision making, where such judgments properly belong.

68. MCA deals with attitudes towards risk and uncertainty at this stage by the use of multiattribute utility functions, which explicitly capture attitudes towards risk. See Keeney and Raiffa (1976) for an application in water resource planning, or Keeney and von Winterfeldt (1987) for application to electric utility planning.

69. We note also that the trade-off analysis and surfaces will be much more complex as the number of attributes increases.

70. Decision analysis distinguishes among several types of dominance – such as strict dominance and significant dominance. See, for example, Meier and Munasinghe (1994) for an application of these concepts to environmental decision making.

71. In general, the trade-off curve may extend into quadrant II, and quadrant III may contain fewer solutions or none at all.

72. However, because of the presence of joint products – each represented by a different dimension in the multidimensional trade-off space – valuations that look only at two dimensions need to be interpreted with some caution.

References

Aaheim, A., 1993: *Benefits and costs of climate measures under joint implementation,* Centre for International Climate and Environment Research, University of Oslo.

Abrahamson, S., *et al.,* 1989: *Health effects models for nuclear power plant accident consequence analysis,* NUREG/CR-4214, Sandia National Laboratories, Washington, DC.

Alfsen, K. H., A. Brendemoen, and S. Glomsrod, 1992. Benefits of climate policies: Some tentative calculations, Discussion Paper, Central Bureau of Statistics, Oslo.

Amman, M., L. Hordijk, G Klaasen, W. Schöpp, and L. Sørensen, 1992: Economic restructuring in Eastern Europe and acid rain abatement strategies, *Energy Policy,* **20,** 1187–1198.

Anderson, D., and R. H. Williams, 1993: The cost-effectiveness of GEF projects, Global Environment Facility, Working Paper 6, Washington, DC.

Arrow, K.J., 1966: Discounting and public investment criteria. In Kneese and Smith, 1966.

Arrow, K.J., 1982: The rate of discount on public investments with imperfect capital markets. In *Discounting for time and risk in energy policy,* R.C. Lind, ed., Resources for the Future, Washington, DC.

Arrow, K.J., and A. Fischer, 1974: Environmental preservation, uncertainty and irreversibility, *Quarterly Journal of Economics,* **88,** 312–319.

Arrow, K.J., and M. Kurtz, 1970: *Public investment, the rate of return and optimal fiscal policy,* Johns Hopkins University Press, Baltimore.

Asian Institute of Technology, 1993: UNEP greenhouse gas abatement cost studies: The case of Thailand, Bangkok, Thailand.

Barker, T., 1993: Secondary benefits of greenhouse gas abatement: The effects of a UK carbon/energy tax on air pollution, Energy Environment Economy Modelling Discussion Paper No. 4, Department of Applied Economics, University of Cambridge.

Baron, W., and P. Hills, 1990: Greenhouse gas emissions: Issues and possible responses in six large lower income Asian nations, Working Paper No. 45, Centre for Urban Planning and Environmental Management, University of Hong Kong.

Bernstein, M., 1992: Costs and greenhouse gas emissions of energy supply and use, Environment Department, Working Paper 1993–40, World Bank, Washington DC.

Blitzer, C., R. Eckhaus, S. Lahiri, and A. Meeraus, 1993: Growth and welfare losses from carbon emissions restrictions: A general equilibrium analysis of Egypt, *Energy Journal,* **14,** 1, 57–81.

Bohm-Bawerk, E. von, 1989: *Kapital und Kapitalzins: Zweite Abteilung: Positive Theorie des Kapitals,* Wagner, Innsbruck.

Braden, J.B. and C.D. Kolstad, eds., 1991: *Measuring the demand for environmental quality,* North Holland/Elsevier, Amsterdam.

Brekke, K. A., 1994: Net national product as a welfare indicator, *Scandinavian Journal of Economics,* **96,** 2, 241–252.

Bromley, D., 1986: *Natural resource economics: Policy problems and contemporary analysis,* Kluwer/Nijhoff Publishing, Boston, MA.

Brown, P. G., 1992: Climate change and the planetary trust, *Energy policy,* **20,** 208–222.

Burgess, J., 1990: The contribution of efficient energy pricing to reducing carbon dioxide emissions, *Energy Policy,* **18,** 449–455.

Butlin, R. (ed.), 1981: *Economics of the environment and natural resource policy,* Westview Press, Boulder CO.

Carraro, C., and D. Siniscalo, 1992: The international dimension of environmental policy, *European Economic Review,* **36,** 379–387.

Carraro, C., and D. Siniscalo, 1993: Strategies for the international protection of the environment, *Journal of Public Economics,* **52,** 309–328.

Center for Energy and Environment, University of Pennsylvania, 1992: Costs and Greenhouse gas emissions of energy supply and use.

Chao, H. P., and Wilson, R., 1993: Option value of emission allowances, *Journal of Regulatory Economics,* **5,** 233–249.

Chichilnisky, G., and G. Heal, 1993. Global environmental risks, *Journal of Economic Perspectives,* **7**(4), 65–86.

Cline, W.R., 1992: *The economics of global warming,* Institute for International Economics, Washington, DC.

Coppel, J. 1991: An analysis of energy policy measures and their impact on CO_2 emissions. In Jones and Wheeler, 1992, pp. 263–286.

Cropper, M.L., and W.E. Oates, 1992: Environmental economics: A survey, *Journal of Economic Literature,* **30** (June), 675–740.

Crousillat, E., 1989: Incorporating risk and uncertainty in power sector planning, Industry and Energy Department Working Paper, Energy Series Paper 17, World Bank, Washington, DC.

Crousillat, E., and S. Martzoukos, 1991: Decision making under uncertainty: An option valuation approach to power planning, Industry and Energy Department Working Paper, Energy Series Paper 39, World Bank, Washington, DC.

Crousillat, E. and H. Merrill, 1992: The trade-off/risk method: A strategic approach to power planning, World Bank, Industry and Energy Department, Washington, DC.

Dasgupta, P., S. Marglin, and A.K. Sen, 1972: *Guidelines for project evaluation,* UNIDO, New York.

Dixit, A.V. and R.S. Pindyck, 1994: *Investment under uncertainty,* Princeton University Press, Princeton, NJ.

Dreze, J.H., and A. Sandmo, 1971: Discount rates for public investment in closed and open economies, *Economica,* **38,** (November), 395–412.

Ebert, C., and A. Karmali 1992: Uncertainties in estimating greenhouse gas emissions, Environment Working Paper 52, World Bank Environment Department, World Bank, Washington, D.C.

Edmonds, J., and J. Reilly, 1985: *Global energy: Assessing the future,* Oxford University Press, New York.

EPA (Environmental Protection Agency), 1989: The potential effects of global climate change on the United States, EPA-230-05-89-050, Washington, DC, December, p. 390.

Fisher, A.C., and M. Hanemann, 1987: Quasi-option value: Some misconceptions dispelled, *Journal of Environmental Economics and Management,* **14,** 183–190.

Fuglestvedt, J., T. Hanisch, I. Isaksen, R. Selrod, J. Strand, and A. Torvanger, 1994: *A review of country case studies on climate change,* Global Environment Facility, Working Paper 7, Washington, DC.

Glieck, P., 1989. Climate change and international policies: Problems facing developing countries. *Ambio,* **18,** 6, 333–339.

Glomsrod, S., H. Vennemo, and T. Johnson, 1990: Stabilization of emissions of CO_2: A computable general equilibrium assessment, Central Bureau of Statistics, Discussion Paper 48, Oslo.

Goto, N., and T. Sawa, 1993: An analysis of the macro-economic costs of various CO_2 emission control policies in Japan, *The Energy Journal,* **14,** 1, 83–110.

Gregory, R., S. Lichtenstein, and P. Slovic, 1993: Valuing environmental resources: A constructive approach, *Journal of Risk and Uncertainty,* **6,** 177–197.

Hanemann, W.M.,1989: Information and the concept of option value, *Journal of Environmental Economics and Management,* **16,** 23–37.

Harberger, A.C., 1976: *Project evaluation: Collected papers,* University of Chicago Press, Chicago.

Heal, G.M. (1981): Economics and resources. In Butlin, 1981.

Hobbs, B.F., 1991: The "most value" test: Economic evaluation of electricity demand-side management considering customer value, *The Energy Journal,* **12,** 2, 67–92.

Hoeller, P., and M. Wallin, 1991: The role and design of a carbon tax in an international climate agreement in climate change. In *Designing a practical tax system,* OECD, Paris.

Hoel, M., and I.S.A. Isaksen, 1993: Efficient abatement of different greenhouse gases, Memorandum No. 5/93 from Department of Economics, University of Oslo, Oslo.

Hoel, M., and I.S.A. Isaksen, 1994: The environmental costs of greenhouse gas emissions, mimeo, Department of Economics, University of Oslo, Oslo.

Hogan, W., 1990: Comments on Manne and Richels: CO_2 emission limits: An economic cost analysis for the USA, *Energy Journal,* **11,** 2, 75–85.

Hossein, J., and C. Sinha, 1993: Limiting CO_2 emissions in the power sector of India, *Energy Policy,* **21,** 1025–1034.

Hueting, R., P. Bosch, and B. de Boer, 1992: Methodology for the calculation of sustainable national income, Statistical Essays, M44, Centraal Bureau for de Statistiek, The Hague.

ICF, Inc., 1990: Preliminary technology cost estimates of measures available to reduce U.S. greenhouse gas emissions by 2010, Report by ICF to the U.S. Environmental Protection Agency.

International Energy Agency, 1988: Emission controls in electricity generation and industry, Paris.

IPCC (Intergovernmental Panel on Climate Change), 1990: *Climate change: The IPCC scientific assessment,* J.T. Houghton, G.J. Jenkins, and J.J. Ephraums, eds., Cambridge University Press, Cambridge, UK.

IPCC (Intergovernmental Panel on Climate Change), 1995: *Climate change 1994: Radiative forcing of climate change and an evaluation of the IS92 emission scenarios,* J.T. Houghton, L.G. Meira Filho, J. Bruce, Hoesung Lee, B.A. Callander, E. Haites, N. Harris, and K. Maskell, eds., Cambridge University Press, Cambridge.

Jackson, T., 1991: Least-cost greenhouse planning: Supply curves for global warming abatement, *Energy Policy,* **19,** 35–46.

Jones, B., and E. Wheeler (eds.), 1992: Greenhouse research initiatives in the ESCAP region, Proceedings of the ESCAP Conference on Greenhouse Research, Bangkok, August 1991, Economic and Social Commission for Asia and the Pacific, United Nations, New York.

Keeney, R.L., and H. Raiffa, 1976: *Decisions with multiple objectives: Preferences and value trade-offs,* Wiley, New York.

Keeney, R.L., and D. von Winterfeldt, 1987: *Optimal procedures to evaluate decisions with multiple objectives,* EPRI EA-5433, Electric Power Research Institute, Palo Alto, CA.

King, K., 1993: Issues to be addressed by the program for measuring incremental costs for the environment, Global Environment Facility, Working Paper 8, Washington, DC.

King, K., and M. Munasinghe, 1992: Global warming: Key issues for the bank, World Bank Environment Department Divisional Working Paper 1992-36, Washington, DC.

Kneese, A., 1984: *Measuring the benefits of clean air and water,* Resources for the Future, Washington DC.

Kneese, A., and S.C. Smith (eds.), 1966: *Water research,* Johns Hopkins University Press for Resources for the Future, Baltimore, MD.

Koopmans, T.C., 1960: Stationary ordinal utility and impatience, *Econometrica,* **28** (2), 287–309.

Kreps, D.M., and E.L. Porteus, 1979: Temporal von Neumann-Morgenstern and induced preferences, *Journal of Economic Theory,* **20,** 81–109.

Larsen, B., 1993: *World fossil fuel subsidies and global carbon emissions in a model with inter-fuel substitution,* World Bank, Washington, DC.

Lave, L., 1991: Are economists relevant? The efficiency of a carbon tax. In *Global warming: Economic policy responses,* Dornbusch and Poterba, eds., MIT Press, Cambridge, MA.

Lave, L., and Gruenspecht, H., 1991: Increasing the efficiency and effectiveness of environmental decisions: Benefit-cost analysis and effluent fees: A critical review, *Journal of the Air Waste Managament Association,* **41**(5), 680–693.

Lave, L., and K. Vickland, 1989: Adjusting to greenhouse effects: The demise of traditional cultures and the cost to the USA, *Risk Analysis,* **9,** 283–291.

Lind, R.C., 1984a: *Discounting for time and risk in energy planning,* Johns Hopkins University Press for Resources for the Future, Baltimore, MD.

Lind, R.C., 1984b: A primer on the major issues relating to the discount rate for evaluating national energy options, in Lind, 1984a.

Little, I.M.D., and J.A. Mirrlees, 1974: *Project appraisal and planning in developing countries,* Heinemann, London.

London Economics, 1992: Economic costs of carbon dioxide reduction strategies, Global Environment Facility Working Paper Series 3, Washington, DC.

Manne, A., 1992: The rate of time preference – Implications for the greenhouse gas debate, mimeo. Stanford University, Stanford, CA.

Manne, A., A. Mendelsson, and R. Richels, 1992: MERGE – A model for evaluating regional and global effects of GHG reduction policies, mimeo. Stanford, Yale, and EPR.

Manne, A., and R. Richels, 1990: CO_2 emission limits: An economic cost analysis for the USA, *Energy Journal,* **11,** 2, 51–74.

Manne, A., and R. Richels, 1992: *Buying greenhouse insurance: The economic costs of carbon dioxide emission limits,* MIT Press, Cambridge, MA.

Marglin, A., 1963: The social rate of discount and the optimal rate of investment, *Quarterly Journal of Economics,* **77,** 95–111.

Markandya, A., 1990: Environmental costs and power system planning. *Utilities Policy,* **1,** 13–27.

Markandya, A., and D. Pearce, 1988: Environmental considerations and the choice of the discount rate in developing countries, Environment Department Working Paper 3, World Bank, Washington, DC.

Meier, P., 1990: Power sector innovation in developing countries: Implementing investment planning under capital and environmental constraints, *Annual Review of Energy,* **15,** 277–306.

Meier, P., and M. Munasinghe, 1994: *Incorporating environmental concerns into power sector decision-making,* World Bank, Washington, DC.

Meier, P., M. Munasinghe, and T. Siyambalapitiya, 1993: *Energy sector policy and the environment: A case study of Sri Lanka,* Environment Department, World Bank, Washington, DC.

Moreira, J.R., A.D. Poole, D. Zylberztajn, and J. Fries, 1992: Brazil country study. In *UNEP Greenhouse Gas Abatement Costing Studies,* UNEP Collaborating Centre on Energy and Environment, RISO National Laboratory, Denmark.

Mubayi, V., L. Neymotin, and V. Sailor, 1991: *Cost-benefit considerations in backfit analysis,* Brookhaven National Laboratory, Upton, NY.

Munasinghe, M., 1990: *Energy analysis and policy,* Butterworths Press, London.

Munasinghe, M., 1991: Sustainable energy options for developing countries, *World Energy Council Journal,* **1,** December, 69–87.

Munasinghe, M., 1993: Environmental economics and sustainable development, *World Bank, Washington, DC.*

Munasinghe, M., D.W. Pearce, K. Hamilton, G. Atkinson, R. Dubourg, and C. Young, 1995: Measuring sustainable development, draft, World Bank, Washington, DC.

Newendorp, P., 1976: Decision analysis for petroleum exploration, Pennwell, Tulsa, OK.

Nordhaus, W., 1991: Economic approaches to greenhouse warming. In *Global warming: Economic policy responses,* Dornbusch and Poterba, eds., pp. 33–69, MIT Press, Cambridge, MA.

Nordhaus, W., 1993: Running the DICE: An optimal transition path for controlling greenhouse gases, *Resources and Energy Economics,* **15,** 27–50.

Nordhaus, W., 1994a: Expert opinion on climatic change, *American Scientist,* **82,** 45–51.

Nordhaus, W., 1994b: Climate change and economic development: Climates past and climate change future. In Proceedings of the World Bank Annual Conference on Development Economics, M. Bruno and B. Pleskovic, eds., pp. 355–376, World Bank, Washington, DC.

Ogawa, Y., and H. Awata, 1991: Simulation study of greenhouse gas emissions due to future energy consumption. In Jones and Wheeler, 1992, pp. 234–262.

OTA (Office of Technology Assessment), 1992: *Fueling develop-ment: Energy technologies for developing countries,* OTA-E-516, Washington, DC.

OTA (Office of Technology Assessment), 1993: *Preparing for an uncertain climate,* OTA-O-568, Washington, DC.

Pachauri, R.K., 1991: Global warming: Impacts and implications for South Asia, *Tata Energy Research Institute, New Delhi.*

Pachauri, R.K., and A. Behi (eds.), 1991: *Global warming: Mitigation strategies and perspectives from Asia and Brazil,* Tata Energy Research Institute/McGraw-Hill for Asian Energy Institute, New Delhi, India.

Page, T., 1977: *Conservation and economic efficiency,* Johns Hopkins University Press, Baltimore.

Pearce, D.W., and R.K. Turner, 1990: *Economics of natural resources and the environment,* Harvester Wheatsheaf, London.

Pearce, D.W., and R.K. Turner, 1991: The role of carbon taxes in adjusting to global warming, *Economic Journal,* **101,** July, 938–948.

Pearce, D.W.., R.K. Turner, and J. Warford, 1993: *World without end: Environment, economics and sustainable development,* Oxford University Press, Oxford.

Proost, S., and D. van Regemorter, 1990: Economic effects of a carbon tax: With a general equilibrium illustration for Belgium, Public Economic Resources Paper No. 11, Katholieke Universiteit, Leuven, Belgium.

Putta, S., 1990: Weighing externalities in New York State, *Electricity Journal,* **3**(6), 42–47.

Ramsey, F., 1928: A mathematical theory of saving, *The Economic Journal,* **38,** 152, 543–559.

Randall, A., 1986: Valuation in a policy context. In Bromley, 1986, pp. 163–200.

Rawls, J., 1972: *A theory of justice,* Oxford University Press. Oxford.

Repetto, R., 1989: *Wasting assets: Natural resources in national income accounts,* World Resource Institute, Washington, DC.

Richels, R., and J. Edmonds, 1995: The economics of stabilizing atmospheric CO_2 concentrations, *Energy Policy,* **23**(3/4), 373–379.

Rose, A., 1990: Reducing conflict in global warming policy: The potential of equity as a unifying principle, *Energy Policy,* **18,** 927–935.

Rutherford, T., 1992: *The welfare effects of carbon restrictions: Results from a recursively dynamic trade model,* Working paper, OECD, Paris.

Ryder, H., and G. Heal, 1973: Optimal growth with intertemporally dependent preferences, *Review of Economic Studies,* **40,** 1–31.

Sen, A., 1967: Isolation, assurance, and the social rate of discount, *Quarterly Journal of Economics,* **81,** 112–124.

Sen, A., 1979: Personal utilities and public judgments: Or what's wrong with welfare economics?, *The Economic Journal,* **89,** 537–558.

Sen, A., 1984: Approaches to the social discount rates for social benefit-cost analysis. In Lind, 1984a.

Sitnicki, S., K. Budzinski, J. Juda, J. Michna, and A. Szpilewicz, 1991: Opportunities for carbon emissions control in Poland, *Energy Policy,* **19,** 995–1002.

Steinberg, M., and H. Cheng, 1985: A systems study for the removal, recovery and disposal of carbon dioxide from fossil fuel power plants in the U.S, Brookhaven National Laboratory, BNL 35666, July.

Sterner, T., 1994: Discounting in a world of limited growth, *Environmental and Natural Resource Economics,* **94,** 4.

TERI (Tata Energy Research Institute), 1991: Report on global warming and associated impacts, Report to the Ministry of Environment and Forests, New Delhi, India.

Turner, R.K., 1991: Environment, economics and ethics. In *Blueprint 2: Greening the world economy,* D. Pearce, ed., Earthscan, London.

UNEP (United Nations Environment Programme), 1988: *Assessment of multiple objective water resources projects: Approaches for developing countries,* New York.

UNEP (United Nations Environment Programme), 1992: UNEP greenhouse gas abatement costing studies, Riso National Laboratory, Denmark.

United Nations, 1993: *Integrated Environmental and Economic Accounting,* Department of Economic and Social Development, Statistical Division, New York.

Weitzman, M., 1976: On the welfare significance of national product in a dynamic economy, *Quarterly Journal of Economics,* **90,** 156–162.

Williams, R.H., 1993: Comment on "Climate and Economic Development" by Nordhaus. In Proceedings of the World Bank Annual Conference on Development Economics, M. Bruno and B. Pleskovic, eds., World Bank, Washington, DC.

Willig, R.D., 1976: Consumer surplus without apology, *American Economic Review,* **66**(4), 589–597.

World Bank, Asia Technical Department, 1993: *Interim report, international collaborative project on acid rain and emissions reduction in Asia,* World Bank, Washington, DC.

World Bank, 1994: *Resettlement and development,* World Bank, Washington, DC.

World Commission on Environment and Development, 1987: *Our common future,* Oxford University Press, Oxford.

Yamagi, K., R. Matsuhashi, Y. Nagata, Y. Kaya, 1993: A study on economic measures for CO_2 reduction in Japan, *Energy Policy,* **21,** 123–132.

6

The Social Costs of Climate Change:
Greenhouse Damage and the Benefits
of Control

D.W. PEARCE, W.R. CLINE, A.N. ACHANTA, S. FANKHAUSER,
R.K. PACHAURI, R.S.J. TOL, P. VELLINGA

CONTENTS

SUMMARY

This chapter is concerned with the socioeconomic assessment of climate change impacts. Estimates of damage related to these impacts can make an important contribution to decision making about climate change responses.

Monetary values reflecting human preferences can provide useful information for decision making. The cost-benefit approach, in particular, requires that the damages from climate change be represented, as far as possible, in terms of money units. To the extent that this is possible, the chapter expresses impacts in these terms; that is, human preferences are expressed by people's willingness to pay (WTP) to secure a benefit or their willingness to accept compensation (WTA) for a cost. Such monetary estimates only measure the impact on individual welfare. Aggregating individual damages to obtain total social welfare impacts requires difficult ethical decisions.

Many of the impacts of climate change will not be revealed directly in the marketplace. These are the so-called nonmarket impacts. In these cases WTP and/or WTA are measured through "surrogate markets" or "hypothetical markets." Surrogate markets are real markets in which environmental change has an influence: A house price or land value may be higher because of an environmental amenity, for example. Hypothetical markets reflect people's responses to questions put to them about their willingness to pay. Monetary estimates are thus able to cover both market and nonmarket impacts, although estimates of the latter are more controversial and less confidence is placed in them.

The level of sophistication in socioeconomic assessments of climate change impacts is still rather modest. Damage estimates are tentative and based on a number of simplifying and often controversial assumptions. Most estimates are for equilibrium climate change associated with a doubling of the preindustrial CO_2-equivalent concentration of all greenhouse gases. Best-guess central estimates of global damage, including nonmarket impacts, are in the order of 1.5=2.0% of world GNP for $2xCO_2$ concentrations and equilibrium climate change. This means that if a doubling of CO_2 occurred now, it would impose this much damage on the world economy now. This chapter stresses the uncertain character of these estimates. The figures are best-guess results, and several impact categories could not be assessed for lack of data. Moreover, the range reflects variations in the best-guess estimates and cannot be interpreted as a confidence interval. Particularly vulnerable sectors include agriculture, the coastal zones, human mortality, and natural ecosystems. The possibility of catastrophes (low probability/high impact events) and surprises cannot be ignored.

The regional variation in damage is substantial. The available studies estimate damages for developed countries at between 1% and 2% of GNP for a $2xCO_2$ climate. Central estimates of the damage in different developing regions range from a minimum of 2% of GNP to a maximum of 9%. For individual nations, or if alternative assumptions are used about the value of a statistical life (see Box 6.1), the figure could be even higher. Small island states and low-lying coastal areas are especially vulnerable. Most impact work is confined to developed nations, however. The confidence in estimates for developing countries is much lower.

The chapter emphasizes the need for a long-term perspective reaching beyond a $2xCO_2$ scenario, even though socioeconomic forecasts over more than a century are highly uncertain. Most models assume a nonlinear (convex) damage-temperature relationship, resulting in damages of 6% or higher for 10°C warming. These figures are illustrative only.

Doubled-CO_2 damage estimates usually form the basis for the calculation of marginal damage – the extra damage done by one extra tonne of carbon emitted. Marginal damage is estimated at $5-$125 per tonne of carbon emitted now. The wide range reflects variations in model assumptions, as well as the high sensitivity of figures to the choice of discount rate. Although estimates based on a social rate of time preference (discount rate) of the order of 5% tend to be about $5-$12, figures assuming a rate of 2% or less can be almost an order of magnitude higher. Current models are simplistic and provide poor representations of dynamic processes. The effect of climate change adaptation in particular is poorly understood.

Marginal climate change damage is equal to the marginal climate change benefits of emission control. However, the benefits of greenhouse gas abatement will not be limited to reduced climate change costs alone. A reduction in CO_2 emissions will often also reduce other environmental problems related to the combustion of fossil fuels. The size of these so-called secondary benefits is strongly site dependent. Studies for Norway, the UK and some other countries indicate that the benefits of reduced air pollution could offset between 30% and 100% of abatement costs.

6.1 Conceptual Framework

6.1.1 Scope and limit of the analysis

This chapter is concerned with the nature of the damage from climate change. Damages here refer to the consequences of climate change for *individual and social welfare* from an economic point of view. That is, climate change damage is defined as the the difference in social welfare between a scenario with and one without anthropogenic climate change. The chapter assesses the possible aggregate scale, the geographical distribution, and the nature of those damages. It raises some issues relating to decision-making rules, since alternative ethical approaches to harm done to future generations have implications for damage assessment (see also Chapters 3 and 5). It should also be borne in mind that

- Damage costs are distinct from the costs of climate change response measures (which are discussed in Chapters 8 and 9).

- Absolute damage levels are not necessarily identical to the benefits of mitigation measures, as will be discussed in Sections 6.6 and 6.7.

- In describing damages in individual countries and regions, no implication is made about the ethical question of who should bear these costs or the costs of measures to avoid them. This issue is dealt with in Chapter 3.

The level of sophistication of climate change damage analysis is comparatively low. Damage estimates are generally tentative and based on several simplifying and often controversial assumptions. The degree of uncertainty is correspondingly high, with respect to both physical impacts and their consequences for social welfare. No attempt has been made to specify confidence intervals. Rather, estimates are *best guesses,* depicting the most likely damages currently associated with a particular climate scenario. Moreover, studies often impose climate change onto the current world, thus ignoring the effect of future economic development and population growth on climate vulnerability.

This low level of sophistication implies that climate change damage analysis is a particularly worthwhile area for further research (see also Chapter 10). Especially needed is a better understanding of regional damage (particularly in developing countries), nonmarket damage, and nonequilibrium (transient) damage. The impact of climate change adaptation is also still poorly understood.

6.1.2 The nature of damage assessment

Two broad anthropocentric perspectives have emerged on ways to analyze decisions which have adverse effects on generations yet to be born: the cost-benefit framework and the sustainability framework. These perspectives initially appear to be very divergent, but they have features in common.

The *cost-benefit framework* requires that the future damages and adaptation costs be weighed, integrated into an overall assessment, and then compared to the costs of mitigation measures undertaken now. In turn, there are two perspectives on the way in which benefits and costs should be compared:

(1) The *judgmental cost-benefit* framework, in which gains and losses are compared but without being reduced to common units. In this approach, the monetary costs of control might be compared with some wide-ranging *environmental impact statement* representing the best state of knowledge about climate change impacts, together with assessments of the *distributional incidence* of those impacts both geographically and across time

(2) The *monetized cost-benefit* framework, in which the common unit of money is used to "reduce" the benefits of climate control to the same units as costs to permit direct comparison, but only as far as "monetization" is credible

Approach (1), which often reduces to multicriteria analysis, is discussed in Chapter 5 (see also the discussion of integrated assessment in Chapter 10). Approach (2) is characterized by the following features:

(a) Benefits and costs are defined in terms of human preferences. A benefit is anything that improves an individual's well-being; a cost is anything that reduces that well-being.

(b) Those preferences are expressed in the marketplace by willingness to pay (WTP) for a benefit and willingness to accept compensation (WTA) for a cost. Although the two concepts are not identical and WTP and WTA estimates may vary by more than a factor of two, they are often used interchangeably (for a comparison of the two concepts, see, e.g., Shogren *et al.,* 1994).

(c) Where markets do not exist – for example, with respect to ecosystem change – WTP and/or WTA are estimated through "surrogate markets" or "hypothetical markets." Surrogate markets are real markets in which environmental change has an influence: a house price or land value may be higher because of an environmental amenity, for example (the hedonic property price method). Hypothetical markets reflect people's responses to questions put to them about their willingness to pay (the contingent valuation method). Although controversial, these approaches are well established in the literature (for an introduction and assessment, see, e.g., Mitchell and Carson, 1989; Braden and Kolstad, 1991).

(d) Future generations' preferences count at least insofar as they are assumed to want what current generations want. If there is evidence that they will want more of the environmental assets affected by climate change, then this "rising relative preference" can be accommodated by cost-benefit approaches by allowing benefits or costs to rise through time. Future generations' preferences may count equally with current generation preferences if the discount rate is set accordingly (see Chapter 4).

(e) Since WTP is constrained by income, it is likely to be less for low income groups than for high income groups. This may appear to give rise to unfairness, since the preferences of low income groups (countries) will carry less weight than the preferences of high income groups. One way to address these ethical issues is to give different weights to different income groups in the aggregation process (see Box 6.2). Another way might be to adopt the valuations of the higher income groups and apply them to all countries (e.g., Hohmeyer and Gärtner, 1992; see also Box 6.1).

(f) Aggregation of damages is also a hallmark of cost-benefit studies. Aggregation poses difficult problems regarding the comparability of individual welfare. Care has also to be exercised in interpreting such aggregate figures since they clearly mask substantial regional variations in impact within a country and between countries as well as redistribution effects between positively and adversely affected regions and sectors.

The second overall perspective – the *sustainability approach* (e.g., Howarth and Monahan, 1992; Spash, 1994) – gives the highest priority to the avoidance of "unacceptable" damage to future generations. Proponents of sustainability would argue that:

(a) There is reasonable evidence to suppose that actions now in emitting greenhouse gases could cause significant damage to future generations, including unborn generations.

(b) Future generations are defenceless against actions taken now in the knowledge that those actions may cause harm.

(c) Current generations are linked to future generations at the very least through parents to children, from children to their children, and so on (Howarth, 1992), or, more generally, current generations have obligations to future generations because future generations have rights, even when those generations are not identifiable and even when their existence is contingent upon actions taken now.

(d) Probable improvements in the well-being of future generations cannot be treated as "compensation" for harm knowingly inflicted on future generations by current generations any more than harming the poor now can be excused by paying them compensation after the event (Spash, 1994).

(e) Hence, doing harm is not reversible by doing good. Benefits and costs cannot be "traded off" in the sense advocated by cost-benefit analysis *regardless of whether the cost-benefit analysis is monetized or not*. There is a *duty* to avoid future harm.

As with the cost-benefit approaches, all the propositions in the sustainability approach are open to dispute. Controversial assumptions include the idea that harm can accrue to individuals whose existence is contingent on actions taken now, and the belief that trade-offs can be avoided, since any action now incurs a cost of abatement, and any abatement cost involves losses for others, which in turn implies that other individuals' rights may be impaired.

Within the sustainability approach there are two contrasting perspectives:

(1) Since the obligation to avoid harm is absolute, the cost of avoiding harm is irrelevant: The benefits of control are so large that inspection of costs is unhelpful. This is the *absolute standards* approach.

(2) Harm should be avoided subject to a constraint that avoiding harm does not itself impose "unacceptable cost" – the *safe minimum standards* approach.

The sustainability approach takes a long-term view and stresses the need to sustain a viable global ecological system. It therefore tends to be characterized by

- the avoidance of unacceptable risk where risks are known
- the "precautionary principle" – whereby actions giving rise to *possible* but quantitatively unknown and potentially very large risks are avoided or corrected
- the view that what is unacceptable is only partly measured by reference to individuals' preferences, since individuals are not well informed about climatic risks, experts are similarly not well informed due to uncertainty about climate change and its effects, and human preferences may not capture other values, for example, the intrinsic value of ecosystems
- very low discount rates of the order of a few percentage points and maybe even zero

Whereas the cost-benefit approach seeks to measure the scale of damage, the sustainability approach assumes that damage will be "significant," so much so that action is warranted regardless of quantification.

The sustainability approach often tends to have as its objective a concern to avoid exceeding some target rate of temperature rise, often quoted as 0.1°C per decade, and some absolute overall rise in temperature, such as 2-3°C. Examples include Krause *et al.* (1989) and Rijsberman and Swart (1990). The costs of achieving these constraints are assumed to be worth incurring to avoid the risks to future generations.

With reference to the cost-benefit approach, it is well known that a timepath in which the present value (i.e., the discounted value) of benefits minus costs is maximized need not be a sustainable path, and that a sustainable path could, in turn, be unacceptable in terms of its implied living standards for each generation (Page, 1977; Pearce *et al.*, 1994; Pezzey, 1994). The choice between cost-benefit approaches and sustainability approaches therefore depends crucially on (a) attitudes to uncertainty, (b) the degree of concern for the well-being of future generations, and (c) beliefs about the damage function, that is, the way in which warming relates to damages.

A third approach is a *consensus viewpoint,* which stresses the following common features of both the sustainability and the cost-benefit approaches. The principal arguments of this viewpoint are that

- The existence of uncertainty cannot justify doing nothing. Action on climate change is justified, because the damage costs could be very high, the "coefficient of concern" for the future is not zero, and there are costs of delayed action since greenhouse gas impacts may not be reversible.

- The right cost-benefit perspective is one that investigates the benefits and costs of taking actions. Such actions could yield benefits of a similar order of magnitude as the costs of the actions for some time to come, since avoided climatic damages are not the only benefits of those actions (Section 6.6). This underlines the difference between *damage estimation* and *abatement* benefits: The latter include the avoided damages estimated by the former concern, but also include other benefits from greenhouse gas abatement (see Section 6.7).

- Neither sustainability nor "maximizing net benefits" is an obviously noncontroversial objective. Sustainability cannot be an overriding objective, independent of the quality of life that is sustained or the costs of achieving it. Maximizing net benefits cannot be an overriding objective, since it may be consistent with an approach that discriminates against future generations (for example, by applying an excessive discount rate; see Chapter 4). This suggests an approach in which the best features of both approaches are taken: a concern for the well-being of future generations and acknowledgment of the limited resources that all societies have at their disposal to tackle global problems. Such approaches come closest to the "safe minimum standards" approach: taking a precautionary approach in favour of the environment, unless the demonstrated costs of so doing are very high. Precaution would have its justification in very high damage costs. Damage costs must therefore be investigated. The acceptability of control costs is dealt with in Chapters 8 and 9.

6.1.3 *The valuation of market and nonmarket impacts*

The cost-benefit approach requires that the damages from climate change be expressed, as far as is possible, in terms of money units reflecting human preferences. Many of the impacts of climate change will not be revealed directly in the marketplace – the so-called nonmarket impacts. The absence of markets does not mean that nonmarket impacts are any less important than market impacts. The point is to ensure that nonmarket impacts are adequately accounted for.

Nonmarket impacts may take a variety of forms. One important example is the impact of warming on human health. Human health care is, generally, publicly provided without full charge. In some cases, there are surrogate markets for risks to life. Occupations are subject to varying degrees of risk of accident and ill-health, and that risk is sometimes compensated for by variations in wages. The "risk premium" in the wage can then be interpreted as a valuation of the risk (see Box 6.1). Valuation of morbidity is more complex: Contingent valuation approaches are best suited to such indicators but, as yet, few studies exist outside the U.S. In their absence, second-best measures such as the costs of treatment tend to be used.

The general approach in cost-benefit studies is to treat market and nonmarket impacts on the same footing. Nonmarket costs can be high, between 30 and 80% of the total, as shown below. Tol (1994a) suggests, however, that the way nonmarket impacts are treated has further significance. The reasons are:

(a) They affect human well-being directly (through the utility function rather than through production.

(b) They are liable to be less substitutable.

(c) Their value will rise relative to other (production) damages.

Market price (in the case of market goods) and elicited WTP (for nonmarket goods) reflect people's appreciation for the *marginal* (last) unit of a good or service consumed. The utility gained from consuming the first unit of a good is usually much higher than that from consuming the last unit, however. Especially for food and other products with price-inelastic demand – where the appreciation of the first, essential units consumed is extremely high – the value lost from a cutback in availability will be understated if the quantity loss is evaluated at the original (*ex ante*) price. Price increases as a consequence of climate change then become important. The correct measure to assess the costs of climate change in such cases is the change in producer and consumer surplus. Consumer surplus is the excess of what consumers would be willing to pay, if necessary, above what they actually pay. Producer surplus is what producers receive in excess of their actual costs of production. This consideration also means that it can be misleading to gauge the potential impact of climate change by the present size of a sector in the economy. In industrial countries agriculture is typically on the order of 3% of GDP, but a reduction of x% in agricultural output could cause far more than $0.03x$% of GDP economic loss because of the induced price increase and loss of consumer surplus.

6.1.4 *Temporal aspects*

Climate change impacts have many temporal aspects. First, ocean thermal lag causes realized equilibrium warming from a given steady-state increase in greenhouse gas concentrations to be delayed by at least two, and perhaps several, decades. Hence, advantages and disadvantages associated with emissions and emission reductions occur immediately, whereas the impacts of climate change occur only after a significant lag. In addition, emissions in a given year typically generate a one-year flow of advantages, but create a stock of climate change that then produces a recurrent annual flow of impacts potentially into the indefinite future. Decision making on climate change thus involves intertemporal issues, of which the question of discounting may be the most crucial one (see Chapters 2, 4, and 5).

Second, the difference between transient and equilibrium climate change is important. The former refers to the transition, the latter to the new stable state of the climate. The large majority of the estimates presented in this chapter refer to equilibrium climate change, particularly the climate associated with an atmospheric concentration of carbon dioxide of approximately 600 ppm ($2 \times CO_2$). This arbitrary and abstract assumption is necessary in order to estimate climate change damage with existing models. In reality, however, society will face a changing (rather than a changed) climate. Sections 6.3 and 6.4 present some preliminary findings on nonequilibrium and long-run aspects. One important implication is that transient warming may not follow a smooth path between the present climate and a future equilibrium climate. Transient climate change in a particular region cannot, therefore, be assumed to be a steadily rising fraction of equilibrium impact; instead, there could be discontinuities and reversals. Damage is likely to be sensitive to such variability. Continual climate shocks would affect the ability of economic and natural systems to adjust and to recover.

Third, the faster climate changes, the greater will be the economic impacts. That is, damage is a function of the rate as well as the magnitude of climate change. Adjustments by natural systems and social institutions are not instantaneous. Species and forests need time for migration, just as agriculture and other climate-sensitive human activities need time for adaptation. Systems are generally more flexible in the long run than in the short run.

Fourth, the world is bound to change profoundly even in the absence of climate change. Sections 6.2 and 6.5 below indicate that the poorer regions are more vulnerable to climate change than the richer (see Box 6.3 for an overview of reasons). Most of these regions are projected to experience rapid economic growth in future years (see Chapter 12). This could mean that their vulnerability to climate change might fall. On the other hand, human activities cause increasing stress to many natural and social systems, which could make them more vulnerable to climate change. At the extreme, climate change could be "the straw that breaks the camel's back." Changes in human prosperity also imply changes in preferences. Intangible impacts constitute a large part of the total impacts, and increased economic wealth could well imply higher human valuation of intangible impacts. In addition, technological change can profoundly alter the options available for low-cost adaptation.

However, most of the estimates presented below concern the impact of an equilibrium climate change on present-day society. This body of information by and large reflects the state of the art of this relatively young research area. Despite its shortcomings, equilibrium climate change analysis can be a useful point of departure for further analysis of this extremely complex issue.

6.1.5 Adaptation

Adaptation offers a means to reduce the possible impacts of future climate change. Measures to adjust to climate change will be taken both on an individual level and by society as a whole. The search for more resilient crops will be intensified, for example, vulnerable coastlines will be defended by sea walls, and improved weather forecasts will permit better preparation for extreme weather events. On an individual

level, farmers will change crops or adjust planting dates, households will increase their demand for air conditioning, people may stop building in or move away from flood plains, and so on. These and other aspects of adaptation are discussed further in Chapter 7 and in Volume 2 of the present report (IPCC, 1996b).

However, the degree of adaptation in developing countries is likely to be far less than in developed countries due to lack of financial resources and lack of institutional capacity (for a discussion of adaptation in developing countries see Jodha, 1989). Where it is feasible, adaptation can potentially be a very powerful option. In a stylized cost-benefit model of climate change policies, Hope *et al.* (1993) found strongly positive cost-benefit ratios for an "aggressive adaptive policy" in Europe, mainly in the form of coastal protection. The benefits of adaptation exceed the costs by more than a factor of 20. Fankhauser (1994a) calculates that in OECD countries it could be economic to protect between 50 and 100% of affected coastlines. In a series of country studies IPCC (1994) found that, through appropriate adaptation measures, the number of people at risk from flooding could on average be reduced by a factor of about 8.

Agricultural studies provide a similar picture. In a case study of the Missouri-Iowa-Nebraska-Kansas (MINK) region, Easterling *et al.* (1993) found that low-cost adaptation measures, like earlier planting or increased irrigation, could succeed in reducing agricultural damages to the region by 30% or more (see also Section 6.2.1). A comparable range was found by Rosenzweig *et al.* (1993) and (using the same yield data) Reilly *et al.* (1994). In the Rosenzweig *et al.* study a change of -1.2 to -7.6% in worldwide cereal production without adaptation is reduced to 0 to -5.0% with moderate farm-level adaptation and $+1.0$ to -2.5% with a more comprehensive managerial adjustment. As a consequence, the global welfare loss reduces from $-\$0.1$ to $-\$61.2$ billion without adaptation to $+\$7.0$ to $-\$37.6$ billion in the case of a moderate response (Reilly *et al.,* 1994).

Adaptation will, in general, not be costless and may require extensive planning. The need for integrated and forward-looking coastal zone management has been identified as an essential prerequisite for the future development of coastal zones (IPCC, 1994). The significance of adaptation will also depend on institutional factors. Limited availability of irrigation water or of sufficient capital to finance increased input requirements (e.g., more fertilizer) may limit the scope of agricultural adaptation in poorer regions (Rosenzweig *et al.,* 1993).

Conceptually, the costs of climate change impacts in the presence of costly adaptation consist of two parts: the costs of adaptation (e.g., for coastal protection) plus the costs of the remaining unmitigated damage (e.g., the loss of unprotected land). The estimates reviewed in Section 6.2 for many categories of climate change damage also incorporate considerations of adaptation. Damage from sea level rise, for example, is partly comprised of the cost of building coastal protection structures. Similarly, the impacts on the electricity sector caused by climate change amount to costs required or induced by adaptation (space cooling). A prime exception is loss of biodiversity, for which there are few adaptation options. Implicitly or explicitly, however, most of the estimates incorporate both effects: costs of plausible adaptation plus the remaining damages of unmitigated impacts.

The question of the optimal level of adaptation is strongly linked to that of optimal mitigation. The most desirable level of adaptation will depend on the amount of greenhouse gas abatement undertaken, and vice versa. Abatement and adaptation policies should therefore be carefully coordinated. For example, if significant abatement can be achieved at reasonably low cost, less action may be needed with respect to adaptation. Conversely, if the consequences of climate change could easily and cheaply be accommodated by adaptation, there would be less need for preventive carbon abatement.

However, as the hybrid damage-cum-adaptation nature of most of the cost estimates suggests, assuming a strict dichotomy between abatement and adaptation would be misleading. Indeed, climate change is likely to require both types of actions. Even with the most ambitious abatement policy, some climate change seems likely to occur. Conversely, even the most extensive adaptation strategy is unlikely to fully mitigate the adverse impacts of climate change. This is particularly the case if warming is subject to discontinuities and reversals. A strategy relying primarily on adaptation rather than abatement could then have significantly higher costs than implied by calculations assuming a smooth warming path. Adaptation is primarily a complement, not an alternative, to greenhouse gas abatement.

6.1.6 Recent scientific evidence

The literature reviewed below largely takes as its point of departure the scientific appraisal of IPCC (1990a). It is important to consider whether changes in the scientific assessment since then provide grounds for altering the economic evaluation of the greenhouse effect. In Volume 1 of the present report (IPCC 1996a), IPCC Working Group I has identified the following changes in the underlying impact science that might affect the measurement of damage.

First, although the climate sensitivity range remains at 1.5-4.5°C for equilibrium 2xCO$_2$, transient realized global mean surface temperature is now expected to rise a further 1.0-3.5°C between 1990 and 2100, based on the full range of IS92 scenarios. This is in addition to the increase observed to 1990 (0.3-0.6°C) but is about one-third lower than the 1990 estimates, mainly due to the inclusion of the "cooling" effect of aerosols.

Second, there is increasing emphasis on regional differences. These stem from the differential impact of ocean thermal lag. In the Northern Hemisphere, where the proportion of land to ocean is greater than in the South, realized warming may be about twice the global mean estimate. In the centres of large land masses the warming rate may be several times the global mean.

Third, taking the effect of sulphate aerosols into account, the central estimate for sea level rise by the year 2100 is now placed at about 50 cm for IS92a, compared to 66 cm in IPCC

(1990a), with other IS92 scenarios giving estimates ranging from 15 to 95 cm.

The broad thrust of these changes is to moderate the expected pace of mean global warming but also perhaps to intensify the role of variability and surprises at the regional level. For mid-continental areas in the Northern Hemisphere, the new estimates would seem to leave even the mean pace of warming close to that in IPCC (1990a). Increasing ocean–land differentials could suggest greater precipitation changes and higher storm damages. This result is controversial, however. Significantly the incorporation of sulphate aerosol effects also carries the implication of potential acceleration from baseline warming if there is greater progress than expected in the reduction of these pollutants (Wigley and Raper, 1992).

These considerations may affect the estimates presented below. In which direction they would change, though, is unclear. On the one hand, the revised IPCC analysis indicates a possibility of increased damage associated with variability and unpredictability, as well as giving greater attention to the high regional warming coefficients for the Northern Hemisphere and for mid-continental areas. On the other hand, damages could be reduced due to the revised timetable of mean warming. The net effect is hard to predict. The revised assessments might imply somewhat later damages for the Southern Hemisphere (and thus, broadly, for developing countries).

6.2 Damage Estimates for Benchmark Warming (2xCO$_2$)

Most available damage estimates are concerned with the impact of an equilibrium climate change associated with a doubling of the pre-industrial carbon dioxide equivalent concentration of all greenhouse gases (referred to here as benchmark warming). Long-run impacts have gained little attention. Nor have the possible impacts of the approximately 0.5°C warming already observed over the past century been studied in much detail.

Monetary values for 2xCO$_2$ damage have been estimated for a number of sectors in the market economy. In addition, there are estimates for some nonmarket damages, which are typically more difficult to quantify (e.g., species loss), and for combined market and nonmarket effects in some sectors (e.g., forest loss in lumber and public use value). Table 6.1 provides an overview of the categories of damages that might be caused by climate change and associated sea level rise. It clearly shows that the estimated damages are not complete. For some categories, monetary estimates of damages have not been attempted. For other categories, the estimated damages only partially reflect the potential welfare loss. In many cases, the preferred measure of welfare impacts – willingness to pay – is approximated by other indicators.

A further source of inaccuracy results from the use of different climate models and scenarios. Although all results reported in this section assume benchmark CO$_2$ warming, estimates may be based on different GCM results. In addition, some authors have "normalized" impacts to a standard warming assumption (usually 2.5°C), others have not. This variety

Table 6.1. *Overview of climate change impacts*

Damages	Market Impacts				Nonmarket Impacts		
	Primary economic sector damage	Other economic sector damage	Property loss	Damage from extreme events	Ecosystem damage	Human impacts	Damage from extreme events
Fully estimated, based on willingness to pay	Agriculture			Dryland loss Coastal protection	Wetland loss		
Fully estimated, using approximations	Forestry	Water supply		Hurricane damage	Forest loss		Hurricane damage
Partially estimated	Fisheries[a]	Energy demand Leisure activity	Urban infrastructure	Damage from droughts[b]	Species loss	Human life Air pollution Water pollution Migration	Damage from droughts[b]
Not estimated		Insurance Construction Transport Energy supply		Nontropical storms River floods Hot/cold spells Other catastrophes	Other ecosystem loss	Morbidity Physical comfort Political stability Human hardship	Nontropical storms River floods Hot/cold spells Other catastrophes

[a]Often included in wetland loss.
[b]Primarily agricultural damage.

in assumptions can be the cause of significant variation in damage estimates (see Smith *et al.,* 1993).

Estimates are predominantly for the U.S. and other OECD countries. Material relating to other countries is sparse although increasing. All the estimates are subject to considerable uncertainty. Furthermore, estimates are usually based on the present-day economy and expressed as a percentage of GDP. Projections then apply these percentage impacts to future world product (e.g., to 2060 for realized benchmark warming). Simply projecting percentage losses to the future is a somewhat unsatisfactory approximation. Future impacts will depend on economic, demographic, and environmental developments. Some of the effects are likely to grow more than proportionately with GDP (e.g., the economic value of nonmarket goods) and others less than proportionately (e.g., agriculture). Future demographic developments may change current vulnerability and migration patterns in an unknown way.

The incomplete nature of the damage estimates presented here must be borne in mind when evaluating the full welfare implications of climate change.

6.2.1 Agriculture

Climate change is expected to damage agriculture in some areas but aid it in others. The principal damage will arise from heat stress, decreased soil moisture, and an increased incidence of pests and diseases. In addition, warmer temperatures could cause the growing cycle of many plants to accelerate, allowing less time for plant development before maturity. Increased rainfall intensity could increase soil erosion in some areas, whereas other regions could be affected by drought. Rind *et al.* (1990) use GCM results to calculate that for many mid-latitude locations (e.g., the U.S.) the incidence of severe droughts that currently occur only 5% of the time would rise to a 50% frequency by the 2050s, based on the difference between precipitation and potential evapotranspiration (E_p). They find that

E_p increases most where the temperature is highest, at low- to mid-latitudes, while precipitation increases most where the air is coolest and easiest to saturate by the additional moisture, at higher latitudes.

The principal beneficial impacts from climate change would be longer growing seasons in some regions and for some crops, and the fertilization effect of greater atmospheric carbon dioxide. Higher atmospheric carbon concentrations are expected to increase photosynthesis, which combines carbon dioxide and water to produce carbohydrates. Laboratory experiments suggest that a doubling of CO_2 from 330 to 660 ppm could raise yields by 34% for C_3 crops (wheat, rice, soybeans, fine grains, legumes, root crops, most trees) and 14%

for C_4 crops (maize, sorghum; see Schneider and Rosenberg, 1989). However, open-field conditions may not necessarily achieve the same yield increases (Parry, 1990; Evans *et al.*, 1991; Körner and Arnone, 1992; Erickson, 1993). Bazzaz and Fajer (1992) note that the laboratory experiments depend on availability of fertilizer and water and conclude that because of "competitive interference and limited nutrients . . . we do not expect that agricultural yields will necessarily improve in a CO_2-rich future." Moreover, the doubled-CO_2 experiment is conceptually wrong for calculating the effects of benchmark (equilibrium) warming. This is because, after taking account of other greenhouse gases, the equilibrium CO_2 concentration in the *2xCO$_2$-equivalent* atmosphere is only 440 ppm (Cline, 1992a, calculated from IPCC, 1990a).

Agriculture is a relatively well-studied area of climate change impact research. Nevertheless, available results are still very diverse and often contradictory. Agricultural models are highly sensitive to a number of key assumptions. This sensitivity can, to a large extent, explain differences in model results. The most important elements of dispute include:

- the effect of CO_2 fertilization, as mentioned above
- the assumed climate scenario (particularly changes in temperature and precipitation)
- the potential and scope for adaptation
- the inclusion of trade effects

The rest of this section summarizes the major contributions to this research so far.

The importance of trade flows is illustrated in a study by Kane *et al.* (1992), which uses the U.S. Department of Agriculture's Static World Policy Simulation (SWOPSIM) model of agricultural trade. Even in a "very adverse" scenario in which yields fall by 5-40% in most developed countries, the former Soviet Union, and China, and remain unchanged in most other developing countries, net global welfare declines by only 0.47% of GDP. World food prices are expected to increase in the order of 40% or more, with repercussions on both producer and consumer welfare. In a food exporting country like the U.S., for example, consumers are expected to lose $40 billion annually in consumer surplus (at 1986 prices), whereas U.S. farmers gain $19 billion annually in producer surplus because price increases more than offset yield reductions. The implication is that a corresponding loss of consumer surplus occurs for importing nations, associated with terms-of-trade gains on U.S. farm exports. The main loser identified in this particular model run is China, with economic losses of more than 5% of GDP. Increased world food prices also heavily affect consumers in the former Soviet Union. In a second, more optimistic scenario, worldwide impacts are practically zero, with negative results in Canada, Japan, and Europe being offset by gains in Australia, and now also the former USSR and China. The Kane *et al.* figures, averaged over all scenarios, are used for the agricultural damage estimates reproduced in Table 6.5.

In another global study, Rosenzweig *et al.* (1993) coordinated research applying crop simulation models in 18 countries to examine the impact of benchmark 2xCO$_2$-equivalent warming on yields for wheat, rice, maize, and soybeans by 2060 (see also Rosenzweig and Parry, 1994). Linking the results in a world trade model (Basic Linked System), they calculated the impact on production levels, prices, and the number of people at risk from hunger. The study found that crop yields would decline in the low latitudes, where they are currently grown near their limits of temperature tolerance. However, yields could increase at middle and high latitudes when carbon fertilization is included. In the case of moderate adaptation, output rises in developed countries by 4-14% but falls by 9-12% in developing countries (which must import). Depending on the GCM used, global output falls by 0-5%, prices rise by 10-100%, and the number of people at risk from hunger rises from a baseline of 640 million to a range of 680-940 million. Although the results are open to the criticism that the climate models used (GISS, GFDL, UKMO) have sensitivity parameters higher than the IPCC's 2.5°C (Reilly and Hohmann, 1993), from another standpoint the results are optimistic. They employ a carbon dioxide concentration of 555 ppm, which is essentially a transient concept for the year 2060, rather than a 2xCO$_2$-equivalent equilibrium concept (where the carbon dioxide concentration is 440 ppm).[1]

The Rosenzweig *et al.* study has been supplemented by Reilly *et al.* (1994), who used the Rosenzweig *et al.* yield data as an input for SWOPSIM. The study estimated global welfare losses at $0.1-$61.2 billion in the scenario without adaptation. In the moderate adaptation scenario welfare changes range between +$7.0 and -$37.6 billion.

The number of additional people suffering from hunger has also been estimated by Hohmeyer and Gärtner (1992). Their estimate of 900 million deaths over a 20-year period up to 2030 is based on a rather ad hoc line of reasoning though. Their figure, which measures actual casualties, also appears rather high, compared to the more sophisticated Rosenzweig *et al.* (1993) estimates of people *at risk*.

On a regional level, a study for the European Union predicts that overall agricultural yields in Europe are likely to increase as a result of increased temperature and precipitation. Welfare gains of ECU 3.2 billion are predicted for 1°C warming and ECU 12.2 billion for 4°C (CRU/ERL, 1992). Gains would mainly occur in the north, whereas the outcome for southern Europe would be more mixed.

For the U.S., Adams *et al.* (1993) estimated the combined economic effects of climate change on agricultural producers and consumers under different scenarios. Impacts were generally negative if based on the 2xCO$_2$ predictions of the UK Meteorological Office. The more benign GISS and GFDL forecasts yielded mostly positive impacts, except for the cases without CO_2 fertilization: Producer gains, particularly in the North, were generally large enough to offset the losses faced by consumers and producers in the South. In their "standard" scenario, with a CO_2 concentration of 550 ppm and no trade and adaptation effects, the estimated agricultural impacts ranged from -$18 billion (UKMO) to +$10 billion (GISS). Substantial economic losses are predicted in all scenarios once temperatures rise by 4°C. Based on earlier estimates by the U.S. EPA (1989), Nordhaus (1991) and Cline (1992a) have estimated U.S. impacts of roughly zero and -$17.5 billion, respectively. The former estimate is based on a 660 ppm scenario, whereas Cline assumes 440 ppm, consistent with CO_2-*equivalent* doubling.

A study of the Missouri-Iowa-Nebraska-Kansas (MINK) area introduces the further influence of farmer adaptation (earlier planting, use of longer season varieties, changes in tillage to conserve water; see Easterling *et al.,* 1993; Rosenberg, 1993). Using actual climate conditions of the 1930s as an analogue for a 2030s climate, the study found that, without carbon fertilization, climate change cuts agricultural production in the area by 17.1%, but only by 12.1% with on-farm adaptations. Adaptation thus reduces losses by about one-third. If 100 ppm carbon fertilization is added as an offsetting factor, output is reduced by only 8.4%. This loss is cut further to 3.3% by adding the influence of adaptation. Considering that regional temperatures were only about 1°C higher than today in the 1930s, and adjusting for commensurate carbon fertilization, the MINK results suggest losses in the order of 10% for benchmark $2xCO_2$ warming, even with farmer adaptation.

Another paper emphasizing adaptation is Mendelsohn *et al.* (1993), who argue that the production function method commonly used in crop models inadequately captures induced producer responses. They suggest that existing cross-section data for different climatic regions can provide a better guide to total effects incorporating these responses. They use county-level U.S. data to regress farmland values on climate and a number of other relevant variables. To simulate the overall impact of climate change on U.S. agriculture, they postulate a rise of 2.8°C (5°F) in mean temperature and 8% in rainfall.[2] Under these assumptions, U.S. agriculture experiences losses of $6 billion to $8 billion annually if county results are weighted by shares in cropland area, but gains of $1 billion to $2 billion if the weights are shares in crop revenue (which gives much more weight to irrigated lands of the West and South). The study shows that agricultural land value is strongly influenced by climate, even after taking adaptation into account. The cross-sectional data used in the study reflect agricultural practices that are highly adapted to the local climate, and yet the authors still find large productivity differences related to climate. The evidence indicates that warmer summers have a negative impact on land values and, by implication, productivity. This is persuasive empirical evidence that adaptation is unlikely to completely offset the effects of climate change on agriculture.

6.2.2 Sea level rise

IPCC Working Group I in Volume 1 of the present report (IPCC, 1996a) estimates a central value for sea level rise by the year 2100 of about 46 cm, compared to 66 cm in IPCC (1990a). The impacts of sea level rise are discussed in detail in Volume 2 (IPCC, 1996b), but the main areas threatened are coastal zones and small islands. These are characterized by highly diverse ecosystems that are important as a source of food and as habitat for many species. They also support a variety of economic activities, some of which put the natural coastal systems under stress.

Both the original and updated IPCC estimates refer to the transient rather than the equilibrium impact of sea level rise. Unlike other damage categories, the equilibrium effects for sea level rise are far greater and take much longer to occur than the point estimates corresponding to the first year of equilibrium warming suggest. For a discrete warming shock, the sea level continues to rise for up to 500 years (Titus, 1992; Manabe and Stouffer, 1993; see also Wigley, 1995). Some studies crudely convert this growing long-term effect into a point estimate of a 1-m sea level rise for a doubling of CO_2, even though realized increases by 2100 are estimated to be lower. Others ignore long-term effects and, in a similar arbitrary fashion, associate a doubling of CO_2 with a sea level rise of about 50 cm.

The literature usually divides the costs of sea level rise into three types: capital costs of protective constructions, the recurrent annual cost of forgone land services, and the costs associated with increased flood frequencies. Coastal protection is a form of adaptation used to avoid land loss and loss from increased flood frequencies. Land loss and flood losses thus depend on the chosen level of protection. Note that the preservation of drylands could imply more rapid loss of wetlands. Other, not fully assessed damages include loss of sovereignty, cultural heritage, and national identity of small island states (see Box 6.2), and the creation of a potentially large refugee population (see Section 6.2.11).

Protection costs. The U.S. EPA estimated that for the U.S., a 1-m rise in sea level by the year 2100 would require $73 billion to $111 billion cumulative capital costs to protect developed areas through the building of bulkheads and levees, pumping sand, and raising barrier islands (U.S. EPA, 1989). Assuming that capital costs would be spread over a 100-year period, Cline (1992a) estimated costs in the order of $1.2 billion for capital construction. Gleick and Maurer (1990) estimated capital construction and maintenance costs for protecting San Francisco Bay from a 1-m rise at $200 million annually, or a sixth of the U.S. total estimated in Cline (1992a). Fankhauser (1995) estimated annuitized costs of coastal protection against a 50-cm rise in the order of $1 billion worldwide, with about half occurring in non-OECD countries.

Land loss. The U.S. EPA (1989) report estimates that under a 1-m rise scenario unprotected dry land amounting to 6,650 square miles would be lost in the U.S., and 49% of today's 13,000 square miles of wetlands would be lost (Titus *et al.,* 1991). Titus *et al.* indicate that wetlands preservation programmes typically cost up to $30,000 per acre. Using a more conservative $10,000 capital cost per acre of wetlands, placing coastal dryland value at $4,000 per acre, and applying a rental opportunity cost of 10%, Cline (1992a) estimates the annual U.S. losses from a 1-m rise at $4.1 billion for wetlands and $1.7 billion for dryland. Assuming only a 50-cm rise in sea levels, but doubling wetland costs to $20,000 per acre, at least in developed countries, Fankhauser (1995) obtains a cost estimate of $45.6 billion annually for forgone land services worldwide, assuming a 33% loss of all remaining wetlands under a 50-cm rise. With over 85% of coastal wetland loss occurring in developing countries, low income regions are by far the most heavily affected areas. For the OECD (excluding

Canada, Australia, and New Zealand) Rijsberman (1991) estimates a loss of coastal wetlands in the order of 48,000-64,000 km² for a 1-m rise, more than 50% of the remaining area of coastal wetland habitats in these countries.

Sea surges. A 1-m rise in sea levels could increase the number of people subject to annual flooding by about 20%, according to estimates in IPCC (1994). Particularly at risk would be coastlines along the Indian Ocean, in the South Mediterranean, and in Africa, as well as small island states. A number of case studies exist which quantify the impact of sea level rise on sea surges in the U.S. (see Titus, 1987). The annual average damages in Charleston, South Carolina, could double due to an 88-cm sea level rise; the damages of a 100-year storm in Galveston, Texas, could triple. (Note, however, that cost-effective protection measures can mitigate this loss.) The U.S. Federal Emergency Management Agency and Federal Insurance Agency (U.S. FEMA-FIA, 1991) estimate that the area inundated by a 100-year flood will increase from 19,500 square miles to 23,000 and 27,000 square miles for a 1-foot and 3-foot sea level rise, respectively, by the year 2100 if no protective measures are taken. The region most significantly affected would be the Louisiana coast. The expected annual flood damage in 2100 increases by 36-58% ($150 million) for a 1-foot rise and by 102-200% ($600 million) for a 3-foot rise in sea level. For the European Union, CRU/ERL (1992) calculate that periodic flooding would increase the costs of sea level rise by as much as a factor of 2.7.

6.2.3 Forests

The impact of climate change on forests is uncertain. Impacts may be beneficial for some regions and species and detrimental for others. IPCC (1996b) identifies three major changes of consequence to the forestry sector. They are

(1)   changes in seasonal climate patterns, which differ with latitude
(2)   water shortages during the growing season
(3)   rate of climate change

The most significant changes over the next fifty years or so are, however, likely to be caused by nonclimate effects, in particular by land use change. No attempt is made within this section to quantify nonclimate impacts (e.g., deforestation and human-caused fires). Although these factors interact with climate, damage estimates here are restricted to impacts attributable to anthropogenic climate change.

Simulations for baseline climate change to 2050 suggest that boreal forests will be more impacted by climate change than tropical forests, which are more affected by changes in land use. It is also anticipated that the impact of climate change on temperate forests would be lessened through ameliorative action. Models neglecting land use effects generally suggest relatively benign impacts. It was estimated that global forest area could increase as much as 9% in this case (IPCC, 1996b).

Furthermore, forests could be adversely affected by an increase in the frequency or intensity of wildfires that may oc-

cur as a consequence of changes in thunderstorm and drought conditions. Estimates by Price and Rind (1994) for the southwestern U.S. suggest that a doubling of CO_2 could lead to a 60% increase in the number of lightning-caused fires. The annual area burned could increase by over 140%.

Current models have a number of deficiencies, however. Current studies of forest responses to climate change are at the ecophysiological level and the ecosystem level. Only the latter impacts, arising from changes in existing forest area, are assessed in this section. Furthermore, models are mostly concerned with equilibrium climate change. It is not currently possible to predict the transient responses at the global scale. Some of the static vegetation models used to estimate potential forest losses under changed climate include the Holdridge vegetation model, IMAGE 2.0, BIOME, and MAPSS (see IPCC, 1996b). In spite of the inability of these models to deal with transient responses, they do provide quantitative estimates of changes in equilibrium vegetation classes under future climate. Because the number of vegetation classes in the models is limited, however, it is believed that they underestimate actual changes.

Most studies concerned with the economic impacts on forests and forestry are based on earlier, perhaps more pessimistic, model runs. A study often used is Sedjo and Solomon (1989), who calculated that steady-state $2xCO_2$ warming could reduce boreal forests by 40% and temperate forests by 1.3% in biomass, but increase tropical forests by 12%. The net change would amount to a decline of 3.7% globally in biomass, and 5.8% in area. Based on these figures, Fankhauser (1995) has estimated annual forestry damages of $1.8 billion in OECD countries and $2 billion worldwide, using forest values of 2000, 400, and 200 $/km², respectively, for high, middle, and low income countries. Because of the positive impact on tropical forests and the use of higher forest values in developed nations, most forestry damage occurs in OECD countries.

Compared to equilibrium estimates, transient effects over as much as three centuries would be much more severe. Whereas the latitudinal borders of potential location for given species would migrate poleward by 600 to 1,000 km over the next century, the actual migration pace could be as low as 100 km (U.S. EPA, 1989). Dieback along low-latitude boundaries would thus exceed additional growth on the poleward boundaries. Over the next 100 years U.S. forests could lose 23-54% of standing biomass in the Great Lakes region and 40% in the West.

On this basis, Cline (1992a) estimates a loss of 40% for U.S. forests. Estimating the value of annual wood extraction at $10 billion, and allowing for some limitation of losses through reforestation, he estimates net U.S. forest loss from benchmark warming at $3.3 billion annually, solely for commercial wood products.

Titus (1992) places U.S. forest damages an order of magnitude higher, at $44 billion annually (central estimate). He estimates a median percentage biomass loss of 34% in the thirty states he examines. His much higher damage estimates thus stem not from greater biomass loss but from a higher valuation of unit forest area. Using a comprehensive measure that includes recreation and other value, Titus values forested area at $45 to $150 per acre per year of forest (above raw land value). The Titus estimates would make U.S. forest loss the largest damage category.

A more recent U.S. study by Callaway *et al.* (1994) provides a more differentiated picture. A decrease in softwood yields in practically all areas (except the Northwest) is partly offset by increased yields in hardwood in most regions. Although producers would gain from price increases, consumers would face substantial losses. On aggregate, annual losses to the U.S. forestry sector are estimated to be between $2.5 billion (for 2.5°C warming, including CO_2 fertilization) and $12 billion (4°C, no CO_2 fertilization). This corresponds to a 4-19% welfare reduction in the U.S. forestry sector.

6.2.4 *Water supply*

Some regions may benefit, but climate change could in many areas put considerable stress on water supply as a result of changes in the timing, regional pattern, and intensity of precipitation events. This, in turn, will affect the magnitude and timing of runoff, while higher temperatures will at the same time lead to changes in evapotranspiration, soil moisture, and infiltration conditions (IPCC, 1996b). In areas and/or periods where precipitation declines or does not rise by enough to compensate for higher evapotranspiration (from warmer temperature), the widening gap would reduce soil moisture and water levels and flows. In coastal regions, saltwater intrusion could affect current freshwater sources. At the same time, the demand for water would tend to rise with warming, because of increased needs for irrigation and for cooling in electric power production (U.S. EPA, 1989) and because of higher residential demand.

Although confidence in projected water runoff is still low, model runs project increased runoff in high latitude regions due to increased precipitation, whereas lower latitudes could experience decreased runoff due to the combined effects of increased evapotranspiration and decreased precipitation. The current arid and semi-arid regions, in particular, could experience some of the largest decreases in runoff (IPCC, 1996b). River basin runoff is very sensitive to small variations in climatic conditions, because runoff is a residual of precipitation on the one hand and soil absorption or evaporation on the other. Consequently, small changes in any of the underlying variables can cause a much larger proportionate impact on runoff.

For the U.S., water basin simulation models show that in a warmer, drier climate (+2°C, −10% precipitation), water supply in 18 major water regions covering the bulk of national supply would decline by approximately one-third (calculated from IPCC, 1990b). The U.S. EPA (1989) predicts that $2 \times CO_2$ warming would reduce annual water deliveries in California's Central Valley basin by 7-16%, in a region where baseline water demand is expected to rise by over 50% as early as 2010 (i.e., before a doubling of CO_2 is likely to occur). For the Sacramento basin, Gleick (1987) estimates that a 4°C increase

in temperature would decrease summer runoff by 55%, even if there were a 10% rise in precipitation. For the Boston area, Kirshen and Fennessey (1992) applied water balance models to GCM projections for benchmark warming and found that reliable water yields of existing water systems could fall by as much as one-third (based on the GISS and GFDL projections) but could also rise (if the UKMO and OSU projections are used). The difference arises because some models predict falling precipitation and others rising. If zero change in precipitation is imposed, the result is a decline of 18% in reliable water yield.

Cline (1992a) sets 10% as a central estimate for water supply reduction in the U.S. from benchmark warming. He estimates national annual withdrawals at 0.4 billion acre-feet and unit price at $250 per acre-foot. The resulting estimate of annual damage is $7 billion. Titus (1992) uses a larger volume impact but lower unit prices to reach a central estimate of $11.4 billion annual losses. For the European Union, CRU/ERL (1992) estimate that the costs of reduced water runoff would amount to ECU 5.8 billion for 1°C warming and ECU 18.8 billion for 4°C.

Fankhauser (1995) uses a somewhat lower percentage volume loss than the U.S. studies but, at least for OECD countries, a unit price more than twice as high as Cline's. He estimates annual losses at $34.8 billion for the OECD, and $46.7 billion worldwide. Again, the high share of OECD damages is due to differences in valuation between regions. In physical units, about three-quarters of water losses occur in non-OECD countries (see Table 6.5).

Adjustments in water management practices can help to ease impacts. More efficient water allocation, for example, with less low-priced water allocated to agriculture and more to urban use could attenuate losses in water supply. On the other hand, it could aggravate prospective agricultural output losses.

Gleick (1992) and Homer-Dixon *et al.* (1993) have emphasized another dimension of water supply effects: the potential for political conflict, such as the dispute over the Jordan River basin that contributed to the 1967 war in the Middle East. From this standpoint, associated damages might appropriately include higher defence costs or, perhaps more appropriately, some unquantified social cost arising from the increased probability of regional wars.

6.2.5 *Space cooling and heating*

Climate change would impose higher air conditioning (space cooling) costs but would reduce heating costs. The net effect on energy costs is ambiguous and will be highly variable across regions. For space cooling costs in Japan, Nishioka *et al.* (1993) report estimates of a 2% increase in electricity demand per °C warming for temperatures between 24°C and 30°C, and a 1% increase for temperatures between 17°C and 24°C. No significant change in demand was found for temperatures above 30°C.

On the basis of detailed energy projections, the U.S. EPA (1989) has estimated that a warming of 3.7°C by 2055 would result in a net increase in U.S. electricity demand of about one-fifth above baseline, requiring additional annual operating costs of $53 billion and cumulative capital cost increases of $224 billion. Scaling back these estimates for consistency with present-day economic size and for a warming of only 2.5°C, Cline (1992a) estimates annualized damages of $11.2 billion from higher net electricity requirements under benchmark warming. Arguably, this estimate is understated, because with global mean warming of 2.5°C, U.S. warming would amount to 3.9°C. Titus (1992) uses the same underlying study to reach a central estimate of $5.6 billion.

More informal estimates of the corresponding savings on space heating are considerably lower, in the range of $1 billion annually (Nordhaus, 1991; Cline, 1992a). A lower gain from reduced heating relative to increased cooling is consistent with the analysis of Loveland and Brown (1992). They find that "annual cooling loads will increase at a much greater rate than heating loads will decrease," and they stress the extra costs of peak cooling loads. However, the expenditure base on which a greater proportionate increase in cooling costs would operate is about 2.3 times lower than the expenditure base of the smaller percentage cut in heating costs (Rosenthal *et al.,* 1994). The net effect is thus ambiguous.

Based on regional GISS warming results reported in Loveland and Brown, Cline (1993a) estimates that for the U.S. the number of cooling degree days would rise by about 100% under $2xCO_2$, compared to a 40% fall in heating degree days. As this ratio for change is approximately the inverse of the ratio for the expenditure base, the overall result would be that the reduction in heating costs would be approximately equal to the increase in cooling costs. On this basis, the Nordhaus (1991) and Cline (1992a) heating cost reductions would appear substantially understated.

Rosenthal *et al.* (1994) go further and estimate that U.S. savings on heating would exceed increased cooling costs by $7.6 billion for 2.5°C U.S. warming. With nearly equiproportionate changes in heating (−14% for 1°C) and cooling (+16%), their results significantly deviate from the Loveland-Brown estimates and stress greater proportionate change in cooling. For higher temperatures, the savings on reduced heating could be expected to fall, and the costs of increased cooling to rise (given the falling base of the former and the rising base of the latter).

For most of the developing countries, because of their location and base climate, savings from reduced heating would tend to be limited even for benchmark warming. Increased cooling costs for these countries would tend to be larger, especially where baselines are already incorporating greater penetration of air conditioning as per capita income rises. Extrapolating the U.S. EPA (1989) data to other geopolitical regions, Fankhauser (1995) assumed an increase in electricity demand of 3.2% in all regions considered, except for the former Soviet Union, where electricity demand was assumed to decrease by 1%. This resulted in increased space cooling costs of about $20 billion in the countries of the OECD and $23 billion worldwide.

Based on estimated changes in heating and cooling degree days, CRU/ERL (1992) calculated significant benefits from reduced heating expenses in the European Union. Gains in Northern Europe were only partly offset by increased costs in Portugal, Spain, and Greece. A 1°C rise in temperature would yield net benefits in the order of ECU 13 billion. For 3°C warming, the figure would rise to about ECU 32 billion. These rather optimistic results appear to be mainly due to the dominance of heating over cooling degree days in the baseline case without warming.

6.2.6 Insurance

The property insurance industry protects other economic sectors from the financial consequences of unexpected or uncertain events, including weather extremes. As such, it is highly exposed to changes in these extremes and thus to climate change (see Section 6.2.14). The basic function of insurance is to transfer financial risk from an individual to a group, that is, to spread a specific individual loss over the entire group of potentially affected people. At present, natural hazard insurance has a capacity to absorb damages of about $100 billion. For comparison, in 1992 Hurricane Andrew alone caused economic damage of $30 billion, about half of which was insured (Dlugolecki *et al.,* 1994). The impact of climate change on other branches of insurance, such as life and liability, and on the wider financial sector (e.g., banking) are discussed in Volume 2 (IPCC, 1996b).

Since 1987, after a relatively quiet period of about twenty years, the insurance industry has been confronted with a large number of major weather-related catastrophes, that is, events that involved insured losses of over $1 billion. This increase was caused by a variety of factors, including population growth, higher standards of living, concentration of people and capital in large conurbations, the development of extremely exposed areas, and environmental changes. Although there is no clear connection to anthropogenic climate change, the reaction of the sector is illustrative of what would happen if climate change were to lead to an increase in the number and intensity of extreme weather events (Berz and Conrad, 1993; Dlugolecki *et al.,* 1994; IPCC, 1996b; Leggett, 1993). Premiums were increased and cover was restricted, but with some delay, as the insurers wanted to make sure that the increase in risk was permanent. In some areas, such as the Southeast states of the U.S., the Caribbean, and the Pacific, insurance and reinsurance supply was withdrawn after a sequence of hurricanes in the late 1980s and early 1990s. As a further reaction, the insurance sector has the following broad options (Dlugolecki *et al.,* 1994; Tol *et al.,* 1994):

(a)   altering the way in which premiums are determined by incorporating more knowledge of the actual risk (so far, premiums have primarily been determined by competition and recent claims)

(b)   changing the product to limit the insurer's exposure

(c)   extending the available funds through further pooling and accumulation, particularly in cooperation with governments and banks. This may require alterations of existing fiscal and institutional regulations

(d)   risk management, that is, getting involved in restricting the damage caused by weather events

(e)   lobbying for and investing in environmentally sound policies and projects

Points (a) and (b) are being implemented, (c) and (d) are under consideration, and (e) awaits further evidence on the relationship between environmental policies and insured risks. Although a number of these points help reduce the overall exposure of society to weather hazards, many merely affect the exposure of the insurance sector itself. Limiting only the insurers' exposure obviously leads to greater exposure for others. Little research has been done to assess the consequences of such a policy on households, industry, and government.

6.2.7   Other market sectors

Construction. Nordhaus (1991) suggests that the construction sector in temperate climates would be favourably affected by climate change because of a longer period of warm weather. However, although construction is adversely affected by frost, it is also inhibited by rainfall, and GCMs typically predict an increase in global mean precipitation by about 8% as the consequence of benchmark $2 \times CO_2$ warming (IPCC, 1990a). The IPCC notes the adverse effect of rainfall on economic activity more generally and points out that "rainfall is responsible for more delays than any other climatic variable" for UK industry (IPCC, 1990b). However, estimates are not available on the net effects of warming and increased precipitation on construction or industry more generally.

Tourism and leisure activities. Studies by the Canadian Climate Program Board suggest that, as a consequence of a shortened period of snow cover, Quebec could lose 40-70% of ski days. In Ontario, the shorter skiing season may cause a loss of up to $50 million in revenue (Canadian Climate Program Board, 1988a, b). Losses would presumably be larger in relative terms in U.S. ski areas, where the temperature base is already higher. Annual U.S. ski activity amounts to an estimated $5.6 billion and 53 million skier visits (Waters, 1990). Assuming a 60% reduction from benchmark warming, and allowing for released productive labour and capital, Cline (1992a) estimates ski industry losses at $1.7 billion annually.

The tourist industry will also be affected by beach erosion and the inundation of beaches (Baan *et al.,* 1993), as well as likely coral reef death and other ecosystem loss (U.S. EPA, 1989).[3] These impacts could be particularly significant for small island states, where tourism frequently accounts for over a third of GDP (see Turner *et al.,* 1994). On the other hand, there could be gains from climate change in other leisure sectors such as camping, boating, and sunbathing. The Ontario case study, for example, predicts an increase in the camping season of up to 40 days in some areas (Canadian Climate Program Board, 1988b). In many regions, however, an increase in summer activities could be hampered by increased rainfall. CRU/ERL (1992) developed a "comfort index," using temperature, sunshine, and rainfall as indicators of the suitability of climates for leisure activities. Using this index and figures on current European tourist revenues, they estimate that 1°C

warming could benefit tourism in the EU by about ECU 4.7 billion. As temperature rises though, the impact, at least in Southern Europe, soon becomes negative.

Urban infrastructure. Adaptation and/or protection of infrastructure from extreme weather events like floods, extreme rainfall, or landslides could cause increased costs under climate change (IPCC, 1996b). The U.S. EPA (1989) has examined the impact of climate change on urban infrastructure costs. For coastal cities, sea level rise or more frequent droughts would increase the salinity of coastal aquifers and tidal surface waters, requiring a response where these are the sources of a metropolitan water supply. In addition, more frequent and intense storms would likely overload existing storm sewer systems. Using U.S. EPA studies for New Orleans, New York, Philadelphia, and Miami, Cline (1992a) suggests that annualized damages of benchmark warming for U.S. urban infrastructure could amount to $100 million yearly.

6.2.8   Health

There are many potential health impacts arising from climate change, some beneficial and some adverse. Although generally difficult to foresee and quantify, these impacts could arise from diverse events, including disturbances in natural or managed ecosystems. They could either be direct, as in the case of heat wave deaths, or indirect, as when caused by changes in the range and transmissibility of vector-borne infectious diseases. Also, it is expected that different populations with varying levels of natural, technical, and social resources would differ in their vulnerability to climate-induced health impacts. The range of health impacts is discussed in more detail in IPCC (1996b).

Given the current state of knowledge, and the influence of environment, socioeconomic circumstances, population density, and nutritional status, among other factors, it is possible to apply only a quantitative cost-assessing approach to a minority of such impacts. In this section monetized health damages reflect only mortality due to heat stress. Other health impacts of climate change are excluded. This is not to imply that these are necessarily the most significant expected health impacts (see IPCC, 1996b).

There is a U-shaped relationship between mortality and outdoor air temperature. Death rates increase as a consequence of both heat waves and very cold weather. The lowest mortality rates are found at temperature levels of about 16-25°C (Kunst *et al.,* 1993; Haines and Parry, 1993). Climate change could lead to an increase of heat-related deaths from coronary disease and stroke, which is likely to more than offset a reduction in winter mortality. Air pollution increases the occurrence of respiratory diseases (such as emphysema and asthma), and longer, warmer summers are expected to increase the severity of air pollution.

Kunst *et al.* (1993) have analyzed the statistical relationship between mortality and air temperature in the Netherlands. They found that a 30-day increase in air temperature of 1°C above its ideal level would lead to a rise in mortality of 1.1-1.9%. A 1°C drop from the optimum would cause a slightly lower rise of about 0.8-1.3%. For the UK, Langford

BOX 6.1: ATTRIBUTING A MONETARY VALUE TO A STATISTICAL LIFE

Attributing a monetary value to a "statistical life" is controversial and raises a number of difficult theoretical and ethical issues. It is important to understand that what is valued is a change in the risk of death, not human life itself. In other words, the issue is how a person's welfare is affected by an increased mortality risk, not what his or her life is worth. If 100,000 people are exposed to an annual mortality risk of 1:100,000, there will, statistically, be one death incidence per year. Removing the risk would thus save one *statistical life*. It is this statistical life that has an economic value. It would make no sense to ask an individual how much he or she is willing to pay to avoid certain death. Nor is that the context of social decision making. But it can make sense to ask what individuals are willing to pay to reduce the risk of death or what they are willing to accept to tolerate an increased risk of death.

The reality is that safety is not "beyond price." If it were, most of the world's wealth would be spent trying to save lives by reducing accidents and preventing disease. Risks are taken every day, both by individuals and by governments in choosing their social and economic expenditures, some of which are specifically directed at protecting and extending human life. For example, if a government introduces a programme of inoculation for childhood diseases that costs $10,000,000 per year and saves an average of 80 lives per year, a statistical life is implicitly valued at $125,000 at a minimum.

Several methods have been applied to calculate the value of a statistical life (VOSL). None of them is without problems.

The prescriptive view

Under a prescriptive or normative approach, the VOSL is not set according to observed behaviour but is based on ethical and political considerations. It poses the question: At how much *ought* a statistical life be valued according to ethical or other criteria? An obvious implication of this approach is that all lives will be treated equally. Each statistical life saved should have the same value. However, the question of what this uniform value should be is difficult to answer from a purely prescriptive point of view. In the context of climate change it has been argued that, since the developed countries have caused the greenhouse problem, OECD VOSLs should be used for all lives under the polluter-pays principle. An example of such an approach is given by Hohmeyer and Gärtner (1992). Based entirely on a "moral imperative," the theoretical economic basis of this approach is weak.

The descriptive view

The alternative is to take a descriptive stance and ask how much people are actually willing to spend to avoid the risk of death. Most studies that attempt to estimate damage due to climate change (see Table 6.4) are based on a descriptive perspective. Two approaches are commonly used in the economic literature: the human capital approach and the willingness-to-pay approach.

The human capital approach: This method involves treating an individual as an economic agent capable of producing an output that is valued in monetary terms. A life lost is then the loss of that output, less any consumption that the individual would have made. One problem with this approach is that it tends to produce extremely low values for those with low earnings, clearly discriminating against the already poor. Another problem is that the approach has no particular relationship to an individual's willingness to pay to reduce his or her risks of mortality. The human capital approach is not properly founded in economic theory.

The willingness-to-pay/willingness-to-accept method: The theoretically preferred approach is to value a statistical life on the basis of what individuals are willing to pay or accept for risk changes. Such values can be based on methods such as "contingent valuation," where individuals are asked directly how much they would be willing to pay to reduce risks. Other measures include finding out how much people are spending on safety and disease-preventing measures, or by how much wages differ between safe and risky jobs (the "hedonic approach"). For example, suppose 100,000 workers are paid an additional $15 each to tolerate an increased risk of mortality of 1/100,000. The increased risk will result in one statistical life lost, valued at $15 \times 100,000 = $1,500,000.

One problem with the willingness-to-pay approach is that it relies on individuals having an adequate perception of the risks undertaken. This will not always be the case, particularly in developing countries. Another disadvantage of the WTP approach is that the resulting figures depend on factors that may be distributed in a way that is not considered just. Most important, the estimates depend on a person's income. Rich people are better able to afford safety expenditures, whereas poorer people's WTP may be constrained by their ability to pay. WTP estimates will reflect this discrimination against the less well off.

BOX 6.1 (*cont.*):

Descriptive VOSL estimates

Studies of the contingent valuation and hedonic wage risk approaches suggest VOSLs in the order of about \$1.8-\$9 million in the developed world, with a best guess of \$3.5 million (Viscusi, 1993; ORNL/RFF, 1994). Few studies exist for the developing world, and it is difficult to say what the results would be for these techniques. Since WTP is constrained by ability to pay (wealth and income), the results are likely to be very much lower. A preliminary study for India places "own valuations" at perhaps \$120,000 (Parikh *et al.,* 1994).

In the absence of developing country studies, various rough and ready approximations have been tried. However, "borrowing" VOSLs from developed economy studies is hazardous. For example, developing country valuations could be estimated as

$$VOSL_{ldc} = VOSL_{dc} \cdot (Y_{ldc}/Y_{dc})^E$$

where *ldc* = less developed or developing economy, *dc* = developed economy, *E* either denotes the income elasticity of demand or the elasticity of the marginal utility of income, and *Y* is income (corrected for purchasing power parity). Developing countries' VOSLs would simply be "scaled down" by the ratio of incomes raised to the power *E*. If *E* = 1, then the scaling down is simply done by the ratio of incomes. The above formula reduces to $VOSL_{ldc}/Y_{ldc} = VOSL_{dc}/Y_{dc}$, that is, the VOSL in a developing country is the same proportion of income as it is in the developed economy. In the light of existing developing country estimates, this seems a reasonable first approximation. But it still leaves the *absolute* VOSL lower in developing countries. Many would therefore argue that interregional comparisons of WTP estimates should be avoided. If comparisons are made, the aggregation of individual damages is crucial.

It can also be argued, however, that the growing international mobility of skills and services will make national differences in VOSL increasingly less relevant over the next half century (R.K. Pachauri, personal communication, 1995).

Aggregation

Aggregation in this context is a political and ethical process, based on rules such as those set out in Chapter 3 (see also Box 6.2). The aggregation process makes it possible to correct for factors not reflected in individual WTP estimates, such as the injustice in the underlying income distribution or different responsibilities for the climate change problem. The process may thus result in changes in monetary values based on income levels as they are computed conventionally. For example, if VOSL is scaled in proportion to income, as suggested above, and aggregation weights are inversely related to income, weighted VOSLs will effectively be equal across countries, as proposed by the prescriptive school.

Sensitivity analysis

Fankhauser (1995) estimates $2 \times CO_2$ damages for all effects at about \$180.5 billion (1.3% of GDP) for OECD countries and about \$89.1 billion (1.6% of GDP) for non-OECD countries (see Table 6.6). Of these damages, human mortality accounts for \$34 billion for OECD countries and \$15 billion for non-OECD countries. If all mortality damages were valued at a uniform average VOSL of \$1 million, OECD mortality damage would fall to \$22.7 billion and the non-OECD figure would rise to \$115 billion. This change would reduce the overall OECD damage only marginally to 1.22% of GDP but would approximately double the non-OECD damage estimate to 3.4% of GDP.

However, because this approach would be premised on a "moral imperative" rather than WTP, for consistency it would seem also to require uniform VOSLs across time as well as across countries. An important implication is that by the time the damage occurred (e.g., in the middle of the next century), a constant life valuation of \$1 million would represent a substantially lower fraction of GDP than it would today, especially for developing countries where per capita income should grow more rapidly. The sensitivity estimate here would thus tend to overstate damage as a percentage of GDP at the time relevant for damage assessment. Similar intertemporal adjustments for moral imperative considerations would generally not be relevant for other damage.

and Bentham (1993) estimate that the higher temperatures predicted for 2050 (about 2.5°C) could result in some 9,000 fewer winter deaths per year.

Most economic studies use Kalkstein (1989) as a basis. Kalkstein has used statistical methods to analyze the mortality impact of changed weather conditions under $2xCO_2$ for 15 major U.S. cities (U.S. EPA, 1989; Kalkstein, 1989). Even after accounting for induced acclimatization (based on mortality statistics for control cities with comparable present climate conditions), he finds that increased summer deaths substantially exceed decreased winter deaths. Cline (1992a) and Fankhauser (1995) weight these results by population and extrapolate that benchmark warming would increase mortality by about 27-40 persons per million population, depending on the warming scenario. Without acclimatization this figure would be about six times higher. For the United States the more conservative figure including acclimatization translates into 6,600 to 9,800 additional deaths annually for the present U.S. population.

Mortality effects could be particularly damaging in the more vulnerable countries of the developing world, where mitigating technologies like air conditioning will be less readily available. In Shanghai, for example, death rates from heat stress are currently over twice those in New York City (Haines and Parry, 1993). Health problems associated with hot spells are also significant in Bangladesh (Asaduzzaman, personal communication, 1994) and Pakistan (Asian Development Bank, 1994). Using uniform mortality rates, Fankhauser estimates that about 115,000 additional casualties per year could occur in non-OECD countries, compared to some 23,000 additional annual deaths in the OECD (for 2.5°C warming; see Table 6.5). Tol (1994b), using the same estimates but without scaling to 2.5°C warming, arrives at a figure of 215,000 additional casualties worldwide.

Expressing the value of this loss in monetary terms is controversial (see Box 6.1). Cline (1992a) uses lifetime earnings to place statistical life valuation at a conservative $595,000 and reaches a corresponding estimate of $5.8 billion in annual U.S. losses from benchmark warming. Using higher values, Titus (1992) obtains annual damages of $9.4 billion for the U.S. Fankhauser uses value-of-statistical-life estimates of $0.1 to 1.5 million, depending on regional income levels, and obtains mortality damages of $10 billion in the U.S., $34 billion for the OECD, and $49 billion worldwide. If an identical value of $1 million for all lives were used worldwide, this latter figure would rise to almost $140 billion. Tol (1995) assesses a statistical life at $250,000 + 175 x (annual income per capita). With worldwide mortality costs of $188 billion, health damages account for more than half his aggregate estimate of warming damages.

In addition to direct heat-related health effects, there could be indirect losses from increased vector-borne diseases like malaria or yellow fever, as their insect vectors adjust to new climate conditions and their risk areas shift (Haines and Parry, 1993). WHO (1990), for example, suggests that the so far disease-free highlands of Ethiopia, Indonesia, and Kenya might be invaded by vectors. A study for Indonesia predicts a fourfold increase in the incidence of dengue fever and a 20-25%

increase in malaria cases (Asian Development Bank, 1994) by the latter half of the next century. Global simulations by Matsuoka *et al.* (1994) indicate a 10-30% increase in the number of people at risk from malaria under $2xCO_2$ conditions. Hohmeyer and Gärtner (1992) estimate that an extra 200 million people could be exposed to malaria worldwide. Martens *et al.* (1994) expect several million additional malaria cases by the year 2100. Vector-borne diseases may also spread into developed countries. Simulations dealing with a possible increase of malaria in the U.S. are, however, inconclusive (U.S. EPA, 1989).

The increased occurrence of flooding could lead to a higher incidence of diseases associated with poor sanitation standards, particularly in developing countries. For Indonesia it has been estimated that, by the year 2070, cases of diarrhoea could rise to over 900 per 10,000 inhabitants, compared to 300 cases per 10,000 people in 1989 (Asian Development Bank, 1994).

The emerging picture of health impacts thus indicates that the indirect effects could by far exceed the direct losses.

6.2.9 Air pollution

A warmer climate could aggravate some urban pollution problems. In the U.S., tropospheric ozone is the most severe problem in terms of the number of persons living in areas with air quality indexes that violate national standards (75 million persons in 1986). Total suspended particulates (TSP) and carbon monoxide are also important (41 million each). TSP is also implicated in widespread health damage in the developing world and the economies in transition.

Numerous studies confirm that ozone concentrations rise with temperature (see, e.g., IPCC, 1990b). The U.S. EPA, after summarizing various U.S. estimates, concluded that a 4°C rise in temperature (about what could be expected for the U.S. under 2.5°C global mean warming) could cause an increase in peak ozone concentrations of 10%. The result would be to double the number of cities in violation of the air quality standards from 68 to 136, causing most midsize and some small cities in the Midwest, South, and East to be added to the list of those presently in violation.

Applying its past models relating ozone concentrations to emissions of volatile organic compounds (VOCs), the U.S. EPA estimates that it would be necessary to reduce VOC emissions in the U.S. by 700,000 tonnes from a year-2000 expected base of 6 million tonnes to offset the effects of $2xCO_2$ warming on ozone formation. At an estimated cost of $5,000 per tonne, the agency calculates that the resulting costs would amount to $3.5 billion annually (U.S. EPA, 1989). Cline (1992a) uses this figure as an estimate of U.S. air pollution damage to be expected from benchmark warming. In comparison, air pollution control expenditure in 1987 was $27 billion (U.S. EPA, 1990).

Titus (1992) estimates U.S. tropospheric ozone damage from benchmark warming at $27.2 billion annually. This is almost an order of magnitude higher than Cline, although both figures are based on the U.S. EPA (1989) estimate. The difference appears to be that Titus projects a high baseline of VOC emissions at 3% growth and reaching an implied 70 million tonnes by 2060, whereas the U.S. EPA estimate is based on the present scale of emissions. Although some allowance for growth in the base is appropriate, it seems unlikely that the pollution baseline would grow faster than GDP and energy output.

Instead of calculating the additional expenditures needed to maintain present air quality standards, Fankhauser (1995) estimates the extra damage occurring if air quality standards were allowed to deteriorate. Assuming increases in NO_x and SO_2 emissions of 5.5% and 2% (based on U.S. EPA, 1989) respectively, he estimates additional air pollution damages of $12 billion in the OECD and $15 billion worldwide (see also Table 6.13 for cost of air pollution estimates).

6.2.10 Water pollution

Titus (1992) and Nishioka *et al.* (1993) have identified water pollution as a category with potentially large damages due to climate change. They predict a likely decline in river flow because of lesser water runoff. Because rivers carry away waste, reduced river flow leaves more waste to be removed by emission controls. In addition, a higher water temperature could affect water quality through a reduction in the level of dissolved oxygen (ECLAC, 1993). Nishioka *et al.* also discuss a possible decrease in the water quality of lakes through increased algae growth.

In assessing the impact of water pollution on the U.S., Titus assumes that the discharge of pollutants would change by the same proportion as river flow. He cites U.S. EPA estimates to establish a base of $64 billion for U.S. water pollution control costs in the year 2000 (at constant 1990 prices). According to his calculations, runoff declines by 2-4% if either precipitation decreases by 1% or temperature increases by 0.4°C. Setting the elasticity of control costs (percent change for a 1% change in river flow) at 1 to 1.7, Titus concludes that water pollution control costs imposed on the U.S. by benchmark climate change would be in the range of $15 billion to $67 billion annually, with a central estimate of $34 billion. Like Titus's estimate for forest damage, this estimate dwarfs typical damage calculations for most other individual categories. Yet water pollution is one of the most underresearched aspects of economic damage.

6.2.11 Migration

Shoreline erosion, river and coastal flooding, and severe drought could displace millions of people. Less dramatically, an accelerated decline in soil quality could also induce additional migration (IPCC, 1996b). Myers (1993) calculates that there will be 150 million additional refugees, or 1.5% of the world population in 2050, as a result of climate change, but many assumptions are behind this estimate. Schelling (1983) views migration as an efficient adaptive response to climate change. However, presumably a cost should be imputed to the utility loss by families compelled to migrate (Jansen, 1993). Indeed, historically, peoples have often fought wars to avoid being forced to leave their homelands. Tol (1995) arbitrarily

puts this loss at three times the average annual income per capita in the region of departure. This is reflected in Table 6.6. There is also a cost to the recipient host country. In many countries, for example, European countries and the U.S., the costs of incorporating immigrants into the social welfare infrastructure are already a politically sensitive source of social and budgetary pressure.

Cline (1992a) reports estimates that 640,000 legal immigrants have entered the U.S. annually in recent years, in addition to perhaps 130,000 illegal immigrants (Goering, 1990). He hypothesizes an increase of 25% in illegal immigration and 10% in legal immigration as a consequence of benchmark warming. Based on total per capita spending by state and local governments, and assuming 18 months before an immigrant's tax payments cover the family's social infrastructure costs, Cline estimates costs at $4,500 per immigrant; Ayres and Walter (1991) similarly cite a figure of $4,000 per refugee accepted by the U.S. Cline (1992a) estimates annual immigration costs to the U.S. from $2xCO_2$ warming at $450 million. The hardship and stress suffered by refugees remain uncounted, though. Fankhauser (1995) has extended the Cline estimates to the world as a whole, assuming that current migration patterns continue to hold (see Table 6.5).

6.2.12 Human amenity

Mearns *et al.* (1984) use statistical distributions of current temperatures to explore the impact of climate change on extreme temperature events for the U.S. In their base case, they increase mean temperature by 3°F (1.7°C) and hold the variance and autocorrelation of daily temperatures constant. Under these assumptions, they calculate that the frequency of heat waves (defined as 5 consecutive days with a maximum temperature of at least 35°C) would multiply threefold (estimated for Des Moines, Iowa). There is reason to believe that people would be willing to pay something to avoid a threefold increase in the incidence of heat waves. At the same time, climate change would reduce the disamenity of severe winters in colder areas. The net balance under benchmark warming is unclear. As suggested below, however, for much higher warming over the very long term, the amenity damages would be more likely to dominate.

Cline (1992a) provides an order of magnitude estimate of the value of the disamenity caused by a sharp increase in the number of hot spells in the U.S. Assuming that people are willing to pay 0.25% of their income to avoid this and other disamenities, a $10 billion loss per year results.[4]

Leary (1994) has surveyed the available empirical evidence on the implicit valuation of amenity due to local climatic variation in the U.S. The results of this study are that:

- Individuals value changes in climate and are willing to accept lower wages, or pay more for property, in return for an improved local climate. Some migratory patterns are explained by climate differences.

- In the U.S. these values appear to be substantial, ranging from hundreds to thousands of U.S. dollars per household (in present value terms).

- Households prefer sunny, mild climates. Increases in winter temperature and the number of sunny days are favoured, whereas higher summer temperatures deter migrants, so that hot summers appear to be a disamenity.

These findings are intuitively appealing, but characterization of "climate preferences" is not robust across the studies evaluated, and it is not possible, therefore, to say whether climate change will, on balance, increase or decrease climate amenity. As with other aspects of the economics of climate change, these results are relevant only to the U.S. and perhaps to other temperate industrialized zones where housing and labour markets respond to differentials in housing and work characteristics.

Amenity effects would vary by region, tending towards damages in presently warm areas and gains in presently cold areas. Potentially, amenity valuations could be large in absolute terms, because they apply to the entire income base of the population. However, the tendency towards neutralization by the differential effects among geographic regions and between seasons would tend to reduce the overall magnitude of amenity effects.

6.2.13 Ecosystem and biodiversity loss

Perhaps the category in which losses from climate change could be among the largest, yet where past research has been the most limited, is that of ecosystem impacts. Uncertainties arise both because of the unknown character of ecosystem impacts, and because of the difficulty of assessing these impacts from a socioeconomic point of view and translating them into welfare costs. Existing figures are all rather speculative. There is a serious need for conceptual and quantitative work in this area.

The U.S. EPA (1989) has noted in general terms the risk of increased species extinction from climate change, because of changes in habitat, predator/prey relationships, and physiological changes. It cites the poleward migration of forests as a major reason to expect stress on species, especially in view of natural and manmade barriers to species migration. Another category of likely species loss is that of coral reefs, as suggested by recent instances of coral death from El Niño warming (Glynn and de Weerdt, 1991).

Economists identify three types of value: direct and indirect *use value* (e.g., plant inputs into medicine and the role of mangrove forests in coastal protection); *option value* (preserving a species to retain the possibility that it may be of economic use in the future); and *existence value* (e.g., the value of knowing that there still are blue whales). Table 6.2 (based on Pearce, 1993) summarizes the results of "contingent valuation" sample survey estimates of what the public would be willing to pay to preserve an endangered animal species. Average values range from $1 to $18 per person per year for preservation of an individual species, with a maximum of $40 to $64 obtained for humpback whales. The willingness to pay figure for the preservation of entire habitats is somewhat higher, with a range of $9 to $107 per person and year.

The economic value of plants is dominated by their potential significance for medicinal purposes. Pearce (1993) notes

Table 6.2. *Preference valuation for endangered species and prized habitats*

Country	Species or Habitat	Value (1990$/year/person)
Norway	Brown bear, wolf, and wolverine	15.0
	Conservation of rivers against hydroelectric development	59.0–107.0
United States	Bald eagle	12.4
	Emerald shiner	4.5
	Grizzly bear	18.5
	Bighorn sheep	8.6
	Whooping crane	1.2
	Blue whale	9.3
	Bottlenose dolphin	7.0
	California sea otter	8.1
	Northern elephant seal	8.1
	Humpback whale[a]	40–48 (without information) 49–64 (with information)
	Grand Canyon (visibility)	27.0
	Colorado wilderness	9.3–21.2
Australia	Nadgee Nature Reserve (NSW)	28.1
	Kakadu Conservation Zone (NT)[b]	40.0 (minor damages) 93.0 (major damages)
UK	Nature reserves[c]	40.0

[a]Respondents divided into two groups, one of which was given video information.
[b]Two scenarios of mining development damage were given to respondents.
[c]Survey of informed individuals only.
Note: People's willingness to pay (WTP) to preserve all listed species is not necessarily identical to the sum of individual WTP estimates, because of the so-called "embedding effect" (WTP estimates elicited in surveys depend on the "bundle of goods" presented to the interviewee; see Mitchell and Carson, 1989).
Source: Pearce (1993).

that in the U.S. about 40 plant species accounted for plant-based prescription sales of some $15-20 billion per year during the 1980s (at 1990 prices). This would imply an economic value of at least $300 million per successful species and year. The figure could rise up to several billion dollars if values were calculated using statistical life valuation on the basis of deaths avoided. These are averages, and some species are clearly more valuable than others. Nevertheless, the figures can provide some indication of the lost pharmaceutical value from disappearing species. Some 60,000 plant species are expected to become extinct over the next fifty years. Given that the probability of a plant species yielding a successful drug is between 1:10,000 and 1:1,000, between 6 and 60 plant species with potential drug value could thereby be lost. With a mean loss of 30 such species, and applying the price range just noted, Pearce (1993) calculates that the annual losses for the U.S. could amount to between $8.8 billion and $180 billion (assuming there are no synthetic substitutes). By implication, if climate change were to increase these expected losses by just 10%, loss of plant species alone could cause annual U.S. losses from climate change in the order of $1 billion to $18 billion in gross value terms (or less, once production costs are allowed for).

Monetary estimates of ecosystem damages through climate change are invariably ad hoc. Fankhauser (1995) cites the Pearce (1993) survey (see Table 6.2) to arrive at a willingness-to-pay estimate of $30 per person per year to avoid species and habitat loss from climate change. Total costs amount to about $40 billion annually for the world as a whole, with about one-third occurring in developing countries. Cline (1992a) arrives at an estimate of about $4 billion annually as a notional value of species loss from benchmark global warming for the U.S., but suggests that the figure could as easily be an order of magnitude higher ($40 billion). The estimate is based on an extrapolation of observed U.S. public expenditures for the preservation of one particular species (the spotted owl).

6.2.14 Extreme weather events

Along with changes in the mean climate, there will most likely also be changes in the extremes. Changes in the extreme values of meteorological variables are not necessarily proportional to changes in the mean (IPCC, 1996b). Also, extremes are of more importance in studying the socioeconomic impacts of climate change. Most societies, in accommodating to the environment they live in, have developed strategies to cope with only a limited range of climatic events. Within this range, the "normal" variability is regarded as a resource, whereas the extremes constitute hazards (Heathcote, 1985). All kinds of adaptation mechanisms, such as dikes, shelters, and insurance, exist to deal with these extremes. However, adaptation requires investment (capital, time, skills) and is, therefore, limited.

Measures to adapt to natural hazards comprise a mix of physical, economic, and societal features. At the physical level, these include building design and protective structures such as dikes. At the economic level, personal savings and insurance help to cover the cost of damage. At the societal level, there are social safety nets, charity, and the government. Different adaptation mechanisms exhibit different degrees of flexibility: An insurance policy is valid for one year; an average Dutch dike has a lifetime of about two centuries. Flexibility to adapt is enhanced through constant changes in population and industrialization, as well as through changes in the legal, economic, financial, and social systems (Berz and Conrad, 1993; see also Section 6.2.6). In addition, climate itself constantly changes.

It is clear from the above that the qualification "extreme event" or "disaster" is a social construct. Such constructs are hard to measure, and the difficulty is increased by continuous change in the mechanisms for coping with natural disasters. Also, although the direct impacts of natural disasters may be harmful, indirect impacts are hardly measurable on a macroscale (Albala-Bertrand, 1993), and the event might set in motion a chain of beneficial changes, compensating for the initial losses. Finally, a disaster is not only local in a social and temporal sense but also in a spatial sense. The current generation of GCMs is not capable of reproducing present extremes very well, certainly not without spatial filtering, nor is it capable of deriving changes in extremes due to the enhanced greenhouse effect (IPCC, 1996b).

Little systematic research into the impact of changes in extreme weather events has been carried out. The natural disaster impact community, structured by the United Nations' International Decade for Natural Disaster Reduction, mainly studies present risks, although it is aware of the enhanced greenhouse effect (Olsthoorn *et al.*, 1994). The climatic change impact community focusses on changes in the mean, although attention is increasingly paid to extremes. So far, most attention has been paid to tropical cyclones, changes in which appear to be one of the most controversial questions of climate change research.

Tropical cyclones. The impact of climate change on tropical storms is still unclear. Hansen *et al.* (1989) concluded that climate change would bring "increased intensity" of "both ordinary thunderstorms and mesoscale tropical storms." Wendland (1977) has provided empirical support for the relationship of hurricanes to ocean surface temperatures. From monthly data for 1971–81, the frequency of hurricanes is closely related to the size of ocean area with temperature over 26.8°C, and the relationship is exponential. Similarly, Emanuel (1987) argues that tropical cyclones are "particularly sensitive to sea surface temperature," and expects a 40-50% increase in the destructive potential of hurricanes under $2xCO_2$. Houghton (1994) estimates an increase in both the frequency and severity of tropical and other storms. Haarsma *et al.* (1993), Ryan *et al.* (1992), and others come to similar conclusions.

Other results, on the other hand, give a less clear-cut picture (see, e.g., Broccoli and Manabe, 1990; Maunder, 1994; Raper, 1993; Lighthill *et al.,* 1994). Idso *et al.* (1990), for example, argue that tropical sea surface temperatures would increase very little with global warming, whereas the tropical tropopause could become more stable, leading to less intense cyclones. Bengtsson *et al.* (1995), using a transient ocean-atmosphere GCM, find a decrease in tropical cyclone numbers under enhanced greenhouse conditions. IPCC (1990a) judges that the impact of climate change on storm intensity is "ambiguous."

The available economic estimates, based on earlier climatological findings, tend to include increased hurricane damage as part of the damage to be expected from climate change (see Tables 6.4 to 6.6). Thus, Cline (1992a) reviews past hurricane damage for the U.S. and applies the 50% increase implied by Emanuel to estimate that benchmark warming would impose average annual damages of $750 million.

In an average year about 70 to 80 tropical cyclones are recorded worldwide, causing damages of about $1.5 billion, with a death toll of 15,000 to 23,000 lives (Smith, 1992; Bryant, 1991). The occurrence of tropical cyclones is distributed unevenly over the globe. Northern regions like the former Soviet Union and Europe, for example, are only marginally

Table 6.3. *Reported fatalities due to weather events, 1989–1992*

Region	Total Number				Average
	1989	1990	1991	1992	
Africa	31	138	> 621	> 360	>288
				>69[a]	>215[a]
Asia	>4,300	>3,280	>142,000	>7,766	>39,337
			>3,000[b]	>2,766[b]	>3,337[b]
South America	>35	26	>117	108	>72
Central and North America	>75	> 57	>86	>131	>87
Southwest Pacific	17	804	6,640	90	1,888
			296		302[c]
Europe	>52	>129	>65	>424	>161
Total:					
All events	>4,483	>4,434	>149,529	>8,879	>41,831
Excluding major disasters	>4,483	>4,434	>4,185	>3,588	>4,173

[a]Excluding the Madagascar 1992 drought and famine.
[b]Excluding the Bangladesh 1991 and Pakistan 1992 floods.
[c]Excluding the Philippines 1991 storm.   *Source:* Limbert (1993).

affected. Based on the natural hazard map of Berz (1990), Fankhauser (1995) has estimated that the U.S. is affected by 7% of all tropical cyclones, while another 7% occur in China, and 29% in OECD nations other than the U.S. (Australia, Japan, New Zealand). Combining this distributional pattern with the Smith and Bryant figures and the Emanuel estimate, Fankhauser (1995) arrives at an additional 8,000 cyclone casualties, practically all in developing countries. Using the same value-of-life estimates as for health damages, and adding in $630 million additional property damages, total worldwide hurricane damage amounts to about $2.7 billion.

These average figures are overshadowed, however, by the disastrous consequences of individual events, with poorer countries, especially small islands, being particularly vulnerable (see Box 6.3). In 1970 a cyclone caused more than 500,000 deaths in what is now Bangladesh. In 1985 a similar disaster in the same region killed another 100,000 people (Bryant, 1991). The $2.7 billion figure may therefore underestimate the true costs.

Extratropical storms. Dlugolecki *et al.* (1994; see also Munich Re, 1993) report that worldwide losses due to major windstorm events have averaged $2.0, $2.9, and $3.4 billion (1990 prices) for the decades of the 1960s, 1970s, and 1980s respectively. For 1990–92 this figure has risen to $20.2 billion. The larger part of these damages is due to tropical cyclones, however. The 1990 storms in Europe (Daria, Herta, Vivian, and Wiebke) resulted in a total loss of DM 25.3 billion (Munich Re, 1993). However, although storm damages have clearly increased over the last decades, it is not clear how much of this rise is attributable to climate change. The results of Changnon and Changnon (1992) and CATMAP (Clark, 1988; 1991) indicate that most of it is probably due to socioeconomic developments.

River floods. Little information is currently available regarding the socioeconomic impact of changes in the frequency and intensity of river floods (see Arnell and Dubourg, 1994).

Droughts. The impact of drought on agriculture has already been treated in Section 6.2.1. As an alternative way to estimate agricultural damage, Cline (1992a) reports an annual loss imposed on U.S. agriculture by increased drought of $18 billion/year. Water supply in general is dealt with in Section 6.2.4. The effect of decreased water runoff is partly dealt with in the section on water pollution (6.2.10). A further impact of drought is land subsidence. The 1975–76 drought in England and Wales, for example, led to a cost of £100 million to the insurance industry (Doornkamp, 1993). By 1979 the costs had amounted to £220 million. Finally, drought has implications for hydropower productivity. The 1987–91 drought in California cost an estimated $3 billion (Gleick and Nash, 1991). Dracup *et al.* (1993) report a potential loss in Northern Californian hydroelectricity of 40%, or $370 million per year, for their drought scenario. Nash and Gleick (1993) highlight the high sensitivity of hydropower production to changes in runoff.

Hot and cold spells. The impacts of hot and cold weather spells on health and human amenity were treated in Sections 6.2.8 and 6.2.12 above.

Total losses. Table 6.3 (after Limbert, 1993) provides total fatalities attributed to weather events between 1989 and 1992 as reported for the six WMO regions. The average number of lives lost to natural hazards is more than 4,000 per year, with the highest proportion occurring in Asia. This is likely to be an underestimate, because some major events are not included. As already mentioned in Section 6.2.6, losses due to natural hazards have increased dramatically over the past decade. Whether anthropogenic climate change has con-

Table 6.4. *Monetized 2xCO$_2$ damage to present U.S. economy
(base year 1990; billion $ of annual damage)*

Damage Category	Cline (2.5°C)	Fankhauser (2.5°C)	Nordhaus (3°C)[a]	Titus (4°C)	Tol (2.5°C)[b]
Agriculture	17.5	8.4	1.1	1.2	10.0
Forest loss	3.3	0.7	small	43.6	—
Species loss	4.0 + a[c]	8.4	c	—	5.0
Sea level rise	7.0	9.0	12.2	5.7	8.5
Electricity	11.2	7.9	1.1	5.6	—
Non-elec. heating	−1.3	—		—	—
Human amenity	+ b[c]	—	⎫	—	12.0
Human morbidity	+ c[c]	—	⎪	—	—
Human life	5.8	11.4	⎪	9.4	37.4
Migration	0.5	0.6	⎪	—	1.0
Hurricanes	0.8	0.2	⎪	—	0.3
Construction	± d[c]	—	⎪	—	—
Leisure activities	1.7	—	⎬ d	—	—
Water supply			⎪		
Availability	7.0	15.6	⎪	11.4	—
Pollution	—	—	⎪	32.6	—
Urban infrastructure	0.1	—	⎪	—	—
Air pollution			⎪		
Trop. O$_3$	3.5	7.3	⎪	27.2	—
Other	+e[c]	—	⎪	—	—
Mobile air cond.	—	—	⎭	2.5	—
Total	61.1	69.5	55.5	139.2	74.2
(% of GDP)	+ a + b + c ± d + e[c] (1.1)	(1.3)	(1.0)	(2.5)	(1.5)[b]

[a]Transformed to 1990 base.
[b]U.S. and Canada, base year 1988.
[c]Costs that have been identified but not estimated.
[d]Not assessed categories, estimated at 0.75% of GDP.
Note: Figures represent *best guesses* of the respective authors. Although none of the studies reports explicit confidence intervals, figures should be seen as reflecting orders of magnitude only.
Sources: Cline (1992a), Fankhauser (1995), Nordhaus (1991), Titus (1992), Tol (1995).

tributed to this death toll and, if so, to what extent, is, however, unclear. The strong trends in losses allow us to display only short-term averages.

6.2.15 Summary of damage estimates

Tables 6.4 to 6.6 summarize the principal existing estimates of climate change damage for major regions of the world. In the U.S., losses from benchmark 2xCO$_2$ equivalent warming reach over 1% of GDP in the Cline, Fankhauser, and Tol compilations, and some 2.5% of GDP in the central Titus estimates. Titus specifies a lower and upper end of his range of estimates, at 0.8% and 5.4% of GDP, respectively. It should also be noted that the Titus estimates are based on GCMs with average warming projections of about 4°C, higher than the IPCC's best guess of 2.5°C. Estimates for other OECD countries are mostly of the same order of magnitude of 1-2% of GDP (see Table 6.6).

Table 6.5. *2xCO₂ damage in physical units: different world regions (2.5°C warming)*

Type of Damage	Damage Indicator	EU	USA	Ex-USSR	China	Non-OECD	OECD	World
Agriculture	Welfare loss (% GNP)	0.21	0.16	0.24	2.10	0.28	0.17	0.23
Forestry	Forest area lost (km²)	52	282	908	121	334	901	1,235
Fishery	Reduced catch (1,000 t)	558	452	814	464	4,326	2,503	6,829
Energy	Rise in electricity demand (TWh)	54.2	92.0	54.6	17.1	142.7	211.2	353.9
Water	Reduced water availability (km³)	15.3	32.7	24.7	32.2	168.5	62.2	230.7
Coastal protection	Annual capital costs (m$/yr)	133	176	51	24	514	493	1,007
Dryland loss	Area lost (1,000 km²)	1.6	10.7	23.9	0	99.5	40.4	139.9
Wetland loss	Area lost (1,000 km²)	9.9	11.1	9.8	11.9	219.1	33.9	253.0
Ecosystem loss	Number of protected habitats lost, assuming 2% loss (Section 3.2.12)	16	8	N/A	4	53	53	106
Health/mortality	Number of deaths (1,000)	8.8	6.6	7.7	29.4	114.8	22.9	137.7
Air pollution Trop. O₃	Equivalent increase in emissions (1,000 t NOₓ)	566	1,073	1,584	227	2,602	1,943	4,545
SO₂	(1,000 t sulphur)	285	422	1,100	258	1,864	873	2,737
Migration	Additional immigrants (1,000)	229	100	153	583	2,279	455	2,734
Hurricanes Casualties	Number of deaths	0	72	44	779	7,687	313	8,000
Damages	m$	0	115	1	13	124	506	630

Source: after Fankhauser (1995).

Less comprehensive estimates by Nordhaus (1991), again for the U.S., arrive at a direct calculation of only 0.26% of GDP, primarily from sea level rise; but Nordhaus also sets 1% of GDP as a reasonable central estimate. The CRU/ERL (1992) estimates for the European Union, on the other hand, are significantly higher, with costs in the order of 1.6% of national income per degree of warming. The principal reason for this is a very high assessment of sea level rise damages, augmented by a factor of 2.7 to account for storm surges. On the other hand, their assessment of non-sea level rise impacts is less pessimistic, with an overall beneficial outcome in these categories.

However, these damage figures are likely to deviate from the "true" impacts, for three main reasons. First, several effects are not adequately quantified (e.g., nontropical storms, droughts, floods, morbidity, transport). Second, adaptation is not fully taken into account. Third, the figures are far from exact, and one should allow for a considerable margin of error. Many are deliberately kept conservative. Species loss valuation in particular could be far higher. The economic figures presented also suffer from the fact that they are based on earlier climate and impact research.

It should also be emphasized that the estimates in Tables 6.4 to 6.6 refer to central warming expectations. The corresponding damages for upper-bound warming would be higher, and more than linearly so. Also, when moving from the question of damage estimation to that of abatement benefits, a number of benefits not related to climate change need to be taken into account (see Section 6.7).

Regional differences can be substantial, as exemplified by the estimates for developing regions and the former USSR (see also Section 6.5). For the former Soviet Union, damage could be significantly below average, or even negative (i.e., climate change would be beneficial). A generally beneficial impact, as, for example, estimated by Tol (1995), mainly stems from large beneficial impacts in the agricultural sector. In the Fankhauser study, on the other hand, possible beneficial yield impacts are more than offset by the adverse impact of increased world prices on food imports. The region will also suffer from particularly high health and air pollution costs. The extremely high estimate for the Asian regions and Africa, on the other hand, are predominantly due to the severe life/morbidity impacts. As explained above, both the quantitative assessment and the underlying value-of-statistical-life estimates are very volatile, and the probability range of total damage is particularly wide for these regions.

Damage is likely to be more severe in developing countries than in developed countries, as is shown in Table 6.6 and discussed in Box 6.3. Table 6.6 reports damages for the non-OECD region of about 1.6-2.7% of GDP, some 50% higher than the OECD average. The main causes for this high estimate are health impacts and the high proportion of natural habitats and wetlands found in developing countries. Although the data for the non-OECD estimates are significantly weaker, they provide a clear indication that climate change will have its worst impacts in the developing world.

In general, the estimates in Tables 6.4 to 6.6 show a relatively narrow band for central damage calculations for the U.S. and for developed countries in general. It is important to recognize, however, that this field of estimates is probably bi-

Table 6.6. *Monetized 2xCO₂ damage in different world regions (annual damages)*

Region	Fankhauser (1995) bn$	Fankhauser (1995) %GDP[a]	Tol (1995) bn$	Tol (1995) %GDP[a]
European Union	63.6	1.4		
United States	61.0	1.3		
Other OECD	55.9	1.4		
OECD America			74.2	1.5
OECD Europe			56.5	1.3
OECD Pacific			59.0	2.8
Total OECD	180.5	1.3	189.5	1.6
E. Europe/ former USSR	18.2[b]	0.7[b]	- 7.9	- 0.3
Centrally planned Asia	16.7[c]	4.7[c]	18.0	5.2
South and Southeast Asia			53.5	8.6
Africa			30.3	8.7
Latin America			31.0	4.3
Middle East			1.3	4.1
Total non-OECD	89.1	1.6	126.2	2.7
World[d]	269.6	1.4	315.7	1.9

[a]Note that the GDP base may differ between the studies.
[b]Former Soviet Union only.
[c]China only.
[d]Percentage of GDP figures are based on market exchange rate GDP. The order of magnitude of estimates does not change if uncorrected damage categories are purchasing-power-parity adjusted and expressed as a fraction of PPP-corrected GDP.
Sources: As shown.

ased towards convergence. The reason is that, with the exception of CRU/ERL (1992), the underlying sources of many of the estimates are the same, particularly U.S. EPA (1989). The convergence tends to become extrapolated to other regions, too, considering that several of the international estimates in Fankhauser (1995) and Tol (1995) are obtained by extrapolation of the U.S. estimates. The similarity of the estimates should therefore not be interpreted as evidence of their robustness. A substantial degree of uncertainty remains. Nevertheless, the relative ranking of regions appears to be reasonably robust, with the most severe impacts to be expected in Asia and Africa, and northern and developed regions suffering less.

The worldwide estimates of Table 6.6 are expressed as the total sum of regional damages relative to the global sum of GDP. As discussed in Box 6.2, this is one of many possible ways to calculate global damages from regional or individual estimates. It can also be argued on equity grounds that there should be greater weights placed on impacts for low income countries than would result from simply applying their shares in the global income base.

It is useful to consider the results of an opinion survey of nineteen climate change experts from both the physical sciences and economics (Nordhaus, 1994a; see Table 6.7). For a 3°C actual warming by 2090 (scenario A), estimated global damages ranged from zero to 21% of gross world product,

with a mean value of 3.6%. For a more rapid and severe warming (scenario C) the mean increases to 10.4% of GDP. Diverging from the estimates of Table 6.4, virtually all the respondents judged that more than half of the damages would occur in the market sectors (such as agriculture) rather than in sectors outside the standard system of national accounts (biodiversity, amenity).

6.3 Damage Estimates for Longer-Term Warming

The benchmark of a doubling of the atmospheric CO_2-equivalent concentration could be reached around the middle of the next century. Yet it is unlikely that greenhouse gas emissions and atmospheric buildup would stop at that point. A long-term view on climate change impacts is therefore important, even though the need for socioeconomic forecasts over more than a century makes this task extremely problematic (see also Section 6.2).

Sundquist (1990) has estimated that there are sufficient fossil fuel reserves to permit emissions of carbon dioxide to rise to the point where its atmospheric concentration might reach 1,600 ppm by the year 2200. After that, the exhaustion of reserves would reduce emissions and allow concentrations to plateau through the year 2300 and then moderate back to about 1,200 ppm over a 400-year period as a consequence of deep-ocean mixing. With preindustrial concentrations of 280 ppm, this scenario amounts to a potential rise of nearly sixfold in carbon dioxide concentrations alone. Cline (1992a) uses a 300-year time horizon to investigate the very long-term impacts of climate change. On the basis of fossil fuel (primarily coal) reserves estimated by Edmonds and Reilly (1985), he concludes that as much as 14,000 Gt of carbon (GtC) could be available for use at reasonable economic cost. He applies three leading energy-carbon models (Reilly *et al.*, 1987; Nordhaus and Yohe, 1983; Manne and Richels, 1992) that project carbon emissions until the year 2100, and extrapolates their growth rates through to 2275. Average emissions rise from 7 GtC today to 20 GtC by 2100 and 56 GtC by 2275.

Under the assumption that atmospheric retention remains at its recent 50% level, Cline estimates carbon concentrations on the order of 2,200 ppm by 2275, or about eight times preindustrial levels. The corresponding increase in radiative forcing would reach about 13 Wm⁻² from carbon alone, or some 19 Wm⁻² from all greenhouse gases. With the central IPCC climate sensitivity of 2.5°C for 4 Wm⁻², the resulting mean warming would be almost 12°C. On this basis, Cline (1992a) sets 10°C (the lowest range from the three emission models) as a tentative benchmark for ultimate warming over a 300-year period.

Warming of this magnitude would take the earth back to the climate of the mid-Cretaceous period 100 million years ago, when mean temperatures were an estimated 6-12°C higher than today (Hoffert and Covey, 1992). If the upper-bound climate sensitivity of 4.5°C is applied instead, very long-term warming could reach approximately 18°C. The corresponding lower-bound estimate would be 6°C if the climate sensitivity were 1.5°C.

Recent simulations using the GFDL (Geophysical Fluid Dynamics Laboratory) general circulation model at Princeton

---

BOX 6.2: DAMAGE AGGREGATION ACROSS COUNTRIES AND INDIVIDUALS

Cost-benefit analysis assumes that the monetary values of damage can be aggregated across (a) individuals and (b) countries. This aggregation process raises an important problem of so-called *interpersonal comparisons* of utility, where utility is simply another word for "welfare" or "well-being." Basically, introspection allows any one individual to assess his or her own preferences, but perhaps not others' preferences. Each individual knows by how much he or she is better off in situation X compared to situation Y, but he or she cannot assess how this change compares to the extent to which someone else is better off for the same change of situation. This inability to assess other minds produces the theorem that it is impossible to make comparisons of well-being between individuals and hence, by extension, between countries. There are, therefore, endless numbers of ways in which individuals' assessments of their own well-being can be aggregated to yield a measure of social welfare change.

Measuring social welfare change requires interpersonal welfare comparisons, comparisons that require ethical judgments about individuals' preferences. Such judgments could include one to the effect that preference measures should not be unduly influenced by income differences. In this case, the individual measures of preference – willingness to pay – could be weighted so as to reflect the preference of someone with an average income. One way to proceed is to indicate the outcomes of the cost-benefit analysis according to differing value judgments about the weights attached to preferences – a kind of "value sensitivity analysis." Thus, whether a change in the state of the world raises or lowers social welfare depends on the impacts on individual welfare as well as on normative criteria for making interpersonal welfare comparisons. Measures of social welfare change are, therefore, not objective but normative. See Chapter 3 for a more extensive discussion of ethical issues in relation to the enhanced greenhouse effect.

The equation below provides an example of "equity-corrected" aggregation: Total damage D is the weighted sum of individual damage d_i; the weights are some power E of the ratio of the reference income Y_a to the individual income Y_i. (Both income terms are purchasing-power-parity corrected.)

$$D = \sum_i d_i * \left(\frac{Y_a}{Y_i}\right)^E$$

In the literature, E is usually either the income elasticity of demand or the elasticity of the marginal utility of income. Other ways of determining equity weights could also be used.

There is also the question of what per capita income figure to use. The obvious choices are those based on purchasing power parity and exchange rate. For low income countries, the former tend to be about three to five times as high as the latter, whereas the two are approximately equal for rich countries.

Consideration of welfare-weighted impacts helps show how extreme effects for some countries might count for relatively more than minor effects summed over a larger block of countries (e.g., welfare effects for an island state expected to be inundated might be larger than would be attributed solely on the basis of its income or even population).

University tend to confirm this range of 6-18°C for baseline warming in the long term. Manabe and Stouffer (1993) examine the effects of a quadrupling of preindustrial carbon dioxide equivalent, a concentration that would be achieved 140 years from now as a result of a 1% annual increase in CO_2-equivalent concentration, consistent with the IPCC "business-as-usual" scenario. The resulting equilibrium mean surface air temperature increase reaches 7°C, whereas the actual transient warming is 5°C. Manabe and Stouffer do not argue that atmospheric concentrations will, in fact, stop rising after 140 years. Instead, their interest is in examining the consequences of this specified century-scale scenario. They find that sea level rise from thermal expansion alone reaches 1.8 m by the 500th year, and would presumably be much greater if the melting of ice sheets were taken into account. Similarly, Wigley (1995) finds that a sea level rise in the order of 2-3 m could occur up to 500 years after greenhouse gas concentrations had been stabilized.

After first conducting an economic analysis of climate change using comparative static techniques with carbon dioxide doubling as the benchmark, Nordhaus (1991; 1993a, b; 1994b) has also adopted a centuries-scale horizon for his dynamic model DICE. However, Nordhaus's baseline involves warming of only 5.5°C by 2275, because of lower emissions (31 GtC by 2275) and lower atmospheric retention. Lower emissions stem from lower growth expectations, as Nordhaus's model assumes a progressive slowdown in technological change that has the effect of limiting growth. Global product by the year 2275 is only 7 times today's output in the DICE baseline but 26 times in Cline (1992b), suggesting that diverging economic assumptions may be even more important than different assumptions about the scientific parameters for such long-term analysis.

If 10°C is used as a guideline to potential warming over a 300-year horizon (or much sooner if the upper end of the climate sensitivity range proves valid), the corresponding economic damages could be very large. The central reason is that, in many categories, damage is likely to rise nonlinearly with warming. As one example, Yohe (1993) finds that, at intermediate sea level rise, a doubling of the rise causes damages to rise about 2½-fold, and above 60 cm, a doubling of the rise causes a tripling of damage.

It seems likely that the damages of high, very long-term warming would be especially pronounced in the nonmarket

---

BOX 6.3: RELATIVE DAMAGE IN DEVELOPING AND INDUSTRIAL COUNTRIES

Various factors influence climate change damage (as a fraction of GDP) in less developed countries (LDCs) as opposed to industrial or developed countries (DCs).

- *Location.* In general, warming is expected to be greater at higher latitudes. Because LDCs tend to be located closer to the equator, one would expect them to be *less* affected than industrial countries.

- *Economic structure.* On the other hand, LDCs have a much higher share of GDP in agriculture and, therefore, a larger share of output directly exposed to climatic influences. Consequently, one would expect a *greater* impact on LDCs than on DCs from this standpoint.

- *Coastal vulnerability.* Coastal vulnerability to sea level rise and the possibility of an increase in tropical cyclone damage are probably *greater* in LDCs than DCs. Although vulnerable areas can be found in industrial countries (for example, Louisiana in the U.S.), vulnerability is particularly high for such LDCs as Bangladesh, Egypt, and China. Low-lying island states tend to be LDCs. Of 50 countries or territories identified as having shore protection costs above 0.5% of GDP annually as a consequence of a 1-m rise in sea level, all but one (New Zealand) are LDCs (IPCC, 1990b).

- *Rigidities.* Capacity to adapt to climate change may be more limited in developing countries. Adaptation requires an investment outlay, and low income communities tend to have lower savings rates and less flexibility to undertake these investments. Thus, Rosenzweig *et al.* (1993) identify *greater* relative damage to agriculture in LDCs, partly because crops are already grown nearer to heat tolerance limits but also because of lesser expected capacity to adapt than in DCs. A possible consideration in the other direction is that the economy that is growing more rapidly has a lower fraction of past fixed investment in its total capital stock, and can thus make an easier adjustment.

- *Human life.* Because of poorer nutrition and health infrastructure, proportionate loss of life from climate change (e.g., from heat waves, increased hunger risk, and a possible increase in tropical storm damage) seems likely to be *greater* in LDCs than in DCs.

- *Valuation.* Monetary damage estimates are based on people's willingness to pay and thus vary according to people's income. People's appreciation for nonmarket goods like ecosystems, for example, is often assumed to rise more than proportionately with income. That is, a given ecosystem's loss affects the welfare of rich people more than that of poor people. However, in many damage estimates, willingness to pay was assumed to be a constant fraction of income. Income differences do not then affect impacts expressed as a fraction of GDP in those estimates.

---

sectors. Human amenity could face major losses, even after accounting for adaptation. For example, in the event of a 10°C global mean warming, Cline (1992a) estimates that among 66 major U.S. cities the number with average daily temperatures exceeding 90°F in July would rise from 18 at present to 62. Sea level rise in the order of 2-3 m (Wigley, 1995) could lead to a loss of cultural heritage and national sovereignty in small island states and multiply the stream of climate refugees.

Attempts to quantify long-term damage are rare and highly speculative. Most often, long-term damage estimates are simple extrapolations of the $2xCO_2$ equilibrium case, using different assumptions about the degree of nonlinearity in the damage-temperature relationship. Cline (1992a) uses a relatively modest degree of nonlinearity for most effects, with an overall damage function exponent of 1.3. Nordhaus (1993a, b) applies a quadratic damage function in his main scenario, whereas Peck and Teisberg (1992) consider several specifications, including quadratic and cubic functions. Table 6.8 reports the central estimates in Cline (1992a) for U.S. damage from 10°C warming. Against the present economic scale, they amount to $335 billion annually, or 6% of GDP. However, the figure is only illustrative.

Assuming a quadratic temperature-damage relationship, U.S. damage from 10°C warming would amount to over 17%

of GDP. If more allowance is made for such categories as species loss, the central range could rise to 2% of GDP for $2xCO_2$ at 2.5°C and 12% of GDP at 10°C. With upper-bound warming, the very long-term damage reaches 20% of GDP. In the long-term scenario in the Nordhaus (1994a) poll of experts, respondents predicted a mean damage of 6.7% of GDP for 6°C warming by 2175. Answers ranged from negligible damages to 35% of GDP (see Table 6.7).

In addition, warming on this scale could trigger some of the catastrophic (low probability/high impact) results sometimes mentioned. These are treated in the following section.

6.4 Climate Catastrophes and Surprises

The discussion so far has focussed on the impact of the best-guess climate change and some reasonable upper and lower bounds in the next century (Section 6.2) and in the longer term (Section 6.3). This section makes some statements on the impact if all the climatic dice roll the wrong way, that is, if the enhanced greenhouse effect leads to rather unlikely but possible changes. Where some information on the physical mechanisms is available, these changes are referred to as climate catastrophes; where the mechanisms are unknown, they are referred to as surprises. Also, the impact of benchmark climate change could be far more dramatic than best-guess esti-

Table 6.7. *Expert opinion on climate change damage*

	Scenario A 3°C Warming by 2090	Scenario B 6°C Warming by 2175	Scenario C 6°C Warming by 2090
Damage relative to world GDP (% of GDP)			
Mean answer	3.6	6.7	10.4
Median answer	1.9	4.1	5.5
Range of answers	0.0–21.0	0.0–35.0	0.8–62.0
Probability of damage > 25% of GNP (%)			
Mean answer	4.8	12.1	17.5
Median answer	0.5	3.0	5.0
Range of answers	0.0–30.0	0.2–75.0	0.3–95.0

Source: Poll of experts, Nordhaus (1994a).

Table 6.8. *Illustrative damages from long-term climate change, present U.S. economy (base year 1990; billion $ of annual damage)*

Damage Category	Long-Term Warming (10°C)
Agriculture	95.0
Forest loss	7.0
Species loss	16.0 + a[†]
Sea level rise	35.0
Electricity	64.1
Nonelectric heating	- 4.0
Human amenity	+ b[†]
Human morbidity	+ c[†]
Human life	33.0
Migration	2.8
Hurricanes	6.4
Construction	± d[†]
Leisure activities	4.0
Water supply	
Availability	56.0
Pollution	—
Urban infrastructure	0.6
Air pollution	
Tropospheric ozone	19.8
Other	+ e[†]
Total	335.7 +a+b+c±d+e[†]
(% of GDP)	(6.1)

[†]Costs that have been identified but not estimated.
Source: Cline (1992a).

mates suggest. The second part of this section briefly discusses impact catastrophes and surprises.

Most attention will be paid to the catastrophes, as surprises are, by definition, unknown. Collard (1988) distinguishes between weakly and strongly catastrophic risks. The former category refers to the case in which the product of the likelihood of an event multiplied by its consequences tends to zero as events become more and more disastrous; in other words, the probability of the event declines more rapidly than its impact grows. A strongly catastrophic risk describes the converse situation, where the severity of an event rises more rapidly than the probability of its occurrence declines. Which category best applies to climate change risks is unclear (Tol, 1995).

Three main types of climate catastrophes are identified in the literature, all associated with strongly nonlinear responses to changed forcing: (1) the runaway greenhouse effect, (2) disintegration of the West Antarctic Ice Sheet, and (3) structural changes in ocean currents. These three categories have a small but unknown probability of occurrence.

A "runaway" greenhouse effect refers to the scenario in which one or more of the positive feedbacks dominate the negative ones such that the climate changes much more and much faster than the common consensus indicates. The main causes are a rapid increase in natural emissions of greenhouse gases (e.g., through methane and carbon dioxide releases from melting permafrost or methane clathrates), a shutdown of major greenhouse gas sinks (e.g., through reduced plankton activity or the reduction of growth or dieback of forests), and changes in atmospheric chemistry.

A "runaway" greenhouse effect not only brings much larger and faster climate change, it also considerably enlarges the possibility of other catastrophes and surprises. The disintegration of the West Antarctic Ice Sheet refers to a rapid melting of this ice sheet, which could be unstable, leading to an additional sea level rise of 5 to 6 m (Revelle, 1983) within the next century, causing large land losses along coastlines and inundating many low-lying islands (Schneider and Chen, 1980).

Changes in ocean currents have regional and global impacts. Potential changes in the thermohaline circulation and Gulf Stream are discussed in Section 4.3.3 of Volume 1. Reconstructions of past climates (Dansgaard *et al.,* 1989) and model experiments (Manabe and Stouffer, 1993) reveal that the energy transport in the North Atlantic can be weakened, start to fluctuate, or even stop completely, possibly as a result of climate change. This would lead to a sharp drop in European mean temperatures. In the $4xCO_2$ experiment of Manabe and Stouffer (1993) circulation almost shuts down, suggesting that under long-term warming without intervention this phenomenon may become the base case rather than a low-probability event.

Rapid climate change brings forward the impact of long-term warming (see Section 6.3), leaving less time to adapt and thereby increasing the impact enormously. Table 6.8 provides an illustration of the possible impacts of a 10°C increase in the global mean temperature (four times the temperature rise associated with $2xCO_2$) and suggests that this might lead to about a sixfold increase in the costs compared to a 2.5°C rise. This figure cannot readily be applied to a 10°C temperature rise in the shorter term, though, since the estimates assume a moderate level of adaptation, and adaptation will be much less successful in the case of rapid climate change. Section 6.1.5 indicates the effectiveness of adaptation in preventing the larger part of the losses due to sea level rise in the OECD and mitigating agricultural losses at a moderate cost. This adaptation cost is likely to be much higher for fast or sudden changes, and some measures might be impossible to implement. The long-term figure stated above thus at best represents the lower bound of the actual costs of a catastrophe.

Unfortunately, it is largely beyond the ability of today's climate science to quantify catastrophic impacts or their likely variation with climate change. Hence, only a few attempts exist. In one example, Tol (1995) splits the damage costs of climate change, *inter alia,* into those due to the rate of temperature change and those due to the magnitude of temperature change. The damage module of his model FUND yields tangible damages of 0.33%, 0.55%, and 1.09% of gross world product for a 2.5°C warming in 2095, 2057, and 2031, respectively. However, no figures for a more rapid temperature increase are reported there. The ICAM2 model (Dowlatabadi and Morgan, 1993) also considers the pace as well as the amount of warming in the damage cost functions, but explicitly adds adaptation. A warming of 8°C at high latitudes and 2.8°C at low latitudes would lead to tangible losses of 6.0% of GDP in the industrialized countries and 6.6% of GDP in the less industrialized countries (Kandlikar, 1994).

Nordhaus's (1994a) poll of experts provides the most direct information: A doubling of the warming by 2090 from 3°C to 6°C almost triples the costs of climate change. The probability of an impact catastrophe (i.e., damages greater than 25% of GNP) could also be considerable. The average probability stated in the survey varies between 4.8% and 17.5%, depending on the underlying warming scenario (see Table 6.7). It is interesting that natural scientists were far more pessimistic in their assessment than economists (see also Section 6.2.15).

Even less is known about impact catastrophes than climate catastrophes. The upper-bound damage in the Nordhaus (1994a) survey is 20% of gross world product for a temperature rise of 3°C by the year 2090. Such a high loss could be brought about by strong nonlinearities in the damage due to climate change, for instance, through rapid deterioration of agriculture due to drought, floods, or pests; rapid loss of species; rapid spread of vector-borne diseases; collapse of the financial sector; large-scale migration; and armed conflict. The last two categories are more likely, should either of the other ones occur. Impact catastrophes are more likely to occur on a regional scale. Their valuation therefore raises questions of equity and damage aggregation (see Box 6.1). Catastrophic impacts, even if limited to certain regions, may be considered undesirable from an equity point of view. The notion of possible catastrophes is particularly profound among advocates of the sustainability approach and the precautionary principle.

6.5 Regional Implications of Climate Change

6.5.1 Regional damage estimates

Section 6.2 surveyed the available literature on comprehensive damage estimates of climate change. These are heavily biased towards the U.S. and, to a lesser extent, Europe, due to the concentration of work on these areas. This section looks at case studies available for other regions. Focussing mostly on agricultural impacts and sea level rise, regional studies predict impacts ranging from slightly beneficial to truly catastrophic. It is difficult to compare studies across regions, though, due to different underlying assumptions in the GCMs used, the nature of the model runs (transient or equilibrium), the models' capabilities for simulating control conditions for the relevant country (e.g., precipitation), and differences in assumptions and scenarios between damage areas. Some scenarios include assumed adaptation measures, others do not. Moreover, studies do not examine interlinkages between areas and thus are not truly integrated assessments.

6.5.2 Japan

In terms of agriculture, research has focussed on the effects of climate change and increased CO_2 on rice production (Nishioka *et al.*, 1993). Laboratory experiments with increased CO_2 concentrations (a rather high 700 ppm) produced grain yield increases of 23-71%.

An increase in effective accumulated temperature in most of the country (except parts of Hokkaido) would reduce the vulnerability of rice to cool summer weather in the north of Japan, perhaps stabilizing production. Rice-producing regions may themselves be extended. Commercial rice cultivation in Japan could shift northward by 200-500 km by the end of the next century. In the north, $2\times CO_2$ conditions could open up most of the land below 500 m elevation for rice cultivation. Currently most cultivated areas are at elevations below 200 m. In light of the altered climate predicted by the GISS model, the transplanting date of rice would be advanced by 20-30 days and the maturing period would be reduced by 25-50 days. For spring wheat and winter wheat respectively, the heading date would advance by 2-5 days and 30 days, and the maturing period would reduce by 3-5 days and 3-6 days. For forests, some replacement of the understory is expected due to the inability of forests to migrate in the short term.

Japan's economic activities are concentrated along the coast. Major cities like Tokyo, Osaka, and Nagoya are all located in the coastal zone. Together the three cities account for more than 50% of Japan's industrial production. Already about 860 km² of coastal land – an area supporting 2 million people and with physical assets worth 54 trillion yen ($450 billion) – is below mean high water level. For a 1-m rise in sea level, this area would expand by a factor of 2.7 to embrace 4.1 million people and assets worth 109 trillion yen ($908 billion). The same sea level rise would expand the flood-prone area from 6,270 km² to 8,900 km², with an additional 3 million people at risk (Nishioka *et al.*, 1993). Coastal protection will be central to the country's response strategy. The costs of adjusting existing protection measures have been estimated at about $80 billion. However, extensive coastal protection would put additional pressure on Japan's remaining natural shorelines (IPCC, 1992b).

6.5.3 Africa

Very little work has been done on impact assessment in Africa. Magadza (1991) predicts reduced precipitation in the rain forests of Zaire and Uganda under doubled CO_2 concentrations. On the basis of general relationships between precipitation and herbivores, Magadza predicts reductions in populations of large herbivores, such as buffaloes and elephants, as savannah productivity is reduced. Shallow lakes, such as Lake Abiata in Ethiopia and Lake Turkana in Kenya, and savannah wetlands are likely to be reduced, affecting resident wildlife and bird migrations.

Ominde and Juma (1991) emphasize the greater vulnerability of Africa to climate change due to high agricultural dependence and limited capacity for adaptation. For Egypt, Rosenzweig *et al.* (1993) predict aggregate yield losses in the order of 25-50% for $2xCO_2$. The implications of agricultural damages on consumer welfare in Africa are expected to be negative even in the most optimistic scenarios (Reilly *et al.*, 1994).

Sea level rise is predicted to affect some 20% of Egypt's 35,000 km^2 of arable land. A 1-m rise in sea level could destroy up to 25% of the Nile Delta's agricultural land and displace about 8 million people. Cotton and rice would be the main crops affected (El-Raey, 1990). Awosika *et al.* (1990) estimate that sea level rise would greatly exacerbate existing erosion at Lagos Beach in Nigeria, and even a modest rise will affect mangroves and wetlands that support timber and fishing industries. In the absence of protection, a 1-m rise in sea level could flood over 18,000 km^2 of Nigeria's land, damaging assets currently worth at least $18 billion, including much of the country's oil industry, which is mostly located near the coast. In addition, over 3 million people would have to be relocated. Protecting at least the highly developed areas would cost $550-700 million (IPCC, 1994, 1992b). A similar case study for Senegal estimates that, in the absence of protection, over 6,000 km^2 of land – some 3% of the country's total area – would be lost under a 1-m rise. Two-thirds of the population and 90% of the industry are located in the coastal zone. Protecting these areas would cost $250-850 million, about three-quarters of which would go towards beach nourishment (IPCC, 1994, 1992b).

6.5.4 Bangladesh

Climate change impacts in Bangladesh were analyzed in detail in the context of a multicountry study on climate change in Asia (Asian Development Bank, 1994). Taking the coastal zone of Bangladesh to be defined by districts with a maritime boundary, the zone is a delta of the combined Ganges-Jamuna-Meghna river system and accounts for some 22% of

Table 6.9. *Losses of rice output due to sea level rise in coastal zones of Bangladesh*

Year	SLR by 2070:45 cm		SLR by 2070:100 cm	
	('000 tonnes)	(% coastal output)	('000 tonnes)	(% coastal output)
2020	125	2	412	4
2050	2,122	28	7,708	49
2070	2,619	28	9,514	50

Source: Asaduzzaman (1994).

the total land area and 16% of the population (Asaduzzaman, 1994). It also accounts for 24% of agricultural value added, 40% of manufacturing fixed assets, and 22% of manufacturing employment. A 45-cm rise in sea levels along the Bay of Bengal coast would submerge some 15,700 km^2 of land (about 11% of the total land area), including some 75% of the Sundarban mangrove forests. Several ports would also be affected. A 1-m rise would affect nearly 30,000 km^2 or about 21% of the land area (Asaduzzaman, 1994). This is somewhat higher than earlier estimates by the Commonwealth Secretariat (1989), which predicted a loss of just under 16% of total land area for a 1-m rise. A 1-m sea level rise would result in all the Sundarban mangrove forests disappearing. Sea level rise will result in saline intrusion further inland than the existing freshwater-saltwater interface. Changes in cyclone frequency appear to be small, and soil erosion effects are unknown.

Impacts on agriculture remain difficult to predict. A study by the Bangladesh Institute of Development Studies suggests losses of rice output due to sea level rise as shown in Table 6.9.

Asaduzzaman (1994) suggests that the overall macroeconomic impact of sea level rise would amount to some 30% of current GNP in the coastal zone, or some 5% of overall Bangladesh GNP. Damages in absolute terms would be perhaps $4.8 billion in terms of "lost" output in 2070.

The loss of the Sundarbans would be particularly severe for the poorest people of Bangladesh, relying as they do on the mangroves for fish, fuel wood, timber, and many other raw materials. Biodiversity losses are incalculable due to the extensive lack of knowledge about the ecological functioning of the Sundarbans. Many shrimp fisheries are likely to disappear, offset to some extent by the emergence of new estuarine fisheries as sea levels rise. Salinization will also affect industries relying on freshwater intakes, including electricity generating plants, raising costs of production. Other industries would have to move and some would be irretrievably lost, such as those relying on shrimp processing. Road and rail links between Dhaka and Chittagong/Khulna would be disrupted by sea level rise, seriously affecting the country's international trade. Asaduzzaman (1994) suggests that a 45-cm sea level rise would affect 195,000 jobs and some 790 km of roads; a 100-cm rise would affect 735,000 jobs and 1,460 km of roads.

Population displacement would be dramatic if impacts are assumed to be "sudden." Some 5% of people would be displaced by a 45-cm rise, and 13.5% by a 1-m rise or, in terms of projected 2070 population, 12 million and 32 million people, respectively. In practice, migration away from the likely-to-be-affected areas is already taking place, partly for weather-related reasons.

6.5.5 India

A broad analysis of climate change impacts in India has been provided as a part of two multicountry studies on climate change in Asia, commissioned by the Asian Development Bank (ADB, 1994) and the South Asian Association for Regional Cooperation (SAARC, 1992). Some 70% of India's annual rainfall occurs in the June-September monsoon season. Chakraborty and Lal (1994) project increases in precipitation in most parts of India due to doubled CO_2 concentrations. The Central Plains and East Coast might expect pronounced increases in annual average precipitation (1 mm/day) and this may be 2 mm/day in the monsoon period in the former region. An increase of 2 mm/day is also estimated for West Bengal in the premonsoon season. Monsoons thus intensify for many regions, and more frequent heavy rainfall events are predicted.

In agriculture, yields may be lowered as a result of enhanced temperatures. Seshu and Cady (cited in SAARC, 1992) suggest a decline in rice yields of 0.7 t/ha for an increase in minimum temperature from 18°C to 19°C, a decrease of 0.4 t/ha for 22°C to 23°C, and 0.04 t/ha for 27°C to 28°C. As for wheat, it has been estimated that each 0.5°C increase in temperature would reduce productivity in Punjab, Haryana, and Uttar Pradesh by about 10%. In Central India, where productivity is lower, the decrease would also be lower (SAARC, 1992). In comparison, the global study by Rosenzweig *et al.* (1993) predicts changes in aggregate agricultural yields for India of +3% to -33% under $2xCO_2$ (including CO_2 fertilization).

Sea level rise will affect many regions, with the Andaman and Nicobar islands and the coral atolls of the Lakshadweep archipelago among the most vulnerable areas. The east coast would be more subject to storm surges than the west coast. The western coastline south of 12° North is likely to become more eroded. No overall estimates of macroeconomic impact are available, though. In a case study of the Orissa and West Bengal region, IPCC (1992b) estimated that in the absence of protection a 1-m sea level rise would inundate an area of 1,700 km², predominantly prime agricultural land, and displace 700,000 people. Protecting the area would require the construction of an additional 4,000 km of dikes and sea walls.

The Asian Development Bank country study for India (ADB, 1994) reports estimates by Asthana (1993) of the costs of a 1-m sea level rise. In the absence of protection, approximately 7 million people would be displaced and some 5,763 km² of land and 4,200 km of roads would be lost. The dominant cost is land loss, which accounts for 83% of all damages. Although the annuitized cost is reported to be 0.18% of GNP,

it appears to be net of the value of land loss. Inclusive of land loss, the correct percentage appears to be 1% of GNP.

6.5.6 Indonesia

Based on calculations by the Commonwealth Scientific and Industrial Research Organisation (CSIRO), the Asian Development Bank's country study for Indonesia (ADB, 1994) predicts an increase in mean annual temperature in Indonesia of 1.5°C (0.4-3.0°C) by 2070 under baseline conditions. Sea levels could rise by 45 cm (with a range of 15-90 cm). In comparison, Parry *et al.* (1991) suggest that $2xCO_2$ could raise the mean annual temperature by as much as 3°C and produce a rise in sea levels of 60 cm. Precipitation under $2xCO_2$ is likely to decline in some regions but might generally increase. Increases in rainfall could lead to a 30% increase in the area under irrigation in the Brantas and Citarum basins in western Java. Soil erosion might increase by 14%, 18%, and 40% in the Citarum, Brantas, and Saddan watersheds respectively, with resulting soybean production losses of 2,000 to 2,700 tonnes in each region.

Enhanced temperatures and, at certain sites, reduced water availability could reduce rice yields, especially for early season rice. This could be offset to some extent by increases in late season rice. Net yields might decline by 4%. Soybean production might decline by around 10%, largely due to lower yields in the early season. Appropriate adaptive measures may lead to productivity increases that could offset potential losses, and overall yields may increase. The biggest yield effects are likely to be on maize output, with declines of 25-65% (Parry *et al.*, 1991).

Indonesia is the world's largest archipelagic state, with nearly 17,000 islands and a shoreline of approximately 81,000 km (ADB, 1994). Sea level rise will affect coastal ecosystems, industrial production, and agriculture alike. In the Krawang and Subang districts, 95% of the predicted reduction in local rice supply (about 300,000 tonnes) and half of the loss in maize output is due to the inundation of coastal land. As a consequence, over 81,000 farmers in the Subang district alone may lose their source of income, and about 43,000 farm labourers could lose their jobs (Parry *et al.*, 1991). Under the baseline assumption of a 15-90 cm sea level rise by 2070, and assuming that the Indonesian population will stabilize between 2030 and 2045, ADB (1994) predicts that about 3.3 million people will be displaced. Some 800,000 households would have to be relocated at a cost of $8 billion. In the absence of protection, a total area of 3.4 million hectares could be inundated.

Indonesia is one of the few countries for which tentative health impact estimates are available. ADB (1994) expects that the incidence of malaria could increase by about 20%, from 2,700 cases per 10,000 people in 1989 to 3,200 cases in 2070. The incidence of diarrhoea and dengue fever could each increase by as much as a factor 3 or 4. The resulting increase in health expenditures for all three diseases could amount to approximately $64.5 billion annually (excluding the costs of death and disrupted livelihoods).

6.5.7 Malaysia

The expected impact of a CO_2 doubling on annual mean temperature in Malaysia ranges from 1-2°C (ADB, 1994) to 3-4°C (Parry *et al.*, 1991). Rainfall increases are likely in January-February in the coastal regions of Sarawak, and in March-May in southwestern peninsular Malaysia. Runs of the CERES rice model predict a 12-22% yield decline for rice in the largest rice growing region of Muda. A 10% reduction in solar radiation in the Serdang region could result in maize production losses of about 20%. Rainfall increases in March-May could increase oil palm productivity in alluvial coastal areas. Temperature and rainfall changes on the eastern coast of Malaysia could make the area too wet for rubber cultivation. Rubber yield is roughly inversely proportional to total annual rainfall. A 10% increase in rainfall could reduce yields by 13%, a figure which could rise to 25-40% due to interference with tapping (ADB, 1994). Marginal rubber cultivations in areas such as the northern states may become uneconomic due to drought. National yield levels could decline by 15%, but improved varieties could more than offset this loss.

The frequency of peak discharges in the Kelantan river basin in northeastern peninsular Malaysia could increase by 9% under $2xCO_2$, implying more flood damage and a 5% increase in the population affected by floods. About 3 million people currently live in flood-prone areas. In terms of flood occurrence, the present once-every-50-years flood would return every 30 years (ADB, 1994). Water deficits in the dry season, on the other hand, are likely to increase due to higher evaporation, reducing water availability for irrigation.

About 70% of the total population of Malaysia live in the coastal zone, which is also the centre of most of the country's economic activities. In addition, important natural ecosystems are located along the coast, with 44 of the 1000 islands designated as marine parks. Sea level rise could have significant consequences for the low-lying coastal plains of Malaysia. Parry *et al.* (1991) report that a 1-m rise could lead to a landward retreat of the shoreline of as much as 2.5 km. Midun and Lee (1989) suggest that such a rise could result in the near total loss of existing mangroves, with little chance of inland migration. Mangrove forests are already under severe stress from human interference. Sea level rise thus aggravates an already urgent environmental problem.

6.5.8 Thailand

Parry *et al.* (1991) suggest that a CO_2 doubling could result in a 3-6°C mean annual temperature change in Thailand, although this appears to be a rather high estimate. GCM predictions for rainfall diverge, but generally show a reduction under the GISS scenario. Northern Thailand will tend to be drier in most months except July. For Ayuthaya Province, the GISS model predicts rainfall reductions in August-September, but other models do not produce this result. With respect to agriculture, preliminary runs using current climate data with the CERES agricultural model for Ayuthaya province showed higher yields for transplanted rice and lower than expected yields for directly seeded rice in comparison to observed val-

ues. Under the $2xCO_2$ climate, rice yields in this province would generally increase. The analysis suggested an 8% increase in cultivation in the province, substantially less than the currently observed year-to-year fluctuations. Off-season rice showed average yield increases of 5%. Model validation is weak, however, and the results should be treated with caution. Results for Chiang Mai, for example, suggest average yield reductions of 5%.

For the Suratthani province in southern Thailand, a case study calculated that 37% of the area would be affected by a 1-m rise in sea level, with losses of over 4,200 ha of agricultural land and many shrimp ponds (Parry *et al.*, 1991).

6.5.9 Latin America

IPCC (1992b, 1994) summarizes sea level rise case studies for several Latin American countries, including Argentina, Uruguay, and Venezuela. In Argentina a 1-m sea level rise would inundate an area of about 3,400 km² (0.1% of the country's total area). Erosion would claim assets and land worth $5 billion. Venezuela could lose about 5,700 km² or 0.6% of its area under a 1-m rise (assuming no protection). Particularly at risk would be the country's low-lying coastal plains and deltas. Although only a small area would be at risk in Uruguay, the coastlines affected would be highly valuable tourist beaches. Uruguay's tourist industry creates over $200 million in revenue per annum, and attracts over 1 million visitors each summer. Protecting the beaches would be expensive. The capital costs of protecting developed areas (mainly beach nourishment) were estimated at $2.9-8.6 billion, or more than five times the costs expected for Venezuela. If spread over 50 years, this would correspond to annual investments of 6-19% above 1987 gross investments (Nicholls *et al.*, 1992).

The impacts of climate change on agriculture are less well studied. Using climate analogues, Magalhães and Glantz (1992) illustrate the consequences of climate extremes on Brazilian agriculture and society. In the semi-arid northeast of Brazil, agriculture is characteristically vulnerable to droughts. Drops in agricultural production cause mass unemployment in the agricultural sector, followed by malnutrition and hunger. Increased migration to urban centres is one of the consequences. In the global agricultural model of Rosenzweig *et al.* (1993), yield impacts in Brazil are among the most severe for all regions. Under $2xCO_2$, yields are expected to fall by 17-33% with CO_2 fertilization, and 38-53% without. Similar reductions are also reported for Uruguay.

6.5.10 Small island states

The small island states (SISs) are clearly among the regions that would be most affected by climate change, not only because of their high vulnerability to sea level rise but also because of their strong dependence on natural resources that may be affected by climate change. Pernetta (1989) suggests a ranking of SISs in the Pacific in terms of their extreme vulnerability to sea level rise. "Profound impacts," including disappearance in the worst cases, could be felt by Tokelau, the Marshall Islands, Tuvalu, the Line Islands, and Kiribati. Severe impacts resulting in major population displacement would be experienced by Micronesia, Palau, Nauru, French Polynesia, the Cook Islands, Niue, and Tonga. Moderate to severe impacts would be felt by Fiji, American Samoa, New Caledonia, the northern Marianas, and the Solomon Islands, while local severe to catastrophic events would be experienced by Vanuatu, Wallis and Futuna, Papua New Guinea, Guam, and Western Samoa.

Pernetta (1992) outlines various impacts that SISs may suffer. These include increased frequency of tropical cyclones in areas not normally affected by them; an altitudinal shift in vegetation zones, affecting alpine grasslands in Papua New Guinea and threatening mid-montane rain forests as they come under pressure for cultivation; some increased capillarity in limestone soils, reducing soil fertility in some areas; some increases in disease due to drier conditions in some countries and longer wet seasons in others; and increases in humidity and hence human discomfort and adaptation costs. Sea level rise will also affect agricultural activity, which now occurs mostly on the coasts, and push it inland onto less suitable soils, thus increasing erosion. In addition, exclusive economic zones based on outlying islands will be affected as some of the islands disappear under the rising ocean waters. Protection measures are limited and expensive. For the Marshall Islands, for example, IPCC (1992b) reports that protecting the Majuro atoll alone would cost 1.5 to 3 times the country's present GDP.

Since many hermatypic corals are growing at their limit of thermal tolerance, any rise in sea water temperature could result in increased coral bleaching, with consequent loss of coral fishery resources and tourism. Although the reactions of corals to such changes appear not to be known in any detail, consequences could be significant. In the Maldives, for example, dependence on the corals is high: they are mined for building materials; fish and marine products account for most of the islands' exports. Tourism based on the corals is also vital. Coral deaths have already occurred, probably due to increased lagoon temperatures, but pollution is also implicated.

6.5.11 Conclusions on regional impacts

Economic analyses of damage to developing countries from climatic change remain limited. Nonetheless, the preceding overview supports the general finding of the broader economic studies by showing that impacts on developing countries are likely to be more severe relative to the wealth of those countries. In some cases, sea level rise alone results in dramatic impacts on the economies and may threaten the existence of whole communities and nation states. The relatively greater vulnerability of the developing countries is relevant to any discussion of the equity case for controlling climate change (see Chapter 3).

Table 6.10 summarizes the results of computing a vulnerability index for different categories of countries. Vulnerability is defined in terms of exposure to foreign economic conditions (export dependence), insularity and remoteness, and proneness to natural disasters. The table illustrates the high vulnerability of developing countries to climate change. In the

Table 6.10. *Vulnerability index for different categories of countries (high vulnerability is indicated by values closer to 1)*

Country Categories	Number of Countries	Index
All countries	113	0.376
Developed countries	22	0.208
Developing countries	91	0.417
Small island developing countries	20	0.590
Other island developing countries	28	0.539

Note: The index is calculated as the average of three variables (i = 1, 2, 3): export dependence, insularity and remoteness, and proneness to natural disasters. Variable i for country j is calculated as

$$V_{ij} = (X_{ij} - minX_i)/(maxX_i - minX_i)$$

where X_{ij} is the value of component i obtained for country j, $maxX_i$ denotes the highest value for component i observed in any country, and $minX_i$ likewise the lowest value observed in the sample. For example, in the case of export dependence, X_{1j} measures the export dependence of country j, $maxX_1$ is the value observed in the country with the highest export dependence, whereas $minX_1$ reflects the lowest export dependence observed. Therefore, in the country with the highest export dependence, V_{1j} is equal to 1. In the country with the lowest export dependence, $V_{1j} = 0$.
Source: Briguglio (1993).

full list for 113 countries, 9 out of the 10 most vulnerable countries are island states (Briguglio, 1993).

6.6 From Greenhouse Damages to Abatement Benefits

There is a distinction between *climate change damage* and the *benefits of policy measures,* although the two concepts are related. In general, *the benefits of greenhouse action are at least equal to the amount of damage avoided,* that is, to the extra damage which would have occurred in the absence of action. In addition, there may also be ancillary benefits that are not related to climate change (see Section 6.7). The principal rule cited above is complicated somewhat by the dynamic character of climate change, however.

Figure 6.1 considers schematically the development of greenhouse gas emissions and damage over time under different scenarios. In the base case, annual emissions are assumed to continue rising over the next 100 years or so (curve labelled "baseline" in the emissions graph). Global mean temperature levels will therefore rise as well, and so too will annual climate change damage (including the influence of adaptation, see Section 6.1.5). The upper curve in the damage chart shows how annual damages may rise in the baseline case. The estimates of benchmark warming damage of Section 6.2 relate to only one particular point on this time path: the point t_{2xCO_2}, when the warming assumptions underlying the estimates are realized. In a baseline scenario, this may, for example, be in the year 2060. If damage levels grow in proportion to GNP

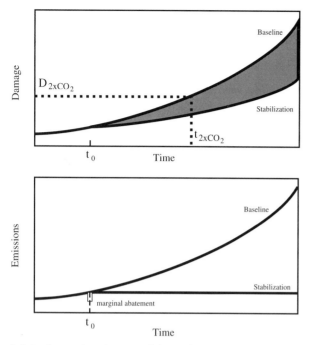

In the baseline, annual greenhouse gas emissions continue to rise over time (upward sloping line in the emissions graph). As a consequence, annual climate change damage rises over time as well (upper line in the damage graph). If instead emissions are stabilized at time t_0 (the horizontal "stabilization" line in the lower graph), the annual damage curve will follow a flatter trajectory (the lower curve in the upper diagram). The benefit of this stabilization policy is represented by the shaded area: the sum of avoided damages in all time periods. Stabilizing emissions constitutes a significant change in policy. Analysts are often interested in the benefits obtained through only a *marginal* deviation from the baseline – the marginal benefit per tonne of carbon abated or the marginal cost of an additional tonne emitted. Marginal changes are conceptually calculated in the same way, i.e., as the difference in the two damage trajectories with and without the abated tonne (not drawn in the figure). The two emission paths with and without marginal abatement would be identical, except for a slight deviation at time t_0.

Figure 6.1:   Doubled CO_2 damage vs abatement benefits.

and assuming the earlier assessment is correct, damage at time t_{2xCO_2} will be in the order of 1.5-2.0 % of gross world product (GWP) (abstracting from possible differences between transient and equilibrium damage). Before this date, annual damage levels (relative to GNP) are lower. In subsequent decades they are higher, due to the rising atmospheric CO_2 concentration.

The $2xCO_2$ benchmark is contrasted with the consequences of a particular greenhouse gas abatement strategy, say a stabilization of emissions at time t_0 (represented by the lower emission and damage curves). Due to the thermal inertia of the climate system, annual damage levels will react only gradually to this policy change, and will initially continue to grow unabated. Eventually, however, damage will start to deviate from the baseline. The level to which the trajectory will converge in the long run is unclear, and depends on the exact character and source of the damages.

What, then, are the benefits of the stabilization policy? They are represented by the shaded area in Figure 6.1, which depicts the difference in annual damage levels between the base case and the control scenario, summed up over the relevant time horizon. In each future time period damage is lower than it would have been otherwise, and the benefit of greenhouse gas abatement is the (discounted) sum of these avoided future damages.

As Figure 6.1 makes clear, there is thus no direct connec-

tion between *damage associated with $2xCO_2$* and the *benefits of greenhouse gas abatement*. The benefits of abatement occur as a *stream* of reduced damage over time, while $2xCO_2$ concerns only one point. Annual damage reduction levels will generally not coincide with the $2xCO_2$ assessment either, for two reasons: first, because even the most stringent abatement scheme is unlikely to succeed in avoiding all damage (since some damage has already been done and may be irreversible); second, because damage reductions will chiefly concern years in which the base case damage would not be that of $2xCO_2$.

Despite this difference, $2xCO_2$ assessments still provide some indication of the size of abatement benefits. For a more precise assessment much more information would be necessary, though. A careful analysis would in particular require a better knowledge about the *slope* of the damage trajectory, that is, about how climate change damage alters over time as a consequence of both climate change and economic and population growth. A further need would be information about the degree of damage that can be avoided through a particular policy measure, that is, the distance between the two damage trajectories of Figure 6.1. The answer here depends chiefly on the exact character of impacts – whether they are reversible or not, whether damage depends on the rate of change or on absolute temperature levels, whether damage is persistent or a transitional adjustment cost, and so on.

Knowledge about damage beyond the $2xCO_2$ benchmark is very limited. The small literature aiming at estimating the benefits of greenhouse gas abatement relies on several rather ad hoc assumptions. In older studies, damage costs were typically specified as a polynomial (usually linear to cubic) function of global mean temperature, calibrated around the $2xCO_2$ estimates. Damage is usually fully reversible and typically assumed to grow with GDP. More recently, studies have started to emerge that explicitly incorporate regionally diversified temperatures and sea levels and that model individual damage categories (e.g., agriculture) separately, or at least distinguish between damages related to absolute temperature level and those related to the rate of change (e.g., Dowlatabadi and Morgan, 1993; Hope *et al.*, 1993; Tol, 1994, 1995).

Table 6.11 provides a list of estimates of the marginal benefits of CO_2 abatement. In terms of Figure 6.1 these numbers represent the shaded area that would occur if emissions were to deviate marginally from the baseline path, say by one tonne in period t_0, but remain unaltered otherwise. The estimates are mostly based on models with simple polynomial damage functions. Several of them are derived from optimal control or dynamic cost-benefit models. In these models, the marginal benefit from abatement, or the "shadow price of carbon," is calculated as the carbon tax necessary to keep emissions on the trajectory that is considered socially optimal by the model. In other models the marginal benefits are calculated directly, as in Figure 6.1, by comparing the present value of the stream of damages associated with a certain emissions scenario to that of an alternative scenario with marginally different emissions in the base period. Estimates vary widely, mainly as a consequence of different assumptions about the discount rate. They rise over time as a consequence of economic growth and increasing concentration levels.

Table 6.11. *The social costs of CO₂ emissions in different decades (in 1990 $/tC)*

Study	Type	1991–2000	2001–2010	2011–2020	2021–2030
Nordhaus (1991)	MC		7.3 (0.3–65.9)		
Ayres and Walter (1991)	MC		30–35		
Nordhaus (1994b)	CBA				
Best guess		5.3	6.8	8.6	10.0
Expected value		12.0	18.0	26.5	n.a.
Cline (1992b, 1993d)	CBA	5.8–124	7.6–154	9.8–186	11.8–221
Peck and Teisberg (1992)	CBA	10–12	12–14	14–18	18–22
Fankhauser (1994b)	MC	20.3 (6.2–45.2)	22.8 (7.4–52.9)	25.3 (8.3–58.4)	27.8 (9.2–64.2)
Maddison (1994)	CBA/MC	5.9–6.1	8.1–8.4	11.1–11.5	14.7–15.2

MC = marginal social cost study.
CBA = shadow value in a cost-benefit study.
Figures in parentheses denote 90% confidence intervals.
Sources: As shown.

Using the base case in DICE, Nordhaus (1993a, b) finds that the shadow price begins at about $5 per tonne of carbon in 1995, rises to about $10 by 2025, and reaches $21 by 2095 (at 1990 prices). Peck and Teisberg (1992, 1993a, b) find values of a similar order of magnitude. Tol's (1995) alternative specification of DICE yields shadow prices of $13 for 1995, rising to $89 for 2095. These model runs assume that parameter values are known with certainty. In the case of DICE, expected shadow prices more than double once uncertainty is added to the model. This result arises because of the skewedness in the damage distribution, which allows for low probability/high impact events (Nordhaus, 1994b). All three authors assume a pure rate of time preference (or utility discount rate, see Chapter 4) of 3%. In contrast, Cline (1992b, 1993d) finds significantly higher shadow prices by using a zero utility discount rate. His reproduction of the DICE model generates a path of shadow prices beginning at about $45 per tonne, reaching about $243 by 2100. Other parameter specifications provide even higher values.

In comparison, Fankhauser (1994b) identifies a lower and flatter trajectory for the shadow price of carbon, rising from $20 per tonne in the decade 1991–2000 to $28 per tonne by 2021–30, with confidence intervals of $6-$45 and $9-$64 respectively. Fankhauser uses a probabilistic approach to the range of discount rates, in which low and high discount rates are given different weights. His sensitivity analysis of the discount rate suggests that moving from high (3%) to low (0%) discounting could increase marginal costs by a factor of 9, from $5.5 to $49 per tonne of carbon emitted now.

6.7 The Secondary Benefits of Abatement Strategies

The benefits of greenhouse gas abatement will not be limited to reduced climate change costs alone but are likely to spill over to other sectors. This is a further reason why the cost of greenhouse damage differs from the benefits of greenhouse gas abatement (see Section 6.6). For example, efforts to halt deforestation to reduce the emission of CO_2 will contribute to the conservation of the world's biological diversity. Other ancillary benefits could occur in the form of local and regional air quality improvements, a reduction in traffic-related externalities like accidents or congestion, and the reduced risk of tanker accidents and oil spills. These problems are tied to climate change in that they are caused by largely the same activities, in particular, the consumption of fossil fuels. Because CO_2-removal technologies are presently not economical, attempts to limit CO_2 emissions currently by and large concentrate on reducing the use of fossil fuels. A reduction in CO_2 emissions will therefore also reduce other environmental problems related to fuel combustion.[5] These effects are often called the *secondary benefits* of carbon abatement.

Secondary benefits from air quality improvements may be quite large. Table 6.12 provides estimates of benefits from reduced air pollution levels as a consequence of carbon abatement (see also Box 6.4). The estimates are not necessarily comparable since the assumed abatement policies differ from study to study, although they all roughly aim at stabilizing CO_2 emissions at about 1990 levels.

What is the value of these emission reductions? Estimates that measure the social costs of each pollutant vary widely between regions, depending on local factors like baseline air quality standards, ecosystem vulnerability, and population at risk. Figures may also differ because impacts caused by a combination of gases are attributed to the initial sources in different ways. A selection of illustrative results is given in Table 6.13. The average secondary benefits implied by these estimates vary widely, from about $2 per tonne of carbon abated to over 500 $/tC in absolute terms. Secondary benefits offset about 30-50% of the initial abatement costs in the case

Table 6.12. *Reduced air emissions due to CO_2 abatement (% reduction from baseline)*

(a) Regional studies

Country	Year	Policy/Scenario	CO_2	SO_x	NO_x	CO	TSP[a]	VOC[b]	Secondary Benefits ($/tC)	Sources
World	2000	World CO_2 emissions stabilized at 1990 level	9	14	10	—	—	—	—	Complainville/ Martins (1994)
U.S.			8	13	8	—	—	—		
Japan			4	4	3	—	—	—		
EU			5	7	4	—	—	—		
Other OECD			10	14	11	—	—	—		
China			18[c]	19	19	—	—	—		
Ex-USSR			17[c]	21	18	—	—	—		
India			17[c]	17	16	—	—	—		
E. Europe			11[c]	11	11	—	—	—		
OECD	2000	CO_2 emissions in OECD stabilized at 1990 level	—	—	—	—	—	—	—	Complainville/ Martins (1994)
U.S.			18	28	25	—	—	—		
Japan			14	12	15	—	—	—		
EU			14	18	15	—	—	—		
Other OECD			21	29	32	—	—	—		
Europe[d]	2000	EU carbon/energy tax								Alfsen *et al.* (1993)
		Current structure	9.4	7.4[e]	6.2	—	—	—	6.1	
		Cost-efficient regime	9.7	9.3[e]	6.4	—	—	—	6.6	

(b) Country studies

Country	Year	Policy/Scenario	CO_2	SO_x	NO_x	CO	TSP[a]	VOC[b]	Secondary Benefits ($/tC)	Sources
Norway	2000	Emission stabilization (at 1989 level)	15.0	20.8[e]	10.8	24.1	4.3	—	40–140[f]	Alfsen *et al.* (1992)
UK	2005	EU carbon/energy tax	12.1	38.3[e]	10.6	9.6	30.3	1.1	40–1,040	Barker (1993)
U.S.	2000	Emission stabilization								Scheraga and Leary (1994)
		Through carbon tax	8.6	1.9	6.6	1.5	1.0/1.8[g]	1.4	2.0–20	
		Through Btu tax	8.6	2.2	6.6	3.4	1.6/2.2[g]	2.7	3.5–28	

[a]Total suspended particles.

[b]Volatile organic compounds.

[c]The study uses the hypothetical scenario of a global carbon tax. Note that the UN Framework Convention on Climate Change does not contain any obligations for developing countries to reduce their CO_2 emissions. Economies in transition are granted a "certain degree of flexibility."

[d]Western and Eastern Europe (UN ECE region). Tax in six EU countries (France, Germany, UK, Italy, Netherlands, Denmark) and three Nordic countries (Norway, Finland, Sweden) only.

[e]SO_2.

[f]Including road traffic benefits (reduced congestion, noise, accident, and road damage costs).

[g]PM_{10} / TSP (PM_{10} refers to fine particles less than 10^{-6} m in diameter).

Sources: As shown.

of Norway (Alfsen *et al.,* 1992) and over 100% in the UK (Barker, 1993). Extending the Alfsen *et al.* study, Amano (1994) has found a similar result for Japan.

The calculations by Complainville and Martins (1994), based on the OECD GREEN model, suggest that secondary air quality improvements may be as significant in developing nations as they are in OECD nations. This is confirmed by Amano (1994), who has calculated secondary benefits for several Asian regions, using the same benefit-abatement ratios (air quality benefits, in percent of GDP, per percentage cut in air pollution) as Alfsen *et al.* (1992). For India, the estimated secondary benefits exceed the primary costs of stabilizing emissions at 1990 levels. For China and the group of "Dynamic Asian Economies" (Hong Kong, Philippines, Singapore, South Korea, Taiwan, and Thailand) secondary benefits are estimated to offset about one-third of the initial abatement costs. The Amano results should be interpreted with caution, however. Using the Norwegian ratios of Alfsen *et al.* for developing economies almost certainly biases the results.

BOX 6.4: VALUATION OF ELECTRICITY-RELATED EXTERNALITIES IN CALIFORNIA

The importance of secondary air quality benefits can be illustrated at least in part from the practices of some regulatory commissions in the U.S. A growing number of these now require that electric utilities take into account the value of air emissions in their cost-benefit analyses of alternative supply- and demand-side resources.

The monetary air quality values used in these assessments do not usually represent actual marginal damages, however. CO_2 values, for example, are mostly based on avoided costs, such as the cost of planting trees on otherwise unforested land. The environmental cost adders for the other pollutants are also mostly based on control costs. The conceptual basis for this method is rather weak: Using abatement costs as a proxy for the damage costs could involve substantial error. The use of damage control costs as a proxy is justified by many regulatory commissions on the basis that damage cost estimates are themselves uncertain.

A study commissioned by the California Energy Commission that does use consistently estimated damage-cost-based externality values is shown below. These estimates for the Los Angeles area imply that for every dollar in benefits resulting from the reduction of CO_2 emissions, there is another $3 benefit to the area resulting from the reduction of conjoint pollutants. It is a reasonable conclusion, therefore, that joint benefits can significantly affect the overall cost-benefit assessment of climate policies.

	Environmental Benefits ($/t)	Emission Rate (lbs/MWh)	Total Benefits ($)
CO_2	26[†]	1,820	23
Joint products			
NO_x	14,483	6	43
SO_2	7,425	6	22
TSP	57,620	0.3	9
Total joint products			74

[†] Benefits in the form of *avoided costs* of carbon sequestration.
Source: Emission rates are based on a notional, new coal-fired plant meeting new source performance standards (NSPS). Externality values for NO_x, SO_2, and particulates (TSP) are in 1989 U.S. dollars, based on Hashem and Haites

---

Table 6.13. *The social costs of air pollution ($/tonne)*

Country	SO_2	NO_x	CO	Particulates	VOC	Source
UK	367	124	15	21,333	n.a.	Pearce (1994)
UN ECE[†]	637	490	n.a.	21,333	n.a.	Pearce (1994)
Norway	500–7,600	1,600–31,400	1–13	2,100–27,700	n.a.	Alfsen *et al.* (1992)
U.S.	4,800	2,000	n.a.	2,700	n.a.	Ottinger *et al.* (1990)
U.S.	300–1,800	10–100	n.a.	400–10,900	360–2,400	Scheraga and Leary (1994)

[†]Damage done by a tonne of UK emissions to Western and Eastern Europe, including UK (UN ECE region).
Sources: As shown.

An alternative way to measure secondary benefits is by estimating the change in the costs of meeting air quality standards. Many industrialized countries are committed to significant cuts in the emission of air pollutants. The Helsinki Protocol of 1985 required a cut in sulphur emissions of 30% by 1993, compared to 1980 levels, for selected European countries. Under the "Second Sulphur Protocol" further reductions will be required up to 2010. The Sofia Protocol on nitrogen dioxide commits signatories to a freeze on emissions at 1987 levels. Greenhouse gas abatement will lower the amount of traditional air pollution abatement needed to meet these targets. Alfsen *et al.* (1993), for example, calculate that the in-troduction of a carbon/energy tax would reduce the cost of traditional SO_2 and NO_x abatement in the nine countries introducing the tax by 25-30% and 12-25 % respectively, implying average secondary benefits of about 2.5 $/tC for NO_x and 4.0 $/tC for SO_2. These results are low compared to those of the previous estimation procedure.

It is important to underline the different character of secondary benefits and primary (warming damage avoidance) benefits. Most critically, secondary benefits do not depend on climate variables but only arise in connection with greenhouse gas *abatement*. They occur locally or regionally and do not share the global character of greenhouse damages. The question of secondary benefits from carbon abatement should also be distinguished from the more comprehensive issue of the optimal abatement mix with respect to all pollutants. The secondary benefit argument is characterized by an implicit primacy of the greenhouse problem, in that improvements in other areas are seen as welcome side effects of a climate change policy, but are not considered or sought in their own right. This is not necessarily the ideal way to proceed. Strictly, each pollutant should be taxed in proportion to the environmental damage it causes. If there are interdependencies between them, as is the case with climate change and air pollution, these would have to be reflected in the relative tax rates. The currently considered abatement strategies may then no longer constitute the optimal approaches. Once secondary benefits are taken into account, *location* will also matter. Although it is not the case for greenhouse gases, for most other air pollutants it matters where they are emitted. Emission reduction measures should therefore be concentrated in those places where the joint benefits of reducing all emissions is highest.

6.8 Conclusions

This chapter has been concerned with the measurement of climate change damage and with the benefit of policies designed to reduce this damage. Climate change impact assessments are an integral input to cost-benefit studies and other decision-making frameworks (see Chapters 2 and 5). Economic models comparing the costs and benefits of greenhouse gas abatement are discussed in Chapter 10.

Social cost estimates are necessarily *uncertain*. Apart from the scientific uncertainty of climate change, there are additional uncertainties associated with

(a) Limited knowledge of regional and local impacts
(b) Difficulties in measuring the economic value of impacts, even where the impacts are known. This is particularly the case for nonmarket impacts and the impacts in developing countries
(c) Difficulties in predicting future technological and socioeconomic developments
(d) The possibility of catastrophic events and surprises

This uncertainty must be emphasized when interpreting the social cost figures in this chapter.

Most impact analysis has been based on equilibrium climate change associated with a doubling of the pre-industrial CO_2 concentration or its equivalent for all greenhouse gases. These studies have usually used the IPCC's 1990 best-guess value for climate sensitivity of 2.5°C and are generally based on the IPCC 1990 impact assessment. The available studies estimate damages for a doubling of CO_2 as follows:

• World impact: 1.5-2.0% of world GDP
• Developed country impact: 1-1.5% of GDP
• Developing country impact: 2-9% of GDP

These are *best-guess* central estimates, including both market and nonmarket impacts, and in some cases also adaptation costs. They are based on a large number of simplifying and of-

ten controversial assumptions. The range does not represent the confidence interval around the estimates, but the spread of the best guesses in existing studies. No attempt has been made to quantify a confidence interval. There will be considerable differences in regional damage figures, with potentially higher impacts for some individual countries, such as small island states. Alternative assumptions about the value of a statistical life (Box 6.1) could further increase regional differences. The regional variability of the social costs underlines the important issues of equity discussed in Chapter 3.

These cost estimates are for $2xCO_2$ concentrations, but concentrations may continue rising above this level. Such *long-term warming* damage may rise more than linearly. The damage associated with 10°C might be 6% of world GDP or more. The probability of a climate catastrophe also increases with the speed and amount of warming.

The *marginal damage,* that is, the extra damage done by one extra tonne of carbon emitted now, is estimated to be in the order of $5-$125 per tonne of carbon. The value will rise for marginal emissions in later periods. The range appears wide but reflects variations in models, discount rates, and other factors. The numbers are particularly sensitive to the choice of the discount rate. Estimates based on a positive social rate of time preference (discount rate) of approximately 5% are usually in the order of about $5-12 per tonne of carbon emitted now, whereas figures assuming a rate of 2% or less are almost an order of magnitude higher. The models on which these estimates are based remain simplistic and are limited representations of the actual processes. But they represent the state of the art at this moment.

The marginal damage of an extra tonne of emissions is not necessarily the same as the marginal benefits of abating an extra tonne. Abatement measures will yield additional benefits besides avoided climate change damage. These are the *secondary benefits,* which occur, for example, in the form of local air quality improvements. The size of secondary benefits depends on local circumstances. Studies for European countries and the U.S. indicate that secondary benefits could offset between 30% and 100% of abatement costs. Whether the benefits of climate policies are high enough to justify the costs of abatement is an issue not addressed here.

Endnotes

1. The eventual equilibrium warming consistent with the 2060 greenhouse gas concentrations would be higher than benchmark warming (because of ocean thermal lag, which would cause a delay of two decades or more between committed and realized warming).

2. These parameters apply to the global mean effects of $2xCO_2$ warming. However, U.S. warming and precipitation may be less favourable than the global mean (IPCC, 1990a). Moreover, the simulations assume that irrigation water will be readily available, whereas its availability is expected to become more constrained in the future, even without climate change (Waggoner *et al.*, 1992).

3. Some of these impacts may be included, as forgone use value, in estimates for ecosystem loss (see Section 6.2.13).

4. As the estimate is purely speculative, it is excluded from the central estimates of that study and from Table 6.4.

5. This is not the only connection between climate change and air pollution. The two issues are heavily intertwined. For example, sulphur abatement, if achieved through the installation of end-of-pipe scrubbers, could lead to a lower system efficiency and thus higher CO_2 emissions. The accumulation of sulphur aerosols in the atmosphere causes a reduction in mean temperature, thus masking the extent of warming. Warmer temperatures, on the other hand, will aggravate photochemical air pollution.

References

Adams, R.M., C. Rosenzweig, R.M. Peart, J.T. Ritchie, B.A. McCarl, J.D. Glyer, R.B. Curry, J.W. Jones, K.J. Boote, and L.H. Allen Jr., 1990: Global Climate Change and U.S. Agriculture, *Nature* **345**(17 May), 219–223.

Adams, R.M, R.A. Fleming, C.C. Chang, B.A. McCarl, and C. Rosenzweig, 1993: A reassessment of the economic effects of global climate change on U.S. agriculture, report prepared for the U.S. Environmental Protection Agency, Washington DC (September).

Adger, N., and S. Fankhauser, 1993: Economic analysis of the greenhouse effect: Optimal abatement level and strategies for mitigation, *International Journal of Environment and Pollution* **3**(1–3), 104–119.

Agarwal, A., and S. Narain, 1991: Global warming in an unequal world. A case of environmental colonialism, mimeo, Centre for Science and Environment, New Delhi.

Albala-Bertrand, J.M., 1993: *The political economy of large natural disasters,* Clarendon Press, Oxford.

Alfsen, K.H., H. Birkelund, and M. Aaserud, 1993: *Secondary benefits of the EC carbon/energy tax,* Research Department Discussion Paper No 104, Norwegian Central Bureau of Statistics, Oslo.

Alfsen, K.H., A. Brendemoen, and S. Glomsrød, 1992: *Benefits of climate policies: Some tentative calculations,* Discussion Paper No. 69, Norwegian Central Bureau of Statistics, Oslo.

Allen, B.D., and R.Y. Anderson, 1993: Evidence from Western North America for rapid shifts in climate during the last glacial maximum, *Science* **260**(25 June), 1920–1923.

Amano, A., 1994: Estimating secondary benefits of limiting CO_2 emissions in the Asian region, mimeo, School of Business Administration, Kobe University.

Anklin, M., *et al.,* 1993: Climate instability during the last interglacial period recorded in the GRIP ice core, *Nature* **364**(15 July), 203–207.

Arnell, N.W., and W.R. Dubourg, 1994: Implications for water supply and water management. In *Economic implications of climate change in Britain,* M.L. Parry and R. Duncan, eds., Earthscan, London.

Asaduzzaman, M., 1994: Economic and social impacts of climate change: A case study of Bangladesh coastal zone, mimeo, Bangladesh Institute of Development Studies, Dhaka.

Asian Development Bank, 1994: *Climate change in Asia,* Asian Development Bank, 8 Country Studies, Manila.

Asthana, V., 1993: Report on impact of sea level rise on the islands and coasts of India, report to the Ministry of Environment and Forests, Government of India, New Delhi.

Awosika, L., A. Ibe, and M. Udo-Aka, 1990: Impact of sea level rise on the Nigerian coastal zone. In *Changing climate and the coast,* Vol. 2, J.G. Titus, ed., U.S. Environmental Protection Agency, Washington DC.

Ayres, R., and J. Walter, 1991: The greenhouse effect: Damages, costs and abatement, *Environmental and Resource Economics* **1**: 237–270.

Baan, P.J.A., H. van der Most, G.B.M. Pedroli, and H. van Zuijlen, 1993: *Systeemverkenning Gevolgen van Klimaatverandering* (in Dutch: *System exploration impact of climatic change*), Delft Hydraulics, Delft.

Barker, T., 1993: Secondary benefits of greenhouse gas abatement: The effects of a UK carbon/energy tax on air pollution, Energy Environment Economy Modelling Discussion Paper No. 4, Department of Applied Economics, University of Cambridge.

Batie, S.S., and H.H. Shugart, 1989: The biological consequences of climate changes: An ecological and economic assessment. In *Greenhouse warming: abatement and adaptation,* N.J. Rosenberg *et al.,* eds., Resources for the Future, Washington, DC.

Bazzaz, F.A., and E.D. Fajer, 1992: Plant life in a CO_2-rich world, *Scientific American,* **266**(1): 68–74.

Bengtsson, L., M. Botzet, and M. Esch, 1995: Will Greenhouse Gas-Induced Warming Over the Next 50 Years Lead to Higher Frequency and Greater Intensity of Hurricanes? Report No. 139. Max Planck Institute, Hamburg.

Berz, G., 1990: Natural disasters and insurance/reinsurance, *UNDRO News,* January/February: 18–19.

Berz, G., and K. Conrad, 1993: Winds of change, *The Review,* June: 32–35.

Bigford, T.E., 1991: Sea level rise, nearshore fisheries, and the fishing industry, *Coastal Zone Management* **19**: 417–437.

Braden, J.B., and C.D. Kolstad, 1991: *Measuring the demand for environmental quality,* Elsevier, Amsterdam.

Briguglio, L., 1993: The economoc vulnerabilities of small island developing states, report to UNCTAD, Geneva.

Broccoli, A.J., and S. Manabe, 1990: Can existing climate models be used to study anthropogenic changes in tropical cyclone climate?, *Geophysical Research Letters* **17** (11): 1917–1920.

Broome, J., 1992: *Counting the costs of global warming,* White Horse Press, Cambridge.

Bryant, E.A., 1991: *Natural hazards,* Cambridge University Press, Cambridge.

Callaway, M., J. Smith, and S. Keefe, 1994: The economic effects of climate change for US forests, report prepared for the U.S. Environmental Protection Agency, RCG/Hagler Bailly, Boulder, CO.

Canadian Climate Program Board, 1988a: Implications of climate change for downhill skiing in Quebec, Climate Change Digest Report, CCD 88-03, Downsview, Ontario.

Canadian Climate Program Board, 1988b: Implications of climate change for tourism and recreation in Ontario, Climate Change Digest Report, CCD 88-05, Downsview, Ontario.

Chakraborty, B., and M. Lal, 1994: Monsoon climate and its change in doubled carbon dioxide atmosphere as simulated by CSIRO 9, mimeo, unpublished.

Changnon, S.A., and J.M. Changnon, 1992: Temporal fluctuations in weather disasters: 1950–1989, *Climatic Change,* **22:** 191–208.

Clark, K.M., 1988: Predicting insurance getting easier, *National Underwriter,* **29,** August.

Clark, K.M., 1991: Catastrophe reinsurers need more, better data, *National Underwriter,* September.

Cline, W.R., 1991: Scientific basis of the greenhouse effect, *Economic Journal* **101**(407): 904–919.

Cline, W.R., 1992a: *The economics of global warming,* Institute for International Economics, Washington, DC.

Cline, W.R., 1992b: Optimal carbon emissions over time: Experiments with the Nordhaus DICE model, mimeo, Institute for International Economics, Washington, DC.

Cline, W.R., 1993a: A note on the impact of global warming on air conditioning and heating costs, mimeo, Institute for International Economics, Washington, DC (January).

Cline, W.R., 1993b: The impact of global warming on the United States: A survey of recent literature, mimeo, Institute for International Economics, Washington, DC (April).

Cline, W.R., 1993c: The costs and benefits of greenhouse abatement: A guide to policy analysis, paper presented at the OECD/IEA Conference on "The Economics of Global Warming," (June), Paris.

Cline, W.R., 1993d: Modelling economically efficient abatement of greenhouse gases, paper presented at the United Nations University Conference on "Global Environment, Energy and Economic Development," (September) Tokyo.

Cline, W.R., and C.T. Hsieh, 1993: Optimal carbon emissions in the Nordhaus DICE model under backstop technology, mimeo, Institute for International Economics, Washington, DC (July).

Complainville, C., and J.O. Martins, 1994: NO_x and SO_x emissions and carbon abatement, Economic Department Working Paper No. 151, OECD, Paris.

Collard, D., 1988: Catastrophic risk: Or the economics of being scared. In *Economics, growth and sustainable environments: Essays in memory of Richard Lecomber,* D. Collard, D.W. Pearce and D. Ulph, eds., Macmillan, New York.

Commonwealth Secretariat, 1989: Climate change: Meeting the challenge, report of a Commonwealth Group of Experts, Commonwealth Secretariat, London.

CRU/ERL, 1992: Development of a framework for the evaluation of policy options to deal with the greenhouse effect: Economic evaluation of impacts and adaptive measures in the European Community, report for the Commission of European Communities, Climate Research Unit (CRU), University of East Anglia, and Environmental Resources Limited (ERL), London.

Daily, G.C., P.R. Ehrlich, H.A. Mooney, and A.H. Ehrlich, 1991: Greenhouse economics: Learn before you leap, *Ecological Economics* **4,** 1–10.

Dansgaard, W., J.W.C. White, and S.J. Johnsen, 1989: The abrupt termination of the younger dryas climate event, *Nature* **339** (15 June), 532–534.

Darmstadter, J., and M.A. Toman (eds.), 1993: *Assessing surprises and nonlinearities in greenhouse warming,* Resources for the Future, Washington, DC.

den Elzen, M.G.J., and J. Rotmans, 1992: The socio-economic impact of sea-level rise on the Netherlands: A study of possible scenarios, *Climatic Change* **20,** 169–195.

Dlugolecki, A., D. Clement, C. Elvy, G. Kirby, J. Palutikoff, R. Salthouse, C. Toomer, S. Turner, and D. Witt, 1994: The impact of changing weather patterns on property insurance, report, Chartered Insurance Institute, London.

Doornkamp, J., 1993: Clay shrinkage induced subsidence, *Geographical Journal* **159**(2), 196–202.

Dowlatabadi, H., and G. Morgan, 1993: A model framework for integrated studies of the climate problem, *Energy Policy* **21,** 209–221.

Dracup, J.A., S.P. Pelmulder, R. Howitt, G. Horner, W.M. Hanemann, C.F. Dumas, R. McCann, J. Loomis, and S. Ise, 1993: *Integrated modeling of drought and global warming – Impacts on selected California resources,* U.S. Environmental Protection Agency, Washington, DC.

Easterling III, W.E., P.R. Crosson, N.J. Rosenberg, M.S. McKenney, L.A. Katz, and K.M. Lemon, 1993: Agricultural impacts of and responses to climate change in the Missouri-Iowa-Nebraska-Kansas (MINK) region, *Climatic Change* **24**(1/2), 23–62.

ECLAC (Economic Commission for Latin America and the Caribbean), 1993: *Climate change and water management in Latin America and the Caribbean,* Economic Commission for Latin America and the Caribbean, Santiago, Chile.

Edmonds, J., and J. Reilly, 1985: *Global energy. Assessing the future,* Institute for Energy Analysis, New York.

El-Raey, M., 1990: Responses to the impacts of greenhouse-induced sea level rise on the Northern coastal regions of Egypt. In *Changing climate and the coast,* Vol. 2, J.G. Titus, ed., U.S. Environmental Protection Agency, Washington DC.

Emanuel, K.A., 1987: The dependence of hurricane intensity on climate, *Nature* **326,** 483–485.

Erickson, J.D., 1993: From ecology to economics: The case against CO_2 fertilization, *Ecological Economics* **8,** 157–175.

Evans, D.J. *et al.,* 1991: *Policy implications of greenhouse warming,* National Academy Press, Washington, DC.

Fankhauser, S., 1994a: Protection vs. retreat. The economic costs of sea level rise, *Environment and Planning A,* **27,** 299–319.

Fankhauser, S., 1994b: The social costs of greenhouse gas emissions: An expected value approach, *Energy Journal* **15**(2), 157–184.

Fankhauser, S., 1995: *Valuing climate change. The economics of the greenhouse,* Earthscan, London.

Fellner, W., 1967: Operational utility: The theoretical background and a measurement. In *Ten economic studies in the tradition of Irving Fisher,* John Wiley & Sons, New York.

Fisher, A.C., and W.M. Hahnemann, 1993: Assessing climage change risks: Valuation of effects. In *Assessing surprises and nonlinearities in greenhouse warming,* J. Darmstadter and M. Toman, eds., Resources for the Future, Washington, DC.

Gleick, P.H., 1987: Regional hydrologic consequences of increases in atmospheric CO_2 and other trace gases, *Climatic Change* **10:** 137–161.

Gleick, P.H., 1992: Water and conflict, project on Environmental Change and Acute Conflict, University of Toronto and the American Academy of Arts and Science, Cambridge, MA.

Gleick, P.H., and E.P. Maurer, 1990: Assessing the costs of adapting to sea level rise. A case study of San Francisco Bay, mimeo, Pacific Institute for Studies in Development, Environment and Security, and Stockholm Environment Institute, Berkeley, CA and Stockholm.

Gleick, P.H., and L.L. Nash, 1991: *The societal and environmental costs of the continuing California drought,* Pacific Institute for Studies in Development, Environment and Security, Oakland CA.

Glynn, P.W., and W.H. de Weerdt, 1991: Elimination of reef-building hydrocorals following the 1982–83 El Niño warming event, *Science* **253** (5 July), 69–71.

Goering, J.M., 1990: The causes of undocumented migration to the United States: A research note, Working Papers No. 52, Commission for the Study of International Migration and Cooperative Economic Development.

Goklany, I.M., 1989: Climate change effects on fish, wildlife and other DOI programs. In *Coping with climate change,* J.C. Topping, Jr., ed., Climate Institute, Washington, DC.

Grubb, M., 1993: The costs of climate change: Critical elements. In *Costs, impacts and benefits of CO_2 mitigation,* Y. Kaya, N. Nakicenovic, W.D. Nordhaus, and F. Toth, eds., IIASA Collaborative Paper Series, CP93-2, Laxenburg, Austria.

Gutowski, W.J., G.F. McMahon, S.S. Schluchter, and P.H. Kirshen, 1992: Effects of global warming on hurricane-induced flooding, paper prepared for U.S. Department of Energy, Washington, DC.

Haarsma, R.J., J.F.B. Mitchell, and C.A. Senior, 1993: Tropical disturbances in a GCM, *Climate Dynamics* **8,** 247–257.

Haines, A., and C. Fuchs, 1991: Potential impacts on health of atmospheric change, *Journal of Public Health Medicine* **13**(2), 69–80.

Haines, A., and M. Parry, 1993: Climate change and human health,

Journal of the Royal Society of Medicine **86**(December), 707–711.

Han, M., J. Hou and L. Wu, 1990: Adverse impact of projected one meter sea level rise on China's coastal environment and cities, mimeo, University of Maryland.

Hansen, J. *et al.,* 1989: Regional greenhouse climate effects. In: *Coping with climate change,* J.C. Topping, Jr., ed., pp. 68–81, Climate Institute, Washington, DC.

Hashem, J., and E. Haites, 1993: *Status report: State requirements for considering environmental externalities in electric utility decision making,* Association of DSM Professionals, Boca Raton, FL.

Heathcote, R.L., 1985: Extreme event analysis. In *Climate impact assessment,* R.W Kates, J.H. Ausubel and M. Berberian, eds., John Wiley and Sons, Chichester.

Hoffert, M.I., and C. Covey, 1992: Deriving global climate sensitivity from paleoclimate reconstructions, *Nature* **360**(10 December), 573–576.

Hohmeyer, O., and M. Gärtner, 1992: *The Costs of Climate Change,* Report to the Commission of the European Communities, Fraunhofer Institut für Systemtechnik und Innovations-Forschung, Karlsruhe, Germany.

Homer-Dixon, T.F., J.H. Boutwell, and G.W. Rathjens, 1993: Environmental change and violent conflict, *Scientific American,* **268**(2), 38–45.

Hope, C., J. Anderson, and P. Wenman, 1993: Policy analysis of the greenhouse effect. An application of the PAGE model, *Energy Policy* **21**(3), 327–338.

Houghton, J., 1994: *Global warming. The complete briefing,* Lion Book, London.

Howarth, R.B., 1992: International justice and the chain of obligation, *Environmental Values,* **1,** 133–140.

Howarth, R.B., and P.A. Monahan, 1992: Economics, ethics, and climate policy, mimeo, Lawrence Berkeley Laboratory, Berkeley, CA.

Idso, S.B., R.C. Balling Jr., and R.S. Cerveny, 1990: Carbon dioxide and hurricanes: Implications of Northern Hemispheric warming for Atlantic/Caribbean storms, *Meteorology and Atmospheric Physics* **42,** 259–263.

IPCC (Intergovernmental Panel on Climate Change), 1990a: Climate change, the IPCC scientific assessment, report from Working Group I, Cambridge University Press, Cambridge.

IPCC (Intergovernmental Panel on Climate Change). 1990b: Climate change, the IPCC impacts assessment, report from Working Group II, Australian Govenment Publishing Service, Canberra.

IPCC (Intergovernmental Panel on Climate Change), 1990c: Strategies for adaptation to sea level rise, report of the Coastal Zone Management Subgroup, The Hague.

IPCC (Intergovernmental Panel on Climate Change), 1992a: Climate change 1992. The supplementary report to the IPCC scientific assessment, Cambridge University Press, Cambridge.

IPCC (Intergovernmental Panel on Climate Change), 1992b: *Global climate change and the rising challenge of the sea. Supporting document for the IPCC update 1992,* WMO and UNEP, Geneva.

IPCC (Intergovernmental Panel on Climate Change), 1994: Preparing to meet the coastal challenges of the 21st century. Conference report of the World Coast Conference, Nordwijk, November 1993, WMO and UNEP, Geneva.

IPCC (Intergovernmental Panel on Climate Change), 1996a: *The science of climate change,* vol. 1 of *Climate change 1995: IPCC second assessment report,* Cambridge University Press, Cambridge.

IPCC (Intergovernmental Panel on Climate Change), 1996b: *Scientific-technical analyses of impacts, adaptations, and mitigation of climate change,* Vol. 2 of *Climate change 1995: IPCC second assessment report,* Cambridge University Press, Cambridge.

Jansen, H.M.A., 1993: Are we underestimating, when valuing the benefits of greenhouse gas reduction? In *Costs, impacts and benefits of CO$_2$ mitigation,* Y. Kaya, N. Nakicenovic, W.D. Nordhaus, and F. Toth, eds., IIASA Collaborative Paper Series, CP932. International Institute for Systems Analysis, Laxenburg, Austria.

Jodha, N.S., 1989: Potential strategies for adapting to greenhouse warming: Perspectives from the developing world. In *Policy options for adaptation to climate change,* N.J. Rosenberg, ed., Resources for the Future Discussion Paper ENR 89-05, RFF, Washingon, DC.

Kaiser, H.M., S.J. Riha, D.S. Wilks, D.G. Rossiter, and R. Sampath, 1992: A farm-level analysis of the economic and agronomic impacts of gradual climate warming, mimeo, Department of Agricultural Economics, Cornell University, Syracuse, NY (May).

Kalkstein, L.S., 1989: The impact of CO$_2$ and trace gas-induced climate changes upon human mortality. In *The potential effects of global climate change on the United States. Appendix G: Health,* J.B. Smith and D.A. Tirpak, eds., EPA, Washington DC.

Kandlikar, M., 1994: Catastrophic climate damage: Results from ICAM-2, mimeo, Carnegie Mellon University, Pittsburgh.

Kane, S., J. Reilly, and J. Tobey, 1992: An empirical study of the economic effects of climate change on world agriculture, *Climatic Change* **21,** 17–35.

Kimball, B., and N.J. Rosenberg, eds., 1990: *The impact of CO$_2$, trace gases, and climate change,* Publication No. 53, American Society of Agronomy, Madison, WI.

Kirshen, P.H., and N.M. Fennessey, 1992: Potential impacts of climate change upon the water supply of the Boston Metropolitan Area, mimeo, U.S. Environmental Protection Agency, Washington, DC (October).

Klinedinst, P.L., D.A. Wilhite, G.L. Hahn, and K.G. Hubbard, 1993: The potential effects of climate change on summer season Dair cattle milk production and reproduction, *Climatic Change,* **23**(1), 21–36.

Knox, L., C.A. and V.K. Smith, 1985: Microeconomic analysis. In *Climate impact assessment,* R.W Kates, J.H. Ausubel and M. Berberian, eds., John Wiley and Sons, Chichester.

Körner, C., and J.A. Arnone III, 1992: Responses to elevated carbon dioxide in artificial tropical ecosystems, *Science* **257** (18 September), 1672–1675.

Krause, F., W. Bach, and J. Kooney, 1989: *Energy policy in the greenhouse,* Earthscan, London.

Kunst, A.E., C.W.N. Looman, and J.P. Mackenbach, 1993: Outdoor air temperature and mortality in the Netherlands: A time-series analysis, *American Journal of Epidemiology* **137**(3), 331–341.

Langford, I.H., and G. Bentham, 1993: The potential effects of climate change on winter mortality in England and Wales, Centre for Social and Economic Research on the Global Environment (CSERGE), Working Paper GEC 93-25, University of East Anglia and University College London.

Leary, N., 1994: The amenity value of climate: A review of empirical evidence from migration, wages and rents, U.S. Environmental Protection Agency, discussion paper (July).

Leggett, J., 1993: *Climate change and the insurance industry. Solidarity among the risk community,* Greenpeace International, Amsterdam.

Lighthill, J., G.J. Holland, W.M. Gray, C. Landsea, G. Craig, J. Evans, Y. Kurihara, and C.P. Guard, 1994: Global change and tropical cyclones, *Bulletin of the American Meteorological Society* **75:** 2147–2157.

Limbert, D.W.S., 1993: The human and economic impacts of weather events in 1992, *WMO Bulletin* **42**(4), 333–341.

Lind, R.C. (ed.), 1982: *Discounting for time and risk in energy policy,* Resources for the Future, Washington, DC.

Loveland, J.E., and G.Z. Brown, 1992: Impacts of climate change on the energy performance of buildings in the United States, mimeo, U.S. Office of Technology Assessment, Washington, DC.

MacDonald, G.J., 1990: Role of methane clathrates in past and future climates, *Climatic Change* **16:** 247–281.

Maddison, D.J., 1994: The shadow price of greenhouse gases and aerosols, mimeo, Centre for Social and Economic Research on the Global Environment (CSERGE), University College London and University of East Anglia, Norwich.

Magadza, C.H.D., 1991: Some possible impacts of climate change on African ecosystems. In *Climate change. Science, impacts and policy. Proceedings of the Second World Climate Conference,* J. Jäger and H.L. Ferguson, eds, Cambridge University Press, Cambridge.

Magalhães, A.R., and M.H. Glantz (eds.), 1992: *Socioeconomic impacts of climate variations and policy responses in Brazil,* United Nations Environmental Programme, Nairobi, and Esquel Brazil Foundation, Brasilia.

Manabe, S., and R.J. Stouffer, 1993: Century-scale effects of increased atmospheric CO_2 on the ocean-atmosphere system, *Nature* **364**(15 July), 215–218.

Manne, A.S., R. Mendelsohn, and R.G. Richels, 1993: MERGE – A model for evaluating regional and global effects of greenhouse gas reduction policies, mimeo, Electric Power Research Institute, Palo Alto, CA.

Manne, A.S., and R.G. Richels, 1992: *Buying greenhouse insurance,* MIT Press, Cambridge MA.

Martens, W.J.M., J. Rotmans, and L.W. Niessen, 1994: *Climate change and malaria risk: An integrated modelling approach,* GLOBO Report No. 3, Dutch National Institute of Public Health and Enviromental Protection (RIVM), Bilthoven, The Netherlands.

Matsuoka, Y., K. Kai, and T. Morita, 1994: An estimation of climate change effects on malaria, AIM Interim Paper, National Institute for Environmental Studies, Tsukuba, Japan.

Maunder, W.J., 1994: *Tropical cyclones in the South Pacific: An historical overview regarding the intensity, tracks, and frequency of tropical cyclones in the South Pacific during the last 100 years, and an analysis of any changes in these factors,* Institute for Environmental Studies, Free University Amsterdam.

McLean, D.M., 1989: A mechanism for greenhouse-induced collapse of mammalian faunas. In *Coping with climate change,* J.C. Topping Jr., ed., pp. 263–267, Climate Institute, Washington, DC.

Mearns, L.O., R.W. Katz, and S.H. Schneider, 1984: Extreme high temperature events: Changes in their probabilities with changes in mean temperature, *Journal of Climate and Applied Meteorology* **23,** 1601–1613.

Mendelsohn, R., W. Nordhaus, and D. Shaw, 1993: The impact of climate on agriculture: A Ricardian approach. In *Costs, impacts and benefits of CO_2 mitigation,* Y. Kaya, N. Nakićenović, W.D. Nordhaus, and F. Toth, eds., IIASA Collaborative Paper Series, CP93-2, Laxenburg, Austria.

Midun, Z., and S. Lee, 1989: Effect of climate change and sea level rise on Bangladesh, report to the Commonwealth Group of Experts, Commonwealth Secretariat, London.

Milliman, J.D., J.M. Broadus, and F. Gable, 1989: Environmental and economic implications of rising sea level and subsiding deltas: The Nile and Bengal examples, *Ambio* **18**(6), 340–345.

Mitchell, R., and R. Carson, 1989: *Using surveys to value public goods: The contingent valuation method,* Resources for the Future, Washington, DC.

Morgenstern, R.D., 1991: Towards a comprehensive approach to global climate change mitigation, *American Economic Review, Papers and Proceedings* **81**(2), 140–145.

Munich Re, 1993: *Winter storms in Europe – Analysis of 1990 losses and future loss potential,* Munich Re, Munich.

Myers, N., 1993: Environmental refugees in a globally warmer world, *BioScience, 43* (11), 752–761.

Nash, L.L., and P.H. Gleick, 1993: The Colorado River Basin and climate change, report No 230R-93-009, U.S. Environmental Protection Agency, Washington, DC.

Nicholls, R., K. Dennis, C. Volonte, and S. Leatherman, 1992: *Methods and problems in assessing the impacts of accelerated sea level rise,* proceedings of the workshop "The World at Risk: Natural Hazards and Climate Change," American Institute for Physics, New York.

Nishioka, S., H. Harasawa, H. Hashimoto, T. Ookita, K. Masuda, and T. Morita, 1993: *The potential effects of climate change in Japan,* Centre for Global Environmental Research, and National Institute for Environmental Studies, Tsukuba (Japan).

Nordhaus, W.D., 1991: To slow or not to slow: The economics of the greenhouse effect, *Economic Journal* **101**(407), 920–937.

Nordhaus, W.D., 1993a: Optimal greenhouse gas reductions and tax policy in the "DICE" model, *American Economic Review, Papers and Proceedings* **83**(2), 313–317.

Nordhaus, W.D., 1993b: Rolling the "DICE": An optimal transition path for controlling greenhouse gases, *Resources and Energy Economics* **15**(1), 27–50.

Nordhaus, W.D., 1993c: Reflections on the economics of climate change, *Journal of Economic Perspectives, 7*(4), 11–25.

Nordhaus, W.D., 1994a: Expert opinion on climate change, *American Scientist* (January-February).

Nordhaus, W.D., 1994b: *Managing the global commons: The economics of climate change,* MIT Press, Cambridge, MA.

Nordhaus, W.D., and G.W. Yohe, 1983: Future carbon dioxide emissions from fossil fuels. In *Changing climate,* National Research Council, National Academy Press, Washington, DC.

Office of Management and Budget (OMB) and U.S. Department of Agriculture (USDA), 1989: Climate impact response functions, report of a workshop held at Coolfont, West Virginia, Sept. 11-14.

Olsthoorn, A.A., P.E. van der Werff, and J. de Boer, 1994: The natural disaster reduction community and climate change policy making, an inquiry among participants of the IDNDR World Conference in Yokohama, 1994, Institute for Environmental Studies W94/14, Vrije Universiteit, Amsterdam.

Ominde, S.H., and C. Juma, 1991: *A change in weather,* African Centre for Technology Studies, ACTS Press, Nairobi.

ORNL/RFF, 1994: Benefits from reducing risk of death fuel cycle externalities, paper No 10 in *Fuel cycle externalities: Analytical methods and issues,* Oak Ridge National Laboratory, Oak Ridge, TN, and Resources for the Future, Washington, DC.

Ottinger, R.L, D.R. Wooley, N.A. Robinson, D.R. Hodas, and S.E. Babb, 1990: *Environmental costs of electricity,* Pace University Center for Environmental and Legal Studies, Oceana Publications, New York.

Page, T., 1977: *Conservation and economic efficiency,* Johns Hopkins University Press, Baltimore.

Parikh, K., J. Parikh, T. Muralidharan, and N. Hadker, 1994: *Valuing air pollution in Bombay,* Indira Gandhi Insitute of Development Research, MEZP Project, The World Bank, Bombay and Washington DC.

Parry, M., 1990: *Climate change and world agriculture,* Earthscan, London.

Parry, M., 1993: Climate change and the future of agriculture, *International Journal of Environment and Pollution* **3**(1-3), 13–30.

Parry, M.L., M. Blantran de Rozari, A.L. Chong, and S. Panich, 1991: *The potential socio-economic effects of climate change in South-East Asia,* United Nations Environmental Programme, Nairobi.

Parry, M.L., T.R. Carter, and N.T. Konijn (eds.), 1988: *The impacts of climate variations on agriculture,* 2 Volumes, Kluwer, Dordrecht.

Pearce, D.W., 1992: The secondary benefits of greenhouse gas control, Global Environmental Change Working Paper GEC 92-12, Centre for Social and Economic Research on the Global Environment, University College London and University of East Anglia, Norwich.

Pearce, D.W., 1993: *Economic values and the natural world,* Earthscan, London.

Pearce, D.W., 1994: Costing the environmental damage from energy production, mimeo, Centre for Social and Economic Research on the Global Environment, University College London and University of East Anglia, Norwich.

Pearce, D.W., G. Atkinson, and R. Dubourg, 1994: What is sustainable development?, *Annual Review of Energy* **19**, 457–474.

Peck, S.C., and T.J. Teisberg, 1992: CETA: A model for carbon emissions trajectory assessment, *Energy Journal* **13**(1), 55–77.

Peck, S.C., and T.J. Teisberg, 1993a: CO_2 emissions control: Comparing policy instruments, *Energy Policy* **21**(3), 222–230.

Peck, S.C., and T.J. Teisberg, 1993b: Global warming uncertainties and the value of information: An analysis using CETA, *Resource and Energy Economics* **15**(1), 71–97.

Penner, J.E., R.E. Dickinson, and C.A. O'Neill, 1992: Effects of aerosol from biomass burning on the global radiation budget, *Science* **256**(5 June), 1432–1433.

Pernetta, J.C., 1989: Projected climate change and sea-level-rise: A relative impact rating for the countries of the Pacific Basin. In *Implications of expected climate changes in the South Pacific: An overview,* J.C. Pernetta and P.J. Hughes, eds., UNEP Regional Seas Reports and Studies 128, United Nations Environmental Programme, Nairobi.

Pernetta, J.C., 1992: Impacts of climate change and sea level rise on small island states, *Global Environmental Change,* **2**(1), 19–31.

Peters, R.I., and T.E. Lovejoy (eds.), 1992: *Global warming and biological diversity,* Yale University Press, New Haven.

Pezzey, J., 1994: The optimal sustainable depletion of non-renewable resources, mimeo, Department of Economics, University College London.

Price, C., and D. Rind, 1994: The impact of a $2 \times CO_2$ climate on lightning-caused fires. *Journal of Climate* **7**, 1484–1494.

Raper, S.C.B., 1993: Observational data on the relationships between climate change and the frequency and magnitude of severe tropical storms. In *Climate and sea level change,* W.W. Warrick, E.M. Barrow, and T.M.L. Wigley, eds., Cambridge University Press, Cambridge.

Reilly, J., J.A. Edmonds, R.H. Gardner, and A. L. Brenkert, 1987: Uncertainty analysis of the IEA/ORAU CO_2 emissions model, *Energy Journal,* **8**(3), 1–29.

Reilly, J., and N. Hohmann, 1993: Climate change and agriculture: The role of international trade, *American Economic Review, Papers and Proceedings* **83**(2), 306–312.

Reilly, J., N. Hohmann, and S. Kane, 1994: Climate change and agricultural trade: Who benefits, who loses?, *Global Environmental Change* **4**(1), 24–36.

Revelle, R.R., 1983: Probable future change in sea level resulting from increased atmospheric carbon dioxide. In *Changing climate,*

National Research Council, National Academy Press, Washington, DC.

Rijkswaterstaat, 1991: *Rising waters. Impacts of the greenhouse effect for the Netherlands,* Dutch Ministry of Transport and Public Works, The Hague.

Rijsberman, F., 1991: Potential costs of adapting to sea level rise in OECD countries. In *Responding to climate change: Selected economic issues,* OECD, Paris.

Rijsberman, F., and R.J. Swart (eds.), 1990: *Targets and indicators of climate change,* Stockholm Environment Institute, Stockholm.

Rind, D., D. Goldberg, J. Hansen, C. Rosenzweig, and R. Ruedy, 1990: Potential evapotranspiration and the likelihood of future drought, *Journal of Geophysical Research* **95**(D7), 9,983–10,004.

Ritchie, J.T., B.D. Baer, and T. Y. Chou, 1989: Effect of global climate change on agriculture: Great Lakes Region. In *The potential effects of global climate change on the United States,* U.S. Environmental Protection Agency, Washington, DC.

Roberts, L., 1991: Costs of a clean environment, *Science* **251** (8 March), 1182.

Rosebrock, J., 1993: Time-weighting emission reductions for global warming projects – A comparison of shadow price and emission discounting approaches, mimeo, The World Bank, Washington, DC.

Rosenberg, N.J., 1992: Adaptation of agriculture to climate change, *Climatic Change* **21**(4), 385–405.

Rosenberg, N.J., 1993: Towards an integrated impact assessment of climate change: The MINK study, *Climatic Change* **24**(1/2).

Rosenberg, N.J., and P.R. Crosson, 1991: The MINK project: A new methodology for identifying regional influences of, and responses to, increasing atmospheric CO_2 and climate change, *Environmental Conservation* **18**(4), 313–322.

Rosenberg, N.J., P. Crosson, W.E. Easterling III, K. Frederick, and R. Sedjo, 1989a: Policy options for adaptation to climate change, Resources for the Future Discussion Paper ENR 89-05, Resources for the Future, Washingon, DC.

Rosenberg, N.J., W.E. Easterling III, P.R. Crosson, and J. Darmstadter (eds.), 1989b: *Greenhouse warming: Abatement and adaptation,* Resources for the Future, Washington, DC.

Rosenthal, D.H., H.K. Gruenspecht, and E.A. Moran, 1994: Effects of global warming on energy use for space heating and cooling in the United States, mimeo, U.S. Department of Energy, Washington, DC (March).

Rosenzweig, C., and D. Hillel, 1993: Agriculture in a greenhouse world, *National Geographic Research & Exploration* **9**(2), 208–221.

Rosenzweig, C., and M.L. Parry, 1994: Potential impact of climate change on world food supply, *Nature* **367**(13 January), 133–138.

Rosenzweig, C., M. Parry, K. Frohberg, and G. Fisher, 1993: *Climate change and world food supply,* Environmental Change Unit, Oxford.

Rothenberg, J., 1993: Economic perspectives on time comparisons. In *Global accord: Environmental challenges and international responses,* C. Nazli, ed., MIT Press, Cambridge, MA.

Ryan, B.F., I.G. Watterson, and J.L. Evans, 1992: Tropical cyclone frequencies inferred from Gray's Yearly Genesis Parameter: Validation of GCM tropical climates, *Geophysical Research Letters,* **19** (18), 1831–1834.

SAARC, 1992: Regional study on greenhouse effect and its impact on the region, report, South Asian Association for Regional Cooperation, Kathmandu.

Schelling, T.C., 1983: Climatic change: Implications for welfare and policy. In *Changing climate,* National Research Council, National Academy Press, Washington, DC.

Schelling, T.C., 1992: Some economics of global warming, *American Economic Review* **82**(1), 1–14.

Schelling, T.C., 1993: Intergenerational discounting, mimeo, University of Maryland, College Park (November).

Scheraga, J.D., and N.A. Leary, 1994: Costs and side benefits of using energy taxes to mitigate global climate change. In *Proceedings of the 86th Annual Conference,* National Tax Association, Washington, DC.

Scheraga, J.D., N.A. Leary, R.J. Goettle, D.W. Jorgenson, and P.J. Wilcoxen, 1993: Macroeconomic modelling and the assessment of climate change impacts. In *Costs, impacts and benefits of CO_2 mitigation,* Y. Kaya, N. Nakićenović, W.D. Nordhaus, and F. Toth, eds., IIASA Collaborative Paper Series, CP93-2, Laxenburg, Austria.

Schlesinger, M.E., and X. Jiang, 1990: Simple model representation of atmosphere–ocean GCMs and estimation of the time scale of CO_2-induced climate change, *Journal of Climate* **3**(December), 12.

Schneider, S.H., and R. S. Chen, 1980: Carbon dioxide warming and coastline flooding: Physical factors and climatic impact, *American Review of Energy* **5**, 107–40.

Schneider, S.H., and N.J. Rosenberg, 1989: The greenhouse effect: Its causes, possible impacts, and associated uncertainties. In *Greenhouse warming: Abatement and adaptation,* Resources for the Future, Washington, DC.

Scott, M.F., 1989: *A new view of economic growth,* Clarendon Press, Oxford.

Sedjo, R.A., and A.M. Solomon, 1989: Climate and forests. In *Greenhouse warming: Abatement and adaptation,* N.J. Rosenberg, W.E. Easterling III, P.R. Crosson, and J. Darmstadter, eds., Resources for the Future, Washington, DC.

Shogren, J.F., S.Y. Shin, D.J. Hayes, and J.B. Kliebenstein, 1994: Resolving differences in willingness to pay and willingness to accept, *American Economic Review* **84**(1), 255–270.

Smith, J.B., S. Ragland, and E. Trabka, 1993: Standardized estimates of climate change damages for the United States, mimeo, RCG/Hagler Bailly, Boulder, CO (December).

Smith, J.B., F.D. Stern, C.A. Yermoli, and W.S. Breffle, 1994: The impact of climate change on thermoelectric power use in Egypt and Poland, mimeo, RCG/Hagler Bailly, Boulder, CO (February).

Smith, K., 1992: *Environmental hazards. Assessing risk and reducing disaster,* Routledge, London.

Spash, C., 1994: Double CO_2 and beyond: Benefits, costs and compensation, *Ecological Economics,* **10,** 27–36.

Sundquist, E.T., 1990: Long-term aspects of future atmospheric CO_2 and sea-level changes. In *Sea-level change,* R.R. Revelle *et al.,* National Research Council, National Academy Press, Washington, DC.

Titus, J.G., 1987: The causes and effects of sea level rise. In *The impact of sea level rise on society,* H.G. Wind, ed., A.A. Balkema, Rotterdam.

Titus, J.G., 1991: Greenhouse effect and coastal wetland policy: How Americans could abandon an area the size of Massachusetts, *Environmental Management Journal* **15**(1), 39–58.

Titus, J.G., 1992: The cost of climate change to the United States. In *Global climate change: Implications challenges and mitigation measures,* S.K. Majumdar, L.S. Kalkstein, B. Yarnal, E.W. Miller, and L.M. Rosenfeld, eds., Pennsylvania Academy of Science, Easton, PA.

Titus, J.G., R.A. Park, S.P. Leatherman, J.R. Weggel, M.S. Greene, P.W. Mausel, S. Brown, C. Gaunt, M. Trehan, and G. Yohe, 1991:

Greenhouse effect and sea level rise: The cost of holding back the sea, *Coastal Zone Management* **19,** 172–204.

Tol, R.S.J., 1994: The damage costs of climate change: A note on tangibles and intangibles applied to DICE, *Energy Policy* **22**(5), 436–438.

Tol, R.S.J., 1995: The damage costs of climate change: Towards more comprehensive calculations, *Environmental and Resource Economics,* **5,** 353–374.

Tol, R.S.J., C. Dorland, E. van der Hul, F.P.M. Leek, A.A. Olsthoorn, P. Velinga, and P.E. van der Werff, 1994: Climate extremes, risk and risk management. In *Climate change and extreme weather events: Scenarios of altered hazards for further research,* T.E. Downing, D.T. Favis-Mortlock, and M.J. Gawith, eds., Environmental Change Unit Research Report 7, University of Oxford, Oxford.

Turner, R.K., S. Subak, N. Adger, and J. Parfitt, 1994: How are socio-economic systems affected by climate-related changes in the coastal zone?, report to IPCC Working Group II, Subgroup B, Centre for Social and Economic Research on the Global Environment (CSERGE), University of East Anglia and University College London.

UK Climate Impact Review Group, 1991: The potential effects of climate change in the United Kingdom, report prepared at the request of the Department of the Environment, HMSO, London.

U.S. Environmental Protection Agency, 1989: *The potential effects of global climate change on the United States,* J.B. Smith and D.Tirpak, eds., EPA, Washington, DC.

U.S. Environmental Protection Agency, 1990: *Environmental investments: The cost of a clean environment,* EPA, Washington DC.

U.S. FEMA-FIA, 1991: *Projected impact of relative sea level rise on the national flood insurance program,* Federal Emergency Management Agency and Federal Insurance Administration, Washington DC.

Viscusi, W.K., 1993: The value of risks to life and health, *Journal of Economic Literature,* **31** (December), 1912–1946.

Waggoner, P.E., 1990: *Climate change and U.S. water resources,* Wiley, New York.

Waggoner, P.E., 1994: How much land can ten billion people spare for nature?, Report No. 121, Council for Agricultural Science and Technology, Ames, IA.

Waggoner, P.E. *et al.,* 1992: Preparing U.S. agriculture for global climate change, report No. 119, Council for Agricultural Science and Technology, Ames, IA.

Waters, S., 1990: *The Travel Industry World Yearbook: The big picture,* World Tourism Organization, Madrid.

Wendland, W.M., 1977: Tropical storm frequencies related to sea surface temperatures, *Journal of Applied Meteorology* **16**(5), 477– 481.

Wigley, T.M.L., 1995: Global mean temperature and sea level consequences of greenhouse gas concentration stabilization, *Geophysical Research Letters* **22**(1): 45–48.

Wigley, T.M.L., and S.C.B. Raper, 1992: Implications for climate and sea level of revised IPCC emissions scenarios, *Nature* **357,** 293–300.

World Health Organisation, 1990: *Potential health effects of climatic change,* WHO, Geneva.

Yohe, G.W., 1993: Sorting out facts and uncertainties in economic response to the physical effects of global climate change. In *Assessing surprises and non-linearities in greenhouse warming,* J. Darmstaedter and M. Toman, eds., Resources for the Future, Washington, DC.

7

A Generic Assessment of Response Options

C. J. JEPMA, M. ASADUZZAMAN, I. MINTZER, R.S. MAYA,
M. AL-MONEEF

Contributors:
*J. Byrne, H. Geller, C.A. Hendriks, M. Jefferson, G. Leach, A. Qureshi, W. Sassin,
R. A. Sedjo, A. van der Veen*

CONTENTS

SUMMARY

In this chapter, current response options for dealing with climate change are assessed on the basis of their feasibility, acceptability, cost-effectiveness, and applicability. As much as possible, specific attention has been given to the applicability of these various options in the developing countries and countries in transition. The chapter does not, however, contain an evaluation of the (macro)economic effects that large-scale applications of the various options might have in different regions of the world.

Conceptually a distinction must be made between mitigation and adaptation options on the one hand, and indirect options – that is, options not designed to have an impact on the greenhouse effect but that nevertheless do – on the other. Indeed, many technological developments and various policies have an impact on energy use and thus on the global climate. An effective climate change response strategy should therefore preferably pay attention to possibilities of joining climate response options with responses to other socioeconomic transition phenomena, as in the application of an integrated systems approach.

The various response options can be assessed in fundamentally different ways. At one extreme is the engineering efficiency approach, which focusses only on costs and how these are related to internal and external economies of scale and learning effects. At the other extreme is the welfare economic approach, which, in addition, considers such welfare aspects as social, political, or environmental resistance to the option's application. Costs associated with the diffusion of technologies, public education, and lifestyle changes are also taken into account.

A number of CO_2 mitigation options have been proposed, including

- Energy conservation and efficiency improvement
- Fossil fuel switching
- Renewable energy technologies
- Nuclear energy
- Capture and disposal technologies
- Enhancing sinks and forestry options

Attention has also been focussed on reducing emissions of methane.

With respect to *energy conservation and efficiency improvement,* reductions in energy intensities during recent decades have varied widely across countries and also within the group of developing countries. Some of this variation, however, reflects differences in how the underlying variables have been measured.

Because reductions in national energy intensities are related to structural changes in national economies, the growth of the secondary sectors in developing countries may give a biassed view of their energy efficiency improvement results. In most industrial countries, in contrast, a trend towards "dematerialization" (i.e., a shift away from the highly energy-intensive secondary towards the less energy-intensive tertiary sector) has favoured lower energy intensities.

There is a broad consensus in the literature in favour of efficiency improvement, because it is seen as directly beneficial irrespective of any impact on greenhouse warming and because it has significant scope for negative net cost (i.e., no-regret) applications. The potential for energy efficiency improvements in production seems promising, especially in the power production, transportation, steel and cement production, and residential sectors. However, because the end use phase is the least efficient part of an energy system, improvements in this area would produce the greatest benefits. The potential for efficiency improvements in the developing countries is roughly similar in magnitude to that in industrialized countries. By contrast, energy conservation may be achieved somewhat more easily in the industrialized countries.

Optimism about the scope for no-regret options with respect to energy efficiency varies considerably and depends to a large extent on the discount rate that is employed. Revealed consumer discount rates for household investments can be very high indeed. Similarly, in developing countries a lack of access to information and limitations of institutional capacity, human skills, and financial resources may cause the revealed time preference to be much higher than commercial interest rates.

The potential for energy savings is estimated at 10-40% for production and 10-50% for residential use. However, to achieve such results, institutional and information factors are crucial. So too is the degree to which the option may help in deriving other environmental benefits.

With respect to *fossil fuel switching,* relatively little information about costs is available, although it is recognized that fossil fuels will remain the dominant energy source for several decades yet. Estimates of the costs of switching vary to a large extent, depending on the type of measure, the fraction of natural gas lost to the atmosphere from leakage during production and distribution, and the opportunity costs of the option (which depend to a large extent on the availability of, for instance, coal reserves).

These opportunity costs may be particularly large in populous countries with massive coal reserves, such as China and India. In fact, in developing countries growth may even result in a transition from less carbon-intensive biomass to more carbon-intensive fossil fuels.

Renewable energy technologies may be sustainable with respect to energy inputs but may not always be socially and environmentally benign in other respects. This is particularly so in the case of large-scale applications (for example, of major hydro or biomass projects) in developing countries.

The technical potential of the renewable options not currently utilized varies from 50% for biomass to 75% for hydro to several thousand per cent for wind. Many renewable technologies, however, tend to be site-specific (i.e., their application is limited to a finite number of specific sites). Other problems include potential environmental risks, technological readiness, and cost-effectiveness.

Though some renewable options are almost mature, others are still in the demonstration stage. Practicable potentials therefore vary to a large extent, although much will depend on the costs of the various options.

Cost estimates diverge widely, mainly due to the time horizon adopted, the discount rate chosen, and the capacity and useful lifetime assumed. Moreover, costs are strongly influenced by site-specificity, variability of supply, and the form of final energy delivered. Other aspects that influence cost behaviours are learning effects, economies of scale, and the need for immediate storage or transport of the energy generated.

The promise of renewables lies mainly in their large potential and modest price on the spot. These factors are particularly relevant for developing countries, which, by using local renewables, could reduce their dependence on imported fossil fuels. Local communities could benefit significantly from small-scale applications and their net positive side effects.

In view of these considerations, the future role of renewables is hard to predict precisely; the share of renewables in the 2020 energy mix will, however, probably not exceed 25%.

Nuclear energy technology is long past the demonstration stage, but the issue of the safe storage of nuclear waste remains unresolved. Because of their long design and construction time (10-15 years) and the enormous investment costs of nuclear power plants, the nuclear option is also rather inflexible.

In view of the waste disposal problem and the consequent lack of public support, the share of nuclear energy in total energy use is expected to increase only to a limited extent during the coming decades.

Capture and disposal have potential in cases where a switch from coal to other fossil fuels is difficult for one reason or another. Some technologies already exist; others are being developed.

The disposal option is ultimately limited not only for technical reasons but also because disposal cannot permanently prevent the reentry of carbon into the atmosphere. This is irrespective of the way in which disposal would take place. The practicability of this option is still a matter of discussion, because in some types of disposal (e.g., in aquifers or oceans) environmental impacts are uncertain.

The scope of *forestry options* is determined by the large expected potential, modest costs, low risk, and positive side effects. However, there is still a large amount of uncertainty with respect to the net carbon release from deforestation and land use changes on the one hand and the long-term carbon absorption capacity of afforestation efforts on the other. Basically, forestry measures, like removal options, are to be seen as an intermediate response policy.

Uncertainties in assessments of the global potential for halting or slowing deforestation and for reforestation are linked to the extent of human encroachment into the forests, the area available for forestry measures, and the annual and cumulative carbon uptake per hectare.

Mitigation policies using forests are generally considered relatively cost-effective, especially if applied in developing countries. With the costs of afforestation, much depends on whether one assumes that the forests can be exploited sustainably or, instead, should be left alone to mature, and on the acceptance of the newly planted forests by the local population.

Halting or slowing deforestation is probably one of the most urgent and cost-effective options. However, social, political, and infrastructural barriers may restrict this option as well as the scope of reforestation.

Estimates of cost-effectiveness of forestry measures depend strongly on whether one takes a static or dynamic point of view. There is a clear tendency to focus increasingly on cost functions rather than point estimates; the former approach seems clearly more relevant in the case of large areas. Moreover, the cost assessment methodology has been increasingly refined (for example, by the inclusion of discounting procedures). Cost estimates, which are now probably more realistic, tend to fall within a range of \$30-\$60/tC for large annual uptakes.

With respect to *methane,* the emission data available reveal wide discrepancies between various regions. Information about methane leakage and distribution is also rather scanty, and some of it is unreliable. The same applies to information about the costs of methane control options.

Information about the cost functions of the various mitigation options is still weak, because the functions are not only time-specific but also region- and context-specific. The weakness of information also relates to the remarkable fact that the scope for no-regret options seems to be significant, especially in developing countries. This apparent scope is most likely due to the high actual time preference rates, lack of information, and limitations of human capacity. All this and the different assessment perspectives mentioned earlier may explain why virtually no studies exist in which the optimal mix of options is designed on the basis of their underlying cost functions and feasibility.

The few studies of this kind that have been done provide only tentative results but do indicate – given present knowledge about the cost functions of the various options – that the pure application of the cost minimization principle would require a significant share (probably more than half) of the emission reduction targets to be achieved via the application of options outside the OECD area. In addition, in terms of the size of the emission reduction, energy conservation and efficiency improvements and the forestry option seem to provide the largest potential from a cost minimization point of view. The potential of the forestry option is widely debated, however, because of the limitations of net absorption in time and because much depends on forest exploitation and local acceptance.

To illustrate how an optimal mix of response options might look, the result of a (linear programming-based) cost minimization simulation using the available cost-function information disaggregated by region is presented in Table 7.13 for a predetermined emission reduction target of 2.4 GtC. In view of the tentative and uncertain character of the underlying data, the outcomes can only be seen as an illustration of what an optimal policy mix might be (recognizing that marginal costs per option per region generally tend to increase to the point where they eventually become prohibitive). Obviously technological or political breakthroughs may significantly affect the optimal mix.

Adaptation options can be surveyed in many ways. One is to consider what should be adapted to and how it should be done. No systematic cost data on the various adaptation options are available, although information about land protection costs against flooding and sea level rise is rapidly increasing. Many efforts are now underway, however, to reduce the vulnerability of agricultural production to climate change through adaptation policies. Especially in developing countries there is an urgent need for both more information and a better infrastructure for the actual implementation of adaptation techniques.

Finally, the point has to be made that when it comes to the introduction and application of the various options, the developing countries occupy a special position. The application and acceptance of these options often crucially depends on the international transfer of technologies as well as the countries' own local institutions and abilities to build their human capacity. Therefore, the conditions needed to ensure the success of these processes, such as joint implementation and technology transfers from developed to developing countries, deserve a high priority on the academic research agenda.

7.1. Introduction

In recent years a host of response options has been proposed to cope with possible climate change. These options can be classified in many ways, including by technology, by sector, by impact, and by strategic approach. This chapter is based on classification by strategic approach, that is, *mitigation, adaptation,* and *indirect policy options.* Many response options are thoroughly discussed in Volume 2 of this report, with a major emphasis on technological feasibility. Some aspects of these options will be taken up here and assessed generically, that is, not only from an engineering efficiency point of view but also from that of welfare economics.[1]

The present chapter surveys the set of options that are feasible from a comparative economic perspective in order to assess the scope and priorities of potential policies. The main purposes are

- To set up a structure so the various options can be put into proper perspective and the assessment to be made can be truly generic (Section 7.2)

- To discuss the various criteria that can be used in assessing the options and the degree to which different criteria can produce different choices in terms of optimal use of the options (Section 7.3)

- To review the various options in terms of (technical and practical) applicability, cost-effectiveness, and social acceptability, both as far as mitigation options (Section 7.4), and adaptation options (Section 7.5) are concerned; special attention will be given to the case of the developing countries and countries in transition, because of their particular circumstances

- To evaluate the scope for integrating response options, in particular, with respect to mitigation options on the basis of information about regional cost functions (Section 7.6)

- To analyze to what extent currently available information about various options might provide a basis for international policy cooperation (Section 7.7)

Sections 7.1 to 7.3 therefore provide the methodological base; Sections 7.4 and 7.5 survey the mitigation and adaptation options, and Sections 7.6 and 7.7 deal with response options and policy application. In this chapter the applicability, feasibility, and cost-effectiveness of the various response options are surveyed; however, a macroeconomic effects assessment of the various options has not been carried out here. (See in this respect also the sections in this report dealing with integrated response options.)

7.2. A Conceptual Framework

Figure 7.1 shows the policy options available to counter greenhouse warming and their possible feedbacks. The diagram may serve to illustrate that one can basically distinguish between three strategic categories of options to deal with the greenhouse issue:

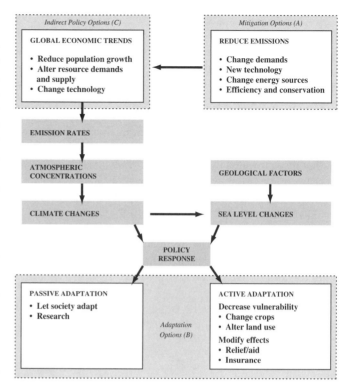

Source: After Viner and Hulme (1994).

Figure 7.1: Schematic overview of available options to counter the greenhouse effect and their possible feedbacks.

(1)   *Mitigation options* (Block A in the figure) are options that, amongst others, strive to prevent climate change, or combat any reinforcement thereof, by reducing the net emissions of greenhouse gases into the atmosphere, either by reducing greenhouse gas emissions (source-oriented measures) or by increasing the sinks for greenhouse gases (effect-oriented measures). See also Chapter 8, Section 8.2.2.2.

(2)   *Adaptation options* (Block B) are options that focus on reducing the expected damages due to rapid climate change by combatting or avoiding their detrimental effects.

(3)   *Indirect policy options* (Block C) are options that are not directly related to the emission or capture of greenhouse gases but that can have a considerable indirect effect on greenhouse gas emissions or greenhouse gas uptake.

Obviously, the various types of options are not mutually exclusive, nor can they be fully separated. Indirect policy options, adaptation options, and mitigation options may even reinforce each other. For example, a population policy, as part of a broader policy mix that slows down population growth in a densely populated country, may contribute to finding cost-effective and acceptable opportunities for mitigation options. Similarly, if policies designed to decrease the intensity of energy and materials use of economic activity are instituted in a country, many technically feasible options for emission reductions may become cost-effective. Technological progress will obviously improve the scope for adaptation and other options. For conceptual reasons, however, the preceding distinction between the various types of options seems a useful starting point. Before moving on to the details, though, it would be only proper to point out what this chapter is not about. Only the broad principles underlying the response options are emphasized here. Their actual application would depend on a host of factors that are very much country-specific and include many economic, social, political, and legal considerations. Thus, they would need to be analyzed on a country-by-country basis for policymaking at national levels.

7.2.1 Mitigation options

In the literature about greenhouse policy options, mitigation options receive by far the most attention. Most commonly the various options are discussed separately and from the engineering perspective. Information about the cost-effectiveness of the various options, for example, in terms of $/tC not released into the atmosphere, is rapidly increasing. The marginal cost-effectiveness of the various options is probably highly dependent on the scale of application, the sector, the country or region of application, and whether or not additional options are applied. Moreover, learning curves, and therefore cumulative application and time, almost invariably play a dominant role in determining the options' economic viability. All these factors point in the same direction, namely, that the mitigation options' cost functions may change in the course of time, sometimes quite rapidly. The same applies with respect to the various options' social and political acceptability. Conclusions about the economic, social, and political viability of various options are therefore highly scale-, time- and location-specific.

In discussing the potential of the various mitigation options a distinction has been made between measures concerning CO_2 and measures concerning other greenhouse gases, because the former are in actual practice largely associated with energy-related activities (i.e., both energy production and consumption) whereas the latter are also associated with other types of activities. Thus, except for some "exotic," mainly effect-oriented options such as geoengineering, orbital shades, iron fertilization, creating algal blooms, and weathering rocks, mitigation options can generally be divided into those affecting CO_2 and those affecting other greenhouse gases.

Measures concerning CO_2 include the following:

(a) Source-oriented measures
    (1) energy conservation and efficiency improvement
    (2) fossil fuel switching
    (3) renewable energy
    (4) nuclear energy
(b) Sink-enhancement measures
    (5) capture and disposal of CO_2
    (6) enhancing forest sinks

Measures concerning other greenhouse gases include phasing out HFCs (in addition to HCFCs, via the Montreal Proto-

col) as well as a variety of measures for reducing emissions of methane (CH_4), nitrous oxide (N_2O), and other greenhouse gases.

Since the energy sector (in terms of both energy production and consumption) is the single largest source of carbon, much of the CO_2 mitigation effort can be concentrated here. Each of the four source-oriented options addresses elements of the energy conversion process, from primary energy production to end-use services.

Both energy conservation and energy efficiency aim to reduce total energy use without changing the current fuel mix or the fundamental structure of the energy conversion process. Energy conservation is used here to mean a reduction in energy needs resulting from a change in the nature or level of energy services (e.g., lighting areas only when they are occupied rather than during specified periods). Energy efficiency means providing the same type and level of energy service with less total energy (e.g., using more efficient lamps to provide the desired lighting level). Since energy conservation is strongly linked to the preferences and behaviour of various economic agents (such as households, firms, and governments), policies aimed at achieving it are more likely to lead to ambiguous conclusions. Consequently, most studies focus on energy efficiency.[2]

A fossil fuel switch alters the mix of fossil fuels in favour of the less carbon-intensive ones such as natural gas (and perhaps oil) and away from coal. Nuclear energy substitutes for fossil fuels as primary energy. Renewable energy is characterized by an extensive natural supply, which is vast compared to current levels of commercial energy use, and by a large long-term potential because of its regeneration capability. Mobilization of this natural supply can in some cases result in severe environmental and societal impacts.

Removal technologies (option 5) extract carbon in one form or another from an energy conversion process even before it has entered the atmosphere. Subsequently, the carbon has to be utilized, stored, or disposed of. Option 6 is in essence outside the energy area. It aims at binding carbon after it is combusted and dispersed throughout the atmosphere by combatting deforestation or by afforestation.[3] It may also refer to activities designed to preserve or enhance carbon uptake by soils.

7.2.2 Adaptation options

Adaptation options have two purposes:

(1) To reduce the damages from climate change
(2) To increase the resilience of societies and ecosystems to the aspects of climate change that cannot be avoided

Clearly, adaptation measures are interlinked with mitigation measures. The more one succeeds in limiting climate change, the easier it will be to adapt to it. This is notwithstanding the fact that there can be reasons for supporting adaptation measures in their own right. Three types of adaptation measures are commonly distinguished: protection, retreat, and accommodation.

As far as the costs of adaptation options are concerned, one can either focus on the opportunity cost, in other words, assess the welfare implications of no-action scenarios, or on the net investment cost involved with adaptation measures. Since Chapter 6 of this report focusses on the former, Section 7.5 of this chapter will mainly consider the latter.

7.2.3 Indirect policy options

Potential climate change is perceived as a problem, mainly because it would interfere with the world's economic, social, and ecological systems, and eventually with its political system. Just as the precise scope and risks of climate change are subject to uncertainty, so is the future development of technology, resources, and the organization and structure of the economic, social, and political systems. However, it seems most likely that the changes in the global climate and the structural changes in the economic and political system differ significantly in at least one respect: the speed or time lag of changes to be expected. Whereas possible severe global climate change generally is expected to take approximately 50 to 100 years (although exceptions can be possible), the economic, social, and political systems may change several times within a similar period.

This difference poses a fundamental dilemma when assessing the various response options to climate change: The changing climate system has to be superimposed on economic, social, and political systems that are in constant flux due to numerous factors, with (potential) climate change being only one of them. This dilemma significantly complicates the assessment process, and even more the process of formulating policy options based thereon. However, recent history has taught that if there is a strong political consensus about the need to take action, such actions can be undertaken vigorously, as in the case of the Montreal Protocol (see Benedick, 1991) and the Convention on International Trade in Endangered Species.

Indeed, climate and ecological change are by no means the only factors that will enforce a deep modification of the present economic situation and that may pose serious problems to society. Other evolutionary trends and structural adjustment processes – driven by such forces as population growth, urbanization, information technologies and their dissemination, the international mobility of labour and capital, the competition for natural resources, and the pattern and speed of technological progress (e.g., in waste management and in redesigning products) – may also be expected to play an important role in shaping the economic, political, and social systems of tomorrow, especially if the policymakers' time horizon is at most a few decades if not shorter. To illustrate, Western nations may well face a combination of problems, such as urban decay, unemployment, massive migration, and changing patterns of economic competitiveness that may easily attract more public and political attention than the climate change issue.

All these problems already call for response options, for instance, in the sphere of consumption and lifestyle policies,

population and migration policies, technology and environmental policies, structural and sectoral adjustments or trade policies, or redistribution policies. Virtually all these policies will also, albeit indirectly, greatly affect energy use and thus the global climate.

An effective climate change response strategy should therefore pay attention to the possibilities of joining climate change response options with responses to other socioeconomic transition phenomena, and thus increase the probabilities of actual implementation.

Examples of this approach can be found in applications of the integrated systems approach. For instance, in many developing countries crop agriculture is at present highly dependent on energy use, both directly and indirectly, and farmers have to depend on outside sources for much of their energy supply. In addition, many of these agricultural systems are based on monocultures (e.g., high-yielding varieties of wheat and rice, which increase soil exhaustion and are more vulnerable to massive infestations of pests and disease). Alternatives like low-external-input sustainable agriculture reportedly lower the need for external and energy-intensive inputs and increase productivity in farming in an ecologically robust way while at the same time reducing concerns for national food security (Reijntjes *et al.,* 1992).

Yet another example of a "multifunctional system" is wave energy. In that case the production of energy is combined with other functions, such as coastal protection or water desalination. However, this technology may also have adverse environmental side effects. All these systems can be particularly promising if applied on a relatively small scale in developing countries.

7.3. Criteria for Assessment

In discussing the assessment of response options, the application possibilities of the options themselves are evaluated rather than the policies that may be expected to cause the various options to be applied or withdrawn. Insofar as the assessment of these policies is concerned, the reader is referred to Chapter 8.

From a methodological point of view one can distinguish between two fundamentally different approaches for assessing response options. These approaches, however, should not be confused with the distinction – which has drawn a lot of attention in the literature – between top-down and bottom-up modelling (see also Chapter 8). On the one hand, the financial costs of the various technologies can be expressed in terms of CO_2 emission reduction/absorption. This could be called the "engineering efficiency" approach. On the other hand, an assessment of the various options could be made in the tradition of welfare economics. According to this line of thinking, determining the costs and benefits of the application of any particular technology should include an assessment of the opportunities forgone by the allocation of the resources. This could be called the "welfare economic" approach. Other categorizations of the assessment approach are also conceivable. In Chapter 8, for instance, the assessment is differentiated according to the level of aggregation (e.g., the aggregate national level or the level of a single project). However, such a differentiation was not considered crucial for the purpose of the present chapter, which is to provide a generic assessment of the various response options.

An afforestation programme may serve to illustrate the differences between these approaches. What investment has to be made to achieve a predetermined target in terms of net CO_2 absorption during some time interval? Using the engineering efficiency approach, one would try to determine the discounted value of the costs of land acquisition, tree planting, maintenance, security, and other needs. Any future (sustainable) harvesting returns would equally be discounted, so that the net levelized costs could be determined in dollars. On the basis of this information, and by comparing this option with other options' cost-efficiencies, one could then decide whether or not to proceed.

However, if the welfare economic approach is taken, the overall assessment may be quite different. By using the land for afforestation purposes, the possibility of using the same land for agricultural purposes is forgone. It therefore matters a great deal if the area has agricultural potential or not. If so, the local population may well be forced to migrate or else to suffer income losses. Moreover, the afforestation programme, if applied on a large scale, may have additional impacts, either positive or negative (e.g., through its effect on local climate and soil fertility, social and cultural life, on infrastructure, tourism, etc.). Ensuring that such side effects are beneficial depends on the establishment of effective monitoring and extension services at the local level. In the assessment, attention can also be paid to the distorting impact of government measures, such as subsidies and taxes, on the efficiency of the forestry option. If all the direct and indirect welfare consequences of the envisaged afforestation programme are going to be assessed, an extensive and complicated social cost-benefit type of analysis may well be called for, because not all aspects can be quantified or monetized (see also Chapter 5).

A priori, there is no reason why the outcomes of the engineering efficiency and welfare economic assessments of the same project would coincide. The costs of the land in monetary terms may not fully reflect the land-use opportunity costs in welfare terms, because in the former no full account is taken of indirect effects, nonmaterial consequences, distributional impacts, and externalities.

In short, the major distinction between the cost assessment methodology in both approaches is that the engineering efficiency approach basically starts from the evaluation of a project from the narrow perspective determined by the project boundaries, whereas the welfare economic approach attempts to account fully for the various interests and impacts inside and outside the societies concerned, including the external effects and the social and political acceptability of the options. A welfare economic approach would therefore imply an assessment based on a general equilibrium model, an exercise conspicuous by its almost total absence in the literature. In this chapter, therefore, response options are evaluated on the basis of important opportunity costs and externalities.

In actual practice, even public agents may not be fully aware of the various externalities and indirect, nonmaterial,

and distributional impacts of the application of response options. For one reason (e.g., pressure from special interest groups), they may not want to take these various aspects into account. For another, the information available for a full welfare assessment may simply be insufficient. What is more, even if all information for assessing the various options is available, obstacles in setting up the institutional machinery can impose serious bottlenecks, so that appropriate action will not follow.

As preceding chapters have already noted extensively, a welfare economic assessment of climate change response options faces some large practical obstacles, particularly in the developing countries. First, the policy priorities, especially with respect to the greenhouse issue, will often differ from those in industrialized countries. Second, information about externalities at the local level may not fully reach the public sector because of limitations in data collection, processing, and communication; on the other hand, policies dealing with externalities may fail to reach part of the local population. Third, most developing countries face a severe lack of institutional and human capacity to deal with these issues.

The general impression also arises that optimism about the potential of technology is larger in the engineering efficiency approach than in the welfare economic approach; in the latter the emphasis is more on the obstacles in society to absorbing and applying new technologies. This distinction can be related to various aspects of the economy-of-scale concept, notably:

(a) Average costs may decrease at a larger scale of application (internal economies of scale).

(b) Costs of a given option may decrease when other options are applied on a larger scale because of positive external effects (external economies of scale).

(c) Costs may decrease as the application time progresses (learning effects).

(d) Costs may increase at a larger scale of application due to increasing resistance and bottlenecks related to social, political, and environmental concerns and to increasing opportunity costs; afforestation projects often provide a clear example.

(e) Costs may increase because achieving the required rate of diffusion of technologies, public education, and lifestyle changes may become increasingly difficult on a larger scale; this problem may be particularly relevant if response technologies require a high level of technical expertise.

If one focusses mainly on items (a) to (c), optimism about the options' economic potential may rise. This is the perspective taken by the engineering efficiency approach. If, however, one focusses instead on items (d) to (e), one might easily take a much more pessimistic view, associated with the welfare economic perspective.

A separate issue in comparing the feasibility of these options is that the various studies differ in the extent to which they take the energy costs and benefits of the options into account. The application of some options, such as capture and disposal, requires significant energy inputs, which are often denoted as energy penalties; other options, such as nuclear or

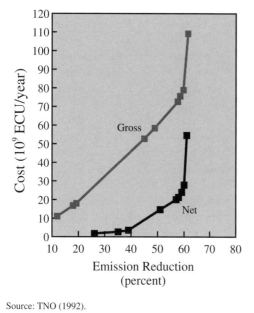

Source: TNO (1992).

Figure 7.2: Options for CO_2 emission reduction in the EU, net and gross costs, and effectiveness.

renewable energy, besides achieving a carbon emission reduction, also produce energy and are therefore substituting for traditional fossil energy resources. This consideration implies that one could distinguish between gross and net energy costs, the latter being gross costs minus the benefits of avoided fossil energy production. In TNO (1992) both cost functions have been derived for the EU (see Figure 7.2). Differences between gross and net costs turned out to be notably relevant for the options of energy saving, renewables, nuclear energy, and energy farming.

A comparable issue is how costs have to be ascribed to the various reductions that are achieved with the help of the investment made. More often than not, investments made for economic and/or environmental reasons have changes in greenhouse gas emission as a side effect. Many no-regrets options belong to this category. The question then becomes how precisely to relate the investment costs to the greenhouse effect.

In any case, from the above it is clear that an assessment based on the engineering efficiency approach alone may easily create a biassed view. A more complete assessment must recognize different priorities within countries, the impact of externalities, the political acceptability at various levels, and a variety of distributional aspects. In this respect it seems that, although both approaches raise analytical concerns that need to be addressed, a high priority item for both should be to pay attention to the special position of developing countries as well as countries in transition.

In other chapters (especially Chapters 8 and 9) the need to reconcile the various types of analyses of the costs of energy-related greenhouse gas mitigation has been underlined. There is indeed a growing convergence of detailed (bottom-up) analyses of technological options and more aggregate (top-down) analyses of economic effects, so that differences in results can increasingly be attributed to differences in input

assumptions rather than to differences in model structure. However, notwithstanding the current progress in greenhouse-related modelling, there are fewer studies for economies in transition or developing country economies. Moreover, where the potential for political, social, and economic change in these economies is great, future predictions are probably more uncertain. In view of the structural changes that are underway in these regions, it is imperative to improve further the understanding of the potential for reducing or absorbing greenhouse gas emissions in these economies, particularly their sensitivities to other important considerations such as economic and technological development.

7.4. Mitigation Options

This section will treat in some detail the mitigation options listed in Section 7.2.1, along with their costs and potential.

7.4.1 Energy conservation and efficiency improvement

In order to put the energy efficiency option into a proper perspective, the Kaya identity (Kaya, 1989) may provide a useful starting point:

$$CO_2 = (CO_2/E) \times (E/GDP) \times (GDP/P) \times P$$

where E = energy consumption; GDP = gross domestic product; P = population.

If population growth is given and the future levels of GDP per capita are predetermined, a given CO_2 emission reduction target can only be achieved by a reduction in carbon intensity (CO_2/E) and/or energy intensity (E/GDP). The need to reduce carbon and energy intensities becomes stronger, the higher the growth rate of population and the more ambitious the targets set with respect to GDP increase. This relationship obviously reinforces the need to pay specific attention to developing countries.

Historically, carbon and energy intensities in most countries have tended to decline due to ongoing technological change and evolution. Energy intensity per unit of value added has been decreasing at a rate of about 1% per year since the 1860s and at about 2% per year (2.6% in IEA member countries during 1980-1984) in most Western countries in the 1970s and much of the 1980s (Nakicenovic *et al.*, 1993). However, the differences between the various countries are enormous, both in terms of the levels of energy intensity and its direction in the course of time. Moreover, the carbon and energy intensity in a number of large rapidly growing developing countries today is much higher than in virtually all presently industrialized countries at a similar stage of technological development (Nakicenovic *et al.*, 1993). Also, in contrast to the postwar trend noticed in industrialized countries, some developing countries have not succeeded in reducing energy intensities.

Indeed, within each group, countries do vary in terms of the capacity, whether potential or realized, to restrict carbon emissions through energy efficiency. Moreover, within a given country, not all sectors have a similar energy efficiency. During 1973-1988, for example, the estimated energy intensity in Japan fell by more than 35% (Ogawa, 1992), with the energy intensity of electric refrigerators falling by nearly 67% between 1973 and 1987 and the efficiency of motorcars increasing from around 9.4 to about 13 km/litre (49%). During the same period, the U.S., (the then) West Germany, and France lowered their energy intensities by 27%, 22%, and 17% respectively, and IEA member countries by 25% (IEA/OECD, 1991). In most cases, changes have been most apparent in the industrial sector. However, low oil prices and economic recession caused a slowdown in energy intensity reduction in the late 1980s and early 1990s.

Over the 1980s, various developing countries managed to lower their industrial energy intensity: China by approximately 30% (Huang, 1993), Taiwan (between 1970 and 1985) by some 40% (Li, Shrestha, and Foell, 1990), and the Republic of Korea by 44% (Park, 1992). However, in other countries, such as Nigeria (Nakicenovic *et al.,* 1993), Egypt (Abdel-Khalek, 1988), and Mexico (Guzman *et al.,* 1987), energy intensity actually increased. In addition, Imran and Barnes (1990) have reported energy intensity increases in Brazil (+20%), Pakistan (+26%), India (+25%), and Malaysia (+48%) for the period 1970-1988.

Changes in aggregate energy intensity must be viewed with caution, however, as they depend on how energy use and economic output are measured. In Brazil, for example, official figures show overall energy intensity remaining roughly constant during 1973-1988. However, if hydropower is counted based on its direct energy content and GDP is corrected to reflect purchasing power parity with the dollar, then overall energy intensity declined 21% during 1973-1988 (Geller and Zylbersztajn, 1991).

Carbon intensity, the other variable in the Kaya identity, also shows a declining trend globally. From 1860 to the present, carbon emissions per unit of primary energy consumed have come down by about 0.3% per year, or from over 0.8 to somewhat over 0.5 tC/kWyr (Nakicenovic *et al.,* 1993). Clearly, decarbonization can be achieved by a variety of options, such as fossil fuel switching and using nuclear and renewable energy as fossil fuel substitutes. However, various projections with respect to developing countries indicate that, without serious policies and changing trends, not only will total emissions increase rapidly but also carbon emissions may increase faster than GDP because demand for energy services is switching from regenerating biofuels to fossil fuels (for India, for instance, see Mongia *et al.,* 1991).

The two factors that underlie reduced energy intensities are improvements to the energy efficiency of individual production processes and structural changes in the economy (in particular, the increasing economic predominance of less energy-intensive sectors, such as many of the service sectors, and the energy efficiency of spatial planning). Only a few studies explicitly incorporate the impact of structural changes. Most focus on energy efficiency measures, which are generally considered to be the most relevant factor.[4] To illustrate, it was estimated that energy efficiency improvements were responsible for about three-quarters of the 26% reduction in U.S. energy intensity during 1973-1986 (Schipper, Howarth, and Geller, 1990).

Disregarding the impact of structural shifts on energy intensity in an intercountry comparison may easily create a biassed view, because the industrialized economies have generally shifted away from the highly energy-intensive secondary towards the less energy-intensive tertiary sector (a process known as "dematerialization"), whereas the developing countries in general are increasingly entering the secondary sector.

Among virtually all studies, there is a broad consensus on the virtue of energy efficiency improvement. Moreover, it is seen as directly beneficial, irrespective of whether greenhouse warming will take place or not, as long as reductions are achieved at a negative net cost (no-regrets policy).

One basic reason why the energy efficiency improvement potential is considered substantial is that the ratio of useful energy (i.e., the amount of energy that provides useful services) to overall primary energy (i.e., the amount of energy recovered or gathered directly from natural sources) is estimated at only 34% globally. It is lowest, at 22%, in the developing countries and highest, at 42%, in the countries in transition (Nakicenovic and Grübler, 1993). This ratio, in turn, is the product of two other ratios:

- The final energy (energy delivered to the point of consumption) to primary energy ratio (with a global average of 74%, a maximum of 80% in the developing countries, and a minimum of 69% in the countries in transition)

- The useful energy to final energy ratio (with an average of 46% globally, 28% in the developing countries, 53% in the industrialized countries, and 60% in the countries in transition)

These numbers suggest that the scope for improving energy efficiency is particularly promising with regard to increasing the useful-to-final energy ratio. Efficiencies are lowered further if seen from an "exergy" point of view, that is, if the actual services (work) supplied by the energy source are related to the corresponding inputs minimally required: The exergy efficiency of primary inputs in the market economies is only a few percent (i.e., of the order of 2.5-5%) if the energy service is fully taken into account.

Indeed, a back-of-the-envelope calculation shows that, if energy efficiencies of the current structure of the OECD technologies were disseminated throughout the world, global primary energy requirements would come down by 17%, from 12 to 10 TWyr/yr. If, instead, the best available technologies instantaneously replaced the current ones, without altering the energy system structure, global annual primary energy requirements would decline to 7.2 TWyr/yr (Nakicenovic and Grübler, 1993). A similar exercise assuming that Japanese industrial efficiency levels would diffuse globally shows an estimated industrial carbon reduction potential of some 730 MtC worldwide, mainly in the steel, chemical, and cement industries (Matsuo, 1991).

Clearly, energy end use is the least efficient part of energy systems, and it is in this area that improvement would bring the greatest benefits. Most studies suggest that a large potential for reducing energy consumption exists in many sectors

Table 7.1. *Energy efficiency potential: summary of opportunities and barriers*

	(A) Estimated Share of Total Final Consumption (%)	(B) Estimated Share of Total CO_2 Emissions (%)	(C) Total Energy Savings Possible[a] (%)	(D) Existing Market/Inst. Barriers[b] (%)	(E) Potential Energy Savings Not Likely to Be Achieved[c] (%)
Residential space heating & conditioning	11.4	11	10–50	Some/Many	Mixed
Residential water heating	3.4	3.6	Mixed	Some/Many	Mixed
Residential refrigeration	1.1	2.1	30–50	Many	10–30
Residential lighting	0.6	1.2	over 50	Many	30–50
Commercial space heating & conditioning	6.1	6.8	Mixed	Some/Many	Mixed
Commercial lighting	1.5	3.4	10–30	Some/Many	Mixed
Industrial motors	4.5	9.0	10–30	Few/Some	0–10
Steel[d]	4.1	4.6	15–25	Few/Some	0–15
Chemicals[d]	8.4	5.9	10–25	Few/Some	0–20
Pulp and paper[d]	2.9	1.2	10–30	Few/Some	0–10
Cement[d]	0.1	0.9	10–40	Few/Some	0–10
Passengers cars	15.2	13.7	30–50	Many	20–30
Goods vehicles	10.1	9.1	20–40	Some	10–20

[a]Based on a comparison of the average efficiency of existing capital stocks to the efficiency of the best available new technology. This estimate includes the savings likely to be achieved in response to current market forces and government policies as well as those potential savings (indicated in Column E) not likely to be achieved by current efforts.

[b]Extent of existing market and institutional barriers to efficiency investments.

[c]Potential savings (reductions per unit) not likely to be achieved in response to current market forces and government policies (part of total indicated in Column C).

[d]Energy use only.

Note: How to read this table: For example, for residential lighting, over 50% per unit savings would result if the best available technology were used to replace the average lighting stock in use today over the next ten to twenty years. Some of these savings would take place under existing market and policy conditions. But due to the many market and institutional barriers, there would remain a 30–50% potential for savings that would not be achieved.

Source: IEA/OECD (1991).

and regions, at the same time acknowledging that institutional, economic, and social barriers may delay or inhibit the achievement of full efficiency potentials in the near future. A review of twelve studies of long-term energy efficiency potential found that in many regions of the world full adoption of cost-effective energy efficiency measures could reduce carbon emissions by 40% or more over the medium to long term, compared to business-as-usual trends (Geller, 1994). An illustrative example that is related to the OECD area is IEA/OECD (1991), as shown in Table 7.1. In this respect it should be mentioned that several policy and regulatory reforms have recently begun to address some of these barriers. In the U.S., for example, more than 30 states have adopted or experimented with regulatory reforms since 1989 to promote demand-side management (DSM) and to encourage integrated resource planning (IRP).

Other studies focus on the energy efficiency improvement potential by analyzing major energy end use (e.g., Blok *et al.*, 1991; OTA, 1991; COSEPUP, 1991; Goldemberg *et al.*, 1988; Kaya *et al.*, 1991; Gupta and Khanna, 1991; and ESCAP, 1991) or focus on specific sectors. To illustrate, recent estimates for the U.S. show energy saving potentials of 45% in buildings, 30% in industries, and 30% in cars (Rubin *et al.*,

1992; DeCicco and Ross, 1993). In rural areas of developing countries, to give another example, the efficiency of wood and charcoal-fuelled cook stoves can be increased from a range of 10-20% to 25-35% using improved stove designs at a capital cost of under $10 per stove. Cooking efficiency can be further increased to the 40-65% range by shifting from biomass-based fuels to kerosene, liquefied petroleum gas (LPG), or electricity, but at a significantly higher capital cost (U.S. Congress, 1992).

Energy efficiency gains may be particularly promising in the following sectors: power production, transportation, steel and cement production, and residential. However, the relative ranking of sectors in terms of energy efficiency improvement potential is highly dependent on whether or not both the direct and indirect requirements of energy are taken into account, in other words, if interindustry demands are included during sectoral comparisons. A comparative study of India (Parikh and Gokarn, 1993) shows, for instance, that if direct carbon emission due to fossil fuel use is considered, then electricity generation tops the list of total emissions (one-third of the total). However, if direct and indirect emissions are taken into consideration, the construction sector emerges as the largest carbon-emitting sector in India (22% of total).

Table 7.2. *Regional potentials for reducing industrial carbon emissions by cost categories (in Mt Carbon)*

	Cost Saving or at Moderate Cost		Cost (< 100 $/t C)		Cost (> 100 $/t C)		Sum[†]	
Market economies								
Efficiency improvement	15	} 116	45	} 47	84	} 207	144	} 370
Structural change/recycling	95		n/a		25		120	
Fuel substitution	6		n/a		n/a		>> 6	
Process technology process	0		2		98		100	
Reforming economies								
Efficiency improvement	48	} 223	53	} 113	n/a	} > 46	> 101	} 382
Structural change/recycling	165		50		n/a		> 215	
Fuel substitution	10		n/a		n/a		>> 10	
Process technology process	0		10		46		56	
Developing countries								
Efficiency improvement	12	} 34	41	} 78	n/a	} > 56	> 53	} 168
Structural change/recycling	19		29		n/a		> 48	
Fuel substitution	3		n/a		n/a		>> 3	
Process technology process	0		8		56		64	
World								
Efficiency improvement	75	} 372	139	} 238	84	} 309	> 298	} 920
Structural change/recycling	279		> 79		> 25		> 383	
Fuel substitution	19		n/a		n/a		>> 19	
Process technology process	0		20		200		220	

[†]Total reduction potential could be higher because not all measures have been assessed.

Note: n/a = not assessed.

Source: Grübler *et al.* (1993a).

The issue of energy conservation and efficiency in the developing countries differs in some respects from the issue in industrialized countries. First, a substantial part of the demand for energy is often met from renewable energy sources like biomass. This is likely to remain so in the short to the medium run, and there are estimates to show that the scope for conservation of biomass is enormous in these countries. One reason is that cooking with traditional biomass fuels is technically very inefficient, although not necessarily from a socio-economic perspective (U.S. Congress, 1992). Second, energy efficiency in industrial activities generally showed little or no improvement (Imran and Barnes, 1990). Third, the demand for electricity is growing at a rate that is often hard to keep up with. There are developing countries that have allocated a quarter to a third of public investment to generation of power, and even this is sometimes inadequate to meet the growing demand (World Bank, 1993). However, due to the presently low level of energy efficiency in the developing countries and the consequently large scope for improvement, the potentials for energy saving in these countries are considered somewhat similar in magnitude to those in industrialized countries at present, notwithstanding adverse factors such as the fast growth in commercial energy use and the increasing weight of the industrial sectors (Ewing, 1985; Levine *et al.*, 1991; U.S. Congress, 1992). Finally, the the energy market in developing countries is often distorted by energy pricing policies.

By contrast, energy conservation may be achieved somewhat more easily in the industrialized countries, insofar as a trend towards lower material and energy consumption appears to be underway. Various indicators, such as the increasing service orientation of the industrial economies, seem to point in this direction.

Much of the discussion seems to focus increasingly on the extent to which improved energy efficiency and conservation can be economically viable in the present while saving energy and reducing CO_2 emissions (a no-regrets option). Optimism about the scope for no-regret options generally is much greater among proponents of the bottom-up approach than amongst those adhering to top-down methodologies.[5]

Various studies have been carried out focussing on both the potential for carbon emission reduction via energy efficiency improvement and the net costs involved. An overview of the potential for emission reductions in the industrial sector is presented by Grübler *et al.* (1993a) in Table 7.2. They argue that a potential reduction of 920 MtC (over 40% of current emissions) could be achieved overall. Of this, 372 MtC could be achieved at net negative or modest positive costs (with about two-thirds of this amount coming from the countries in transition). These estimates disregard the potential for fuel switching and for decarbonizing the electricity supply and assume an annuity rate of 10% throughout the lifetime of the investment.

The choice of a financial discount rate is an important factor in evaluating the cost-effective energy efficiency potential in a particular sector or region. Studies that have tried to assess the implicit consumer discount rates of household investments in energy efficiency reveal ranges that vary (depending on income classes and other factors) from only a few percent to well over 50%. Train (1985) found a range of 10-32% for improvements to the thermal integrity of buildings, 4-36% for space heating and fuel type, 3-29% for air conditioning, 39-100% for refrigerators, and 18-67% for other home appliances.

Thus, it is clear that the scope estimated for no-regrets options is crucially dependent on the discount factor employed. If one were to use an interest rate (whether based on market or normative considerations) that was considerably lower than that applied by the actual investor or consumer, a no-regrets option would not materialize, even if access to information and the availability of human capacity and financial resources did not provide any serious bottlenecks. However, the practical situation, especially at the grassroots level in developing countries and countries in transition, is such that even the latter conditions are seldom fulfilled.

Consider, for example, the problem of how to increase energy efficiency in the consumption of wood fuels in the developing countries. Here institutional measures and proper distribution (keeping in view local societal and cultural factors) are probably quite important. Popularizing energy-efficient cooking stoves among hundreds of thousands of households would necessitate efforts at many levels. Suitably designed credits and, if necessary, subsidies or tax breaks may help in manufacturing the new stoves in large numbers, but dissemination may be difficult (Hurst, 1990). Nongovernment efforts in this area may go a long way towards solving the problem (Asaduzzaman, 1995).

As a general remark with respect to the above, it should be noted that a high implicit discount rate does not mean that substantial energy efficiency improvements and consequent benefits for the economy are not possible. Rather it suggests that significant policy intervention will be required to achieve such improvements. For example, in spite of a high implicit discount rate, the average energy efficiency of new refrigerators sold in the U.S. nearly tripled between 1972 and 1993. This large and steady improvement was due primarily to the adoption of minimum efficiency standards, first at the state level and then at the national level (Geller and Nadel, 1994).

The choice of a discount rate can affect the overall magnitude of energy efficiency improvements that are considered economical. Meier (1991) argued that by assuming an annual discount rate of 10% more than a quarter of U.S. electricity demand for refrigerators could be reduced by cost-efficient measures; using a 30% rate results in positive costs for all these measures. Similarly, the Committee on Science, Engineering, and Public Policy (COSEPUP, 1991) has shown how the percentage savings in electricity, at the point where the costs of conserved electricity equal the typical operating costs for an existing U.S. power plant, vary according to the discount rate: At a 3% rate the electricity saving potential is almost 45%; at a 10% rate, it is about 30%; and at a 30% rate, it is about 20%.

Notwithstanding the above, a host of studies has emerged suggesting a considerable scope for no-regrets options, especially in the household and tertiary sector (e.g., Springmann, 1991; Mills *et al.*, 1991; Rubin *et al.*, 1992; Jackson, 1991; Blok *et al.*, 1993; UNEP, 1993; Robinson *et al.*, 1993).

Table 7.3. *Energy mix: Annual past and future global fuel use (Gt oil equivalent)*

	1960	1990	in 2020			
			A	B1	B	C
Coal	1.4	2.3	4.9	3.8	3.0	2.1
Oil	1.0	2.8	4.6	4.5	3.8	2.9
Natural gas	0.4	1.7	3.6	3.6	3.0	2.5
Nuclear	—	0.4	1.0	1.0	0.8	0.7
Large hydro	0.15	0.5	1.0	1.0	0.9	0.7
Renewables						
"Traditional"	0.5	0.9	1.3	1.3	1.3	1.1
"New"	—	0.2	0.8	0.8	0.6	1.3
Total	3.3	8.8	17.2	16.0	13.4	11.3

Source: WEC Commission (1993).

Finally, in addition to the potential for energy efficiency improvement, there clearly is also considerable scope for conservation options, even if their assessment often can only be somewhat qualitative and impressionistic. There seems to be ample opportunity for increasing energy conservation in the industrialized countries through the imposition of stricter standards with respect to energy and materials use and, most of all, through alterations and adjustments in lifestyles.

7.4.2 Fossil fuel switching

According to most studies, the present dominance of fossil fuels in global (primary and noncommercial) energy consumption will continue to exist in the decades to come. According to recent authoritative World Energy Council scenarios[6] (WEC Commission, 1993) (Table 7.3), fossil fuels will account for between 66% (scenario C, where renewables are fully explored) and 76% (scenario A, where fossil fuels remain dominant) of world energy consumption in 2020, compared to 77% in 1990.

All the scenarios reflected in the table show that:

- Fossil energy remains dominant

- The share of natural gas, environmentally the least damaging of the fossil fuels, increases from the present quarter to one-third at most

- The share of nuclear remains modest

- The relative potential of the presently modest "new" renewables is not insignificant, as opposed to the limited size of the projected shifts for large hydro and "traditional" energy sources (in this respect, see also Chapter 9 and, for a different point of view, Kassler, 1994).

The remaining dominance of fossil fuels is due to the large resource base,[7] the strongly vested position of the current vintage of technologies, and price distortions that externalize the environmental costs. Estimates point out that total identified fossil fuel reserves will suffice to provide for current (1990) levels of energy consumption for the next 130 years.[8] This time span may become considerably shorter, as energy use in the developing countries will increase rapidly.

Of the three fossil fuels, natural gas is the least and coal the most carbon-intensive.[9] Natural gas also produces minimal sulphur emissions and virtually no airborne particulates (World Resources Institute, 1994). Therefore, a switch from coal and/or oil to natural gas is seen as a response option with multiple benefits. Current estimates of the natural gas resource base, which will likely be revised upwards in the future, allow for a massive switch-over for the next century or so to come. If so, the entailed transition of the current vintage of energy technology would, as an additional beneficial side effect, pave the way for a broad diffusion of gas from biomass or coal gasification, or of hydrogen, a potentially massive renewable energy source for later in the next century.[10]

The costs of this fuel stem from retrofitting or replacing the current vintage of energy technology and, in some cases, building additional transport grids to connect more remote urban areas with gas fields. Estimates of the costs of switching, even without extending the existing networks, depend to a large extent on the type of measure. For example, switching building heating from electric to natural gas (improving overall efficiency by 60-70%) would, according to Rubin *et al.* (1992), yield a net benefit of $90/tCO$_2$ in constant 1989 dollars (assuming a 6% real discount rate). According to the same source, however, switching coal consumption in industrial plants to natural gas or oil, where technically feasible, would involve net direct implementation costs of some $60/tCO$_2$ in constant 1989 dollars.

Ettinger *et al.* (1991) have estimated the investment costs of exploration and extraction for a fuel switch scenario involving a natural gas supply growth rate of 3.3% per year between 1988 and 2005 plus the costs of extending the existing supply network into a global gas distribution system (based on 1989 data from the Dutch Gas Union and an average transport distance of 2500 km). They calculate that total costs would be in the order of $70 billion gross per year, corresponding to $70/tC on average.

However, two caveats should be mentioned. First, much of the attractiveness of natural gas as a less carbon-intensive fossil fuel is lost if a sizable fraction evaporates into the air by leakage during production and distribution. This is due to the substantially higher global warming potential of methane (CH$_4$), which is about 24.5 times that of CO$_2$.[11] Estimates of common current leakage rates range from 0.3% to 4% for distribution and from 0.13% to 6% for production (Simpson and Anastasi, 1993).[12] The break-even point, that is, the rate at which the reduced total warming potential is just offset by leakage of methane, occurs at 7%[13] for switching from coal to gas and at 3% for switching from oil to gas (adopting a global warming potential index for CH$_4$ of 24.5 for a 100-year time horizon). These figures point to the need for strict control of leakage rates.[14] Additional questions revolve around what happens to leakage rates in the case of a large-scale fuel switch and whether leakage rates of newly built and/or additional grids (i.e., marginal leakages) can be reduced.

Second, the costs of the fuel switch option can also be approached on the basis of the opportunity cost concept. For countries such as China and India that dispose of massive coal reserves and that may contribute increasingly in an ab-

solute sense to the global greenhouse problem, the opportunity costs of fossil fuel switching may be considered large, especially if the environmental costs of coal are not taken into account.

7.4.3 Renewable energy technologies

Today many technologies have been developed to provide energy on a sustainable basis, in the sense that they harness energy resources that are practically unlimited and require relatively little additional energy input. Moreover, exploitation of renewable energy resources with appropriate technologies has the advantage of releasing relatively little carbon in net terms.[15] Consequently, a switch from fossil fuels to renewables will result in reduced absolute greenhouse gas emissions.

However, renewable technologies are not always sustainable in the sense of being socially and environmentally benign. Particularly in the case of large-scale applications in developing countries, notably of hydropower and biomass, adverse effects may arise for the local population. Moreover, adverse environmental side effects may occur, such as smog from the use of traditional biomass fuels (fuelwood, dung, and crop residues) or changes in biological habitats and local climate.

The following classes of renewable energy resources are commonly distinguished: solar, wind, hydro, geothermal, ocean, and traditional and modern biomass.[16] To understand the main factors that underlie the costs and energy potential of renewables as a group, a detailed treatment of their diversity is required.[17] Most of them, with the exception of biomass, are variable in supply, and some of them (especially traditional biomass, wind, and solar) are relatively more cost-competitive with fossil sources when they are produced on a small scale and near the spot of consumption. These latter aspects make them a potentially attractive option in remote and underdeveloped areas.

Further, large differences exist in the technical and economical readiness of these options. Hydro, wind, and traditional biomass are relatively well-developed, whereas some ocean technologies are still in a demonstration stage, although tidal and wave technologies may soon become more practical economically. Solar, modern biomass, and geothermal are in between, and photovoltaics may become competitive with fossil-fuel power plants within a decade or so (Mills *et al.*, 1991).

Table 7.4 breaks down the contribution of the various technologies to renewable energy production in 1990 and makes clear that traditional biomass and large hydro are presently the most prominent renewable energy sources.

Some estimates of "practicable"[18] potentials (relative to current use) are given in Table 7.5. It clearly shows how small current use is when related to various estimates of practicable potential, whatever discrepancies may exist in estimates of that concept. Notable exceptions are large hydro and traditional biomass, which, according to the data presented, are exploited at about a quarter to half of probable capacity. Judging by these figures only, the potential contribution for renewables is promising. However, a truly comprehensive assessment must also consider the costs involved.

Table 7.4. *Contribution of various technologies to renewable energy production in 1990*

Energy Technology	Mt Oil Equivalent	% of Total
New renewables		
Solar	12	0.8
Wind	1	0.1
Geothermal	12	0.8
Modern biomass	121	7.8
Ocean	0	0.0
Small hydro	18	1.2
Total new renewables	164	10.5
Traditional biomass[†]	930	59.6
Large hydro	465	29.8
Total	1559	100

[†]Includes fuelwood and dung.
Source: WEC, 1993.

Table 7.6 gives a selective overview of cost estimates of renewable energy technologies. As usual, figures diverge widely. This variation is mainly due either to (1) the calculation method used or (2) the inherent peculiarities of the technology. As for (1), the time horizon adopted, the level of discount rate chosen, and the assumed capacity and useful lifetime are important factors. As for (2), costs are strongly influenced not only by the site specificity and temporal variability of supply as mentioned above but also by the form of final energy delivered.[19]

Other aspects relevant to cost behaviour are learning effects, economies of scale, and the need for immediate storage or transport of the energy generated (the costs of which are very difficult to assess with any precision). Immediate storage or transport needs occur not only when the timing of supply and demand fail to coincide, as is commonly the case with solar and wind, but even more when sources and points of end use are far apart. Preferably, generated electricity should be fed into a linked distribution system of sufficient capacity to handle its intermittent supply. Different, but equally difficult to assess, are the problems of location and transportation associated with storable biofuel.

On the basis of the prices in Table 7.6, it has been concluded that hydro, wind, and some solar and biomass technologies are already becoming more competitive with conventional sources. Although many wind and solar power applications are still subsidized or legislatively supported, substantial cost reductions are to be expected within the next few decades.[20] Whether these technologies actually become competitive, however, will also depend on local conditions that shape a renewable's attractiveness and complementarity between renewables and nonrenewables.

In contrast to fossil fuels, renewable energy at the moment is less portable: Consumption currently seems to be more strongly bound to the production location. Whereas fossil fuels can be relatively easily stored or transported with the existing infrastructure, similar exploitation of the new re-

newables would in most cases require new investment. The competition between renewables is generally more complex than that between fossil fuels. Solar and geothermal energy, for example, can only be produced on the basis of complementarity by using temporal variation of supply.

Conversely, what often makes up the main part of a renewable's promise are its large potential and modest price on the spot relative to the availability and prices of conventional sources. Moreover, by using local renewables, countries could reduce their dependence on imported fossil fuels and also reduce foreign exchange constraints. In addition, in the case of biomass, local communities could significantly benefit from small-scale applications and their net positive side effects. In this respect, local renewables, like energy efficiency measures, offer a basis for no-regrets policies.

There is some reason to believe that a new generation of renewable energy technologies now under development could well become commercially viable in the near future. For example, a variety of promising photovoltaic technologies designed to shave commercial building demand during peak load periods is under active consideration in the U.S. and elsewhere and might become commercially feasible in the foreseeable future (Byrne *et al.*, 1994; Wenger *et al.* 1992).

As the preceding discussion implies, the future role of renewables is hard to predict precisely. Although some scenarios are more optimistic than others, the share of renewables in the 2020 energy mix will probably not exceed 25%.[21] However, most studies agree that the new renewable mix will tend to be a hybrid that will exploit a variety of renewable energy sources backed up by fossil fuels, which will remain dominant for decades to come.

7.4.4 Nuclear energy[22]

Nuclear energy now accounts for about 5% of all primary energy production or 17% of the world's electricity generation. Its production, like that of renewables, emits relatively little CO_2.[23] Moreover, its technology has passed the demonstration stage, except for the large but still unresolved issue of nuclear waste storage. On the other hand, further dissemination could be strongly prohibited by lack of public acceptance due to major concerns about reactor safety, the risk of theft of nuclear technologies or materials, the proliferation of nuclear weapon capabilities, and the final treatment and disposal of fission products.

Barring these limitations, nuclear energy, if evaluated on the basis of the engineering efficiency approach, can be competitively applied, and in various countries it is, albeit to a largely different degree. (For comparison with gross costs, see Figure 7.2; for an estimate of the UK cost-effective potential, see Jackson, 1991, who used data from the mid-1980s.) Because of the long design/construction time (up to 10-15 years) and the enormous per plant investment costs, the nuclear option is rather inflexible now. According to Table 7.7 (note that the figures in the table are based on averages from existing plants rather than new plants), costs to produce electricity with nuclear energy ($/kWh) appear to fall within the range of renewable options, though nuclear costs seem to have been rising and not falling (MacKerron, 1992).

Table 7.5. *Current use and practicable potentials of renewable energy technologies (TWh/yr)*

	Current use	Practicable potential estimate			
		Johansson *et al.* 2020	Swisher *et al.* 2030	WEC 2020	Read average to 2050
Solar	54		1395	489–1592	
Wind	3.2		4931	20148	
Hydro	2281.2[a]	6000–9000	7077	8295[b]	
Geothermal	37–57	>53	1499	178–405	
Ocean	0.6		247	48–240	
Traditional biomass	4170		8003	7031–7269	
Modern biomass[c]	543				about 35,000[d]

[a]Includes 81.7 TWh/yr for small hydro.

[b]Includes 211-308 TWh/yr for small hydro.

[c]Modern biomass refers to the use of biomass (e.g., timber or sugar cane) for the production of electricity, liquid fuels, and heat using modern technology.

[d]Assumes 740 million hectares become available for biofuel production by 2050 (proportionately less according to technical progress with biofuel productivity per hectare) with a slow start and more rapid build up after 2010. The 35,000 TWh would yield about 18,000 TWh of electricity given advanced generating technology expected to be in use next century.

Source: Johansson *et al.* (1993), WEC (1993), Swisher *et al.* (1993), Read (1994b).

Table 7.6. *Estimates of current[1] and future costs of renewable energy technologies (U.S.¢ per kWh)*

						Biomass	
Source	Solar	Wind	Hydro	Geothermal	Ocean	Electric	Fuel ($/GJ)
IEA[2]	0–14[a] 7.6–41.9 (15–174)[b] 5.2–26 (22.61)[c] 5(50)[g]	3.5–4.2[h] (4.48–7.62) (20)[i]		(3.6–9.2)	5–20[m,3] (11.5–50) 6.7–8[n]		7.58–12.80 (1.85–16.68)[q] 12.70–20.85[t] 15.64–23.70[u]
Johansson *et al.*[2]	4.5–11.7 (7.5–32.8)[c] 4.9–9 (8.5–28)[g]	3.13–4.46 (4.29–8.4)		3–12[k] 0.15–2.5[l]	(5–30)[m] 12–25[n] 22–30[o]		1.86–2.73 (2.73–3.86)
Swisher *et al.*	5–10[d](12) 4–8[e](25) 4–8[g](30)	3–6(7)	5–10(5)[j]		8–10[m]	5	6[r](8) 7[s](13) 10[t](15) 1[v]
WEC	no storage[w] 0.4–2.5(0.5–10)[c] 1–11 (1.2–28)[f] 4–14 (28–45)[g]	3–9 (5–10)		n/a	(5–12)[m] (5–7)[n] (12)[o] (10–14)[p]	n/a	n/a

Note: n/a = not assessed.

[1]current costs in parentheses.

[2]1984 cents.

[3]UK pence per kWh.

[a]passive solar; [b]active solar; [c]solar thermal (line focus); [d]solar thermal (line focus); [e]solar thermal (point focus); [f]solar thermal-electric; [g]photovoltaic; [h]small/medium wind energy conversion systems; [i]large wind energy conversion systems; [j]small hydro; [k]electric; [l]direct heat; [m]tidal; [n]wave; [o]salt gradient; [p]ocean thermal; [q]ethanol from corn; [r]ethanol from sugar; [s]ethanol from wood; [t]methanol from wood; [u]methanol from herbage; [v]methanol from biomass; [w]costs exclude storage systems.

Sources: IEA/OECD (1987); Johansson *et al.* (1993); Swisher *et al.* (1993); and WEC (1993).

Table 7.7. *Examples of avoided emissions and their costs: Electricity*

Electricity[a] (Cost of avoided resource (coal):$0.44/kWhe)	Measure Resource Cost ($/kWh)	Avoided Emissions (g Carbon-eq/kWh)	%	Cost of Avoided Carbon-equivalent (CaCeq) ($/tonne)
End-use efficiency[b]				
Available technologies				
Lighting (incandescent to compact fluorescent)	-0.011	318	100	-171
Lighting (efficient fluorescent tube)	-0.007	318	100	-159
Lighting (lamps, ballasts, reflectors)	0.013	318	100	- 96
Refrigerator/freezer, no CFCs	0.018	318	100	- 79
Freezer, automatic defrost, no CFCs	0.022	318	100	- 67
Heat pump water heaters	0.034	318	100	- 30
Variable-speed motor drive	0.011	318	100	-102
U.S. field data, multifamily, leaking retrofits	0.038	318	100	- 19
Retrofits in 450 U.S. commercial buildings	0.026	318	100	- 54
No-cost or behavioural measures	0	318	100	-137
Electricity production (busbar costs)				
Available technologies				
Biomass steam-electric (woodfuel)	0.041	318	100	- 9
STIG[c] (gasified coal)	0.041	9	3	-313
STIG[c] (natural gas)	0.027	163	51	-103
Wind (1988)	0.054	318	100	- 33
Solar thermal electric (1988)	0.114	318	100	-221
Solar photovaltaics (1988)	0.231	318	100	-588
Nuclear	0.057	318	100	- 41
Emerging technologies				
ISTIG[d] (gasified coal)	0.034	57	18	-176
ISTIG[d] (natural gas)	0.024	187	59	-106
Chemically recuperated gas turbine	0.029	204	64	- 73
Solar thermal electric				
(2000)	0.043	318	100	- 1
(2010)	0.036	318	100	- 24
(2020)	0.031	318	100	- 40
Solar photovoltaics				
(2000)	0.072	318	100	89
(2010)	0.050	318	100	22
(2020)	0.036	318	100	- 24
Wind				
(2000)	0.033	318	100	- 33
(2010)	0.027	318	100	- 51
Nuclear – industry target for U.S.	0.040	318	100	- 11
Fuel choice (STIG[c] technology in all cases)				
Avoided resource cost (gasified coal:$0.071/kWh)				
Gasified coal to natural gas (1990)	0.027	155	50	- 91
Gasified coal to biomass (sugar)(~ 2000)	0.033	309	100	- 25

[a]Unless noted, the annualized costs of efficiency and supply measures are calculated with a 6% real discount rate and no taxes. For details on the other assumptions, see source.

[b]Lighting and refrigeration measures calculated using a 7% real discount rate.

[c]Steam-injected gas turbine.

[d]Intercooled steam-injected gas turbine.

Source: Mills *et al.,* 1991.

Social opportunity costs will remain high until a full and credible investigation of the safety aspects of nuclear power plants is completed. However, if the nuclear option is assessed from the welfare economic point of view, the final assessment becomes much more uncertain because the lack of public acceptance and the various risks, advantages, and uncertainties now also have to be taken into account explicitly. This holds not only in the industrialized countries, but also in the developing countries and the countries in transition. In addition, any future use of nuclear energy, like any switch from fossil to nonfossil fuels, will depend on the underlying cross-price elasticities and energy price assumptions, inflation, public policy, and technological progress. Taking these complicating factors into account – namely, that there is no established technology for decommissioning nuclear plants, that there are hidden external costs regarding nuclear power-related damage, and that efforts are being made to develop intrinsically safe nuclear reactors – the IEA projects the share of nuclear energy in total energy use at 6.1% by 2010; the WEC C-scenario (see also note 6) projects the share of nuclear at 6.2% in 2020.

7.4.5 Capture and disposal

CO_2 capture and disposal is understood as any sequence of processes in which carbon is recovered in one form or another from an energy conversion process and disposed of at sites other than the atmosphere. It should be noted though, that disposal capacity is ultimately limited, both for technical reasons and because not all forms of disposal ensure a permanent prevention of carbon reentering the atmosphere. However, assuming sufficient and feasible disposal, the further development of these technologies in combination with coal gasification is thought to have significant intermediate potential, especially for coal-rich countries such as China, India, the U.S., or the Russian Federation (see also Nakicenovic and Victor, 1993, and the outcomes of the OECD Model Comparison Project as discussed in Chapter 8).

Since places of recovery do not generally coincide with places of disposal, transport of the recovered carbon is required as an additional process. In principle, carbon can be recovered from each fossil fuel conversion process. However, recovery is most attractive at energy-intensive stationary point sources, such as steel manufacturing, fertilizer, and power plants.[24] To date, most research effort has been spent on power plants. For these, two types of recovering technologies exist:[25] those that combine separation of the CO_2 from the flue gases (scrubbing) with modifications to the energy conversion process and those that rely on CO_2 scrubbing only. Modifications to the energy conversion process, which are now in experimental use, include an Integrated Coal Gasifier Combined Cycle (ICGCC) system, modification of boilers, and modification of gas turbines.[26] The main separation options are chemical or physical absorption, the use of membranes, and cryogenic fractionation. Of these, chemical and physical absorption are most developed and membrane separation and cold distillation least.[27]

Depending on the place of disposal, transport will take place onshore or offshore. Onshore, pipelines are most economical. Estimated transport costs vary between $1 and $4/t$CO_2$ over 100 km, depending on the flow rate (Hendriks, 1994). Offshore, tankers compete with pipelines. For larger distances, tanker transport is likely to be cheaper. Pipeline transport costs are more or less proportional with distance and decrease with increasing flow rate of the gas and decreasing ambient temperatures. Estimates of costs offshore therefore vary between somewhat more than one-half to three times the costs onshore (Hendriks, 1994; TNO, 1992).

After the carbon is recovered, it has to be handled so that reentry into the atmosphere is prevented or at least delayed as much as possible, that is, so that the mean retention time is large compared to the residence time of CO_2 in the atmosphere (since not all applications ensure entire or long-term storage of the carbon).[28]

Disposal can occur in two ways: The gas can be utilized for the production of long-lived materials,[29] or it can be stored underground, either in aquifers (which, technically, have almost unlimited storage potential), or in the ocean.[30] Environmental risks seem to be involved, however, especially in the latter cases.

7.4.6 Enhancing sinks: Forestry options[31]

Unlike removal options, options that enhance sinks remove carbon after it has been dispersed into the atmosphere. All sources seem to agree that much more carbon is stored in soils than in forests. This would suggest that significant attention be given to measures that promote soil conservation, reduce carbon mobilization from soils to air, and increase soil storage of atmospheric carbon through the action of soil microorganisms. Nevertheless, the main option for enhancing carbon sinks – except for iron fertilization and weathering rocks, which are both still in their experimental stage – relates to forestry measures. Their importance is due to their expected large storage potential and relatively modest costs. The enhancement of forest sinks is also one of the lowest-risk options and offers substantial positive side effects in the environmental and sometimes also in the socioeconomic sphere.

Just as the potential for forestry measures in enhancing sinks is probably sizable, so is the contribution of deforestation to greenhouse gas emissions. After fossil energy-related activities, deforestation and other land use changes are the second-largest source of carbon emissions. The net annual flux of carbon to the atmosphere as a result of land use changes and deforestation ranged between 0.6 and 2.8 GtC during the early and probably the rest of the 1980s, compared with global emissions of slightly less than 6 GtC from burning of fossil fuels, manufacture of cement, and flaring of natural gas (Grübler *et al.*, 1993a; see also Houghton, 1990). The large amount of uncertainty about the net quantity of carbon released by deforestation and land use changes relates to the extent of the area undergoing land use change, the carbon content of biota and soils in the deforested land, and the dynamic release profile of biotic and soil carbon after disturbance.

The following subclasses of forestry measures are commonly distinguished:

(1)   Halting or slowing deforestation

(2)   Reforestation and afforestation[32]

(3)   Adoption of agroforestry practices

(4)   Establishment of short-rotation woody biomass plantations

(5)   Lengthening forest rotation cycles

(6)   Adoption of low-impact harvesting methods and other management methods that maintain and increase carbon stored in forest lands

(7)   Sustainable forest exploitation cum sequestration of carbon in long-lived forest products[33]

The first six measures sequester carbon by increasing the standing inventory of biomass or by preventing a decrease thereof. This amounts to a once-for-all uptake of carbon. In contrast, the seventh measure aims at continuing to break the carbon cycle, thereby, in principle, enabling its permanent application. This option becomes even more efficient and attractive if the timber is used to substitute on a large scale for products such as bricks, concrete, steel, and plastics whose manufacture releases much greater quantities of CO_2.

However, in practice all forestry measures are ultimately limited: The first six by the area available in competition with other potential land uses and the seventh by saturation of demand for timber and other long-lived wood products and the eventual decay of the wood. Therefore, forestry measures, like removal options, are to be seen as an intermediate response policy. In this respect it is worth mentioning that trees grown on fairly short rotations (harvested at maximum Mean Annual Increment) are more effective carbon sinks than trees that are allowed to mature in old-growth forests. This fact has large implications, especially for developing countries, where by far the largest demand for wood is for fuelwood and small construction poles that can be grown on short rotations.

In assessing the global potential for halting or slowing down deforestation and for reforestation, there are three main sources of uncertainty:

(1)   The potential for slowing deforestation depends on resolving complex problems that are linked to societal and economic pressures, such as large-scale settlement on forest lands and the sale of timber for export earnings in tropical countries or policy distortions (e.g., below-cost sales of timber on government lands) in industrialized countries.

(2)   The potential of the option depends on the amount of area globally available for some kind of forestry measure (see also Volume 2, Chapter 24).

(3)   The incremental (i.e., annual) and net cumulative carbon uptake per hectare[34] for the main forest species[35] have yet to be reliably determined.

In addition, it should be noted that large-scale monoculture forestry may not be acceptable to many environmentalists; moreover, local ecosystems may be destabilized.

With respect to the first of these uncertainties, there is a near consensus in the literature that most deforestation in tropical countries occurs because standing forests are converted to crop and pasture land. This happens because those encroaching on the forests consider them to have lower economic value than crop and pasture land. The potential for slowing deforestation is therefore hard to estimate. Furthermore, slowing deforestation requires the application of effective solutions to highly politicized problems such as inequitable land distribution and lack of secure land tenure. It also requires effective means of increasing the per hectare productivity of crops and livestock. The solutions to these problems are partly technical but mostly economic in nature and include improved price structures for farmers (e.g., higher crop and livestock prices versus lower prices for inputs such as fertilizer) and better access to markets.

As for tropical deforestation rates, estimates of these vary widely among the various sources, partly due to different definitions of both tropical forests and deforestation (for a discussion, see Jepma, 1994). According to FAO (1991), annual tropical deforestation for the late 1980s amounted to some 17 million ha; other estimates vary between 3 and 20 million ha. Estimates of global annual biotic carbon fluxes from closed forests during the late 1980s show an equally large variety, ranging between 600 MtC (IPCC, 1992) and 2800 MtC (WRI, 1990). For an overview, see Grübler *et al.* (1993b).

A similar discussion has arisen on the issue of the global land area that would be suitable and available for carbon sequestering plantations. One study of the maximum worldwide potential of this approach, Sedjo and Solomon (1989), suggests that 2.9 Gt of atmospheric carbon could be sequestered annually by approximately 465 million ha of fast-growing plantation forests at a cost of about $186-372 billion. Without employing fast-growing species the area needed would be several times larger, but many factors will determine whether such high rates of uptake can be achieved.

Clearly, a large potential for the enhancement of forest sinks exists in the tropics. A recent survey, carried out under the auspices of the Asian Development Bank in eight Asian countries (Pakistan, India, Sri Lanka, Bangladesh, Indonesia, Malaysia, Vietnam, and the Philippines) indicates not only that climate change is likely to have large and generally adverse impacts on forests and forest ecosystems in the Asia-Pacific region, but also that "forest conservation and afforestation can often be judged to be cost-effective and excellent opportunities for limiting net greenhouse emissions" (Qureshi and Sherer, 1994). However, it should be emphasized once again that much of the land availability will depend on the willingness of the local population to cooperate, given their perceptions of the most appropriate land use.

Keeping these limitations in mind, one can compare the preceding findings with the estimate of Grübler *et al.* (1993b) that at present 265 million ha globally would be available and suitable for forest plantations and 85 million ha for agroforestry. Other sources (Winjum *et al.*, 1992) suggest significantly larger areas (some 400-1200 million ha). These figures are in sharp contrast with the potential in the OECD countries, amounting to 15-50 million ha in future in the EU, mainly due to redundancy of farmlands, and 30-60 million ha in the U.S.

Carbon sequestered in global annual wood production is currently estimated at some 1 GtC (TNO, 1992). Such a high rate may not represent the actual "net" addition to the wood products pool and may not be sustainable in the future, but it gives some indication of the potential of carbon sequestration through wood products, depending on the price, product lifetime, and the trend in timber demand (which is commonly projected to rise). Some authors argue that a considerably higher demand for wood could be achieved if it were used to produce electricity (and liquid fuels such as methanol). This would replace fossil fuel CO_2 emissions with a system in which net emissions are zero, provided the wood is grown on a sustainable basis. (For a feasibility study, see BTG, 1994; for a discussion of the institution building needed to monitor and account for the global realization of this concept, see Read, 1994a.)

With respect to the problem of estimating carbon uptake, only rough estimates exist for the carbon content of biomass and soils in disturbed areas. Estimates of average annual carbon uptake vary considerably, depending on, among other factors, plantation age (for a correlation, see Cannell, 1982), timber species, soil/climate conditions, and management practices (see also Houghton, 1991). However, most estimates are in the likely range of 1-8 tC/ha/yr. Cumulative carbon uptake would as a maximum be somewhere between 100 and 150 tC/ha for the main forest types. (Note that the vegetation and soils of undisturbed forests can hold 20-100 times more carbon than agricultural systems.)

7.4.6.1 Costs

Mitigation policies using forests are generally considered relatively cost-effective, especially if applied in developing countries. An early U.S. high cost estimate of $100/tC (Nordhaus, 1990) now seems to have ignored changes in soil carbon through tree planting and to have underestimated the carrying capacity and length of productivity of forest plantations. Richards *et al.* (1993a, 1993b) estimate that the overall costs of stabilizing U.S. carbon emissions could be reduced by as much as 80% by forestry options.

Most of the studies, which deal with the costs of afforestation or halting or slowing down deforestation, take the engineering efficiency approach rather than the welfare economic approach. With respect to afforestation, the assumption is commonly made that the forests would not be harvested but would be left alone to mature (the so-called carbon cemetery forests). The emphasis, therefore, is on the assessment of the costs of afforestation (plantations), including maintenance and protection, and of land requirement. Sometimes, however, land required for establishing carbon plantations may be considered free, therefore implying that opportunity costs would be zero (e.g., Winjum *et al.,* 1992).

One generally recognizes that the second option, halting or slowing down deforestation, is probably one of the most urgent and most cost-effective options (Grübler *et al.,* 1993b). However, experience with the closest alternative, reforestation, has so far produced mixed results. On the one hand, some success stories can be told about reforestation projects in Sweden, Finland, and parts of Canada. On the other hand,

large losses have occurred in Angola, Nigeria, Morocco, and several other countries, and in China the rate of survival of reforestation efforts is estimated to be not higher than 20% (Nakicenovic and John, 1991).

There are two crucial factors that appear to determine the feasibility of the afforestation option in actual practice. First, it matters a great deal whether the forest can be harvested sustainably and forest products sold at commercial rates, or whether it instead should be left alone. Second, much depends on the acceptance of the newly planted forests by the local population, as might be expected if they were to derive an economic benefit from it.

If, for instance, the forest can be harvested through the exploitation of timber and nontimber products, and if, in addition, carbon sequestration credit can be given to trees that are harvested, then the net costs of afforestation could easily become negative. In that case, afforestation would become a no-regrets option, and initiatives could allow the payback period to be left to the market. However, not all externalities can be incorporated in the prices of the timber and nontimber products (e.g., trees may provide environmental benefits but may also contribute to productive losses by shading adjacent field crops or competing with them for water).

Even if afforestation with sustainable exploitation offers a net positive return, many other factors may still form an obstacle to its implementation. Actual experience has made abundantly clear that, even if environmental quality and economic productivity in a certain area are both low, those who use the land may still be unwilling to convert it to forest. Some even argue that for at least a decade social, political, and infrastructural barriers will keep reforestation rates very modest (Trexler, 1991). Indeed,

> [T]ropical forestry programmes undertaken with global climate change mitigation in mind will need to be integrated into the social, environmental, and economic contexts and needs of the countries [and local communities] in which they are undertaken. Failure to understand this has brought about the failure of many tropical forestry efforts intended to solve fuelwood and other problems. The same could easily occur with forestry efforts intended to mitigate global climate change. (Brown *et al.,* 1993)

Estimates of the cost-effectiveness of forestry measures in the engineering efficiency approach are also subject to uncertainties about the availability of land area, carbon uptake per hectare, and costs of establishment and maintenance per hectare. In addition, figures diverge depending on the methodology adopted. In this respect, two problems should be discussed: (1) the derivation of point estimates or of cost *functions,* and (2) forestry cost function methodology.

With respect to (1), most effort so far has been spent on deriving point estimates from average costs. Some selected cases are given in Tables 7.8 and 7.9. Note that the estimates generally assume a "tree cemetery" approach. Tables 7.8 and 7.9 show relatively low carbon sequestering costs through tree planting, in many cases under $10/tC and rarely over $30/tC. Other studies with similar results, stressing various aspects of the problem, include Trexler *et al.* (1989), Swisher (1991), Winjum and Lewis (1993), and Faeth *et al.* (1993). For a de-

Table 7.8. *Costs of sequestering carbon through forest projects: Some selected cases ($/tC)*

| Source | Tropical | | Temperate Plantation | Boreal | |
	Agroforestry	Plantation		Plantation	Protection
Andrasko (1993)	3–5	3–6	0–2		
Dixon *et al.* (1993)	4–16	6–60	2–50	3–27	1–4
Krankina and Dixon (1993)			1–7	1–8	1–3
Houghton et al. (1991)	3–12	4–37			

Source: Adapted from Dixon *et al.* (1993).

Table 7.9. *Establishment costs of cost-efficient practices*

Forest Type/Practices	Median $/tC[†]	Median $/ha[†]
Boreal		
Natural regeneration	5	93
	(4–11)	(83–126)
Reforestation	8	324
	(3–27)	(127–455)
Temperate		
Natural regeneration	1	9
	(< 1–1)	(9–100)
Afforestation	2	259
	(<1–5)	(41–444)
Reforestation	6	357
	(3–29)	(257–911)
Tropical		
Natural regeneration	1	178
	(<1–2)	(106–238)
Agroforestry	5	454
	(2–11)	(254–699)
Reforestation	7	450
	(3–26)	(303–1183)

[†]The numbers in parentheses are interquartile ranges (middle 50% of observations).
Source: Turner *et al.*, 1993.

tailed assessment see Turner *et al.* (1993) and Volume 2, Chapter 24, of this report.

Though these point estimates may give a satisfactorily accurate description of cost effectiveness for small areas and single plantation programmes, they are bound to lack validity in the case of large areas. From a global perspective, the costs in terms of economic welfare are likely to rise with the scale of the effort. Four forces underlie this cost pattern:

(1) Diminishing uptakes as less suitable or less well-managed land is forested, resulting in a lower carbon uptake per hectare

(2) Increasing public resistance and social and legal objections by the local population against interference with present land use

(3) Rising opportunity costs as fallow land is used up and plantations move on to land suitable for alternative uses[36]

(4) No or negligible economies of scale in operating and maintenance costs

Together these factors generally mean that marginal costs will rise as the area being forested increases. Exceptions to this rule might only occur if the amount of land needed for agriculture shows a declining trend. Clearly, this is almost nowhere the case in developing countries, but it might hold for parts of the Western world. Only recently have a number of somewhat more sophisticated studies begun to appear that do take increasing marginal costs explicitly into account. Such studies also do more justice to the welfare economic point of view by explicitly recognizing that an expansion of the area forested will most likely increasingly interfere with the expanding domestic demand for agricultural land. Table 7.10 sketches this feature.

When comparing Tables 7.8 and 7.9 with Table 7.10, it is apparent that the figures correspond roughly only for low levels of sequestering effort. For higher levels, the divergence grows rapidly. Therefore, the conclusion seems justified that point estimates, though valid for small areas, seriously fail to describe actual costs for larger areas.

With respect to the second issue, a number of more sophisticated studies have recently begun to appear. These include Moulton and Richards (1990), Adams *et al.* (1993), Parks and Hardie (1992), Richards *et al.* (1993a), and Read (1994b). These studies refine the approach to estimating the cost of establishing carbon sequestering tree plantations in three ways. First, they estimate a cost function, not a point. Second, they refine the cost estimates for establishing tree plantations by recognizing differences associated with location and site considerations. Third, they build discounting procedures into the methodology – a common practice in the assessment of other options, but until recently one that was virtually ignored with respect to this option (see also Richards, 1993). Keeping all this in mind, it is clear that both the methodology and the empirical estimates of the various studies are still amenable to further revision. It is probably a justified generalization to state that the newer research is tending to find a somewhat steeper increase in costs than did the earlier studies, with the

Table 7.10. *Estimates of cost of carbon sequestered by tree planting: some comparative results for the U.S.*

	Total Carbon Sequestered (Mt)			
	140	280	420	700
Study	Costs ($/tC)			
Moulton/Richards (1990)	16.57	20.69	23.24	34.73
Adams *et al.* (1993)	18.50	25.11	37.21	95.06
Parks/Hardie (1992)	175.00	n/a	n/a	n/a

Note: n/a = not assessed.
Sources: As shown.

marginal costs per tonne of carbon roughly doubling, from about $30 to $60, for large annual uptakes.

Finally, it is increasingly recognized that there are probably limits to the extent to which the global system can maintain forest stocks. Nevertheless, sustainable forest management can make an important long-term contribution to providing a continuous flow of substitutes for net-emitting energy sources such as coal.

7.4.7 Methane

Methane currently accounts for about 20% of expected warming from climate change. This contribution is a result of methane's potency as a greenhouse gas and dramatically increased anthropogenic emissions. Currently, about 70% of global methane emissions are associated with human-related activities such as energy production and use (coal mining, oil and natural gas systems, and fossil fuel combustion); waste management (landfills and wastewater treatment); livestock management (ruminants and wastes); biomass burning; and rice cultivation.

Technologies and practices for reducing methane emissions from their major anthropogenic sources have been identified and reviewed through a number of expert meetings and studies, many under the IPCC. Many of the technological options currently available are cost-effective in many regions of the world and have been implemented to a limited extent. The available options represent different levels of technical complexity and capital needs and therefore should be adaptable to a wide variety of country situations. In total, it appears to be technically feasible to reduce methane emissions by about 120 Tg (75 to 170 Tg) per year through reductions in emissions from the following methane sources.

Coal mining. Techniques for removing methane from gassy underground mine workings have been developed primarily for safety reasons, because methane is highly explosive in air in concentrations between 5% and 15% and is the cause of mining accidents. Some of these same techniques can be adapted to recover methane in concentrations of 30% or more, so the energy value of this fuel can be put to use. Methane emissions into the atmosphere can be reduced by up to 50-70% at gassy mines using available techniques such as gob gas recovery (IPCC, 1990a, 1990b, 1990c; U.S. EPA, 1993; IPCC, 1993).

Oil and natural gas systems. Methane is the primary constituent of natural gas, and significant quantities of methane can be emitted to the atmosphere from components and operations throughout a country's natural gas system. The technical nature of emissions from natural gas systems is well understood, and emissions are largely amenable to technological solutions through enhanced inspection and preventative maintenance, replacement of equipment with newer designs, improved rehabilitation and repair, and other changes in routine operations. Reductions in emissions in the order of 10 to 80% are possible at particular sites, depending on site-specific conditions (IPCC, 1990b; U.S. EPA, 1993; IPCC, 1993).

Landfills. The methane generated in landfills as a direct result of the anaerobic decomposition of solid waste can be reduced by recovering this medium-BTU gas for use in electricity generation equipment or for direct use in heating or cooking equipment. At many sites reductions of up to 90% are possible. Additional benefits that result from landfill methane recovery include improved air and water quality and reduced risk of fire and explosion (IPCC, 1990b; U.S. EPA, 1993; IPCC, 1993).

Ruminant livestock. Many opportunities exist for reducing methane emissions from ruminant animals by improving animal productivity and reducing methane emissions per unit of product (e.g., methane emissions per kilogram of milk produced). In general, a greater portion of the energy in the animals' feed can be directed to useful products instead of wasted in the form of methane. As a result, herd size can be reduced while productivity remains the same. Current technologies and management practices can reduce methane emissions per unit product by 25% or more in many animal management systems (IPCC, 1990a, 1990c; U.S. EPA, 1993; IPCC, 1993).

Livestock manure. Methane emissions from anaerobic digestion of animal manures constitute a wasted energy resource which can be recovered by adapting manure management and treatment practices to facilitate methane (biogas) collection. This biogas can be used directly for on-farm energy or to generate electricity for on-farm use or for sale. The other products of anaerobic digestion, contained in the slurry effluent, can be used as animal feed and aquaculture supplements and as a crop fertilizer. Additionally, managed anaerobic decomposition is an effective method of reducing the environmental and human health problems associated with manure management. Current reduction options can reduce methane emissions by as much as 25-80% at particular sites (IPCC, 1990c; U.S. EPA, 1993; IPCC, 1993).

7.5. Adaptation Options

There are no comprehensive surveys of the various adaptation options and their costs, probably because adaptation covers such a broad range of potential action and also because of the large uncertainties surrounding these options. The literature on the subject is limited but growing.[37] In any case, it is clear

Table 7.11. *Agricultural yield changes under a* $2 \times CO_2$ *climate (percentage of gross agricultural product)*[a]

Region/Scenario[c]	UKMO Model[b]			GISS Model[b]			GFDL Model[b]		
	1	2	3	1	2	3	1	2	3
OECD America	-20.0	-5.0	-5.0	10.0	10.0	10.0	-5.0	10.0	10.0
OECD Europe	5.0	5.0	5.0	10.0	10.0	10.0	-5.0	-5.0	-5.0
OECD Pacific	7.5	7.5	7.5	7.5	7.5	7.5	7.5	7.5	7.5
Central and Eastern Europe and former Soviet Union	-7.5	-7.5	-7.5	22.5	22.5	22.5	7.5	7.5	7.5
Middle East	-22.5	-22.5	-7.5	-7.5	-7.5	7.5	-7.5	-7.5	7.5
Latin America	-22.5	-22.5	-8.5	-15.0	-15.0	-1.0	-10.0	-10.0	4.0
South and Southeast Asia	-20.0	-20.0	-10.0	-10.0	-10.0	0.0	-10.0	-10.0	0.0
Centrally Planned Asia	-7.5	7.5	7.5	7.5	22.5	22.5	7.5	22.5	22.5
Africa	-20.0	-20.0	-20.0	-7.5	-7.5	7.5	-15.0	-15.0	0.0

[a]After Rosenzweig *et al.* (1993); cf. also Fischer *et al.* (1993), Rosenzweig and Parry (1994), and Reilly *et al.* (1994).

[b]The climate change scenarios are based on equilibrium $2 \times CO_2$ experiments using the General Circulation Models of the UK Meteorological Office (UKMO), the Goddard Institute for Space Studies (GISS), and the Geophysical Fluid Dynamics Laboratory (GFDL).

[c]The scenarios are (1) no adaptation, (2) minor shifts, and (3) major shifts in behaviour.

Source: Tol (1994).

that society now already incurs large costs in adapting to climate extremes; climate change will just increase these costs.

When talking about adaptation, the central questions relate to (1) what impacts to adapt to, (2) how to adapt, and (3) when to adapt. In this section only the first two questions will be considered; no attention will be given to the aspect of insurance, which could be viewed as an adaptation option in its own right (see also Chapter 6). The question of when to adapt is one of implementing no-regrets adaptation options now (possibly developing drought-resistant cultivars and techniques) and of weighing the implementation of mitigation options now against adaptation options in the future. In the literature hardly any attention has been paid to any possible trade-off between both types of options. The section concludes with some remarks on the modelling of adaptation.

7.5.1 Adaptation to what?

Adaptation in various degrees and in some form or other may be necessary to cope with ecosystem changes that have interfaces with human (economic, social, political, legal, and cultural) activities (for a more detailed assessment of adaptation options, see Volume 2). The extent of these changes and their subsequent impact on human affairs will depend on the sequence, severity, and characteristics of the climatic changes that initiated them. Changes in temperature and associated rainfall regimes may lead to more droughts in some localities and heavier rainfall in others, thus affecting worldwide surface and groundwater availability, which in turn will affect agronomic practices and yields in agriculture. Fisheries and forestry will be affected by changes in temperature and the availability and quality of water (e.g., salinity). Temperature rise may also affect livestock populations and output through

heat stress and climate-related influences on infestations of parasites, insects, and disease.

Climate change may cause accelerated sea level rise, possibly attended by increased flooding, changes in regional temperature, increases in the frequency of storms and hurricanes, and changes in surface runoff and river discharges resulting from changes in the mean value and variability of precipitation. Impact scenarios differ considerably, however, as a result of differences in their starting assumptions: IPCC (1994), for example, assumes a 1-m sea level rise over 100 years, whereas other scenarios are based on a 50-cm rise. In addition, the response options that are considered adequate or appropriate differ significantly from study to study.

Global research on sea level rise is increasingly being carried out (Tol, 1994; Nordhaus, 1993; Cline, 1992a; Fankhauser, 1992, 1993, 1994a, and 1994b). The World Coast Conference 1993 (IPCC, 1994) has pointed to the need to integrate responses to long-term threats such as climate change and associated sea level rise with existing planning and management efforts to arrive at Integrated Coastal Zone Management. On the impacts of changes in river discharges, only some scattered information is available. As an example, the discharge of the Rhine in the Netherlands is predicted to fall by 10-15% due to an assumed temperature increase of 4°C in the Alpine part of the basin (Kwadijk, 1991, as cited in Penning-Rowsell and Fordham, 1994). However, little research has yet been carried out to combine the effects of changes in precipitation with the effects of temperature rise. A major conclusion in IPCC (1994) is that it is very difficult to differentiate between sea level rise and nonclimate-related factors, such as subsidence and excessive groundwater withdrawal, which may be equally important determinants in *relative* sea level rise.

Table 7.12. *Annual costs of sea level rise*[†]

Region	Wetland (mln. $)	Drylands (mln. $)	Protection (mln. $)	Total (mln. $)	Total (% of GDP)
OECD America	5,000	2,000	1,500	8,500	0.15
OECD Europe	4,000	500	1,700	6,200	0.14
OECD Pacific	4,500	4,000	1,800	10,300	0.45
Central and Eastern Europe and former Soviet Union	1,250	1,250	500	3,000	0.07
Middle East	0	0	0	0	0
Latin America	1,500	500	1,000	3,000	0.35
South and Southeast Asia	1,500	1,000	2,000	4,500	0.65
Centrally Planned Asia	500	0	500	1,000	0.29
Africa	500	500	500	1,500	0.43

[†]All estimates ± 50%.
Source: Tol, 1993.

Quite obviously, countries where sea level rise may become prominent may face challenges beyond what only a climate change would have entailed. Effects on agriculture may be caused by regional changes in temperature and by sea level rise. A sea level rise of 1 m would affect the supply of rice of more than 200 million people in Asia (IPCC, 1994). Changes in temperature would have mixed regional effects (Tol, 1994, based on Rosenzweig *et al.,* 1993). Effects on agriculture would depend on the full range of possible impacts of climate change (as well as CO_2 fertilization) and not just temperature (see Table 7.11).

Similarly, a sea level change would pose problems or present opportunities for numerous other activities, including fishing and mangrove forestry. Human habitation would also be affected (through changes in water quality) as would industry and trade (through relocation of industries and loss of infrastructure). Both these factors may also affect human health and nutrition.

Human adjustment, however, will be affected by a complex array of factors over time. Thus, a study by the Asian Development Bank (1994) shows that, whereas the agronomic yield of rice may increase, the "realized" increase may be lower than the agronomic potential due to the interplay of demand and supply factors.

7.5.2 How to adapt

Options for adapting to sea level rise can generally be categorized as retreat, accommodation, or protection.

Retreat will cause loss of dry land and loss of wetlands. IPCC (1994) computes that a 1-m sea level rise could threaten 170,000 km^2 (or 56%) of the world's coastal wetlands. Loss of dry land means losses in agriculture, in forestry, in species, and in physical assets and implies migration of people. Attempts to estimate these losses can be found in Ayres and Walter (1991), Rijsbergen (1991), Fankhauser (1994a), Cline (1992a and 1992b), Suliman (1990), Nordhaus (1991b, 1993), and Tol (1994). In some of these sources estimates have been made of the land protection costs insofar as land loss is pre-

vented on economic grounds by such factors as coastal infrastructure. An overview of costs (on an annual basis) for both wetland and dryland losses is given in Table 7.12.

Not all estimates include the side effects of resettling people that used to live on the lost land. These costs involve the costs of taking up refugees on the one hand and of people leaving (and of the hardships they may endure) on the other. By combining various sources of information, Tol estimated the global annual costs of relocating due to sea level rise at some $14 billion. These costs vary between 0.01 and 0.03% of GDP for the OECD and the countries in transition, and between some 0.3 and 0.8% for the developing regions (Tol, 1993).

Accommodation to sea level rise involves not only the adaptation of existing structures to a higher sea level but also a variety of other responses, such as the elimination of subsidized insurance in industrialized countries for building new structures along sea shores. In a state of transition it may also involve the need to respond to inundations causing loss of life and damage to assets, agriculture, and the environment (Penning-Rowsell and Fordham, 1994).

Protection against sea level rise would involve major costs, but estimates of these differ. Tol (1994), assuming a 0.5-m rise, estimates annual global coastal defence costs at $9.5 billion in constant 1988 dollars. IPCC (1994), assuming a 1-m rise, computes costs of $10.0 billion per year in constant 1994 dollars, whereas Ayres and Walter (1991, as cited in Winpenny, 1994) derive a figure of $50-100 billion in constant 1981 dollars. Protection costs for an increased intensity of storms are not available on a global scale.

Adaptation to changes in river water discharge involves the same choice of options as adaptation to sea level rise: retreat, accommodation, or protection. Unfortunately, no global costs are available for any of these. For a European example, however, see Penning-Rowsell and Fordham (1994).

Adaptation to changing temperatures involves adjustments in health care, heating and cooling facilities, and household activities, and the adaptation of agriculture and fisheries. Improvements to infrastructure, including urban buildings and construction as well as water control and storage systems (such as dams, drainage and sewer systems, dikes, and locks), would also be needed.

In agriculture various types of technical responses are available. These include changes in farming strategies and crop management as well as changes in crop variety, irrigation, fertilizer, and drainage. Some salt-tolerant crops, to give an example, can be very successfully grown along the shoreline of coastal deserts when irrigated with ocean water. Global and regional estimates for different levels of adaptation in agriculture are presented in Table 7.11.

Given our still limited understanding of climate change, extending the range of policy options rather than refining technical responses seems to be the most logical approach at the moment. The following options deserve special attention:

- Capacity building, in both industrialized and developing countries, to educate people in the former about the effects of their activities on carbon-trapping biota and people in the latter about responses to the effects of natural climatic variability and of potential future climate changes.

- Changes in land use allocation, including developing the potential of tropical plant species. Since most of the world's plant food comes from only 20 species, the potential of the vast majority of plant species is still to be developed.

- Improvements in food security policies and reduction of postharvest losses. Given that postharvest losses – due to deficient systems of storage and transport – amount in many developing countries to 50% of production, or more, major scope for improvement does seem to exist.

- Conversion to "controlled environment agriculture." Massive introduction of integrated "controlled environment agriculture" in developing countries might easily require an investment of several tens of billion of dollars, or billions of dollars per annum if introduced over some decades.

- Aquaculture. Climate change affects ocean circulation in the upper layers, upwelling, and ice extent, all of which affect marine biological production and, hence, marine fisheries. One way to adapt is to intensify efforts to develop aquaculture. Integrating aquaculture with "controlled environment agriculture" has a great potential, given recent dramatic advances in marine biotechnology. The almost sterile, nutrient-rich bottom water from Ocean Thermal Energy Conversion (OTEC) systems holds considerable promise as a culture medium for kelp, abalone, oysters, and a range of fish species.

It should also be mentioned that for marginal groups the risks of damages due to climate change will become larger the more unequal the land distribution system is. Changes in land tenure may, therefore, as a side effect, reduce these risks and can be viewed as an indirect adaptation option in themselves. As a final remark, it may be pointed out that patterns of scarcity and surplus will change across regions and over time, presenting new opportunities for trade between nations as they respond to stabilize supply.

7.5.3 Adaptation measures in developing countries

In developing countries, as elsewhere, adaptation depends on the type and intensity of the impacts of climate change that may occur. Depending on these impacts, adaptation may be applied immediately or may be delayed. In the case of the African countries, however, no real adaptation studies have yet been carried out. Current bilateral and multilateral activities are expected to lead to a more systematic assessment of adaptation options and their costs.

The quest for adaptation options, however, already existed long before the global debate on climate change began. Countries in arid and semi-arid zones have tried to find long- and short-term responses to recurrent droughts for some time now, while countries in heavy rainfall regions and those affected by storms and cyclones in their coastal areas have tried to find both structural (engineering) and nonstructural (institutional) means for dealing with recurrent floods.

Short-term responses to recurrent droughts include improvements in drought preparedness and focus primarily on drought relief and drought recovery activities. Drought relief typically includes supplementary food programmes and programmes to protect and replenish livestock. Drought recovery entails such activities as the provision of seed and land preparation supplements to farmers after a period of drought. However, even for these short-term responses no systematic studies have been carried out.

Long-term measures include regional and national research efforts to develop drought-resistant crops and breed hardy livestock. The incorporation of drought and salt resistance in crop varieties is thus already a major item on the research agenda in some developing countries. Further activities, particularly those strengthening research capacity and financial support for research, are necessary and will almost certainly prove to be cost-effective. In areas where water resource management will become crucial because of large changes in rainfall regimes, an improved and more environmentally sound infrastructure will be necessary, while policies encouraging water conservation (e.g., pricing mechanisms in which prices reflect social scarcity) will need to be introduced.

The electricity generation sector, which will also be heavily affected by changes in climatic patterns, has already had to develop adaptation responses to problems outside the context of climate change. Facing massive river and dam silting and below-average precipitation to replenish hydroelectric installations, some nations have sought to develop alternative base load systems, such as coal thermal. A more systematic assessment of these responses will prove to be crucial, particularly in the light of the indicated importance of decarbonizing the fuel base to reduce emissions.

7.5.4 Modelling adaptation

Climate change adaptation models have been developed for sea level rise, storminess, and changes in river discharges. A methodology for assessing damages can be found in Howe *et al.* (1991) and in Green *et al.* (1994). Penning-Rowsell and Fordham (1994) present a general methodology for adaptation, whereas models for flood hazard assessment and management can be found in Klaus *et al.* (1994). Correia *et al.* (1994) present a framework for the analysis of river zone management, including the institution setting.

Two important problems can be mentioned with respect to the modelling of adaptation. The first of these includes the general set of greenhouse assessment problems, such as the handling of time, uncertainty, and discount rate. The second is specific to adaptation and involves the valuation of intangibles, such as wetlands and species. A valuation in dollars per person for protecting threatened species, for example, cannot be compared with a valuation in dollars per kilometre for protecting threatened coasts.

7.6 An Integrating Approach

A major part of the literature on response options focusses on the various technologies and their cost-effectiveness within a specific option. The options themselves, however, are not assessed on the basis of broader comparisons. The main explanation for this "partial" approach is probably the limited availability of reliable and accepted data about the options' costs and benefits. Moreover, one increasingly recognizes that the costs of the various options critically depend on the assumptions employed about the efficiency of the baseline scenario used in the analysis (see also Chapters 8 and 9).

A truly generic assessment, however, requires an integrating framework that allows a simultaneous evaluation of the various technologies. If emission reduction targets are to be achieved in an optimal way, not only economically but also in terms of flexibility and spreading of risks, a full picture of all the alternatives should be available, so an integrated portfolio of options can be determined that minimizes the costs of a given level of carbon reduction. (The integrating approach in this chapter should not be confused with the integration of costs of a given option, or with the integrated modelling approach treated elsewhere in this report.) One option, drawing on economics, is to apply a cost-benefit or cost-effectiveness criterion for decision making. That approach is highlighted here. Other approaches to decision making are also possible. One could rely, for example, on the concept of safe minimum standards (which may be particularly important in evaluating investments in nuclear power plants).

The need for an integrating approach is reinforced by the fact that many of the options' cost functions appear to show internal diseconomies of scale (for some evidence with respect to forestry options see, e.g., Moulton and Richards, 1990; Adams *et al.*, 1993; Parks and Hardie, 1992; and Qureshi and Sherer, 1994; with respect to energy technologies, see, for example, Kram, 1994b, and Southern Centre/ Risö, 1993; for a broader analysis, see, e.g., TNO, 1992). The implication is, therefore, that, after reaching a certain scale of application, the most efficient option will become more costly than another option, and this, in turn, may eventually become more costly than yet another option. The discussion of the marginal costs of CO_2 abatement in Chapters 8 and 9 is relevant in this respect.

The need for an integrating approach is further reinforced by evidence that cost functions per option also differ from place to place because of regional variations in supply conditions, levels of technology, infrastructure, and other factors. Evidence suggests that even within a relatively homogeneous area, such as the European Union, marginal emission reduction cost curves differ significantly (COHERENCE, 1991); *a fortiori,* one can hypothesize that some options can also be significantly more cost-effective in one place than in another (McKinsey & Company, 1989).

A number of integrating studies have been carried out. Those using a "top-down" methodology attempt to provide a comprehensive analysis based on generalized estimates of the cost functions of the various options (McKinsey & Company, 1989; Nordhaus, 1991a; Jepma and Lee, 1995). Others using a "bottom-up" approach commonly pursue a greater level of detail (Jackson, 1991; Rubin *et al.,* 1992; Mills *et al.,* 1991; Kram, 1994b).[38]

In the "top-down" studies, the regional differences between the cost functions of the various options provide a strong case for their joint implementation, if the ultimate CO_2 reduction target is to be achieved with the least cost (see also Article 4.2.B of the Framework Convention on Climate Change). This result is valid, irrespective of which parties take the main responsibility for financing the options.

However, a number of other considerations may affect these conclusions. Often the various sources are not completely clear as to the degree to which opportunity costs, social and institutional barriers, and other environmental side effects have been included in the cost functions employed. Furthermore, as was explained in Section 7.3, cost functions may differ depending on whether they have been designed according to the engineering efficiency approach or the welfare economic approach. These differences will obviously have a strong impact on the outcome of integrated assessments. To the extent that welfare considerations will cause cost functions to shift upward (especially for countries in transition and developing countries) in comparison with those calculated on the basis of engineering efficiency, the anticipated scope for joint action may be reduced.

In addition, the global costs of achieving ambitious long-term emission reduction targets (such as reducing annual emissions to half the present level) – commonly estimated at several hundreds of billions of dollars per annum – turn out to be rather sensitive to the degree to which one assumes scope for no-regrets policies, especially in energy conservation, efficiency improvement, and fossil fuel switching.

The top-down studies also indicate that in the optimal case all options must be applied at the same time and in all regions, albeit to different degrees. The largest potential in overall emission reduction at current cost estimates seems to be in forestry (especially in developing countries) and energy conservation and efficiency improvement (especially in the OECD and Eastern Europe). Renewable energy (particularly in developing countries) and fuel switching (especially in Eastern Europe if methane leakages can be limited) are also important, though to a lesser extent. Needless to say, the optimal mix may easily change as a result of future technological progress.

To illustrate, the results of a linear programming optimization procedure have been presented in Table 7.13 (Jepma and

Table 7.13. *Base case simulation: Optimal mix of options for a global emission reduction of 2.4GtC (marginal costs: $50/tC)*

Option	Level of emission reduction (MtC)			
	OECD	Eastern Europe	Rest of the World	Total
1. Energy conservation and efficiency improvement	250 (250; 250)	250 (250; 250)	100 (100; 100)	600 (600; 600)
2. Fuel switching	50 (50; 50)	50 (50; 50)	50 (50; 50)	150 (150; 150)
3. Removal and disposal	100 (100; 150)	50 (50; 100)	0 (0; 50)	150 (150; 300)
4. Nuclear energy	50 (50; 50)	50 (50; 50)	0 (0; 50)	100 (100; 150)
5. Renewable energy	50 (100; 50)	50 (100; 100)	100 (150; 150)	200 (350; 300)
6. Forestry	250 (250; 250)	250 (250; 250)	700 (550; 400)	1200 (1050; 900)
Total	750 (800; 800)	700 (750; 800)	950 (850; 800)	2400 (2400; 2400)

Note: Figures in parentheses give the results for a 50% reduction in the marginal costs of renewables and for a doubling of the marginal costs of forestry respectively over all intervals.
Source: Jepma and Lee (1995).

Lee, 1995). The procedure starts from a predetermined emission reduction target and is applied to the cost functions of the various options per region, featuring stepwise increasing marginal costs and based on data from a combination of sources (McKinsey & Company, 1989; Jackson, 1991; Mills *et al.,* 1991; and Rubin *et al.,* 1992). The table shows the optimal mix of options both in terms of types of options and of regions of application if a medium-term emission reduction target of -2.4 GtC is to be achieved. (The figures in parentheses show the outcomes if the marginal costs of the renewable option are assumed to be 50% of those in the base case and if the marginal costs of the forestry option are doubled compared to the base case. This sensitivity test suggests that the outcomes are fairly robust. Obviously, various other sensitivity tests, e.g., on the impact of changing lifestyles, could be carried out.)

Kram (1994b) is a detailed integrating response study in the bottom-up tradition. Here an overall assessment was made on the basis of long-term bottom-up country models (MARKAL) for nine Western countries, which integrate more than 70 technologies (including more than 30 supply technologies and more than 40 end-use technologies). A range of targets for CO_2-emission reductions by 2020 was tested to determine the mix of energy technologies that would produce the reductions at the least total energy system cost. The results revealed considerable diversity in the optimal paths of the different countries in terms of the mix of energy technologies and the cost and amounts of reductions that could be achieved. This diversity resulted from the future energy needs of the various countries, as well as their existing energy systems, natural resources, technology options, and energy policies (especially with regard to hydroelectric and nuclear power).

Second, the study calculated the marginal costs of CO_2 reduction for the 1990-2020 period for several countries (Figure 7.3). The results clearly show that the marginal costs vary greatly among the countries according to their circumstances. If one accepts that the most efficient allocation of emission reductions would be at the point of equal marginal costs, these results provide further justification for implementing options on a joint or cooperative basis.

There appear to be no similar detailed integrating response studies dealing with the developing countries. Given that the

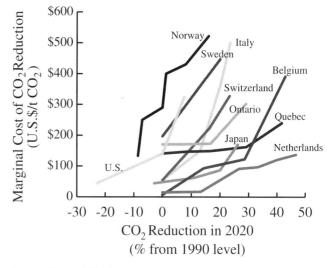

Source: Kram (1994a).

Figure 7.3: Marginal costs of CO_2 emission reduction.

level of economic development and other circumstances vary greatly among developing countries, the marginal costs of emission reduction for them are likely to be very context-specific.

7.7. Regional Differences and International Cooperation

Though the industrialized countries constitute only 25% of the world's population, they account for 72% of the current global energy-related carbon emissions and some 80-85% of cumulative historical carbon emissions (Fujii, 1990). Clearly, such numbers require the industrialized countries to assume their historic responsibility, which has been translated into the concept of "common but differentiated responsibilities" mentioned in Article 3.1 of the FCCC. This has also been elaborated in Principle 7 of the Rio Declaration, which states: "The developed countries acknowledge the responsibility that they bear in the international pursuit of sustainable development in view of the pressures their societies place on the global envi-

Table 7.14. *Comparison of the 1990 development and energy situations in the main (sub-)regions*

Region	Tr (%)	Fossil Fuels (%)				Al (%)	ΣEner (Mtoe)	Popul (million)	GNP (billion US$)	GNP ($ per capita)	Toe per capita	Toe per thousand US$
		Co	Oi	NG	ΣF							
West	1	22	41	19	82	16	4296	849	14840	17473	5.06	0.29
North America	2	22	38	24	83	15	2289	276	5738	20809	8.30	0.40
Western Europe	1	22	43	16	80	19	1456	430	5901	13726	3.39	0.25
JANZ	1	21	51	12	84	16	551	144	3202	22279	3.83	0.17
East	1	25	29	36	91	8	1745	414	3047	7367	4.22	0.57
Eastern Europe	1	46	25	19	90	9	352	125	388	3101	2.81	0.91
ex-USSR	1	20	31	40	91	8	1393	289	2660	9215	4.83	0.52
North	1	23	38	24	85	14	6040	1263	17888	14163	4.78	0.34
Africa	37	21	27	9	57	6	360	642	385	600	0.56	0.93
North and South Africa	5	na	na	na	92	3	154	150	227	1507	1.02	0.68
sub-Sahara	61	na	na	na	30	9	206	492	158	322	0.42	1.30
Asia	15	49	25	4	78	7	1475	2810	1237	440	0.53	1.19
China	6	72	15	2	89	5	718	1139	393	345	0.63	1.83
India	25	41	23	4	68	7	253	853	287	337	0.30	0.88
Other	23	19	39	8	66	11	504	817	557	681	0.62	0.90
Latin America	15	4	45	14	63	22	556	448	842	1880	1.24	0.66
Brazil	30	5	34	2	41	29	185	150	375	2495	1.23	0.49
Mexico	5	5	49	22	75	20	142	89	170	1920	1.60	0.83
Other	9	4	51	20	74	17	228	209	297	1421	1.09	0.77
Middle East	1	1	64	32	98	1	233	129	370	2860	1.81	0.63
South	17	31	33	10	74	10	2624	4029	2835	704	0.65	0.93
World	6	25	36	20	81	13	8664	5292	20723	3916	1.64	0.42

Note: Tr = traditional (woodfuel, crop residues, and animal dung); Co = coal; Oi = Oil; NG = natural gas; ΣF = total fossils; Al = alternatives (nuclear, hydro, wind, geothermal, etc) (all % of total energy); ΣEner = total energy; (M)toe = (million) tons of oil equivalent; popul = population; GNP = gross national product in 1989; na = not available. Western Europe = OECD Europe (includes Turkey); Eastern Europe = non-OECD Europe (includes Cyprus, Gibraltar, and Malta); JANZ = Japan, Australia, and New Zealand.
Source: Ettinger (1994), based on *BP Statistical Review 1992* and *World Resources 1992-93.*

ronment and of the technologies and financial resources they command."

Table 7.14 compares economic development and energy use in the main regions of the world as of 1990. This comparison highlights the following striking differences between and within regions and may serve to clarify why the involvement of the developing countries in greenhouse policy formulation and implementation is imperative:

- GNP per capita varies from an average of $440 for Asia to $17,473 for the West, a ratio of 1:40. The differences can also be large within regions: sub-Saharan Africa has a per capita GDP of $322 as against $1,507 for North and South Africa together (a ratio of 1:5).

- The relative use of traditional energy (woodfuel, crop residues, and animal dung) varies from 1% for the East to 37% for Africa. Within Africa traditional fuels account for 5% of energy use for all Africa and 61% for sub-Saharan Africa (1:12).

- The relative use of fossil fuels varies from 57% of total energy use for Africa to 98% for the Middle East; within Africa it amounts to 30% for sub-Saharan Africa and 92% for all Africa (1:3).

- The relative use of nuclear plus renewables varies from 1% for the Middle East to 22% for Latin America. Within Africa it amounts to 3% for all Africa and 9% for sub-Saharan Africa.

- The share of the main energy source as a percentage of total energy use varies from 36% for the East (natural gas) to 64% for the Middle East (oil). For Africa the main source is traditional fuels, for Asia coal, and for the West and Latin America oil.

- In terms of energy use per capita, the regions vary from 0.53 tonnes of oil equivalent (toe) for Asia to 5.06 toe for the West, or by a ratio of 1:10. Between Western subregions it still varies from 3.39 for Western Europe to 8.30 toe for Northern America.

- Energy intensity (in toe/$1000 GNP) varies from 0.29 for the West to 1.19 for Asia (1:4). Between Western subregions, it varies from 0.17 for Japan/Australia/New Zealand to 0.4 for Northern America. However, energy intensity in toe/$1000 GNP is an unreliable yardstick for comparisons between regions, especially between developed and developing countries, because of differences between nominal GNP and real GDP in purchasing power parities (PPP). If the energy intensity data in the table were expressed in toe/$1000 PPP (correction based on UNDP Human Development Report, 1993 data), the energy intensity of the developing countries would become 0.35 but that of the developed countries would remain 0.34. However, this would still ignore the relatively higher energy content of the developing countries' imports and the lower energy contents of its exports. If these two factors are taken into account, energy intensity in the developing countries would be higher than that in toe/$1000 PPP but probably lower than that in nominal terms (Ettinger, 1994).

Most scenarios suggest that during the next decades the growth in carbon emissions will increasingly take place in the developing countries. According to the data summarized in Alcamo *et al.* (1995), the mean world annual growth rate of CO_2 emissions for over twenty scenarios is 1.56%, the corresponding mean rate for China is 2.83%, for Eastern Europe and ex-USSR 0.76%, and for Africa 3.85%. Consequently, according to one of the scenarios in that study, ECS 92 (dynamics as usual), the share of developing countries in global CO_2 emissions is projected to reach 46% by 2020 (as compared to 34% for the OECD and 20% for countries in transition). However, according to the various World Energy Council (WEC) scenarios, the developing countries' share in 2020 would be over 60%. In any case, it seems most likely that the developing countries as a group will start to become the major CO_2 emitters within a few decades. This picture is reinforced if the emissions of CH_4 from wetland rice cultivation and from enteric fermentation are also taken into account.

At the same time it is clear that, although the scope for effectively applying policy options in the developing countries seems to be significant (for a recent evaluation of various technical options at the country level, see UNEP, 1994), so are the obstacles to be encountered. Indeed, the availability of technical options for higher energy efficiency, to give just one example, does not guarantee their adoption on a large scale. There may need to be a significant stimulus to achieve widespread efficiency improvements, particularly in markets characterized by high implicit discount rates. But a combination of education, financial incentives, and minimum efficiency standards coupled with freedom from distortionary policies can effectively transform energy use markets so that large energy savings and emission reductions are achieved along with net economic savings (Geller and Nadel, 1994).

The literature on the adoption and diffusion of technology clearly indicates that while profitability is probably the most straightforward determinant of the adoption of a new idea, a new technology, or new equipment, various other factors may also be important. A review of recent research into the diffusion of energy technologies in developing countries shows that there are many financial, institutional, and other factors that influence the successful adoption of these technologies (Barnett, 1990; Ghai, 1994). In Africa, for example, social resistance has impeded the diffusion of drought-tolerant crop varieties, and such resistance could also inhibit the adoption of new energy technologies. Often the initial awareness of benefits and new opportunities may be contingent on such

factors as winning the support of women for more energy-efficient cooking stoves.

Moreover, one necessary ingredient for the adoption of new technology, namely a pool of local skills to draw on, may be lacking or inadequate in many cases, so that even proven technologies may spread rather slowly in these countries. For all these reasons an adequate and timely process of energy efficiency institution building seems imperative, especially in developing countries. There is evidence that the existence of such separate institutions has helped Indonesia, South Korea, and Thailand, for instance, make greater headway in the scope and coverage of their energy efficiency policies and programmes (Byrne *et al.*, 1991).

Endnotes

1. The "engineering efficiency" approach determines the financial costs and benefits of various options to an individual agency or other entity in terms of CO_2 emission reduction/absorption; in the "welfare economics" approach the broadly defined costs and benefits of options to society are determined. These two approaches will be further discussed in Section 7.3.

2. For an example of energy conservation, see, for example, Rubin *et al.* (1992). Here 25% of employer-provided parking places are eliminated and the remainder taxed to reduce solo commuting by 15-20% in the U.S. Net costs are estimated to be -$22/tC (a negative cost is the same as a saving).

3. It may seem that, although the above categories are conceptually distinct, in real life they are not strictly mutually exclusive; that is, measures are conceivable that can be classified in more than one category. An example would be the plantation of forests or biomass used for energy purposes. These measures seem to fall into both category 3 (renewable energy) and category 6 (enhancing carbon sinks). However, this is not the case. The measures are an example of how easily markedly different processes that underlie the measures can be confused.

In the case of forests, broadly three types of measures are conceivable to fix carbon: (1) to afforest new lands to let the forest simply mature; (2) to plant forest and sequester the timber derived from it; and (3) to use the wood for energy purposes on a sustainable basis, thereby avoiding the alternative use of fossil fuels. In the following, (1) and (2) are discussed in Section 7.4.6 (forestry options), whereas (3) belongs to the renewable/biomass category. With respect to (3), it should be borne in mind that sometimes a significant amount of additional energy may be required to turn the biomass into energy. This is, for instance, the case for the production of ethanol from corn, where additional energy requirements are of the order of the energy content of the produced ethanol itself (Swisher *et al.*, 1993).

Another example would be to classify an Integrated Gasification Combined Cycle (IGCC) or the hydrocarb process in both category 1 (energy saving and efficiency) and category 3 (clean fossil technologies). In the present chapter, both are considered primarily clean technologies and both change the energy conversion process to the extent of violating the definition for the energy saving and efficiency category. However, ultimately no clear distinction can be made as modifications in the energy conversion process become minor (due to further technological progress).

4. For instance, studies for Poland, Hungary, and the former USSR indicate that a combination of energy efficiency improvements, fuel substitution, and structural change (Chandler, 1990), could reduce carbon emissions by 40-60% from base case projected levels by 2030. In the case of Poland, Sitnicki *et al.* (1990) suggest that base-

line emissions of 260 Mt could be reduced to 117 Mt by 2030; for the former USSR, Makarov and Bashmakov (1990) suggest that a reduction of 40% would be feasible.

5. For a more detailed discussion of top-down versus bottom-up modelling, see Chapters 8 and 9.

6. The WEC distinguishes four scenarios for the energy mix in 2020: Scenario A assumes high annual world economic growth (especially in developing countries), high annual energy intensity reduction, and very high total energy demand; scenario B1 assumes moderate annual world economic growth rates, moderate annual energy intensity reduction, and high possible total energy demand; scenario B, the reference scenario, assumes high annual energy intensity reduction; scenario C assumes moderate annual economic growth, very high energy intensity reductions, and relatively low total energy demand in 2020.

7. Here a set of definitions taken from Rogner *et al.* (1993) is used to distinguish between different levels of geological certainty and economical and technical feasibility. The *resource base* is defined to consist of (proven) *reserves* and *resources*. Reserves are those occurrences that are identified, measured, and known to be economically and technically recoverable at current prices and using current technologies. Resources are the remainder of occurrences with less certain geological and economic characteristics. Additional quantities with unknown certainty of occurrence or with unknown or no economic significance at present are referred to simply as *occurrences*.

8. Total global energy consumption amounted to 10 TWyr in 1990, whereas identified fossil energy reserves are estimated at 1,280 TWyr (Rogner *et al.* 1993). Obviously, the use of an aggregate figure for fossil fuels (mostly coal reserves) should not obscure the fact that the corresponding time span for the individual fossil fuels differs widely. The ratio of proven reserves to annual production (R/P) is estimated at about 55 years for natural gas, 45 years for oil, and 235 years for coal.

9. IPCC carbon emission rates are 15.3, 20.0, and 25.8 kg of carbon per GJ for natural gas, crude oil, and (bituminous) coal respectively (IPCC, 1995).

10. This is because transport and combustion technologies are roughly the same for natural gas and hydrogen (H_2).

11. Assuming a 100-year time horizon. For the various ways the GWP measure for methane could be calculated, see, for example, Reilly and Richards (1993).

12. This would imply that 3-41% (for distribution) and 1-63% (for production) of the carbon reduction from a 100% coal-to-natural-gas fuel switch would be offset by the detrimental effects of leakage.

13. $BEP = A/[(MER \times GWP) + A]$, with BEP = break-even point, $A = (25.8 - 15.3) \times 3.67$, MER = mass:energy ratio for methane = 22 Tg CH_4/EJ, GWP = global warming potential index of methane = 24.5. The term A is the additional mass of carbon dioxide released by coal compared to methane per GJ of energy and is composed of the difference between the carbon emission rates of coal and methane (25.8 and 15.3 kgC/GJ respectively; see note 9) times the mass ratio of CO_2:C (3.67). The calculation assumes a zero leakage rate of methane in coal production.

14. See, for example, Jackson (1991) for an analysis of cost-effectiveness in the UK, explicitly incorporating CH_4 leakage.

15. In this respect, electricity and hydrogen appear as ideal intermittent energy carriers from a technological point of view.

16. See, for example, IPCC (1991), Johansson *et al.* (1993), WEC Commission (1993), or WEC (1994). With respect to the classification of renewables, it should be noted that the classification of geothermal as a renewable resource is technically not correct, as the Earth's core will slowly but surely cool down.

17. Solar can broadly be subdivided into solar thermal, solar architecture, solar thermal-electric, photovoltaic systems, and thermo-

chemical and photochemical systems. Wind and hydro are relatively homogeneous energy technologies, the largest differences stemming from scale of operation. Here, a distinction is made between small/medium-scale and large-scale conversion systems. In contrast, biomass appears to be the most complex of all technologies. A wide range of conversion technologies exists, depending on the type of feedstock used and the form of energy output required. Geothermal consists of hydrothermal, hot dry rock, geopressured, and magma resources technologies. Current ocean technologies encompass tidal, wave, biomass, and salt and thermal gradient technologies.

18. Different approaches describe the concept of "practicable," i.e., realizable, potential. The most common categorizations are physical, technical, and economic, in that order, with each ensuing category being a subset of the earlier mentioned one. The physical potential would denote the maximum potential that is constrained by geological, geophysical, and meteorological factors only. Technical potential would refer to that part of physical potential that can be exploited given the state of technology at hand. Finally, the remainder of technical potential after excluding what is not deemed feasible due to prevailing economic constraints (such as a prohibitive level of costs, institutional constraints in the energy markets, etc.) would pass for economic potential. Notice that, for the present purpose, the former of the three can be considered constant in time, whereas the others prevail only at a certain moment.

Practicable potential would now be defined as somewhere between technical and economic potential. This is because the two do not hold independently but are interlinked in time; e.g., technical potential is enlarged by investments that stimulate technological progress. Conversely, the impact of improvements of, say, silicon films in photovoltaic systems on the price of solar energy is obvious.

19. For example, wind energy costs depend heavily on wind speed and solar energy costs on solar irradiance, features that are not equally favourable for all locations, seasons, or times of day.

20. It is not possible to derive cost developments for individual subclasses of technologies from the listed figures, as they are aggregated into ranges of similar technologies. The same holds for disparities stemming from differences among sites. It should be realized that these limitations significantly hamper direct comparison. However, greater detail was avoided for the purpose of clarity.

21. Estimates are 21.3-29.6% in 2020 (WEC, 1993), 15% (6% of which comes from hydro) in 2020 (Grübler *et al.*, 1993a), and close to 43% in 2025 (Johansson *et al.*, 1993). According to Grübler *et al.* (1993a) the latter estimate is most likely too high. It would imply an unprecedented rate of change of technology and infrastructure. For comparison, it took about 80 years for the market share of oil to grow to 40% of global primary energy supply (Grübler *et al.* 1993a). In the past the mean interval for replacing most technological systems has been about 30 to 40 years.

22. For an extensive discussion of the nuclear option, see Volume 2 of this report.

23. Relatively minor fossil fuel inputs are used to support the overall functioning of breeder reactors.

24. Recovery of carbon at power plants has the advantage of removing carbon from energy before it is distributed to highly dispersed end users.

25. As the use of coal and natural gas is predominant in power plants, virtually no technologies are based on oil.

26. In an ICGCC coal is converted prior to combustion. After some intermediate steps, CO_2 and H_2 are obtained. The former can be extracted by absorption at a 98% rate and the latter can be used either directly in the power plant to generate electricity or as a carbon-lean fuel to be distributed to end-user sectors, like households, industry, or transport. Modifying a conventional gas- or coal-fired boiler involves changing the oxidant from air to pure oxygen. The gas turbine of an ICGCC or a STIG (steam-injected gas turbine) can be modified by changing the combustion medium into an O_2/CO_2 medium.

27. Cost information suggests that absorption and oxyfuel combustion are the most attractive. It appears that absorption is cheaper for conventional coal-derived flue gases than for natural gas flue gases. An ICGCC is promising, though it is not clear yet whether it will replace proven conventional pulverized coal-fired installations.

28. In enhanced oil recovery, part of the injected CO_2 reenters the atmosphere, and in food packaging CO_2 is released within days or weeks. Obviously, insofar as CO_2 is released into the atmosphere, these applications, though perhaps commercially interesting, are of no significant long-term interest from an abatement point of view.

29. Applications of carbon (dioxide) storage exist in the food, chemical manufacturing, metal processing, and oil industries. Enhanced oil recovery, in which carbon dioxide is pumped into the production well to increase recovery rates, has the highest potential.

30. The combined potential of the other applications is limited to several hundreds of MtC per year. Storage capacity of the ocean is very uncertain, as it already contains nearly 40,000 GtC as (dissolved) CO_2 (compared with some 750 GtC in the atmosphere). Moreover, most of the injected carbon will come out after fifty to several hundreds of years, depending on the depth and method of injection. There may also be objections to ocean storage because of potential environmental impacts from the methods used.

31. Here, forestry measures are distinguished from the use of biomass as a renewable energy resource. Forests, like biomass, could be classified as a renewable energy resource if harnessed for energy purposes and harvested in such a way that supply is practically unlimited and no additional energy is required. This means that the plantation is rotated after harvest and *net* energy inputs for the energy extraction and conversion process are negative, or *gross* energy inputs are easily paid out of the extracted energy.

Inherently, net carbon emissions (removal) will be zero for such applications of forests or biomass, since the carbon emitted by combustion is exactly offset by the carbon removed in the next generation of plantations. Net carbon emissions are reduced *only if* the forest or biomass energy substitutes for fossil energy.

32. Afforestation is defined to apply to lands that have not been covered by forests for the last 50 years. In contrast, reforestation applies to lands that were cleared no longer than 50 years ago.

33. See Volume 2, Chapter 24, Management of Forests for Greenhouse Gas Emissions, for a detailed assessment of the subclasses of forestry measures, the potential quantity of carbon that could be conserved and sequestered by forestry measures, the effects of climatic and demographic changes on the potential amount of carbon conservation and sequestration, and the new research directions needed to improve the assessment and development of practical forestry strategies.

34. Besides carbon stored in the forest wood itself, soil carbon and carbon in other biomass growing in the forest are included.

35. Cumulative uptake refers to the total amount of carbon stored after a certain period, usually after the forest has reached maturity and no (net) carbon is absorbed any more; in other words, the incremental uptake is nil. Annual uptake is usually described by one figure only, this being an average annual uptake rate. However, for plantations logged before maturity, it should be noted that the absorption rate is dependent on the age of the forest. In contrast to widespread belief, annual uptake of a newly planted forest is in general not greatest in the first years, but only after the forest has reached an intermediate age. More precisely, accumulated uptake is an S-shaped growth function of time (Nilsson, 1982; Cooper, 1983).

36. Opportunity costs for land would in theory largely be reflected in land market rents.

37. For an overview of the costs of greenhouse damages, see also Chapter 6 of this report.

38. Note that the distinction between bottom-up and top-down modelling employed here does not coincide with a similar distinction elsewhere in the literature, where top-down approaches are associated with macroeconomic modelling techniques assuming fixed behaviourial patterns, and bottom-up approaches with identifying the (technical) opportunities presented by a changeable world.

References

Abdel-Khalek, G., 1988: Income and price elasticities of energy consumption in Egypt: A time series analysis, *Energy Economics,* **10**(1), 47–57.

Adams, R.M., D.M. Adams, C.C. Chang, B.A. McCarl, and J.M. Callaway, 1993: Sequestering carbon on agricultural land: A preliminary analysis of social cost and impacts on timber markets, *Contemporary Policy Issues,* **11**(1), 76–87.

Alcamo, J., A. Bouwman, J. Edmonds, A. Grübler, T. Morita, and A. Sugandhy, 1995: An evaluation of the IPCC IS92 emission scenarios. In Intergovernmental Panel on Climate Change, *Climate Change 1994,* J.T. Houghton, L.G. Meira Filho, J. Bruce, Hoesung Lee, B.A. Callander, E. Haites, N. Harris, and K. Maskell, eds., pp. 247–304, Cambridge University Press, Cambridge.

Asaduzzaman, M., 1995: Energy savings potentials, issues and constraints, paper presented at the international conference on Joint Implementation, Groningen, 1–3 June 1994, *The Feasibility of Joint Implementation,* C.J. Jepma, ed., Kluwer Academic Publishers, Dordrecht, The Netherlands.

Asian Development Bank, 1994: *Climate change in Asia: Bangladesh, country report,* Manila.

Ayres, R.U., and J. Walter, 1991: The greenhouse effect: Damages, costs and abatement, *Environmental and Resource Economics,* **1,** 237–270.

Baldwin, S.F., 1986: *Biomass stoves: Engineering design, development, and dissemination,* VITA, Arlington, VA.

Barnett, A., 1990: The diffusion of energy technology in the rural areas of developing countries: A synthesis of recent experience, *World Development,* **18**(4), 539–553.

Benedick, R.E., 1991: *Ozone diplomacy,* Harvard University Press, Cambridge, MA.

Blok, K., C. Hendriks, and W.C. Turkenburg, 1989: The role of carbon dioxide removal in the reduction of the greenhouse effect, *Energy technologies for reducing emissions of greenhouse gases – Proceedings of an experts' seminar,* Volume 2, 135–156, IEA/OECD, Paris.

Blok, K., J. Farla, C. Hendriks, and W. Turkenburg, 1991: Carbon dioxide removal: A review, paper presented at the International Symposium on Environmentally Sound Energy Technologies and their Transfer to Developing Countries and European Economies in Transition (ESETT'91), October 1991, ENI – San Donata Milanese, Italy.

Blok, K., E. Worrell, R. Culenaere, and W. Turkenburg, 1993: The cost-effectiveness of CO_2 emission reduction achieved by energy conservation, *Energy Policy,* **21,** 656–667.

Brown, S., C.A.S. Hall, W. Knabe, J. Raich, M.C. Trexler, and P. Woomer, 1993: Tropical forests: Their past, present, and potential future role in the terrestrial carbon budget, *Water, Air, and Soil Pollution,* **70,** 71–94.

BTG (Biomass Technology Group), 1994: Potential woodfuel production in developing countries for power generation in The Netherlands, University of Twente, Enschede, The Netherlands.

Byrne, J., S. Letendre, C. Govindarajalu, and Y-D. Wang, 1996: Evaluating the economics of photovoltaics in a demand-side management role, *Energy Policy,* forthcoming.

Byrne, J., Y-D. Wang, K. Ham, I. Han, J. Kim, and R. Wykoff, 1991: Energy and environmental sustainability in East and Southeast Asia, *IEEE Technology and Society Magazine,* Winter 1991/1992, 21–29.

Byrne, J., S. Letendre, R. Nigro, and Y-D. Wang, 1994: PV-DSM as a green investment strategy, *Proceedings of the Fifth National Conference on Integrated Resource Planning,* pp. 272–285, National Association of Regulatory Utility Commissioners, Washington, DC.

Cannell, M., 1982: *World forest biomass and primary production data,* Academic Press, London.

Chandler, W.U., 1990: *Carbon emissions control strategies,* World Wildlife Fund, Washington, DC.

Cline, W.R., 1992a: *The economics of global warming,* Institute for International Economics, Washington, DC.

Cline, W.R., 1992b: *Global warming – The benefits of emission abatement,* OECD, Paris.

COHERENCE, 1991: *Cost-effectiveness analysis of CO_2 reduction options, Synthesis Report,* Report for the CEC CO_2 Crash Programme, Commission of the European Community, Brussels.

Cooper, C.F., 1983: Carbon storage in managed forests, *Canadian Journal of Forestry Research,* **13,** 155–166.

Correia, F.N., M. da Graca Saraiva, J. Rocha, M. Fordham, F. Bernardo, I. Ramos, Z. Marques, and L. Soczka, 1994: The planning of flood alleviation measures: Interface with the public. In *Floods across Europe: Flood hazard assessment, modelling and management,* E.C. Penning-Rowsell and M.H. Fordham, eds., Middlesex University Press, London.

COSEPUP (Committee on Science, Engineering and Public Policy), 1991: *Policy implications of greenhouse warming,* Report of the Mitigation Panel, U.S. National Academy of Sciences, U.S. National Academy of Engineering and Institute of Medicine, U.S. National Academy Press, Washington, DC.

DeCicco, J., and M. Ross, 1993: *An updated assessment of the near-term potential for improving automotive fuel ecomony,* American Council for an Energy-Efficiency Economy, Washington, DC.

Dessus, B., B. Devin, and F. Pharabod, 1992: World potential of renewable energies. In *Houille blanche,* **1,** 21–70.

Dixon, R.K., K.J. Andrasko, F.A. Sussman, M.A. Livinson, M.C. Trexler, and T.S. Vinson, 1993: Forest sector carbon offset projects: Near-term opportunities to mitigate greenhouse gas emissions, *Water, Air, and Soil Pollution,* **70,** 561–577.

Dronkers, J., R. Boeye, and R. Misdorp, 1990. Socio-economic, legal, institutional, cultural and environmental aspects of measures for the adaptation of coastal zones at risk to sea level rise. In *Changing climate and the coast,* vol. 1: *Adaptive responses and their economic, environmental, amd institutional implications,* J.G. Titus, ed., U.S. Environmental Protection Agency, Washington, DC.

Economic and Social Commmission for Asia and the Pacific (ESCAP), 1991: *Energy policy implications of the climatic effects of fossil fuel use in the Asia-Pacific region,* Bangkok.

Ettinger, J. van, T.H. Jansen, and C.J. Jepma, 1991: Climate, environment and development, *The European Journal of Development Research,* **3**(1), 108–133.

Ettinger, J. van, 1994: Sustainable use of energy, a normative energy scenario: 1990–2050. *Energy Policy,* **22,** 111–118.

Ewing, A.J., 1985: *Energy efficiency in the pulp and paper industry with emphasis on developing countries,* World Bank Technical Paper No. 34, Washington, DC.

Faeth, P., C. Cort, and R. Livernash, 1993: *Evaluating the carbon sequestration benefits of sustainable forest projects in developing countries,* World Resources Institute Report, Washington, DC.

Fankhauser, S., 1992: Global warming damage costs: Some monetary estimates, CSERGE Working Paper GEC 92-29, Centre for Social and Economic Research on the Global Environment, London and Norwich.

Fankhauser, S., 1993: The economic costs of global warming: Some monetary estimates. In *Costs, impacts and benefits of CO_2 mitigation,* Y. Kaya, N. Nakicenovic, W.D. Nordhaus, and F.L. Toth, eds., Proceedings of the IIASA Workshop on the Costs, Impacts, and Benefits of CO_2 Mitigation, IIASA, Laxenburg, Austria.

Fankhauser, S., 1994a: *Global warming damage costs: Some monetary estimates* (revised version), Centre for Social and Economic Research on the Global Environment, London and Norwich.

Fankhauser, S., 1994b: *The economic costs of global warming damage: A survey,* Centre for Social and Economic Research on the Global Environment, London and Norwich.

Fankhauser, S., 1995: *Valuing climate change: The economics of the greenhouse effect,* Earthscan, London.

FAO (Food and Agriculture Organization), 1991: FAO's 1990 re-assessment of tropical forest cover, *Nature and Resources,* 27(2), 21–26.

Fischer, G., K. Frohberg, M.L. Parry, and C. Rosenzweig, 1993: Climate change, world food supply, demand and trade, *Global Environmental Change,* **4**(1), 7–23.

Fujii, Y., 1990: *Assessment of CO_2 emission reduction technologies and building of a global energy balance model,* Kaya Laboratory, Department of Electrical Engineering, University of Tokyo, Japan.

Geller, H.S., 1994: *Review of long term energy efficiency potential throughout the world,* American Council for an Energy Efficient Economy, Washington, DC.

Geller, H.S., and S. Nadel, 1994: Market transformation strategies to promote end-use efficiency, *Annual Review of Energy and Environment* **19:** 301–346, Annual Reviews Inc., Palo Alto, CA.

Geller, H.S., and D. Zylbersztajn, 1991: Energy intensity trends in Brazil, *Annual Review of Energy and Environment,* **16:**179–203, Annual Reviews Inc., Palo Alto, CA.

Ghai, D., 1994: Environment, livelihood and empowerment, *Development and Change,* **25** (1) (January), 1–11.

Goldemberg, J., T.B. Johansson, A.K.N. Reddy, and R.H. Williams, 1988: *Energy for a sustainable world,* Wiley Eastern Limited, New Delhi, India.

Green, C.H., A. van der Veen, E. Wierstra, and E.C. Penning-Rowsell, 1994: Vulnerability refined: Analysing full flood impacts. In *Floods across Europe: Flood hazard assessment, modelling and management,* E.C. Penning-Rowsell and M.H. Fordham, eds., Middlesex University Press, London.

Grübler, A., and N. Nakicenovic, 1992: International burden sharing in greenhouse gas reduction, Environment Working Paper No. 55, World Bank Environment Department, Washington, DC.

Grübler, A., S. Messner, L. Schrattenholzer, and A. Schäfer, 1993a: Emission reduction at the global level. In *Long term strategies for mitigating global warming,* N. Nakicenovic, ed., Special Issue of *Energy – The International Journal,* **18**(5), 539–581.

Grübler, A., S. Nilsson, and N. Nakicenovic, 1993b: Enhancing carbon sinks. In *Long term strategies for mitigating global warming,* N. Nakicenovic, ed., Special Issue of *Energy – The International Journal,* **18**(5), 499–522.

Gupta, S., and N. Khanna, 1991: India country paper. In *Collaborative study on strategies to limit CO_2 emissions in Asia and Brazil,* P. Ghosh, A. Achanta, P. Bhandari, M. Damodaran, and N. Khanna, eds., Asian Energy Institute, New Delhi, India.

Guzman, O., A. Yunez-Naude, and M.S. Wionczek, 1987: *Energy efficiency and conservation in Mexico,* Westview Press, Boulder and London.

Hendriks, C.A., 1994: *Carbon dioxide removal from coal-fired power plants,* Kluwer, Dordrecht, The Netherlands.

Houghton, R.A., 1990: The future role of tropical forests in affecting the carbon dioxide concentration of the atmosphere, *Ambio,* **18,** 204–209.

Houghton, R.A., 1991: Tropical deforestation and atmospheric carbon dioxide, *Climatic Change,* **19,** 99–118.

Houghton, R.A., J. Unruh, and P.A. LeFèbvre, 1991: *Current land use in the tropics and its potential for sequestering carbon,* Proceedings of the Technical Workshop to Explore Options for Global Forestry Management, 24–29 April 1991, Bangkok, Thailand, pp. 279–310, International Institute for Environment and Development (IIED), London.

Howe, C.H., H.C. Cochrane, J.E. Bunin, and R.W. King, 1991: *Natural hazard damage handbook,* University of Colorado, Boulder.

Huang, J-P., 1993: Industry energy use and structural change: A case study of the People's Republic of China, Energy Economics, **15**(2), 131–136.

Hurst, C., 1990: Establishing new markets for mature energy equipment in developing countries: Experience with windmills, hydropowered mills and solar water heaters, *World Development,* **18**(4), 605–615.

IEA/OECD, 1987: *Renewable sources of energy,* IEA/OECD, Paris.

IEA/OECD, 1991: *Energy efficiency and the environment,* IEA/OECD, Paris.

Imran, M., and P. Barnes, 1990: Energy demand in the developing countries: Prospects for the future, World Bank Staff Working Paper No. 23, World Bank, Washington, DC.

IPCC (Intergovernmental Panel on Climate Change), 1990a: *Methane emissions and opportunities for control,* Subgroup for Methane Emissions and Opportunities for Control, Workshop results of the Response Strategies Working Group, U.S. Environmental Protection Agency, Washington, DC.

IPCC (Intergovernmental Panel on Climate Change), 1990b: *Energy and industry subgroup report,* Energy and Industry Subgroup, Workshop results of the Response Strategies Working Group, U.S. Environmental Protection Agency, Washington, DC.

IPCC (Intergovernmental Panel on Climate Change), 1990c: *Greenhouse gas emissions from agricultural systems,* Agriculture, Forestry, and Other Human Activities Subgroup, Workshop results of the Response Strategies Working Group, U.S. Environmental Protection Agency, Washington, DC.

IPCC (Intergovernmental Panel on Climate Change), 1991: *Climate change, the IPCC response strategies,* Island Press, Washington, DC.

IPCC (Intergovernmental Panel on Climate Change), 1992: *Climate change 1992,* J.T. Houghton, B.A. Callander, and S.K. Varney, eds., Cambridge University Press, Cambridge.

IPCC (Intergovernmental Panel on Climate Change), 1993: *Methods in national emissions inventories and options for control,* Proceedings of International IPCC Workshop on Methane and Nitrous Oxide, A.R. van Amstel, ed., Rijksinstitut voor Volksgezondheid en Milieuhygiene (RIVM), Bilthoven, The Netherlands.

IPCC (Intergovernmental Panel on Climate Change), 1994: Preparing to meet the coastal challenges of the 21st century, conference report, World Coast Conference 1993, National Institute for Coastal and Marine Management, The Hague.

IPCC (Intergovernmental Panel on Climate Change), 1995: *Greenhouse gas inventory: IPCC guidelines for national greenhouse gas inventories,* vol. 3, Reference Manual, United Kingdom Meteorological Office, Bracknell, England.

Jackson, T., 1991: Least-cost greenhouse planning: Supply curves for global warming abatement, *Energy Policy,* **19**(1) (Jan/Feb), 35–46.

Jepma, C.J., 1995: *Tropical deforestation: A socio-economic approach,* Earthscan, London.

Jepma, C.J., and C.W. Lee, 1995: Carbon dioxide emissions: A cost-effective approach. In *The feasibility of joint implementation,* C.J. Jepma, ed., pp. 57–68, Kluwer, Dordrecht, The Netherlands.

Johansson, T.B., H. Kelly, A. Reddy, R.H. Williams, and L. Burnham (executive editor), 1993: *Renewable energy: Sources for fuels and electricity,* Island Press, Washington, DC, Covelo, CA.

Kassler, P., 1994: *Energy for development,* Shell Selected Paper, Shell International Petroleum Company Ltd., London.

Kaya, Y., 1989: Impact of carbon dioxide emission control on GNP growth: Interpretation of proposed scenarios, Intergovernmental Panel on Climate Change, Response Strategies Working Group.

Kaya, Y., Y. Fujii, R. Matsuhashi, K. Yamaji, Y. Shindo, H. Saiki, I. Furugaki, and O. Kobayashi, 1991: Assessment of the technological options for mitigating global warming, paper presented at the IPCC EIS Group Meeting, 6–7 August 1991, Geneva, Switzerland.

Klaus, J., W. Pflügner, R. Schmidtke, H. Wind, and C.H. Green, 1994: Models for flood hazard assessment and management. In *Floods across Europe: Flood hazard assessment, modelling and management,* E.C. Penning-Rowsell, and M.H. Fordham, eds. Middlesex University Press, London.

Kram, T., 1994a: Boundaries of future carbon dioxide emission reduction in nine industrial countries (executive summary), Energy Technology Systems Analysis Programme, Annex IV: Greenhouse Gases and National Energy Options, Netherlands Energy Research Foundation ECN/International Energy Agency.

Kram, T., 1994b: *National energy options for reducing CO_2 emissions,* vol. 1, *The international connection,* a report of the Energy Technology Systems Analysis Programme/Annex 4 (1990–1993): Greenhouse Gases and National Energy Options, Netherlands Energy Research Foundation ECN/International Energy Agency.

Krankina, O.N., and R.K. Dixon, 1993: Forest management options to conserve and sequester terrestrial carbon in the Russian Federation, *World Resources Review,* **6**(1), 88–101.

Kwadijk, J.C.J., 1991: Sensitivity of the river Rhine discharge to environmental change: A first tentative assessment, *Earth Surface Process and Landreforms,* **16,** 627–637.

Ledbetter, M., and M. Ross, 1989: *Supply curves of conserved energy for automobiles,* American Council for an Energy Efficient Economy, Washington, DC.

Levine, M.D., A. Gadgil, S. Meyers, J. Sathaye, and T. Wilbanks, 1991: *Energy efficiency, developing nations, and Eastern Europe,* report to the U.S. Working Group on Global Efficiency, Washington, DC.

Li, J., R.M. Shreshtha, and W.K. Foell, 1990: Structural change and energy use: The case of the manufacturing sector in Taiwan, *Energy Economics,* **12**(2), 109–115.

MacKerron, G., 1992: Nuclear costs, why do they keep rising? *Energy Policy,* **20,** 641–652.

Makarov, A.A., and I. Bashmakov, 1990: The Soviet Union. In *Carbon emissions control strategies,* W.U. Chandler, ed., World Wildlife Fund, Washington, DC.

Mathur, A., 1991: Energy and CO_2 scenarios developed within Asian Energy Institute's collaborative study, paper presented at the Workshop on CO_2 reduction and removal: Measures for the next century, 19–21 March 1991, International Institute for Applied Systems Analysis, Laxenburg, Austria.

Matsuo, N., 1991: Japan country paper. *Collaborative study on strategies to limit CO_2 emissions in Asia and Brazil,* Asian Energy Institute (AEI), AEI, New Delhi, India.

McKinsey & Company, 1989: Protecting the global environment, findings and conclusions, appendices, Ministerial conference on atmospheric pollution and climate change, Noordwijk, The Netherlands.

Meier, A., 1991: Supply curves of conserved energy, *Proceedings of the IEA International Conference on Technology Responses to Global Environmental Challenges: Energy Collaboration for the 21st Century,* 6–8 November 1991, vol. 1, Inter Group Corporation, Kyoto, Japan.

Mills, E., D. Wilson, and T.B. Johansson, 1991: Getting started – No-regrets strategies for reducing greenhouse gas emissions, *Energy Policy,* **19,** 527–542.

Mongia, N., R. Bhatia, J. Sathaye, and P. Mongia, 1991: Costs of reducing CO_2 emissions from India, *Energy Policy,* **19**(10), 978–986.

Moulton, R.J., and K.R. Richards, 1990: *Costs of sequestering carbon through tree planting and forest management in the United States,* USDA Forest Service General Technical Report WO-58, Washington, DC.

Nakicenovic, N., and A. John, 1991: CO_2 reduction and removal measures for the next century, *Energy – The International Journal,* **16**(11/12), 1347–1377.

Nakicenovic, N., and A. Grübler, 1993: Energy conversion, conservation and efficiency. In *Long term strategies for mitigating global warming,* N. Nakicenovic, ed., Special Issue of *Energy – The International Journal,* **18**(5), 421–435.

Nakicenovic, N., and D. Victor, 1993: Technology transfer to developing countries. In *Long term strategies for mitigating global warming,* N. Nakicenovic, ed., Special Issue of *Energy – The International Journal,* **18**(5), 523–538.

Nakicenovic, N., D. Victor, A. Grübler, and L. Schrattenholzer, 1993: Introduction. In *Long term strategies for mitigating global warming,* N. Nakicenovic, ed., Special Issue of *Energy – The International Journal,* **18**(5), 401–419.

Nilsson, N.-E., 1982: An alley model for forest resources planning. In *Statistics in theory and practice: Essays in honour of Bertil Matrn,* B. Ranneby, Ed., Section of Forest Biometry, Swedish University of Agricultural Sciences, Umea, Sweden.

Nordhaus, W.D., 1990: An intertemporal general-equilibrium model of economic growth and climate change, *Proceedings of the workshop on economic/energy/environmental modeling for climate analysis,* D.O. Wood, Y. Kaya, eds., 22–23 October, 1990, pp. 415–433, MIT Center for Energy Policy Research, Cambridge, MA.

Nordhaus, W.D., 1991a: The cost of slowing climate change: A survey, *The Energy Journal,* **12,** 37–65.

Nordhaus, W.D., 1991b: To slow or not to slow: The economics of the greenhouse effect, *The Economics Journal,* **101** (July) 920– 937.

Nordhaus, W.D., 1993: *Managing the global commons: The economics of climate change,* MIT Press, Cambridge, MA.

Ogawa, Y., 1992: Analysis of factors affecting carbon dioxide emissions due to past energy consumption around the world, *Green house research initiatives in the ESCAP region: Energy,* ESCAP, Bangkok.

OTA (Office of Technology Assessment), 1991: *Changing by degrees: Steps to reduce greenhouse gases,* OTA-0-842, Congress of the United States, U.S. Government Printing Office, Washington, DC.

Parikh, J., and S. Gokarn, 1993: Climate change and India's energy policy options: New perspectives on sectoral CO_2 emissions and incremental costs, *Global Environmental Change* (September) 276–291.

Park, S-H., 1992: Decomposition of the industrial energy consumption: An alternative method. *Energy Economics,* **14**(4), 265–270.

Parks, P.J., and I.W. Hardie, 1992: *Least-cost forest carbon reserves: Cost-effective subsidies to convert marginal agricultural land to forest,* American Forests, Washington, DC.

Penning-Rowsell, E.C., B. Peerbolte, F.N. Correia, M. Fordham, C.H. Green, W. Pflügner, J. Rocha, M. Saraiva, R. Schmidtke, J. Torterotot, and A. van der Veen, 1992: Flood vulnerability analysis and climate change: toward a European methodology. In *Floods and flood management,* A.J. Saul, ed., Kluwer, Dordrecht, The Netherlands.

Penning-Rowsell, E.C., and M.H. Fordham, 1994: *Floods across Europe: Flood hazard assessment, modelling and management,* Middlesex University Press, London.

Qureshi, A., and S. Sherer, 1994: *Climate change in Asia: Forestry and land use,* Asian Development Bank's Regional Study on Global Environment Issues, Climate Institute, Washington, DC.

Read, P., 1994a: *Responding to global warming: The technology, economics and politics of sustainable energy,* ZED Books, London and Atlantic Highlands, NJ.

Read, P., 1994b: Biofuel as the core technology in an effective response strategy, economics discussion paper, Massey University, Palmerston North, New Zealand.

Reijntjes, C., B. Haverkort, and A. Waters-Bayer, 1992: *Farming for the future: An introduction to low-external-input and sustainable development,* Macmillan, London.

Reilly, J., and K.R. Richards, 1993: Climate change damage and the trace gas index issue, *Environmental and Resources Economics,* **3,** 41–61.

Reilly, J., N. Hohmann, and S. Kane, 1994: Climate change and agricultural trade: Who benefits, who loses? *Global Environmental Change,* **4**(1), 24–36.

Richards, K.R., 1993: Valuation of temporary and future greenhouse gas reductions, paper presented at Western Economic Association Annual Meeting, Lake Tahoe, NV.

Richards, K.R., R. Moulton, and R.A. Birdsey, 1993a: Costs of creating carbon sinks in the U.S., *Energy Conservation and Management,* **34**(9-11), 905–912.

Richards, K.R., D.H. Rosenthal, J.A. Edmonds, and M. Wise, 1993b: The carbon dioxide emissions game: Playing the net, paper presented at Western Economic Association Annual Meeting, Lake Tahoe, NV.

Rijsbergen, F., 1991: Potential costs of adapting to sea level rise in OECD countries. In *Responding to climate change: Selected economic issues,* OECD, Paris.

Robinson, J., M. Fraser, E. Haites, D. Harvey, M. Jaccard, A. Reinsch, and R. Torrie, 1993: *Canadian options for greenhouse gas emission reduction (COGGER),* final report of the COGGER Panel to the Canadian Global Change Program and the Canadian Climate Program Board, Royal Society of Canada, Ottawa.

Rogner, H., N. Nakicenovic, and A. Grübler, 1993: Second- and third-generation energy technologies, *Energy – The International Journal,* Special Issue, Long term strategies for mitigating global warming, **18**(5), 461–484.

Rosenzweig, C., M.L. Parry, G. Fischer, and K. Fröberg, 1993: *Climate change and world food supply,* Environmental Change Research Unit, Number 3, University of Oxford, Oxford.

Rosenzweig, C., and M.L. Parry, 1994: Potential impact of climate change on world food supply, *Nature,* **367,** 133–138.

Rosenzweig, C., M.L. Parry, and G. Fischer, 1996: Climate change and world food supply. In K.M. Strzepek and J.B. Smith, eds., *As climate changes: International impacts and implications,* Cambridge University Press, Cambridge.

Rubin, E.S., R.N. Cooper, R.A. Frosch, T.H. Lee, G. Marland, A.H. Rosenfeld, and D.D. Stine, 1992: Realistic mitigation options for global warming, *Science,* **257** (July) 148–149, 261–266.

Schipper, L., R.B. Howarth, and H. Geller, 1990: United States energy use from 1973 to 1987: The impacts of improved efficiency, *Annual Review of Energy,* **15,** 455–504.

Sedjo, R.A., and A.M. Solomon, 1989: Climate and forests. In

Greenhouse warming: Abatement and adaptation, N.J. Rosenberg, W.E. Easterling III, P.R. Crosson, and J. Darmstadter, eds., Resources for the Future, Washington, DC.

Simpson, V.J., and C. Anastasi, 1993: Communication, future emissions of CH_4 from the natural gas and coal industries, *Energy Policy,* **21,** 827–830.

Sitnicki, S., K. Budzinski, J. Juda, J. Michna, and A. Szpilewica, 1990: Poland. In *Carbon emissions control strategies,* W.U. Chandler, ed., pp. 55–80, World Wildlife Fund, Washington, DC.

Southern Centre/Risö, 1993: UNEP Greenhouse gas abatement costing studies, *Zimbabwe country study, phase two,* Southern Centre for Energy and the Environment, Harare, Zimbabwe, and Risö National Laboratory Systems, Analysis Department, Copenhagen, Denmark.

Springmann, F., 1991: *Analysis of the ecological impact of demonstration projects in the field of rational use of energy: Development of evaluation criteria,* study on behalf of the Commission of the European Communities, Directorate General for Energy (DG XVII), Regio-Tec GmbH, Starnberg, Germany.

Suliman, M., 1990: Introduction: Africa in the IPCC report. In *Greenhouse effect and its impact on Africa,* M. Suliman, ed., Institute for African Alternatives, London.

Swisher, J.N., 1991: Cost and performance of CO_2 storage in forestry projects, *Biomass and Energy,* **1**(6), 317–328.

Swisher, J., D. Wilson, and L. Schrattenholzer, 1993: Renewable energy potentials. In *Long term strategies for mitigating global warming,* N. Nakicenovic, ed., Special Issue of *Energy – The International Journal,* **18**(5), 437–459.

Task Force on Energy, 1991: Report for the Bangladesh government, vol. 3, University Press Limited, Dhaka.

TNO (Netherlands' Organization for Applied Scientific Research), 1992: *Confining and abating CO_2 from fossil fuel burning – A feasible option?* rapport voor de EG, TNO, Apeldoorn.

Tol, R.S.J., 1993: *The climate fund: Survey of literature on costs and benefits,* Institute for Environmental Studies (IVM), Free University Amsterdam.

Tol, R.S.J., 1994: *The climate fund: Optimal greenhouse gas emission abatement,* Institute for environmental studies, W-94-08, Free University Amsterdam.

Train, K., 1985: Discount rates in consumers' energy-related decisions: A review of the literature, *Energy,* **16**(12):1243–1253.

Trexler, M.C., 1991: Estimating tropical biomass futures: A tentative scenario, *Proceedings of the Technical Workshop to Explore Options for Global Forestry Management,* 24–29 April 1991, Bangkok, Thailand, International Institute for Environment and Development, UK.

Trexler, M.C., P.E. Faeth, and J.M. Kramer, 1989: *Forestry as a response to global warming: An analysis of the Guatemala agroforestry and carbon sequestration project,* World Resources Institute, Washington, DC.

Turner, D.P., J.J. Lee, G.J. Koperper, and J.R. Barker (eds.), 1993: *The forest sector carbon budget of the United States: Carbon pools and flux under alternative policy options,* U.S. EPA, ERL, Corvallis, OR.

UNEP, 1993: *UNEP Greenhouse gas abatement costing studies, analysis of abatement costing issues and preparation of a methodology to undertake national greenhouse gas abatement costing studies,* Phase Two Report, UNEP Collaborating Centre on Energy and Environment, Risö National Laboratory, Denmark.

UNEP, 1994: *UNEP Greenhouse gas abatement costing studies,* Part

1: *Main report,* and Part 2: *Country summaries,* UNEP Collaborating Centre on Energy and Environment, May, Risö National Laboratory, Denmark.

U.S. Congress, 1992: Office of Technology Assessment (OTA), *Fueling development: Energy techologies in developing countries,* OTA-E-516, U.S. Government Printing Office, Washington DC., April.

U.S. EPA (U.S. Environmental Protection Agency), 1990: Policy options for stabilizing global climate, report to Congress: technical appendices, EPA Washington, DC.

U.S. EPA (U.S. Environmental Protection Agency), 1993: Options for reducing methane emissions internationally, vol. I, Technological options for reducing methane emissions, report to Congress, K.B. Hogan, ed., Office of Air and Radiation, Washington, DC.

Viner, D., and M. Hulme, 1994: *The Climate Impacts LINK Project, providing climate change scenarios for impacts assessments in the UK,* report for the UK Department of the Environment, Climate Research Unit, University of East Anglia, Norwich.

Watson, R.T., L.G. Meira Filho, E. Samhueza, and A. Janetos, 1992: Sources and sinks. In *Climate change 1992: The supplementary report to the IPCC scientific assessment,* J.T. Houghton, B.A. Callander, and S.K. Varney, eds., pp. 25–46, Cambridge University Press, Cambridge.

WEC (World Energy Council) Commission, 1993: *Energy for tomorrow's world – The realities, the real options and the agenda for achievement,* World Energy Council, Kogan Page Ltd., London, St. Martin's Press, New York.

WEC (World Energy Council), 1993: *Renewable energy resources: Opportunities and constraints 1990–2020,* Kogan Page, London.

WEC (World Energy Council), 1994: *New renewable energy resources: A guide to the future,* Kogan Page, London.

Wenger, H., T. Hoff, and R. Perez, 1992: Photovoltaics as a demand-side management option: Benefits of a utility-customer partnership. In *Proceedings of the Fifteenth World Energy Engineering Conference,* October 1992, Atlanta, GA.

Winjum, J.K., R.K. Dixon, and P.E. Schroeder, 1992: Estimation of the global potential of forest and agroforest management practices to sequester carbon, *Water, Air, and Soil Pollution* **62,** 213–227.

Winjum, J.K., and D.K. Lewis, 1993: Forest management and the economics of carbon storage: The nonfinancial component, Research paper, U.S. EPA, Environmental Research Laboratory, Corvallis, OR.

Winpenny, J.T., 1994: The relevance of global climatic effects to project appraisal, In *The economics of project appraisal and the environment,* J. Weiss, ed., Edward Elgar, Aldershot, UK.

World Bank, 1993: *Energy efficiency and conservation in the developing world,* A World Bank Policy Paper, Washington, DC.

WRI (World Resources Institute), 1990: *World resources 1990-91: A guide to the global environment,* WRI, Oxford University Press, New York.

WRI (World Resources Institute), 1994: *World resources 1994-95: A guide to the global environment.* Oxford University Press, Oxford.

Yamaji, K., R. Matsuhashi, Y. Nagata, and Y. Kaya, 1991: An integrated system for CO_2/energy/GNP analysis: Case studies on economic measures for CO_2 reduction in Japan, paper presented at the Workshop on CO_2 Reduction and Removal: Measures for the Next Century, 19-21 March 1991, International Institute for Applied Systems Analysis, Laxenburg, Austria.

8

Estimating the Costs of Mitigating Greenhouse Gases

Convening Author:
J.C. HOURCADE

Principal Lead Authors:
R. RICHELS, J. ROBINSON

Lead Authors:
W. Chandler, O. Davidson, J. Edmonds, D. Finon, M. Grubb, K. Halsnaes,
K. Hogan, M. Jaccard, F. Krause, E. La Rovere, W.D. Montgomery, P. Nastari,
A. Pegov, K. Richards, L. Schrattenholzer, D. Siniscalco, P.R. Shukla, Y. Sokona,
P. Sturm, A. Tudini

CONTENTS

SUMMARY

- It is important to be clear what types of costs (e.g., direct, sectoral, macroeconomic, or welfare) are included in the analysis. What matters from the point of view of policy are, not the total costs of emission reduction, but the net costs (i.e., the total costs minus any positive side effects of mitigation).

- The insights generated by modelling analyses (e.g., the areas of greatest potential for cost-effective emission reduction or the directional effect of tax recycling on emission reduction costs), not the specific numerical results of any one analysis, are what matter. Although researchers attempt to incorporate their best understanding of development processes into the studies, neither the baseline nor the intervention scenarios should be interpreted as representing likely future conditions, especially more than a decade into the future.

- The size of the costs of mitigation depends critically on assumptions about the efficiency of the baseline scenario used in the analysis. The higher the underlying economic growth assumed in the baseline scenario, the greater the estimated costs of mitigation. The more emission reduction built into the baseline, the higher the estimated costs of further reduction.

- Whether a no-regrets potential for emission reduction exists depends on whether the economy under consideration is on or below its theoretical production frontier. The existence of a no-regrets potential (where the society is below the frontier) implies, first, that significant market failures exist that give rise to increased greenhouse gas emissions and, second, that policies can be designed and implemented that correct those market failures.

- Mitigation costs will be affected by a wide range of factors, including population growth, consumption patterns, resource and technology availability, geographical distribution of activity, land use and transportation patterns, and trade. As a result, there exists a range of quite different socioeconomic and technological development paths that would give rise to quite different emission scenarios and costs of mitigation. Existing energy and emissions models do not address these underlying factors in a very effective way. Future analysis should make use of multiple baseline scenarios to help capture the differences in these factors.

- Infrastructure decisions are critical because they can allow or restrict future options, and different infrastructure decisions can lead to very different cost outcomes. This issue is of particular importance to developing countries, where major infrastructure decisions will be made over the next 25 years.

- Mitigation costing analyses reveal the costs of mitigation relative to a given baseline. The results of different studies cannot easily be compared to each other unless differences in baselines are taken into account. Neither the baseline nor the intervention scenarios should be interpreted as representing likely future conditions.

- There is growing integration of bottom-up and top-down analyses of the costs of energy-related greenhouse gas mitigation. This convergence means that differences in results are increasingly driven by differences in input assumptions rather than by differences in model structure. Nevertheless, differences in structure remain important, since each type of model is best suited to answer particular kinds of questions. Models with more detailed representation of technology are better suited to identify technical potentials and financial costs and savings; models with more detailed representation of broader economic activity are better suited to identify costs in terms of higher or lower economic growth.

- Substantial disagreement still exists over the existence and size of a significant no-regrets potential for greenhouse gas emission reduction. Bottom-up approaches, which assume that there exist substantial correctable market imperfections, show significant no-regrets potential; top-down studies, which assume that existing markets are relatively efficient, show little no-regrets potential.

8.1 Introduction

Although most of the available literature on the costs of greenhouse gas mitigation has been written in the period since 1988, interest in this issue began more than a decade earlier with Nordhaus (1977, 1979), whose contribution was followed by about a dozen other pioneering studies. The picture has evolved since then, but it retains two features from this earlier period:

- First, the focus is on CO_2 emissions, and the other greenhouse gases are either ignored or treated separately in an ad hoc manner. This emphasis has arisen not only because of the importance of CO_2 emissions relative to those of other gases, but also because the long-term demand and supply models[1] that energy analysts had begun to develop by the mid-1970s in response to the instability of energy markets at that time could also be used to analyze CO_2 emissions associated with the energy system. Such a modelling capability was not available with respect to other greenhouse gases.

- Second, the structure of the debate was determined very early by the wide range of numerical results about the potential for, and costs of, mitigation. Some studies argued that mitigation policies were likely to entail very substantial costs (Nordhaus, 1977, 1979), others that the costs would be relatively small (Edmonds and Reilly, 1983), and still others that these costs might in fact be negative and bring an overall benefit (Lovins *et al.*, 1981).[2]

This debate became more heated and the demand for this type of study accelerated after the drafting of the UN Framework Convention on Climate Change in 1992. At that time, the climate issue appeared on the international negotiation agenda and countries and international organizations became interested in obtaining information on the economic impacts of climate policies.

The resultant controversies about the overall costs of greenhouse gas mitigation strategies made it clear that statesmen, policymakers, business people, journalists, and the public stand in an ambiguous posture in the face of figures coming from economic modelling exercises. On the one hand, they expect a clear-cut answer, for example, about the effectiveness of a carbon tax to curb greenhouse gas emissions and about the economic consequences of this tax, preferably in the form of an aggregated figure such as a loss of GDP. On the other hand, they are very sceptical about the results of such analyses and, above all, about the capacity of economic models to provide reliable predictions and cost estimates.[3]

Much of this scepticism stems from the well-documented failure of energy forecasters in the past to provide reliable predictions of future energy demand after the oil shock disturbances beginning in the early 1970s,[4] as well as the more general problems of errors in economic forecasting even over short time horizons (Ascher, 1978, 1990). Part, however, stems from misunderstandings about what the economic and energy models are able to capture and about how to use their results. There is a gap between the answers expected by policymakers and the type of information economic models can supply. As discussed below, the lessons of modelling analysis are rather complex and cannot easily be translated into clear-cut policies.

Another difficulty stems from the fact that, in the first phase, many agencies have employed models not initially designed to shed light on the cost of emission reductions. Examples would be models with energy sectors so aggregated that it was impossible to describe a substitution between fossil and nonfossil energies, or the use of short-term models to simulate medium-term impacts. Some of these models are no longer in use; others have evolved through several versions, further complicating the picture. Thus, some of the apparent diversity of results is simply due to the development of research itself.

These circumstances have created an opportunity for misusing models and their results in public debates. To help avoid either uncritical acceptance or total distrust of current analyses of mitigation costs, this chapter will present a discussion of the critical determinants likely to influence the overall cost of climate policies and of the main methodologies employed to account for them. The purpose is not to have a detailed comparative analysis of the models, but to concentrate on the possible sources of misunderstanding and misinterpretations of the results presented in Chapter 9 so as to facilitate the discussion of their policy implications. We will examine successively

- the various concepts of costs used in the literature (Section 8.2)
- the relationships between economic cost assessments and assumptions about development patterns and technical change which (explicitly or implicitly) underlie any economic scenario used to assess mitigation costs (Section 8.3)
- the main methodological approaches for costing assessments, the key assumptions likely to determine the numerical results, and the lessons derived from modelling debates in the energy field and in the forestry sector (Section 8.4)

We will not in this chapter discuss methodological issues having to do with integrating cost analyses with analyses of the benefits of mitigation, or issues having to do with the costs of adaptation to climate change. Discussions of estimates of benefits (i.e., of the costs of climate change) are contained in Chapter 6, whereas Chapter 10 on integrated assessment discusses issues related to the integration of costs and benefits. The purpose of this chapter is to provide a basis for a better understanding of the mitigation cost estimates summarized in Chapter 9. These cost estimates need to be compared with the estimates of the benefits of mitigation (i.e., the damages or costs of not reducing net emissions) summarized in Chapter 6 in order to provide the basis for an integrated assessment of the merits of climate control policies (Chapter 10). But the research agenda for the new generation of integrated models must address the methodological issues discussed in this chapter.

8.2 Costs: Definitions and Determinants

8.2.1 *How is cost measured?*

It is important to begin with a clear definition of costs that can be found in the literature because many misunderstandings stem from using very different cost concepts when presenting and comparing the results of existing studies. There are two crucial elements in the definition: how costs are measured and whether costs are reported on a gross basis or net of some of the ancillary benefits of policies to reduce emissions.

8.2.2 *Taxonomy of mitigation cost concepts*

In any economic analysis, the cost of mitigation is calculated as a difference in costs (defined in monetary units) between a reference situation and a new one characterized by lower emissions. Beyond this common feature, however, the concept of cost is not unique, meaning that the measurement criteria chosen to represent costs will depend on the level and the purpose of the measurement.

First, we have to recall here a classic distinction in economics between marginal costs (e.g., the incremental cost of removing an additional tonne of carbon or its equivalent), total costs (the sum of all marginal costs), and average costs (total cost divided by the quantity removed). Confusion can easily result from neglecting these distinctions. For example, the marginal cost of reducing emissions by an additional tonne may be very high in a given scenario, whereas the total and average cost of the relevant emissions reduction policy may be very low or even negative.

Before comparing any results, attention must be given to the way reduction targets are set, because this affects their meaning. Some models calculate reduction from a benchmark date in the recent past; this is the case for most of the U.S. studies in the energy field and for the studies carried out in the other OECD countries on how to stabilize greenhouse gas emissions over the next decade. Others (usually in studies of developing countries and very often in long-term studies in Europe) set the target in terms of a percentage reduction from an emission level at a specified future date. Given expected growth in baseline emissions in the future, a particular percentage reduction from a base year typically implies a much greater total reduction than the same percentage reduction from the future baseline scenario.

Second, to understand many critical debates about the costs of climate policies, it is necessary to distinguish four types of cost concepts used in costing analysis:

(1) *The direct engineering and financial costs of specific technical measures.* Examples include the cost of switching from coal to gas in electric production, of improving the thermal efficiency of existing homes, or of planting trees in reforestation programmes. Costs are normally reported in present-value terms and can represent the life-cycle cost of the technique used or of the project (the up-front cost of the measures considered plus annual energy and operating costs, all reduced to a

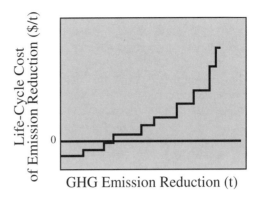

Figure 8.1: Energy technology cost curve.

net present value or levelized costs). Technical costs can show negative net costs because a given technology may yield enough energy cost savings to more than offset the costs of adopting and using the technology. These costs depend on both technico-economic data and a given interest rate and can be used to construct technology cost curves of the kind shown in Figure 8.1. They can be calculated in the absence of any global scenario but they do not then provide a macroeconomic cost assessment unless they are fed into coherent technical and economic frameworks. They are the only source of information available for mitigation measures in sectors for which no comprehensive sectoral model has so far been developed.

(2) *Economic costs for a given sector.* Here sectoral models are used to integrate sets of measures to provide consistent pictures of a given sector, and to compare the relative costs of different scenarios. These sectoral scenarios take some macroeconomic indicators, such as the overall rate of growth, as input parameters. They provide what is referred to as "partial equilibrium" analysis, in the sense that these sectoral models do not capture the feedback effects between the behaviour of a sector and that of the overall economy. To date, most sectoral analyses of mitigation costs have used energy sector models and forestry models. Other sectoral models, such as transportation or agriculture models, exist but have not yet been used so extensively for analysis of greenhouse gas mitigation.

(3) *Macroeconomic costs.* These measure the impact of a given strategy on the level of the gross domestic product (GDP) and its components (household consumption, investment, etc.). This aggregate index measures the monetary value added of goods and services produced in a single year and provides an index of the scale of human activities that pass through markets plus, by convention, the imputed value of some nonmarket activities (such as the value of services provided by public administration). At this level of cost analysis one tries to account for the interrelationships between a specific sector and the overall economy. This requires the use of either pure macroeconomic models or modelling frameworks cou-

pling sectoral models and macroeconomic models in order to capture the changes throughout an economy caused by policies in a given sector (what is commonly labelled the "general equilibrium" effects, which are to be distinguished from the partial equilibrium – sectoral – effects).

(4) *Welfare costs.* GDP variations do not provide direct measures of human welfare. There are many reasons for this. First a climate policy may change the composition of GDP in the direction of higher investments and lower consumption. These changes are invisible to an analysis looking only at the level of GDP. Second, human welfare may not increase linearly with consumption, so consumption changes do not necessarily indicate commensurate changes in welfare. Third, changes in the level of GDP do not account for the relationship between distribution of income and overall welfare. Finally, environmental degradation reduces welfare but does not result in a corresponding reduction in GDP. Even GDP, therefore, does not include all costs of interest, and even general equilibrium models that attempt to measure welfare costs do not include all costs that matter.

The four types of costs outlined here represent increasing levels of generality as one moves from direct financial costs to welfare costs. However, the different levels of cost cannot be aggregated or correlated in any systematic way. For example, ranking efficient technologies in terms of their individual prices in the marketplace (direct engineering and financial costs) usually does not reveal the way these technologies will actually be adopted and combined in an economically consistent productive system. For instance, a given energy efficiency improvement may not be rationally selected if this improvement occurs in a period of time when the electric supply is in excess. Similarly, a sectoral optimization of a very capital-intensive sector may not give results consistent with some form of overall macroeconomic optimum. For example, the amount of money spent in optimized energy supply or transportation systems may prevent investment in other sectors such as education and health. On the other hand, aggregate macroeconomic analysis of costs may not capture important changes to the extent that these changes remain below the level of "noise" of analysis and are not captured by historical variation in the aggregate variables. This typically occurs with respect to the agricultural sector, which represents a small part of overall GDP but often plays a decisive role in the social and spatial equilibrium of a society.

Existing studies use a variety of methodological approaches and definitions of cost to assess the costs of mitigating greenhouse gas emissions. No existing study provides a complete evaluation of full social costs. Instead, studies provide a range of cost estimates: Some provide estimates of the direct financial costs of specific technological options; others provide estimates of the effect of broad policies on aggregate economic activity; still others attempt to estimate welfare costs. It is important to be clear about which type of cost is being estimated in any specific study.

Despite these differences, progress has been made in incorporating successively broader concepts of costs in greenhouse gas mitigation studies and in accounting for the possibility of significant technological and behavioural change. Two more general problems remain, however.

The first lies in the fact that, in a number of developing countries and Central and Eastern European countries, the level of government intervention or market distortion (which exists also in OECD countries but not to the same degree) may interfere with the absolute and relative prices of goods and services. This results in distortions between observed market prices of technologies and their "shadow prices," assessed "at factor costs," which would reflect the actual balance between human capabilities, technical potentials, natural resource scarcity, and final needs within a given economy. The use of market prices is relevant for assessing the likeliness of the adoption of some technical alternatives for a given institutional context,[5] but if these costs are interpreted as capturing the overall social costs of a given measure, they conceal some intrinsic costs of goods and services and can be very misleading.

The second problem is that it is more difficult, empirically and theoretically, to pass from GDP costs to welfare costs. In most empirical models, welfare cost measures are calculated as the income required to leave a typical household no worse off after a tax than before. In models devoted to analyzing the short-term impact of taxation policies, welfare costs are calculated on the basis of the so-called "Harberger triangle," which demonstrates that welfare losses grow at a higher rate than increases in taxes. These results, however, are not complete measures of welfare costs. In the first place, they generally do not account for the dynamic effects of important policy measures on technology and consumer preferences. In the second place, they depend on market product calculations and do not account for the goods and services produced by nonmarket and informal economies.

In fact, much work remains to be done in attempting to incorporate broader conceptions of human welfare (such as those that are suggested in the United Nations Development Programme's Human Development Index), and most of the models reported hereafter ignore such factors. The cost figures provided by current economic modelling studies must therefore be strictly interpreted as estimates of losses (or gains) in the value of new final goods and services forgone as a consequence of the policy, not as estimates of impacts on overall welfare.

The different kinds of cost estimates discussed here are of great value; the point is simply that, before drawing policy-relevant conclusions from even the most general analyses of macroeconomic costs, it is important also to examine other indicators (e.g., unemployment, distribution of income, security, and political stability).

8.2.2.1 Gross costs, net costs, and the overall cost-benefit balance of mitigation strategies

In assessing the costs of mitigation, it is first important to make a distinction between what economists call the gross costs and the net costs of mitigation strategies. The two are different to the extent that there are possible positive side effects of mitigation strategies that would offset some of the gross costs. These positive side effects can be divided into three categories:

(1) The *negative cost potential,* namely mitigation caused by technologies whose costs are lower than the technologies currently in use. As discussed further in Section 8.4 below, this issue is controversial, since it implies that there are cost-effective mitigation strategies not now in use and that will not be adopted in the absence of new policies. Such measures would have a negative net cost and obviously lower the gross cost of greenhouse gas mitigation for a given target.

(2) An *economic double dividend,* such as the possible positive effects on growth or employment of the recycling of carbon tax revenues or of the technological externalities (i.e., side effects) associated with fostering research and development programmes.

(3) An *environmental double dividend,* namely the synergy between greenhouse gas mitigation strategies and the mitigation of other environmental nuisances such as local air pollution, urban congestion, or land and natural resource degradation, such that greenhouse gas mitigation also contributes to reducing these other problems.

The existence of such positive side effects within an array of mitigation measures would result in lowering the gross cost of these measures. To the extent that such positive side effects may totally offset the gross costs of a specific emission strategy, they represent what has been called a "no-regrets potential": measures that are worth undertaking whether or not there are climate-related reasons for doing so.[6] However, as discussed below, there is much controversy about the existence and magnitude of these positive side effects. The point here is simply that, from the point of view of greenhouse gas mitigation policy, what matters is the net costs of mitigation strategies, that is, gross costs minus any positive side effects.

To avoid any possible misunderstanding, it must be emphasized that the double dividend positive side effects described here are not the same as, and should not be mistaken for, the benefits of mitigation policies (or costs of climate change), which are discussed in Chapter 6. Instead, these positive side effects represent items that lower the total cost of mitigation policies, and it is the resulting net cost figure (gross cost minus positive side effects) which is to be compared with the benefits of mitigation policies (whether these benefits are accounted for in terms of explicit monetary values or purely normative mitigation targets).[7]

Many current economic models account for the first and/or the second category of positive side effects; very few account for positive environmental externalities.[8] This is mainly due to practical reasons, and one can expect quick improvement in this direction in the near future. In the meantime, it is useful to remember that the studies discussed in this chapter do not include this category of secondary benefit.

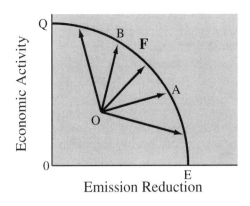

Figure 8.2: Relationship between economic activity and emission reduction.

8.2.3 Key factors affecting the magnitude of costs: Costs as a function of baselines and policy strategies

The above taxonomy suggests that assessing the costs of greenhouse gas mitigation strategies is not equivalent to adding up the direct costs of individual measures or policies. The cost of mitigation is always a net incremental cost (or a marginal cost) relative to a given scenario – usually called a baseline scenario. This means that the calculation of these net costs is determined in large part by both the assumptions underlying such baseline scenarios and the assumptions about mitigation policies.

8.2.3.1 Baselines and magnitude of the "no-regret" potentials

The most sensitive issue in the debates about how to interpret the results of the models is the way assumptions about the existence and the size of potentials for so-called "no-regret strategies" are conveyed in specific modelling frameworks and baseline scenarios.

The discussion of "no-regrets" potential has triggered a sensitive policy debate which can be summarized rather simply, though rather abstractly, in graphical form (see Figure 8.2). To begin, we represent the whole economy as producing two sets of goods and services: (1) a composite good Q, namely an aggregate of all existing goods and services, and (2) a given level of environmental quality E, represented in this case by a certain amount of emission reductions. Given such an assumption, it is possible to construct a curve F(Q,E), called a theoretical production frontier by economists, which represents the trade-off between economic activity (Q) and emission reduction (E). For a given economy at a given time, each point on this curve shows the maximum size of the economy for each level of emission reduction; put another way, it shows the maximum emission reduction for each level of economic activity. If the economy is at a size and level of emission reduction below and to the left of this curve (e.g., point O in Figure 8.2), it is possible for that economy to move upwards (e.g., from O to B), producing more goods without increased emissions, or to move to the right (e.g., from O to A), reducing emissions without reducing the size of the economy, or to move somewhere in between A and B, increasing both economic activity and emissions reduction.

From the point of view of cost analysis, a key consideration is what is assumed about the location of the reference or baseline scenario with respect to this curve. If the baseline scenario assumes the economy to be located somewhere on the theoretical production frontier (curve F), it is clear that there is a direct and unavoidable trade-off between economic activity and the level of emissions. In effect, all increases in emission reduction (moving down the surface of the curve to the right) will decrease economic activity (i.e., increase costs). That is, there is no no-regrets potential: Moving up to the left on the curve will increase economic activity but also increase emissions. In such a context, an appropriate policy mix can minimize the net cost of lower emissions but can never offset it totally.[9] Conversely, in a baseline scenario that describes an economy below the production frontier represented by curve F, no-regret strategies are possible, by moving from O to any point between A and B on curve F. Under these conditions, emissions can be reduced without reducing the size of the economy (i.e., without increasing overall costs) and possibly with some enhancement of economic activity.[10]

The critical question is, then, whether the reference or baseline scenario to which mitigation scenarios are compared is on this frontier or not. Assuming that a no-regrets potential exists suggests implicitly that any baseline scenario is below the frontier and that appropriate policies would move the economy up towards that frontier. The counterargument is that, if such a potential had existed, it would already have been adopted by the marketplace at least as long as there were no institutional failures preventing market forces from operating.[11] This line of reasoning leads many analysts to assume that any cost-effective emission reduction is already embodied in any baseline scenario and to locate their baseline scenario on the frontier.[12] In this sense the economic debate is as much about the location and characteristics of the baseline scenarios as it is about the nature and costs of specific mitigation measures.[13]

In fact, the existence of a no-regrets potential implies (1) that markets and institutions do not behave perfectly because of market failures (lack of information, distorted price signals, lack of competition, etc.) and/or institutional failures (inadequate regulation, inadequate delineation of property rights, distortion-inducing fiscal systems, etc.); (2) that it is possible to identify policies that have the ability to correct these market and institutional failures without incurring implementation costs larger than the benefits gained; and (3) that a policy decision is made to eliminate selectively those failures that give rise to increased greenhouse gas emissions (since there may exist other market failures whose removal might increase these emissions).

In other words, the existence of market and institutional failures that give rise to a no-regrets potential is a necessary but not a sufficient condition for the development of strategies to realize that potential. The latter depends on the existence of significant political desire to realize the potential. In practice, in many fields of public policymaking, countries will consider climate policies in a multiobjective decision-making frame-

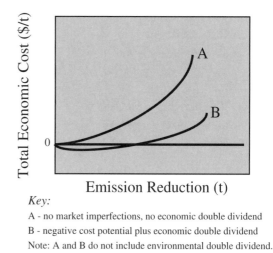

Key:

A - no market imperfections, no economic double dividend

B - negative cost potential plus economic double dividend

Note: A and B do not include environmental double dividend.

Figure 8.3: Alternative views on cost of emission reduction.

work, whereby greenhouse gas mitigation policies are likely to be a by-product or joint product of policies developed in part for other reasons. Few costing studies address these complexities; however, some bottom-up studies examine the additional environmental benefits of greenhouse gas mitigation policies, whereas some top-down studies examine the benefits of using carbon taxes to reduce other tax distortions in the economy.

The existence of these issues means that the results of any analysis of the relationship between cost and emission reduction are largely determined by a set of underlying assumptions about negative cost potentials and the economic double dividend. Figure 8.3 illustrates two different views of this relationship that underlie the top-down and bottom-up debate. Both curves in Figure 8.3 show how total costs would increase for higher levels of emission reduction, given different underlying assumptions about the efficiency of existing energy markets. Both curves assume that there exist no environmental double dividends.

Curve A (traditionally associated with a top-down perspective) assumes that there exist no reducible market imperfections (i.e., no negative cost potential), or that reducible market imperfections are already incorporated in the base case, or that the costs of reducing market imperfections outweigh the benefits. It also assumes there is no economic double dividend. As a result, the greater the level of emission reduction, the higher the costs. In this perspective, the net costs may be even higher than the gross costs because of inefficient recycling of carbon tax revenues.

Curve B (traditionally associated with a bottom-up perspective) starts below the *x*-axis because it assumes that there exists some combination of (1) market failures in the energy, transportation, or agricultural systems that can be corrected by (or are corrected in parallel with) emission reduction policies at negative cost, and (2) economic double dividends that offset the costs of emission reduction policies. Thus Curve B shows the existence of some no-regrets or "worth doing anyway" potential below the *x*-axis.

The differences between Curves A and B represent different underlying views of the efficiency of the economy. Since many current models can adopt either view of the economy, such underlying assumptions are often the main reason for the differences in quantitative results among different analyses.

8.2.3.2 *Target setting: Level and timing*

The growth rate of CO_2 emissions is determined by the growth rate of GDP, the ratio between GDP and the required level of end-use services (energy, transport, food), and the level of greenhouse gas emission per unit of each of these services.[14]

When targets are set for levels of emissions calculated from a given benchmark year in the recent past, the level of baseline emissions is critical because higher rates of economic growth increase the gap between baseline and target emissions, thus making any given target more costly to achieve. The higher the rate of growth, the bigger the emission reductions required to meet the target. This tendency is obviously less strong if the baseline incorporates some decoupling between economic growth and emissions and assumes many flexibilities in the system (e.g., high responsiveness of consumption to price and nonprice signals, or availability of low carbon-intensive techniques).

With emissions growing over time, a target of constant emissions implies that larger emission reductions are required in every future time period, requiring recourse to more and more expensive measures. Conversely, if a model embeds optimistic assumptions about technological progress in the long run, this tendency may be counterbalanced and the costs in the short term may be higher than over the long run.

For given assumptions about technical progress the time profile of the abatement may be a key determinant of differences in cost estimates. In discussing these timing issues, there is a key distinction between the transition period and the backstop period.

The transition period comes first and is characterized by an existing capital stock and limited technological options for replacing existing techniques with less carbon-intensive or carbon-free techniques. In effect, much of the infrastructure and technology is fixed. The backstop period is entered after sufficient time elapses to allow the entire capital stock to be replaced and for carbon-free backstop technologies to become available, in other words, technologies available for widespread adoption at the end of the economic life of existing equipment.

One of the most important determinants of costs during the transition period is the turnover of the capital stock. Over the backstop period the cost of carbon-free technologies places an upper limit on how great the costs of reducing carbon emissions can be. Successful research and development that accelerate the availability of less carbon-intensive and carbon-free technologies can reduce costs in the backstop period.

8.2.3.3 *Policy instruments – the tax recycling issue*

What policy instruments are used to trigger modifications in consumption and technical adoption behaviours and how they are accounted for in the models can also affect the models' results. The types of policy instruments that have been studied in detail are energy taxes and quotas on the one hand and

---

BOX 8.1: BASIC PRINCIPLES FOR THE ASSESSMENT OF THE WELFARE COST OF A TAX

Make D the demand curve for a given good (part (a) of the diagram). Then Q_0 is the quantity of this good purchased for a market price P_0. The increase in welfare for the consumer who purchases Q_0 of the good is the area of the triangle $P_mP_0D_0$, namely the sum of the differences between the maximum price the consumer would be willing to pay for each quantity of the good below Q_0 and the price actually paid (the consumer would have been willing to pay P_m for the first unit of the good but got all Q_0 units for P_0).

When a tax is levied, the new price of the good is P_1. The consumer surplus is now the area of the triangle $P_mP_1D_1$, and the tax revenue is the area of the rectangle R. The net loss in welfare is then equal to the triangle A (i.e., A is the part of the triangle $P_mP_0D_0$ that does not accrue to the consumer, who gets $P_mP_1D_1$, or to the government, which gets R). If a tax is levied on top of an existing tax (part (b) of the diagram), then the loss of welfare is A plus A′ plus B. The higher the preexisting tax, the greater the loss of welfare.

Added burden from tax

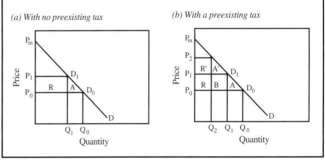

constant a collection of regulatory programmes, efficiency standards, incentives, information programmes, and voluntary programmes that are intended to bring about adoption of specific technical measures to reduce energy use on the other. Significant differences exist among different models as to which instruments are considered and how they are treated.

To date the focus of macroeconomic models has been on carbon or energy taxes (the focus of new generations of sectoral technico-economic analysis has been on the impact of other types of incentive instruments), and significant differences result from the way the revenues of a carbon tax are recycled in an economy. In the earlier models, many simulations were made that assumed no tax recycling. Such an assumption amounts to treating a carbon tax as an external shock such as an oil shock and places an upper bound on the macroeconomic costs. In later analyses, most of the models represented recycling in the form of a lump-sum process, namely, without modifying the rest of the fiscal structure. This method makes comparison easier but does not describe the recycling techniques that have the highest probabilities of being implemented. In a third stage, models tried to exemplify other ways of recycling a tax by changing the level of payroll taxes, income taxes, and corporate taxes, or simply by reduc-

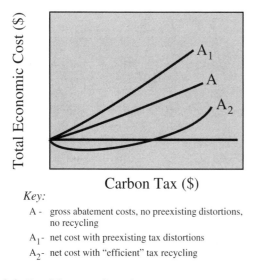

Key:

A - gross abatement costs, no preexisting distortions, no recycling

A_1 - net cost with preexisting tax distortions

A_2 - net cost with "efficient" tax recycling

Figure 8.4: Possible costs of a carbon tax.

ing public deficits. This methodological development complicates comparisons, because the outcome depends on the ways the existing distortions of fiscal structures (or subsidies) are accounted for in the baseline, but it is more meaningful from a policymaking viewpoint.

Theoretically, the recycling of a carbon tax may result in either an economic double dividend or an added tax burden (Bovenberg and Van der Mooij, 1994; Goulder, 1994). Rectangle B in the right-hand diagram in Box 8.1 represents the additional cost of a tax when it is levied on top of existing distorting taxes. In these circumstances, emission taxes would yield an economic double dividend if the added tax burden they cause is lower than the decreased tax burden they make possible by reduced taxes on other factors of production (labour, capital, rent). Otherwise, emission taxes would increase the costs to the economy.

These alternative possibilities are shown graphically in Figure 8.4. Curve A shows the gross abatement costs. Curve A_1 represents the net costs of emission taxes when there is no tax recycling and when there are preexisting tax distortions. The difference between A and A_1 is the added tax burden of the emission tax. (A_1 would be lower, but still above A if the emission tax revenues were used to reduce taxes that had less distorting effects than the emission taxes.) Curve A_2 shows the result if the emission tax revenues are recycled to reduce taxes that are more distorting than the emission taxes. Under these circumstances there exists an economic double dividend.

8.2.3.4 *International dimensions of climate policies*

The last factor affecting the cost figures provided by the models is the nature of the assumptions made about the international context of climate policy.

On the one hand, unilateral reduction of greenhouse gases may negatively affect the competitiveness of national industries and create a "leakage" of emissions from one country to another. This leakage occurs if mitigation policies in a given country cause firms to relocate their polluting plants and production processes to countries in which no such policies exist, or if firms in countries with no mitigation policies gain a com-

parative advantage due to lower costs, thus increasing their output and thereby their emissions relative to countries with mitigation policies.

On the other hand, international coordination of mitigation policies may not only prevent these outcomes but may result in lowering the total global cost of reducing emissions. To the extent that mitigation costs vary among countries, the costs of a given level of global mitigation would be lower if international coordination resulted in the most cost-effective mitigation policies being adopted first than if each country were simply to reduce emissions by an equal amount.[15] Although a growing theoretical literature is currently devoted to this issue, it has been addressed only in a few empirical modelling works to date. Most analyses have assumed either that national polices are adopted unilaterally or that the effects of national mitigation policies are neutral with respect to international competition (i.e., that all countries take the same action or that the effects of mitigation policies on international competitiveness are small). So far no model has been able to account for the gains to be expected from international cooperation in research and development and technology diffusion.

8.3 Patterns of Development and Technological Change

For any country, both the level of greenhouse gas emissions and the costs of mitigation of those emissions depend on a series of factors, including the range of technologies used and the underlying technological and socioeconomic conditions that give rise to final demands for land, energy services, or transportation. Most models explore options and alternatives as well as possible future economic conditions, based on particular assumptions about the nature of technological change and long-term development paths. This section of the chapter will explore this latter question in more detail, because these assumptions are very often only implicit and because current modelling methodologies and current data do not enable modellers to treat explicitly some of these critical parameters.

8.3.1 *Links between development patterns, technical change, and mitigation costs*

8.3.1.1 *The importance of the socioeconomic assumptions underlying scenarios*

Much of the discussion about the costs of greenhouse gas emission reduction – and indeed about energy issues in general – naturally focusses on economic concepts and variables such as income, prices, and growth in GDP. We therefore begin with a rather trite observation: Damages to the environment are not caused by a given amount of dollars, yen, pounds, Deutschmarks, or francs, but by the material content of the consumption or production activities that are the counterparts of these monetary values. To put it in another way, greenhouse gas emissions over the long run depend not only on the rate of economic growth but also on the structure and physical content of this growth.

It is well known that countries with rather similar development levels may have very different energy consumption per capita or very different transportation requirements. Some of these disparities are obviously due to natural and geographical characteristics (temperature, population density, etc.), but many stem from differences between the development patterns of these countries. Comparative studies aiming at explaining these differences (Martin, 1992; Darmstadter *et al.*, 1977) suggest the importance of five considerations that will influence the amount of greenhouse gas emissions, given a certain overall rate of economic growth:

(1) *Technological patterns* in sectors such as energy, transport, heavy industry, construction, agriculture, and forestry. As discussed below, these patterns encompass individual technological choices and options but also include overall technological systems, with their particular internal consistencies and dynamics.

(2) *Consumption patterns.* For a given per capita income, parameters such as housing patterns, leisure styles, or the durability and the rate of obsolescence of consumption goods will have a critical influence on long-run emission profiles. Beyond their purely technical aspects, these patterns are also related to the level of education, distribution of income, and degree of dualism in an economy.

(3) *The geographical distribution* of activities, which encompasses the distribution of human settlements in a given territory, climate impacts on energy demand, and the nature of urban form within a given settlement. The impact of this parameter is threefold: first on the evolution of land uses, second on mobility needs and transportation requirements, and third on the energy used for heating and cooling.

(4) *Structural changes* in the production system and, in particular, the role of high or low energy-intensive industries and services. The energy content of industries such as steel, nonferrous metals, chemicals, or pulp and paper is between four and six times that of other industries. At the other end of the spectrum, a simple phone call on a given commodity futures market can generate substantial economic gains (and e-mail connections can increase the efficiency of researchers) for a negligible energy content. A shift in the relative size of primary production and service industries in an economy may or may not affect the overall level of economic activity but will have significant implications for energy use.

(5) *Trade patterns.* It is generally argued in the economic literature that removing tariff and nontariff trade barriers enhances overall economic efficiencies. But, because historical experience demonstrates that some form of protectionism was considered necessary to many countries at the early stages of industrialization, and because of the transition problems for removing these barriers (risks of social and economic disruptions), free trade will be implemented only gradually after the Uruguay Round. In the meantime, the world is apparently moving towards the creation of regional trading blocs (European Union, NAFTA, Mercosul). The future of these arrangements is very hard to predict and will alter significantly the access to the best available technologies, the capacity of developing countries to generate high enough internal capital accumulation to finance infrastructures and education, the location of industrial activities, and future land uses (because of the impact on agricultural markets).

These factors are not ignored by current economic models. They are in some way captured by changes in economic parameters such as the structure of household expenses devoted to heating, transportation, or food, the share of each activity in the total value added, the share of energy costs and transportation costs in the production function of industrial sectors, or import-export elasticities. This type of treatment is convenient for addressing the requirements traditionally posed by policymakers since the beginning of economic modelling just before the Second World War: to provide information on the consequences of economic policies (e.g., a monetary devaluation, a fiscal policy, an incentive to final demand through public investment programmes, etc.) over the short term (one to three years), or to develop consistent economic scenarios to frame sectoral planning and policy (mainly in energy and transportation) over the medium term (four to ten years). For these time horizons and objectives it is logical to assume a continuation of historical trends in the main characteristics of development patterns and in the speed and direction of the transformation of these characteristics.

For the longer-term periods under consideration in greenhouse debates, these assumptions cannot easily be maintained, and economic parameters cannot easily be viewed as the sole command variables needed to predict the future of our production and consumption systems. For example, a given amount of added value produced by the steel or chemical industries may correspond to very different levels of material production (and thus energy demand), depending on the level of sophistication of the final product; in the same way, the differences in household budgets devoted to transportation may not fully express the differences in mobility and transportation patterns prevailing between towns with or without public rail transport systems, or between car uses in towns with very different levels of congestion.

More fundamentally, the dynamics of long-term technological development cannot be fully captured by changes in the capital output ratio (the aggregate amount of economic capital used per unit of output) or by the impact of the rate of investment on overall productivity. These parameters are important, but the outcome in terms of greenhouse gas emissions will also depend on dynamic linkages between technology, consumption patterns (mainly with respect to energy requirements), transportation, urban infrastructure, and the rural–urban distribution of population. We will return to the attempts of existing models to take these parameters into account, but the lack of knowledge available about their dynamic linkages and about their interactions with economic policies and economic signals over the long run must be underlined at this stage, together with the intrinsic difficulty of predicting innovations and transformations of lifestyles over the long run. Many fields of social science address these issues, but such information is not typically available in a form easy to process in a numerical model.

In principle, alternative configurations of the factors determining development patterns could be formally combined to give internally consistent scenarios characterized by various physical and technical characteristics and economic equilibria for a given rate of economic growth. But this does not mean that all these possible scenarios are viable and that it is possible to achieve a transition towards these long-term pictures without entailing high social and economic costs. What matters here is that these underlying technological and consumption factors are critical not only for the definition of the baseline scenarios but also for the assessment of actual mitigation costs for a given mitigation policy or objective.

As explained earlier, the mitigation costs attached to each possible baseline scenario depend not only on the absolute level of the required mitigation and on the array of available technologies (energy efficiency, fuel switching, biomass planting, other renewable energy development, modal choices in transport) but also on the timing of this mitigation (Grubb *et al.*, 1994; Hourcade and Chapuis, 1993; Manne and Richels, 1992). In this connection, three groups of partly related issues become very important:

- *The flexibility (or inertia) of consumption patterns* underlying the activity of greenhouse-gas-emitting sectors such as energy, transportation, or cement. This flexibility parameter determines the speed of adaptation to a given economic or noneconomic signal and encompasses two aspects. The first is the rate of renewal of existing end-use equipment, which is likely to be an important factor in the case of buildings, which have longer lifetimes than most capital stock. A more critical inflexibility of this kind stems from the systemic linkages between consumption patterns, technology, and the spatial distribution of activities. To take an extreme illustration, it would be far more costly to move away from oil-based automobile fuels in big conurbations where urban structure makes the use of cars almost essential than in, say, a European town of about 40,000 inhabitants, where it would be easier to satisfy a significant proportion of intra-urban personal transportation with electric buses and bicycles. The second aspect is the length of time required for turnover of the energy supply system. In this connection, the flexibility of policy response is associated with the size and lead times associated with new energy supply technologies.

- *Behavioural characteristics* that determine technical change and the evolution of lifestyles; this point will be elaborated further with regard to technical adoption mechanisms. With regard to consumption patterns, the roots of inertia are not only technical. Anthropology and social psychology demonstrate how far they are embedded in cultures and habits and, more generally, how individual consumption behaviours are shaped by social determinants that are hard to change overnight (Robinson, 1991a; Lutzenhiser, 1993; Schipper, 1995).

- *Interactive effects* due to feedbacks between the use of certain options and the rest of the economy. Some sectors have a pervasive effect on the rest of the economy,

and drastic adaptations would trigger strong structural shocks on the entire productive system.

As a whole, the critical role of these factors comes from the fact that the bigger the inertia of the production and consumption systems the more the mitigation costs will be determined by the timing of the required mitigation. This is easy to understand as far as the adoption of new technologies is concerned: To accelerate the replacement of old equipment could be a major source of the costs of mitigation. But this inertia also determines the magnitude of the loss of consumer welfare associated with mitigation. If the range of available alternatives is restricted by the material and spatial features of one's living conditions, the consumer will tend to suffer from welfare losses during the transition period towards consumption and production systems that emit lesser quantities of greenhouse gases.

A special case for inertia in development patterns occurs when such inertia creates irreversible processes of technological change. To a large extent such irreversibilities emerge out of the factors traditionally discussed in the literature on technical innovation: learning curves, economies of scale, increasing informational returns, positive network externalities, barriers to entry, and others. Irreversibility occurs when these factors combine in such a way that a particular trajectory of technological change and development is created that effectively makes impossible the alternative choices that were available earlier. This gives rise to a time dependence of technical choices and the occurrence of "lock-in" effects (Arthur, 1988). Beyond such a bifurcation point, market forces will reinforce the first choice in a self-fulfilling process. To give a single example, given the high research and development costs involved in a new automobile engine, it is unlikely that research and development risks will be incurred simultaneously on electric cars, 2L/100 km gasoline engines, and biofuel engines. As with gasoline engines in the early days of the automobile, the choice of a particular technology, beyond a certain point, will essentially prevent the development of alternatives.

More generally, beyond technical considerations themselves, the self-reinforcing loops among technical choices, consumer demand, and the geographical distribution of activities and human settlements explain the fact that particular sets of technological and behavioural options can be clustered into consistent packages which, at least for a rather long period of time, foreclose other options in technology and innovation. These clusters are rather systematic in industries relying on network structures, such as energy, transportation, or telecommunications because of the characteristics of their production function (e.g., discontinuities and economies of scale), the need for technical harmonization across the network, and their dynamic interactions with markets. This is particularly clear for transportation systems because of their linkages with urban and spatial dynamics. For example, trends observed in Western Europe over the past several decades could be expected to lead to a doubling of road freight on highways in the next fifteen years under the influence of the Single European Market. However, if Austria and Switzerland maintain their policies of limiting international truck freight transport across their countries for reasons of security, avoidance of congestion, and local environmental protection, Europe will over the next several decades create two very different transport systems (combined rail–road versus road-dominant) and will certainly have gone beyond a bifurcation point (Hourcade, 1992). A similar choice will be faced in the near future in many large developing countries with far more dramatic consequences, as decisions are made about how to expand road, rail, and air networks to respond to large increases in the demand for personal mobility and freight transport.

8.3.1.2 Current and future socioeconomic development patterns

Combining the issues and considerations discussed above suggests that four factors are likely to give rise to differences in the development paths underlying different baseline emission levels and differences in the cost curves for emission reduction or adaptation. We will describe briefly the nature of these factors and point out to what extent they are amenable or not to conscious control or choice. A critical issue is that many of the decisions that are apt to control the long-term paths will be taken for reasons that have nothing to do with energy policy or climate change and are indeed very often taken "in passing" rather than as an explicit component of public policy.

8.3.1.2.1 Material and energy content of development in industrialized countries

An important determinant of greenhouse gas emissions in any economy is the raw material intensity of that economy: the amount of matter and energy used per unit of economic activity. Broadly speaking, for any given level of economic activity, the lower the raw material intensity, the lower the greenhouse gas emissions. Over the past several decades, the raw material intensity of industrialized countries has dropped significantly (Bernardini and Galli, 1993; Williams, 1987). An important question is whether such reductions in industrialized economies simply reflect the shifting of energy- and matter-intensive activities to other countries, in which case the net effect may be to hold constant or even increase global raw material intensities, or whether there is a net reduction in the energy- and matter-intensive activities themselves (IAEE, 1993).

Some of the major socioeconomic factors that will affect the future raw material intensity of industrialized economies are

- structural shifts in the economy towards services
- increases in the "information intensity" of industrial processes, goods, and services (Chen, 1994)
- the effects of telecommunications on travel and transportation energy use (Selvanathan and Selvanathan, 1994)
- saturation in the consumption of some goods and services and the emergence of less energy- and material-intensive goods and services

Clearly, quite different configurations of these factors are possible, resulting in potentially large differences in raw material intensities, independent of the size of the economy. It is also clear that mitigation costs per unit of economic activity

(e.g., dollars per unit GDP) will depend on both the mix and level of such intensities and will necessarily vary among economies with different structures.

8.3.1.2.2 Links among energy, transport, and urban planning

Transportation energy use accounts for a significant proportion of greenhouse gas emissions,[16] and its growth rate is typically higher than for other categories of energy demand. More critically it is the only form of energy use not drastically decoupled from economic growth after the oil shocks. This does not demonstrate that price effects are nonexistent in this sector but that they interact with other structural determinants. In the U.S., for example, energy efficiency gains in private automobiles over the 1970s and early 1980s were almost exactly offset by increases in distances driven per vehicle, so that total energy use remained flat (Schipper and Howarth, 1990). In the transport sector, the types and quantities of emissions are a function of the demand for transportation services (i.e., the amount of travel), the mode chosen (auto, air, bus, rail), the efficiency of the vehicles, and the types of fuel used. The first two of these, in turn, are greatly influenced by, and influence, the size and configuration of cities and towns (Newman and Kenworthy, 1989). Important factors are:

- the location of housing relative to jobs, schools, and retail outlets
- the distribution of retail trade and industrial activities within the region
- the road and rail network within and among different cities and towns
- the investment in, and choice of, public transit systems

Beyond these specific urban planning issues are a whole set of questions related to the development of new transportation systems and technologies (Ross, 1989). The growth in personal mobility in industrialized countries over the past several centuries has been closely correlated with technological advances in the types of vehicles used, the availability of fuels, and the transportation infrastructure created. In turn, such factors interact in complex ways with urban form and development and directly affect both the prospects for, and the costs of, greenhouse gas mitigation. For example, the level of future automobile emissions will depend in part on the degree of saturation of automobile densities in urban areas, due to the availability of effective air and high-speed ground transportation alternatives (Grübler *et al.,* 1993). In the metropolitan area of Sao Paulo, for instance, most of the increase in atmospheric pollution has been attributed to increased congestion, as indicated by a reduction in the average vehicle speed from 28 km/h during the 1980s to 20-24 km/h in 1993 (CETESB, 1994).

Clearly, quite different configurations of these factors can be envisaged, and transportation is a field in which bifurcations towards contrasted paths may occur in any region of the world. These alternatives would have major impacts on mitigation options and costs. As suggested earlier, the costs of transportation sector mitigation for a population that depends on automobile transport in highly dispersed suburban developments would be very different than for a population located at the core of dense urban agglomerations with a well-developed urban transit infrastructure. Since towns and cities take shape, and change, over periods measured in decades and centuries, the full effect of alternative transportation developments will be manifest only in the relatively long term, and there is high uncertainty about the long-run impact of short-term decisions (or nondecisions).

8.3.1.2.3 Land use and human settlements

This issue concerns all regions of the world, but its quantitative impact is more impressive in developing countries. Land use and human settlement pattern changes derived from agricultural and forestry activities as well as rural–rural and rural–urban migrations are among the main sources of greenhouse gas emissions in these countries. Deforestation in most developing countries is a complex phenomenon, mainly caused by the expansion of the road network, logging and cattle raising activities, agricultural production, and population growth fostered by rural-rural migrations due in some cases to the absence of guaranteed access to land (agrarian reform) in other regions.

These examples show that mitigation cost estimates that don't take account of such institutional factors can be very misleading. Hidden costs can be very important when one considers the institutional capacity building needed, on the one hand, to prevent the harvesting of a very low-cost natural resource such as a native forest, and, on the other, to enforce environmental protection, reforestation, and forest management practices. Moreover, even macroeconomic cost assessments don't tell the whole story: For instance, the whole GDP loss resulting from completely halting the development of the Amazon region would be very low compared with overall Brazilian economic output, but this scenario is clearly inconceivable for social, political, and cultural reasons. Deep structural changes in these underlying factors will often be required in order to make mitigation strategies feasible.

In this field it is even more important than elsewhere to look at the "greenhouse gas mitigation component" of more general development strategies instead of considering greenhouse gas mitigation measures on their own. Mitigation costs are only one part of a larger set of factors in such cases. For example, a first step towards increasing the relevance of a carbon sequestration objective might be to link it with the preservation of biodiversity. This could increase its political and social relevance, though it would make the task of quantifying economic costs and benefits much more difficult.

8.3.1.2.4 Development patterns in developing countries

The importance of development pattern assumptions for the assessment of mitigation costs is of particular importance in the case of developing countries. Since a major part of the infrastructure needed for development is still to be built, the spectrum of future options is considerably wider than in industrialized countries. The traditional approach of using "business-as-usual" assumptions as the baseline is then particularly problematical.

Nor can one assume that developing countries will automatically follow the past development paths of industrialized countries. The significant transformations recently evident in the international economy and energy markets highlight the dangers of such an assumption. In this respect, we need to consider the structure of GDP in developing countries and how that structure might evolve given such global transformations. A crucial question concerns the developing countries' share of the world production of highly energy- and pollution-intensive goods, such as steel and aluminum. For example, as the recent shift of heavy industries from the industrialized countries towards the developing countries begins to slow, economic output may increasingly come from services and other less energy-intensive activities.

Moreover, technological choices in both the production and consumption sectors can substantially change critical parameters such as the elasticity of energy demand/GDP. Experience demonstrates that countries entering the path to development have had generally lower energy profiles than countries that developed sooner. This does not mean that the energy/GDP ratio will not increase in many of these countries and that increases in consumer purchasing power in many developing countries will not drive up energy demand growth rates. It does mean, however, that these countries are not bound to reach the same levels of energy intensity that countries such as the UK and the U.S. have; they are not bound, for example, to adopt the same models of energy-wasteful refrigerators still in operation in developed countries or to reject up-to-date technological improvements for factories, many of which will improve energy efficiency. More generally a growing interest in the preservation of cultural diversity might also favour less energy-intensive housing, transportation, leisure, and consumption patterns, and less resource consumption per capita than is currently the case in the developed countries. One among other possible examples is related to avoiding low urban population densities that are coupled with long daily trips to work and to large shopping centres by car.

Finally, the spatial distribution of the population and of economic activities is still not settled in most developing countries, a situation that offers the possibility of adopting urban/regional planning and industrial policies directed towards rural development and a stronger role for small and medium-sized cities, thus reducing the rural exodus and the concentration of populations in large cities.[17] But, given current heavy trends towards huge conurbations, it will be difficult to shift them over the next century. We are faced here with a strong socioeconomic inertia that can be overcome only gradually through steady long-term policies. Nevertheless, more decentralized development patterns than those typical of industrialized countries could allow developing countries to use modern technology (biotechnology, solar energy, wind, and small-scale hydro power) to tap their large reserves of natural resources. In the same ways, it could change their needs for transportation.

These examples show that developing countries might be in a position to adopt anticipative strategies that could prevent in the long term some of the problems faced today by industrial societies. For example, in industrialized countries, energy demand/GDP elasticities first increased with successive stages of industrialization (with an acceleration during the 1950s and 1960s), but have sharply decreased since then, due to a number of different factors, such as the relative growth of services in the share of GDP and technical progress and energy conservation induced by higher oil prices (Martin, 1988; Darmstadter, *et al.,* 1977). Developing countries could skip these intermediate stages and follow a path leading directly to less energy-intensive development patterns, thus avoiding a large increase in energy/GDP intensities in the short and medium terms – the so-called "tunnel effect" (Berrah, 1983).

Such possibilities for alternative development patterns highlight the technical feasibility of low carbon futures in developing countries that are compatible with national objectives. But, in the opposite direction, past and present experience suggest that these countries might not be in a position to switch to low greenhouse gas emitting profiles. The recent experience of economic stagnation or recession in a substantial set of developing countries demonstrates, for example, that access to superior technologies may be limited by the increase of the share of second-hand equipment or the import of obsolete technology and products. In the same way, the prevailing indebtedness of many countries could prohibit the adoption of interurban transportation infrastructures or of the urban infrastructure policies necessary to reach balanced human settlement patterns. As a whole, barriers to more sustainable development in the developing countries can hardly be underestimated. These include insufficient capital stock (thus preventing the use of the same technologies as in developed countries), tariff and nontariff trade barriers, the organization of international trade and the international financial environment, the distribution of national income, and very often the lack of appropriate institutions for using the indigenous technical and economic capabilities of these countries.

This means that switching to very different development patterns from those that have customarily been expected will depend on the removal of these barriers, the setting up of an appropriate context by national public policies, and the evolution of international economic relationships. Development theory has not yet provided the analytical tools needed to enable modellers to discriminate among such different development patterns.

8.3.2 *Modelling development paths and mitigation costs*

It is clear from the preceding discussion that the underlying socioeconomic conditions will have an important effect on future greenhouse gas emissions and the costs of mitigation and that great uncertainty exists about how those conditions may evolve over the next century. The extent of these uncertainties is such as to cast some doubt on the ability of models to predict long-term development patterns. On the other hand, it has become more and more important to have a scientific understanding of the implication of possible long trends. It is in the context of this tension between uncertainty and the need to act that we have to assess the meaning and lessons of modelling analyses.

8.3.2.1 *Prediction and simulation: The need for multiple baselines*

In coming to grips with the uncertainty of long-term projections, it is important to recognize that it goes beyond the uncertainty associated with any given emissions forecast or scenario. The point is not that individual base case forecasts or scenarios are uncertain with regard to exogenous parameters such as economic or population growth rates. It is that there exists a range of quite different underlying socioeconomic development paths that would give rise to different emission scenarios and costs of mitigation. Each of these development paths can be thought of as a base case, and each one will vary, given different assumptions about economic or population growth rates (i.e., each is uncertain in the familiar sense). Moreover, to the extent that these different development paths contain the kinds of bifurcations and irreversibilities discussed above, it would be very difficult, and indeed misleading, to estimate a probability distribution for the set of long-term development paths. Such bifurcations can be triggered by short-term decisions and actions that are not predictable.

The logical implication of this is the necessity to use multiple scenarios in order to account for uncertainties with regard to development patterns and the possibility of what economists call "multiple equilibria" in the long term. At time "*t*" there are several possible market equilibria that could arise at a future time "*t+n*" (i.e., several possible states of the world characterized by different technical circumstances), and these are not easily predictable from current trends. From a purely analytical perspective, it is possible to resort to some subjective or logical probability distribution in order to narrow these uncertainties among different scenarios and provide some kind of average scenario. But to do so could be misleading in a decision-making context. The point is that these alternative base cases represent quite different, and internally consistent, patterns of development. Many of the uncertainties about such scenarios are due to the long-term consequences of short-term decisions or behaviours, and to the collective expectations prevailing at time "*t*" regarding the future time "*t+n*." In turn, these decisions, behaviours, and expectations are part of a particular sequence of events and cannot simply be combined with the components of other such sequences to produce an "average" or "most likely" sequence.

These concerns about the predictive value of models and the existence of possible multiple economic equilibria and their logical implications for policymaking are not new in various fields of socioeconomic modelling. The origin of these concerns lies in the recognition of the limits of empirical forecasting methods that appeared by the end of the 1960s. This led to the development of approaches that tend to emphasize the unpredictability of alternative futures, the degree to which they are subject to partial choice, and the need for analysis of multiple alternatives.[18] In the energy field, approaches based on the concept of multiple scenarios were pioneered by Shell International Petroleum Company in the early 1970s (Jefferson, 1982, 1983; Wack, 1985a, 1985b) and have been adopted in various forms in the electric utility industry in North America (Southern California Edison Company, 1988; Northwest Power Planning Council, 1991). More recently, building in part on some of the methodological insights of the Shell scenario analysis approach, there has emerged a set of more general analyses of long-term futures focussing on questions of surprise and alternative development paths defined in very broad terms (FRN, 1987; Stockholm Environment Institute/ Greenpeace, 1993; Gal and Frick, 1987).

In parallel to methodological concerns about the inadequacy of traditional forecasting techniques, there emerged over the same period a series of critiques of conventional "business-as-usual" views of the future, usually based on an explicitly "environmental" perspective. One of the most prominent of these early critiques was the Club of Rome's 1972 book *The Limits to Growth* (Meadows and Meadows, 1972), the publication of which stimulated an extensive debate about alternative development paths. This debate was fuelled by the publication of a number of subsequent studies (Mesarovic and Pestel, 1974), including the Leontieff report launched to illustrate the International Development Strategy adopted by the United Nations in 1974 (Tinbergen, 1976). An attempt to articulate a developing country perspective on these issues led to the publication of the Bariloche report, which presented a rather different development path for these countries (Herrera *et al.,* 1976).

These studies provided fertile ground for the development of alternative energy modelling analyses, which were also triggered in part by the oil shock of 1973–74. These studies expressed concerns about possible long-range limits in fossil fuel resources and about the risks likely to be involved in nuclear power. Following a major study by the Ford Foundation in 1974 (Ford Foundation, 1974), Amory Lovins' provocative "soft energy path" analysis (Lovins, 1977) helped to launch many exploratory works in North America (Craig *et al.,* 1978; Sant, 1979; Brooks and Ginzton, 1980; Stobaugh and Yergin, 1979; Ross and Williams, 1981) and in many European countries (Johansson and Steen, 1977; Leach, 1979; Norgaard, 1979; Krause, 1981; Messenger, 1981; Olivier *et al.,* 1982) about the possibility of avoiding possible physical limits and nuclear risks. These studies were based on what came to be called "bottom-up" methods of analysis and produced energy system projections that were significantly at odds with the "top-down" analyses prepared by governments, the energy industry, and energy economists relying on extrapolatory approaches.[19] In the early 1980s, the publication of the results of the energy study by the International Institute of Applied Systems Analysis (Häfele, 1981) sparked a vigorous debate both about modelling methods and about alternative energy futures (Keepin and Wynne, 1987; Thompson, 1984). Subsequent analyses continued to reflect the split between top-down and bottom-up analyses, but revealed a growing interest in global environmental issues such as global climate change, influenced in part by the publication of the final report of the World Commission on Environment and Development (1987) and the subsequent UN Conference on Environment and Development in 1992.[20] Throughout this period, a major bone of contention has been how best to analyze alternative patterns of development and technological change that represent significant departures from past trends.

These debates are not restricted to applied analyses. They have taken place in many fields of social science theory concerning such issues as the plurality of economic equilibria generated by different sets of expectations – such as the works about the so-called "sunspot theory" (Azariadis and Guesnerie (1986); self-fulfilling or self-defeating prophecies (Hen- shell, 1982, 1993); "common knowledge" and "conventions" (Lewis, 1969; Dupuy, 1989); the coordination game (Schelling, 1960; Aumann, 1987), and, more generally, the outcome of repeated games in game theory (Fudenberg and Tirole, 1991) and chaotic behaviour and surprises. There is thus a rich history of approaches to modelling that emphasize the importance of incorporating alternative development paths and multiple base cases. As yet such approaches have made only a modest impact on greenhouse gas costing studies.

8.3.2.2 *Economic modelling and development trends: Some limits*

There is perhaps an unavoidable gap between advanced thinking in social science theory and the empirical tools available, given computational constraints, data availability, and the priorities adopted by research-funding institutions. Nevertheless, over the past two decades the state of the art of long-term, policy-oriented energy system modelling has evolved very rapidly, due in large part to significant progress in energy-economy modelling. This progress has been driven by questions raised by the oil shocks of the 1970s, perceived risks of resource exhaustion, the nuclear debate, and climate change issues.

Progress has also been made in modelling other sectors such as transportation or agriculture. However, less attention has been paid in this work to the linkage between sectoral analysis and a more comprehensive economic framework than in energy sector analyses, where policymakers have been very interested in the macroeconomic imbalances (e.g., external debt, inflation, and unemployment) caused by oil imports and price increases. Moreover, the initial focus on energy-related greenhouse gas emissions has meant that less attention has been paid to the emissions of other sectors. As a result, the state of the art of modelling in the energy field serves as a useful indicator of the difficulties of representing alternative development paths in current models.

Many econometrically driven "top-down" models of the whole economy, initially built for the purpose of economic policy analysis, have contained aggregate or quasi-aggregate sectoral demand functions (e.g., one or two energy services or one composite food product). They consequently capture future development and technological trends in the econometric relationships among standard econometric indicators. Using them in long-term analyses amounts to extrapolating current trends in the relationships among key variables, on the assumption that historical changes in these relationships capture the essence of likely future responses to changes in input variables such as the future price of energy, food, or wages. Such analysis provides very useful "counterfactual" information on, say, the aggregate impact of a carbon tax relative to some assumed base case, but it does not have a very high predictive value for the very long term or with regard to, say, bifurcations in the transportation sector.

"Bottom-up" models have historically been designed for the specific needs of a given sector (e.g., energy, transportation, forestry) and in the energy field are sometimes linked to standard macroeconomic models. They give a disaggregated picture of demand and supply and point out potential gains in efficiency from specific technologies or the potential for substitution of carbon-free technologies. Such models rely on rather precise descriptions of end-use and production technologies (e.g., appliance penetration and usage rates, building heat-loss coefficients, tonne-kilometres of freight transportation), but typically do not contain much treatment of feedbacks between these parameters and underlying economic variables (e.g., feedbacks between energy efficiency savings and employment, or consumption behaviour for other goods and services). The lack of macroeconomic behavioural feedback means that such models are better suited to "what if" simulation analyses rather than to prediction.

Virtually all the modelling analyses of greenhouse gas mitigation undertaken to date have used some form of these two approaches to energy/economy modelling or coupled models that combine them. We will discuss the difference between these approaches in more detail in Section 8.4. The point here is simply that, to date, neither set of models has been used to address the issues of alternative development patterns and technological change that have just been identified. Instead, the models have been used to analyze the effect of certain policies (typically tax policies in the case of top-down models and the implementation of specific energy-using technologies in the case of bottom-up models), given implicit assumptions about development paths.

This suggests that existing mitigation cost studies are meaningful primarily at the margin of a given development path, which, in turn, means that they are valid under the following conditions:

(1) As long as historical development patterns and relationships among key underlying variables hold constant for the projection period (top-down analyses)

(2) If there are no important feedbacks between the structural evolution of a particular sector in a mitigation strategy and the overall development pattern (bottom-up analyses)

8.3.3 *Multiple baselines, uncertainty, and long-term mitigation costs*

The preceding discussion explains why there is a growing consensus among experts on the necessity to work with multiple baseline scenarios when long-term horizons are under consideration. Such scenarios would represent different, internally consistent sets of assumptions concerning the factors discussed above rather than simple variants of a base case generated by altering the input parameters of a given model. The point is not to predict what long-term outcomes are most likely, an exercise at which few have been successful (Ascher, 1978, 1990), but to explore the economic and technical feasibility and the costs associated with quite different development paths. This would make it possible, for example, to respond to the frequent request for the generation and analysis of "sustainable development" or other scenarios that represent very different assumptions about economic and technological development paths and about "lifestyle change" from those contained in traditional analyses.

The limitations of current methodologies heighten the need for making clear the types of structural determinants that are explicitly accounted for in each baseline scenario and those that are considered fixed. There is a need for more sensitivity testing of modelling analyses, together with viability testing to ensure the overall consistency of long-term projected trends, given financial, political, and institutional constraints. However, it is also important to avoid the common mistake of using arguments about the inadequacies of existing models to project wishful thinking about future development options and mitigation strategies.

Beyond these technical difficulties we should like to emphasize two major issues stemming from the existence of several baseline scenarios. These are covered in the following two sections.

8.3.3.1 *Multiple baselines and the noncomparability of cost assessments*

Given that different underlying assumptions about development patterns can give rise to a number of quite different types and levels of greenhouse gas emissions as well as a number of mitigation costs and potentials, a comparison between cost assessments theoretically requires an evaluation of baseline and mitigation scenario assumptions, including technology data.[21] Such an evaluation should identify key determinants and assumptions behind the cost estimates in order to explain differences. Without a comprehensive and transparent documentation of the full set of scenario assumptions, any comparison of cost estimates coming from studies with very different baseline scenarios will be misleading. For example, a mitigation scenario derived from baseline assumptions that incorporate substantial air quality improvements for reasons unrelated to climate change is likely to give rise to greenhouse gas mitigation cost estimates that are higher per unit of reduction than a mitigation scenario derived from baseline assumptions without such improvements.[22] Similarly, mitigation cost estimates would vary depending on the transportation infrastructure assumed in the baseline scenario.

Moreover, given the difficulties of incorporating the feedbacks between development patterns and economic variables discussed above, a baseline scenario that gives rise to lower estimates of mitigation costs cannot be interpreted as being economically superior to other baselines. Such lower cost estimates simply mean that the incremental costs of mitigation relative to that baseline are lower than the incremental costs relative to some other baseline. However, this leaves unanswered the question of the relative costs (e.g., transaction or political costs) of achieving these different baseline scenarios. Since different baselines represent different development paths, each of which is potentially an "efficient" scenario, they cannot be directly compared and no overall relative cost assessment is possible. Two historical examples help to illustrate that this problem is far from being a simple intellectual fancy.

If the climate debate had emerged in 1973, just before the decision to launch the nuclear programme in France, any assessment of mitigation costs should have considered two possible baseline scenarios. The first would have excluded the nuclear programme and, consequently, would have had a higher CO_2 emission level. In this case, the cost of the nuclear programme could have been included in the mitigation costs.

The second would have included the nuclear programme, and, as a result, emissions and mitigation requirements would have been lower. Paradoxically, however, the costs of an incremental emission reduction of, say, 20% would have been far higher.

Beyond the fact that assessing the relative costs of two totally different energy systems is technically difficult (for example, in the nonnuclear case France would not have developed electrical heating to the degree it did), the critical point is that the nuclear choice would not have been made, in practice, purely for climate-related reasons, but in any negotiation the French administration would have tended to argue the contrary so that this programme could have been considered a specific contribution to a collective climate policy.

Turning to the second example, baseline and mitigation scenarios for Brazil differ considerably, depending on assumptions about the future of the country's biomass ethanol programme. Even though no environmental factors were taken into account to justify it when it was first implemented in 1975, the existence of this programme can be shown to lead to both a lower emission baseline and lower mitigation costs. So far, many questions are still pending with regard to the overall costs and benefits of this programme in terms of macroeconomic impacts, distributional effects, refining structure of oil, and innovation, and until these are answered it is difficult to predict the likely future of the programme. Furthermore, the choice of a baseline with or without the ethanol programme would entail many types of transaction and political costs and would have many external effects, all of which are very hard to assess.

In both cases, the best analytical strategy is to recognize that different baselines are possible and that any mitigation cost estimates are only relevant at the margin of each baseline and not in absolute terms.

8.3.3.2 *The meaning of the baseline*

The function of baseline scenarios in cost studies is to provide a basis of comparison for calculating mitigation costs. It is important to bear in mind that such baselines pose a somewhat artificial distinction between a notional "business-as-usual" case (i.e., what would happen if no mitigation policies were instituted) and a "policy intervention" case (what would happen if they were). Although such a procedure is required to obtain a basis of comparison and thus an estimate of the costs of intervention, in principle it does not imply anything about the likelihood or relative economic efficiency of the baseline compared to the intervention case.

A similar problem emerges with respect to the conventional distinction between adaptation to, and mitigation of, climate change. In the short term, the meaning of that distinction is clear: Mitigation means reducing sources of emissions and/or increasing sinks (i.e., reducing the causes of climate change), whereas adaptation means improving our capability to withstand changes in the global climate system (i.e., responding to the effects of climate change). In the longer term, however, the distinction between mitigation and adaptation begins to blur. Not only will measures adopted to serve one goal have significant consequences for the other (e.g., some energy efficiency measures adopted to reduce emissions may make energy systems more resilient to climate variabil-

ity) but many policies adopted for other reasons will have both adaptive and mitigative effects (e.g., urban planning decisions).

Indeed, in the long term, both "mitigation" and "adaptation" measures are part of the adaptive responses of societies to a whole host of conditions and perceived problems, only a small number of which have to do with climate change. From this point of view, the distinction between baseline and intervention scenarios is particularly artificial. Their importance lies in the difference between them, not in the plausibility or likelihood of either as a forecast.

8.4 Differences among Models and their Results

8.4.1 *General methodological considerations*

We have suggested that existing models do not generally address questions of alternative paths of technological development in an effective way. Yet even without the variation that consideration of such factors might be expected to induce, the range of results in greenhouse gas mitigation costing studies is very large. It is difficult to disentangle the various reasons for these disparities in modelling results, given the diversity of tools used to calculate cost estimates, the many and varied assumptions employed, and the disparities in the geographical coverage of different studies.

Historically the debate in the energy field has been framed by the distinction between "top-down" and "bottom-up" studies, a distinction that can be applied in other fields as well. Basically, top-down models analyze aggregated behaviours based on economic indices of prices and elasticities. These models began mainly as macroeconomic models that tried to capture the overall economic impact of a climate policy, which, because of the difficulty of assessing other types of policy instruments, was usually in the form of a carbon tax or, more rarely, tradable permits. Bottom-up models, on the other hand, rely on the detailed analysis of technical potential, focussing on the integration of technology costs and performance data.

Not all models fall neatly into one of these two categories, and several "hybrid models" are now available in which analysts have attempted to merge top-down and bottom-up model characteristics. As a result, differences in findings are increasingly the effect of differences in input assumptions rather than differences in model structure. However, the discussion in this text will maintain the dichotomy between these models to the extent that this distinction remains meaningful for understanding some critical policy issues.

The top-down/bottom-up categorization has been portrayed as opposing the optimism of the "engineering paradigm" to the pessimism of the "economic paradigm" (Grubb *et al.*, 1993). From an engineering standpoint the evidence is that the best available technologies have not been adopted so far; this "efficiency gap" is the gap between the energy efficiency of equipment actually chosen by consumers and the energy efficiency of the technology that could theoretically minimize the costs entailed in providing a given amount of energy service. Bottom-up models are able to demonstrate the existence of such an "efficiency gap," and thus they suggest that, thanks to "negative cost measures," substantial emission reductions could be achieved with low taxes and low costs or even net savings.

In response, the professional reflex of many economists has been to call attention to the reasons why consumers do not adopt technologies that appear to be optimal, and to suggest that accounting for these reasons, together with the economic feedbacks of a given policy, would reduce the magnitude of, or eliminate, the efficiency gap that is actually achievable. Top-down macroeconomic models concluded, at least in early analyses, that relatively large carbon taxes resulting in significant economic costs would be required to counter current emission trends.

The methodological difficulties lying behind these differences revolve around how to describe the processes of technology adoption, the decision-making behaviour of economic agents, and the feedbacks between any public policy measures and the overall economy, and how markets and economic institutions actually operate over a given period of time. From this viewpoint the opposition between top-down and bottom-up methodologies does not fully represent the whole spectrum of critical issues. That is why, before discussing the lessons of the top-down and bottom-up debate, we will sketch the dimensions of a typology of existing models. Given the prominence of energy/economy models in the field, we will concentrate on them. However, many of the methodological issues are quite general and similar arguments could be applied to models of other sectors, such as forestry.

8.4.2 Critical dimensions of a typology of existing models

Various attempts have been made to categorize the large variety of models that have been used to analyze the costs of reducing greenhouse gas emissions.[23] Instead of providing a new and necessarily arbitrary typology, we will try here to provide a description of the main characteristics that differentiate these models, in order to make clear what each type of model describes and what kind of policy question it can best address. To accomplish this, we will focus on those characteristics that have to do with the purposes of the models (i.e., the questions they are meant to address), their structure (i.e., those assumptions that are embedded in the equation system), and their external assumptions (i.e., assumptions expressed in terms of inputs to the models). Because few, if any, individual models represent pure types within these categories, we have not referenced specific models. Instead the goal here is to provide a general framework to aid in interpretation of the specific modelling results presented in Chapter 9.

8.4.2.1 Diversity of models, diversity of purposes
Significant misinterpretations of the results of modelling studies can arise from overlooking the purpose of the analysis they were used for. The meaning of the numerical results of a model will differ depending on whether a given scenario is used to predict (forecast) the future or to explore it as a tool for "backcasting" exercises.

Many models are used to try to *"predict"* the future and to provide an estimate of the most likely set of future events. This purpose imposes very strict methodological constraints on the modeller. He or she must produce a base case forecast, which amounts to a best-guess projection of the most likely future; to do this requires an endogenous representation of economic behaviour and general growth patterns. This type of predictive exercise attempts to extrapolate the interactions of historical trends into the future, with a minimum of exogenous parameters. This approach has been typical of government and sectoral forecasting activities (e.g., in energy, transport, and heavy industries such as steel) and early climate change scenario analyses. It remains both necessary and convenient for analyzing the short-term impacts of climate policies, since a number of critical underlying development variables can reasonably be assumed to remain constant for these time periods. Most short-term, econometrically driven macroeconomic models adopt this approach. For the long-term (middle of the next century), the Jorgenson-Wilcoxen, McKibbin-Wilcoxen, and Goulder models are the only models in this category. It is noteworthy that none of these models tries to work on the basis of forecasts of explicit technological trends in the engineering sense.

Because of the difficulty of extrapolating past trends over the long run, the purpose of some modellers is to *"explore"* the future rather than to predict it and, in so doing, to provide potentially counterintuitive assessments. This leads to a scenario analysis approach, which involves building up different coherent visions of the future (each of which can undergo sensitivity testing) based on different values for key assumptions about economic behaviour, physical resource endowments, or technical progress, together with assumptions about economic or population growth. The first step in such analyses is the generation of a "reference" or "nonintervention" scenario. This is then contrasted with alternative cases involving an array of policy measures, such as carbon taxes or energy efficiency regulations, giving rise to one or several "policy" or "intervention" scenarios, but the policy analysis is relevant only in the context of each baseline scenario. This approach is increasingly used for climate change analysis and was the basis for the 1992 IPCC scenarios (Legget *et al.,* 1992). It is shared by both bottom-up and top-down models[24] and indeed tends to favour the development of hybrid models (which are discussed at greater length below). In terms of representing and simulating the behaviour of economic agents, two methodologies predominate. One involves an assumption of least-cost optimization, in which society maximizes the utility of consumption over the long run. The other involves the effort to simulate real-world behaviour predictively in terms of technology adoption.

Finally, another possible purpose of models is to assess the feasibility of alternative futures, often defined in terms of desirability rather than likelihood. This contrasts with the two previous approaches, insofar as it involves the development of a vision of a future state of the system being studied and then an analysis of how that future system might be realized. This *"backcasting"* methodology[25] allows for identification of major changes as well as discontinuities in present trends that might be required if a desirable future is to be attained (Robinson, 1988, 1990). Two types of research can be carried

Table 8.1. *Key structural characteristics in energy/economy models*

Structural Characteristics	Policy Issues
1. Degree of endogenization (the extent to which behavioural relationships are endogenized in the model equations or left to be supplied as exogenous assumptions)	Models that endogenize behaviour are suited to predict actual outcomes; those that exogenize it are more suited to simulate the effects of changes in historical patterns
2. Extent of description of the nonenergy sector components of the economy (investment, trade, consumption of nonenergy goods and services, income distribution, etc.)	Models that describe these sectors in more detail are more suitable for analyzing the wider economic effects of energy policy measures
3. Extent of description of energy end uses	Models that describe these end uses in more detail are more suitable for analyzing the technological potential for energy efficiency
4. Extent of description of energy supply technologies	Models that describe these end uses in more detail are more suitable for analyzing the technological potential for fuel substitution and new supplies

Table 8.2. *Key structural distinctions between bottom-up and top-down models*

Structural Dimension	Early Models		More Recent Models	
	Top-Down	Bottom-Up	Top-Down	Bottom-Up
1. Endogenization of behaviour	High	Low	High	Increasing
2. Detail on nonenergy sectors	High	Low	High	Increasing
3. Detail on energy end-uses	Low	High	Increasing	High
4. Detail on energy supply technologies	Low	High	Increasing	High
5. Predictive orientation	High	Low	Decreasing	Increasing

out in this approach. Most studies involve generating and analyzing normative scenarios about desired futures. Many alternative energy studies, for example, belong to this category. But it is also possible to use a backcasting methodology as a purely analytical tool by simply linking bottom-up analyses about the long-term evolution of technology and development patterns to a macroeconomic framework (Hourcade, 1993) so as to be able to assess the economic consistency of different and competing views about the long run.[26]

These three different purposes have implications for the types of models required, the analytical questions being asked, and the meaning of the results. As discussed in Section 8.3.2.1 above, there has been a historical correlation between attempts to move away from predictive modelling approaches and the early development of bottom-up models; these models were in many cases built precisely to undertake simulation and backcasting analyses not possible with the current generation of top-down models (Baumgartner and Middtun, 1987; Robinson, 1982). The greatest emphasis in current climate modelling efforts is on "exploratory" analyses, using a combination of top-down and bottom-up methods, but there is continuing interest in "backcasting" analyses aimed at exploring quite different future scenarios than would otherwise be examined (FRN, 1987; Goldemberg *et al.*, 1988; Jäger *et al.*, 1991; Rothman and Coppock, 1996; Robinson *et al.*, 1996).

8.4.2.2 *The structure of existing models*
A second basis for distinguishing among different models is the nature of the model itself (i.e., those assumptions embed-

ded in the mathematical structure of the model). At a very general level, it is possible to characterize some of the main structural differences among existing energy and emissions models in terms of four main dimensions. This description abstracts from a number of more detailed distinctions among models but captures the points that differentiate the models in a way that helps to show the connection between model structure and policy questions (see Section 8.4.3 for a more detailed discussion of the top-down/bottom-up modelling debate). These four dimensions, and some of the related policy questions, are sketched in Table 8.1 for energy/economy models. An equivalent typology could be applied to models focussing on other sectors (e.g., forestry).

Each of the four structural characteristics shown in Table 8.1 represents a spectrum from more to less, and individual models can be located on that spectrum for each dimension. This means that individual models are more or less suited to answering particular policy questions, depending on where they are located on the spectrum for each dimension. Traditionally, top-down models have represented one end of the spectrum on each of these dimensions, and bottom-up models the other, as illustrated in Table 8.2.

It is clear from Table 8.2 that the early top-down and bottom-up models represented virtual mirror images of each other, with respect to the four characteristics shown in Table 8.1. Bottom-up models tended to describe the energy system in great detail, with little endogenization of behaviour or description of other parts of the economy. Top-down models tended to have very little detail on the energy sector but explicit treatment of behaviour and larger economic relationships. As suggested in the previous section, these characteristics caused the two types of model to be useful for answering somewhat different questions. Bottom-up models were better at simulating detailed technological substitution potentials ("exploration") and top-down models were better at predicting wider economic effects ("prediction").

Table 8.2 also shows that this simple characterization of the differences between top-down and bottom-up approaches is increasingly misleading, since more recent versions of each approach have tended to move in the direction of greater detail in those dimensions that were relatively less developed in the past. This is possible because the four dimensions shown in Table 8.1 are independent of each other. Thus any particular model can be located at virtually any point on the spectrum represented by each dimension. It is this independence of key structural characteristics that makes it so hard to classify the large population of existing models on any single spectrum, whether bottom-up to top-down, or any other. Instead we increasingly have a wide range of models, which, in terms of their structure, occupy different places on each of the four dimensions shown here. Thus, although the differences represented by these four dimensions remain important, no simple classification scheme is adequate.

Tables 8.1 and 8.2 describe the differences among different types of models in very general terms. Coming closer to the structure of actual models, we can distinguish several kinds of modelling procedures.

Among bottom-up models two approaches are usually distinguished: (1) spreadsheet models that solve a simultaneous set of equations to describe the way a given set of technologies is (or could be) adopted throughout the economy; and (2) simulation or optimization models, which simulate investment decisions endogenously. Each of these two approaches can be used in two different ways: prescriptively or descriptively.

A prescriptive model examines the effect of acquiring only the most efficient technologies available or of minimizing explicit costs for a given service at a system level (e.g., electric supply, urban transportation, or land use). A descriptive model, in contrast, would try to estimate the technology mix that would result from actual decisions, based on factors such as more complex preferences (people preferring private cars even if the cost per kilometre is higher than railway transportation), intangible costs (differences in cost of acquisition of technologies), capital constraints, attitudes to risk (via higher discount rates for some agents) and uncertainty (actual performance of new technologies), or any kind of market barriers. Such analyses will typically tend to be less optimistic than prescriptive studies about mitigation, unless appropriate policies are assumed to remove existing barriers to the adoption of the best available technologies. Considerable disagreement about the potential for such policies now exists in the literature.

Top-down models resort to two main methodologies: (i) neo-Keynesian macroeconomic models and (ii) computable general equilibrium models and models simulating very long-term growth paths. Each of these approaches can be coupled to process energy models.

Neo-Keynesian macroeconomic models incorporate econometrically estimated sets of equations that trace the short- and middle-run dynamics of the national economic aggregates and related components of economic activity (labour, savings, consumption). They usually simulate aggregate potential output as a function of aggregate inputs of capital and labour, and some of them include energy and materials as production factors. They use input-output tables to describe the transactions among economic sectors and lag equations to model inertia in the adjustment processes and to allow for unemployment in the short run in response to shocks. Some of them also allow for structural unemployment due to inadequate demand for labour in the long run. The fact that they assume a nonperfect equilibrium economy explains, as seen in Chapter 9, why there is a gap between their findings in the case of nonrecycling or lump-sum recycling of tax revenues (which show high costs) and their findings in the case of efficient recycling (some of which show an overall economic benefit from a carbon tax). For longer time periods, they fail to account for the effects of intertemporal preferences and expectations and capture technical change in a rather static fashion.

Computable general equilibrium (CGE) models or optimal growth models (for the very long term) focus mainly on a long-term analysis of the effect of climate policies in the period after adjustment of the economy to short-term effects. They rely on a resource allocation principle (maximization of utility and cost minimization) and a market clearing mechanism for all goods (which functions by equating prices with marginal costs). The dynamics of these models are produced by capital accumulation and/or by the exogenous growth of the factors of production and productivity. In contrast to the neo-Keynesian models, they do not rely systematically on econometric relationships. They are, instead, frequently benchmarked on a given year in order to guarantee the consistency of the parameters. This allows for a greater flexibility in using information coming from other models (or expert judgments) about possible shifts in current trends (technological breakthroughs, for example) that would not be picked up in econometric relationships but that would be dangerous to neglect when very long time periods are being examined.

Given an exogenous perturbation such as a change in the tax system, CGE models produce a price-dependent general equilibrium response due to the behaviour of economic agents. This has two impacts on cost assessments. First, because of their formal structure, these models calculate a new equilibrium but do not provide an accurate picture of the time path towards this equilibrium; they tend as a result to understate transition costs. Second, because they rely on a perfect market equilibrium assumption, they do not allow for structural unemployment over the long run and in this and other ways reduce the room for economic double dividends.

8.4.2.3 The role of key input assumptions

A third important factor to consider in evaluating the meaning of the specific modelling studies is the nature and value of the assumptions that make up the external inputs to those models. Changes in the values of these assumptions will have a significant impact on the results of the study.

Table 8.3. *Key input assumptions in greenhouse gas mitigation costing studies*

Assumption	Meaning and Relevance
1. Population	Other things being equal, population growth increases greenhouse gas emissions
2. Economic growth	Increased economic growth increases energy-using activities and also increases the turnover of energy-using equipment (e.g., allowing penetration of more efficient equipment)
3. Energy demand	
a) Structural change	Degree to which the structure of the economy changes. Since different sectors of the economy have different energy intensities, this will have a significant impact on overall energy use
b) Technological change/choice	Energy intensity of energy-using equipment and processes. This "energy efficiency" variable influences overall energy demand
4. Energy supply	
a) Short-term availability of alternative supplies	This determines the potential for fuel substitution
b) Backstop technology	The cost at which an infinite alternative supply of energy becomes available. This sets an upper bound on the cost of mitigation
5. Price and income elasticities of energy demand	These elasticities measure the relative change in energy demand, given relative changes in energy prices and in incomes. Higher elasticities cause larger changes in energy use
6. Existing tax system and tax recycling	Whether carbon tax revenues are used to reduce the distortionary effects of existing taxes. This has a large impact on the overall cost of carbon taxes

The distinction between input assumptions and assumptions embedded in the structure of the models changes as models evolve and incorporate more assumptions in the model structure. Moreover, input assumptions can be based on implicit mental models of the way the world works that have not yet been formalized in explicit terms and that may or may not be consistent with the formal model. For example, assumptions about the cost and availability of biomass energy supplies in the future imply certain assumptions about future land use that may or may not be consistent with the rest of the analysis.

The assumptions listed in Table 8.3 represent assumptions that are important for any findings regarding costs of GHG mitigation.

8.4.3 The top-down versus bottom-up modelling controversy: Some lessons from the energy field

As discussed earlier, many recent analyses incorporate features of both bottom-up and top-down models, and consequently the distinction between these two approaches is becoming blurred. Despite this convergence, we have reported the results of costing studies separately in Chapter 9 for bottom-up and for top-down analyses. We turn here to a discussion of the bottom-up/top-down controversy. There are several reasons for doing so. First, the prevalence of this modelling dichotomy in the energy field (most studies in other fields are bottom-up) means that the bulk of studies reviewed in Chapter 9 fall naturally into these two camps. Second, the differences between the results that emerged from these two approaches – the bottom-up models tended to suggest that the costs of greenhouse gas mitigation might be low to negative and top-down models tended to suggest the opposite – has led to a significant level of controversy organized around this distinction. Finally, the fundamental questions raised by these differences in approach need to be understood if the results of these studies are not to be misinterpreted. Such questions also raise important policy issues that are relevant beyond the energy field.

The terms "top" and "bottom," in the usual jargon of the economic modeller, are commonly linked to the distinction between aggregate models and disaggregated models. The top-down label is derived from the way in which the first developers of these models applied econometric techniques (i.e., statistical analyses of past trends) to historical data on consumption, prices, and incomes to estimate price and income elasticities for final demand goods and services such as energy, transportation, food, and industrial products. These aggregate models were criticized for not providing a complex enough description of the determinants underlying sectoral demand dynamics. Simulation-oriented technico-economic models were then developed to explore the potential for a possible decoupling of energy demand from economic growth. This required "bottom-up" or disaggregated analysis of technical alternatives and demand for specific services.

Many macroeconomic models are also very detailed, but not in the same way as bottom-up models: They account for economic activities at a two-digit SIC level[27] and can break down consumer demands into many household types. In many models the so-called aggregate demand functions of consumer expenditure are constructed by summing "individual demand functions," which allows for testing the income distribution effects of various types of climate change policies.

A second methodological distinction, partly correlated with but not identical to the aggregation level, is the degree to which behaviours are endogenized in the model equations (i.e., predicted by the model) and extrapolated over the long run. Because econometric relationships among aggregated variables are generally more reliable than among disaggregated variables, and the behaviour of the model is more stable with such variables, it is common to adopt higher levels of aggregation (two to ten goods and services) in top-down econometric models to make them more robust over longer periods. This means that, with a few notable exceptions, as longer periods are under consideration, the aggregation gap between top-down and bottom-up models tends to increase.

Top-down models attempt to examine a broad equilibrium framework, which means that they are interested in feedbacks between the energy system and other sectors of the economy, between all sectors of the economy and the macroeconomic performance of the economy, between national energy markets and global energy markets, and, in some cases, between the national economy and the global economy. Given all of these feedbacks, early top-down models generally included little detail of the energy-consuming side of the economy, especially at a technology-specific level. In contrast, bottom-up models attempt to examine in detail the technological options, especially on the energy-consuming side of the economy. This detailed representation of different energy end uses and the technologies serving those end uses generally meant that greater feedbacks were ignored. In a sense, bottom-up models hold a magnifying glass up to the energy demand side of the economy, and other connections and feedbacks tend to be obscured or lost from the picture.

Another way to look at the difference between these approaches is by returning to the first three types of mitigation costs introduced in Section 8.2.2. We can say that top-down models focus on financial flows across the whole economy and provide rather sophisticated analytical tools at this level, whereas sectoral models, which have tended also to be top-down but sometimes with fairly detailed representation of technology, focus on the market dynamics stemming from a given industrial structure but with little consideration devoted to overall feedbacks with the rest of the economy. Finally, bottom-up models focus on the technical margins of freedom likely to be evident at a microeconomic level and provide a detailed analysis of the technical and economic dimensions of specific policy options. Historically, this difference in focus has led to significant differences in results, as the larger economic feedback loops originally examined in top-down models (higher production costs, lower investments in nonenergy sectors) tended to increase the macroeconomic costs of emission reductions relative to bottom-up results. More recently, other feedbacks that tend to reduce overall costs (efficient tax recycling) have been examined in top-down analyses, while bottom-up methods have increasingly incorporated detailed examination of the behavioural dimensions of the policy options examined. The effect of these developments has been to narrow the range of difference in mitigation costs between these two approaches.

But behind this difference in the level of aggregation lie two different ways of representing technology. One side tries to capture technology in the engineering sense: a given technique (coke oven steel versus electric-arc steel, an incandescent light bulb versus a fluorescent light bulb) with a given technical performance at a given direct cost. On the other

side, the technology term in macroeconomic models, whatever their level of disaggregation, is represented by the share of the purchase of a given input in intermediate consumption (e.g., steel products to make cars) and by the allocation of the sales revenue among the cost of intermediary inputs, returns to labour, and returns to capital. These shares constitute the basic ingredients of the economic description of a technology in which, depending on the choice of production function, the share elasticities represent the degree of substitutability among inputs.

This characterization of technology makes it difficult to establish explicit links between production functions in economic models, on the one hand, and technology projections, on the other. For example, there is a gap between the findings of Jorgensen (1984) and Hogan and Jorgensen (1990) with respect to the negative correlation between technological change and energy prices in the U.S. economy and the observation of accelerated innovations at a technical level. The reason is that that technological innovation can have two different meanings. For an engineer, it implies increased technical efficiency and therefore greater productivity. For an economist, however, productivity growth is correlated with many other factors than mere technology characteristics; a depressed demand or growing uncertainties can result in lower economic productivity, even if the technical efficiency of equipment is very high. Consequently, both descriptions can be internally consistent, but each may represent a different definition of technological change. Put another way, both descriptions of technological change make sense, but each captures a different aspect of that change.

Bearing in mind these fundamental differences in approach, it is clear that there is no *a priori* reason that the two modelling approaches must give different results. Top-down models account for the technological changes so valued by bottom-up analysis via two parameters: (1) the autonomous energy efficiency index (AEEI), and (2) the elasticity of substitution between the aggregate inputs to households and firms. AEEI is a function of time (a proxy for all the long-run reasons for equipment turnover) and suggests the rate at which the penetration of new technologies may change the energy intensity of the economy. The elasticity of substitution is a function of relative prices of inputs and allows measurement of the degree to which capital or labour can be substituted for energy (i.e., energy intensity can be reduced) as energy prices rise relative to these other inputs. The values of AEEI and elasticity of substitution can be adjusted to provide results that match those frequently suggested by bottom-up analysts.

Recently, top-down modellers have proven to be more willing to abandon historically derived parameter values in favor of other values – derived, for example, from detailed bottom-up analysis – both for AEEI and the elasticity of substitution (benchmarking). They have also worked to develop model forms that achieve greater disaggregation of energy end-use activities, if not a detailed representation of technologies. Thus a relatively high value for AEEI, say a 2% decline per year, and a relatively high value for the elasticity of substitution between fossil fuels and other factors of production (labour, capital) could lead to top-down simulation results in which the costs of carbon emission stabilization, and even of a 15-30% reduction, were below or close to zero. This way of incorporating bottom-up information is relatively easy to achieve in the generation of top-down models that use "benchmarking" approaches to calibration; in the case of econometrically calibrated models, it requires the adoption of some ad hoc hypotheses and a departure from strict econometric consistency.

Conversely, some of the recent refinements to bottom-up modelling have tended to use more sophisticated economic indicators than the technical costs per toe or kWh. They seek to be more descriptive of the actual energy-using behaviours of firms, households, and institutions and to account for actual experience as regards responses to price signals and energy efficiency programmes as well as the administrative costs of policy programmes. In particular, bottom-up studies have become increasingly sophisticated with regard to

- *Technical analysis:* This is especially so with respect to demand-side efficiency improvements. On the one hand, more detailed studies will tend to arrive at larger cost-effective efficiency potentials; on the other, analysts may disagree about the actual performance of various types of end-use equipment.

- *Administrative costs:* Most earlier bottom-up analysis assumed that energy efficiency standards – with minimal administrative costs – would be the main policy tool. More recent analysis has increasingly taken stock of the administrative costs involved in actual utility demand-side management programmes. These costs are estimated to range from negligible (Krause, 1994; Eto *et al.,* 1994) to substantial (Joskow and Marron, 1993), depending primarily on the type of incentive programme involved.

- *Policy effectiveness:* The results of bottom-up studies vary as a function of the way they account for transaction costs and transition costs. At one end of the spectrum, it is assumed that a complete shift to efficient equipment will be achieved within one cycle of capital stock turnover (ten to twenty years); at the other end, political and institutional constraints inhibit the use of the most energy-efficient technology over periods of forty years or more.

In addition to differences in their approach to technology, bottom-up and top-down models typically contain important differences in their assumptions about consumer surpluses, so-called intangible costs, and the role of market barriers (Sutherland, 1991).

An example of a consumer surplus would be the extra satisfaction or value that a consumer derives from a particular automobile that may not be reflected in its capital or operating costs. Common examples of intangible costs are (1) the cost of becoming sufficiently informed about a new technology in order to consider it seriously as an option, (2) the perceived risks associated with the capital or operating costs of a technology, (3) the various transaction costs associated with finding, ordering, shipping, installing, operating, and maintaining a technology, and (4) externally or internally imposed restrictive investment criteria that differ from the social time preference of consumers and the opportunity cost of capital of firms. A difference in one or more of these intangible cost factors or consumer surpluses between two goods could be sufficient to offset differences in their tangible costs.

Top-down modellers, who tend to be economists, are generally reluctant to doubt the economic efficiency of business and household consumption choices, although they recognize the classical imperfections in markets (oligopoly, natural monopoly, subsidies, etc.). They therefore tend to assume that if a technology does not penetrate the market to the extent that a bottom-up engineering/economic analysis (which focuses only on tangible costs) suggests that it could, it is probably at least in part because the intangible cost and/or consumer surplus differences are at least large enough to make the investment unattractive. As a consequence of this general assumption, many economists are more likely to put their faith in values for AEEI and elasticity of substitution that have emerged from studies of the economy using statistical regression of historical aggregate data sets. Since the historical data are based on actual behaviour, behavioural parameters estimated from them may represent a synthesis of all tangible and intangible costs, including differences in consumer surplus.

Sweeping generalizations about bottom-up modelling are as hazardous as those about the top-down aproach. However, bottom-up analysts in the past were more likely to be engineers and physicists. Their detailed knowledge of thermodynamic potentials and the energy-using characteristics of new and emerging technologies has led them to be aware of the apparent economic potential for society to adopt new technologies more quickly than in the past, at rates that, in aggregate, would imply values for AEEI and elasticity of substitution far higher than the values emerging from the historical data relied on originally by top-down modellers. This issue is particularly important when faced by the very long time frame required by the climate change issue.

In addition, bottom-up modellers point out that econometrically derived relationships may incorporate market imperfections. Although many top-down modellers assume that these imperfections are either negligible or very costly to correct, bottom-up modellers have noted the significant successes in fostering technological innovation and greater energy efficiency resulting from government initiatives during the energy crises of the 1970s and utility and government initiatives in the 1980s, and have pointed to these experiences as evidence that well-designed research and development and policy programmes could significantly affect the evolution of the AEEI term and even the elasticity of substitution. Thus, insofar as institutional reforms could lower transaction costs and remove barriers to the adoption of technologies whose life cycle costs are advantageous, there would be room for no-regrets improvements.

The empirical foundations of this position come from evaluation studies of the impacts of demand-side management programmes on end-use markets, product choices, consumers, manufacturers, and trade allies (DeCanio, 1993; Howarth and

Anderson, 1993; Howarth and Winslow, 1994; Koomey and Sanstad, 1994; Krause *et al.,* 1989; Levine and Sonnenblick, 1994; Nadel, 1992). Though the argument has been challenged by various critics (Sutherland, 1991; Joskow and Marron, 1993), the authors of these studies suggest that there exists significant potential to capture the "negative cost potential" that can be realized by well-designed demand-side management programmes.

Given these results, bottom-up modellers tend to argue that appropriate policy will cause energy-efficient products to have lower transaction costs and risks than are generally assumed by firms and households, and perhaps even higher consumer surpluses. From this perspective the market as currently configured simply isn't delivering levels of energy efficiency that would be economically advantageous, neither at the microlevel of direct financial costs nor at more aggregate and inclusive levels of cost. They suggest that significant potentials for emission reductions can be realized at net economic savings through a combination of mandatory energy efficiency standards, labelling and auditing programmes, least-cost planning-oriented utility regulatory reforms, profit incentives for utility demand-side management programmes, market-pull incentives provided through innovative utility or government procurement programmes, and increased research and development and commercialization efforts for small-scale modular technologies with short gestation periods, notably energy efficiency improvements and renewables.

The differences in results between top-down and bottom-up modelling analyses are thus rooted in a complex interplay among differences in purpose, model structure, and input assumptions. The growing tendency for the development of hybrid modelling approaches means that differences of model structure are becoming a less important factor in many climate modelling studies, but the overall difference in perspective and thinking about energy markets represented by these two modelling approaches remains. Hybrid modelling approaches allow the exploration of the relative importance and implications of different input assumptions and are likely to narrow the difference in results between bottom-up and top-down studies. They will not by themselves, however, resolve the underlying question of whether energy markets are efficient with respect to the delivery of energy efficiency gains or not. Moreover, it remains to be seen whether these new approaches will be equally fruitful in fulfilling the range of purposes discussed in Section 8.4.2.1. In particular, it seems clear that we need to improve our ability to explore long-term development paths or configurations of technology that are very different from those typical of experience in past decades (i.e., the alternative future scenarios that are often the subject of backcasting analyses). Much work remains to be done to address these larger issues in a satisfactory way.

8.4.4 Beyond energy: Carbon sinks and nonenergy greenhouse gas emissions

The focus of the literature on the potential for controlling CO_2 from energy sources has developed in part because of the ready availability and adaptability of models designed to ana-

lyze energy markets. However, because of the lack of similar ready-made models that could be adapted to analyses of carbon sequestration and reductions in emissions of methane, nitrous oxide, halogenated substances, and other greenhouse gases, debates in these areas have not been structured around the bottom-up and top-down modelling approaches. Nevertheless, controversies in these fields are related to fundamental issues similar to the ones underlying the bottom-up versus top-down division in the energy field, namely, the reasons for a wedge between the direct cost of technical alternatives from an engineering viewpoint and the overall costs of their adoption and implementation if transaction costs and economic general equilibrium effects are included in the accounting.

8.4.4.1 Carbon sequestration studies
Carbon sequestration cost studies fall into four general categories:

(1)   studies of the cost of removal and storage of carbon dioxide from emission sources such as power plants
(2)   studies of biomass energy technologies that allow displaced fossil carbon to remain undisturbed
(3)   studies of practices to maintain and expand the biological carbon sink, particularly in forests
(4)   studies of technologies to expand the storage of carbon in wood products

Technologies for the removal and storage of carbon dioxide are prohibitively expensive at this time (see e.g., Riemer, 1993). Biomass energy is treated elsewhere in this report as a renewable energy technology that lowers net emissions of carbon dioxide. The analysis of the costs of increasing the use of long-lived wood products is not well developed. Consequently, this review of carbon sequestration cost studies will concentrate on the expansion of forest and agricultural carbon sinks.

Most carbon sink cost analyses have examined the direct costs of specific technical measures.[28] Under this approach, forestry or agricultural practices are matched with appropriate geographic regions. The pairs are defined as unique technologies with specific production functions. Then, in a very simple analysis, three key variables are identified for each region/practice combination:

(1)   the suitable land area for that practice (e.g., hectares)
(2)   the treatment cost and land cost for the practice (e.g., annualized costs per hectare per year)
(3)   the annual carbon yield (e.g., tonnes of carbon per hectare per year)

Two critical results can be derived from these three pieces of data. The potential yield of carbon in tonnes per year, from a given region and practice can be derived by multiplying the first and third factors. The unit cost of carbon sequestration, in dollars per tonne, can be derived by dividing the second factor by the third. These results can then be combined across practices within a region to develop a supply curve for carbon sequestration.

The simple analysis described above has been employed by some studies (see, e.g., Moulton and Richards, 1990). However, estimation of the three variables listed above – land

BOX 8.2: COST CONCEPTS IN CARBON SINK ANALYSIS

Most carbon sink cost analyses have employed the first cost concept described in this chapter, namely, the direct costs of specific technical measures. However, there is no consensus on the definition of the summary statistic "dollars per tonne of carbon sequestration." A review of the literature reveals at least three distinct definitions of this measure of cost-effectiveness. The difference among the definitions revolves around how each deals with the relative value of carbon flows that occur at different points in time. Because carbon-sink-enhancing activities result in very uneven flows of carbon – flows that in some cases occur over several decades – the treatment of the time value of carbon is important.

Flow summation method. The simplest approach to summarizing the cost-effectiveness of carbon sequestration projects is simply to sum the total tonnes of carbon captured, regardless of when the capture takes place. This approach treats early capture (or release) of carbon equally with later capture (or release).

Average storage method. This approach involves dividing the present value of the sum of all implementation costs over a specified period (e.g., Dixon *et al.,* 1991, use 50 years) by the mean standing carbon storage averaged over several rotation periods (see the discussion of carbon flow treatment in Chapter 9, Section 2).

Levelization/discounting method. This method differentiates costs and accomplishments according to when the carbon is captured. There are actually two approaches that yield identical results but are conceptually distinct. The first, which is similar to the approach used in many bottom-up energy studies, annualizes (levelizes) the present value of costs over the period of carbon flows and divides the annualized costs by the annual carbon capture rate. The second approach, developed to address uneven flows of carbon, applies the social discount rate to discount the tonnes of carbon captured back to a summary statistic – the present tonnes equivalent (PTE) – and divides that figure into the present value of costs.

The choice of an appropriate method from among these three candidates depends on the underlying policy for which the cost analysis is being conducted. If the underlying policy is one that is concerned only with capping the total accumulation of carbon and is not concerned with when the reductions occur, then the flow summation method may be appropriate.[†] Alternatively, if the damages from greenhouse gas accumulation are continuous so that interim damages matter, the levelization discounting method, which recognizes the relatively higher value of reductions that are achieved sooner, is the appropriate choice.

Choice of method: An illustration

The importance of the choice of summary statistic can be illustrated by calculating the dollars per tonne associated with the three stylized tree planting examples described below. For purposes of this illustration, a social discount rate of 5% is assumed. For a discussion of the appropriate discount rate, see Chapter 4.

Example 1: A new one-hectare forest plantation is established to sequester carbon. The net present value of all costs (land, establishment, maintenance, and administration) is $1000. The forest captures 2 tonnes of carbon per year for 50 years. Thereafter, the plantation captures no additional carbon and stands permanently without being harvested.

Example 2: The forest plantation is established as above, but the species is slower growing initially. During the first 25 years it captures only 1 tonne of carbon per year and then increases to 3 tonnes per year for 25 years. Thereafter, the plantation captures no additional carbon and stands permanently without being harvested. The net present value of costs is still $1000.

Example 3: The forest plantation is established and grown as in Example 1, but in the fiftieth year it is harvested. All accumulated carbon is released to the atmosphere upon harvest. The area is replanted and the process repeats itself. The net present value of all costs, including perpetual replanting is $1100.

Flow summation method. Using this approach, the first and second examples each have a cost of $10 per tonne. The third example is more difficult to analyze. Although the costs are clearly identified, the accomplishments are unclear. The cumulative tonnage sequestered is cyclical, rising to 100 tonnes just prior to harvest but falling to zero immediately following harvest. This suggests that under this analytical approach there is no value to capturing carbon that is eventually released, no matter how long the storage period.

Average storage method. Examples 1 and 2 are difficult to analyze using this approach, since the rotation length is not clear. As the rotation length approaches infinity, the mean carbon storage asymptotically approaches 100 tonnes. This would suggest a cost of $10 per tonne. If, instead, a 50-year rotation length is artificially imposed on the analysis, the mean carbon storage is 51 tonnes in Example 1 and 38.5 in Example 2. In this case the carbon costs in the two examples would be $19.61 per tonne and $25.97 per tonne respectively. In Example 3, where the rotation length is clearly 50 years, the mean carbon storage is 51 tonnes, giving a carbon cost of $21.57 per tonne.

Levelization/discounting method. In Example 1 the present value of costs, $1000, can be levelized over a 50-year period at 5% to $54.78 per year. Dividing by the annual carbon yield of 2 tonnes per year leads to a cost of $27.39 per tonne. Alternatively, discounting the number of tonnes of carbon at a 5% social discount rate yields 36.5 PTEs, and dividing by the present cost also gives $27.39 per tonne. The irregular carbon flow in Example 2 makes it more difficult to apply the levelization approach in that case, but the carbon discounting approach can easily be applied to derive a yield of 26.58

Box 8.2 (*cont.*)

PTEs and a carbon cost of $37.62 per tonne. In Example 3, the stream of flows is also irregular, so the discounting approach is easier to apply than the levelization method. The continuous rotations yield approximately 30.49 PTEs for a cost of $36.08 per tonne.

The accompanying table summarizes the cost figures that the three approaches yield for each of the examples. Note that the flow summation method and the average storage method yield the same result in the first two examples, where the rotation period approaches infinity. It is only by imposing an artifical rotation period that the average storage method can be forced to differentiate between these examples. In contrast, the levelization/discounting approach differentiates between Examples 1 and 2 purely on the basis of when the flows occur. The flow summation method provides no results with respect to Example 3, where continuous rotations occur. The result in that example is that the cost oscillates between 10 and infinity dollars per tonne of carbon sequestered. For both the average storage and levelization/discounting methods the costs rise slightly in Example 3 relative to Example 1.

Carbon sequestration costs for examples, by method

Method	Costs ($/tonne)		
	Example 1	Example 2	Example 3
Summation	10	10	Indeterminate
Average storage	10 [19.61]†	10 [25.97]†	21.57
Levelization/ Discounting	27.39	37.62	32.80

†Term in brackets is derived by imposing a 50-year rotation period.

†Implicit in this policy is the assumption that greenhouse gases cause no damages until they reach some critical level, after which their impacts are catastrophic. A corollary is that, prior to reaching the critical level, temporary removal of carbon from the atmosphere has no value.

area, land and treatment costs, and carbon yield – is not simple in practice. As discussed in Box 8.2, there is no consensus on the definition of the summary statistic "dollars per tonne of carbon sequestration."

As with energy studies, differences among the studies that give rise to a wedge between direct and social costs occur at each stage of a carbon sequestration cost analysis. These stem from different assumptions about the suitability of forestry and agricultural practices, different treatments of land and forestry practice implementation costs, and different estimates of the expected accomplishments of the forestry practices and valuation of those accomplishments in terms of greenhouse gas emission benefits as well as double dividend GDP and natural resource conservation benefits.

As Table 8.4 indicates, a variety of forestry practices may contribute to increasing the size of forest and agricultural carbon sinks. Just as in the case of the best available technologies in the energy field, the appropriate forestry and agricultural practice for carbon sequestration is, in theory, the one that promises to be a cost-effective way to capture and store carbon (see, e.g., Parks and Hardie, 1995). The analysis of "intangible" factor costs or of "double dividends" has been hampered by the fact that many studies have imposed constraints on the forestry or agricultural practices or types of land that they consider. These have included requirements that the adopted practices decrease soil erosion (Moulton and

Table 8.4. *Examples of forestry and agricultural practices to increase carbon sequestration*

Afforestation of agricultural land
Reforestation of harvested or burned timberland
Preservation of forestland from conversion
Protection of forests from destructive wildfire, pests, and diseases
Adoption of agroforestry practices
Establishment of short-rotation woody biomass plantations
Lengthening forest rotation cycles
Modification of forestry management practices to emphasize carbon storage
Adoption of low-impact harvesting methods to decrease carbon release
Adoption of carbon-enhancing agricultural practices (crop rotation, no-till, etc.)

Richards, 1990), meet local needs (Andrasko *et al.,* 1991), or provide other environmental benefits such as habitat preservation. Ideally, these factors should be accounted for as beneficial secondary effects, which in many cases appear to be greater than those associated with energy efficiency.

Significant differences in the cost assessments of sink studies can be traced back to disagreements about the assumptions and data related to the many factors that determine total costs:

land costs, first treatment and maintenance costs, transaction costs, the valuation of wood and agricultural products, and the discount rate applied to expenditures.

With respect to land costs, studies of carbon sequestration costs in the United States have, for example, employed several methods to identify the social costs associated with conversion of land to forest. These have included the use of land rental rates derived from surveys (Moulton and Richards, 1990), the use of market prices adjusted for the elasticity of demand for agricultural land (Richards *et al.*, 1993), the use of the estimated lost profits from removing the land from agricultural production (Parks and Hardie, 1995), and the use of consumer surplus loss from increases in food prices due to the constriction of agricultural land availability (Adams *et al.*, 1993). In the absence of extensive experience, estimating land costs has proven difficult because of the low reliability of data on the elasticities of demand for agricultural land. Also, uncertainties about the future of government subsidies for agriculture raise questions about land costs. These subsidies tend to drive a wedge between the market prices and the social cost of land.

These accounting difficulties are obviously far more difficult in countries that do not have well-established land markets, that have land tenure laws that do not allow permanent transfer of land at all, or where the government owns a significant portion of the land. Further, even where land markets do function, the lack of data about market activities often renders land cost estimates speculative at best.

The transaction costs associated with establishing forestry and agricultural programmes can be significant. Even in well-developed market economies, the costs of programme administration can rise to as much as 15% of the total costs of land rental, establishment, and maintenance (Richards *et al.*, 1993). It is reasonable to expect that where land and labour markets are not so well developed or when a public administration is not well established, the costs of administration and information gathering might rise to a much higher level, even surpassing the direct costs of land acquisition and tree planting. In the context of developing countries where the pressure on land comes from agricultural purposes or from wood exports, these transaction costs should, theoretically, encompass policies required to slacken this pressure. This requirement could narrow the range of feasible low-cost carbon sequestration opportunities.

Some carbon sink cost studies have assumed that carbon sequestration project lands would be taken out of agricultural or timber production permanently (Nordhaus, 1991; Richards *et al.*, 1993), but most have allowed for some kind of derivative benefit in the form of forestry products such as pulpwood, timber, firewood, and biomass energy. Such derivative benefits raise several additional issues. First, harvesting of forestry products suggests the need to modify the flows of carbon associated with the project to reflect the removal of carbon from the site. This in turn raises the question of the rate of release of the carbon back to the atmosphere after harvest. Second, in a cost-effectiveness study, the costs of the project should be reduced to reflect the noncarbon benefits of the wood and agricultural products. Furthermore, to the extent that sequestration timber drives down the price of timber products, additional markets will develop for wood substitutes for energy-intensive materials such as concrete, aluminum, and steel, thus yielding further reductions in emissions.

As the discussion in Box 8.2 illustrates, harvesting can have a significant impact on the carbon benefits of a project. Further, the measure of that impact depends very much on the choice of summary statistics. At the same time, the economic benefits of timber harvesting can be significant. Studies that do not quantify either of these two effects will overstate both the costs of carbon sink projects and the carbon benefits. Although the effect on carbon costs of ignoring timber harvesting is indeterminate, it is likely that inclusion of forestry products in the analysis would generally lower unit costs. Studies that include the effects of harvesting on carbon flows but do not incorporate its economic benefits will almost certainly overstate the unit costs of carbon sequestration. At the extreme, some forestry practices may pay for themselves in the form of forestry products and provide the carbon benefits as a costless (i.e., no-regrets) bonus.[29] For example, Xu (1994) suggests that there may be negative costs associated with some carbon sequestration practices in China. Conversely, those studies that only consider the benefits of forestry products but do not adjust the carbon flows to reflect increased releases of carbon back to the atmosphere will understate the costs of carbon sequestration.

It is apparent that, as with energy efficiency, there are opportunities in the carbon sink area to achieve double dividends. These become evident under an analysis that provides full accounting for the wood and agricultural product benefits of projects as well as for the less tangible benefits such as habitat and watershed preservation, improved local self-reliance, and soil erosion control. However, full accounting must also include those factors that tend to increase social costs relative to financial costs, such as the effect of removing nonmarginal quantities of land from agricultural production, transaction costs associated with establishing new land use patterns, administrative costs of implementing a large-scale carbon sequestration programme, and decreases in carbon benefits associated with timber harvesting. Perhaps the most important factor, one commonly ignored by carbon sink cost studies, is the potential for "leakage" of the carbon sequestration gains. If the selective subsidization of tree planting and other forestry practices creates a new supply of timber, owners of existing forests may try to avoid competition in the timber market by accelerating harvest of their stocks, decreasing the amount of postharvest replanting, and avoiding expansion of the holdings.

Endnotes

1. However, in most of the earlier modelling work, "long-term" generally meant a time horizon of 10-20 years, not the 100-plus horizons used in many of the current climate change scenarios.
2. For a discussion of the wide range of views on future energy demand in the early 1980s, see Caputo (1984) and Thompson (1984).
3. For a discussion of the sometimes vexed relationship between energy modellers and energy policymakers, see Robinson (1992).

4. See, for example, the various national case studies in Baumgartner and Middtun (1987).

5. Early work in this area is summarized in Little and Mirrless (1974) and in Squire and Van der Tak (1976). A good illustration of the interest in making a distinction between market prices and factor costs is given in the debates around the biomass ethanol program in Brazil, where the cost of ethanol is far lower if one utilizes some form of social costs (Nastari, 1991).

6. Most discussions of the issue of "no-regrets" potential have centred on the first category of positive side effects described here: Whether there exist "negative cost" measures such as some types of energy efficiency programmes. But whatever positive side effects are included, the "no-regrets" concept should not be taken to imply that undertaking all such measures will guarantee no regrets with regard to the effects of climate change, if these effects ultimately are proven to be very significant. A better term might be "worth doing anyway" measures.

7. See also the discussion in Chapter 7.

8. See section 6.7 for a discussion of some estimates of the magnitude of the environmental double dividend.

9. For a more general presentation of the different meanings of concepts such as "no regrets" and "double dividend," see Goulder (1994).

10. It would also be possible to move from O to a point B′ above and to the left of point B (increasing economic growth and also increasing emissions). This means that the economic surplus gained, thanks to the removal of inefficiencies (i.e., moving from O to curve F), will be devoted to improving environmental quality only if there is a collective preference and political will to do so. It could also be possible to move to a point A′ below and to the right of A (reducing both emissions and economic activity) if the surplus is devoted to very high investments with a low return and a very low efficiency in terms of environmental quality improvement. This could occur in the case of misallocation of efforts for a given level of concern for environmental quality.

11. Another counterargument acknowledges that a suboptimal baseline may be the most realistic assumption, even over the long run, but suggests that the cost of greenhouse gas abatement measures should be calculated net of the effect of any measures taken to move the economy towards the production frontier (e.g., the effect of fiscal distortions are assumed to be removed before calculating the cost of a carbon tax). By suggesting that no economic double dividends should be included in the cost of abatement, such arguments reduce the estimate of cost-effective abatement potential, but they also imply that the adoption of a more economically efficient baseline scenario may result in lower emissions for reasons unrelated to climate policies.

12. All this assumes a static production frontier. In reality, as a result of technological change and other factors, the frontier moves, usually to the right, over time. Bottom-up analyses often compare a point below the *current* production frontier with a *future* production frontier. Bottom-up analysts typically also argue that due to market imperfections, actual future production is likely to lie below the future production frontier. Since bottom-up analyses tend to calculate the cost savings due to the adoption of future technologies relative to existing ones, it is not surprising that they usually show net savings. Top-down analyses tend to focus on the costs of moving from the current to a future production frontier. Since that entails investment, it is not surprising that the costs are always positive.

13. The same debate surrounds the analysis of fiscal reforms linked to a carbon (or energy) tax. Some analysts consider that it is more legitimate to separate the gains from reducing existing fiscal distortions from the incremental impact of a carbon tax. Others consider that both such effects should be accredited to the carbon tax, since reducing fiscal distortions is associated with, and may be made more politically palatable by, the imposition of the carbon tax. This latter position, which can reduce or more than offset the costs of the carbon tax, is the position taken in most of the empirical modelling work to date; the former assumption is more typical of more theoretical analyses.

14. This is a logical identity and does not imply the independence of these three terms.

15. To give a simple example, if there were two countries in the world and the cost of emission reduction were always twice as expensive in country A as in B, then it would be cheaper to use up all the reduction potential in country B before reducing emissions in A.

16. In 1989, the range for Europe and North America was 20-35% (World Resources Institute, 1993).

17. Alternatively, countries might adopt strategies encouraging the development of very energy-efficient urban design.

18. For a discussion and examples of such approaches see Jantsch (1967); Harman (1976); Gault *et al.* (1987); Gal and Fric (1987); Glimel and Laestadius (1987); Godet (1986); and Robinson (1988).

19. See the country case studies in Baumgartner and Middtun (1987). See also Caputo (1984). For a discussion of the bases of disagreement, see Robinson (1982).

20. For examples of more recent bottom-up work, see Goldemberg *et al.* (1988) and Johansson *et al.* (1993).

21. In Chapter 9 we provide the findings of some attempts to provide such comparisons.

22. Since the former baseline already incorporates such improvements, the cost per unit of further reduction is likely to be higher than it would be relative to a baseline without such improvements.

23. For a recent example, which characterizes individual bottom-up and top-down models, see Grubb *et al.* (1993).

24. For example, Environmental Protection Agency (1990); Edmonds *et al.* (1986); Manne and Schrattenholzer (1993); and Johansson *et al.* (1993).

25. The term "backcasting" is also used in the economic modelling literature to refer to the process of simulating a model over historical time in order to compare such a historical "projection" with actual historical data.

26. For a discussion of the implications of backcasting approaches for modelling, see Gault *et al.* (1987) and Robinson (1991b).

27. The SIC (Standard Industrial Classification) is an internationally recognized classification system for industrial activity. The two-digit SIC level is a very highly aggregated level that breaks industrial activity into less than ten sectors.

28. Apparently, the only carbon sequestration cost study to employ an econometric approach (in contrast to the least-cost analysis of most other studies) is currently being prepared by Robert Stavins of the Kennedy School of Government, Harvard University, U.S. The econometric analysis not only incorporates data from direct financial cost, but accounts for behavioural considerations regarding how landowners respond to economic incentives.

29. As in the case of energy-related emissions, the assertion that opportunities exist to enhance sinks at a negative cost raises the obvious question of why these activities are not already being undertaken if their costs truly are negative.

References

Adams, R., D. Adams, J. Callaway, C. Chang, and B. McCarl, 1993: Sequestering carbon on agricultural land: Social costs and impacts on timber markets, *Contemporary Policy Issues,* **11**(1), 76–87.

Andrasko, K., K. Heaton, and S. Winnett, 1991: Evaluating the costs and efficiency of options to manage global forests: A cost-curve approach, Proceedings of the Technical Workshop to Explore Options for Global Forestry Management, Bangkok, 24-30 April, pp. 216–233, International Institute for Environment and Development, London.

Arthur, W.B., 1988: Self-reinforcing mechanisms in economics. In *The economy as an evolving complex system,* P.W. Anderson, K.J. Arrow, and D. Pines, eds., pp. 9–31, Addison-Wesley, Redwood City.

Ascher, W., 1978: *Forecasting – An appraisal for policy-makers and planners,* Johns Hopkins University Press, Baltimore.

Ascher, W., 1990: Beyond accuracy, *International Journal of Forecasting,* **5,** 469–484.

Aumann, R.J., 1987: Correlated equilibrium as an expression of Bayesian rationality, *Econometrica,* **55**(1), 1–18.

Azariadis, C., and R. Guesnerie, 1986: Sunspots and cycles, *Review of Economic Studies,* **53,** 725–737.

Baumgartner, T., and A. Middtun (eds.), 1987: *The politics of energy forecasting,* Oxford University Press, New York.

Bernardini, O., and R. Galli, 1993: Dematerialization: Long-term trends in the intensity of use of material and energy, *Futures,* **25**(4), 431–448.

Berrah, N.E., 1983: Energy and development: The tunnel effect, *Revue de l'Energie,* no. **356** (August/September), Paris.

Bovenberg, L. and R. Van der Mooij, 1994: Environmental levies and distortionary taxation. *American Economic Review,* **84**(84), pp. 1085–1089.

Brooks, D., and J. Robinson, 1983/84: 2025: *Soft energy futures for Canada,* study prepared by Friends of the Earth Canada, 12 volumes, Department of Energy, Mines, and Resources, Ottawa.

Brooks, H., and E. Ginzton (eds.), 1980: *Energy in Transition 1985–2010.* Final report of the Committee on Nuclear Alternative Energy Systems of the National Research Council, W.H. Freeman & Sons, San Francisco, CA.

Caputo, R., 1984: Worlds in collision, *Futures,* **16,** 233–259.

CETESB (Companhia de Saneamento Basico do Estado de Sao Paulo), 1994: *Relatorio de qualidade do ar em Sao Paulo,* Secretaria de Saneamento e do Meio Ambiente do Estado de Sao Paulo, Sao Paulo.

Chen, X., 1994: Substitutions of information for energy, *Energy policy,* **22**(1), 15–27.

Craig, P., M. Christensen, M. Levine, D. Muhamed, and M. Simmons (eds.), 1978: Distributed energy systems in California's future, report prepared for the U.S. Department of Energy. U.S. Government Printing Office, Washington, DC.

Darmstadter, J., J. Dunkerely, and J. Alterman, 1977: *How industrial societies use energy,* Johns Hopkins University Press, Baltimore.

DeCanio, S., 1993: Barriers within firms to energy efficiency investments, *Energy Policy,* **21**(9), 901–914.

Dixon, R., P. Schroeder, and J. Winjum (eds.), 1991: *Assessment of promising forest management practices and technologies for enhancing the conservation and sequestration of atmospheric carbon and their costs at the site level,* Report of the U.S. Environmental Protection Agency, #EPA/600/3-91/067, Environmental Research Laboratory, Corvallis, OR.

Dupuy, J.-P., 1989: Convention et common knowledge, *Revue économique,* **40**(2), 361–400.

Edmonds, J.A., and J. Reilly, 1983: A global energy-economic model of carbon dioxide release, *Energy Economics,* **5**(2), 74–88.

Edmonds, J. A., *et al.* 1986: *Future atmospheric carbon dioxide scenarios and limitation strategies,* Noyes Publications, Park Ridge, NJ.

Environmental Protection Agency, 1990: Policy options for stabilizing global climate, report to Congress, D. Lashof and D. Tirpak, eds., EPA, Washington, DC.

Eto, J., E. Vine, L. Shown, R. Sonnenblick, and C. Payne, 1994: *The cost and performance of utility commercial lighting programs,* LBL-34967, Energy and Environment Division, Lawrence Berkeley Laboratory, Berkeley, CA.

Ford Foundation, 1974: *A time to choose: America's energy future,* Energy Policy Project, Ballinger, Cambridge, MA.

FRN (Forskningsrådsnamden), 1987: Surprising futures, notes from an International Workshop on Long-Term World Development, Forskningsrådsnamden (Swedish Council for Planning and Coordination of Research), Report 87:1, Stockholm.

Fudenberg, D., and J. Tirole, 1991: *Theory game,* MIT Press, Cambridge, MA.

Gal, F., and P. Fric, 1987: Problem-oriented participative forecasting. *Futures,* **19**(6), 678–685.

Gault, F., K. Hamilton, R. Hoffman, and B. McInnis, 1987: The design approach to socio-economic forecasting, *Futures,* **19**(1), 3–25.

Glimel, H., and S. Laestadius, 1987: Swedish futures studies in transition, *Futures,* **19,** 635–650.

Godet, M., 1986: Introduction to La Prospective, *Futures,* **18**(2), 134–157.

Goldemberg, J., A. Reddy, T. Johansson and R. Williams, 1988: *Energy for a sustainable world,* Wiley-Eastern Ltd., New Delhi.

Goulder, L.H., 1994: Environmental taxation and the double dividend: A reader's guide, paper presented at the 50th Congress of the International Institute of Public Finance, (August), Session 1, Green taxes and the rest of the tax system, Harvard University, Cambridge, MA.

Grubb, M., J. Edmonds, P. ten Brink, and M. Morrison, 1993: The costs of limiting fossil-fuel CO_2 emissions: A survey and analysis, *Annual Review of Energy and the Environment,* **18,** 397–478.

Grubb, M., M. Ha Duong, and T. Chapuis, 1994: Optimising climate change abatement responses: On inertia and induced technology development. In *Integrative assessment of mitigation impacts and adapation to climate change,* N. Nakicenovic, W.D. Nordhaus, R. Richels, eds., pp. 513–534, Proceedings of an IIASA workshop held on 13-15 October 1993 at Laxenburg, Austria, International Institute for Applied Systems Analysis, Laxenburg.

Grübler, A., N. Nakicenovic, A. Schäfer, 1993: Dynamics of transport and energy systems, Report RR-93-19, International Institute for Applied Systems Analysis, Laxenburg, Austria.

Häfele, W., 1981: *Energy in a finite world: A global systems analysis,* Ballinger, Cambridge, MA.

Harman, W., 1976: *An incomplete guide to the future,* San Francisco Books, San Francisco.

Henshell, R, 1982: The boundary of self-fulfilling prophecy and the dilemma of prediction, *British Journal of Sociology,* **33,** 511–528.

Henshell, R., 1993: Do self-fulfilling prophecies improve or degrade predictive accuracy? How sociology and economics can disagree and both be right, *Journal of Socio-Economics,* **22**(2), 85–104.

Herrera, O., H. Scolnik, G. Chichilnisky, G. Gallopin, J. Hardoy, D. Moscovich, E. Oteiza, G. de Romero Brest, C. Suarez, and L. Talavera, 1976: *Catastrophe or new society? – A Latin American world model,* International Development Research Centre, Ottawa.

Hogan, W., and D. Jorgensen, 1990: Productivity trends and the cost of reducing CO_2 emissions, Global Environmental Policy Project, Discussion Paper E-90-07, Harvard University, Cambridge, MA.

Hourcade, J.-C., 1992: Modelling long run scenarios: Methodology lessons from a prospective study on a low CO_2 intensive country, *Energy Policy,* **21**(3), 309–311.

Hourcade, J.-C., and T. Chapuis, 1993: No regret potentials and technical innovation: A viability approach to integrative assessment of climate policies. In *Integrative assessment of mitigation impacts and adaptation to climate change,* Proceedings of an IIASA workshop held on 13-15 October 1993 at Laxenburg, Austria, N. Nakicenovic, W.D. Nordhaus, R. Richels, eds., pp. 535–558, International Institute for Applied Systems Analysis, Laxenburg.

Howarth, R., and B. Anderson, 1993: Market barriers to energy efficiency, *Energy Economics, 15*(4), 262–272.

Howarth, R., and M. Winslow, 1994: Energy use and climate stabilisation: Integrating pricing and regulatory policies, *Energy – The International Journal, 19,* 855–867.

IAEE (International Association for Energy Economics), 1993: *Energy, environment, and sustainable development: Challenges for the 21st century,* 16th Annual International Conference of the International Association for Energy Economics, Nusa Dua, Bali, Indonesia, July.

Jäger, J., N. Sonntag, D. Bernard, and W. Kurz, 1991: The challenge of sustainable development in a greenhouse world: Some visions of the future, report of a Policy Exercise held in Bad Bleibert, Austria, Sept. 2-7, 1990, Stockholm Environment Institute, Stockholm.

Jantsch, E., 1967: *Techological forecasting,* OECD, Paris.

Jefferson, M., 1982: Historical perspectives of societal change and the use of scenarios in shell. In *Social forecasting for company planning,* B. Twiss, ed., Macmillan, London.

Jefferson, M., 1983: Economic uncertainty and business decision-making. In *Beyond positive economics,* J. Wiseman, ed., Macmillan, London.

Johansson, T., and P. Steen, 1977: *Solar Sweden,* Secretariat for Futures Studies, Stockholm.

Johansson, T., H. Kelly, A. Reddy, and R. Williams, 1993: *Renewable energy: Sources for fuels and electricity,* Island Press, Washington, DC.

Jorgensen, D., 1984: Economic effects of the rise in energy prices: What have we learned in ten years? The role of energy in productivity growth, *The American Economic Review, 74*(2), 26–29.

Joskow, P., and D. Marron, 1993: What does a negawatt really cost? Evidence from utility conservation programs, *The Energy Journal, 13*(4), 41–74.

Keepin, B., and B. Wynne, 1987: The roles of models – What can we expect from science? A study of the IIASA World Energy Model. In. *The politics of energy forecasting,* T. Baumgartner, A. Middtun, eds., pp. 33–57, Clarendon Press, Oxford.

Koomey, J., and A. Sanstad, 1994: Technical evidence for assessing the performance of markets affecting energy efficiency, *Energy Policy, 22*(10), 826–832.

Krause, F., 1981: The industrial economy – An energy barrel without a bottom, paper presented at the Second International Conference on Soft Energy Paths, ENI, Rome.

Krause, F., R. Howarth, J. Koomey, and A. Stanstad, 1994: Top-down and bottom-up methods for calculating the cost of cutting carbon emissions: An economic assessment, invited contribution to the 1995 IPCC Scientific Assessment, Working Group III, Intergovernmental Panel on Climate Change, Geneva.

Krause, F., E. Vine, and S. Gandhi, 1989: *Program experience and its regulatory implications: A case study of utility lighting efficiency programs,* LBL-28268, Lawrence Berkeley Laboratory, Berkeley, CA.

Leach, G., 1979: *A low energy strategy for the United Kingdom,* The International Institute for the Environment and Development, Science Reviews Ltd., London.

Legget, J., W.J. Pepper, and R.J. Swart, 1992: Emissions scenarios for the IPCC: An update. In Intergovernmental panel on climate change, *Climate change 1992: The supplementary report to the IPCC Scientific Assessment,* J.T. Houghton, B.A. Callander, S.K. Varney, eds., pp. 69–95, Cambridge University Press, Cambridge.

Levine, M., and R. Sonnenblick, 1994: On the assessment of utility demand-side management programs, *Energy Policy, 22*(10), 848–856.

Lewis, D. K., 1969: *Convention: A philosophical study,* Harvard University Press, Cambridge, MA.

Little, I., and J. Mirrless, 1974: *Project approach and planning for developing countries,* Basic Books, New York.

Lovins, A.B., 1977: *Soft energy paths: Towards a durable peace,* Penguin Books, Harmondsworth.

Lovins, A.B., L. Lovins, F. Krause, and W. Bach, 1981: *Least-cost energy – solving the CO_2 problem,* Brick House Publishing, Andover, MA.

Lutzenhiser, L., 1993: Social and behavioural aspects of energy use, *Annual Review of Energy and the Environment, 18,* 247–289.

Manne, A., and R. Richels, 1992: Buying greenhouse insurance. The economic costs of carbon dioxide emission limits, The MIT Press, Cambridge, MA.

Manne, A., and L. Schrattenholzer, 1993: Global scenarios for carbon dioxide emissions, *Energy, 18*(12), 1207–1222.

Martin, J.M., 1988: L'intensité énergétique de l'activité économique dans les pays industrialisés: les évolutions de longue période livrent-elles des enseignements utiles? *Série Economie de l'énergie, 4,* 9–27.

Martin, J.M., 1992: *Economie et politique de l'énergie,* Armand Collin, coll. Cursus Economie, Paris.

Meadows, D.H., and D.L. Meadows, 1972: *The limits to growth,* Universe Books, New York.

Mesarovic, M., and E. Pestel, 1974: *Mankind at the turning point,* E.P. Dutton, New York.

Messenger, M., 1981: A high technology–low energy demand for Western Europe, *Energy–the International Journal, 6,* 1481– 1503.

Moulton, R., and K. Richards, 1990: Costs of sequestering carbon through tree planting and forest management in the United States, General Technical Report WO-58, U.S. Department of Agriculture, Washington, DC.

Nadel, S., 1992: Utility demand-side management experience and potential–A critical review, *Annual Review of Energy and the Environment, 17,* 507–535.

Nastari, P.M., 1991: Economic viability of sugar cane ethanol in Brazil considering its social cost of production, Proceedings of the IX international symposium on alcohol fuels, Florence, Italy, pp. 981–983.

Newman, P., and J. Kenworthy, 1989: Cities and automobile dependence: An international sourcebook, Gower Technical, Aldershot, UK.

Nordhaus, W.D., 1977: Strategies of the control of carbon dioxide, Cowles Foundation Discussion Paper No. 443, Yale University, New Haven, CT.

Nordhaus, W.D., 1979: *The efficient use of energy resources,* Yale University Press, New Haven, CT.

Nordhaus, W.D., 1991: The cost of slowing climate change: A survey, *The Energy Journal, 12*(1), 37–65.

Norgaard, J., 1979: Household and energy, report No. 4 from DEMO-Project, Physics Laboratory III, Danish Institute of Technology, Lyngby, Denmark.

Northwest Power Planning Council, 1991: *1991 Northwest conservation and electric power plan, Volume II,* Portland, Oregon.

Olivier, D., H. Miall, F. Nectoux, and M. Opperman, 1982: *Energy-efficient futures: Opening the solar option,* Earth Resources Research, London.

Parks, P., and I. Hardie, 1995: Least cost forest carbon reserves: Cost-effective subsidies to convert marginal agricultural land to forests, *Land Economics,* **71**(1), 122–136.

Richards, K., R. Moulton, and R. Birdsey, 1993: Costs of creating carbon sinks in the U.S. In *Proceedings of the International Energy Agency Carbon Dioxide Symposium,* P. Riemer, ed., pp. 905–912, Pergamon Press, Oxford.

Riemer, P. (ed.), 1993: Proceedings of the International Energy Agency Carbon Dioxide Symposium, *Energy Conversion and Management,* **34,** 711–1227.

Robinson, J., 1982: Bottom-up methods and low-down results: Changes in the estimation of future energy demands, *Energy – The International Journal,* **7,** 627–635.

Robinson, J., 1988: Unlearning and backcasting: Rethinking some of the questions we ask about the future, *Technological Forecasting and Social Change,* **33**(4), 325–338.

Robinson, J., 1990: Futures under glass: A recipe for people who hate to predict, *Futures,* **22**(9), 820–843.

Robinson, J., 1991a: The proof of the pudding: Making energy efficiency work, *Energy Policy,* **19**(7), 631–645.

Robinson, J., 1991b: Modelling the interactions between human and natural systems, *International Social Sciences Journal,* **130,** 629–647.

Robinson, J., 1992: Of maps and territories: The use and abuse of socio-economic modelling in support of decision-making, *Technological Forecasting and Social Change,* **42**(3), 147–164.

Robinson, J., D. Biggs, G. Francis, R. Legge, S. Lerner, S. Slocombe, and C. Van Bers, 1996: *Life in 2030: Exploring a sustainable future for Canada,* UBC Press, Vancouver, forthcoming.

Ross, M., 1989: Energy and transportation in the United States, *Annual Review of Energy,* **14,** 131–171.

Ross, M., and R. Williams, 1981: *Our energy – Regaining control,* McGraw Hill, New York.

Rothman, D., and R. Coppock, 1996: Scenarios of sustainability : The challenges of describing desirable futures. In *Climate change and world food security,* Thomas E. Downing, ed., Springer-Verlag, London.

Sachs, I., 1993: *Ecodéveloppement: stratégies de transition vers le XXIe siècle,* Collection Alternatives économiques, Syros, Paris.

Sant, R., 1979: *The least-cost energy strategy,* The Energy Productivity Center, Carnegie Mellon Institute, Arlington, VA.

Schelling, T.C., 1960: *The strategy of conflict,* Harvard University Press, Cambridge, MA.

Schipper, L., 1995: Determinants of automobile use and energy consumption in OECD countries, *Annual Review of Energy and the Environment,* **20,** 325–386.

Schipper, L., and R. Howarth, 1990: United States energy use from 1973 to 1987: The impacts of improved efficiency, *Annual Review of Energy,* **15,** 455–504.

Selvanathan, E., and S. Selvanathan, 1994: The demand for transport and communications in the United Kingdom and Australia, *Transportation Research B,* **28B**(1), 1–9.

Southern California Edison Company, 1988: *Strategies for an uncertain future,* Southern California Edison Company, Rosemead, CA.

Squire, L., and H. Van der Tak, 1976: *Economic analysis of projects,* Johns Hopkins University Press, Baltimore.

Stavins, R., 1995: *The costs of carbon sequestration, A revealed preference approach,* J.F. Kennedy School of Government (Harvard University) and Resources for the Future, Cambridge, MA, and Washington, DC.

Stobaugh, R., and D. Yergin, 1979: *Energy future – Report of the energy project at the Harvard Business School,* Random House, New York.

Stockholm Environment Institute/Greenpeace, 1993: Towards a fossil free energy future, the next energy transition, a technical analysis for Greenpeace International, SEI, Boston.

Sutherland, R., 1991: Market barriers to energy-efficient investments, *The Energy Journal,* **12**(3), 15–34.

Thompson, M., 1984: Among the energy tribes: A cultural framework for the analysis and design of energy policy, *Policy Sciences,* **17**(4), 321–339.

Tinbergen, J., 1976: *Reshaping the international order,* report to the Club de Rome, E.P. Dutton and Co., New York.

Wack, P., 1985a: Scenarios: Uncharted waters ahead, *Harvard Business Review,* **63**(5), 72–89.

Wack, P., 1985b: Scenarios: Shooting the rapids, *Harvard Business Review,* **63**(6), 139–150.

Williams, R., 1987: Materials, affluence and energy use, *Annual Review of Energy,* **12,** 99–144.

World Commission on Environment and Development, 1987: *Our common future,* Oxford University Press, Oxford.

World Resources Institute, 1993: *World resources, 1992-3,* Oxford University Press, New York.

Xu, P., 1994: The potential for reducing atmospheric carbon by large-scale afforestation in China and related cost/benefit analysis, *Biomass and Bioenergy,* forthcoming.

9

A Review of Mitigation Cost Studies

Convening Author:
J.C. HOURCADE

Principal Lead Authors:
K. HALSNAES, M. JACCARD, W D. MONTGOMERY, R. RICHELS,
J. ROBINSON, P.R. SHUKLA, P. STURM

Lead Authors:
W. CHANDLER, O. DAVIDSON, J. EDMONDS, D. FINON, K. HOGAN,
F. KRAUSE, A. KOLESOV, E. LA ROVERE, P. NASTARI, A. PEGOV,
K. RICHARDS, L. SCHRATTENHOLZER, R. SHACKLETON, Y. SOKONA,
A. TUDINI, J. WEYANT

CONTENTS

SUMMARY

- Estimates of the cost of greenhouse gas emission reduction are sensitive to assumptions about appropriate model structure, demographic and economic growth, the cost and availability of both demand-side and supply-side energy options, the desired level and timing of abatement, and the choice of policy instruments. Different assumptions have led to a wide range of emission reduction cost estimates.

- Despite significant differences in views, there is agreement that some energy efficiency improvements (perhaps 10-30% of current consumption, depending on baseline assumptions and the implementation time frame) can be realized at negative to slightly positive costs. The existence of such a no-regrets potential depends on the existence of substantial market or institutional imperfections that prevent cost-effective emission reduction measures from being taken. The key question is whether such imperfections can be removed cost effectively by policy measures.

- Energy-related emissions can be reduced through both demand-side (energy efficiency) and supply-side (alternative sources of supply) options. In the short term, demand-side options are cheapest in most countries.

- Estimates of the costs of stabilizing CO_2 emissions vary widely as a result of differences in both the baseline scenarios used (e.g., how much energy efficiency or what rate of economic growth is contained in the baseline scenario) and the calculated costs of various policy measures.

- There has been much more analysis to date of emission reduction potentials and costs for industrialized countries than for other parts of the world. Moreover, many existing models are not well suited to the study of economies in transition or developing country economies. Much more work is needed to develop and apply models outside of developed countries.

- The overall cost of abatement programmes will ultimately be determined by the rate of capital replacement, the discount rate, and the effect of research and development. Appropriate long-run signals to encourage research and development will reduce long-run costs. The implementation of any no-regrets potential will increase the time available to learn about climate risks and to bring new technologies to the marketplace.

- Infrastructure decisions are critical because they can enhance or restrict the number and types of future options, and different infrastructure decisions can lead to very different cost outcomes. This issue is of particular importance to developing countries, where major infrastructure decisions will be made in the near term.

- If the climate issue is formulated in terms of stabilizing atmospheric concentrations, the choice of emission time path is critical in determining the overall price tag. It is important to identify those paths that minimize the costs of achieving a particular concentration target.

- The overall impact of a carbon tax will depend not only on the size of the tax, but also on the uses to which the revenues are put. If carbon tax revenues are used to reduce more distortionary taxes, overall emission reduction costs will be reduced. Inefficient tax recycling could, however, increase costs.

- There is need for both detailed (bottom-up) analysis of technological options and also more aggregate (top-down) analysis of economic effects. Whereas, historically, these two approaches have been associated with very different cost estimates, this need not always be the case. Indeed, widely differing methods can produce quite similar results when calibrated to the same set of input assumptions.

- Given differences in marginal emission reduction costs among countries, international cooperation can significantly reduce the global price tag for emission reduction. Economic efficiency would be enhanced by carrying out emission reduction where it is cheapest to do so.

- Estimates of the costs of stabilizing CO_2 emissions at 1990 levels in *OECD countries* vary widely. Many bottom-up studies suggest that the costs of achieving this target over the next few decades may be negligible. Given rising baseline emissions over the longer term, however, many top-down studies suggest that the annual costs of stabilizing emissions may ultimately exceed 1-2% of GDP.

- The costs of stabilizing emissions for *economies in transition* may be small relative to OECD countries. The potential for cost-effective reductions in energy use is apt to be considerable, but the realizable potential will depend on what economic and technological development path is chosen. A critical issue is the future of structural changes in these countries that may drastically change the level of baseline emissions and the emission reduction costs.

- Analyses suggest that there may be substantial low-cost emission reduction opportunities for developing countries. However, these are likely to be insufficient to offset rapidly increasing emission baselines associated with, for example, increased economic growth. In the absence of a highly favourable allocation of carbon emission rights, the likely magnitude of emission reduction will be particularly costly for developing countries.

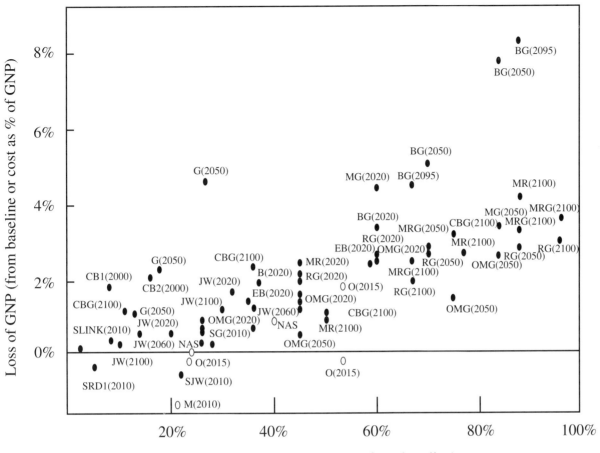

Note: to avoid crowding, labels are not attached to all points. For key to studies used, see Table 9.1.
Source: Grubb *et al.* (1993).

Figure 9.1: U.S. Studies: Cost of CO_2 abatement relative to baseline projection.

9.1 Introduction

A variety of options can be adopted in response to the prospect of increasing concentrations of greenhouse gases in the earth's atmosphere. These include (1) options that eliminate or reduce greenhouse gas emissions; (2) options that offset emissions, for example, through the enhancement of sinks; and (3) options that help human and ecological systems adjust or adapt to new climate conditions and events. The first and second types of interventions take effect prior to climate change and are intended to slow its pace. As such, they are referred to as "mitigation" measures. The third category of intervention takes place after warming occurs and falls under the classification of "adaptation." This chapter reviews existing studies of the economic costs of mitigating greenhouse gas emissions.

A caveat is in order. Sensible greenhouse policy should involve careful consideration of the costs of mitigation and adaptation options together with what these options may buy

in terms of reduced environmental impacts. This chapter focusses exclusively on mitigation costs. The benefits from adopting such options are discussed in Chapter 6.

9.2 Review of Existing Studies of the Costs of Reducing CO_2 Emissions

In recent years, there have been numerous studies of the costs of reducing CO_2 emissions. Unfortunately, estimates have spanned such a wide range that they have been of limited value to policy making. Figure 9.1 and Table 9.1, which summarize recent projections of the costs of reducing U.S. CO_2 emissions, are illustrative of the current lack of consensus about the costs of emission reduction. Whereas some studies place losses at several percent of GDP, others question whether there will be any losses at all. An essential first step in narrowing the range of disagreement is to determine why estimates differ so widely.

Table 9.1. *U.S. CO_2 abatement cost modelling studies*

Author (year)	Key[a]	CO_2 Reduction from Baseline (%)	Cost/GNP Reduction from Baseline (%)
Barns *et al.* (1992)	BG(2020)	26, 45, 60	0.6, 2.0, 3.2
Barns *et al.* (1992)	BG (2050)	45, 70, 84	1.9, 4.9, 7.5
Barns *et al.* (1992)	BG(2095)	67, 88, 96	4.3, 8.0, 10.9
DRI (1992)	B(2020)	37	1.8
CBO-PCAEO, DRI (1990)	CB1(2000), CB2(2000)	8, 16	1.9, 2.0
CBO-DGEM (1990)	CB3(2000)	36	0.6
CBO-IEA-ORAU (1990)	CBG(2100)	11, 36, 50, 75	1.1, 2.2, 0.9, 3.0[b]
Edmonds & Barns (1991)	EB(2020), EB(2100)	35, 59	1.3, 2.3
Goulder (1991)	G(2050)	13, 18, 27	1.0, 2.2, 4.5
Jackson (1991)	J(2005)	43, 40, 46	-0.2, -0.1, 0.1
Jorgenson & Wilcoxen (1990a)	JW(2060)	20, 36	0.5, 1.1
Jorgenson & Wilcoxen (1990a)	JW(2100)	10, 20, 30	0.2, 0.5, 1.1
Jorgenson & Wilcoxen (1990b)	JW(2020)	8, 14, 32	0.3, 0.5, 1.6
Manne & Richels (1990a)	MR(2020)	45	2.2
Manne & Richels (1990a)	MR(2100)	50, 77, 88	0.8, 2.5, 4.0[c]
Manne (1992)	MRG(2020)	26, 45, 60	0.8, 2.2, 4.2
Manne (1992)	MRG(2050)	45, 70, 84	1.4, 2.7, 3.3
Manne (1992)	MRG(2100)	67, 88, 96	2.3, 3.1, 3.4
Mills *et al.* (1991)	M(2010)	21	-1.2[d]
NAS (1991)	N	24, 40	0, 0.8
Oliveira-Martins *et al.* (1992b)	OMG(2020)	26, 45, 60	0.2, 1.1, 2.4
Oliveira-Martins *et al.* (1992b)	OMG(2050)	45, 75, 84	0.4, 1.3, 2.4
OTA (1991)	O(2015)	23, 53, 53	-0.2,[e] -0.2,[e] 1.8
Rutherford (1992)	RG(2020)	26, 45, 60	0.5, 1.3, 2.5
Rutherford (1992)	RG(2050)	45, 70, 84	1.2, 2.4, 2.5
Rutherford (1992)	RG(2100)	67, 88, 96	1.8, 2.6, 2.8
Shackleton *et al.* (1992)	SJW(2010), SLINK(2010)	22, 2	-0.6, 0.1
Shackleton *et al.* (1992)	SDRI(2010), SG(2010)	5, 28	-0.4, 0.2
U.S. Energy Choices (1991)	USEC(2030)	67.5	-0.6

[a]If the model used is a global model, the key includes the letter "G" before the date.

[b]The first two results use multilateral taxes, the others use unilateral taxes; taxes are flat only in the first and third estimates, rising in the other estimates.

[c]Values represent different assumptions for technological developments: an optimistic, an intermediate, and a pessimistic view.

[d]Arising from 11 specified regulatory changes; estimated from claimed savings of $85 billion per year.

[e]The benefit shown in the OTA cost estimates is only an indicative value; no explicit modelling value was calculated.

Source: Grubb *et al.* (1993).

There are many possible explanations for the disagreement – choice of methodologies, underlying assumptions, emission scenarios, policy instruments, reporting year, and others. In Chapter 8, we explored some of these reasons, paying particular attention to the modelling of technological change. In this chapter, we provide a more systematic examination of the published literature on the costs of reducing CO_2 emissions. In doing so, we will show that, despite widely varying and often contradictory findings, the impressive accumulation of modelling results provides a number of useful insights for climate policy making.

The review is organized by region, with results presented successively for

(1)   The U.S.
(2)   Other OECD countries
(3)   The transitional economies of Eastern Europe and the former Soviet Union
(4)   Developing countries
(5)   The globe as a whole

For each region, we first present results from top-down studies, then from bottom-up studies. In the last section, we attempt to draw some general conclusions based on our interpretation of these studies.

9.2.1 Studies of the costs of reducing CO_2 emissions in the U.S.

9.2.1.1 A review of top-down studies
There have been several attempts in the U.S. to systematically compare the results of modelling analyses of greenhouse gas emission reduction costs. One of the best known is a study conducted by the Energy Modeling Forum of Stanford University (EMF, 1993). A diverse group of economic models, employing common assumptions for selected numerical in-

Table 9.2. *Models used in Energy Modeling Forum (EMF) study*

Model/Modellers	Model Type	Time Horizon	Regions
CETA (Peck and Teisburg)	Aggregate economic equilibrium	2100	1 global
CRTM (Rutherford)	Disaggregated economic equilibrium	2100	5 global
DGEM (Jorgenson and Wilcoxen)	Disaggregated economic equilibrium	2050	U.S
ERM (Edmonds and Reilly	Energy-sector equilibrium	2095	9 global)
Fossil 2 (Belanger and Naill)	Energy-sector equilibrium	2030	U.S.
Gemini (Cohan and Scheraga)	Energy-sector equilibrium	2030	U.S.
Global 2100 (Manne and Richels)	Aggregate economic equilibrium	2100	5 global
Global Macro-economy (Pepper)	Energy-sector equilibrium	2100	9 global
Goulder	Disaggregated economic equilibrium	2030	U.S.
GREEN (Martins and Burniaux)	Disaggregated economic equilibrium	2050	8 global
IEA (Vouyoukas and Kouvaritakis)	Energy-sector equilibrium	2005	10 global
MARKAL (Morris)	Energy-sector optimization	2025	U.S.
MWC (Mintzer)	Energy-sector equilibrium	2095	9 global
T-GAS (Kaufmann)	Energy demand simulation	2010	14 global

Source: EMF (1993).

puts, was used to analyze a standardized set of emission reduction scenarios. In all, 14 top-down models participated in the study (Table 9.2).

The EMF exercise provides the most comprehensive application of top-down methodologies to date. The harmonization of key exogenous inputs makes it possible to verify to what extent the disagreements were due to differences in methodologies and to understand better the economic meaning of these differences. Although the focus was primarily on the U.S., many of the insights are applicable to developed countries in general. It is instructive to examine its results in some detail.

9.2.1.1.1 Key assumptions

In selecting parameters for standardization, the EMF study focussed on what were felt to be the most influential determinants of emission reduction costs. These included GDP, population, the fossil-fuel resource base, and the cost and availability of long-term supply options.

Given its importance as a determinant of future carbon emissions,[1] the GDP growth rate was a key parameter for harmonization across models. For its reference case, EMF adopted the average of the IPCC high and low economic growth cases (Report of the Response Strategies Working Group, 1990). Also, for consistency with the IPCC, the study adopted the population growth projections of Zachariah and Vu (1988).

Although the EMF models differed considerably in their technology representation, the study attempted to impose uniformity with regard to world oil prices, the oil and gas resource base, and the cost of backstop technologies. For the reference case, world oil prices were specified exogenously and assumed to be $24/barrel in 1990 and to increase at a real rate of $6.50/barrel each decade until 2030. (This trajectory was selected for its consistency with those from general equilibrium models which incorporate international trade in oil.) Oil and gas resources were based on the optimistic 95th percentile estimates of Masters *et al.* (1987).

To facilitate model comparisons, the study assumed that three types of backstop technologies would ultimately become available:

(1) a liquid synthetic fuel derived from coal or shale at $50/barrel of oil equivalent

(2) a noncarbon-based liquid fuel at $100/barrel of crude oil equivalent

(3) a noncarbon based electric option at 75 mills/kWh.

Although these technologies were assumed to be initially available in 2010, the models imposed constraints on the rate at which they could enter the marketplace. The models also included a variety of carbon- and noncarbon-based supply options that are available at rising marginal costs.

Table 9.3. *Description of EMF scenarios*

Scenario	Description
Reference	No controls
Stabilization	Stabilize CO_2 emissions at their 1990 levels by 2000
20% reduction	Stabilize CO_2 emissions at their 1990 levels by 2000, reduce by 20% below 1990 levels by 2010
50% reduction	Stabilize CO_2 emissions at their 1990 levels by 2000, reduce by 20% below 1990 levels by 2010, reduce by 50% below 1990 levels by 2050

Source: EMF (1993).

Table 9.4. *Comparison of carbon taxes and GDP losses in 2010*

Model	Carbon Tax ($ per tonne of C)		GDP Loss (% of GDP)	
	Stabilization	20% Reduction	Stabilization	20% Reduction
CRTM	150	260	.2	1.0
DGEM	20	50	.6	1.7
ERM	70	160	.4	1.1
Fossil 2	80	250	.2	1.4
Gemini	120	330		
Global 2100	110	240	.7	1.5
Global Macro	20	130		
Goulder	20	50	.3	1.2
GREEN	80	170	.2	.9
MWC	70	180	.5	1.1

Source: EMF (1993).

9.2.1.1.2 Emission scenarios

Table 9.3 describes the principal EMF emission scenarios, and the reference case in Figure 9.2 represents the average of the model results. The reference case projects future emissions in the absence of control measures. Hence, it indicates the amount of carbon that must be removed from the energy system in order to meet a given target. Figure 9.3 compares emission projections for the U.S. In all models, emissions increase well beyond 1990 levels. There are, however, substantial differences in the rate of increase.

The following identity (Kaya, 1989) helps explain why projections differ:

The growth rate in CO_2 = the growth rate in GDP
　　　　　　　　－ the rate of decline of energy
　　　　　　　　　use per unit of output
　　　　　　　　－ the rate of decline of CO_2
　　　　　　　　　emissions per unit of energy use

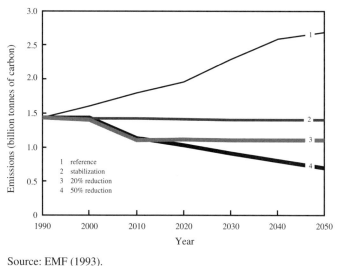

Source: EMF (1993).

Figure 9.2: EMF reference and emission reduction scenarios.

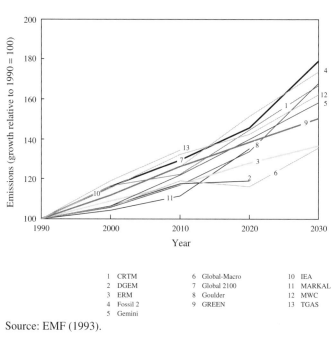

Source: EMF (1993).

Figure 9.3: U.S. carbon emissions.

The models employed common assumptions about GDP growth, but they differed with regard to the last two terms in the identity. The more optimistic the models were about the prospects for reducing energy intensity or the availability of low-cost substitutes, the lower their CO_2 growth rates.

9.2.1.1.3 Carbon taxes

The modellers generally used taxes based on the carbon content of fossil fuels in order to achieve a prescribed emission reduction. The magnitude of the tax provides a rough estimate of the degree of market intervention that would be required to achieve a given emission target.

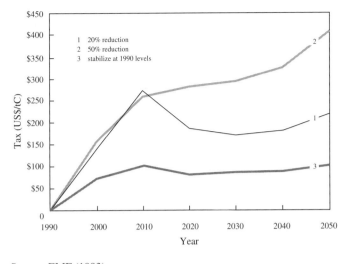

Source: EMF (1993).

Figure 9.4: U.S. carbon taxes – average of all models.

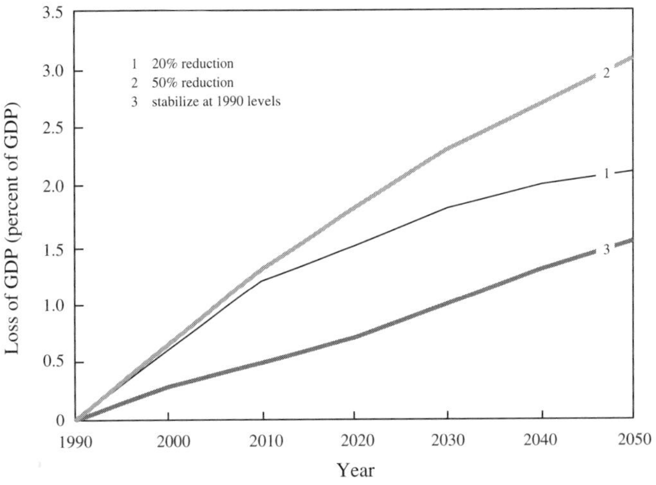

Source: EMF (1993).

Figure 9.5: U.S. GDP loss projections – average of all models.

Table 9.4 compares model results for the stabilization and 20% reduction scenarios. Estimates range from $20 to $150 per tonne for the carbon taxes required to hold emissions at 1990 levels in 2010. Estimates of the carbon taxes required to reduce emissions by 20% below 1990 levels in 2010 range from $50 to $330 per tonne.

Two parameters are particularly important in explaining the differences in tax projections: the price elasticity of energy demand and the speed with which the capital stock adjusts to higher energy prices. Neither was controlled in the EMF experiments but rather they were left to the choice of the modellers. Not surprisingly, those models using lower price elasticities required higher taxes to achieve the same emission goal. Those models which assumed greater malleability of capital required lower taxes.

The wide spread in the numerical results should not, however, obscure the fact that the models were in agreement on a number of important points. First, despite the inclusion of improved technologies and improved energy efficiencies in the reference case, all the models projected that intervention would be required to achieve the emission targets.

Second, the size of the required tax increases with the stringency of the carbon limit. The tax paths shown in Figure 9.4 represent averages of the model results. Note that the size of the required tax doubles as the limit is tightened to 20% below 1990 levels and nearly doubles again for a 50% reduction. This suggests that the economic efforts to be devoted are non-linear with respect to the level of controls in any given year. That is, incremental reductions are apt to cost more as the absolute level of allowable emissions is reduced.

Figure 9.4 also suggests that the size of the carbon tax is apt to vary over time, even for the same target. Recall that the purpose of the tax is to raise the price of carbon-intensive fuels to the point where consumers will turn to price-induced conservation and less carbon-intensive supply alternatives. Hence, the size of the tax is determined by the cost differential

between the marginal sources of supply with and without a carbon constraint. This cost differential is apt to change over time.

9.2.1.1.4 GDP losses

Table 9.4 shows a marked variation in GDP losses across models. Stabilizing emissions at their 1990 levels is estimated to reduce GDP by 0.2-0.7% in the year 2010 – roughly a $20 billion to $70 billion loss for that year. Estimates of the costs of reducing emissions by 20% below 1990 levels in the year 2010 range from 0.9 to 1.7% of GDP.

In making these calculations, the modellers assumed a lump sum redistribution of tax revenues. That is, tax revenues are used to replace other tax payments by individuals and corporations without affecting marginal tax rates or total tax revenues. The GDP losses calculated in this manner measure the cost of the distortions to the economy caused by the imposition of the carbon tax. There are no credits or penalties for lump sum recycling of tax revenues. The assumption of lump sum recycling avoids confusing the economic impacts of carbon taxes with costs or benefits attributable to potential uses of the revenues.

Figure 9.5 shows the average GDP losses across models. Holding emissions at 1990 levels results in a 0.3% loss in 2000. This rises to 1.5% by the middle of the century. That GDP losses increase over time should come as no surprise. Increasing amounts of carbon must be removed from the energy system in order to hold emissions to a particular target. GDP losses also increase with the stringency of the target. With a 20% reduction, losses exceed 2% of GDP in 2050. With a 50% reduction, losses exceed 3% of GDP.

It is interesting that GDP losses for a given year are influenced by expectations about limits in future years. All three emission reduction scenarios require that emissions be stabilized at 1990 levels by the year 2000. Yet losses are higher in 2000 for the scenarios that eventually require further reductions. The reason is that energy investments are typically long

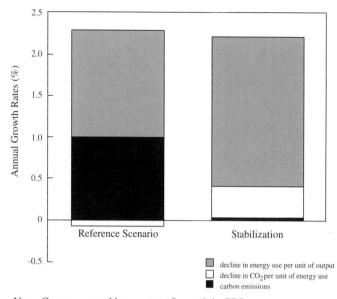

Note: Components add up to rate of growth in GDP.
Source: EMF (1993).

Figure 9.6: Decoupling between carbon emissions and GDP annual growth rates (1990–2010).

lived. Consumers and producers will be influenced by their expectations about the future.

9.2.1.1.5 *The nature of the GDP losses*

Figure 9.6 is helpful in understanding the nature of the GDP losses. Losses occur when the carbon taxes lead to investments that are more expensive than those that would take place in the absence of the taxes. Economists refer to these as deadweight losses. The higher the carbon taxes, the greater the investment in price-induced conservation and the more fuel switching towards the less carbon-intensive substitutes. Hence, the higher the deadweight losses.

In the reference case, carbon emissions and GDP grow at average annual rates of 1.0% and 2.2% respectively. In the absence of carbon taxes, there is some minor fuel switching towards more carbon-intensive fuels, but this is more than offset by the decline in energy use per unit of economic activity. (Note that the top-down models, on average, project sizable reductions in energy intensity – even in the absence of explicit measures to reduce emissions.)

In the stabilization case (the right-hand bar in Figure 9.6), the carbon taxes induce sufficient investment in supply-side substitutes and conservation to eliminate growth in carbon emissions. That is, carbon emissions and GDP are virtually decoupled. Note that this is done without significantly reducing the growth of the economy. Whereas the losses may be considered large in absolute terms, they are relatively small when measured in terms of a reduction in the GDP growth rate.

9.2.1.1.6 *The relationship between timing, emission reduction costs, and research and development*

The EMF study also addressed the issue of the timing of the carbon constraint. An important question is whether the costs

of CO_2 emission reduction will increase or decrease if emission reductions are deferred to the future. This is a complex issue. If marginal costs increase with the level of emission reduction in each period, deferring a tonne of emission control from the near term to a more distant time could raise costs, all else being equal. But all else is not equal.

There are three reasons why deferring emission reductions may reduce costs (Leary and Scheraga, 1994). First, large emission reductions in the near term will require accelerated replacement of the existing capital stock. This is apt to be costly. There will be more opportunities for reducing emissions cheaply once the current capital equipment turns over. Second, the availability and cost of technologies for fuel switching will improve over time. As a result, the costs of reducing a tonne of carbon emissions will decline. Finally, even if the costs of removing a tonne of carbon were the same in all periods, a positive discount rate will favour deferred reductions. Model simulations by Cohan *et al.* (1994) suggest that deferring emission reductions can reduce costs substantially. For example, they find that delaying the target date from 2010 to 2015 for stabilizing U.S. emissions at 20% below 1990 levels can reduce control costs by as much as 20%.

To bound the potential benefits from successful research and development, the EMF modellers were asked to examine an "accelerated R&D" scenario in which the cost of the non-electric backstop is reduced from $100 to $50 per barrel and the cost of the electric backstop is reduced from 75 to 50 mills/kWh. In such a scenario, the carbon-free backstops are economical in their own right. That is, they will enter the marketplace even in the absence of a carbon constraint.

Four models (ERM, Fossil 2, Global 2100, and CRTM) were used to examine the accelerated R&D scenario. The models were remarkably consistent in their estimates of the potential savings. With the more optimistic technology assumptions, GDP losses for the 20% emission reduction scenario were reduced by approximately 65%. This implies that approximately 35% of the discounted losses are incurred during the transition. The remaining costs accrue once we enter into the backstop phase, that is, once the backstops are available in unlimited quantities.

9.2.1.1.7 *The issue of revenue recycling*

The overall impact of a carbon tax will depend not only on the size of the tax but also on the uses to which the revenues are put. In the standard EMF scenarios, it was assumed that tax revenues would be redistributed in a neutral manner (i.e., without affecting the marginal tax rates). There are, of course, numerous ways in which tax revenues can be used. These include reducing budget deficits; reducing marginal rates of income, payroll, corporate, or other taxes; granting tax incentives to preferred activities; and increasing the level of government expenditures. The costs of the tax will vary widely, depending on how the revenues are recycled.

Table 9.5 shows the range of GDP losses associated with a carbon tax rising from $15 per tonne in 1990 to $40 per tonne in 2010, with alternative methods of recycling the revenues. The analysis is performed using four different economic mod-

Table 9.5. *GNP loss, 1990–2010 (percent of discounted constant price GNP)*

	DRI (%)	LINK (%)	DGEM (%)	Goulder (%)
Lump sum tax cuts	-0.58	-0.46	-0.62	-0.24
Revenue raising	-0.40	-1.02		-0.24
Personal income tax cuts	-0.56	-0.53	-0.16	-0.16
Corporate income tax cuts	0.40	-0.11	0.60	-0.17
Payroll tax cuts				-0.18
(Employee only)	-0.58	-0.53		
(Employer only)	0.19	-0.25		
Investment tax credit	1.55	1.67		0.00

Source: Shackleton *et al.* (1992).

els of the U.S. economy (two macroeconomic models and two general equilibrium models). The first alternative recycling method, lump sum tax cuts, is in the neutral manner described above.

Note that the GDP costs of the carbon tax vary considerably depending on how the revenues are recycled. In some models, the costs are more than offset by tax policies that encourage investment. On the other hand, one model suggests that the GDP costs of the tax would be increased over the neutral case if the revenues were used simply to reduce the government budget deficit. Although the alternative of increasing government expenditures was not examined, it is possible that such a policy would increase the GDP costs of the tax over the neutral case (see Nordhaus, 1994).

9.2.1.2 A review of bottom-up studies
Bottom-up research in the U.S. and Canada has tended to suggest, as elsewhere in the world, that significant decreases in CO_2 emissions are possible without great cost to the economy. In this sense, their results differ from those of most top-down research, the latter suggesting substantial economic costs to CO_2 emission abatement. The methodology discussions in Chapter 8 reviewed some of the key reasons for these differences.

9.2.1.2.1 Variations in bottom-up technology and policy assumptions
Like top-down studies, bottom-up studies result in a wide range of reduction cost estimates. Compared with top-down studies, the structure of formal models in bottom-up analysis is generally less important for the results. Instead, input assumptions are dominant. Inspection of available studies shows that for a given time horizon and geographic region, divergent results arise mainly from differences in two factors:

- The quality of the technical analysis from which supply curves for energy efficiency and supplies from cogeneration and renewables are derived
- The assumed effectiveness of policy instruments in mobilizing the economically cost-effective resource potential

Differences in technical analysis. Of the technical factors, the most important ones appear to be the level of detail in analyzing and representing supply curves for technology options, notably those for demand-side efficiency improvements. Bottom-up studies that rely on more detailed and comprehensive assessments of these options will tend to arrive at larger efficiency potentials and lower costs of saved energy than less detailed studies. Even then, uncertain baseline data for equipment use and efficiency in individual end uses leave some room for disagreement among technologists.

The treatment of administrative policy costs can also be important. Most bottom-up analyses assume that energy efficiency standards would be the main policy tool. These have negligible administrative costs, but such costs can be more substantial in some types of incentive programmes. State-of-the-art studies give a differentiated treatment.

Depending on the scope of the study, additional no-regrets opportunities may be quantified. These are lower generating costs from utility regulatory reforms, savings in acid rain and other pollution control expenditures from reduced fossil fuel use, the cost-reducing effects of monetizing environmental externalities (other than climate change), and increased cost-effectiveness of nonfossil options when removing fossil fuel price subsidies.

Most bottom-up studies do not quantify the feedback effect from lower energy demand on fuel prices or the further effect of lower fuel prices and energy service costs on cost-effective energy efficiency levels and energy demand. As a result, bottom-up studies may at once underestimate feasible economic savings from carbon reductions and overestimate the amount of emission reductions that market transformation policies can bring.

Differences in assumed policy effectiveness. All bottom-up scenarios of future energy demand assume that policy intervention can at least partially shift the investment behaviours of consumers and firms from historically suboptimal patterns to economically optimal choices. At one end of the spectrum, it is assumed that policies will shift every replacement purchase or expansion of end-use equipment that will occur over the time period studied. For short time horizons (10-20 years), this assumption implies that a complete shift to efficient equipment will be achieved within one cycle of capital turnover. This assumption is almost certainly too optimistic.

At the other end of the spectrum, some bottom-up studies take a very pessimistic view in which current political difficulties are assumed to limit the more widespread application of market transformation policies. As a result, the savings potential is estimated to be low.

A compromise position is found in studies using longer time horizons (30-40 years or more). Over these long periods, most capital goods will be replaced more than once, and many several times. Here, least-cost efficiency levels can be achieved within the time horizon, even if policies do not shift all or most investments the first time around. The assumptions about the effectiveness of policies are thus more realistic in these studies.

Table 9.6. *Bottom-up studies of U.S. emission mitigation costs – major assumptions (percent annual growth rates)*

Study	Base Year	Forecast Year	GDP	Discount Rate	Energy Prices			
					Oil	Coal	Gas	Elect.
Alliance to Save Energy	1988	2000	2.4	3	6.5	0.9	7.8	
(1991)		2010	2.4	3	6.0	1.4	10.0	
		2030	1.8	3	4.3	1.8	7.1	
National Academy of Sciences (1991)	1990	n/s	—	6				
Office of Technology Assessment (1991)	1987	2015	2.3	—				
U.S. EPA (1990)	1988	2005	2.5	7				
		2010	2.5	7				
SEI/Greenpeace (1993)	1988	2030	2.1	8				
Carlsmith *et al.* (1990)	1988	2010	2.5	7	5	1	5	0.2
Chandler & Kolar (1990)	1985	2030	2.5	7	2.5	1	2.5	1
Chandler & Nicholls	1989	2000	—	—	2.4	8.8	2.5	
(1990)		2000	—	—	6.9			
Chandler (1990)	1989	2010	2.5	7	4.8	0.8		0.2
Lovins & Lovins (1991)	1988	n/s	—	5				
Mills *et al.* (1991)	1988	2000	2.5	6				
Rubin *et al.* (1992)	1989	n/s	—	6				

Note: n/s = not specified.

9.2.1.2.2 *Key study assumptions and results*

In this section we survey some of the key assumptions and results in the studies for the U.S. and Canada. Unfortunately, there has not yet been an effort to undertake a harmonized comparative analysis of U.S. bottom-up models, as has been the case with U.S. top-down models. However, a survey of the approaches at least allows for some tentative observations.

An overview of the main assumptions and results of the major U.S. bottom-up studies is provided in Tables 9.6, 9.7, and 9.8. Most studies test the sensitivity of their results to changes in GDP growth rates. Energy prices are generally assumed exogenously, although some studies have attempted to incorporate the feedback effects of efficiency measures on energy prices. RIGES, for example, assumes low-cost fossil energy in 2030 as a consequence of successful implementation of energy efficiency. In contrast, Carlsmith *et al.* (1990) assume high rates of near-term energy price increases.

Table 9.8 reconfigures the reported results from Tables 9.6 and 9.7 in a manner that facilitates interpretation for policy-making. Results are now reported in terms of the percentage reduction below the level of base year emissions that was found to have zero net cost. The base year variously ranged from 1985 to 1990. The reduction cost is reported as average cost. For studies where negative net average costs were reported, feasible zero-net cost reductions are shown with a "greater than" sign to indicate that the results reported do not include reduction options up to the intersection point with the x-axis.

The following reductions were found to result in zero net costs:

- By the turn of the century: 0-21% (median of 11%)

- By about 2005–2010: >0-26% (median of 13%)

- By about 2015–2020: >23-58% (median of 41%)

- By about 2025–2030: >61-82% (median of 72%)

The studies were not normalized in terms of such factors as economic growth, structural change, fuel prices, technology costs, or policy effectiveness. As a result, the above reduction ranges are wider than a coordinated analysis based on uniform assumptions would have found.

Despite these wide variations, the figures show a consistent pattern: Within the 1990–2030 time frame, progressively larger reductions become feasible at zero net cost as time horizons grow longer. And for any given reduction target, mitigation costs decline as longer adjustment periods are allowed.

This pattern reflects two major factors: First, bottom-up studies identify large energy efficiency potentials that are not exploited in the reference case, due to the high transaction costs. The studies assume that these high transaction costs will be reduced through suitable policy interventions, such as efficiency standards, least-cost utility planning, and other information and financial incentive programmes.

Second, most energy efficiency improvements are introduced at the "economic optimal" rate of capital stock turnover, with the more long-lived capital goods lasting twenty to thirty years or more. Due to this synchronicity, cost-effective efficiency improvements continue to occur. As a result, bottom-up studies estimate "transition benefits" rather than "transaction costs" for such technology shifts.

Third, cheap cogeneration opportunities are assumed in a number of studies to make an important further negative-cost contribution. Compared to separate generation of electricity and heat, this supply-side technology is found to have significant cost-effective resource potentials. They are typically not

Table 9.7. *Results of U.S. emission mitigation studies*

Study	Forecast Year	CO$_2$ Reduction (% from Base Year)	Cost of Reduction (% of GNP)	Average Cost ($/tC)
Alliance to Save	2000	26	-0.4	
Energy *et al.* (1991)	2010	53	-0.5	
	2030	82	-0.6	
National Academy	n/s	24	0	0
of Sciences (1991)		40	0.8	9
Office of Technology	2015	23	-0.2	
Assessment (1991)		53	-0.2/1.8	
U.S. EPA (1990)	2005	3	0	0
	2010	0		
SEI/Greenpeace (1993)	2030	74[†]	0	0
Carlsmith *et al.* (1990)	2010	0	0	0
		20	0.5	
Chandler & Kolar (1990)	2005	0	0	0
		20	0.5	
Chandler & Nicholls (1990)	2000	7		
		20		82
Chandler (1990)	2010	0		0
		20		92
Lovins & Lovins (1991)	n/s	58[†]	0	0
Mills *et al.* (1991)	2000	21	-1.2	-231
Nordhaus (1990)	n/s	6		13
Rubin *et al.* (1992)	n/s	35	0	0 – <0
RIGES (Johansson *et al.*, 1993)	2025	61	<0	<0
FFES (SEI/Greenpeace 1993)	2030	75	<0	<0

[†]Scenario ultimately yields 100% reduction.

Table 9.8. *Zero average net cost reduction potential in U.S. bottom-up studies (estimated reduction potential below base year achieved with zero net average cost)*

Study	Country	2000 (%)	2005/10 (%)	2015/20 (%)	2025/30 (%)
Alliance to Save Energy *et al.* (1991)	U.S.		>26	>53	>82
Carlsmith *et al.* (1990)	U.S.		0		
Chandler and Nicholls (1990)	U.S.	0			
Chandler and Kolar (1990)	U.S.		>0		
DPA (1989)	Canada		>13		
Lovins and Lovins (1991)	U.S.			58	
Mills *et al.* (1991)	U.S.	21			
NAS (1991)	U.S.			>24	
OTA (1991)	U.S.			>23	
Rubin *et al.* (1992)	U.S.			35	
RIGES (Johansson *et al.*, 1993)	U.S.				>61
FFES (SEI/Greenpeace, 1993)	U.S.				>75
Range		0–21	0–26	>23–58	>61–82
Midpoint		11	> 13	>41	>72

Note: A "greater than" sign indicates that the study did not include options up to the intersection point of the x-axis. This means that larger reductions could be achieved for negative or zero average cost.

included in the reference case, since business-as-usual regulatory regimes do not sufficiently control monopsonistic utility buy-back practices. Like most energy efficiency improvements, cogeneration technologies are already commercially available and thus add to the near- to medium-term no-regrets potential.

When cost estimates are compared for a given reduction target, the results of the bottom-up studies listed in Tables 9.6,

9.7, and 9.8 lie within a reasonably close range, despite considerable differences in assumptions. Most cost estimates for a 20% reduction in U.S. CO_2 emissions by the year 2010, for example, range between -0.6 and +0.5% of GDP (see Table 9.7). As noted in Section 9.2.1.2.1, this difference is explained at least in part by the nature and level of detail of the input assumptions used in the specific bottom-up analysis. The RIGES study, for instance, assumes 100% penetration of markets by all technologies shown to be cost-effective in an engineering/economic analysis. In contrast, Carlsmith *et al.* constrained their model to link all market penetration of efficient technologies to consumer behaviour parameters (expressed as price elasticities) and policy constraints (expressed as efficiency standards). This latter approach did not lead to 100% market penetration by efficient technologies.

Bottom-up studies for Canada were compiled and reviewed by Robinson and colleagues in the study *Canadian Options for Greenhouse Gas Emission Reduction* (Robinson *et al.,* 1993). Estimates of cost-effective CO_2 emission reduction potential by 2010, relative to a reference (or baseline) scenario, ranged from 20% to 40%, with a median of about 23%. Relative to 1988 or 1990, many studies showed savings in energy use or emissions of between 10 and 30%, with a median of about 16%.

9.2.2 Studies of the costs of reducing CO_2 emissions in other OECD countries

Much of the early work on the costs of CO_2 emission reduction was U.S.-based and, as a result, tended to be U.S.-focussed. More recently, there has been a flurry of analytical activity elsewhere in the OECD, mainly in Western Europe. In general, these country-specific studies have had relatively shorter time horizons than in the U.S., focussing on the costs of stabilizing emissions in 2000 or a 20% reduction by 2005 (the "Toronto target").

9.2.2.1 A review of top-down studies

Top-down studies of non-U.S. OECD countries have been of two types: those focussing on an entire region or a subset of countries and those focussing on individual nations. In this section, we review the results of both. As with the U.S. studies, the types of policies tested have been limited for the most part to taxes on CO_2 and energy under alternative domestic fiscal recycling schemes.

9.2.2.1.1 Regional studies

One notable attempt at a systematic comparison of non-U.S. OECD models was conducted by the OECD in the early 1990s (Dean and Hoeller, 1992). The exercise was patterned after the parallel study being undertaken by the Energy Modeling Forum. The two studies used many of the same models and shared a common set of input assumptions. For purposes of the model comparison, the OECD was divided into two regions: the U.S. and "other OECD countries."

The OECD analysis encompassed both the transition and backstop phases. This long-term perspective is useful when examining issues related to the timing of the transition from

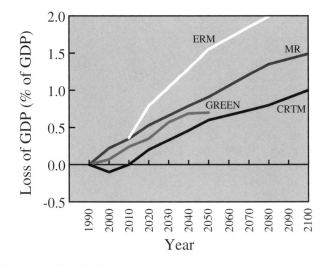

Source: OECD (1993).

Figure 9.7: GDP losses in the stabilization scenario (other OECD countries).

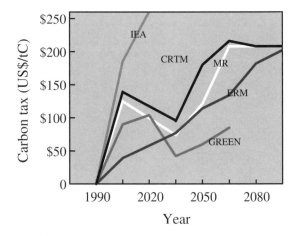

Source: OECD (1993).

Figure 9.8: Carbon taxes in the stabilization scenario (other OECD countries).

fossil fuels and the potential role of technical innovation in lowering overall costs of reducing emissions.

The OECD study examined a range of emission reduction scenarios. Among the more interesting is one in which emissions are permanently held to 1990 levels. Figure 9.7 compares annual GDP losses for four models: ERM, GREEN, Global 2100 (MR), and CRTM. The models are fairly consistent in their projections of losses in 2010 – between 0.3% and 0.5% of GDP. The convergence is due in part to the standardization of key input parameters, but it is also important to note the aggregation effect. Combining all non-U.S. OECD countries into a single region masks important intercountry differences. As we will see in the next section, the variance around the mean is likely to be large.

The models are less in agreement as we move beyond 2010. The differences can be traced to several uncontrolled parameters. Global 2100, for example, employs a lower rate of autonomous energy efficiency improvements than ERM

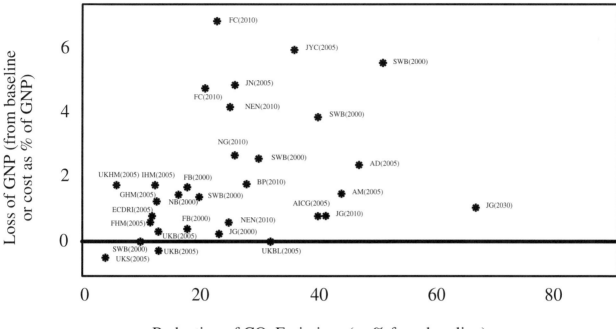

Note: See Table 9.9 for key.

Figure 9.9: Non-U.S. OECD studies: Cost of CO_2 abatement relative to baseline projection.

and GREEN. As a result, its baseline emissions path is somewhat higher. More carbon must be removed from the energy system to maintain emissions at 1990 levels. Hence annual abatement costs are apt to be higher.

Assumptions about the rate at which backstop technologies can enter the energy system also tend to be important. GREEN, for example, assumes a more rapid introduction rate than the other models. Hence its marginal cost of emission abatement is lower for each tonne of carbon removed. This, coupled with the lower baseline, easily explains the ranking of cost estimates among the three models.

Figure 9.8 shows the taxes that would be required to stabilize emissions at 1990 levels. The results from ERM, Global 2100 (MR), and GREEN are consistent with the above discussion. In addition, the OECD report included results from the IEA model. This econometric model is relatively pessimistic about the prospects for energy conservation and fuel switching. Hence, high energy prices are needed to bring down emissions.

9.2.2.1.2 National studies

Figure 9.9 and Table 9.9 report the results of a number of non-U.S. OECD national studies. Annual GDP losses are plotted as a function of emission abatement, where reductions are measured relative to the model's baseline or business-as-usual emission path. Note that the story is quite similar to that for the U.S. The costs of emission abatement tend to be positive, but there is considerable variation in the magnitude of the losses.

Some of the differences are to be expected, particularly when they are intercountry in nature. Nations differ widely in indigenous resources, supply infrastructure, and energy-use

patterns. Those with plentiful natural gas supplies, large existing investments in carbon-free alternatives, and/or low energy use per unit of economic activity should see less growth in carbon emissions. Hence, less carbon will have to be removed from the energy system to meet a specific target.

Intercountry differences do not explain, however, the wide variety of results for an individual country. Here, as in the case of the U.S., we must turn to the models and their underlying assumptions. Unfortunately, sorting out the reasons for the disagreement is more difficult. Unlike the EMF exercise, these studies were conducted independently. There was no opportunity to standardize key inputs. Nevertheless, based on the discussions of the previous sections, it is possible to identify several probable causes for the lack of consensus.

The national studies encompass the period in which we are apt to be most constrained on the supply side, namely, the next ten to fifteen years. During this time, the fuel switching capability will be mostly restricted to switching among fossil fuels. Consequently, much of the burden for the emission reductions will fall on the demand side of the energy sector.

The potential for end-use improvements remains one of the more controversial issues in the greenhouse debate. The wide range of opinion is reflected in the models. Those which are optimistic about the prospects for price- and nonprice-induced conservation show relatively low abatement costs. Conversely, the more pessimistic models tend to report considerably higher losses.

Assumptions relating to what is going on in other countries are also important. In the absence of an explicit model of global world trade,[2] the common practice is to adopt one of two hypotheses: (1) unilateral initiatives – which place an upper bound on trade losses, or (2) initiatives coordinated at an

Table 9.9. *Non-U.S. OECD CO_2 abatement cost modelling studies*

Country	Author (year)	Key[a]	CO_2 Reduction from Baseline (%)	Cost/GDP Reduction from Baseline (%)
Australia	Dixon *et al.* (1989)	AD (2005)	47	2.4
Australia	Industry Commission (1991)	AICG (2005)	40	0.8
Australia	Marks *et al.* (1990)	AM (2005)	44	1.5[b]
Belgium	Proost & van Regemorter (1990)	BP (2010)	28	1.8
EC	DRI (1992)	ECDRI (2005)	12	0.8
Finland	Christensen (1991)	FC (2010)	23, 21	6.9, 4.8[c]
France	Hermès-Midas (1992: Karadeloglou)	FHM (2005)	11	0.7
Germany	Hermès-Midas (1992: Karadeloglou)	GHM (2005)	13	1.3
Italy	Hermès-Midas (1992: Karadeloglou)	IHM (2005)	13	1.9
Japan	Ban (1991)	JB (2000)	18, 18	0.4, 1.7[d]
Japan	Goto (1991)	JG (2000, 2010, 2030)	23, 41, 66	0.2, 0.8, 1
Japan	Nagata *et al.* (1991)	JN (2005)	26	4.9
Japan	Yamaji (1990)	JYC (2005)	36	6[e]
Netherlands	NEPP (1989)	NEN (2010)	25, 25	4.2, 0.6[f]
Norway	Bye, Bye, & Lorentson (1989)	NB (2000)	16	1.5[g]
Norway	Glomsrod *et al.* (1990)	NG (2010)	26	2.7
Sweden	Bergman (1990)	SWB (2000)	10, 20, 30, 40, 51	0, 1.4, 2.6, 3.9, 5.6
UK	Barker (1993)	UKB (2005)	12	-0.2, +0.4[h]
UK	Barker & Lewney (1991)	UKBL (2005)	32	0
UK	Sondheimer (1990)	UKS (2000)	4	-0.5
UK	Hermès-Midas (1992: Karadeloglou)	UKHM (2005)	7	1.9

[a]The letters in the key refer to the country and author.

[b]Study combines technology with macroeconomic assessment of GDP impact.

[c]Unilateral action and global action.

[d]Tax case and regulation case.

[e]Values of both 5 and 6% have been given.

[f]National policy scenario and global policy scenario.

[g]GDP costs for OECD range from 1 to 2%.

[h]GDP gain when OECD tax levied with VAT reduced to maintain revenue neutrality; GDP loss when tax used to reduce the public sector borrowing requirement.

OECD or European Community level – which place a lower bound on trade losses. The impact of a tax at the national level will also vary depending on industrial composition (particularly, the existence of large energy-intensive industries) and a country's dependence on international trade.

How a model handles preexisting energy taxes is also important. Many countries rely on energy taxes as a major source of revenue. If a country already has high energy taxes, it will require a larger carbon tax to achieve a given percentage increase in retail energy price levels. This means that the deadweight losses are apt to be higher. Unfortunately, models differ widely in their treatment of preexisting taxes, adding yet another reason why results are apt to differ. (This was not an issue in our review of U.S. analyses since U.S. energy taxes are much lower than those elsewhere in the OECD. They are more like user fees, barely covering the costs of highway construction and maintenance.)

Still another reason why estimates of GDP losses differ so widely relates to the revenue recycling issue. The national models embrace an even wider variation of recycling options than those reported for the U.S. Indeed, some modellers assume that the tax revenues are removed entirely from the

economy, likening an imposition of a carbon tax to the oil import price shocks of the 1970s. As one would expect, this leads to much higher GDP losses than would be the case if the revenues were used to stimulate investments or consumption elsewhere in the economy. In the next section, we look at the issue of revenue recycling in some detail.

9.2.2.1.3. The effects of revenue recycling on GDP and employment

It has recently been advocated that, independently of their intrinsic environmental merits, carbon or energy taxes could be used to reduce nonwage labour costs and thus increase employment in Europe. Many European countries finance not only their public administration but also their health system, social security, and teaching system by raising funds from taxes levied directly or indirectly on wages. This way of meeting these financial requirements raises total labour costs and causes structural unemployment.

In some countries, taxes on labour represent nearly half of total hourly labour costs. These taxes create a wedge between what an employer must pay for an hour of labour and what a worker receives, and thus tend to reduce both labour supply

Table 9.10. *Effects of different types of tax revenue recycling at the community level: GDP losses in the last year (energy tax of 10$/bbl)*

	HERMES[a] 4 countries (%)	QUEST[b] EUR 12 (1995) (%)	DRI[c] EUR 11 (2005) (%)	HERMES-LINK[d] 6 countries (2001) (%)
Without recycling (revenue raising)				
GDP losses in the last year	-1.6	-1.2		
Employment	-0.9	-0.4		
Payroll tax cuts (employer only)				
GDP losses in the last year	-0.2	-0.7		+0.15
Employment	+0.3	-0.3		+0.64
Personal income tax cut				
GDP losses in the last year	-0.3	-1.1		
Employment	0	0		
VAT cut with mixed policy			-0.36[e]	

[a]Rate of change after seven years with reference to the baseline (Donni *et al.*, 1993).
[b]Rate of change after seven years (DG II-CEC, 1992).
[c]Tax revenue is shared between reduction in personal income tax (30%), employers' social security taxes (30%), reduction in corporate taxes (10%), and incentives for energy conservation (30%) (DRI, 1994).
[d]Bureau du Plan-Erasme (1993).
[e]Calculation from results of the average reduction of the rate of growth.

and the demand for workers. There is considerable evidence to suggest that this tax wedge, in combination with existing legal or contractual minimum wage arrangements, entails particularly detrimental employment effects for low-skilled or unskilled labour. It is not surprising, therefore, that the revenue recycling debate in Europe has tended to focus on shifting the tax burden from labour to energy.

Table 9.10 compares the results from several studies which suggest that the GDP losses of an energy tax could be at least partially offset through effective revenue recycling that reduces particularly burdensome taxes. Where the agreement among these studies breaks down is in whether recycling through appropriate cuts in other taxes can actually increase employment. This type of recycling can be compared either with a lump-sum recycling measure or with the baseline scenario. The first comparison aims at isolating the gains due to corrections of fiscal distortions from the specific dividend yielded by the carbon/energy tax; the second comparison assesses the overall result of the reform and assumes implicitly that concern about climate change facilitates its acceptance. Some studies suggest that the negative impact on employment and growth is only slightly reduced, whereas others suggest that an energy tax recycled through a payroll tax cut will increase employment.

The differences among modelling results can be explained in part by choice of methodology. Because of the focus on short- to medium-term time horizons, most studies for Europe have been carried out using neo-Keynesian models. Some studies employed general equilibrium models to examine the period 2000–2005: namely, Glomsrod *et al.* (1990) for Nor-

way, Proost and van Regemorter (1991) for Belgium, Bergman (1990) for Sweden, and Conrad and Schröder (1991) for Germany. In these studies, carbon taxes were generally used to maintain a given level of CO_2 emissions, assuming lump-sum recycling. The results are not surprising. The loss in GDP is relatively low for reductions of up to 10-20% from the baseline (below 0.2% of GDP) but rises dramatically as the target is increased. The reason for this is that these studies are pessimistic about the tax level required to achieve a 25-30% reduction from baseline emissions in 2000–2005 (e.g., 380 $/t in Belgium, 100 $/t in Norway, 195 $/t in Sweden). GDP reductions approximating 2.5% are found in some studies. It is important to note, however, that such studies typically ignore the effects of correcting distortions in the existing tax system.

Neo-Keynesian models embed more rigidities and allow for unemployment in the short run in response to shocks to the economy and for structural unemployment due to inadequate demand for labour. Recycling through payroll tax reductions can increase demand for labour and reduce both cyclical and structural unemployment. In both types of models, other factors may also be responsible for increased employment. For example, energy taxes may cause labour to be substituted for energy and technical progress to be biassed to save energy and use labour.

Many studies have been carried out at a national level, and it is impossible to analyze them in detail. Beyond the specifics of the national context, comparison is made difficult by the variety of ways in which the models handle preexisting energy and nonenergy taxes and in which a carbon or energy tax is levied. Nevertheless, some useful insights can be derived.

Table 9.11. *Differences in the macroeffects of the energy tax between EC countries in 2001 or 2005 (difference in % from the baseline)*

		FRG	F	UK	IT	NL	B	EUR6
GDP	Hermès[a]	0.22	0.06	-0.72	0.72	-0.16	0.57	0.15
	DRI[b]	-0.26	-0.13	-0.52	-0.39	-0.39	-0.65	
Private consumption	Hermès[a]	0.27	0.03	-0.57	0.75	0.34	0.23	0.15
	DRI[b]	-0.78	-0.39	-0.78	-1.04	-1.18	-0.78	
Employment	Hermès[a]	0.79	0.44	0.56	0.79	0.30	0.88	0.64
	DRI[b]	-0.39	-0.13	-0.39	-0.39	-0.26	-0.39	
Inflation	Hermès[a]	0.45	0.80	2.12	0.86	0.80	0.20	0.95
	DRI[b]	1.96	1.30	2.22	2.10	2.10	2.00	

[a]Recycling by reduction of social charges in HERMES Model (1993). The reference year is 2001.
[b]Mixed policy test in DRI Model (1994). The tax revenue is recycled as follows: 30% for reduction of employers' social charges, 30% for reduction of personal direct taxes, 10% for reduction of corporate taxes, 30% for incentives to energy conservation. The reference year is 2005.
Note: FRG = Germany; F = France; UK = United Kingdom; IT = Italy; NL = Netherlands; B = Belgium; EUR6 = the six European countries listed.

For example, a study by Erasme originally placed the costs of a carbon and energy tax in France at 1.5% of GDP in 2000, assuming lump-sum recycling of tax revenues (Beaumais, 1992). In a subsequent study, applying a complex mix of recycling measures (decreased payroll taxes, incentives to energy efficiency), the same team calculated GDP gains (of 0.5 to 0.93%) from an identical tax (Godard and Beaumais, 1993). In the same way, connecting a bottom-up approach with a macroeconometric long-term model, Walz et al. (1994) link a 40% CO_2 reduction up to 2020 in Germany with an increase in GDP of between 0.2 and 0.7%.

In the English context, Barker (1994) and Barker, Baylis, and Bryden (1994) point out that results differ according to the product on which the excise tax is levied. Indeed, they show that a road duty increase in the UK is likely to reduce inflation (when the revenues are recycled via reductions in employment taxes) and raise employment.

Difficulties in comparative analysis are less important when the same measure is analyzed, which was the case with studies of a proposal for limiting CO_2 emissions put forth by the European Commission (1992). This proposal, which called for levying new taxes on most major sources of energy, was perhaps the most widely discussed emission-control initiative in Europe. The taxes would be based partly on carbon and partly on energy content. Using oil as a point of reference, the tax would level off at $10 per barrel in the year 2000. Nuclear energy and large-scale hydroelectric projects would be taxed at a lower rate – approximately 50% of that of carbon-intensive fuels.

Several studies have attempted to quantify the macroeconomic impacts of the EC proposal. Table 9.10 reports results from the HERMES, QUEST, DRI, and HERMES-LINK models. It compares a case without tax recycling with different recycling schemes. In all cases lump-sum recycling entails net costs. The payroll tax cut is demonstrated to be more efficient than a personal income tax cut in the European context. It is interesting that recent HERMES-LINK (Bureau du Plan-Erasme, 1993) and QUEST (European Commission, 1994)

simulation results arrive at a net positive impact. The employment effects tend to be particularly favourable when the energy taxes are simultaneously introduced by several countries and when the compensatory reduction in social security contributions is targeted to the low-skilled.

A more detailed picture of effects on GDP, private consumption, employment, and inflation is given in Table 9.11, which reports results based on the HERMES and DRI models for six countries. Note that in the case of HERMES, the carbon/energy tax would result in a net positive effect on employment for all the countries and a slight increase in GDP for four of them. In the case of the DRI model the results are less optimistic and do not yield a double dividend that would totally offset the costs of a carbon tax; this is due in part to differences in model structure. Nevertheless, the comparison captures the likely range of impacts.

9.2.2.1.4 The effectiveness of energy and carbon taxes
If the taxes were restricted to carbon-based fuels, the impacts would be unambiguous. Carbon taxes reduce CO_2 emissions through both their effects on energy consumption and fuel choice. However, the EC proposal represents a departure from this pure case by not only aiming to limit CO_2 emissions but also to achieve other objectives such as energy efficiency in general. Coupling an energy tax to a carbon tax will lead to additional conservation, but it will also have an impact on fuel choice. By raising the price of carbon-free substitutes, the proposal will reduce incentives for fuel switching.

Several analysts have looked at this issue and found that extending a carbon tax to other forms of energy reduces the effectiveness of measures to reduce CO_2 emissions because taxing noncarbon-based fuels provides a disincentive for fuel switching. Table 9.12 compares the results from four such studies.

These analyses could be criticized from a bottom-up perspective because they rely on econometric assumptions and neglect the role of accompanying measures apt to accelerate the adoption of energy-efficient equipment. Moreover, such

Table 9.12. *Comparison of emission reductions from an energy tax and a carbon tax*

Study	Geographical Scope	Horizon	Carbon Tax (%)	Energy Tax (%)
Global 2100 (Tax: 10 $/bbl)	Europe	2030	-50	-46.5
Hermès-Midas (Tax: 10 $/bbl)	France	2005	-10.83	-9.27
	Germany	2005	-12.95	-7.12
	UK	2005	-6.55	-4.40
	Italy	2005	-12.74	-3.10
Melodie (Tax: 200 $/t)	France	2010	-15.7	-12.6
Midas (Tax: 10 $/bbl)	Germany	2005	-11.63	-7.52
	UK	2005	-3.98	-3.29

stylized model simulations neglect important institutional characteristics of the European energy market (for example, natural gas prices are effectively linked to mineral oil prices, as there is no gas-to-gas competition, and the decision to build nuclear power stations is significantly influenced by political choices). They therefore tend to overstate the likely actual difference between carbon and energy taxes.

These analyses, however, provide important insights into the impact of other environmental concerns on CO_2 abatement and illuminate the importance of the trade-off between reductions of CO_2 emissions and broader energy efficiency goals. This loss in terms of CO_2 reduction seems more important in Germany (between 35 and 49%) and in the UK (32.5% for the HERMES-MIDAS model) than in France (20%), where the shift of electric supply towards nuclear energy has been completed during the past two decades.

9.2.2.2 A review of bottom-up studies

This section principally summarizes and compares the results of major coordinated bottom-up studies for Europe. It also accounts for studies in other regions such as the Pacific area, Australia, and New Zealand. Studies for other countries are less available. The main differences in results are discussed in relation to different input and baseline scenario assumptions and to modelling methodology. The studies considered are:

- The CEC DG XII JOULE (1991a, b) "Cost-effectiveness analysis of CO_2 reduction options" studies (sometimes also called the Crash Programme), covering nine European Union countries and carried out by the CO-HERENCE research team and a group of other European teams

- The United Nations Environment Programme (UNEP) International Project for Sustainable Energy Paths: country studies for Denmark, France, and the Netherlands (UNEP, 1994b)

- The five-country International Project on Sustainable Energy Paths (IPSEP, 1993)

- The nine-country Energy Technology Systems Analysis Project (ETSAP) (Kram, 1993)

- Johansson *et al.* (1989) on Sweden

- The Australian Bureau of Agricultural Resource Economics (ABARE) MENSA study (Naughten *et al.*, 1994) on Australia

- The German Parliamentary Enquete Commission (Enquete Kommission des deutschen Bundestages, 1991; BMU, 1993)

9.2.2.2.1 Methodology of the studies

The CEC study adopted a uniform modelling methodology, the EFOM linear optimization model developed by the research programme of the European Community. In the UNEP studies, a common methodological framework was applied across countries, comprising uniform assumptions on technology input data, time horizons, and cost definitions. The Netherlands study used the MARKAL linear optimization model, whereas the French study relied on a bottom-up simulation framework combined with a static general equilibrium model. The Danish study used an integrated energy system simulation model.

The IPSEP studies defined a common methodological framework and uniform assumptions like those of the UNEP studies. The customized scenario modelling in the IPSEP project involved a combination of detailed assessment of potential, cost, and behaviour related to energy end-use demand and supply options. The participating countries were the UK, the Netherlands, Italy, Germany, and France.

The ETSAP project was a comparable assessment for nine countries using the MARKAL model with consistent cost and technology assumptions. The participating countries were Belgium, Italy, the Netherlands, Norway, Sweden, and Switzerland, together with non-European countries (the U.S., Japan, and the provinces of Ontario and Quebec in Canada). Extension of this MARKAL-based comparison to Australia was undertaken independently (Dickson *et al.*, 1994; Naughten *et al.*, 1994), using the MENSA model of the Australian Bureau of Agricultural and Resource Economics. An innovation in this study was the incorporation of greenhouse gases other than CO_2 that are emitted by the energy system.

The study by Johansson and Bodlund did not use any formalized modelling framework but focussed on the assessment of technologies and their emission reduction potential.

Table 9.13. *Zero average net cost emission reduction potential in non-U.S. OECD bottom-up studies (estimated reduction potential achieved with zero net average cost)*

Study	Country	2005/10 (%)	2015/20 (%)	2025/30 (%)
IPSEP (1993)	EC-5		>26-58	>60
FRG Enquete (1992)	FRG	30		
Mills *et al.* (1991)	Sweden	>35		
Crash Programme (1991)	UK	10		
UNEP (1994a,b)	Denmark	>21		>45
Range		10-35	>26–58	>45–60
Median		>17	>42	>53

Note: A "greater than" sign indicates that the study did not include options up to the intersection point of the x-axis. This means that larger reductions could be achieved for negative or zero average cost. The Mills *et al.* study uses the same input assumptions as the Johansson *et al.* study of 1989.

9.2.2.2.2 Main objectives and definitions

The primary aims of the European bottom-up studies have been to demonstrate the existence of an energy savings potential with negative or low economic cost and to analyze the cost-effectiveness of energy supply and demand options for achieving a certain emission-reduction goal. An important outcome of such a cost-effectiveness analysis is to enable intercountry comparison of reduction costs.

The study results are highly dependent on the assumptions about energy efficiency measures included in the baseline scenario. The CEC report and the ETSAP project assumed that all profitable efficiency measures were included in the baseline, which meant that all emission reduction options, by definition, would have positive costs in these analyses. However, in the MARKAL-based study, the baseline was designed to exclude energy efficiency devices and policy responses, both in road passenger transport (Naughten *et al.,* 1993) and in the residential sector (Naughten *et al.,* 1994; Naughten and Dickson, 1995).

In the UNEP studies, Denmark, the Netherlands, and France followed different principles for baseline definition. Denmark defined its baseline to include the persistence of some major inefficiencies, implying that a large potential for no-regrets emission reduction options was found. The French study accounted for transaction costs for adopting more efficient technologies, and the Dutch study assumed the baseline to be efficient.

The IPSEP and Johansson *et al.* studies included major inefficiencies in their baselines and found, consequently, a significant potential for no-regrets options.

9.2.2.2.3 The results

Despite differences in baseline definition, all the studies demonstrated the existence of an economic potential for the reduction of CO_2 emissions at negative or low costs. Table 9.13, for example, shows the zero-cost emission reduction potential as estimated by five of these studies for three different time horizons. At the same time, the results demonstrated that the basic assumptions have an important bearing on the existence and possible implementation of no-regret options. This can have a significant effect on the total, average, and marginal costs of reductions.

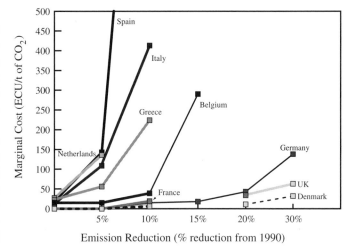

Figure 9.10: Marginal costs of emission reductions in 2010 relative to 1990 emissions in the Crash Programme study.

Table 9.14. *Marginal costs for CO_2 emission reductions from 1988 levels in 2010 as estimated by the Crash Programme study (ECU/t of CO_2)*

	Stabilization	5%	10%	15%	20%	30%
Belgium	15	15	39	290	—	—
Denmark	0	0	6	—	11	32
France	0	0	1	—	43	—
Germany	0	0	15	18	43	138
Greece	23	56	224	—	—	—
Italy	22	112	413	—	—	—
Netherlands	27	136	—	—	—	—
Spain	18	143	1,556	—	—	—
United Kingdom	0	0	12	—	35	63

The main results of the European CO_2 Crash Programme (CEC 1991a, b) study are shown in Figure 9.10 and Table 9.14. The study found that it is less costly for the UK, Germany, and Denmark to reduce emissions from current levels than for Greece and Spain. The study estimated neglible costs

Table 9.15. *Main categories of CO_2 reduction options and related marginal abatement costs for Denmark (reductions in relation to baseline in 2030)*

CO_2 Reduction Option	Reduction (%)	Marginal Abatement Cost (U.S. $/t CO_2)
Connection to natural gas network	0.5	-222.7
Connection to district heating network	2.0	-100.0
Electricity conservation in households	4.8	-70.0
Increased use of combined heat and power	1.5	-62.9
Electricity conservation in services	2.8	-55.7
Conservation in industry	1.6	-40.0
Conservation in agriculture	0.4	-38.6
Combined cycle – natural gas (1,000 MW)	3.8	-8.6
Biogasification (1,500 MW)	11.6	-4.3
Wind turbines (1,800 MW)	3.0	0.0
Central natural gas comb. (4,000 MW)	2.6	1.0
Decentralized biomass combustion	9.4	12.9
Solar collectors for hot water	0.3	30.0
Insulation in office buildings	0.7	81.4
Photovoltaics (800 MW)	2.9	185.7
Fuel cells (1,000 MW)	1.0	614.3
Total CO_2 reductions	48.9	

associated with 10% reductions in emissions from present levels for the former three EC members. For the latter two, even to keep emissions constant at today's levels would require substantial investment in energy conservation schemes or investment in nonfossil fuel technologies.

Baseline CO_2 emissions range from a drop of 3% by 2010 in Luxembourg, to stabilization for Germany, to increases of 72 and 86% for Greece and Portugal respectively (all from current levels). Some of this variation can certainly be explained by intercountry variations in industrial structure, capital stock constraints, and different national energy systems. However, the national baseline projections and cost assessments still show larger differences than would be expected from "real" differences in energy systems and economic development patterns.

One of the main explanations behind the widely varying national cost assessments is the heterogeneity of the national technology data used in the analyses. Differences in economic growth, for instance, for Spain and Greece, could partly explain the high marginal emission reduction cost in these countries but cannot fully explain all intercountry differences. It seems surprising, for example, that France, with an energy system characterized by a large installed nuclear capacity, arrives at an abatement cost very similar to that of Britain and Germany, whereas Italy arrives at a lower cost. This is in contrast to the fact that the scope for manoeuvre in the power sector is much higher in the latter three countries, owing to the dominance of conventional power production, than in France.

The CEC study also assessed the potential for negative cost energy efficiency improvements for the countries. This was done by the construction of a MURE scenario in which the end-use savings measures were embedded. It was estimated that the implementation of these negative cost options would lead to a 6% decrease in CO_2 emissions in 2010 from the 1988

level for Germany, stabilization in Denmark, and increases of 52 and 82% for Greece and Portugal respectively.

The CEC study was a pioneering study in the development of the methodology of comparative cost assessment between countries and produced important methodological lessons on the requirements for uniformity in national scenario and technology assumptions. This experience has induced CEC to organize a homogeneous new data collection on efficient energy technologies.

Differences in assessed reduction potential and costs are also seen in the UNEP studies for Denmark, the Netherlands, and France. These differences are consistent with "real" variations between the countries in the carbon content of their energy systems and in baseline definitions. Denmark and, to a lesser degree, the Netherlands have high emission-reduction potentials as a consequence of carbon-intensive power production systems. In the French country study, certain more costly transportation sector measures are included because reductions in electricity supply are limited and end-use demand options are assumed to have positive costs.

The Danish study assumed that the emission reduction case included further profitable efficiency improvements and energy end-use savings compared with baseline development. These measures required introduction of legislation in the form of norms, efficiency standards, and obligatory connection to already established natural gas and district heating grids, which at present have a low utilization rate. Other options with positive, but low costs were wind turbines and biomass fuels. The Danish options are listed in Table 9.15.

The study for the Netherlands assumed that all possible efficiency improvements in the energy system would be implemented in the reference case. A consequence of the "efficient" definition of the baseline case for the far-reaching reduction targets includes energy end-use savings amounting to 3-4% followed by the use of costly advanced technologies such as

Table 9.16. *Main categories of emission reduction options in the UNEP study for the Netherlands*

Sector	CO$_2$ Reduction Option
Industry	Fuel cell combined heat and power More efficient industrial processes Gas-fired and electric heat pumps Solar heated and biomass- and hydrogen-fired equipment
Transport	More efficient engines, brake energy recuperation Biofuel, hydrogen, electric, liquefied petroleum gas, compressed natural gas, and methanol engines
Households and services	Condensing gas boiler Gas-fired and electric heat pumps Improved building insulation More efficient electric and lighting appliances Solar boilers, biomass- and hydrogen-fired equipment District heating Greenhouse insulation
Electricity and heat production	Gas combined cycle and advanced coal combined cycle plants Fuel cell industrial cogeneration and district heat Electricity storage, hydrogen fuel cells CO$_2$ removal at gas-fired and coal-fired plants Wind turbines, biomass power plants, solar photovoltaic
Other processes	Methanol production from natural gas and coal Hydrogen production from natural gas and imports from Sahara Liquid and solid biofuel production (from sugar beet, wheat, straw, wood) Combined methanol and electricity production Transport of CO$_2$ to depleted natural gas fields and to aquifers

Note: The nature of the energy system optimization model used in the Netherlands study does not allow percentage reductions and costs to be assigned to individual options. The options here are part of an integrated package of technologies that leads to a reduction in CO$_2$ emissions of 50% from baseline in 2030 at a total cost of U.S. $3.8 billion.

CO$_2$-removal technologies and the importation of hydrogen from the Sahara. The emission reduction options for the Netherlands are shown in Table 9.16.

The French study estimated a tax rate for emission reductions in a macroeconomic model. In one scenario variant, it was assumed that carbon taxes would be combined with complementary measures in the energy sector, such as information, efficiency standards, research and development, and grants, as well as with measures in other sectors. The tax was estimated to be $187.5 per tonne of CO$_2$ for a 12.5% reduction in 2005, $75 per tonne of CO$_2$ for a 25% reduction in 2030,

and $1,425 per tonne of CO$_2$ for a 50% reduction in 2030. The lower tax rates in 2030 compared with 2005 are due to assumptions relating to technical progress.

Despite the fact that the UNEP country studies followed a common methodological framework (UNEP 1994a, b, c,) the results still show differences as a consequence of different baseline approaches and emission-reduction technology assumptions. This means that the comparison of national results requires a detailed and transparent documentation of assumptions and national modelling frameworks and that the marginal emission reduction costs cannot be compared directly between the countries.

Yet another approach to country differences is taken in the IPSEP study (IPSEP, 1994/1995). Though the analysis started from baseline data and demand growth projections unique to each of the five countries, emphasis was placed on the aggregate carbon reduction potentials and costs in the five-country region. This approach reflects current plans for a common internal market within the European Union, including a homogenization of the policy frameworks for the energy sector.

The costs and potentials for conventional fossil supplies, nuclear power, demand-side efficiency, cogeneration, and renewables were developed in several substudies. The costs of efficiency resources include estimates of administrative costs developed from a review of market transformation policies (standards, financial incentives to consumers, financial incentives to manufacturers of efficiency equipment, and utility regulatory reforms) in Europe and the U.S. High/low sensitivity ranges were modelled not only for fuel prices but also for technology and administrative programme costs. In the behavioural module of the analysis, assumptions about implementation rates were validated on the basis of empirical investigations of consumer responses to market transformation policies (standards, financial incentives to consumers, financial incentives to manufacturers of efficiency equipment, and utility regulatory reforms). The effectiveness of these policies was expressed in terms of the fraction of the resource potentials assumed to be mobilized by 2020. Scenario results were then given for 25%, 50%, 75%, and 100% effectiveness.

Figure 9.11 shows results for the least cost (*a*) and minimum risk (*b*) cases for the five countries combined, based on four permutations of cost assumptions. Because of feedback effects, a combination of low fuel prices and low technology costs is considered most plausible. For this combination, emissions drop by 40% relative to 1985 in the least-cost case, and energy service costs drop by 23% relative to the CEC reference case. In the minimum risk case, the same combination of cost assumptions results in emissions dropping by 60%, while costs decline 20%.

The country-to-country variations of the IPSEP study are smaller than in other studies. The large no-regrets efficiency potentials found in all five countries make the differences on the supply side relatively less important.

Although the Swedish exercise of 1989 included a detailed assessment of end-use demand options similarly to the IPSEP study, a higher abatement cost was estimated. This is because Swedish CO$_2$ emissions are already very low as a result of the present use of hydro power and nuclear power. In the future these two supply sources will be, respectively, supplemented and

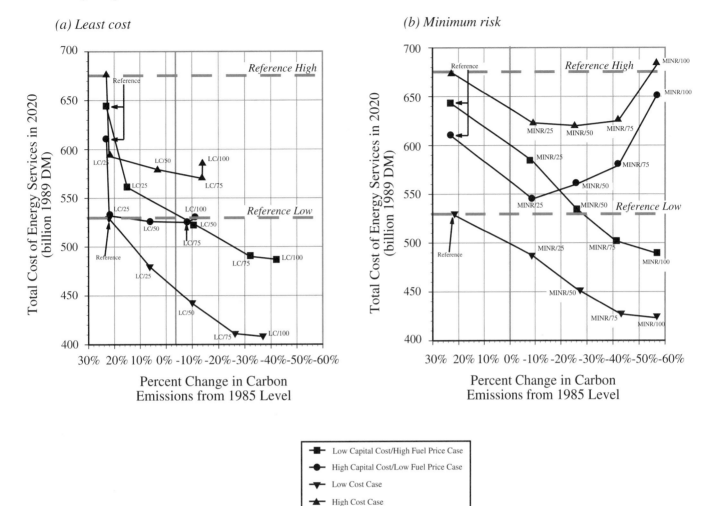

(a) Least cost

(b) Minimum risk

Low Capital Cost/High Fuel Price Case

High Capital Cost/Low Fuel Price Case

Low Cost Case

High Cost Case

1985 EC-5 Carbon emissions = 532 MtC

Source: Krause and Koomey (1992).

Figure 9.11: Sensitivity analysis of the cost of carbon reductions in the IPSEP EC-5 study, all sectors.

substituted for by fossil fuels because of the implementation of regulations concerned with environmental problems other than CO_2 emissions.

The results of the ETSAP study are illustrated in Figure 9.12, showing the marginal costs in the year 2020 of successive carbon emission reductions from the 1990 level. There are wide variations among the countries. Norway, Sweden, and Switzerland – which now emit the least CO_2 per capita – would be among those measuring the highest cost to achieve a specific further reduction in CO_2 emissions. The Netherlands, with the assumed availability of CO_2 removal and storage options, appears able to make the greatest reductions at the least cost.

In addition to the results of these comparative exercises, it is worth recalling that one of the largest studies of energy end-use sectors, the German Enquete Commission study (involving 150 studies by 50 institutes), identified a zero cost potential of 16.5% in Germany compared with 1981 levels (Enquete Kommission, 1991; BMU, 1993).

Time dependence of abatement costs. In all assessed European bottom-up exercises, the slope of marginal cost curves, notably after a certain level of reduction of CO_2 emissions, is highly dependent on the package of technologies considered.

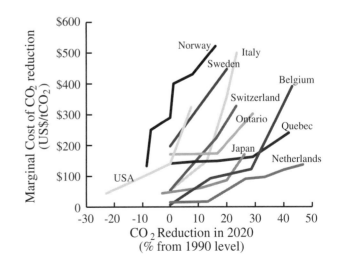

Source: Kram (1993).

Figure 9.12: Marginal costs of CO_2 emission reductions in 2020 in the ETSAP study.

The slope may be very steep after a certain level: For instance, in the Danish UNEP study, abatement costs increase from 10 $/t to 200 $/t if the 2030 reduction goal is switched from 25 to 50% (baseline reference). The introduction of more technologies into the analysis, however, tends to extend the segment of the cost curve where marginal costs increase slowly. This tendency has also been seen when short-term reduction targets are compared with long-term targets. The marginal emission reduction cost decreases with time for similar reduction targets (from baseline) in the Danish and Dutch UNEP studies, simply because more technologies are available.

9.2.2.2.4 *The dependence of the results on input assumptions and modelling approach*

Though the choice of models in bottom-up analysis may have some effect on the calculated carbon reduction costs, these results cannot explain the large differences in findings between the CEC, UNEP, IPSEP, and ETSAP studies. Differences appear to be explained mainly by two factors: the assumptions in some studies that the reference case is already more or less economically efficient, and the use of a limited number of possible efficiency improvements.

More systematic research on the importance of technology assumptions for the results of bottom-up studies would be useful. Optimization models may encourage the use of more historically estimated aggregate data on end-use demand, and in some cases this may imply a more pessimistic assessment of the savings potential. However, studies have also shown that when identical data on efficiency potentials are used, optimization and simulation results appear to converge. Also, optimization models are being continuously refined. A large contribution is currently being made to provide a range of standardized algorithms for modelling alternative policy programmes.

Overall, the divergent results suggest the need for further empirical research on how end-use market efficiency could improve the implementation of end-use options. Also, comparative bottom-up modelling exercises should be done in which identical national data sets for efficiency potential and costs are combined with standardized assumptions about implemented fractions in the reference case. Such an exercise has already been done for Denmark. It was shown that uniform input assumptions in the EFOM model for Denmark and the Danish UNEP study produced very similar results (Morthorst, 1993).

The French study, conducted by CIRED using the NEXUS-IMACLIM model (Hourcade *et al.,* 1993), shows the sensitivity of the results to the assumptions about the industrial and socioeconomic content of the reference scenario. New urban policies could determine a change of fuel consumption growth between now and 2030 just as effectively as a huge carbon tax of 280 $/t.

9.2.3 *Studies of the costs of reducing CO$_2$ emissions in transitional economies*

The term *economies in transition* is often applied to the countries of the former Soviet Union and Eastern Europe. The implication is that market reforms are underway. The stage of the transition varies dramatically, however, by country, and there is no certainty about the types of economic institutions likely to emerge at the end of this process and the type of development pattern which will ultimately win out.

Modelling emission reduction costs for the formerly planned economies is particularly challenging. The region has a long history of highly subsidized energy prices and other inefficiencies in the structure of incentives. Models that assume the existence of market mechanisms will miss many of the important features of these economies. Applying equilibrium models to planned economies assumes that planned economies mimic the behaviour of market economies. This assumption is impossible to accept empirically, particularly with regard to energy use.

The problem is further complicated by the fact that the transition process *per se* generates a state of crisis. The Russian GDP, for example, declined by 18.5% in 1992. Economic activity in 1993 was lower than in 1985 (Bashmakov, 1994). Consequently, standard econometric techniques derived from historical records cannot be applied without caution. Until these economies achieve a more stable footing, predictions of future growth will be tenuous at best.

9.2.3.1 *A review of top-down studies*

The problems of adapting top-down models for use in transitional economies explains the dearth of applications for this region of the world, particularly when compared to the growing body of analysis being amassed on developed economies. The few studies that do exist were conducted by Western economists, using models that are best suited to application within the OECD. This is not to imply that the studies are without merit. Indeed, as we will see shortly, they have yielded some useful insights about the costs of emission reduction. Nevertheless, as we move beyond the OECD, we need to be increasingly circumspect in the interpretation of results.

A key issue in extending top-down models to emerging economies is the treatment of energy subsidies. Many of the early studies showed relatively high emission reduction costs (see, for example, Dean and Hoeller, 1992). Critics claim that the high numbers reflect "the difficulties top-down studies have with economies undergoing restructuring" (Grubb *et al.,* 1993). They argue that the high costs are in large part due to a failure to adequately capture the demand-side impacts of subsidy removal.

There is some validity to this criticism. The removal of subsidies will result in costless conservation in addition to that induced by the AEEI factor. The early studies may have indeed underestimated the potential for demand-side savings. (It is important to note, however that the criticism applies more to the choice of input assumptions than to the models themselves. The models are not inherently limited to scenarios in which markets clear.)

The importance of existing market distortions is underscored in a modelling experiment by Manne and Schratten-holzer (1993) using the Global 2100 model. They examined the sensitivity of emission reduction costs for emerging

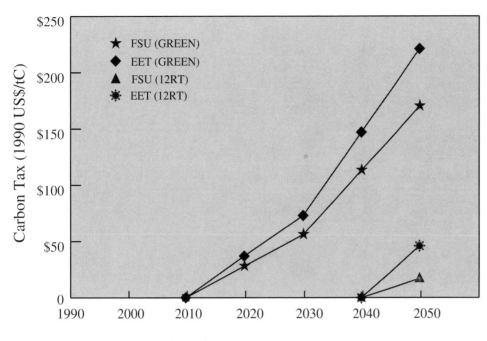

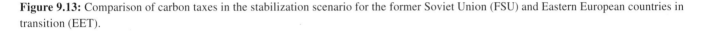

Source: Manne and Oliveira-Martins (1994).

Figure 9.13: Comparison of carbon taxes in the stabilization scenario for the former Soviet Union (FSU) and Eastern European countries in transition (EET).

economies to the presumed size of the subsidy. They found that the removal of artificial barriers that insulate domestic consumers from international price movements may well lead to a drop in carbon emissions over the next several decades.

The same result was confirmed in a subsequent study by Manne and Oliveira-Martins (1994). They used the GREEN and 12RT models to estimate the taxes required to stabilize emissions at 1990 levels. It is important to note that price reforms and the removal of energy subsidies were included in the "baseline" scenario. This enabled the modellers to separate the impacts of restructuring from those of a carbon constraint.

For purposes of the analysis, the transition economies were divided into two regions: the countries of the former Soviet Union (FSU) and the Eastern European countries in transition (EET). Figure 9.13 displays the carbon taxes required to achieve stabilization in each region. The former Soviet Union and Eastern European countries have zero taxes before 2010. This is because the stabilization constraint is nonbinding. With price reforms and the removal of energy subsidies, there is a decline in emissions relative to 1990 levels. For the GREEN model, the emission constraint begins to bite shortly after 2010. For 12RT, emissions remain below the 1990 level until 2040.

Despite the intrinsic difficulties of applying top-down models to the countries of this region, the implications of these analyses are important. The costs of stabilizing emissions may be relatively small for economies in transition. Market failures have undoubtedly limited investment in energy efficiency. The potential for cost-effective reductions in energy use is apt to be considerable. The key is a successful transition towards a new institutional context in which, under whatever form, prices actually reflect the full cost of energy.

9.2.3.2 *A review of bottom-up studies*

This section summarizes selected studies of the potential and cost of carbon emission reduction strategies in the post-planned economies. Bottom-up methods offer some advantages for dealing with transitional economies in that they focus on the physical stock of equipment and apply scenarios for its evolution, with less concern for anticipating macroeconomic equilibrium conditions. However, just as top-down models are constrained by lack of reliable data for these economies, the data on current equipment end-use energy efficiencies are severely limited.

9.2.3.2.1 *Summary of studies*

A comparison of emission reduction cost studies reveals striking differences with the results of some top-down models (see Table 9.17). On the bottom-up side, a number of studies have been completed by Eastern European and Russian experts since the transitions of 1989. Some of this work was conducted using the U.S. EPA End-Use Energy Model, whereas others applied indigenous country-specific models. The EPA model projected future energy demand to the year 2030 in five-year increments, giving results for the major fuel types and future aggregate industrial energy intensity. The model estimates energy demand on the basis of economic growth, structural change, price response, and technical energy-efficiency improvements not attributed to price response. Like that of Makarov and Bashmakov (1991), most of the bottom-up studies incorporated economic restructuring as a driving force for determining the energy content of economic growth.

A relationship can be seen between the magnitude of the emission reductions and costs in the above scenarios, but this factor cannot explain the significant variation in cost esti-

Table 9.17. *Cost of carbon dioxide emission reductions*

Country	Study	Type[†]	Forecast Year	CO$_2$ Reduction from Baseline (%)	Cost of Reduction (% of GNP)
Former Soviet Union	Burniaux *et al.* (1992)	TD	Bur-2020	45	0.9
	Burniaux *et al.* (1992)	TD	Bur-2050	70	2.3
	Burniaux *et al.* (1992)	TD	Bur-2100	88	3.7
	Kononov (1993)	BU	Kon-2005	50	0.3
	Makarov and Bashmakov (1991)	BU	Mak-2005	23	0.5
	Makarov and Bashmakov (1991)	BU	Mak-2020	44	1
	Manne (1992)	TD	Man-2020	45	3.1
	Manne (1992)	TD	Man-2050	70	6.4
	Manne (1992)	TD	Man-2100	88	5.6
	Oliveira-Martins *et al.* (1992b)	TD	OM-2020	45	1.7
	Oliveira-Martins *et al.* (1992b)	TD	OM-2020	70	3.7
	Rutherford (1992)	TD	Rut-2020	45	1.5
	Rutherford (1992)	TD	Rut-2050	70	5.8
	Rutherford (1992)	TD	Rut-2100	88	4.1
Hungary	Jaszay (1990)	BU	Jas-2005	17	-0.1
Poland	Leach & Nowak (1991)	BU	L&N-2005	37	-0.1
	Leach & Nowak (1991)	BU	L&N-2005	53	-0.1
	Sitnicki *et al.* (1991)	BU	Sit-2005	44	0
	Sitnicki *et al.* (1991)	BU	Sit-2030	62	0.3
	Radwanski *et al.* (1993)	TD	Rad-2010	27	0
	Radwanski *et al.* (1993)	TD	Rad-2030	39	0
Czechoslovakia	Kostalova *et al.* (1991)	BU	Kos-2005	20	0
	Kostalova *et al.* (1991)	BU	Kos-2030	29	0

[†]TD indicates a top-down study and BU indicates a bottom-up study.

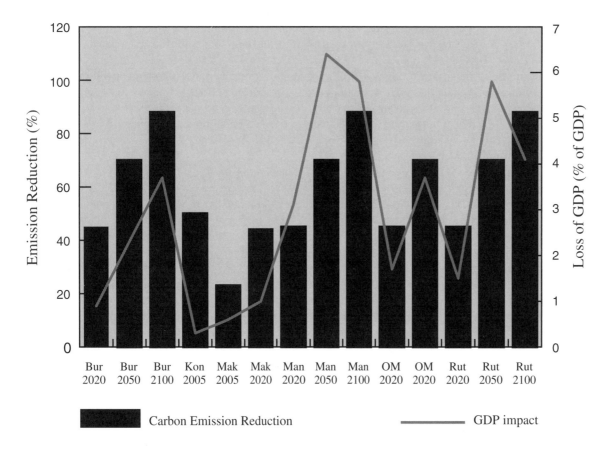

Note: See Table 9.17 for key to studies.

Figure 9.14: Cost studies: Emission reduction in the former Soviet Union.

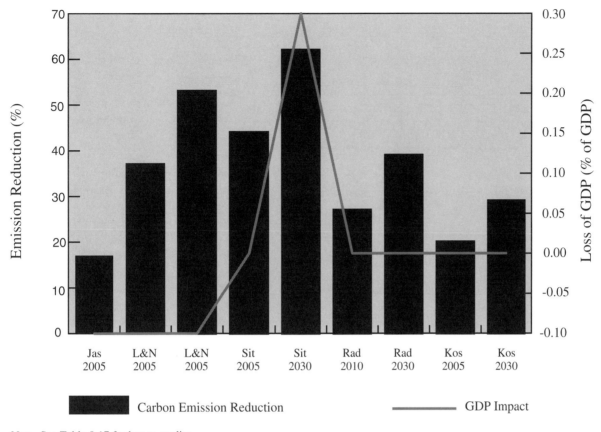

Note: See Table 9.17 for key to studies.

Figure 9.15: Cost studies: Emission reduction in central Europe.

mates across the scenarios. And although the discount rate varied among the studies, there seems to be little correlation between this variation and variations in cost estimates. Indeed, the one major difference in cost estimates in the studies reviewed appears to be that of methodology, specifically the choice of top-down versus bottom-up models (see Figures 9.14 and 9.15, and Table 9.17). This result is not surprising, given the special nature of planned and postplanned economies.

9.2.3.2.2 Evaluation of cost estimates

The most energy-intensive economies in the world are the planned or postplanned economies (Chandler *et al.,* 1990). However, a short digression on the meaning and importance of energy intensity in the formerly planned economies may be in order. Energy intensity (energy consumed per unit of economic output) is an important, if controversial, indicator of future emissions. This value is difficult to measure accurately because GNP is difficult to quantify in comparable units. Also, differences in economic structures may be significant.

If energy intensity is interpreted simply as a surrogate for energy efficiency, then the measure can be misleading. However, a vast literature has verified that the planned economies use energy inefficiently (Baron, 1992), both because they use technology that is not optimized and because the structures of their economies – that is, the remarkable reliance on heavy industry – is not economically justified. Yet, most models of future energy use in this region imply both continued high

energy intensity and high economic growth. The results are sometimes startling when one realizes that they indicate that the formerly planned economies would be twice as energy intensive in the year 2050 as Japan is today. This outcome may be unlikely, given the constraints of international competition.

9.2.4 Studies of the costs of reducing CO_2 emissions in developing countries

Climate analysts have long recognized that greenhouse gas emissions and consequent climate change are a global phenomenon and need global responses. Yet in most global abatement studies, the analysis for developing nations is usually piggybacked on the developed country studies. The fact that the developed nations at present account for nearly two-thirds of greenhouse gas emissions has continued to overshadow the results of the very same models that show that within two or three decades emissions from the developing nations will surpass those from the present developed nations. Greenhouse gas abatement studies for developing countries are therefore not numerous. The relatively few that do exist are based on methodologies designed for use in developed nations. These models tend to treat the dynamics of developing nations as a caricature of industrialized economies.

9.2.4.1 A review of top-down studies

This tendency is particularly marked in the case of top-down models, where the dynamics of each geographical region

Table 9.18. *Comparative results from five models for carbon emission and GDP losses for China and the world*

Model[†]	Average Annual Carbon Emission Growth (%) BAU Scenario		GDP Loss in Year 2100 over BAU (%) Scenario I		Scenario II		Major Policy Option for Carbon Emission Abatement
	World	China	World	China	World	China	
CRTM	1.6	2.5	4.0	3.5	3.5	5.0	—
ERM	1.2	2.5	7.0	6.0	4.5	13.0	Fuel substitution
GREEN	2.0	3.8	3.6	1.5	3.0	5.5	Energy conservation
IEA	2.0	3.2	—	—	—	—	Energy conservation
MR	1.7	2.2	5.2	5.0	4.5	5.0	Fuel substitution

[†]Terminal years for ERM, GREEN, and IEA models are 2095, 2050, and 2005 respectively, and 2100 for the other two models.

tends to mirror the market-based economies of the Western world. The models are market-driven and assume the existence of future markets, perfect information, competitive economic dynamics on the demand and supply side, and optimizing behaviour on the part of producers, consumers, and government. These assumptions are often found to be invalid in developing countries.

In addition, top-down models tend to underestimate the contributions from the informal sectors of developing economies. In many countries, these sectors account for the overwhelming proportion of agriculture and land-use activities, employment, and household energy consumption. Activities in these sectors, such as traditional biomass use, deforestation, rice cultivation, and animal husbandry, account for significant greenhouse gas emissions. The models also have difficulty capturing the patterns of change whereby the development process transforms the traditional activities into modern activities through a myriad of simultaneous processes such as monetization, market development, technology penetration, institution building, and education.

Despite these limitations, top-down models can still provide useful insights into a number of important issues. These include taxes and subsidies, revenue recycling, international trade, allocations for research and development, and backstop technologies. But until more effort is devoted to broadening the frameworks to deal better with the unique characteristics of developing countries, one must be especially cautious in interpreting their results. With these caveats in mind, we now turn to a review of several recent applications.

9.2.4.1.1 The costs of emission reduction in China
Of the developing countries, China has received the greatest attention from top-down modellers. This is not surprising, since it is currently the third-largest emitter of CO_2 and there is every expectation that its share will continue to increase. Because of its importance, global models usually include China as a single geographical region. For example, the OECD Model Comparison Project divided the globe into five geopolitical regions: the U.S., other OECD countries, the former Soviet Union, China, and ROW (a catch-all category for the rest of the world). This study provides some interesting

perspectives on how China's emissions are likely to evolve in the future and the potential costs of emission reduction.

Besides business as usual (BAU), the following two scenarios were examined. Scenario I postulated a reduction in the rate of growth of emissions in each region by 2% per annum, and Scenario II a stabilization of emissions at 1990 levels in each region. Table 9.18 summarizes the percent growth rate of carbon emissions for BAU. In addition, it shows GDP losses for China and the world under the two control scenarios.

Note that all models report very large GDP losses for China under the stabilization scenario. This is because the emissions growth under business as usual is much higher for China compared to the world as a whole. Hence, China requires especially large emission cuts. In Scenario I, absolute cuts in emissions will be required in the industrialized countries, whereas a low growth in emissions is allowed for China. Hence, the GDP losses for China (in percentage terms) are somewhat lower. Although figures are not reported for ROW due to the heterogeneous nature of this region, the study indicates that the GDP losses are also very high for ROW under both scenarios. Among other things, the OECD study underscores the importance of the emission allocation scheme in determining costs to individual countries. This issue will be addressed in greater detail in the section on global models.

9.2.4.1.2 The impacts of energy subsidy removal
One area in which macroeconomic models are particularly useful relates to the removal of market imperfections. In a recent study, the OECD used the GREEN model to analyze the impact of subsidy removal in China and India (OECD, 1994). The "distortion removal" scenario assumes phasing out of subsidies on the sale price of oil by 2000 and on coal and gas by the year 2010. Under the "no distortion removal" scenario, the energy subsidies observed in the base year continue.

The study finds that the removal of energy subsidies has a major impact in reducing energy consumption and carbon emissions. Under the no distortion removal scenario, energy consumption and carbon emissions increase by a factor of 14 between 1985 and 2050 in both countries. Under the distortion removal scenario, energy consumption and emissions in China and India are reduced to nearly 40 and 60% respec-

Table 9.19. *Bottom-up studies of developing countries*

Energy and Environmental Division, LBL (1991)
CO_2 *Emissions from Developing Countries.* India, Indonesia, China, Argentina, Brazil, Mexico, Venezuela. High and low emission scenarios, 1985–2025. Country-specific reductions between 13 and 54% in 2025. Energy system model, STAIRS.

Asian Development Bank (1993)
National Response Strategy for Global Climate Change: People's Republic of China. Business as usual with high and low economic growth and policy scenario, 1990–2050. CO_2 reduction of 23% from baseline in 2050. Energy system model combined with macroeconomic assessment.

UNEP Collaborating Centre on Energy and Environment (1994)
UNEP Greenhouse Gas Abatement Costing Studies. Brazil, India, Egypt, Senegal, Thailand, Venezuela, and Zimbabwe. Reference and abatement scenario. Emission reduction targets for 2005/10 and 2020/30 covering 12.5 to 50% reductions from baseline. Construction of abatement cost curves using energy system models.

Davidson (1993)
Carbon Abatement Potential in Western Africa. Ghana, Sierra Leone, and Nigeria. High and low emission scenarios, 1985–2025. Reductions from baseline in 2025: Ghana 38%, Sierra Leone 25%, Nigeria 24%. Energy system model, STAIRS.

La Rovere, Legey, and Miguez (1994)
Alternative Energy Strategies for Abatement of Carbon Emissions in Brazil. Three alternative economic growth scenarios, abatement scenario using high economic growth scenario as baseline. Reductions in 2010: 8 and 50% from baseline; reductions in 2025: 5 and 50% from baseline. Input-output model used to forecast energy requirements, scenarios taking Japan, Spain, and Colombia as focal points. Energy system analysis.

Asian Energy Institute (1992)
Collaborative Study on Strategies to Limit CO_2 Emissions in Asia and Brazil. Brazil, China, India. The Islamic Republic of Iran, South Korea, and Thailand. Emission reduction scenarios for 1988/90 to 2000/2015. Calculation of investment cost for individual abatement technologies.

SEI/Greenpeace (1993)
Towards a Fossil Free Energy Future. Global study including regional assessment for Africa, centrally planned Asia, Latin America, and South and East Asia. Reduction scenario 1988–2100. Global reductions from 1988 level: 50% in 2030, 100% in 2100. Energy system model and macroeconomic assessment of the impact of carbon taxes on final energy prices.

---

tively in 2050 (relative to the no distortion removal case). The reductions are due to higher energy efficiency and the use of backstops promoted by higher market prices for carbon-based fuels. Moreover, a Chinese study using a World Bank model suggested that joint reforms of energy prices and the exchange rate can offset the effects of energy price increases on the exchange rate, thus reducing the cost of imports (Peng and Hanslow, 1993).

9.2.4.1.3 The impacts of carbon taxes on noncommercial energy consumption
Even if this conclusion relies on incontestable price mechanisms, however, the informal economy could offset part of the expected gain. That was pointed out by, among others, researchers from the OPEC Secretariat who used a macroeconomic model to investigate the impacts of OECD-type carbon taxes on developing countries (Walker and Birol, 1992) and explicitly considered the impact of income and price levels on the consumption of non-commercial energies (NCE).

The study finds that the income elasticity of NCE is negative, meaning that consumers shift from NCE to commercial energy as income rises. But more important, the study also reports an inverse relationship between noncommercial and commercial energy demand in response to price changes. That is, consumers switch from commercial to NCE in the case of price hikes. Both these results point to behaviour that is obvi-

ous and well known in the developing nations and yet is most often forgotten or ignored in top-down analyses. Based on the above findings, the study concludes that in the absence of effective controls on deforestation, carbon taxes would create incentives to deplete forests for energy use.

9.2.4.2 A review of bottom-up studies
Most bottom-up studies of developing countries focus on the national level and typically cover scenario periods from 1988/90 to 2020/30. This contrasts with top-down studies, which tend to treat developing countries in groups, with the exception of China. Some of the principal studies are listed in Table 9.19.

Emission reduction targets in developing-country bottom-up studies typically formulate reduction targets, either as percentage changes from a baseline (or reference) case reflecting the reference case's economic development and energy requirements (UNEP, 1994a) or as percentage changes related to a high emission case for the energy sector (Sathaye and Goldman, 1991). This contrasts with most of the studies carried out for developed countries, where emission reductions targets generally are calculated for each scenario in terms of stabilization or abatement from a given benchmark date.

Two major multicountry studies have been conducted by the Lawrence Berkeley Laboratory (LBL) and by the UNEP Collaborating Centre on Energy and Environment (UCCEE).

Table 9.20. *Growth rates of population, GDP, primary energy consumption, and CO_2 emissions in reference and abatement scenarios for 1990 to 2020/30 (in percentages)*

Country	Population	GDP	Primary Energy		CO_2 Emissions	
			Reference	Abatement	Reference	Abatement
Argentina[a]	1.1	2.0	1.8	1.1	1.8	1.1
Brazil	1.4	4.7	3.5	3.7	5.3	3.2
Brazil[a]	1.2	3.2	2.5	1.0	2.8	0.8
China[b]	0.8	6.0	3.1	2.7	2.6	2.1
Egypt	1.7	5.0	3.2	1.4	3.8	1.4
Ghana[c]	2.9	4.0	3.6	3.0	5.4	5.0
India	1.8	6.3	4.2	3.8	3.8	2.9
India[a]	2.0	5.0	4.6	4.3	4.6	4.3
Indonesia[a]	1.2	3.0	3.6	3.2	4.0	3.5
South Korea[d]	0.4	5.0		3.1		2.8
Mexico[a]	1.7	4.4	2.7	2.1	2.8	1.9
Nigeria[c]	3.0	4.0	2.6	1.9	4.6	3.8
Senegal	3.0	3.2	3.2	2.4	3.5	2.6
Sierra Leone[c]	2.3	3.0	2.2	1.7	3.5	3.0
Thailand	0.9	4.8	4.8	4.2	5.5	4.6
Thailand[d]	2.0	8.0		7.8		7.8
Venezuela	2.1	3.8	2.5	1.9	3.1	2.1
Venezuela[a]	1.7	4.0	3.4	3.2	2.5	1.9
Zimbabwe	2.4	4.2	2.8	1.9	3.2	1.5
Latin America[e]	1.5	3.3	2.7		3.2	
Latin America[f]	1.3	3.2		0.9		-0.6
Southeast Asia[e]	1.4	4.6	3.9		5.1	
Southeast Asia[f]	1.6	4.1		1.5		-0.2

[a]Energy and Environmental Division, LBL (1991).

[b]Asian Development Bank (1993).

[c]Davidson (1993).

[d]Asian Energy Institute (1992).

[e]IPCC (1992a).

[f]SEI/Greenpeace (1993).

Note: Unreferenced studies were part of the UNEP greenhouse gas abatement costing project, UNEP (1994).

These two research centres facilitate comparative analysis in a way similar to that employed for the studies carried out by the Energy Modeling Forum in the U.S. The results and methodological framework of the UNEP and LBL multicountry studies are reviewed in this section along with similar country studies for China, West Africa, and South East Asia.

Only some of the studies have estimated the costs of achieving emission reduction. These are the UNEP studies (UNEP, 1994a), the LBL study for India (Mongia, 1991), and the ADB study for China (Asian Development Bank, 1993). The UNEP studies, for example, have estimated emission reduction costs for a range of target reductions running from 12.5 to 25% in 2005/10 and from 25 to 50% in 2020/30 (UNEP 1994a, b).

The coordinated country study programmes conducted by UNEP and LBL defined an analytical framework for mitigation analysis comprising uniform assumptions about analytical structure that allow the use of different national modelling tools. The recommended framework for each country consists of the following analytical steps:

- Construction of reference scenario
- Assessment and ranking of greenhouse gas reduction options
- Construction of greenhouse gas reduction scenarios
- Macroeconomic impact assessment (to the extent possible)

The studies have in this way tried to combine traditional elements from bottom-up models with macroeconomic assessment. In practice this has meant that the baseline has been constructed to reflect general macroeconomic and energy system development trends. Similarly, after mitigation options have been assessed, the most important macroeconomic impacts of implementing specific options have been considered either qualitatively or quantitatively.

9.2.4.2.1 Main assumptions and findings: Trends and margins of freedom for curbing greenhouse gas emissions
The country studies considered here exhibit a striking similarity to the expected development trends in energy and carbon intensity for the economies of developing nations. This has

Table 9.21. *Key elasticities in reference and abatement scenarios for 1990 to 2020/30*

Country	Reference			Abatement		
	Energy/GDP	CO_2/Energy	CO_2/G	Energy/GDP	CO_2/Energy	CO_2/GD
Argentina[a]	0.9	1.0	0.9	0.6	1.0	0.5
Brazil	0.8	1.5	1.1	0.8	0.9	0.7
Brazil[a]	0.8	1.1	0.9	0.3	0.8	0.3
China[b]	0.5	0.8	0.4	0.5	0.8	0.4
Egypt	0.6	1.2	0.8	0.3	1.0	0.3
Ghana[c]	0.9	1.5	1.4	0.8	1.7	1.3
India	0.7	0.9	0.6	0.6	0.8	0.5
India[a]	0.9	1.0	0.9	0.9	1.0	0.9
Indonesia[a]	1.2	1.1	1.3	1.1	1.1	1.2
South Korea[d]				0.6	0.9	0.6
Mexico[a]	0.6	1.0	0.6	0.5	0.9	0.4
Nigeria[c]	0.7	1.7	1.1	0.5	2.0	1.0
Senegal	1.0	1.1	1.1	0.8	1.1	0.8
Sierra Leone[c]	0.7	1.6	1.2	0.6	1.8	1.0
Thailand	1.0	1.2	1.1	0.9	1.1	1.0
Thailand[d]				1.0	1.0	1.0
Venezuela	0.7	1.3	0.8	0.5	1.1	0.6
Venezuela[a]	0.9	0.7	0.6	0.8	0.6	0.5
Zimbabwe	0.7	1.2	0.8	0.4	0.8	0.4
Latin America[e]	0.8	1.2	1.0			
Latin America[f]				0.3	-0.7	-0.2
Southeast Asia[e]	0.8	1.3	1.1			
Southeast Asia[f]				0.4	-0.1	0.0

[a]Energy and Environmental Division, LBL (1991).
[b]Asian Development Bank (1993).
[c]Davidson (1993).
[d]Asian Energy Institute (1992).
[e]IPCC (1992a).
[f]SEI/Greenpeace (1993).
Note: Unreferenced studies were part of the UNEP greenhouse gas abatement costing project, UNEP (1994).

important implications for CO_2 reduction potential and related costs in these countries. The studies indicate a tendency for the energy intensity of economic growth to decrease, primarily as a consequence of structural economic change and technological development. In contrast, however, the studies generally expect the CO_2 intensity of primary energy consumption to increase more or less as a direct outcome of the introduction of commercial fossil fuels as a major source of energy supply in future baseline development. The overall result is that the CO_2/GDP intensity is close to unity for most of the studies. The policy implication of these trends is that the studies indicate a potential for implementing low-carbon energy technologies, but a special effort seems to be necessary to reverse the strong tendency for fossil fuel energy technologies to become the major source of supply in the future.

The actual assumptions for GDP growth rates, primary energy consumption, and CO_2 emissions are shown in Table 9.20 for the reference and emission reduction scenarios of the country studies. The annual economic growth rates range from 3.0 to 8.0%, with most projections below 5.0%.

In some cases economic growth rate forecasts for the same country differ between studies. Generally the growth rates of

the UNEP studies are higher than for the LBL studies. This is understandable, given the uncertainties surrounding such long-term forecasts.

The elasticities of primary energy to GDP, CO_2 emissions to primary energy, and CO_2 emissions to GDP are shown in Table 9.21 for the reference or baseline scenarios. Many of the countries have a primary energy/GDP elasticity between 0.6 and 0.8, implying that a 1% increase in GDP leads to a 0.6-0.8% increase in energy consumption. This value, which is low if one considers current trends in developing countries, is explained in the individual countries in part by national economic development plans that include structural changes and technical efficiency improvements. However, the CO_2/GDP elasticity is close to unity for most of the country studies as a consequence of high expected CO_2/energy elasticities. This latter elasticity reflects development processes in which traditional local biomass and also hydropower resources are "squeezed out" while commercial fossil fuels are expected to play a larger role. A remarkably high CO_2/energy elasticity is projected in the UNEP Brazil study and in the studies by Davidson for Ghana, Nigeria, and Sierra Leone, where biomass is substantially replaced by fossil fuels in the reference scenario.

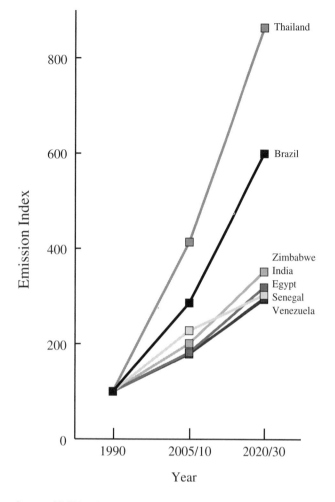

Source: UNEP (1994a).

Figure 9.16: UNEP baseline projections.

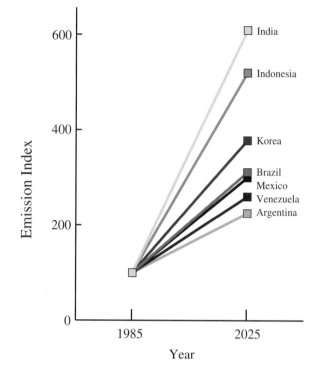

Source: Davidson (1993) and Sathaye and Goldman (1991).

Figure 9.17: LBL baseline projections.

The different assumptions for development in GDP, primary energy, and CO_2 emissions in the baseline case imply that the imposition of reduction targets related to future emissions has quite different consequences for the absolute level of emissions in the reduction scenarios.

The baseline emission projections range from a six- to eightfold increase for Thailand and Brazil to a two- to threefold increase for other developing countries in the UNEP study. In the LBL study, the projection is for a five- to sixfold increase for Indonesia and India and a two- to threefold increase for China, Argentina, Brazil, Mexico, and Venezuela. The emission projections of the UNEP and LBL studies are illustrated in Figures 9.16 and 9.17.

We later discuss the implications of such a wide range of results. Despite these discrepancies, however, the studies converge in suggesting a steep increase in CO_2 emissions from developing countries over the next decades. A second important convergence relates to the estimate of a significant potential for reducing the CO_2 intensity of economic growth in these countries.

In the UNEP study, the emission reduction scenario for most of the countries shows a 40-50% emission reduction from the reference scenario by 2020/30. However, this still

implies up to a doubling of emissions, relative to the present. Exceptions are the studies for Thailand and Brazil, where 30 and 50% reductions from the reference scenario still lead to a sixfold and threefold emission increase respectively from the present.

The emissions develop more slowly in the low emission scenarios of the LBL study than in the UNEP study, even though reductions in the scenarios are less extensive than in the UNEP studies. The LBL studies are typically focussed around a 20-30% reduction from the reference scenario in 2025. This reflects the lower economic growth rate projections of the LBL study. The lowest projections are seen for Brazil and Argentina, where the emissions are projected to increase by only about 50% from present levels. Another group of countries, including China, Mexico, and Venezuela, is projected to have emission levels two to three times higher than present levels after reductions of 20-30% from the baseline scenario. India and Indonesia are projected to have persistently high emission levels.

The reduction scenarios of the UNEP and LBL studies are shown in Figures 9.18 and 9.19.

Although these potentials for CO_2 emission reduction in developing countries are insufficient to offset the overall tendency towards increased CO_2 emissions, they are large enough to contradict the notion that, given current low levels of energy consumption in these countries, there is little utility in efforts to control their emissions.

A similar conclusion can be drawn from a study by the Asian Energy Institute that included Bangladesh, Brazil, China, India, the Islamic Republic of Iran, Japan, South Korea, and Thailand. The study assessed the potential for emis-

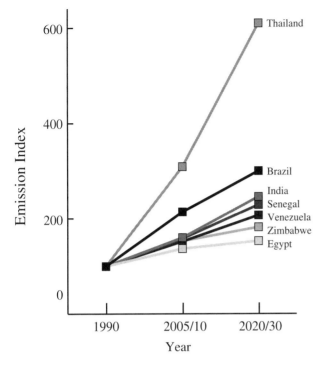

Source: UNEP (1994a).

Figure 9.18: UNEP reduction estimates.

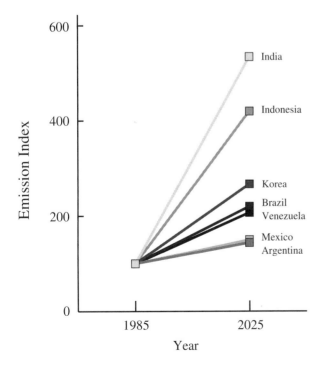

Source: Davidson (1993) and Sathaye and Goldman (1991).

Figure 9.19: LBL reduction estimates.

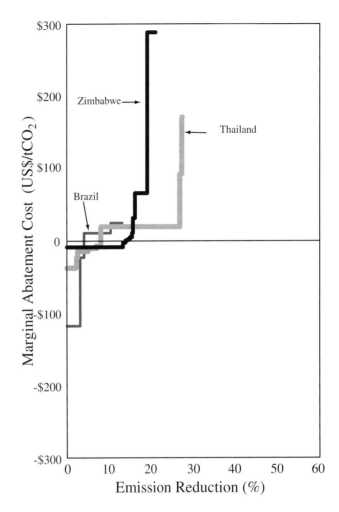

Figure 9.20: UNEP studies: Marginal abatement costs for 2005/2010.

sion reduction in these countries through energy efficiency improvements and a shift towards supply technologies with lower carbon intensity.

9.2.4.2.2 Emission reduction costs

The UNEP country studies include an estimation of costs for emission reductions of between 12.5 and 25% from the baseline scenario in the short term (2005/10) and between 25 and 50% in the long term (2020/30). Cost is defined as financial cost at the energy sector level including investment, operation and maintenance, and fuel costs (UNEP, 1994c). Cost curves, containing a number of target reductions in one year, should be regarded as snapshot pictures of the levelized cost of achieving a given reduction in that year. Figures 9.20 and 9.21 show the cost curves for a number of participating countries for both the short-term and long-term target years.

The short-term cost curves for Brazil, Thailand, and Zimbabwe (Figure 9.20) exhibit a number of similarities. A particular feature is the large potential for negative-cost or low-cost emission reduction options of up to 10-15% from baseline emissions. The curves are also similar in shape. The first part of the cost curves, up to about the 5% reduction level, indicates very low emission reduction costs. It is followed by a long interval up to about the 25% level in which emission

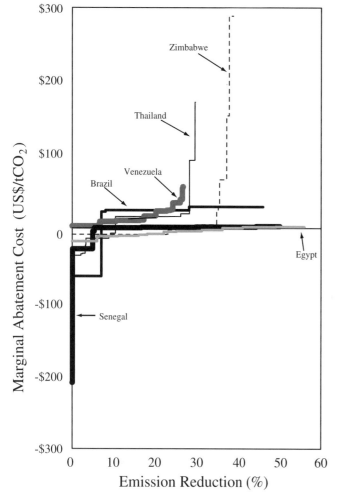

Figure 9.21: UNEP studies: Marginal abatement costs for 2020/2030.

and +$37 per tonne of CO_2. Total emission reduction costs are negative for Egypt and Senegal.

The study of India (Mongia *et al.,* 1991) estimated emission reduction costs for two emission reduction scenarios, one reflecting a least-cost emission reduction case and the other a "capital-constraint" case with more emphasis on domestic resources. The least-cost case requires foreign capital and resources. The other emission reduction scenario assumes a more expensive supply system with greater reliance on domestic renewable energy sources. The first emission reduction case estimates average emission reduction costs in 2025 to be -$9.5 per tonne of CO_2 for a 45% reduction from baseline. The second case, based on domestic resources, is significantly more costly for achieving the same level of emission reduction.

9.2.4.2.3 The special case of China

Because of the expected magnitude of its future CO_2 emissions, China warrants separate treatment. The Asian Development Bank has, in collaboration with the Chinese government, conducted a comprehensive technology assessment study for the period 1990 to 2050 (ADB 1993). The study assumes very low energy/GDP and CO_2/energy elasticities due to expectations of structural economic development away from heavy energy-intensive industries and fast technological change. The baseline furthermore assumes a decreasing share for coal in primary energy consumption and includes the implementation of large efficiency improvements, especially in the power production sector.

The GDP growth rate over the period is assumed to be 6.0%, primary energy growth 3.1%, and the CO_2 emission growth rate 2.6%. The resulting elasticities are 0.5 for energy/GDP, 0.8 for CO_2/energy, and 0.4 for CO_2/GDP. The share of coal in the primary energy supply is projected to decrease from 75.5% in 1990 to 58.0% in 2050. Coal is replaced by an increasing amount of natural gas and nuclear power, as well as some hydro power.

It is assumed that efficiency gains in power production of 3.6% annually can be achieved until 2000, decreasing to 1.8% for the rest of the period to 2050. These high savings are related to the current low power-production efficiency of 30% combined with the expected high investment level in new efficient electricity generation technologies.

The reduction scenario results in a 23% emission reduction from baseline in 2050. The main reduction options in the supply system are expanded nuclear capacity, increasing from 9.9% of primary energy supply in the baseline to 13.9% in the reduction case. The share of hydro power and natural gas is also increased slightly. Total primary energy consumption is decreased by 14% from the baseline level.

9.2.4.2.4 Comparing numerical results: Critical assumptions about technical options

The country studies have estimated emission reduction potential as a function of technical options, either related to individual technologies or to comprehensive packages of technologies. The costs and emission reduction potential of technical options have been assessed using energy system

reduction costs fall within a relatively narrow range between -$10 and +$30 marginal cost per tonne of CO_2 reduced. Few higher cost options were included in the analysis. Those that were included tend to be expensive supply-side options, implying a steep increase in marginal costs.

The long-term marginal cost curves shown in Figure 9.21 show many similarities for the developing countries. Venezuela, with a completely positive cost curve, may be regarded as an exception. This is the result of the methodological decision to include most no-regret options in the baseline.

Senegal and Thailand have a negative marginal cost potential up to about a 15% reduction from baseline emissions in 2020/30. This potential is expanded to about the 40% reduction level for Egypt and to about the 30% reduction level for Zimbabwe. The marginal cost curve for Brazil cuts the horizontal axis around the 10% reduction level. As with the short-term curves, the long-term marginal cost for most countries falls within an interval of -$10 to +$25 per tonne of CO_2 for emission reductions of about 5-25%.

Table 9.22 shows marginal, average, and total emission reduction costs for the short- and long-term reduction targets. In the long term, the average emission reduction costs for reductions from baseline of between 25 and 50% range between -$5

Table 9.22. *Marginal (MAC), average (AAC), and total abatement costs (TAC) for maximum reductions and comparison with GDP*

Country	Year	Baseline Emissions (Mt CO_2)	Reduction (Mt CO_2)	Reduction (%)	MAC[†] ($/t CO_2)	TAC ($mil.)	AAC ($/t CO_2)	GDP ($bn.)	TAC/GDP (%)
Short term									
Brazil	2010	767.8	101.3	13.2	25	-1881	-18.6	848	-0.22
Thailand	2010	359.8	99.3	27.6	171	1027	10.3	250	0.41
Zimbabwe	2010	32.7	6.9	21.1	289	221	32.0	35	0.64
Long term									
Brazil	2025	1611.2	741.2	46.0	29	9339	12.6	1740	0.54
Egypt	2020	253.0	141.7	56.0	2	-732	-5.2	118	-0.62
Senegal	2020	15.4	7.7	50.0	3	-16	-2.0	14	-0.12
Thailand	2030	751.4	221.7	29.5	171	2089	9.4	538	0.39
Venezuela	2025	189.4	50.4	26.6	56	685	13.6	177	0.39
Zimbabwe	2030	57.4	22.2	38.7	289	205	9.3	72	0.28

[†]The marginal abatement cost (MAC) here corresponds to the level at the maximum reduction achieved for the country.

models that evaluate end-use efficiency improvements, fuel substitution, and new supply technologies in an integrated way. This analysis has been conducted with different degrees of sophistication, depending on the individual country and the particular study.

As an example, a summary of the main technical emission reduction options in the UNEP country studies is presented in Table 9.23. The national options are listed in aggregated form and thus represent classes of options rather than individual technologies. The cost curve may be considered in three segments, representing negative and low-cost options, intermediate-cost options, and high-cost options. One general similarity among the country studies is that the least expensive part of the cost curve contains energy end-use savings in households and/or industry. Another is that electricity supply options first appear in the intermediate-cost part of the reduction potential cost curves.

On the supply side, most of the studies focussed on traditional energy-supply technologies and few included more advanced technologies and/or renewable energy technologies. Consequently, the cost curve either increases very sharply or simply does not include any further reduction after the exhaustion of these options.

9.2.4.2.5 Comparability of the national studies

The emission reduction costing studies for developing countries considered here exhibit similarities with regard to the assessed potential for negative or low-cost emission reduction. In general, these options comprise end-use efficiency improvements, energy supply efficiency improvements, and the introduction of fuels with lower carbon intensity. These technologies cover the first and cheapest part of the emission reduction potential and can, especially in the longer term, be supplemented with renewable energy technologies, more far-reaching end-use savings, and advanced combustion technologies. The long-term emission reduction potential will consequently be extended and is also likely to be cheaper than currently estimated.

The individual country studies are difficult to compare quantitatively because of differences in methodological approach and in scenario assumptions for economic growth, energy requirements, and emission reduction costs. There is an important difference in the baseline scenario assumptions used in the different studies.

In the UNEP studies, the national research teams took official macroeconomic forecasts as the starting point for energy demand projections. In contrast, the LBL studies used a broader-level international perspective to estimate an economic structure and income distribution which could be achieved in a developing country within a given time horizon. Thus, for example, it could be assumed that a country like Brazil would approach an economic structure and income distribution comparable to that of Spain in a given time frame.

The advantage of using national macroeconomic projections is that they reflect national views on development. However, national forecasts may be only partially consistent and realistic, whereas international economic studies may be of help in establishing a consistent data set across countries. Studies using a common, well-documented background can also be easier to compare than studies that use different national forecasts.

The degree of optimism of experts from different countries with regard to the penetration of energy efficiency or of carbon-free energy supply options may differ dramatically. This can lead to different critical assumptions in the baseline as well as in the emission reduction scenarios. One approach is to assume that all possible efficiency improvements will be implemented as part of the baseline scenario, implying that only positive-cost options remain. If the existing energy system – as in many developing countries at present – is relatively inefficient, the above-mentioned approach implicitly assumes large investment programmes to implement the "efficient options" in parallel to any emission reduction effort. If, instead, it is assumed that major inefficiencies persist in the

Table 9.23. *Technical emission reduction options in the UNEP study*

Country	CO_2 Reduction Option	Reduction (%)	MAC (US$)
Brazil	Electricity savings (industry, services, and residential)	7	-66
	Solar uses in agriculture	1	22
	Fuelwood and charcoal for afforestation programmes	21	24
	Ethanol, bagasse, and electricity generation from bagasse	19	29
	Total CO_2 reduction	48	
Egypt	Fuel switching in households	6	-21
	Efficient industrial equipment and maintenance	10	-12
	Transportation	2	-12
	Heat recovery and new industrial processes	9	-8
	New raw materials	5	-3
	Efficient household appliances	5	-3
	Electricity generation	8	-1
	Efficient stoves	7	2
	Total CO_2 reduction	52	
Senegal	Early hydropower implementation	0.1	-210
	Agriculture intensification	5	-28
	Energy conservation in industry	0.4	-4
	Dissemination of improved stoves	11	0
	Improved carbonization efficiency	13	1
	Liquefied petroleum gas substitution for charcoal	15	2
	Biomass from afforestation	6	3
	Total CO_2 reduction	50	
Thailand	Efficient air conditioners	2	-36
	Electronic ballast	1	-27
	Compact fluorescent lamps (service sector)	6	-14
	Compact fluorescent lamps (residential sector)	2	-9
	Nuclear electricity	18	20
	Highly efficient gasoline cars	1	92
	Total CO_2 reduction	30	
Venezuela	Reduced flaring and leakage of methane	7	3
	Efficient boilers and kilns	10	10
	Freight transport	1	13
	Efficient electric motors in the industrial sector	2	17
	Passenger transport	4	21
	Electric sector	0.6	35
	Other energy savings in the industrial sector	2	39
	Efficient electric appliances	0.4	52
	Total CO_2 reduction	27	
Zimbabwe	Efficient boilers	23	-9
	Energy savings in the industrial sector	4	-2
	Efficient motors and power factor correction	2	-2
	Increased hydropower	5	5
	Efficient furnaces	2	66
	Central photovoltaic power	1	153
	Coal for ammonia	1	289
	Total CO_2 reduction	38	

energy system in the baseline scenario, there will be an inter-relationship between emission reduction measures and the general effort to overcome barriers for efficiency improvements in the energy system. In the UNEP study, for example, where country research teams were free to make judgments, the team for Venezuela assumed that all profitable efficiency improvements would be implemented in the reference scenario, whereas the teams for Brazil and Thailand assumed a relatively inefficient reference scenario.

Another difference in reference scenarios between countries relates to difficult assumptions about structural change in the economy, about fuel supply, and about the overall level of development. The striking difference between the UNEP and LBL studies in the case of Brazil is an enlightening example of the consequence of these differing assumptions. Part of the difference between the two studies for Brazil can be explained by different assumptions for the GDP growth rate, but another key difference is a consequence of the low fuel price increase projected in the UNEP study during a period long enough to make the existing alcohol fuels programme unprofitable. Consequently, the reference case assumes replacement of present biomass use, including ethanol, with fossil fuels. The LBL studies, in contrast, defined reference energy scenarios as a continuation of historical trends, implying that the alcohol fuels programme would be sustained in Brazil. A new Brazilian study (La Rovere *et al.*, 1994) has been carried out as a compromise between the assumptions of the UNEP and LBL studies.

9.2.4.2.6 Conclusion

The bottom-up CO_2 emission reduction costing studies carried out for the energy sector for developing countries exhibit, despite differences in methodological approach, some common empirical results, namely:

- The 30–40-year reference scenario projections show a tendency to decreasing energy/GDP intensity but increasing CO_2/energy intensity.

- The potential for a 30-40% emission reduction from baseline over a 40-year time frame has been estimated. However, even after such a reduction, emissions will, on average, be two or three times more than present levels, because of economic growth.

- The emission reduction potential includes low- or negative-cost options relating to end use and conventional supply technologies in the short to medium term. In the 30–40-year time frame, the UNEP country studies have estimated average emission reduction costs to be below $14/tonne of CO_2.

9.2.5 Global studies of the costs of reducing CO_2 emissions

The review of existing studies up to this point has focussed on country or regional analyses. In addition, there are a growing number of studies that attempt to provide a global perspective on the assessment of abatement costs. These studies are important for several reasons. The enhanced greenhouse effect is inherently a global issue. If significant reductions in emissions are required, they can be accomplished only through international accords and cooperation. It will be helpful to have some sense of the overall costs before confronting the difficult issue of burden sharing.

A global perspective is also important in assessing the costs to individual countries. Actions taken in one region are apt to have "spillover" effects into other regions. Partial equilibrium analyses ignore potentially significant linkages (e.g., trade in oil, gas, and carbon-intensive basic materials) that could substantially alter the economic impacts of a carbon constraint.

Finally, and perhaps most important, a global perspective is necessary if we are to identify economically efficient strategies for achieving emission targets. The Framework Convention on Climate Change states that "policies and measures to deal with climate change should be cost-effective so as to ensure global benefits at the lowest possible cost." This means

that emission reductions should be carried out where it is cheapest to do so. Analysis on a global scale is needed in order to construct a "least-cost" global abatement supply curve.

All the caveats expressed above about the limitations of country-specific or region-specific models should be borne in mind when evaluating the results of global studies. In addition, the global analyses confront aggregation issues which further complicate the interpretation of results. Not surprisingly, most of the global studies employ top-down methodologies. Nevertheless, there have been several bottom-up studies that illustrate the overall technical potential for curbing greenhouse emissions at a world level.

9.2.5.1 A review of top-down studies: The importance of international cooperation

Several of the models used for regional analyses provide the capability for analyzing emission abatement costs at a global level. These models typically divide the globe into five or more geopolitical groupings. By necessity, they are highly aggregate in their treatment of macroeconomic and technology issues. In order to represent regional differences and trade effects, local economic and technological detail must be sacrificed.

Figure 9.22 and Table 9.24 summarize the results of recent studies using these models. Not surprisingly, there is considerable disagreement concerning the costs of emission abatement. Consistent with the regional studies, the top-down global analyses indicate that emission abatement will involve positive costs, but the size of the model-based cost estimates varies from study to study. In this section, we will try to understand why these results differ so widely. In doing so, we hope to gain additional insights into the costs of emission abatement at the regional and global level.

9.2.5.1.1 The costs of stabilizing global emissions

A systematic comparison of global models was undertaken by the OECD in its 1992 Model Comparison Project (OECD, 1993). The study included all available global models with the capability of simulating regional carbon tax rates required to achieve specific emission abatement targets and the resulting output losses. The models were calibrated to the same set of input assumptions employed by the parallel study being conducted by the Energy Modeling Forum (see earlier discussion). Results were reported for a business-as-usual scenario and for four scenarios involving various levels of emission reduction. We begin by examining the results from the emission stabilization scenario.

For emission stabilization, it was assumed that global emissions would be permanently held at 1990 levels. The global costs of meeting such a target will depend, in part, upon how emission reductions are allocated among regions and whether there is scope for international cooperation. In the analysis that follows, each region is required to hold emissions at 1990 levels without the opportunity of shifting emission abatement from high to low marginal abatement cost regions. That is, there is no trade in carbon emission rights. Later on, we will explore the potential benefits from relaxing this constraint.

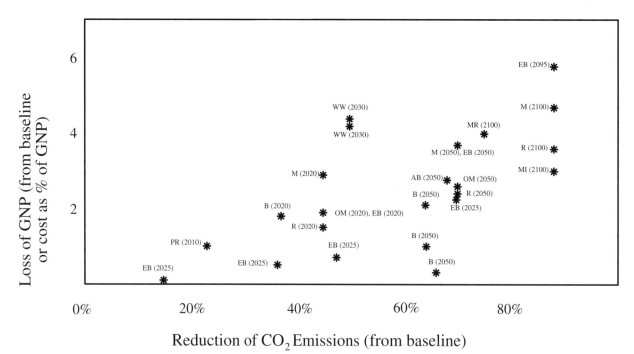

Note: See Table 9.24 for key to studies.

Figure 9.22: Global studies of CO_2 abatement costs relative to baseline projection.

Table 9.24. *Global CO_2 abatement cost modelling studies*

Author (year)	Key	CO_2 reduction from baseline (%)	CO_2 reduction from reference year[a] (%)	GNP impact/cost (reduction) from baseline (%)
Anderson and Bird (1992)	AB (2050)	68	-17	2.8
Burniaux *et al.* (1990)	B (2020)	37	17	1.8
Burniaux *et al.* (1992)	B (2050)	64, 64, 66	-18, -18, -11	2.1, 1.0, 0.3[b]
Edmonds and Barns (1990)	EB (2025)	14, 36, 47, 70		0.1, 0.5, 0.7 2.2[c]
Edmonds and Barns (1992)	EB (2020, 2050, 2095)	45, 70, 88	22, 41, 53	1.9, 3.7, 5.7
Manne and Richels (1990b)	MR (2100)	75	-16	4.0
Manne (1992)	M (2020, 2050, 2100)	45, 70, 88	13, 25, 21	2.9, 2.7, 4.7
Mintzer (1987)	Mi (2075)	88	67	3.0
Oliveira-Martins *et al.* (1992)	OM (2020, 2050)	45, 70	-2, 2	1.9, 2.6
Perroni and Rutherford (1991)	PR (2010)	23		1.0
Rutherford (1992)	R (2020, 2050, 2100)	45, 70, 88	15, 28, 43	1.5, 2.4, 3.6
Whalley and Wigle (1990)	WW (2030)	50		4.4, 4.4, 4.2[d]

[a]Negative values imply an increase in CO_2.

[b]Toronto-type agreement in all three cases, with tradable permits in the second and third cases and removal of energy subsidies in the third case.

[c]Costs as estimated from consumer + producer surplus.

[d]The three numbers refer to three different tax forms: a national producer tax, a national consumer tax, and a global tax with per capita redistribution of revenues.

Source: Grubb *et al.* (1993).

Figure 9.23 compares output losses entailed by stabilizing emissions in the year 2020. It also shows the reduction in the average annual growth rate of CO_2 emissions required to achieve the desired target. Note that losses vary by a factor of nearly three – from 0.8 to 2.2% of gross world product. Al-though it is difficult to isolate all the reasons for the differences in abatement cost estimates, the discussion of the preceding sections points to several possible causes.

The higher the baseline emission level, the more stringently carbon emissions have to be curtailed to achieve a

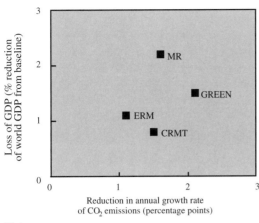

Notes:

1 No emission trading between regions

2 CRMT (Rutherford) model, ERM (Edwards-Reilly) model, GREEN model, and MR (Manne-Richels Global 2100) model as described and referenced in OECD (1993)

Source: OECD (1993).

Figure 9.23: Estimated CO_2 abatement cost in 2020.

given target, and hence the higher the overall abatement costs. The global baseline, however, is not the sole determinant of global abatement costs: The GREEN baseline projects the highest emissions for the year 2020; hence it requires the largest annual reductions to stabilize emissions at their 1990 level. Yet this does not lead to the highest cost estimates. This is because GREEN is the most optimistic of the four models concerning the speed at which backstop technologies can be introduced into the energy system.

Conversely, ERM projects the lowest baseline emissions for 2020. As a result, less carbon must be removed from the energy system to achieve emission stabilization. Even so, estimated costs (in terms of resulting output losses) are higher than those projected by CRTM since ERM does not include backstop technologies, thus producing higher marginal costs of emission abatement.

In a separate experiment, it was determined that ERM and MR project essentially the same baseline emissions when employing identical rates of autonomous energy efficiency improvements. However, for the OECD study, MR adopted an average annual rate of 0.5%, whereas ERM assumed that non-price-induced efficiency improvements occur at twice this rate. Standardization for this key parameter brings results of these two models much closer together with respect to their projections of GDP losses from emission abatement.

The OECD Model Comparison Project suggests that the principal reason why model-based studies differ with respect to estimates of both baseline emissions and abatement costs is alternative views about the future characteristics of the energy system embodied in the models. In an attempt to place greater reliance on expert knowledge in this area, Manne and Richels (1994) polled a group of individuals on their beliefs about key parameters to which abatement costs are particularly sensitive.

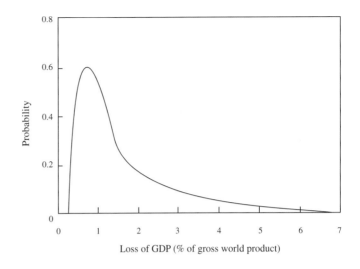

Source: Manne and Richels (1993).

Figure 9.24: Costs of stabilizing global carbon emissions.

For each parameter, expert beliefs were encoded in the form of probability distributions, which were, in turn, combined to form a group probability distribution. Using the poll responses to describe the uncertainty surrounding critical parameters, a probability distribution was then constructed for the costs of stabilizing global CO_2 emissions at 1990 levels, using the MR model.

Figure 9.24 presents the results of this analysis. The spread of the distribution is quite broad, ranging from 0.2 to 6.8% of gross world product (GWP). The median (that is, the fiftieth percentile) is located at approximately 1.0% of GWP. The distribution is also quite skewed. Because of the long right-hand tail, the mean value lies to the right of the median, at approximately 1.5% of GWP.

9.2.5.1.2 The costs of meeting alternative emission targets
In addition to the emission stabilization scenario, the OECD project examined three alternative scenarios in which the growth rate of emissions in each region is reduced below that of the corresponding baseline by one, two, and three percentage points respectively. Figure 9.25 plots global GDP losses as a function of worldwide emissions. Although there are significant variations in cost estimates across models, the results yield some important insights.

First, notice that the costs of stabilizing emissions at 1990 levels (approximately 6 billion tonnes) are lower in 2000 than in later years. The increase in emissions over the present decade is relatively small and there are ample quantities of low-cost substitutes (e.g., natural gas and demand-side management) to achieve the objective.

This situation does not persist as we move out in time. Although technical progress may eventually lower the marginal cost of emission abatement, more and more carbon must be removed from the energy system in order to maintain a particular target. As a result, annual losses are apt to increase.

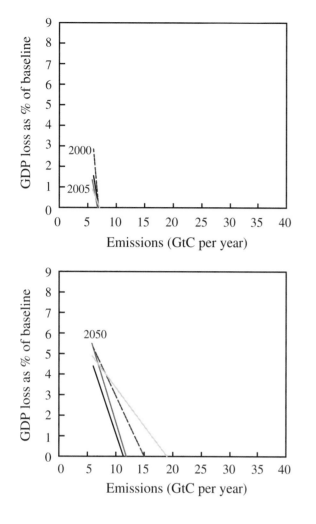

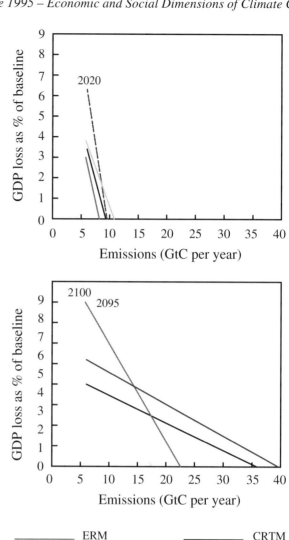

Sources: Dean and Hoeller (1992) and OECD (1993).

Figure 9.25: World CO_2 emissions and GDP losses.

As one would expect, output losses and macroeconomic impacts in general rise with the stringency of the abatement target, but it is important to note the nonlinear nature of the loss function. Successive emission reductions are increasingly expensive. As cheaper options to reduce emissions are used up, it becomes increasingly difficult to substitute for, or to economize on, fossil fuels.

For the models that incorporate backstop technologies, the marginal cost of emission abatement will eventually level off. This accounts for the linear nature of the total cost curves for CRTM and MR in the year 2100. ERM, on the other hand, excludes backstop technologies. Hence costs continue to rise nonlinearly as a function of emission reduction.

9.2.5.1.3 The potential gains from international cooperation
Emission reduction actions affect global emissions and atmospheric concentrations equally, irrespective of the geographical and sectoral origin of the emissions. The least-cost global abatement strategy requires reducing emissions where

it is cheapest to do so. If the costs of reducing emissions by a tonne were constant and the same in all regions, the location of emission reduction would have no effect on global costs. To the extent that marginal costs vary among regions, there are opportunities for efficiency gains through international cooperation.

These opportunities may be exploited through a system of international trade in carbon emission rights or a global carbon tax. Emission rights trading would allow for a more efficient allocation of emission reductions across regions by letting countries trade to the point where the marginal cost of emission reduction is the same in all places and activities. A global carbon tax would also result in the marginal cost of emission reduction being equal for all countries.[3]

Figure 9.26 shows the carbon taxes that would be required, on a region-by-region basis, to achieve a given level of emission reduction (OECD, 1994). Although these "emission reduction cost curves" are from a single study, virtually all analysts agree that the marginal costs of emission reduction

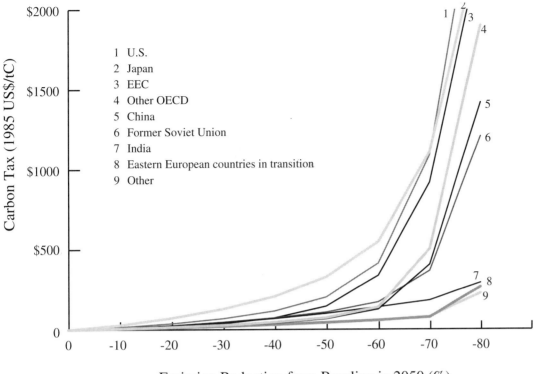

Source: Van der Mensbrugghe (1995).

Figure 9.26: Regional carbon tax required for equiproportional CO_2 emission reductions from baseline.

Table 9.25. *Cost differences for emission trading (global aggregates based on a 2% reduction in emissions from the baseline)*

		ERM[a]		GREEN		MR	
		Tax ($/tC)	GDP loss (%)	Tax ($/tC)	GDP loss (%)	Tax ($/tC)	Welfare loss[b]
2020	No trade	283	1.9	149	1.9	325	—
	Trade	238	1.6	106	1.0	308	—
2050	No trade	680	3.7	230	2.6	448	—
	Trade	498	3.3	182	1.9	374	—
2100	No trade	1304	5.7	—	—	242	8.0
	Trade	919	5.1	—	—	208	7.5

[a]End year is 2095 for ERM.
[b]Consumption losses through 2100, discounted to 1990 at 5% per year, in trillions of 1990 dollars.
Source: OECD (1993).

are apt to vary widely among regions. Efficiency gains can be reaped from shifting abatement from high- to low-marginal-cost regions. If the resulting cost savings are shared between participating regions, all can be made better off.

The potential gains from international cooperation can be demonstrated by comparing the total costs of achieving a global emission target alternatively with and without trade in carbon emission rights. Table 9.25 reports the results of such an experiment. From the viewpoint of economic efficiency, trade in emission rights is clearly worthwhile. A reduction

strategy in which each region reduces its emissions by the same percentage will not be globally cost-effective.

9.2.5.1.4 Allocation of emission rights

The establishment of international trade in carbon emission rights requires a decision on the allocation of emission rights among nations, which, in turn, has major implications for the international distribution of wealth. The decision on how emission rights are distributed among regions does not affect global abatement costs significantly,[4] but it will have a major

a) Permit allocation (% of total)

b) Income received from sales of excess emission rights (billions of 1990 US$)

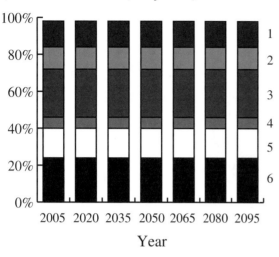
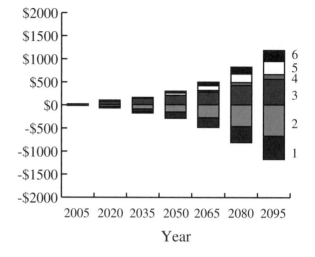

c) Total cost of emission reductions plus net wealth transfers from sales of emission rights (as % of GDP)

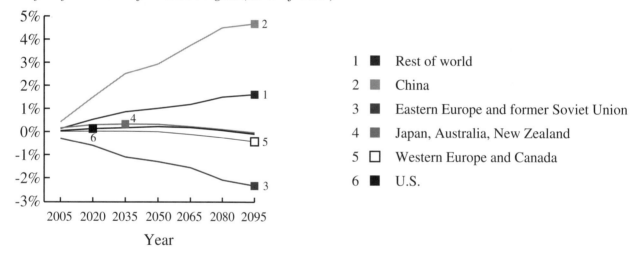

1	■	Rest of world
2	■	China
3	■	Eastern Europe and former Soviet Union
4	■	Japan, Australia, New Zealand
5	□	Western Europe and Canada
6	■	U.S.

Figure 9.27: Grandfathered emission rights principle.

impact on the distribution of net gains and losses and hence on the perceived equity of the agreement. This is a political rather than an economic issue, as there is no unique or objective definition of fairness (see Chapter 3).

Edmonds *et al.* (1993) studied a variety of schemes for allocating emission rights. At one end of the spectrum they considered a "grandfathered" emissions principle in which future emission rights are allocated on the basis of the share in global emissions at the time of joining a global abatement agreement. At the other end·of the spectrum, they examined an "equal per capita emissions" principle where emission rights are allocated on the basis of regions' shares in adult population.

Figures 9.27 and 9.28 are based on a scenario in which global emissions are permanently held at 1990 levels. They show results obtained by grandfathering emission rights and by granting quotas on the basis of equal per capita emis-

sions respectively. The allocation of emission rights differs markedly under these two extreme quota allocation schemes.[5] OECD countries hold on to approximately 50% of the emission rights when the status quo is maintained under a grandfathering scheme, whereas under the equal per capita emissions scheme, their share drops to about 20%.

The implications for regional net costs incurred are as expected: The OECD is much better off with the status quo. Indeed, the OECD region would be a seller of emission rights, and the resulting wealth transfers would be sufficient to reduce the region's GDP losses from abatement to a negligible level. Under such a scenario, the economic burden on developing countries would be substantial.

An equal per capita quota allocation scheme would substantially shift the distribution of net benefits and losses: The OECD countries would incur significant net income losses, and the region labelled "rest of world" would be a winner. Un-

a) Permit allocation (% of global total)

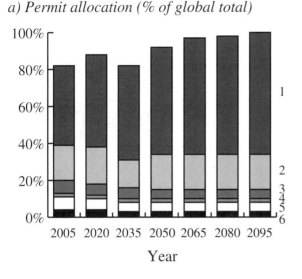

b) Income received from sales of excess emission rights (billions of 1990 US$)

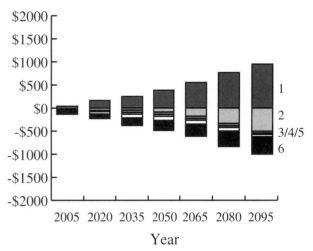

c) Total cost of emission reductions plus net wealth transfers from sales of emission rights (as % of GDP)

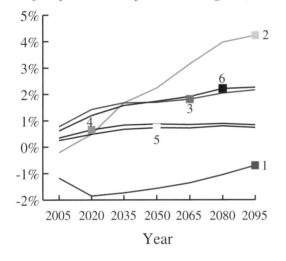

1 ■ Rest of world

2 ▨ China

3 ▨ Eastern Europe and former Soviet Union

4 ▨ Japan, Australia, New Zealand

5 □ Western Europe and Canada

6 ■ U.S.

Figure 9.28: Equal per capita emission rights principle.

der either scheme, China incurs substantial net income losses, except for the very early years. Unless it can greatly increase its energy efficiency, China's rapidly rising demand for energy would far outstrip its allocation of emission rights, even under an equal per capita quota allocation rule.

In a third scenario, also based on emission stabilization, Edmonds *et al.* (1993) explore what they refer to as a "no harm to developing nations" principle. Here developing nations receive sufficient emission rights to cover their own emissions and to generate sufficient revenue from excess quota sales to cover the economic cost of participating in the agreement. This leaves developing countries no worse off (in terms of total production plus income transfers) than had they not participated in the agreement. The results under such a quota allocation rule in terms of regional wealth transfers and net income changes are shown in Figure 9.29.[6]

9.2.5.1.5 Carbon "leakage"

Some proposals for limiting CO_2 emissions call for high-income countries to take the lead in reducing emissions. When abatement actions are limited to a subset of regions, it is important to consider so-called carbon "leakage" effects, which represent the impact of the emission policies of the abating regions on the emission levels of nonabating regions.

Leakage can occur through a number of channels, including

- The relocation of the production of energy-intensive products to nonabating regions
- Energy market effects, including increased energy consumption in nonabating regions and interfuel substitution between fuels of differing carbon contents, due to the differential decline in fossil fuel prices in response to reduced demand in abating regions

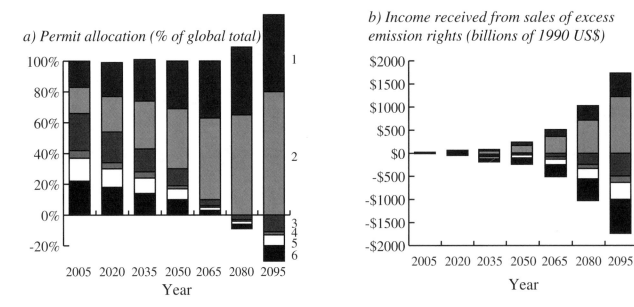

a) Permit allocation (% of global total)

b) Income received from sales of excess emission rights (billions of 1990 US$)

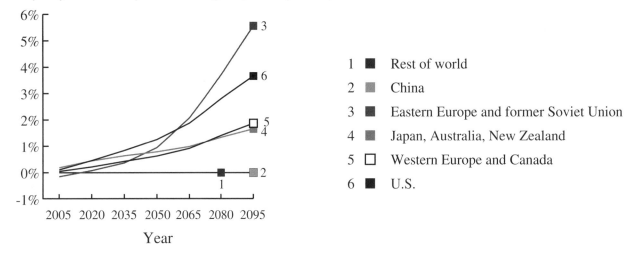

c) Total cost of emission reductions plus net wealth transfers from sales of emission rights (as % of GDP)

1 ■ Rest of world
2 ▨ China
3 ■ Eastern Europe and former Soviet Union
4 ▨ Japan, Australia, New Zealand
5 □ Western Europe and Canada
6 ■ U.S.

Figure 9.29: No net cost to developing nations emission rights principle.

- Changes in regional incomes (and thus energy demand) due to terms-of-trade changes

Such leakage effects can be positive or negative (negative leakage reduces carbon emissions in nonparticipating regions). Negative carbon leakage occurs, for example, in the case of a reduction in the incomes of energy-exporting regions when major oil consuming regions abate carbon emissions.

There have been several attempts to estimate leakage (Barrett, 1994). Unfortunately, estimates have varied so widely that they provide little guidance to policy makers. Pezzey (1992), using the model of Whalley and Wigle, examined a scenario in which the European Union acted unilaterally to reduce emissions by 20% below baseline. He found that for every 10 tonnes of carbon abated, global emissions fell by only 2 tonnes. Leakage rates for the OECD as a whole were of the order of 70%.

Horton, Rollo, and Ulph (1992) estimated even higher leakage rates. They found that for some industries (e.g., fertilizer production), unilateral abatement action could lead to substantial relocation. They argue that if the shift is to countries where energy use is more carbon-intensive, leakage rates may exceed 100%.

In contrast, Oliveira-Martins, Burniaux, and Martin (1992), using the OECD's GREEN model, estimate relatively small leakage rates. They examined a scenario in which emissions are held to 1990 levels and found leakage rates of 11.9% and 3.5% respectively for the European Union and the OECD as a whole.

Using still another model (12RT), Manne (1993) estimated leakage rates that fall between those of Pezzey (1992) and Oliveira-Martins, Burniaux, and Martin (1992). He examined a case in which the OECD acts unilaterally to reduce emissions by 20% below 1990 levels and found that 25% of the

OECD region's reductions could be offset indirectly through changes in international trade patterns.

The reasons why results differ so widely are not entirely clear. Manne and Oliveira-Martins (1994) have attempted a systematic model comparison of results from 12RT and GREEN. They find that much of the differences in leakage rates reported by the two models can be explained by different assumptions concerning the response of trade flows to changes in comparative advantage and competitiveness entailed by a regional carbon tax. Other key determinants of the extent of carbon leakage are the size and composition of the region undertaking unilateral abatement action, the supply elasticities of different fossil fuels, and the elasticity of substitution between energy and primary inputs of labour and capital in the production process in participating and nonparticipating regions. The model comparison needs to be extended to include other models, particularly those that show extremely high rates of leakage, before firm conclusions concerning the leakage effects of unilateral abatement actions can be drawn.

9.2.5.1.6 The costs of stabilizing atmospheric CO_2 concentrations

The United Nations Framework Convention on Climate Change has as its ultimate objective the "stabilization of greenhouse gas concentrations in the atmosphere at a level that would prevent dangerous anthropogenic interference with the climate system." The question of what constitutes an appropriate concentration level is likely to remain the subject of intense discussion for some time. But there is little disagreement over a subsequent clause in the Convention – that "policies and measures to deal with climate change should be cost-effective so as to ensure global benefits at the lowest possible cost."

The issue of cost-effectiveness was addressed in a study by Richels and Edmonds (1995). Although a particular concentration target can be achieved in a variety of ways, some will be more costly than others. Using a reduced-form carbon cycle model to construct alternative emission paths, the authors attempted to identify those paths that would minimize the costs of achieving alternative concentration levels. Global abatement costs were calculated using the Global 2100 and Edmonds-Reilly models.

The analysis suggests that the emission time path may be as important as the concentration level itself in determining the costs of emission abatement. Time is needed both for an economical turnover of the existing capital stock and to develop and deploy low-cost carbon-free alternatives.[7] The most cost-effective emission time paths are those which provide the greatest flexibility in managing the transition away from fossil fuels. Shifting emission reductions into the outer years can reduce costs substantially while preserving both the concentration target and the date at which the target is achieved.

As an example, the authors calculated the economic costs associated with two alternative scenarios for limiting atmospheric CO_2 concentrations to 500 ppmv in the year 2100 (see Figure 9.30). One involved stabilizing global emissions at 1990 levels through the end of the next century. The second

provided for some increase in emissions during the early years when the costs of emission abatement are highest, followed by sharp reductions in the later years when it is cheapest to do so. For this example, shifting emission reductions into the outer years reduced costs by as much as 50%.

The fact that deferring reductions may lead to lower costs should not be interpreted as supporting a "wait and see" or "do nothing" strategy. Part of the savings stems from not having to turn over the existing carbon-intensive capital stock prematurely. If we are to depart from the business-as-usual path, however, it is important that the new capital stock be less carbon-intensive. This means that energy sector decision makers must commit to a less carbon-intensive infrastructure when making new investments. Second, new supply options typically take many years to enter the marketplace. To have sufficient quantities of low-cost substitutes in the future will require a sustained commitment to research and development today.

9.2.5.2 A review of bottom-up studies: The critical role of innovation

Several bottom-up studies have been carried out with the aim of providing a normative view of the very long-term future. Following the work of Lovins in the 70s and early 80s, *Energy for a Sustainable World* by Goldemberg *et al.* (1987) played a key role in emphasizing the differences between the developed and developing countries.[8]

A typical global bottom-up analysis has been carried out for Greenpeace International by the Stockholm Environment Institute, Boston (SEI/Greenpeace, 1993) for the period 1985-2100. The study includes a "fossil-free energy future" (FFES) scenario with the following objectives and constraints:

- Meeting ambitious global CO_2 emission reduction targets
- Phasing out nuclear power by the year 2100
- Considering alternative scenarios for infrastructure, population, and GDP, with the aim of achieving a greater degree of economic equity between different regions of the world

For comparison, two reference scenarios were used as a high and low projection case for CO_2 emissions, namely the IPCC 1991 scenario as the high alternative and an average of the U.S. EPA's "Rapidly and Slowly Changing World" cases as the low alternative (IPCC 1992a; Lashof and Tirpak, 1990). Emissions for these reference scenarios for the year 2100 range between 22.6 PgC for IPCC 91 and 17.7 (PgC) for the EPA. These reference scenarios are shown in Table 9.26, together with the policy scenarios of the FFES study and the EPA "Rapid Reductions" policy scenario. CO_2 emissions decrease to zero in the year 2100 in both these intervention cases.

The total primary energy consumption is relatively similar in the FFES and the EPA policy scenarios. The scenarios differ significantly, however, in their assumptions about the structure of the energy supply system. The EPA scenario as-

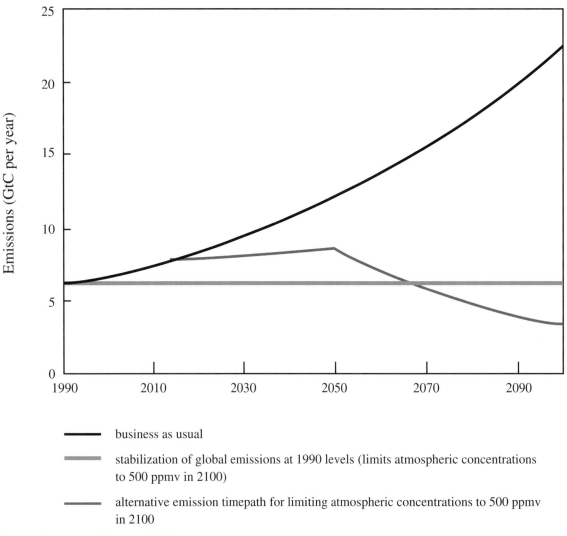

Source: Richels and Edmonds (1995).

Figure 9.30: Global carbon emissions: Three emission scenarios.

sumes that renewables will account for 77% of total primary energy consumption in 2100, whereas the FFES assumes 100%. Furthermore, the EPA scenario assumes a higher share of biomass, amounting to 58% of primary energy consumption in 2100, compared with only 18% in the FFES. The FFES instead expects solar and wind resources to cover as much as 79% of primary energy consumption in 2100, compared with only 9% in the EPA policy scenario.

The high expectations for the contribution of solar and wind in the FFES scenario are linked to the assumption that a breakthrough in solar photovoltaic electricity production costs will occur during the period 2010-2030. In the same way, a breakthrough is expected for electricity production costs for wind power, providing an opportunity for utilizing the technology in areas with weaker wind resources. Finally, the development of advanced storage facilities would enable intermittent solar and wind resources to service a greater part of the electricity system's load. All these assumptions about technical progress explain why the FFES concludes that the

costs of implementing the scenario are small compared with the baseline.

The high contribution of renewables in the FFES scenario is in contrast to the conclusion of an extensive analysis of renewable energy potential by the World Energy Council (1994). This analysis estimates that renewable energy resources would supply about 50% of total energy by 2100 in the so-called "ecologically driven scenario," which is the most far-reaching CO_2 reduction case in the study. It is argued that the total potential for renewable energy could be larger at that time, but if renewables make a major contribution and energy growth is low over a long period, as assumed in the ecologically driven scenario, supplies of petroleum and natural gas will still be available and it will be beneficial to mix different sources in the supply system.

An integrated global top-down/bottom-up analysis has been carried out in relation to the IPCC Working Group II Second Assessment Report (Volume 2 of this report) and has resulted in the construction of scenarios for a Low Emissions

Table 9.26. *Results of the IPCC 91 and U.S. EPA reference scenarios and the FEES and U.S. EPA policy scenarios*

	1988	2000	2010	2030	2100
IPCC 91					
CO_2 (PgC)	5.9	7.3	9.4	12.8	22.6
Primary energy (EJ)	349	460	471	797	1641
Renewables %					
Solar/Wind %					
Biomass %					
EPA Reference					
CO_2 (PgC)	5.1	6.6	7.7	9.9	17.7
Primary energy (EJ)	302	384	451	585	1067
Renewables %	7	8	10	13	19
Solar/Wind %	0	0	1	2	7
Biomass %	0	0	1	3	5
FFES					
CO_2 (PgC)	5.3	5.7	5.6	2.6	0.0
Primary energy (EJ)	338	396	400	384	987
Renewables %	13	21	29	62	100
Solar/Wind %	0	5	9	31	79
Biomass %	7	10	13	24	18
EPA "Rapid" Reductions"					
CO_2 (PgC)	5.1	5.5	4.5	2.5	1.5
Primary energy (EJ)	302	334	408	545	799
Renewables %	7	10	38	68	77
Solar/Wind %	0	1	2	4	9
Biomass %	0	0	26	54	58

Note: For the EPA and IPCC studies, the 1988, 2010, and 2030 values were interpolated from the 1985, 2000, and 2050 results. Results shown for IPCC for 1988 are actually 1990 values.
Source: SEI/Greenpeace (1993).

Supply System (LESS). The scenario results are shown in Figure 9.31.

The LESS scenarios provide estimates of the potential for greenhouse gas emission abatement using data developed from the detailed technology assessments in the Working Group IIa assessment process (IPCC, 1994). The energy supply systems were constructed from work by both bottom-up (Lashof and Tirpak, 1990; Johansson *et al.,* 1993) and top-down modellers (Edmonds, Wise, and MacCracken, 1994). In the bottom-up variant, the starting point is energy demand projections by world regions for 2025, 2050, 2075, and 2100 developed by the Response Strategies Working Group (RSWG, 1990) as part of the IPCC 1990 Assessment Report. A high economic growth version of this scenario demonstrated the importance of technology assumptions in the LESS analysis. Global GDP grows eight-fold by 2050 relative to 1985 and twenty-eight-fold by 2100, with primary energy consumption growing from 323 EJ in 1985 to 559 EJ in 2050 and 664 EJ in 2100. Biomass, mostly for fuels used directly, plays a major role. It accounts for 31% of primary energy in 2025 and rises to 50% in 2100 in one version of the scenario.

As a comparison with the LESS bottom-up analysis, a top-down analysis was carried out (Edmonds, Wise, and Mac-Cracken, 1994), incorporating performance and cost parameters for some of the key energy technologies used in the construction of the base case. The following six technology cases were modelled:

(1)   A reference scenario very similar to IPCC IS92a (IPCC, 1992b; the exogenous end-use energy intensity improvement rate is 0.5% by 2005, rising to 1.0% by 2035 and reaching 1.5% by the year 2065)

(2)   Similar to Case 1, but with an emphasis on energy-efficient power generation from fossil fuels (efficiency reaches 66% by 2095)

(3)   Similar to Case 1, but hydrogen, solar, and wind power become more competitive

(4)   Similar to Case 3, but compressed hydrogen is used instead of liquefied hydrogen

(5)   Similar to Case 4, but biomass prices are more competitive

(6)   Similar to Case 5, but the autonomous rate of energy efficiency improvements increases to 2.0% per year in 2050.

The results of the scenario analysis are given in Figure 9.32, showing global annual fossil fuel CO_2 emissions and energy production and use.

Cases 5 and 6 show that if the assumed technological characteristics are realized, significant reductions in CO_2 emissions could be achieved without economic penalty, as the technologies embodied in the low emissions scenario become competitive under market conditions with traditional fossil fuel technologies.

These LESS scenario results support the suggestion that the disagreement in the numerical outcomes of the models, in the very long run, is due less to the modelling structure than to the exogenous hypotheses. From a decision-making point of view, however, the most sensitive issue that remains to be addressed is the plausibility of this long-term transition, if one accounts for all the general equilibrium effects (for example, the implications of shifting agricultural activities to fuel production) and for all the transaction costs involved in the transition. It is worth noting, however, that the World Energy Council's ecologically driven case, which includes less voluntaristic assumptions than the LESS scenarios, points to a 60% reduction in global energy-related CO2 emissions from 1990 levels in 2100. This scenario is buttressed by a very comprehensive study of renewable energy prospects out to the year 2100.

9.3 Studies of the Costs of Carbon Sequestration

There are many difficulties inherent in developing and comparing estimates of carbon sequestration costs (see Chapter 8, Section 8.3.4). Those difficulties notwithstanding, this section attempts to summarize, compare, and critique the results of national, regional, and global carbon sequestration studies.

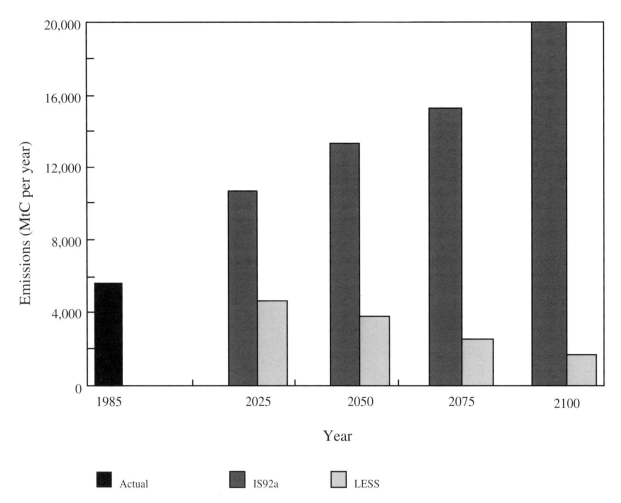

Figure 9.31: Global CO_2 emissions from fossil fuel burning for the LESS base case compared with actual emissions and the IPCC IS92a scenario.

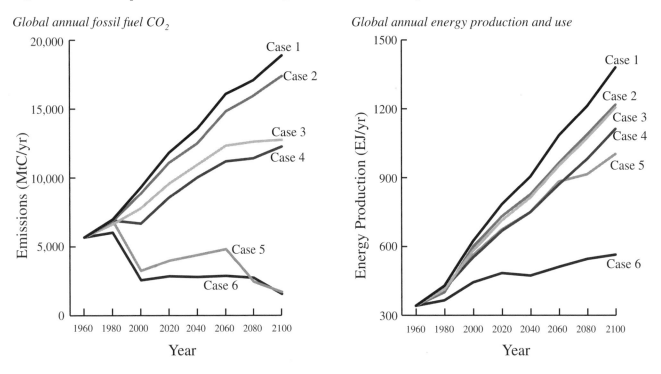

Source: Edmonds, Wise, and MacCracken (1994).

Figure 9.32: Global annual CO_2 emissions from fossil fuel burning and global annual primary energy production and use for six alternative cases constructed by IPCC WG IIa.

Table 9.27. *Land area availability*

Study	Region	Area Available for Practice (ha x 10⁶)		
		Forest Plantation	Forest Management	Agroforestry
Sedjo and Solomon (1989)[a]	Global	465	—	—
Grainger (1988)	Tropics	621	137	—
Barson and Gifford (1990)	Australia	1.6	—	—
Moulton and Richards (1990)	United States	112	31	—
Nordhaus (1991b)	Global	510	—	—
Dixon, Schroeder, and Winjum (1991)[b]	Boreal	Not specified	Not specified	Not specified
	Temperate	Not specified	Not specified	Not specified
	Tropical	Not specified	Not specified	Not specified
Dixon, Winjum, and Krankina (1991)	Latin America	214	—	372
	Africa	222	—	305
	Asia	115	—	159
	North America	42	—	7
New York State (1991)	New York State	0.4	0.2	—
Van Kooten *et al.* (1992)	Canada	8.6	19.7	—
Adams *et al.* (1993)[c]	United States	114	—	—
Houghton *et al.* (1993)	Latin America	44	634	737
	Africa	178	26	888
	Asia	29	393	270
Richards *et al.* (1993)	United States	102	—	—
Dixon *et al.* (1994)	South America	—	—	65–380
	Africa	—	—	300–440
	South Asia	—	—	130–225
	North America	—	—	90–140
Tasman Institute (1994)	New Zealand	5	—	—
Masera *et al.* (1994)	Mexico	18.2–19.6	18.7	—
Ravindranath and Somashekhar (1994)[d]	India	41.3	36.9	96
Xu (1994)[e]	China	110.5	19.2	70.5
Parks and Hardie (1995)[f]	United States	10.4	—	—

[a]The estimate by Sedjo and Solomon (1989) was based on a carbon budget goal of 2.9 x 10⁹ tonnes per year. It is not based on an assessment of land actually available, but rather land needed to meet the goal.

[b]Land area availability was not specified by practice. Therefore, costs and yields could not be matched with land areas available. The land area for dryland ecoregions was split between temperate and tropical categories.

[c]The land area was constrained by the carbon budget goal, not by availability.

[d]Community woodlots, timber forestry, and softwood forestry are included in the forest plantation category; natural regeneration and enhanced natural regeneration are included in the forest management category.

[e]The planting areas and sandy waste areas are included in the forest plantation category; open forest is included in the forest management category; dry cropland is included in the agroforestry category.

[f]The land area was constrained by budget, not by availability.

Many factors affect estimates of carbon sequestration costs and potential, including assumptions, methods, and data used with respect to land area, land costs, treatment costs, discount rates, carbon capture rates and patterns, ecosystem components included in the analysis, and the treatment of forest products. The data and assumptions employed by various studies are summarized in Tables 9.27 to 9.34.

As an illustration of the variation in data employed by sequestration studies, consider the estimates of land availability by region and practice listed in Table 9.27. For the tropics, Grainger (1988) has estimated that 621 x 10⁶ ha of land may be suitable for establishing forest plantations. Dixon, Winjum, and Krankina (1991) appear to be in relatively close agreement with this figure, estimating that there are 551 x 10⁶ ha suitable for forest plantations in Latin America, Africa, and Asia. In contrast, for the same three regions, Houghton *et al.* (1993) arrive at an estimate of less than half as much – 251 x 10⁶ ha. The more optimistic figures are lent support by the studies of Ravindranath and Somashekhar (1994) and Xu (1994), which, taken together, suggest that there are more than 150 x 10⁶ ha suitable for forestry plantations in India and China. The estimates for tropical agroforestry show similar disagreement. Dixon, Winjum, and Krankina (1991) estimate that there are 835 x 10⁶ ha suitable for agroforestry, whereas Houghton *et al.* (1993) suggest that 1895 x 10⁶ ha are available.

Table 9.28. *Land costs*

Study	Region	Land Costs		
		Forest Plantation	Forest Management	Agroforestry
Sedjo and Solomon (1989)	Global	400 U.S.$/ha		
Moulton and Richards (1990)	United States	360–8400 U.S.$/ha	120–1440 US$/ha	—
Nordhaus (1991b)	Global	20–200 U.S.$/ha	—	—
Dixon, Schroeder, and Winjum (1991)	Boreal	0	0	0
	Temperate	0	0	0
	Tropical	0	0	0
Dixon, Winjum, and Krankina (1991)	Latin America	0	—	0
	Africa	0	—	0
	Asia	0	—	0
	North America	0	—	0
New York State (1991)	New York State	0–1200 U.S.$/ha	0	—
Van Kooten *et al.* (1992)	Canada	0	0	—
Adams *et al.* (1993)[a]	United States	Not reported separately	—	—
Richards *et al.* (1993)	United States	275–5135 U.S.$/ha	—	—
Dixon *et al.* (1994)	South America	0	—	0
	Africa	0	—	0
	South Asia	0	—	0
	North America	0	—	0
Masera *et al.* (1994)	Mexico	0	0	—
Ravindranath and Somashekar (1994)[b]	India	16 U.S.$/ha	16 US$/ha	0
Xu (1994)[c]	China	0	0	0
Parks and Hardie (1995)[d]	United States	40–650 U.S.$/ha/yr	—	—

[a]The opportunity cost of land was derived within a mathematical programming model as the consumer welfare loss associated with the withdrawal of agricultural land from production.

[b]These figures are referred to as land rent but are included in the total investment costs, which appear to be initial costs only. It is not clear whether these figures represent a one-time cost or an annual rent.

[c]It is not clear whether land costs are included in investment costs (see Table 9.29).

[d]The land costs were expressed as annual rental payments to landowners within a subsidy programme.

The range of estimates of land availability for the temperate areas is only slightly narrower. Adams *et al.* (1993), Moulton and Richards (1990), and Richards *et al.* (1993) suggest that there may be 100×10^6 to 115×10^6 ha of marginal agricultural land suitable for afforestation in the U.S. alone. However, Dixon, Winjum, and Krankina (1991) identify only 42×10^6 ha in all of North America as suitable for forest plantations. Parks and Hardie (1995) consider only 10.4×10^6 ha in the United States for their analysis, though the land area constraint in their study is determined by an assumed programmatic budget and not by land availability.

Studies have also shown a wide range of estimates of land costs. As discussed in Chapter 8, this factor has proved particularly difficult because of the many nonmarket considerations associated with the social cost of converting between land uses. Because of the difficulty of determining the appropriate figures, some studies simply have not included land costs as an element of the cost analysis (e.g., Dixon, Schroeder, and Winjum, 1991). Others have apparently assumed that the use of land is costless because it is either public land (New York State, 1991) or because the wood products will eventually pay for the land (Van Kooten *et al.*, 1992; Xu, 1994). As might be expected, those that have included land costs have arrived at a wide range of estimates for that variable. Table 9.28 provides a summary of the land cost data employed in the various studies. Note that several of the studies that provide estimates of land availability do not include cost figures.

Initial treatment costs (see Table 9.29) are generally expressed as a capital outlay, whereas the maintenance costs, if included, are expressed as annual costs. Land costs may be expressed as either annual costs (rent) or capital costs. The cost analysis is facilitated by summarizing these costs as either a net present value equivalent or an equivalent annual cost. The key factor for this operation is the discount rate applied to these costs. Table 9.30 summarizes the discount rates used by various sequestration cost studies.

The importance of the choice of discount rate depends critically on the specific structure of the analysis. For example,

Table 9.29. *Treatment cost*

Study	Region	Treatment Costs (U.S.$/ha)		
		Forest Plantation	Forest Management	Agroforestry
Sedjo and Solomon (1989)	Global	400	—	—
Moulton and Richards (1990)	United States	140–520	—	
Nordhaus (1991b)	Global	400–450	—	—
Dixon, Schroeder, and Winjum (1991)	Boreal	125–450	50–250	—
	Temperate	25–800	0–1600	1000
	Tropical	250–320	50–500	250–750
Dixon, Winjum, and Krankina (1991)	Latin America	150–800	10–85	—
	Africa	30–1400	3–60	—
	Asia	150–375	15–30	—
	North America	50–1200	10–400	—
New York State (1991)	New York State	660	288	—
Van Kooten *et al.* (1992)	Canada	300–500	650–1000	—
Adams *et al.* (1993)	United States	140–520	—	—
Richards *et al.* (1993)	United States	190–690	—	—
Dixon *et al.* (1994)	South America	—	—	500–3500
	Africa	—	—	500–3500
	South Asia	—	—	500–3500
	North America	—	—	500–3500
Masera *et al.* (1994)	Mexico	387–700	NA[a]	—
Ravindranath and Somashekhar (1994)	India	367–550	77–205	39
Xu (1994)	China	46–828	11–31	14–240
Parks and Hardie (1995)[b]	United States	350	—	—

[a]Although there are costs associated with forest management in Masera *et al.* (1994), it is not clear how to interpret the figures.
[b]Parks and Hardie categorize treatment according to hardwood and softwood treatments. The figure listed here is for hardwood treatment. The costs for softwood are cited but not listed.

Table 9.30. *Discount rate applied to financial outlays*

Study	Annual Discount Rate (%)
Sedjo and Solomon (1989)	Not specified
Moulton and Richards (1990)	10
Nordhaus (1991b)	8
Dixon, Schroeder, and Winjum (1991)	5
Dixon, Winjum, and Krankina (1991)	Not specified
New York State (1991)	10
Van Kooten *et al.* (1992)	10
Adams *et al.* (1993)	10
Richards *et al.* (1993)	5
Dixon *et al.* (1994)	Not specified
Masera *et al.* (1994)	10
Ravindranath and Somashekhar (1994)	12–17.25
Xu (1994)	Not specified
Parks and Hardie (1995)	4

Moulton and Richards (1990) defined the cost per tonne of carbon as a ratio of land rent plus annualized establishment costs to average annual carbon capture. Since establishment costs are such a small part of the total costs in that analysis,

the difference between applying a 4% and a 10% discount rate was minor. However, in Richards *et al.* (1993), which used land purchase costs and time-dependent carbon yield curves, an increase in the discount rate from 3 to 7% nearly doubled the unit cost of carbon sequestration.

One of the most significant differences among studies is how they have addressed the irregular flows of carbon inherent in carbon sequestration and the differences in patterns among activities. For example, as Figure 9.33 illustrates, planting Loblolly pine on agricultural land in the Southeast region of the U.S. leads to carbon uptake rates that peak during the second decade after planting and taper off during the next four decades. In contrast, the carbon uptake rates associated with planting Ponderosa pine in the mountain states do not peak until the sixth decade after planting. Other carbon sequestration activities have similar variations in their flows over time.

Carbon sequestration studies have dealt with carbon flows in one of several ways. Some, such as Adams *et al.* (1993) and Moulton and Richards (1990), have used average carbon yields, expressed in tonnes per acre per year over the first forty years after tree stand establishment. Others have used yield curves such as those illustrated in Figure 9.33 to describe expected carbon flows on a year-by-year basis (Nordhaus, 1991b; Richards *et al.*, 1993). A third approach,

Table 9.31. *Treatment of carbon flows*

Study	Region	Yield Method	Estimate of Carbon Yield		
			Forest Plantation	Forest Management	Agroforestry
Sedjo and Solomon (1989)	Global	Average flow	6.24 tonnes/ha/yr		
Moulton and Richards (1990)	United States	Average flow	2.0–10.9 tonnes/ha/yr	0.0–7.6 tonnes/ha/yr	—
Nordhaus (1991b)[a]	Global	Yield curve	0.8–1.6 tonnes/ha/yr	—	—
Dixon, Schroeder, and Winjum (1991)	Boreal	MCS	15–40 tonnes/ha	4–20 tonnes/ha	—
	Temperate	MCS	30–175 tonnes/ha	10–125 tonnes/ha	15–160 tonnes/ha
	Tropical	MCS	25–125 tonnes/ha	20–200 tonnes/ha	50–150 tonnes/ha
Dixon, Winjum, and Krankina (1991)[b]	Not specified				
New York State (1991)	New York State	Average flow	2.1 tonnes/ha/yr	1.1 tonnes/ha/yr	—
Van Kooten *et al.* (1992)	Canada	Average flow	0.6–0.8 tonnes/ha/yr	0.6–2.1 tonnes/ha/yr	—
Adams *et al.* (1993)	United States	Average flow	2.0–10.9 tonnes/ha/yr	—	—
Richards *et al.* (1993)[c]	United States	Yield curve	0.0–9.4 tonnes/ha/yr	—	—
Dixon *et al.* (1994)	South America	MCS	—	—	39–195 tonnes/ha
	Africa	MCS	—	—	29–53 tonnes/ha
	South Asia	MCS	—	—	12–228 tonnes/ha
	North America	MCS	—	—	90–198 tonnes/ha
Masera *et al.* (1994)[d]	Mexico	MCS	25–150 tonnes/ha	121–134 tonnes/ha	—
Ravindranath and Somashekhar (1994)[e]	India	Standing carbon	76–121 tonnes/ha	62–87 tonnes/ha	25 tonnes/ha
Xu (1994)[f]	China	MCS	22–146 tonnes/ha/yr	9–15 tonnes/ha	6–33 tonnes/ha
Parks and Hardie (1995)	United States	Average flow	0.4–0.8 tonnes/ha/yr	—	—

[a]Nordhaus (1991b) develops logistic yield curves based on estimated carrying capacity, average flows, and time to maturity.

[b]Dixon, Winjum, and Krankina (1991) concentrated on forestry activities and their costs rather than on carbon yields.

[c]The range represents the lowest yield year for the slowest growing species to the highest yield year for the highest yield species.

[d]Carbon yield for forest management is estimated as avoided emissions from deforestation.

[e]These figures appear to count total carbon standing at 50 years, but may use MCS since rotations are considered for some practices.

[f]In addition to the components included in MCS, carbon stored in wood products is included in the carbon flow figures.

Table 9.32. *Treatment of forest products in carbon sequestration studies*

Study	Forest Product Treatment
Barson and Gifford (1990)	Outlines three scenarios for decay rates of wood products described, but effect on carbon accounting unclear.
Moulton and Richards (1990)	Does not consider forest products.
Nordhaus (1991b)	Discusses land purchased and turned into permanent forest cover; does not consider harvest.
Dixon, Schroeder, and Winjum (1991)	Assumes products are harvested; 100% release of stored carbon at time of harvest; no accounting for value of forest products.
Dixon, Winjum, and Krankina (1991)	Does not address carbon flows or stocks; benefits of harvest are implicitly considered in forestry practice costs analysis via derivation of internal rate of return on investment.
New York State (1991)	Subtracts value of timber products from costs of carbon sequestration.
Van Kooten *et al.* (1992)	Does not consider forest products.
Adams *et al.* (1993)	Explicitly includes timber harvest but does not include value of products in the benefits accounting; effect of release of carbon on carbon accounting not clear.
Richards *et al.* (1993)	Discusses land purchased and turned into permanent forest cover; does not consider harvest.
Dixon *et al.* (1994)	Discusses effects of harvest qualitatively; does not quantify.
Tasman Institute (1994)	Accounts for carbon loss at harvest; assumes 100% loss at harvest.
Masera *et al.* (1994)	Discusses harvest of timber; carbon and cost accounting methods unclear.
Ravindranath and Somashekhar (1994)	Includes value of forest products in cost accounting.
Xu (1994)	Forest products are explicitly included as a component of carbon storage and net costs.
Parks and Hardie (1995)	Does not consider forest products.

Table 9.33. *Ecosystem components included in carbon sequestration studies*

Study	Ecosystem Carbon Components Included
Sedjo and Solomon (1989)	Above- and below-ground tree
Barson and Gifford (1990)	Above- and below-ground tree, soil, understory
Moulton and Richards (1990)	Above- and below-ground tree, soil, understory, litter
Nordhaus (1991b)	Not specified
Dixon, Schroeder, and Winjum (1991)	Above- and below-ground tree
Dixon, Winjum, and Krankina (1991)	Above- and below-ground tree
New York State (1991)	Above- and below-ground tree
Van Kooten *et al.* (1992)	Tree bole only[†]
Adams *et al.* (1993)	Above- and below-ground tree, soil, understory, litter
Houghton *et al.* (1993)	Above- and below-ground biomass
Richards *et al.* (1993)	Above- and below-ground tree, soil, understory, litter
Dixon *et al.* (1994)	Above- and below-ground tree[†]
Tasman Institute (1994)	Above- and below-ground tree, understory, litter
Masera *et al.* (1994)	Above- and below-ground tree, soils
Ravindranath and Somashekhar (1994)	Above- and below-ground tree, understory, soils
Xu (1994)	Above- and below-ground tree, litter, and wood products
Parks and Hardie (1995)	Above- and below-ground tree

[†]Implied in text, but not explicitly stated.

Table 9.34. *Summary statistic applied to studies*

Study	Summary Statistic Used
Moulton and Richards (1990)	Levelized costs
Nordhaus (1991b)	Levelized costs
Dixon, Schroeder, and Winjum (1991)	Average storage method
Dixon, Winjum, and Krankina (1991)	Average storage method
New York State (1991)	Levelized costs
Van Kooten *et al.* (1992)[a]	Flow summation
Adams *et al.* (1993)[b]	Levelized costs
Richards *et al.* (1993)	Levelized costs/discount method
Dixon *et al.* (1994)	Average storage method
Masera *et al.* (1994)	Average storage method
Ravindranath and Somashekhar (1994)[c]	Flow summation
Xu (1994)	Average storage method
Parks and Hardie (1995)	Levelized costs

[a]Van Kooten used a flow summation method in the text and expresses a clear preference for this approach. Levelized costs are provided in an appendix.
[b]Levelized costs are not explicitly derived but are implicit in the form of the model.
[c]Because the carbon accounting method is unclear (see Table 9.31) it is uncertain whether the costs are derived using the flow summation method (if carbon accounting is based on standing carbon) or the average storage method (if carbon accounting is based on MCS).

introduced by Schroeder (1992), expresses programme effects on carbon in terms of storage rather than flows. The method, called mean carbon storage (MCS), assumes that once a practice is implemented, the forest system is sustained in the same use over time. Thus the carbon changes are expressed in terms of the change in the amount of carbon storage on site, averaged over one full rotation. This is expressed as

$$MCS = \frac{\sum_{i=1}^{n} C_i}{n}$$

where C_i is the standing carbon (tonnes) in year i, and n is the rotation length. Finally, the standing carbon method, a variation on the MCS method, expresses accomplishments in terms of the carbon standing at the end of the analysis period, say fifty years. This is the method used by Ravindranath and Somashekhar (1994). Table 9.31 provides a summary of the methods employed by sequestration cost studies.

Carbon flows into forests can also be reversed by harvesting. Those studies that have concentrated on plantation establishment have dealt with this issue in one of three ways (Table 9.32). The group of studies that employ the MCS method as-

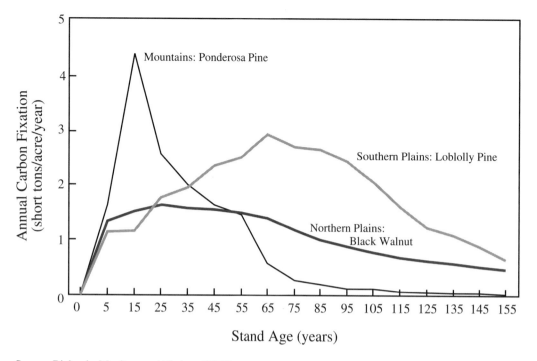

Source: Richards, Moulton, and Birdsey (1993).

Figure 9.33: Rate of annual carbon fixation as a function of forest-stand age for three region/species combinations.

sume that all carbon is released on harvest, but that the forestry practices are repeated in continuous rotations (Dixon, Schroeder, and Winjum, 1991; Dixon, Winjum, and Krankina, 1991; Dixon *et al.*, 1994). Hence, the concept of carbon release is built into the analysis. Another group of studies assumes that the land planted with trees is permanently withdrawn from other uses, including harvest of wood products, so that there is no release of carbon (e.g., Nordhaus, 1991b; Richards *et al.*, 1993). This assumption must be reflected in the calculation of land costs. Finally, some studies simply do not address the release of carbon on harvest, implicitly assuming that either the forest area will not be harvested, or that the harvest will occur so far in the future as not to be a concern (e.g., Moulton and Richards, 1990).

Several components of a forest ecosystem store carbon, including tree trunks, branches, leaves, and coarse and fine roots, as well as soils, litter, and understory. Studies have varied significantly with respect to how they address these various components. Some have included all components in their carbon accounting (e.g., Moulton and Richards, 1990). Others have limited their analysis to tree carbon only (e.g., Dixon, Schroeder, and Winjum, 1991). Table 9.33 provides a summary of which carbon components are included in each of the studies reviewed.

Box 8.2 in Chapter 8 provided a discussion and sample calculations of the various summary statistics used in carbon sequestration cost studies to capture the concept of "dollars per tonne of carbon sequestration." Van Kooten *et al.* (1992) demonstrate the importance of the choice of summary statistics in their analysis of the cost-effectiveness of carbon sequestration in Canada. Their analysis of costs employs the flow summation method, whereas an appendix provides calculations using the levelization approach. The costs in the lat-

ter case rise by a factor of five to ten relative to the former case. Table 9.34 provides a summary of the approaches employed by the carbon sequestration cost studies.

9.3.1 Costs of carbon sequestration

Table 9.35 summarizes the estimates of unit costs of carbon sequestration provided by the studies reviewed here. The studies fall into four general categories. One group concentrates on the potential of North America to sequester carbon (Adams *et al.*, 1993; Moulton and Richards, 1990; Ottinger *et al.*, 1990; Van Kooten *et al.*, 1992; Parks and Hardie, 1995; Richards *et al.*, 1993). This group predominantly uses a cost levelization/discounting approach. The one exception is Van Kooten *et al.* (1992), and they complement their flow summation approach by providing cost levelization results in their appendix. The second group considers the carbon sequestration potential of major ecological regions of the world using the average storage method (Dixon, Schroeder, and Winjum, 1991; Dixon, Winjum and Krankina, 1991; and Dixon *et al.*, 1994). The third group comprises studies of the global potential and cost of carbon sequestration (Sedjo and Solomon, 1989; Nordhaus, 1991b). Sedjo and Solomon (1989) do not provide unit cost calculations of carbon sequestration, whereas Nordhaus applies a discounting method. The fourth group, a set of recent studies, examines the potential for carbon sequestration in individual developing countries (Masera *et al.*, 1994; Ravindranath and Somashekhar, 1994; Xu, 1994). These studies use either the average storage method or the flow summation method.

Among the group of studies that concentrate on North America, the estimates of carbon costs fall into a relatively

Table 9.35. *Unit costs of carbon sequestration*

Study	Region	Method	Cost of Carbon Sequestration ($/tonne)		
			Forest Plantation	Forest Management	Agroforestry
Sedjo and Solomon (1989)[a]	Global	Levelized	7	—	—
		Flow summation	3	—	—
Moulton and Richards (1990)[b]	United States	Levelized	9–41	6–47	—
		Flow summation	2–9	2–9	—
Nordhaus (1991b)	Global	Levelized	42–114	—	—
Dixon, Schroeder, and Winjum (1991)	Boreal	Average storage	5–8	7	—
	Temperate	Average storage	2–6	1–13	23
	Tropical	Average storage	7	1–9	5
New York State (1991)	New York State	Levelized	14–54	12	—
Van Kooten *et al.* (1992)	Canada	Flow summation	6–18	8–23	—
		Levelized	66–187	39–108	—
Adams *et al.* (1993)	United States	Levelized	20–61	—	—
Richards *et al.* (1993)[b]	United States	Levelized	9–66	—	—
		Flow summation	2–9		
Dixon *et al.* (1994)	South America	Average storage	—	—	4–41
	Africa	Average storage	—	—	4–69
	South Asia	Average storage	—	—	2–66
	North America	Average storage	—	—	1–6
Masera *et al.* (1994)	Mexico	Average storage	5–11	0.3–3	—
Ravindranath and Somashekhar (1994)[c]	India	Flow summation	0.13–1.06	0.09–1.22	0.95–2.78
Xu (1994)[d]	China	Average storage	(12)–2	(2)–1	(13)–(1)
Parks and Hardie (1995)	United States	Levelized	5–90	—	—

[a]Sedjo and Solomon do not provide a unit cost figure for carbon sequestration. The figures presented here are based on their cost and yield estimates treated over 40 years with a 5% discount rate on financial outlays.

[b]The flow summation method is not used in these reports. The figures are supplied here for purposes of comparison with other studies.

[c]The interpretation of these figures is unclear. See discussion in text.

[d]Figures in parentheses indicate negative costs.

narrow range. After accounting for the differences attributable to cost analysis methods, Moulton and Richards (1990) provide the lowest cost range ($9 per tonne to $41 per tonne of carbon captured). These estimates were subsequently revised to reflect refined (lower) carbon yield estimates, the elasticity of demand for agricultural land, administrative costs, and failure rates (Richards *et al.* 1993). The analysis by Adams *et al.* (1993), which is based on a method of imputing land values through consumer welfare loss derived within a mathematical programming model of the agricultural sector, tends to confirm the revised results. Both studies suggest that the marginal cost of carbon sequestration would range from $9 to about $65 per tonne of carbon captured (levelized cost basis). The New York State (1991) study is in close agreement with the previous two studies.

Parks and Hardie (1995) present a very different picture. They suggest a similar lower range on costs but a very rapid increase that approaches $90 per tonne of carbon. Several factors contribute to the difference in costs. First, Parks and Hardie recognize much less land availability than either Adams *et al.* (1993) or Richards *et al.* (1993). This means that they move into more expensive land very quickly. Also, as indicated by Tables 9.28 and 9.30, their annual land costs are estimated at $40–$650/ha/yr and their discount rate is 4%. This suggests a capitalized land cost over the ten-year rental con-

tracts of $320/ha to $5300/ha, a rental cost that is higher than that used by Richards *et al.* (1993) for the outright purchase of land. Also, Parks and Hardie (1995) use lower carbon yield estimates and include only tree carbon in their calculations. Finally, and perhaps most important, their costs are annualized over only a 10-year contract period. Although this may be appropriate for their analysis of a specific hypothetical government programme, it almost certainly overstates the costs of carbon sequestration in a broader context, since carbon capture continues for several decades into the future, even after the end of government land rental payments.

In their presentation of levelized costs, Van Kooten *et al.* (1992) also provide higher estimates of carbon costs than either Adams *et al.* (1993) or Richards *et al.* (1993). This might be surprising in light of the fact that they do not include land costs in their estimates. The difference can be attributed in part to higher initial establishment costs and the low growth rates expected in the Canadian forests. Their carbon capture rates are also low because they consider only the carbon in the tree trunks and not whole ecosystem carbon.

The two studies of broad geographic/climate regions suggest that the costs of carbon sequestration may be relatively low for all three types of practices – forest plantations, forest management, and agroforestry (Dixon, Schroeder, and Winjum, 1991; Dixon *et al.*, 1994). Although these costs are

Table 9.36. *Potential carbon sequestration*

Study	Region	Carbon Sequestration Potential		
		Forest Plantation	Forest Management	Agroforestry
Sedjo and Solomon (1989)	Global	2900 x 10^6 tonnes/year	—	—
Barson and Gifford (1990)[a]	Australia	7 x 10^6 tonnes/yr	—	—
Moulton and Richards (1990)[b]	United States	630 x 10^6 tonnes/yr	110 x 10^6 tonnes/yr	—
Nordhaus (1991b)[c]	Global	280 x 10^6 tonnes/yr	—	—
Dixon, Schroeder, and Winjum (1991)[d]	Boreal	2 x 10^9 tonnes		
	Temperate	20 x 10^9 tonnes		
	Tropical	53 x 10^9 tonnes		
New York State (1991)[b]	New York State	0.8 x 10^6 tonnes/yr	0.2 x 10^6 tonnes/yr	—
Van Kooten *et al.* (1992)[e]	Canada	6 x 10^6 tonnes/yr	13 x 10^6 tonnes/yr	—
Houghton *et al.* (1993)[f]	Latin America	2.3 x 10^9 tonnes	13.2 x 10^9 tonnes	49.1 x 10^9 tonnes
	Africa	13.6 x 10^9 tonnes	0.4 x 10^9 tonnes	52.6 x 10^9 tonnes
	Asia	1.9 x 10^9 tonnes	15.0 x 10^9 tonnes	18.7 x 10^9 tonnes
Adams *et al.* (1993)[g]	United States	640 x 10^6 tonnes/yr	—	—
Richards *et al.* (1993)[h]	United States	49 x 10^9 tonnes	—	—
Dixon *et al.* (1994)[i]	Global	—	—	1100-2200 x 10^6 tonnes/yr
Tasman Institute (1994)[j]	New Zealand	5 x 10^6 tonnes/yr	—	
Masera *et al.* (1994)	Mexico	1.4-2.0 x 10^9 tonnes	1.5-2.3 x 10^9 tonnes	—
Ravindranath and Somashekhar (1994)[k]	India	3.7 x 10^9 tonnes	2.6 x 10^9 tonnes	2.4 x 10^9 tonnes
Xu (1994)	China	8.5 x 10^9 tonnes	0.2 x 10^9 tonnes	1.1 x 10^9 tonnes
Parks and Hardie (1995)[l]	United States	150 x 10^6 tonnes/yr	—	—

[a]This carbon capture rate is maintained over 25 years.

[b]This carbon capture rate is maintained over 40 years.

[c]This is the average carbon capture rate over an assumed 40-year growing period with planting spread over 35 yers, so that actual carbon capture is spread over 75 years.

[d]Dixon, Schroeder, and Winjum (1991) do not differentiate total potential yield on the basis of forestry practice. This is MCS for all practices averaged over a 50-year period.

[e]Rates are maintained over 60 years for plantation forests and 80 years for forest management.

[f]This is MCS. Period of time over which MCS is calculated is not specified.

[g]Adams *et al.* (1990) do not specify the period over which this carbon capture rate could be maintained.

[h]Richards *et al.* (1993) report potential cumulative carbon yield of 49 x 10^9 tonnes over a 160-year period. Of that 40 x 10^9 tonnes occur in the first 100 years. Hence, there is a potential average carbon capture rate of 400 x 10^6 tonnes per year for 100 years.

[i]This is MCS over 50 years.

[j]This carbon capture rate is maintained over 30 years.

[k]This is apparently total standing carbon at 50 years.

[l]Calculations are based on 10-year period only. However, sequestration is likely to continue over a longer period.

calculated using the average storage method, the costs presented here suggest lower estimates than those derived in the North American studies. It is interesting to note that Dixon, Schroeder, and Winjum (1991) find relatively little difference among the boreal, temperate, and tropical regions with respect to the carbon sequestration costs associated with forest plantations and forest management, which range from $2 per tonne to $8 per tonne and $1 per tonne to $13 per tonne respectively. In contrast, the carbon sequestration costs associated with agroforestry are considerably higher in the temperate region ($23 per tonne) than in the tropics ($5 per tonne). In the second study, however, that relation seems to be reversed: the cost in the tropical areas is $2 to $69 per tonne

and the cost in North America is $1 to $6 per tonne (Dixon *et al.*, 1994). The reversal appears to have occurred because the relative costs of initial treatment in the temperate zone were lowered in the second study and the MCS capacity of land was raised.

The two studies of global cost estimates differ significantly. The Nordhaus (1991b) estimate of $42-$114 per tonne for the unit costs of global carbon sequestration through afforestation is much higher than those from the other two groups. This is a bit surprising, given the fact that Nordhaus's land and treatment cost figures are similar to those used by other studies. The difference in results is almost entirely attributable to how the Nordhaus study treats carbon yields.

First, Nordhaus uses average carbon yield factors derived from the review of greenhouse gas policy options conducted by the U.S. EPA (1989). These figures are for carbon yields on average commercial timber land and probably substantially underestimate yields expected from conversions of marginal agricultural land to forestry plantations (see Table 9.31 for a comparison with other studies). Second, the analysis limits the total cumulative carbon to a range of 30-50 tonnes/ha and assumes that this amount occurs over a forty-year period following plantation establishment. These figures are certainly at the low end of the expected carrying capacity of forest plantations (see, e.g., Dixon, Schroeder, and Winjum, 1991). Finally, to portray the timing of carbon capture, Nordhaus applies a logistic growth curve that has the effect of delaying carbon uptake relative to the rate given by the average-flow approach or the MCS approach. Combined with a levelization approach to costs, this delay in carbon uptake contributes to an increased unit cost of carbon capture. (Richards *et al.*, 1993, also captures this effect.)

At the other extreme, Sedjo and Solomon (1989) provide land and treatment cost figures that would suggest a cost of carbon sequestration of $7 per tonne on a cost levelization basis and $3 per tonne on a flow summation basis. This relatively low cost estimate is due to their optimistic assumption regarding carbon yield, which is based on growth rates in the Pacific Northwest and Southeast regions of the U.S. Applying these rates to a global analysis is probably unrealistic, but it does suggest that, at least in some regions, carbon sequestration should be relatively inexpensive.

The three studies of individual developing countries provide an interesting contrast. Whereas Masera *et al.* (1994) estimate that carbon sequestration on forest plantations in Mexico would cost $5-$11 per tonne, Xu (1994) calculates that carbon could be stored on plantations in China at a negative cost. The latter result stems from the fact that Xu includes revenues from the sale of forestry products in the cost calculations, and that China has a largely unmet demand for timber. The interpretation of cost figures for Ravindranath and Somashekhar (1994) is unclear. Although they apparently use a flow summation approach in their cost calculations, they do discuss the application of discounting, at zero and 1%, to the carbon flow, which would suggest a levelized cost method. Their costs range from $0.09 to $2.78 per tonne.

9.3.2 Potential quantities of carbon

The studies show a wide range of estimates of potential for carbon sequestration (Table 9.36). At one extreme, Sedjo and Solomon (1989) have estimated that if 465 million hectares of land can be secured, 2.9 Gt of carbon per year can be removed from the atmosphere in forest plantations. This is nearly one-half of current annual global levels of carbon emissions. At the other extreme, Nordhaus (1991b) suggests that an average of only 0.28 GtC per year can be captured over a period of 75 years, even in the presence of a global effort. The difference between these two estimates is almost entirely due to the estimates of carbon yields, since their assumptions on land availability are very similar. At the same time, Dixon *et al.* (1994)

estimate that globally 1.1-2.2 GtC can be captured annually using expanded agroforestry practices alone.

Opportunities in the most northern latitudes appear somewhat limited. Dixon, Schroeder, and Winjum (1991), summing across all practices, suggest that only 2 GtC can be accumulated in the boreal regions. Averaged over their 50-year period, this yields 0.04 GtC per year. However, Van Kooten *et al.* (1992) suggest that forestry opportunities in Western Canada alone may provide as much as 0.13 Gt of carbon capture per year.

In the temperate regions there appear to be significant opportunities. Richards *et al.* (1993) suggest that in the U.S. alone, an aggressive tree planting programme could yield an average of 0.4 GtC per year for 100 years and a cumulative total of 49 GtC if the plantations are undisturbed for 160 years. Dixon, Schroeder, and Winjum (1991) are not so optimistic. That study suggests that across all forestry practices a total of 20 GtC could be accumulated in the entire temperate zone. Over their 50-year analysis period this averages to 0.4 GtC per year. Because the estimate of land area availability in Dixon, Schroeder, and Winjum (1991) is much higher than in Richards *et al.* (1993), the difference in the estimates must be due to the fact that the latter study uses much higher estimates of potential carbon accumulation per hectare.

The outlook in the tropics is even better than that in the temperate region. Dixon, Schroeder, and Winjum (1991) suggest that a cumulative total of 53 GtC could be captured across all forestry practices. Houghton *et al.* (1993) provide an even more optimistic estimate of the potential in the tropics, 167 GtC, though they provide no cost figures.

Two other studies that provide estimates of carbon sequestration potential without analyzing costs suggest that Australia and New Zealand could capture carbon at a rate of 0.007 Gt per year and 0.005 Gt per year respectively for 25 to 30 years (Barson and Gifford, 1990; Tasman Institute, 1994). For Mexico, India, and China it is estimated that approximately 3.5 GtC, 8.7 GtC, and 9.8 GtC respectively could be accumulated (Masera *et al.*, 1994; Ravindranath and Somashekhar, 1994; Xu, 1994).

9.3.3 Cost curves

This discussion has suggested that there is significant variation among the studies with respect to their estimates of the unit costs and potential amounts of carbon sequestration. The studies also differ in how they present the results. Several of the reports (Sedjo and Solomon, 1989; Nordhaus, 1991b; New York State, 1991; Van Kooten *et al.*, 1992) have presented their results as point estimates, that is, as estimates of the costs of achieving a specified amount of carbon sequestration.[9] Other studies have developed cost curves that illustrate the increasing marginal cost of carbon sequestration as a function of the level of sequestration (Adams *et al.*, 1993; Moulton and Richards, 1990; Dixon, Schroeder, and Winjum, 1991; Parks and Hardie, 1995; Richards *et al.*, 1993). The cost curves provide information that is not conveyed in either point estimates or in the ranges of unit costs listed in Table 9.35.

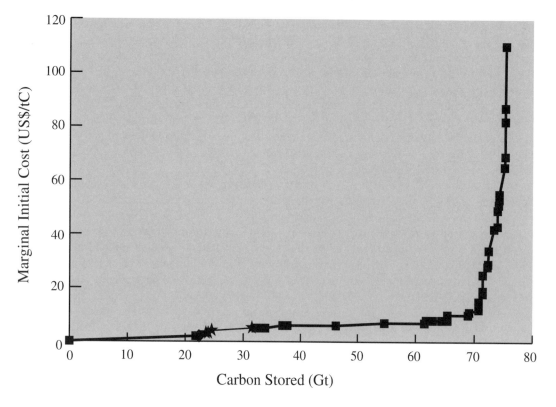

Source: Dixon, Winjum, and Schroeder (1991).

Figure 9.34: Marginal initial costs of sequestering carbon in forest systems employing forestation and forest management practices.

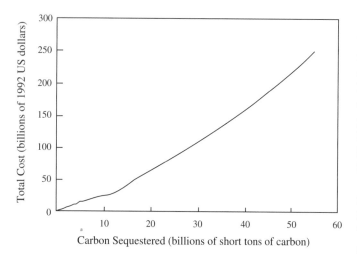

Source: Richards, Moulton, and Birdsey (1993).

Figure 9.35: Total cost curve for carbon sequestration under a 160-year programme.

An example of a cost curve for global carbon sequestration is shown in Figure 9.34. The dependent variable is the marginal cost of carbon capture expressed in dollars per tonne of carbon. Costs rise only slightly from 0 to 70 GtC, but climb sharply for higher levels of carbon sequestration. The MCS method, which expresses a stock of carbon over several rotations rather than total accumulation of carbon, is used for the figure.

Other studies report the marginal cost of carbon sequestration differently. Marginal costs are sometimes expressed in nominal dollars, that is, the dollars of the year in which the sequestration occurs. To arrive at this figure, current costs are levelized (inflated) at the discount rate. Carbon sequestration can also be measured as tonnes of carbon accumulated and permanently stored. Marginal cost curves that use either or both of these procedures tend to have a shape that is very similar to the curve in Figure 9.34 despite the significant underlying differences in the meaning of the values.

Figure 9.35 illustrates the total cost for carbon sequestration under a 160-year programme. The total cost rises almost linearly over the range of 12 to 55 billion short tons of carbon. This implies that the average cost per ton of carbon stored is roughly constant over this range. A constant average cost means that the marginal cost per ton of carbon stored is also constant and is equal to the average cost. Thus, Figure 9.35 confirms the relatively constant marginal cost over this range shown in Figure 9.34.

9.4 Studies of the Costs of Reducing Nonenergy Greenhouse Gas Emissions

Although much of the work on greenhouse gas emission reduction has focussed on energy-related emissions, a growing literature is emerging on nonenergy emissions. Table 9.37 summarizes some of the major options and their reduction potential.

Table 9.37. *Summary of economically viable options for reducing methane, nitrous oxide, PFC, and HFC emissions*

Source	Global Emissions (Tg/yr)[c]	Available Options for Reducing Emissions Profitably	Reductions Possible at Site-Specific Projects[b] (%)	Near-term Profitable Reductions Possible Worldwide (Tg/yr)[a]
Methane				
Natural gas systems	30–65	Yes	20–80	4–16
Coal mining	24–40	Yes	30–90	4–7
Waste disposal	20–40	Yes	up to 90	9–14
Ruminant livestock	65–100	Yes	5–60	4–10
Animal manures	10–18	Yes	up to0	over 1
Wastewater treatment	30–40	NA[d]	—	—
Rice cultivation	60–100	under development	—	—
Biomass burning	29–68	under development	—	—
Nitrous Oxide				
Adipic acid production	0.3–0.6	under development	up to 90	not examined
Nitric acid production	0.1–0.3	not examined	—	not examined
Fertilizer use	0.3–3.0	under development	up to 30	not examined
PFCs				
Aluminum production	0.03	Yes	up to 90	not examined
HFCs				
Intentional manufacture	NA	not examined	—	not examined
Manufacturing by-product	NA	Yes	up to 50	not examined
Applications	NA	not examined	—	not examined

[a]U.S. EPA (1994) for methane emissions; see text for other greenhouse gases.

[b]U.S. EPA (1993a).

[c]U.S. EPA (1993b).

[d]Methane emissions from wastewater treatment are largely from lesser developed countries, and large capital investments may be necessary to construct effective wastewater management facilities. In these cases the captured methane is not expected to substantially offset the investments.

9.4.1 Costs of reducing methane emissions

Methane is emitted from a diverse set of human-related activities which currently represent about 70% of global emissions annually (U.S. EPA, 1994). It is important to note that methane emissions from these systems represent the waste of a valuable fuel, and the methane can often be recovered where the saved fuel justifies the investment. Recent reports suggest that substantial reductions in methane emissions can be achieved profitably or at low cost while providing a number of other benefits (RIVM, 1993; U.S. EPA, 1993a; U.S. EPA, 1993b).[10] These studies indicate that it may be possible to employ existing technologies and practices profitably and reduce methane emissions from landfills, oil and natural gas systems, coal mining, and ruminant livestock by about 18 to 37 Tg per year over the next 5 to 10 years. These reductions represent about 5-10% of current anthropogenic emissions. Emission reductions of about 36 to 97 Tg per year (10-30% of current anthropogenic emissions) may be possible in the longer term. This section summarizes the results of these reports for each major methane source.

9.4.1.1 Natural gas and oil systems

Cost studies of available technologies and practices for reducing methane emissions from natural gas systems indicate that reductions in emissions in the order of 20-80% are possible at particular sites for particular types of emissions, depending on site-specific conditions (U.S. EPA, 1993a). The major opportunities for profitably achieving such reductions are in countries or regions where (1) gas prices are at world levels and reasonably small incremental investments can modify current operations and practices and save sufficient gas; (2) there are old or stressed systems handling large quantities of gas; and (3) work is underway to reduce emissions of volatile organic compounds or air toxics. For example, recent analyses in the U.S. show that, through incremental investments in a small set of best management practices, emissions can be profitably reduced over the next decade by about 30% or 1 Tg (U.S. EPA, 1993c). Globally it may be possible to reduce emissions by 4-16 Tg per year over the next decade through profitable ventures (U.S. EPA, 1993b) .

9.4.1.2 Coal mining

With available techniques, methane emissions into the atmosphere can be profitably reduced by up to 50-70% at gassy mines, depending on site-specific conditions (U.S. EPA, 1993a). The best opportunities for profitably achieving such reductions are for mines that are classified as "gassy," that produce substantial quantities of coal, and that are close to demand points for a medium quality fuel. However, with appropriate technical expertise, high quality gas can be produced and sold to pipeline systems. Recent analyses in the U.S. show that emissions from coal mining can be profitably reduced over the next decade by about 35% or 1.5 Tg with intensified efforts at about twenty-five mines (U.S. EPA, 1993c). Globally it may be possible to reduce emissions by 4-7 Tg per year over the next decade by profitable expansion of methane recovery techniques (U.S. EPA, 1993b). Reductions of 1.5-2.5 Tg per year could be achieved just by using the medium quality gas that is currently recovered in major coal regions around the world but then vented to the atmosphere (U.S. EPA, 1993b).

9.4.1.3 Waste disposal
Methane emissions from landfill sites can be profitably reduced by up to 90% and provide additional benefits in the form of improved air and water quality and reduced risk of fire and explosion (U.S. EPA, 1993a). Landfills that provide the best opportunities for profitably using captured gas fall into two categories: (1) those that are currently receiving wastes, that are expected to have more than 0.5 to 1 million tonnes of waste in place, and that have a clearly identifiable energy user or buyer nearby, and (2) those likely to be subject to air pollutant emission rules. Recent analyses in the U.S. show that methane emissions from landfills may be reduced by as much as 7 Tg per year as a by-product of new air emission regulations. Some of these landfills will be able to comply through revenue-generating recovery projects. Globally it may be possible to reduce emissions by 9-14 Tg per year over the next decade through methane recovery projects (U.S. EPA, 1993b).

9.4.1.4 Ruminant animals
Analyses show that available technologies and management practices can reduce methane emissions per unit of product by 5-60% in many animal management systems (U.S. EPA, 1993a). The greatest near-term potential for the profitable use of better management practices is in countries or regions where animals have not experienced large changes in productivity in the preceding decades, the animals are currently eating poor quality forage, product markets and physical infrastructure exist, and traditions and customs will support the changing practices. In particular, dairy herds in many developing countries offer great potential because the current diets are nutrient-deficient, there is daily access to animals so that feed supplementation and other management practices are possible, and the increased production of milk is quickly seen and marketable. Globally, emission reductions of 4-10 Tg per year may be achieved over the next decade by promoting available practices in these regions (U.S. EPA, 1993b).

9.4.1.5 Animal manures
Available technologies can profitably reduce methane emissions by as much as 80% at particular sites (U.S. EPA, 1993a). The best opportunities for profitably using recovered biogas from animal manures are in fairly warm regions with farms with large numbers of animals. For example, in the U.S. only

farms in the Southern states that have over 500 head of dairy cattle and over 1500 head of swine and that lagoon their animal manures and purchase fairly expensive electricity are expected to recover methane profitably. However, these farms represent about 30% of total U.S. emissions and could profitably reduce these by about 1 Tg per year.

9.4.2 Costs of reducing emissions of other greenhouse gases

Nitrous oxide and halogenated substances are the other greenhouse gases that make sizable contributions to the increasing radiative forcing in the earth's atmosphere. Although there remains much uncertainty about the magnitude of emissions of these gases from different sources, a number of activities are underway around the world to further assess and implement technologies and practices for reducing the emissions. This section briefly describes these activities. In general, the technologies and practices already in use are economically viable or were pursued for reasons other than the reduction of greenhouse gas emissions.

9.4.2.1 Nitrous oxide
Nitrous oxide is emitted from a variety of industrial and agricultural activities. These human-related activities represent about 20% of global emissions annually (IPCC, 1992b). Technologies and practices for reducing nitrous oxide emissions from some of the key sources have been identified and are at different stages of use and development as discussed below (RIVM, 1993).

9.4.2.1.1 Adipic and nitric acid production
Nitrous oxide is produced as a waste gas during the production of adipic acid, which is used primarily in the manufacture of nylon. The production of nitric acid – mostly for fertilizer production, with a small percentage used as an input into adipic acid production – also produces nitrous oxide. These sources are estimated to emit about 0.4 to 0.6 TgN and 0.1 to 0.3 TgN as N_2O per year, respectively (IPCC, 1992b). Over 80% of these emissions are from the industrialized countries. Emissions of nitrous oxide from adipic acid production already reflect a 30% emission abatement from the use of reductive furnaces to reduce emissions of NO_x (RIVM, 1993). Reductive furnaces reduce emissions by about 98%. A number of additional options for reducing these emissions with similar reduction efficiencies are under study by the adipic acid producers, some of whom expect to have these technologies in place by 1996. Emissions of nitrous oxide from nitric acid production are not currently abated, and technologies for reducing these emissions do not appear to be under discussion.

9.4.2.1.2 Fertilizer use
Cultivated soils are the principal anthropogenic source of nitrous oxide emissions. Nitrogen-based fertilizers contribute a large proportion of these emissions, primarily through denitrification processes (RIVM, 1993). Emissions from cultivated soils are estimated to be 0.3 to 3.0 TgN per year (IPCC, 1992),

with about 80% of these emissions from the developing countries and countries with economies in transition. Approaches for reducing nitrous oxide emissions from cultivated soils concentrate on the more efficient use of fertilizers, since these approaches can result in sizable cost savings. More efficient fertilizer use can be achieved by testing soils for nutrient deficiencies, applying fertilizers to meet specific plant needs, and using application methods that reduce the loss of nitrogen. Although these alternatives need substantial development and demonstration before becoming widely available, the potential for reductions in fertilizer use and greenhouse gas emissions is substantial. For example, the U.S. expects to reduce fertilizer use by 10-30% by 2000 by demonstrating such alternatives and educating farmers through agricultural extension networks about the cost savings associated with more strategic and efficient fertilizer use (Clinton and Gore, 1993).

9.4.2.1.3 Other sources

Nitrous oxide is also produced from a variety of other sources, including biomass burning and mobile and stationary combustion. Technologies and practices for controlling emissions from these sources are in the research phase.

9.4.2.2 Halogenated substances

Halogenated substances are emitted by a variety of manufacturing processes and numerous types of equipment and appliances. The halogenated substances of greatest interest to the IPCC are perfluorocarbons (PFCs) and hydrofluorocarbons (HFCs), which are highly potent and long-lived greenhouse gases. They are not ozone depleters and therefore are not covered by the Montreal Protocol. Activities are underway to determine low-cost, if not profitable, ways to reduce emissions of these gases.

9.4.2.2.1 Perfluorocarbons (PFCs)

The production of aluminum is thought to be the largest source of emissions of two PFCs, CF_4 and C_2F_6. These emissions are produced primarily by anode events, which occur during the reduction of alumina in the primary smelting process. World emissions are highly uncertain and are estimated to be in the order of 30,000 tonnes per year and 3000 tonnes per year for CF_4 and C_2F_6, respectively (Cook, 1995). Approaches for reducing emissions focus on technological upgrades for highly inefficient smelters and practices for reducing the frequency and duration of anode events at more modern smelters. These practices include improved algorithms for controlling automated processes, better management of alumina additions, and improved training for personnel. Because aluminum smelters are large consumers of energy, the costs of these modifications can be largely offset by the costs of saved energy. The U.S. estimates that it can reduce PFC emissions from aluminum smelting by 30-60% by 2000 by promoting profitable changes in current practices (Clinton and Gore, 1993).

9.4.2.2.2 Hydrofluorocarbons (HFCs)

HFCs occur as by-products of manufacturing, are intentionally produced as CFC and HCFC substitutes, and are used in a variety of applications. Emissions are difficult to estimate on an annual basis because, even though almost 100% of manufactured HFCs will be emitted to the atmosphere eventually, emission rates can vary from low but long-term (as in the case of insulating materials) to instantaneous (as in the case of aerosol propellants). Currently, emissions of these gases are relatively small, but emissions can be expected to grow substantially over the next decades as HFCs are increasingly used as substitutes for CFCs and HCFCs. The U.S., for example, expects more than a doubling in these emissions between 1990 and 2000. A number of options for reducing emissions of these gases are being investigated. Options include process changes, substitution of other substances, recycling, and destruction. The U.S. expects to reduce emissions of HFCs by about 25% by 2000 by encouraging product stewardship, by implementing regulations under the Clean Air Act, and by reducing HFC emissions as a by-product of HCFC-22 production.

9.4.2.2.3 Other halogenated substances

Other halogenated substances include sulphur hexafluoride and fluoroiodocarbons. Sulphur hexafluoride is produced for use as insulation for electrical equipment, to degas molten reactive metals, and as a tracer gas. Fluoroiodocarbons may be produced for intentional use as halon alternatives or refrigerants. These substances are emitted in relatively small quantities worldwide. Technologies or practices for reducing or limiting emissions of these gases do not appear to be under discussion.

Endnotes

1. See, for example, Nordhaus and Yohe (1983), Edmonds and Reilly (1985), and Manne and Richels (1994).

2. In fact, with the exception of the Quest model in the European Community, no model was able to assess satisfactorily the effects on trade balances or on delocation of big consuming industries because of the lack of a model of international trade.

3. The Pareto efficiency of abatement strategies based on a uniform carbon tax or tradable emission quotas (regardless of the quota allocation) has recently been contested at a theoretical level by Chichilnisky and Heal (1994). Their paper has triggered various responses; however, the ongoing debate on this issue is not pursued further here.

4. There may be second order general equilibrium effects on global abatement costs from changes in regional demand patterns, terms of trade, and investment paths, due to region-specific income and price elasticities of demand and to investment propensities.

5. These extreme quota allocation rules, neither of which is likely to be an acceptable basis for a global abatement coalition, were chosen for purely illustrative purposes.

6. Towards the end of the 110-year simulation period, even allocating all emission rights to developing countries is insufficient to compensate them for income losses from a global emission stabilization agreement. As a result, it is necessary to allocate "negative emission rights" to some developed regions under this allocation rule.

7. Recall from Section 9.2.1.1.6 that a positive discount rate also favours deferring reductions.

8. See, for example, the "Jérémie" and "Noé" scenarios by B. Dessus (1991).

9. Nordhaus (1991) presented a range for the cost estimates associated with a target level.

10. RIVM 1993 is a report of the proceedings of an international IPCC workshop on methane and nitrous oxide held in Amersfoort, the Netherlands, 3-5 February, 1993; U.S. EPA 1993a is a Report to Congress developed by the U.S. Environmental Protection Agency (U.S. EPA) building on efforts of the U.S./Japan Working Group on Methane for the Response Strategies Working Group of the IPCC, which solicited methane-related information from all IPCC country representatives. U.S. EPA 1993b is a follow-on to the first U.S. EPA report, which examines the applicability of available technologies and practices in specific countries. These reports are based on numerous supporting studies, some of which are listed at the end of this section.

References

Adams, R., D. Adams, J. Callaway, C. Chang, and B. McCarl, 1993: Sequestering carbon on agricultural land: Social cost and impacts on timber markets, *Contemporary Policy Issues,* **11**(1), 76–87.

Alliance to Save Energy, American Council for an Energy-Efficient Economy, Natural Resources Defense Council, Union of Concerned Scientists, 1991: *America's energy choices: Investing in a strong economy and a clean environment,* Union of Concerned Scientists, Cambridge MA.

Anderson, D. and C.D. Bird, 1992: Carbon accumulations and technical progress – a simulation study of costs, previously released as mimeo (1990), *Oxford Bulletin of Economics and Statistics* **42**(1).

Asian Development Bank, 1993: *National response strategy for global climate change: Peoples Republic of China,* Asian Development Bank, Beijing.

Asian Energy Institute, 1992: *Collaborative study on strategies to limit CO_2 emissions in Asia and Brazil,* Asian Energy Institute, New Delhi.

Ban, K., 1991: Energy conservation and economic performance in Japan – An economic approach to CO_2 emissions, Faculty of Economics, Discussion Paper Series, No. 112, Osaka University, Japan.

Barker, T., 1993: A UK carbon/energy tax: The macroeconomic effects, *Energy Policy,* **21**(3), 296–308.

Barker, T., 1994: Taxing pollution instead of employment: Greenhouse gas abatement through fiscal policy in the UK, Energy-Environment-Economy Modelling Paper No. 9, ESRC (Economic and Social Research Council), Department of Applied Economics, University of Cambridge, Cambridge, UK.

Barker, T., and R. Lewney, 1991: A green scenario for the UK economy. In *Green futures for economic growth: Britain in 2010,* T. Barker, ed., Cambridge Econometrics, Cambridge, UK.

Barker, T., S. Baylis, and C. Bryden, 1994: Achieving the Rio target: CO_2 abatement through fiscal policy in the UK, Energy-Environment-Economy Discussion Paper No. 9, ESRC (Economic and Social Research Council), Department of Applied Economics, University of Cambridge, UK.

Barns, D.W., J. A. Edmonds, and J. M. Reilly, 1992: The use of the Edmonds-Reilly models to model energy related greenhouse gas emissions, OECD, Economics Dept. Working Paper No. 113, Paris.

Baron, R., 1992: *Dynamic cost estimates of carbon dioxide emissions reduction in Eastern Europe and the former Soviet Union:* *An evaluation,* Pacific Northwest Laboratories, Advanced International Studies Unit, Global Studies Program, Richland, WA.

Barrett, S., 1994: Climate change policy and international trade, *Climate change: Policy instruments and their implications,* Proceedings of the Tsukuba Workshop of IPCC Working Group III. Tsukuba, Japan, 17–20 January 1994.

Barson, M., and R. Gifford, 1990: Carbon dioxide sinks: The potential role of tree planting in Australia. In *Greenhouse and energy,* D. Swaine, ed., pp. 433–443, Commonwealth Scientific and Industrial Research Organisation (CSIRO), Canberra.

Bashmakov, I., 1992: World energy development and CO_2 emission, *Perspectives in Energy,* **2,** 1–12.

Bashmakov, I., 1993: *Russia: Energy related greenhouse gas emissions, present and future,* International Workshop on Integrated Assessment of Mitigation, Impacts and Adaptation to Climate Change, International Institute for Applied Systems Analysis, Laxenburg, Austria, October 1993.

Bashmakov, I., 1994: Russia – Energy-related greenhouse gas emissions: Present and future. In *Integrative assessment of mitigation, impacts, and adaptation to climate change,* N. Nakicenovic, W.D. Nordhaus, R. Richels, and F.L. Toth, eds., pp. 379–401, International Institute for Applied Systems Analysis, Laxenburg, Austria.

Bashmakov, I., and V. Chupiatov, 1991: *Energy conservation: The main factor for reducing greenhouse gas emission in the former Soviet Union,* Pacific Northwest Laboratories, Advanced International Studies Unit, Global Studies Program, Richland, WA.

Beaumais, D., 1992: Variations sur l'écotaxe: évaluation des effets macroéconomiques et énergétiques de court-moyen terme induits par l'instauration d'une taxe carbone énergie sur les énergies fossiles, École centrale de Paris et Université de Paris I.

Bergman, L., 1990: General equilibrium effects of environmental policy: A CGE-modelling approach, Research Paper No. 6415, Handelshegskolan, Stockholm, Sweden.

Berry, L., 1989: *The administrative costs of energy conservation programs,* ORNL/CON-294, Oak Ridge National Laboratory, Oak Ridge, TN.

Bhide, A.D., S.A. Gaikwad, and B.Z. Alone, 1990: Methane from land disposal sites in India. In *International workshop on methane emissions from natural gas systems, coal mining, and waste management systems,* funded by Environment Agency of Japan, U.S. Agency for International Development, and U.S. Environmental Protection Agency, Washington, DC (April).

Bibler, C.J., J.S. Marshall, and R.C. Pilcher, 1992: *Assessment of the potential for economic development and utilization of coalbed methane in Czechoslovakia,* prepared for U.S. Environmental Protection Agency, Washington, DC.

BMU (Federal Environment Ministry of Germany), 1993: *Synopsis of CO_2 reduction measures and potentials in Germany,* Federal Environment Ministry of Germany, Bonn.

Bodlund, B., E. Miller, T. Karlsson, and T.B. Johansson, 1989: The challenge of choice: Technology options for the Swedish electricity sector. In *Electricity: Efficient end use and new generations of technologies, their planning implications,* T.B. Johansson, B. Bodlund, and R.H. Williams, eds., Lund University Press, Lund, Sweden.

Boero, G., R. Clarke, and L.A. Winters, 1991: *The macroeconomic consequences of controlling greenhouse gases: A survey,* Department of the Environment, Environmental Economics Research Series, HMSO, London.

Bowman, R.L., 1992: *Reducing methane emissions from ruminant livestock: India prefeasibility study,* prepared for U.S. Environmental Protection Agency, Washington, DC.

Bowman, R.L., T.D. Robertson, and S.A. Ribeiro, 1993: *Reducing methane emissions from ruminant livestock: Tanzania prefeasibility study,* prepared for U.S. Environmental Protection Agency, Washington, DC.

Brandsma, A., 1992: Les effets macroéconomiques d'une introduction d'une taxe CO_2/énergie: simulation sur la base du modèle Quest. In *Economie européenne* (Le Défi climatique: aspects économiques de la stratégie communautaire), **51** (May).

Bureau du Plan-Erasme, 1993: Un redéploiement fiscal au service de l'emploi: réduction du coût salarial financée par la taxe CO_2/énergie, rapport à la DG XI de la Commission Economique Européenne, Brussels.

Burniaux, J-M., J.P. Martin, G. Nicoletti, and J.O. Martins, 1990: *The costs of policies to reduce global emissions of CO_2: Initial simulation results with GREEN,* OECD Dept. of Economic Statistics, Resource Allocation Division, Working Paper No. 103, OECD/GD(91), Paris.

Burniaux, J-M., G. Nicoletti, and J.O. Martins, 1992: GREEN: A global model for quantifying the costs of policies to curb CO_2 emissions, *OECD Economic Studies,* **19** (Winter), 49–92.

Bye, B., T. Bye, and L. Lorentson, 1989: SIMEN: Studies in industry, environment and energy towards 2000, Central Bureau of Statistics Discussion Paper No. 44, Oslo, Norway.

Carlsmith, R., W.U. Chandler, J.E. McMahon, and D.J. Santini, 1990: *Energy efficiency: How far can we go?,* Oak Ridge National Laboratory, ORNL/TM - 11441, Oak Ridge, TN.

CBO (Congressional Budget Office), 1990: *Carbon charges as a response to global warming: The effects of taxing fossil fuel,* U.S. Congress, Washington, DC.

CEC DG XII JOULE, 1991a: *Cost-effectiveness analysis of CO_2-reduction options: Synthesis report,* Report for the CEC CO_2 Crash Programme, COHERENCE, Commission of the European Communities, Brussels.

CEC DG XII JOULE, 1991b: *Increase of taxes on energy as a way to reduce CO_2 emissions: Problems and accompanying measures,* Report for the CEC CO_2 Crash Programme prepared by V. Detemmerman, E. Donni, and P. Zagamé, Commission of the European Communities, Brussels.

Chandler, W., 1988: Carbon Emissions Control Strategies: The Case of China, *Climatic Change,* **13**(3), 241–265.

Chandler, W. (ed.), 1990: *Carbon emissions control strategies: Case studies in international cooperation,* Conservation Foundation, Washington DC; Carbon Dioxide Information Center, Oak Ridge National Laboratory, Oak Ridge, TN.

Chandler, W., 1994: Confidential report to the international finance corporation on energy-efficiency investment opportunities and constraints in Poland, draft, February.

Chandler, W. (rapporteur), W.B. Ashton, I.A. Bashmakov, S. Buttner, L.L. Jenney, E.K. Jochem, A.B. Lovins, R.G. Richels, L. Schipper, and K. Yamaji, 1992: Group Report: What are the options available for reducing energy use per unit of GDP? In *Limiting greenhouse effects: Controlling carbon dioxide emissions,* G.I. Pearman, ed., Wiley, West Sussex, UK.

Chandler, W., and S. Kolar, 1990: *Carbon emissions futures for eight industrialized countries,* The Fridtjof Nansen Institute, Oslo.

Chandler, W., and S. Kolar, 1991: *Energy and energy conservation in Eastern Europe: Two scenarios for the future,* Battelle, Pacific Northwest Laboratories, prepared for the U.S. Agency for International Development Global Energy Efficiency Initiative. Washington, DC.

Chandler, W. and A.K. Nicholls, 1990: Assessing carbon emission control strategies: A carbon tax or a gasoline tax?, Policy Paper No. 3, American Council for an Energy-Efficient Economy (ACEEE), Washington, DC.

Chandler, W.U., A. Makarov and Z. Dadi, 1990: Energy for the Soviet Union, Eastern Europe, and China, *Scientific American,* **263**(3), 120–127.

Chichilnisky, G., and G. Heal, 1994: Who should abate carbon emissions: An international perspective, *Economics Letters* **44,** 443–449.

Christensen, A., 1991: Stabilisation of CO_2 emissions – Economic effects for Finland, Ministry of Finance, Paper No. 29, Helsinki, Finland.

Cline, W.R., 1992: *The economics of global warming,* Institute for International Economics, Washington, DC.

Clinton, W.J., and A. Gore, Jr., 1993: *The climate change action plan.* The White House, Washington, DC.

Cohan, D., R.K. Stafford, J.D. Scheraga, and S. Herrod, 1994: The global climate policy evaluation framework. In *Proceedings of the 1994 Air and Waste Management Association Global Climate Change Conference,* Phoenix, 15–18 April, Air and Waste Management Association, Pittsburgh, PA.

Conrad, K., and M. Schröder, 1991: The control of CO_2 emissions and its economic impact, *Environmental and Resource Economics,* **1,** 289–312.

Cook, E., 1995; *Lifetime commitments: Why policymakers can't afford to overlook fluorinated compounds.* World Resources Institute, Washington, DC.

Davidson, O., 1993: Carbon abatement potential in Western Africa. In *The global greenhouse regime: Who pays?* P. Hayes and K. Smith, eds., United Nations University, Tokyo.

Dean, A., and P. Hoeller, 1992: Costs of reducing CO_2 emissions: Evidence from six global models, *OECD Economic Studies,* **19** (Winter), 15–47.

Dessus, B., 1991: Jérémie et Noé: deux scénarios énergétiques mondiaux à long terme, *Revue de l'énergie,* **421,** 291–307.

DG II-CEC, 1992. : *The economics of limiting CO_2 emissions,* Special edition of *European Economy,* Commission of the European Communities, Office for Official Publications, Luxembourg.

Dickson, A., K. Noble, Z. Peng, J. Dlugosz, and R. Stuart, 1994: Meeting greenhouse targets in Australia: Implications for coal use, ABARE paper presented at Greenhouse 94; An Australia-New Zealand Conference on Climate Change, Wellington, New Zealand (October).

Dixon, P.B., R.E. Marks, P. McLennan, R. Schodde, and P.L. Swan, 1989: *The feasibility and implications for Australia of the adoption of the Toronto Proposal for CO_2 emissions,* Report to Charles River & Assoc., Sydney, Australia.

Dixon, R., J. Winjum, K. Andrasko, J. Lee, and P. Schroeder, 1994: Integrated land-use systems: Assessment of promising agroforest and alternative land-use practices to enhance carbon conservation and sequestration, *Climatic Change,* **30,** 1–23.

Dixon, R., J. Winjum, and O. Krankina, 1991: Afforestation and forest management options and their costs at the site level. In *Technical workshop to explore options for global forestry management,* International Institute for Environment and Development, eds., pp. 319–328, Proceedings of a conference held 24–30 April, 1991, Bangkok, Thailand.

Dixon, R., P. Schroeder, J. Winjum (eds.), 1991: *Assessment of promising forest management practices and technologies for enhancing the conservation and sequestration of atmospheric carbon and their costs at the site level,* Report of the U.S. Environmental Protection Agency, #EPA/600/3-91/067, Environmental Research Laboratory, Corvallis, OR.

Donni, E., P. Valette, and P. Zagamé, 1993: *HERMES, Harmonised econometric research for the modelling of economic systems,* North Holland, Amsterdam.

The DPA Group Inc., 1989: Study on the reduction of energy-related greenhouse gas emissions, Ministry of Energy, Toronto, Ontario.

DRI (Data Resources Inc.), 1992: *The economic effects of using carbon taxes to reduce carbon dioxide emissions in major OECD countries,* prepared for the U.S. Department of Commerce by DRI/McGraw-Hill, Lexington, MA.

DRI (Data Resources Inc.), 1994: *Potential benefits of integrating environmental and economic policies,* Report prepared for the European Commission, Office for the Official Publications of the European Union, Luxembourg.

ECN 1994: *Boundaries of future carbon dioxide emission reduction in nine industrial countries,* Netherlands Energy Research Foundation, ECN-C-94-025, Amsterdam.

Edmonds, J.A., and J.M. Reilly, 1985: *Global energy: Assessing the future,* Oxford University Press, New York.

Edmonds, J.A., and J.M. Reilly, 1986: *Uncertainty in future global energy use and fossil fuel CO$_2$ emissions 1975 to 2075,* prepared for the U.S. Department of Energy, DOE/NBB-0081, Department of Energy, Washington, DC.

Edmonds, J.A., and D.W. Barns, 1990: *Estimating the marginal cost of reducing global fossil fuel CO$_2$ emissions,* Pacific Northwest Laboratories, PNL-SA-18361, Washington, DC.

Edmonds, J.A., and D.W. Barns, 1991: *Use of the Edmonds-Reilly model to model energy-sector impacts of greenhouse emissions control strategies,* prepared for the U.S. Dept. of Energy, by Pacific Northwest Laboratories, Washington, DC.

Edmonds, J.A., and D.W. Barns, 1992: Factors affecting the long-term cost of global fossil fuel CO$_2$ emissions reductions, *International Journal of Global Energy Issues,* **4**(3), 140–166.

Edmonds, J.A., D.W. Barns, and M. Ton, 1993: Carbon coalitions – The cost and effectiveness of energy agreements to alter trajectories of atmospheric carbon dioxide emissions, draft, Pacific Northwest Laboratories, Washington, DC.

Edmonds, J.A., M. Wise, and C. MacCracken, 1994: *Advanced energy technologies and climate change: An analysis using the global change assessment model (GCAM),* report prepared for the IPCC Second Assessment Report, Working Group IIa, Energy Supply Mitigation Options.

Edmonds, J.A., M. Wise, H. Pitcher, R. Richels, T. Wigley, and C. MacCracken, 1994: *The accelerated introduction of advanced energy technologies and climate change: An analysis using the global change assessment model (GCAM),* Battelle Pacific Northwest Laboratories, draft, Washington, DC.

EMF (Energy Modeling Forum), 1993: *Reducing global carbon emissions – Costs and policy options,* EMF-12, Stanford University, Stanford, CA.

Energy and Environmental Division, LBL (Lawrence Berkeley Laboratories), 1991: *CO$_2$ emissions from developing countries: Better understanding the role of energy in the long term,* Lawrence Berkeley Laboratories, Berkeley, CA.

Enquete Kommission des deutschen Bundestages (German Parliamentary Enquete Commission), 1991: *Schutz der Erde: eine Bestandsaufnahme mit Vorschläge zu einer neuen Energiepolitik,* Bundestag, Bonn.

European Commission, 1992: A community strategy to limit carbon dioxide emissions and to improve energy efficiency, communication from the Commission to the Council, European Commission, Brussels.

European Commission, 1994: Taxation, employment and environment: Fiscal reform for reducing unemployment, *European Economy,* **53,** 137–177.

Faruqui, A., M. Maulden, S. Schick, K. Seiden, G. Wikler, and C. Gellings, 1990: *Efficient electricity use: Estimates of maximum energy savings,* EPRI CU-6746, Electric Power Research Institute, Palo Alto, CA.

Fickett, A.P., C.W. Gellings, and A.B. Lovins, 1990: Efficient use of electricity, *Scientific American,* **263**(3), 64–75.

Glomsrod, S., H. Vennemo, and T. Johnson, 1990: Stabilization of emissions of CO$_2$: A computable general equilibrium assessment, Central Bureau of Statistics, Discussion Paper No. 48, Oslo, Norway.

Godard, O., and O. Beaumais, 1993: L'Économie face à l'écologie, La Découverte/La Documentation française, Paris.

Goldemberg, J., T.B. Johansson, A. Reddy, and R. Williams, 1987: *Energy for a sustainable world.* report, Wiley Eastern, New Delhi, India; World Resources Institute, Washington, DC.

Goto, U., 1991: A study of the impacts of carbon taxes using the Edmonds-Reilly model and related sub-models. In *Global warming and economic growth,* A. Amano, ed., Centre for Global Environmental Research, Tsukuba, Japan.

Goulder, L.H., 1991: Effects of carbon taxes in an economy with prior tax distortion: An intertemporal general equilibrium analysis for the U.S., working paper, Stanford University, Palo Alto, CA.

Grainger, A., 1988: Estimating areas of degraded tropical lands requiring replenishment of forest cover, *The International Tree Crops Journal,* **5,** 31–61.

Grubb, M., J. Edmonds, P. ten Brink, and M. Morrison, 1993: The costs of limiting fossil-fuel CO$_2$ emissions: A survey and analysis, *Annual Review of Energy and Environment,* **18,** 397–478.

Hanslow, K., M. Hinchy, J. Small, B. Fisher, and D. Gunasekera, 1994a: Trade and welfare effects of policies to address climate change, paper presented at IPCC Workshop on Equity and Social Considerations Related to Climate Change, Nairobi, July, ABARE Conference paper 94.24, Australian Bureau of Agricultural and Resource Economics, Canberra.

Hanslow, K., M. Hinchy, J. Small, B. Fisher, and D. Gunasekera, 1994b: Climate change: Trade and welfare effects, *Australian Commodities,* **1**(3), 344–354.

Hanslow, K., M. Hinchy, J. Small, B. Fisher, 1994c: International greenhouse economic modelling, paper presented to Greenhouse 94 Conference, Wellington (October), ABARE Conference paper 94.34, Australian Bureau of Agricultural and Resource Economics, Canberra.

Hirst, E. and M. Brown, 1990: Closing the efficiency gap: Barriers to the efficient use of energy, *Resources, Conservation and Recycling,* **3**(4), 267–281.

Hoeller, P., A. Dean, and J. Nicolaisen, 1991: Macroeconomic implications of reducing greenhouse gas emissions: A survey of empirical studies, *OECD Economic Studies,* **16** (Spring), 45–78.

Horton, G.R., J.M.C. Rollo, and A. Ulph, 1992: The Implications for trade of greenhouse tax emission control policies, working paper prepared for UK Department of Trade and Industry, London.

Houghton, R., J. Unruh, and P. Lefebvre. 1993: Current land cover in the tropics and its potential for sequestering carbon, *Global Biogeochemical Cycles,* **7**(2), 305–320.

Hourcade, J-C., R. Baron, and N.B. Chaabane, 1991: *Politique énergétique et l'effet de serre: Une esquisse des marges de manoeuvre à 2030,* rapport pour l'Atelier Prospective du Commissariat Général du Plan, Centre International de Recherche sur l'Environnement et le Développement (CIRED), Montrouge, France.

Hourcade, J.-C., N.B. Chaabane, K. Helioui, V. Journe, and P.

Mabire, 1993: *UNEP GHG abatement costing studies: France country study,* Centre International de Recherche sur l'Environnement et le Développement (CIRED), Montrouge, France.

Industry Commission, 1991: *Costs and benefits of reducing greenhouse gas emissions,* report 15, vol. 1, Industry Commission, Canberra.

IPCC, 1992a: *Emission scenarios for the IPCC, an update: Assumptions, methodology, and results,* W. Pepper, J. Leggett, R. Swart, and J. Wasson, prepared for The Intergovernmental Panel on Climate Change, Working Group 1, U.S. Environmental Protection Agency, Washington, DC.

IPCC, 1992b: *Climate change 1992. The supplementary report to the IPCC scientific assessment,* J.T. Houghton, B.A. Callendar, and S.K. Varney, eds., Cambridge University Press, Cambridge.

IPCC, 1994: *Energy supply mitigation options,* IPCC Working Group IIa (draft 18), July. Intergovernmental Panel on Climate Change, Geneva.

IPSEP (International Project for Sustainable Energy Paths), 1993: *Energy policy in the greenhouse: The economics of energy tax and non-price policies,* International Project for Sustainable Energy Paths, Lawrence Berkeley Laboratories, Berkeley, CA.

IPSEP, 1994/1995: *The cost of cutting carbon emissions in Western Europe: Energy policy in the greenhouse,* Volume II, Parts 2-6, Report to the Dutch Ministry of Environment, International Project for Sustainable Energy Paths (IPSEP), El Cerrito, CA.

Jackson, T., 1991: Least-cost greenhouse planning: supply curves for global warming abatement, *Energy Policy,* 19(1), 35–46.

Jaszay, T., 1990: Hungary, In: *Carbon emission control strategies: case studies in international cooperation,* W. Chandler, ed., Conservation Foundation, Washington, DC; Carbon Dioxide Information Center, Oak Ridge National Laboratory, Oak Ridge, TN.

Johansson, T.B., B. Bodlund, and R.H. Williams (eds.), 1989: *Electricity: Efficient end-use and new generation technologies and their planning implications,* Lund University Press, Lund, Sweden.

Johansson, T.B., H. Kelly, A.K.N. Reddy, and R.H. Williams, 1993: A renewables-intensive global energy scenario (RIGES). In: *Renewable energy: Sources for fuels and electricity,* T. Johansson, *et al.,* eds., Island Press, Washington, DC.

Jones, B., Z. Peng, and B. Naughten, 1994: Reducing Australian energy sector greenhouse gas emissions, *Energy Policy,* 22(4), 270–286.

Jorgenson, D.W., and P.J. Wilcoxen, 1990a: *The cost of controlling carbon dioxide emissions,* presented at Workshop on Economy-Energy-Environment Modelling and Climate Policy Analysis, Washington, DC; Harvard University, Cambridge, MA.

Jorgenson, D.W., and P.J. Wilcoxen, 1990b: *Reducing U.S. carbon dioxide emissions: The cost of different goals,* Harvard Institute for Economics and Resources, Harvard University.

Karadeloglou, P., 1992: Energy tax versus carbon tax: A quantitative macro-economic analysis with the Hermès/Midas models, *European Economy,* Special Edition no. 1., CEC DG-II, Brussels, Belgium.

Kaya, Y., 1989: Impact of carbon dioxide emissions on GNP growth: Interpretation of proposed scenarios, Intergovernmental Panel on Climate Change/Response Strategies Working Group, IPCC, Geneva.

Kononov, Y., 1991: *The cost of reducing Soviet CO$_2$ emissions,* Siberian Energy Institute, Irkutsk, Russia.

Kononov, Y., 1993: *Impact of the economic reforms in Russia on greenhouse gases emissions, mitigation and adaptation,* International Workshop on Integrated Assessment of Mitigation, Impacts and Adaptation to Climate Change, October, International Institute for Applied Systems Analysis, Laxenburg, Austria.

Koomey, J., C. Atkinson, A. Meier, J. McMahon, S. Boghosian, B. Atkinson, I. Turiel, M. Levine, B. Nordman, and P. Chan. 1991: *The potential for electricity efficiency improvements in the US residential sector,* Report LBL-30477, Lawrence Berkeley Laboratory, Berkeley, CA.

Kostalova, M., J. Suk, and S. Kolar, 1991: *Reducing greenhouse gas emissions in Czechoslovakia,* Pacific Northwest Laboratories, Advanced International Studies Unit, Global Studies Program, Richland, WA.

Kram, T., 1993: *National energy options for reducing CO$_2$ emissions,* vol. 1, Netherlands Energy Research Foundation (ECN), Amsterdam.

Krause, F., and J. Koomey, 1992: The greenhouse abatement dividend, *Electricity Journal,* 5(7).

La Rovere, E., L.F.L. Legey, and J.D.G. Miguez, 1994: Alternative energy strategies for abatement of carbon emissions in Brazil, *Energy Policy,* 22(11), 914–924.

Lashof, D.A., and D.A. Tirpak, 1990: *Policy options for stabilizing global climate,* Appendices, Report to Congress from the Office of Policy, Planning and Evaluation. U.S. Environmental Protection Agency, Washington, DC.

Leach, G., and Z. Nowak, 1991: Cutting carbon dioxide emissions from Poland and the UK, *Energy Policy,* 19(10), 918–925.

Leary, N.A., and J.D. Scheraga, 1994: Policies for the efficient reduction of carbon dioxide emissions, *International Journal of Global Energy Issues,* 6(1/2), 102–111.

Leng, R.A., 1991: *Improving ruminant production and reducing methane emissions from ruminants by strategic supplementation,* U.S. Environmental Protection Agency, Washington, DC.

Levine, M., A. Gadgil, S. Meyers, J. Sathaye, J. Stafurik, and T. Wilbanks, 1991: *Energy efficiency, developing nations, and Eastern Europe,* report to the U.S. Working Group on Global Energy Efficiency, International Institute for Energy Conservation, Washington, DC.

Lovins, A.B., and L.H. Lovins, 1991: Least cost climate stabilization, *Annual Review of Energy and the Environment,* 16, 433– 531.

Mahin, D.B., 1988: Prospects in developing countries for energy from urban solid wastes. In *Bioenergy Systems Report,* U.S. Agency for International Development, Washington, DC.

Makarov, A., and I. Bashmakov, 1991: An energy development strategy for the USSR: Minimizing greenhouse gas emissions, *Energy Policy,* 19(10), 987–994.

Manne, A.S., 1992: Global 2100: Alternative scenarios for reducing emissions, OECD Working Paper 111, OECD, Paris.

Manne, A.S., 1993: International trade: The impacts of unilateral carbon emission limits, paper presented at the International Conference on the Economics of Climate Change, OECD, Paris.

Manne, A.S., and R.G. Richels, 1990a: CO$_2$ emission limits: An economic cost analysis for the USA, *The Energy Journal,* 22(2), 51–74.

Manne, A.S., and R.G. Richels, 1990b: Global CO$_2$ emissions reductions – the impacts of rising energy costs, *The Energy Journal,* 12(1), 87–102.

Manne, A.S., and C. Wene, 1990: *MARKAL-MACRO: A Linked model for energy-economy Analysis,* Brookhaven National Laboratory, Upton, NY.

Manne, A.S., and R.G. Richels, 1991: Buying greenhouse insurance, *Energy Policy,* 19(6), 543–552.

Manne, A.S., and R.G. Richels, 1992: *Buying greenhouse insurance – the economic costs of CO$_2$ emission limits,* MIT Press, Cambridge, MA.

Manne, A.S., and R.G. Richels, 1993: The EC proposal for combining carbon and energy taxes: The implications for future CO_2 emissions, *Energy Policy,* **21**(1), 5–13.

Manne, A.S., and L. Schrattenholzer, 1993: Global scenarios for carbon dioxide emissions, *Energy,* **18**(12), pp. 1207–1222.

Manne, A.S., and J. Oliveira-Martins, 1994: Comparisons of model structure and policy scenarios: GREEN and 12RT, draft, Annex to the WP1 Paper on Policy Response to the Threat of Global Warming, OECD Model Comparison Project (II). OECD, Paris.

Manne, A.S., and R.G. Richels, 1994: The costs of stabilizing global CO_2 emissions: A probabilistic analysis based on expert judgments, *The Energy Journal,* **15**(1), 31–56.

Marks, R.E., P.L. Swan, P. McLennan, P.B. Dixon, and R. Schodde, *et al.,* 1990: The cost of Australian carbon dioxide abatement, *The Energy Journal,* **12**(2), 135–152.

Marshall, J.S., R.C. Pilcher, C.J. Bibler, and D.K. Kruger, 1993: *Reducing methane emissions from coal mines in Russia and Ukraine: The potential for coalbed methane development,* prepared for the U.S. Environmental Protection Agency, Washington, DC.

Masera, O., M. Bellon, and G. Segura. 1994: Forest management options for sequestering carbon in Mexico, *Biomass and Bioenergy* (forthcoming).

Masters, C.D., E.D. Attanasi, W.D. Dietzman, R.F. Meyer, R.W. Mitchell, and D.H. Root, 1987: World resources of crude oil, natural gas, natural bitumen, and shale oil, *Proceedings of the 12th World Petroleum Congress,* vol. 5, John Wiley, London.

McDougall, R.A., 1993: Energy taxes and greenhouse gas emissions in Australia, Monash Centre for Policy Studies, General Paper No. G-104, December, Clayton, Australia.

McDougall, R.A., and G.A. Meagher, 1994: Distributional effects of a carbon tax in Australia, paper presented to the 50th Congress of the International Institute Public Finance (IIPF), August, Cambridge, MA.

Meredydd, E., 1994: The status of joint implementation programs in the countries of Central and Eastern Europe and the former Soviet Union, draft, Battelle Pacific Northwest Laboratories, Washington, DC.

Michalik, J., S. Pasierb, J. Piszczek, M. Pyka, and J. Surówka, 1992: *Evaluation of the feasibility and profitability of implementing new energy conservation technologies in Poland,* Polish Foundation for Energy Efficiency (October), Warsaw.

Mills, E., D. Wilson, and T.B. Johansson, 1991: Getting started: No-regrets strategies for reducing greenhouse gas emissions, *Energy Policy,* **19**(6), 526–542.

Mintzer, I., 1987: *A matter of degrees: The potential for controlling the greenhouse effect,* Research Report #5, World Resources Institute, Washington, DC.

Mongia, N., R. Bhatia, J. Sathaye, and P. Mongia, 1991: Cost of reducing CO_2 emissions from India: Imperatives for international transfer of resources and technologies, *Energy Policy,* **19**(10), 978–986.

Morthorst, P.E., 1993: The cost of CO_2 reduction in Denmark – methodology and results, Risø National Laboratory, Systems Analysis Department, Denmark (December).

Morthorst, P.E., and P.E. Grohnheit, 1992: *UNEP greenhouse gas abatement costing studies – Danish Country Report, Phase 1,* Risø National Laboratory, Denmark.

Moulton, R., and K. Richards, 1990: Costs of sequestering carbon through tree planting and forest management in the United States. In *General Technical Report WO-58,* U.S. Dept. of Agriculture, Washington, DC.

Nadel, S., 1992: Utility demand-side management experience and

potential – A critical Review, *Annual Review of Energy and the Environment,* **17,** 507–535.

Nagata, Y., K. Yamaji, and N. Sakarai, 1991: *CO_2 reduction by carbon taxation and its economic impact,* Central Research Institute of the Electric Power Industry, Economic Research Institute Report No. 491002, Tokyo.

NAS (National Academy of Sciences), 1991: *Policy implications of greenhouse warming,* National Academy Press, Washington, DC.

Naughten, B., and A. Dickson, 1995: Greenhouse, energy efficiency and cost effectiveness. In *Outlook 95,* vol. 3, *Minerals and energy,* pp. 289–301, Proceedings of the National Agricultural and Resources Outlook Conference, Australian Bureau of Agricultural and Resource Economics, Canberra (February).

Naughten, B., B. Bowen, and A. Beck, 1993: Energy market failure in road transport: Is there scope for "no regrets" greenhouse gas reduction? *Climatic Change,* **25,** 271–288.

Naughten, B., P. Pakravan, J. Dlugosz, and A. Dixon, 1994: *Reductions in greenhouse gas emissions from the Australian energy system: A report on modelling experiments using ABARE's MENSA model,* report prepared for the Australian Department of the Environment, Sport, and Territories, Canberra.

NEPP (National Environmental Policy Plan), 1989: To choose or to lose, Second Chamber of the States General, The Hague, Netherlands.

New York State, 1991: Analysis of carbon reduction in New York State, Report of the New York State Energy Office in consultation with NYS Department of Environmental Conservation and NYS Department of Public Service, NYS Energy Office, Albany, NY.

Nordhaus, W.D., 1991a: To slow or not to slow: The economics of the greenhouse effect, *Economic Journal,* **101,** 920–937.

Nordhaus, W.D., 1991b: The cost of slowing climate change: A survey, *The Energy Journal,* **12**(1), 37–65.

Nordhaus, W.D., 1994: *Managing the global commons: The economics of the greenhouse effect,* MIT Press, Cambridge, MA.

Nordhaus, W.D., and G. Yohe, 1983: Future carbon dioxide emissions from fossil fuels. In *Changing climate,* National Research Council, Carbon Dioxide Assessment Committee, National Academy Press, Washington, DC.

Northwest Power Planning Council (NWPPC), 1991: Northwest conservation and electric power plan, NWPPC, Portland, OR.

OECD, 1993: The costs of cutting carbon emissions: Results from global models, OECD, Paris.

OECD, 1994: *GREEN: The reference manual,* OECD Department of Economics, Working Paper No. 143, OECD, Paris.

OECD, 1995: *Global warming: Economic dimensions and policy responses,* OECD, Paris.

Oliviera-Martins, J., 1995: Unilateral emissions control, energy-intensive industries, and carbon leakages. In OECD. *Global warming: Economic dimensions and policy responses,* pp. 105–124, OECD Monograph OECD, Paris.

Oliveira-Martins, J., J-M. Burniaux, and J.P. Martin, 1992a: Trade and the effectiveness of unilateral CO_2 -abatement policies: Evidence from GREEN, *OECD Economic Studies,* **19,** 123–140.

Oliveira-Martins, J., J.-M. Burniaux, J.P. Martin, and G. Nicoletti, 1992b: The cost of reducing CO_2 emissions: A comparison of carbon tax curves with GREEN, OECD Economics Working Paper No. 118, OECD, Paris.

Orlic, M.A., and R.A. Leng, 1992: *Preliminary proposal to assist Bangladesh to improve ruminant livestock productivity and reduce methane emissions,* Prepared for U.S. Environmental Protection Agency, Washington, DC.

OTA (Office of Technology Assessment), 1991: *Changing by de-*

grees: Steps to reduce greenhouse gases: Summary, OTA-0-482, U.S. Government Printing Office, Washington, DC.

Ottinger, R.L., D.R. Wooley, N.A. Robinson, D.R. Hodas, and S.E. Babb, 1990: *Environmental costs of electricity.* Prepared by Pace University Center for Environmental Legal Studies for the New York State Energy and Developmental Authority and the U.S. Department of Energy, Oceana Publications, New York, NY.

Parks, P., and I. Hardie, 1995: Least cost forest carbon reserves: Cost-effective subsidies to convert marginal agricultural land to forests, *Land Economics,* **71**(1), 122–136.

Peng, Z., and K. Hanslow, 1993: China's greenhouse policy options: A general equilibrium analysis, ABARE paper presented to the International Congress on Modelling and Simulation, Modelling change in Environmental and Socioeconomic Systems, University of Western Australia, Perth (December).

Perroni, C., and T. Rutherford, 1991: *International trade in carbon emissions rights and basic materials: General equilibrium calculations for 2020,* Dept. of Economics, Wilfred Laurier University, Waterloo, Canada, and Dept. of Economics, University of Western Ontario, London, Canada.

Petersen, E., S. Belanger, D. Cohan, A. Diener, J. M. Drozd, and A. Gjerde, 1993: *The transition to reduced levels of carbon emissions,* WP 12.6, Energy Modeling Forum, Stanford University, Stanford, CA.

Pezzey, J., 1992: Analysis of unilateral CO_2 control in the European Community and OECD, *The Energy Journal,* **13**, 159–171.

Pilcher, R.C., C.J. Bibler, R. Glickert, *et al.,* 1991: *Assessment of the potential for economic development and utilization of coalbed methane in Poland,* prepared for U.S. Environmental Protection Agency, Washington, DC.

Proost, S., and D. van Regemorter, 1990: Economic effects of a carbon tax – With a general equilibrium illustration for Belgium, Public Economics Research Paper No. 11, Katholieke Universiteit, Leuven, Belgium.

Radwanski, E., A. Gromadzinski, E. Hille, P. Skowronski, and S. Szukalski, 1993: *Case study of greenhouse gas emission in Poland: Final report* (Polish Foundation for Energy Efficiency), prepared for the Pacific Northwest Laboratories, Washington, DC.

Ravindranath, N., and B. Somashekhar, 1994: Potential and economics of forestry options for carbon sequestration in India, *Biomass and Bioenergy* (forthcoming).

Richards, K.M., 1989: Landfill gas: Working with Gaia. In *Biodeterioration extracts,* no. 4, Energy Technology Support Unit, Harwell Laboratory, Oxfordshire, UK.

Richards, K.M., R. Moulton, and R. Birdsey, 1993: Costs of creating carbon sinks in the U.S. In *Proceedings of the International Energy Agency Carbon Dioxide Disposal Symposium,* P. Riemer, ed., pp. 905–912, Pergamon Press, Oxford.

Richels, R., and J. Edmonds, 1995: The economics of stabilizing atmospheric CO_2 concentrations, *Energy Policy,* **23**(3/4), 373–379.

RIVM (Rijksinstitut voor Volksgezondheid en Milieuhygiene), 1993: *Proceedings of International IPCC Workshop on Methane and Nitrous Oxide: Methods in national emissions inventories and options for control,* A.R. van Amstel, ed., RIVM, Amersfoort, the Netherlands.

Robinson, J., M. Fraser, E. Haites, D. Harvey, M. Jaccard, A. Reisch, and R. Torrie, 1993: *Canadian options for greenhouse gas emission reduction: Final report of the COGGER panel to the Canadian Global Change Program and the Canadian Climate Program Board,* Canadian Global Change Program Technical Report Series No. 93-1. The Royal Society of Canada, Ottawa.

Roos, K., 1991: Profitable alternatives for regulatory impacts on livestock waste management. In *Proceedings of American Society of Agricultural Engineers 1991 National Livestock, Poultry, and Aquaculture Waste Management Meeting.*

RSWG (Report of the Response Strategies Working Group), 1990: *Formulation of response strategies,* Intergovernmental Panel on Climate Change, Island Press, Washington, DC.

Rubin, E., R. Cooper, R. Frosch, T. Lee, G. Marland, A. Rosenfeld, and D. Stine, 1992: Realistic mitigation options for global warming, *Science,* **257,** 148ff.

Rutherford, T., 1992: The welfare effects of fossil carbon reductions: Results from a recursively dynamic trade model, Working Papers, No. 112, OECD/GD(92)89, OECD, Paris.

Saadullah, M., 1991: Importance of urea molasses block lick and bypass protein on animal production (Bangladesh), paper presented at International Atomic Energy Agency, International Symposium on Nuclear and Related Techniques in Animal Production and Health, Vienna, Austria (April).

Sathaye, J., and N. Goldman (eds.), 1991: *CO$_2$ emissions from developing countries: Better understanding the role of energy in the long term,* vol. 2, *Argentina, Brazil, Mexico and Venezuela,* and vol. 3: *China, India, Indonesia and South Korea,* Energy & Environment Division, Lawrence Berkeley Laboratory, Berkeley, CA.

Schipper, L., and S. Meyers, 1992: The potential for energy conservation. In *World energy: Building a Sustainable Future,* Stockholm Environmental Institute, Stockholm.

Schroeder, P., 1992: Carbon storage potential of short rotation tropical tree plantations, *Forest Ecology and Management,* **50,** 31–41.

Sedjo, R., and A. Solomon, 1989: Climate and forests. In *Greenhouse warming: Abatement and adaptation,* N. Rosenberg, W. Easterling, P. Crosson, and J. Darmstadter, eds., pp. 110–119, Resources for the Future, Washington, DC.

SEI (Stockholm Environment Institute)/Greenpeace, 1993: *Towards a fossil free energy future, The next energy transition,* a technical analysis for Greenpeace International, Tellus Institute, Boston.

Shackleton, R., M. Shelby, A. Cristofar, R. Brinner, and J. Yancher, 1992: *The efficiency value of carbon tax revenues,* U.S. Environmental Protection Agency, Washington, DC.

Shackleton, R., and M. Shelby, 1992: The efficiency value of carbon tax revenues. Programme de recherche interdisciplinaire sur société, technique, et énergie, Centre national de recherche scientifique (PRISTE-CNRS), Oct. 1992: Paris.

Sinyak, Y., and K. Nagano, 1992: *Global energy strategies to control future carbon dioxide emissions,* Status Report, International Institute for Applied Systems Analysis, Laxenburg, Austria.

Sitnicki, S., K. Budzinski, J. Juda, J. Michna, and A. Spilewicz, 1991: Poland: Opportunities for carbon emissions control, *Energy Policy,* **19,** 995–1002.

Sollid, A.E., and M.J. Walters, 1992: *Reducing ruminant methane emissions in China: Findings of a prefeasibility site visit,* prepared for the U.S. Environmental Protection Agency, Washington, DC.

Sondheimer, J., 1990: Energy policy and the environment: Energy policy and the Green 1990s – macroeconomic effect of a carbon tax, paper presented at The Economy and the Green 1990s, Robinson College, Cambridge, UK (July), seminar sponsored by Cambridge Econometrics and UKCEED.

Tasman Institute, 1994: *A framework for trading carbon credits from New Zealand's forests,* Report C6, Tasman Institute, Melbourne, Australia.

Tonkal, V., N. Gnedoy, M. Kulik, M. Mints, and M. Raptsoun, 1992: *Case study and the potential of energy conservation in Ukraine,* Battelle Pacific Northwest Laboratories, Richland, WA.

UNEP, 1994a: *UNEP greenhouse gas abatement costing studies, Phase Two Report, Part 1: Main Report,* UNEP Collaborating Centre on Energy and Environment, Risø National Laboratory, Denmark.

UNEP, 1994b: *UNEP greenhouse gas abatement costing studies, Part Two: Country summaries,* UNEP Collaborating Centre on Energy and Environment, Risø National Laboratory, Denmark.

UNEP, 1994c: *UNEP greenhouse gas abatement costing studies, Annex: Guidelines,* UNEP Collaborating Centre on Energy and Environment, Risø National Laboratory, Denmark.

U.S. Department of Energy, 1991: *Limiting net greenhouse gas emissions in the United States,* Washington, DC.

U.S. EPA (Environmental Protection Agency), 1989: *Policy options for stabilizing global climate – Report to Congress,* vol. 1: Chapters 1-6, United States Environmental Protection Agency, Washington, DC.

U.S. EPA (Environmental Protection Agency), 1990: *Preliminary technology cost estimates of measures available to reduce U.S. greenhouse gas emissions by 2010,* Washington, DC.

U.S. EPA, 1993a: *Options for reducing methane emissions internationally,* vol. 1: *Technological options for reducing methane emissions,* report to Congress, K.B. Hogan, ed., EPA 430-R-93-006 A, Office of Air and Radiation, Washington, DC.

U.S. EPA, 1993b: *Options for reducing methane emissions internationally,* vol. 2: *International opportunities for reducing methane emissions,* report to Congress, K.B. Hogan, ed., EPA 430-R-93-006 B, Office of Air and Radiation, Washington, DC.

U.S. EPA, 1993c: *Opportunities to reduce anthropogenic methane emissions in the United States,* report to Congress, K.B. Hogan, ed., EPA 430-R-93-012, Office of Air and Radiation, Washington, DC.

U.S. EPA, 1994: *International anthropogenic methane emissions: Estimates for 1990,* report to Congress, M.J. Adler, ed., Office of Policy Planning and Evaluation, Washington, DC.

Van der Mensbrugghe, D., 1995: Regional carbon tax required for equiproportional CO_2 emission reduction from baseline, Interval Note, OECD, Paris.

Van Kooten, G., L. Arthur, and W. Wilson. 1992: Potential to sequester carbon in Canadian forests: Some economic considerations, *Canadian Public Policy,* **18**(2), 127–138.

Walker, I.O., and F. Birol, 1992: Analysing the cost of an OPEC environmental tax to the developing countries, *Energy Policy,* **20**, 559–567.

Walz, R., M. Schon, *et al.,* 1994: Gesamtwirtschaftliche Auswirkungen von Emissionsminderungsstrategien, Study for the Enquete Commission of the German Bundestag, Karlsruhe/Berlin.

Whalley, J., and R. Wigle, 1990: Cutting CO_2 emissions: The effects of alternative policy approaches, *The Energy Journal,* **12**(1), 109–124.

Williams, R.H., 1990: Low-cost strategies for coping with CO_2 emission limits, The Energy Journal, **11**(3), 35–59.

Wilson, D., and J. Swisher, 1992: *Exploring the gap: Top-down vs. bottom-up analyses of the cost of mitigating global warming,* Summer School on Science and World Affairs, Shanghai, China.

World Energy Council, 1994: *New renewable energy resources: A guide to the future,* World Energy Council, London.

Xu, D., 1994: The potential for reducing atmospheric carbon by large-scale afforestation in China and related cost/benefit analysis, *Biomass and Bioenergy* (forthcoming).

Yamaji, K., 1990: Japan. In *Carbon emissions control strategies: Case studies in international cooperation,* W.U. Chandler, ed., pp. 155–172, Conservation Foundation, Washington, DC; Carbon Dioxide Information Center, Oak Ridge National Laboratory, Oak Ridge, TN.

Ybeme, J.R., P.A. Okken, T. Kram, and P. Lako, 1993: *CO$_2$ abatement in the Netherlands: A study following the UNEP guidelines,* ECN Policy Studies, Amsterdam.

Zachariah, K.C., and M.T. Vu, 1988: *World population projections, 1987-1988 edition,* World Bank, Johns Hopkins University Press, Baltimore, MD.

10

Integrated Assessment of Climate Change: An Overview and Comparison of Approaches and Results

Convening Lead Author:
J. WEYANT

Principal Lead Authors:
O. DAVIDSON, H. DOWLATABADI, J. EDMONDS, M. GRUBB, E.A. PARSON,
R. RICHELS, J. ROTMANS, P.R. SHUKLA, R.S.J. TOL

Lead Authors:
W. CLINE, S. FANKHAUSER

CONTENTS

SUMMARY

Integrated assessments are convenient frameworks for combining knowledge from a wide range of disciplines. These efforts address three goals:

(1) Coordinated exploration of possible future trajectories of human and natural systems
(2) Development of insights into key questions of policy formation
(3) Prioritization of research needs in order to enhance our ability to identify robust policy options

The integration process helps the analyst coordinate assumptions from different disciplines and introduce feedbacks absent in conclusions available from individual disciplinary fields.

Historically, the most common approach to integrated assessment has been the attempt by individual researchers or research teams to integrate the information available from the relevant disciplines and provide policy advice in books and reports. Although this has typically been accomplished via informed qualitative linkages, Integrated Assessment Models (IAMs) use a computer program to link an array of component models based on mathematical representations of information from the various contributing disciplines. This approach makes it easier to ensure consistency among the assumptions input to the various components of the models, but may tend to constrain the type of information that can be used to what is explicitly represented in the model.

IAMs can be divided into two broad classes: *policy optimization models* and *policy evaluation models*. Policy optimization models optimize key policy control variables such as carbon emission control rates or carbon taxes, given formulated policy goals (e.g., maximizing welfare or minimizing the cost of meeting a carbon emission or concentration target). Policy evaluation models, on the other hand, project the physical, ecological, economic, and social consequences of specific policies.

Policy optimization models can be divided into three principal types:

(1) Cost-benefit models, which attempt to balance the costs and benefits of climate policies
(2) Target-based models, which optimize responses, given targets for emissions or climate change impacts
(3) Uncerainty-based models, which deal with decision making under conditions of uncertainty

Policy evaluation models are of two types:

(1) Deterministic projection models, in which each input and output takes on a single value

(2) Stochastic projection models, in which at least some inputs and outputs are treated stochastically

Each approach has strengths and weaknesses and produces particular insights regarding climate change and potential policy responses to it. Some of the more advanced models can be used for several purposes.

Cost-benefit models

Cost-benefit IAMs balance the marginal costs of controlling greenhouse gas emissions against those of adapting to any climate change. In this approach any constraint on human activities is explicitly represented and costed out. At present, models of this type include highly aggregated representations of climate damages, generally representing economic losses as a function of mean global surface temperature but sometimes disaggregating total damages into market and nonmarket damage components.

Keeping in mind the uncertainties and limitations inherent in these models, they can nevertheless be used to compute optimal control strategies. Specifically, results relating to optimal CO_2 emission control rates (percentage reductions in emissions relative to baseline emissions) and carbon taxes (equivalent to the marginal cost of efficient carbon emission reductions) over the next century vary widely, in part because of debates about the nature and valuation of climate impacts and in part because of debates about how to represent the dynamics of energy systems and technology development processes. However, the models do agree that higher control costs, lower damage estimates, and higher discount rates lead to lower initial optimal control rates, whereas lower control costs, higher damage estimates, and lower discount rates lead to higher initial control rates. For example, if new technology development is highly responsive to the level of control, lower control costs will result over time and a higher initial optimal control rate will be implied. Conversely, breakthroughs in biotechnology that would be expected to reduce the damages resulting from climate change on agriculture would (other things being equal) reduce the optimal initial control rate.

Target-based models

In target-based IAMs, targets for greenhouse gas emissions, atmospheric concentrations of greenhouse gases, climate change, or climate impacts can be set to avoid certain types of risks, perhaps according to the "precautionary principle." As a result, the guiding principle of the cost-benefit models, economic optimization (i.e., the marginal cost of implementing the mitigation and adoption measures resulting from the individual targets should equal the marginal economic benefits of the impacts avoided), is replaced by an emphasis on precautionary targets, risk aversion, and physical criteria.

Several integrated assessment efforts have attempted to identify the cost-effective emission timepath for reaching a particular CO_2 concentration target, that is, to identify the emission profile that minimizes abatement costs. The initial path depends on assumptions about the current availability of low-cost measures and the inertia of the system, but after taking account of these factors the least-cost path tends to remain close to the reference path initially and to diverge at a rate that depends on the concentration target (among other things). Factors that tend to favour deferral of reductions include

(1)   A positive marginal product of capital
(2)   The prospect of autonomous reductions in the cost of carbon-free substitutes
(3)   More time to achieve an optimal configuration of the capital stock in anticipation of emission constrictions
(4)   The carbon cycle (the earlier the release, the more time for removal from the atmosphere)

Conversely, factors that make greater early action optimal include

(1)   Lower marginal product of capital
(2)   The prospect of inducing further cost reductions through abatement action
(3)   The prospect of avoiding being locked in to more carbonintensive patterns of development
(4)   The extent to which inertia may amplify the costs of having to make more rapid emission reductions later

It is important to note that these analyses were conducted with top-down models of the global energy-economic system. Although the models incorporate opportunities for "no-regrets" measures, they assume that such options are in insufficient supply to displace fossil fuels altogether. Hence, they show emissions continuing to grow under a wide range of assumptions about population and economic growth. IAMs that include the full range of factors that bear on the optimal timing of emission reductions have not yet been developed.

Uncertainty-based models

As a result of the high level of uncertainty about the future evolution of socioeconomic and natural systems, some researchers have put the analysis of climate change into explicit frameworks of the kind discussed in Chapter 2 for analyzing decision making under uncertainty. Generally, this has been done either by including an uncertainty representation of all key parameters within simplified models of the types discussed above or by adding a limited number of alternative states to full cost-benefit models. In addition, many of these models allow policies to be changed as uncertainties are resolved through time, although the process by which uncertainty is resolved is usually repre-

sented quite simplistically, perhaps even unrealistically. The uncertainty-based cost-benefit assessments completed thus far find higher optimal rates of abatement than do the deterministic cost-benefit models. Uncertainty analyses with target-oriented IAMs have also been used to calculate the likelihood of certain key physical thresholds being exceeded in the future.

Policy evaluation models

Policy evaluation IAMs are comprehensive, process-based models that attempt to provide a thorough description of the complex, long-term dynamics of the biosphere-climate system. The dynamic description often includes a description of atmospheric chemistry, climate, and ecological impact processes as well as a number of geophysical and biogeochemical feedbacks within the system. Some of the models even deal with biosphere-climate dynamics at a geograpically explicit level. On the other hand, the socioeconomic system in these models is usually poorly represented. The larger models usually do not serve the purpose of performing cost-benefit or cost-effectiveness analyses, but they can provide insights into the intricate interrelationships between the various components of the human system and the biosphere-climate system. Ideally, such insights can lead to new priority setting in the analysis of the climate change process. Policy evaluation IAMs provide a useful framework for identifying, illuminating, and clarifying current uncertainties. The most important uncertainties can be compared and ranked, and then the model can show how they propagate through the whole human/climate/biosphere system.

Policy Evaluation IAMs have helped identify critical knowledge gaps in several areas. Some of the most important findings from these models relate to the balancing of the carbon cycle, integrated land-use analysis, and sulphur aerosols.

The carbon cycle. IAM assessments of the impact of feedback mechanisms within the global carbon cycle have demonstrated that there are a large number of representations of the cycle that balance the past and present carbon budget, each of which can lead to very different atmospheric concentration levels for a specific projection of future carbon emissions.

Land-use analysis. The integration of geographically explicit representations of agriculture and land cover with climate change calculations has already provided new insights into climate-related shifts in agricultural areas and the influence of changing land cover on climate. Preliminary results suggest that regional demands for land can serve as a surrogate for the regional and local forces that are driving local land cover changes. These results also show the vulnerability of protected areas under shifting vegetation zones, and the consequences for biodiversity and nature conservation.

Sulphur aerosols. The first integrated assessments incorporating both SO_2 and CO_2 emissions show that it is conceivable that reductions in radiative forcing resulting from rapid reductions in coal use in some regions could be more than offset for a decade or two by increased radiative forcing from the associated reductions in SO_2 emissions. However, spatial and temporal differences in sulphur emissions and the local nature of the changes in radiative forcing due to sulphur aerosols mean that the effects cannot be considered to cancel each other out in terms of impacts on regional climate patterns.

Although integrated assessment of climate change is a rapidly evolving field, the following additional preliminary conclusions can be made from the work completed thus far:

(1) Integrated assessments are no stronger than the underlying natural and economic science that supports them. Nevertheless, by bringing many components of the climate change problem into a common framework, they offer potentially useful insights that would be unavailable from a purely disciplinary research programme. In applying these assessments to climate policy design, two critical factors should be noted. First, researchers should provide a measure of the confidence with which such policy assessments can be made; and, second, the models should indicate the distribution across countries and income levels of impacts associated with particular policy goals and implementations.

(2) Recent refinements to Integrated Assessment Models show increased diversity in the distribution of regional costs and benefits. This implies potentially greater difficulty in reaching agreements but also opens up the possibility of greater gains in global welfare from achieving them.

(3) From the integrated assessment perspective, there are important gaps in disciplinary research and inconsistencies between the information produced by the various disciplines whose reconciliation would lead to improved integrated assessments. Much of the underlying fundamental science needed to develop coordinated integrated assessments is not in a form suitable for immediate use. Different disciplinary experts, for example, have held different factors constant. This contributes to the difficulty of developing, calibrating, and validating the models. In addition, some of the underlying fundamental research has not been performed. For example, models of adaptive decision making do not yet explicitly consider how social goals and progress towards them are measured over time or how global change processes are detected. Finally, there are some highly

uncertain components in the current set of integrated assessments, including the sensitivity of the climate system to changes in greenhouse gas concentrations, the physical and economic impacts of any climate change that may occur, and the applicability and choice of discount rates. One of the main values of the integrated assessment approach to the study of climate change lies in the identification of gaps and inconsistencies in our knowledge of the underlying phenonema and their implications for future research.

(4) Although it is difficult confidently to choose one policy in preference to others based on current knowledge about the climate system and human interactions with it, it has been demonstrated that the policy objective, discount rate, and timing of compliance can be critical to short-term policy formulation and the overall cost of action.

(5) Given the considerable uncertainties associated with how the climate system will evolve and interact with human activities, policies that enhance the flexibility of nations and individuals to respond to any impacts that do emerge tend to have high value. Because they can be focussed directly on the impacts of climate change, research and development activities related to technologies and institutions that facilitate the process of adaptation to climate change generally have a high payoff. Research and development activities directed towards technologies that lower material use in economic activity are also a good bet.

(6) Most current models do not match the social and economic organization of the developing economies well. For example, none of the existing models can incorporate hierarchical decision structures or represent the operation of the informal economies that are important in many developing countries. This can lead to biases in global assessments when impacts in developing countries are valued as if these countries were no different from developed countries.

(7) Finally, climate change is but one dimension of global change. For example, integrated assessments suggest that ecosystem impacts from projected climate change, agriculture management, and urbanization could well be of similar magnitudes.

10.1 Introduction

The historic Framework Convention on Climate Change (FCCC), signed by 154 countries at the UN Conference on Environment and Development (UNCED) in Brazil in June 1992, had as its central objective the stabilization of atmospheric greenhouse gas concentrations at a level that would prevent dangerous anthropogenic interference with the climate system. It also stated that this goal should be realized soon enough that ecosystems could adapt naturally to climate change, that food production would not be threatened, and that sustainable economic development could proceed (FCCC, Article 2). The text does not specify, though, what the operational meaning of "dangerous anthropogenic interference" is, how its occurrence or the risk of its occurrence could be detected, or what measures, applied at what level of stringency, would be justified in avoiding it. The other central concepts in the objective – natural adaptation of ecosystems, threats to food production, and sustainable economic development – are also not articulated precisely. Nor could they be.

Rendering the Convention's objective into operational specifics will require further deliberations, informed by the best available synthesis of current scientific, technical, economic, and sociopolitical knowledge. Such a synthesis can help define and assess the risks associated with climate change, ecosystem responses, and human adaptive responses as well as the feasibility, effectiveness, cost, and side effects of potential response measures. Synthesizing and communicating such knowledge in support of policy deliberations is the function of assessment.

Integrated assessment is distinguished from disciplinary research by its purpose, which is to inform policy and decision making rather than to advance knowledge for its intrinsic value. Integrated assessment is identified by the breadth of knowledge sources and the variety of disciplines from which it draws. It is to be distinguished from those (infrequent) instances in which a significant policy issue can be well informed by clear presentation of a body of knowledge held within a single discipline.

The broader the set of knowledge domains that must be synthesized to inform a policy or decision, the greater the intellectual and managerial problems that must be overcome to do the assessment well and make it useful to its audience. How integrated any particular assessment must be depends on the issue or decision to be informed. Perhaps more than any other policy issue, global climate change requires integrated assessment. Making rational, informed social decisions on climate change potentially requires knowledge of a large number of interrelated processes, beginning with the human activities that affect greenhouse gas emissions and extending to the atmospheric, oceanic, and biological processes that link emissions to atmospheric concentrations, the climatic and radiative processes that link atmospheric concentrations to global and regional climate, the ecological, economic, and sociopolitical processes that link changed climate to valued impacts, and the processes by which such evaluations are made. Any progress in understanding and responding to an issue of such complexity will require the capacity to interpret, integrate, reconcile, organize, and communicate knowledge across domains – that is, to do integrated assessment. This need has been widely recognized in calls to advance methods of integrated assessment and in the large number of projects now underway. Although there have been past examples of integrated assessments of major environmental issues – for example, the American CIAP Project (Grobecker *et al.,* 1974) and the European acid rain studies integrated in the RAINS model (Alcamo *et al.,* 1990) – the current level of integrated assessment activity on global climate change is unprecedented.

10.1.1 *Purposes of integrated assessment*

Integrated assessment can in principle serve three purposes. First, integrated assessment can help assess potential responses to climate change, by (1) representing physical, ecological, economic, and social processes to project the consequences of climate change and of particular policy responses to it, (2) using a cost-benefit formulation to compare costs of responses to the severity of the impacts they are intended to prevent, or (3) using a cost-effectiveness formulation to compare the relative effectiveness and cost of different responses to meet a specified target. Whichever of these formulations is employed, integrated assessment performs this function by making consistent, contingent, appropriately qualified projections of the likely cost and effect of specified responses.

Second, by providing a coherent, systematic framework to structure present knowledge, integrated assessment can bring two important benefits: It can promote a broad view of the climate issue that may facilitate more systematic searching for possible responses and avoid prematurely settling on one or a few proposed responses; and it can provide a consistent representation of current uncertainties, permitting identification and prioritization of those that are most important in practical terms – that is, those uncertainties that are most important to reduce in order to understand what should be done. Since the most important uncertainties from the perspective of policy relevance will not necessarily be the most important for advancing basic understanding, this function of integrated assessment can be of the highest importance.

Third, integrated assessment can help to address the most fundamental policy question about global climate change: How important is it relative to other matters of human concern? Gaining insight into this question will require comparing the aggregate social effect of climate change and potential responses to it with the aggregate social effect likely to arise from other changes and risks over the same period of time.

In fulfilling these purposes, integrated assessment supplements disciplinary research but does not replace it. A disciplinary research programme in the natural or social sciences, even one including components representing every relevant discipline that could contribute to informing policy choice, will not normally emphasize a synthesis of knowledge across domains and so cannot typically do the jobs of assessing the consequences of potential responses or prioritizing decision-relevant uncertainties and research needs. Current experience suggests that cross-disciplinary integration to fill these needs does not happen spontaneously and can be both difficult and costly. But this does not mean that integrated assessment can replace disciplinary research, even for providing policy-relevant knowledge. Although integrated assessment is needed to identify and prioritize policy-relevant and scientific gaps in knowledge, the gaps so identified can normally only be filled by disciplinary research, whether in the natural or social sciences.

10.2 Approaches to Integrated Assessment

Integrated assessments can be integrated over different dimensions and to different degrees. In contributing to general policy-relevant understanding, studies using many different dimensions and degrees of integration, from the broadest to the narrowest, may make important contributions. In informing the deliberations or decisions of a particular policy audience, though, the appropriate form and extent of integration are determined by the needs of the audience. When there is a specific audience or decision to inform, a useful assessment will seek to represent the kinds of policies and decisions they are concerned with, at a resolution that corresponds to their responsibilities and concerns, while taking appropriate opportunities to help broaden their understanding of the issue.

An area of active current exploration is the development of global-scale climate assessments with "end-to-end" integration that combines assessment of emissions and abatement measures with impacts and adaptation measures. Such projects often (implicitly or explicitly) pursue a cost-benefit framing of the climate issue to shed light on decisions about optimal global emission abatements and efficient means of achieving them. The implied audience for such assessments consists of those decision makers with the authority to balance the extent and form of abatement measures, adaptive and compensatory measures, and possibly geoengineering measures. Assessments of this kind are likely to help improve the general understanding of appropriate responses to the climate issue. Broad balancing of abatement and adaptation measures will be done, implicitly or explicitly, and it is clearly desirable that whatever knowledge is available to illuminate such broad trade-offs be presented to those involved in such choices.

Such assessments, though, are not the only kind that can be useful, or necessarily the most appropriate for informing many specific policy decisions. Certain international deliberations and negotiations, for example, may need assessments that are integrated even more broadly. At this level, it may not be possible to address the climate issue without making judgments of its significance relative to other environmental and policy issues. To engage choices of this breadth, assessments may be required that facilitate comparing the potential impacts of climate change and other issues. End-to-end assessment of climate change may be a necessary component of such assessments, but it may not by itself be sufficient. For questions of such breadth, the most useful assessments may be those that focus not on a single environmental issue but on basic policy choices and long-term technological trends in

areas of human activity that affect a variety of environmental and other issues, such as agriculture or energy.

On the other hand, most policy audiences are likely to need less broadly integrated assessments. This may be so even for assessments to inform international negotiations, if, as often happens, a preexisting political commitment either to a simple heuristic principle like the precautionary principle or to a specific numerical policy target truncates the consideration of responses (Parson and Zeckhauser, 1995; Levy, 1993). In such cases the most useful assessment may be to adopt a cost-effectiveness approach, comparing emission types, sources, gases, and regions, to determine feasible, low-cost ways to meet specified abatement goals (see, e.g., Read, 1994b). However, integrated assessment models can also be used to test the cost-effectiveness or welfare implications of those principles.

Integrated assessment may also be of interest to small countries or regions that may suffer climate impacts but have little or no influence over global emissions. For authorities in such jurisdictions, the crucial dimension of integration will be across dimensions of impact – sector, location, group, and time – under an illustrative set of climate change scenarios. This type of assessment could inform their decisions about long-term climate-dependent investment, emergency response measures, zoning, and insurance and compensation schemes that form the bulk of adaptation response. Recent empirical study suggests that assessments conducted at relatively fine levels of spatial or sectoral aggregation and initiated by decision makers with direct responsibility for making such decisions or responding to such impacts tend to be more immediately useful and more directly used than assessments with national or international scope (Clark and van Eijndhoven, 1996).

10.2.1 Integrated modelling and other methods of integration

A variety of methods to conduct integrated assessment are possible. Current projects on global climate have largely, but not exclusively, pursued integration through a formal integrating model, though the centrality and manner of use of the model vary among projects (Rotmans *et al.,* 1995). Other integration methods that have been tried include special senior commissions or panels whose members span the required range of expertise and integrate knowledge judgmentally through their deliberations; formal models of problem subcomponents, linked through an external, judgmental combination of results rather than through a formal integrating model; collaborative interdisciplinary research teams whose continuing interactions develop collective skills at exchanging and sharing knowledge across their fields; and individual essays by authors with sufficient multidisciplinary competence to encompass the policy problems (see Box 10.1). Other integrating devices are not yet thoroughly developed but may hold substantial promise. These include simulations or policy exercises – devices for joint deliberation by researchers and policymakers in a hypothetical policy setting employing knowledge available from a variety of sources, including existing literatures, formal models, and expert judgment (Brewer, 1986; Parson, 1995b).

BOX 10.1: HISTORY OF INTEGRATED ASSESSMENT

Integrated assessment is neither a new concept nor an activity restricted to climate change. This box provides an illustrative review of landmarks in the history of integrated assessment of global environmental issues.

The first integrated assessment of a global environmental issue was the Climatic Impacts Assessment Program (CIAP), conducted by the U.S. Department of Transportation to assess the environmental impacts of stratospheric flight by supersonic aircraft (Grobecker *et al.*, 1974). Six separate interdisciplinary expert teams examined one link in a causal chain stretching from human activities (scenarios of supersonic flight and jet engine design) through atmospheric chemistry and radiation to biological, economic, and social impacts. The teams exchanged numerical estimates of key quantities, ultimately yielding quantitative estimates of the environmental and economic impacts of specific scenarios of stratospheric flight.

Through the 1970s and early 1980s, several other major integrated assessments were conducted using a similar structure of interdisciplinary expert panels. Early assessments of global climate change that helped lay the groundwork for the present IPCC approach included Clark (1982) and the U.S. National Research Council (1983). This comprehensive, interdisciplinary approach, which does not centrally depend on formal modelling, continues to the present in such bodies as the Assessment Panels of the Montreal Protocol and the IPCC itself. Since CIAP, however, no assessment has attempted such a precise, comprehensive integration of processes, from human activities to valued consequences, without using a formal integrating model.

Formally modelled integrated assessment studies trace their inspiration, if not their precise methods, to the global models of the 1970s, such as Meadows *et al.* (1972) and Mesarovic and Pestel (1974). (This field was reviewed in Meadows *et al.*, 1982.) These highly aggregated dynamic models of world development included generalized representations of pollution and resource depletion but did not address any particular environmental issue.

Formal integrated assessment models of climate change emerged in the late 1970s from earlier economic and technical models of energy policy. Nordhaus (1979) presented the first model that combined energy conversion, emissions, and atmospheric CO_2 concentration. Subsequent efforts in integrated assessment of climate change that stressed formal modelling included the IIASA energy project (Häfele *et al.*, 1981), Nordhaus and Yohe (1983), which added uncertainty to modelled projections of future CO_2 concentrations, and Edmonds and Reilly (1985).

Through the 1980s, climate assessment studies using formal integrated modelling were narrower in scope than those using interdisciplinary expert panels. The modelled assessments normally extended no further than atmospheric CO_2 concentration, excluding both non-CO_2 greenhouse gases and resultant changes in climate and impacts. A separate line of work, beginning with the MINK project (Rosenberg and Crosson, 1991; Rosenberg, 1993) focussed specifically on climate impacts, combining detailed sectoral models of agriculture, forests, energy, and water resources.

The first integrated assessment model to extend fully from emissions to impacts did not address climate change but the more analytically tractable issue of acid rain. The RAINS model of acidification in Europe was developed at IIASA beginning in the early 1980s (Alcamo, Shaw, and Hordijk, 1990). RAINS integrates models of acid emissions, atmospheric transport and deposition, and effects. The RAINS project also pioneered a close relationship between the modelling team and policymakers, arguably leading to a more policy-relevant model and a more useful contribution to negotiations and policy making than has yet been attained on other issues.

The first steps to extend formal integrated modelling of climate change were taken by Mintzer (1987), who added non-CO_2 gases and global temperature change, and subsequently by Lashof and Tirpak (1989) in their Atmospheric Stabilization Framework. The first model to attempt a fully integrated representation of climate from sources to impacts was IMAGE 1.0 (Rotmans, 1990), which subsequently became the basis for the integrated European model ESCAPE (Hulme *et al.*, 1995).

Since 1990, the number of projects in integrated assessment modelling of global climate change has expanded rapidly. The idea that useful models could be developed to span the full range of the climate issue has gained increasing acceptance, as advances in computing power and in the disciplinary understanding and sectoral modelling efforts on which such integrated modelling projects depend have made projects of this kind increasingly feasible. A landmark of the maturation of integrated assessment modelling of climate change was the first conference to assess activity in the field (Nakicenovic *et al.*, 1994). Since then, as discussed in the text of this chapter, the field has continued to expand and develop rapidly.

Doing integrated assessment by building an integrating model has several evident advantages. Constructing a model imposes common standards of coherent, precise communication on project participants. It also imposes common data definitions and standards of consistency and scale on problem components and can facilitate the incorporation of new knowledge in an assessment. Attendant disadvantages are that the modelling may force more precise representation than the underlying knowledge in particular domains allows, may impose inappropriate restrictions, and may direct excessive project effort toward technical problems of model convergence, hence giving aggregate results that say as much about algorithmic artifacts as they do about component understanding. Integration through integrated modelling may be particularly

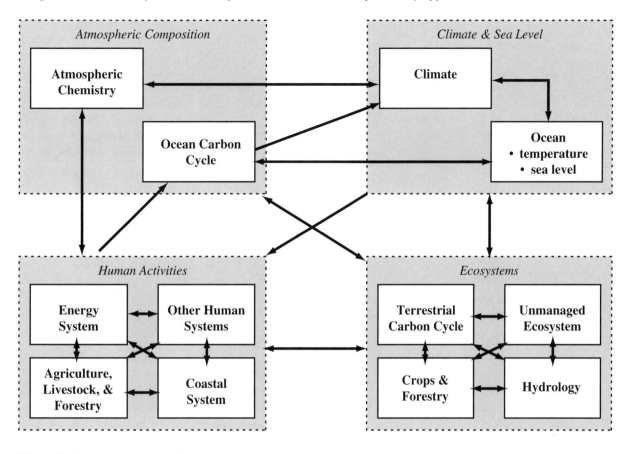

Figure 10.1: Key components of full scale IAMs.

weak in representing policies and decisions realistically and in reflecting knowledge of relevant social, political, institutional, and negotiation processes.

10.2.2 The current state of integrated assessment activity

Most current integrated assessment projects focus principally on building integrated models, although this is only one of various possible approaches to integrated assessment. In addition, most current projects are characterized by a national to global scale, a rather coarse spatial and sectoral resolution, and weak representation of policies and political processes. The balance of this chapter focusses on recent and current work in integrated assessment modelling (IAM), discussing the structure, modelling approaches, and major weaknesses of present projects and reviewing preliminary results.

Despite its importance, the field of integrated assessment is relatively immature and lacks a shared body of professional knowledge and standards of "best practice." Such knowledge will require more experience to develop; in its absence, it would be ill-advised to pursue a single, authoritative vision of integrated assessment. On many intellectual and managerial dimensions, there are many plausible ways of meeting the challenges of integrated assessment, but there is no evident single right way. Consequently, there is much to be gained from the parallel pursuit of diverse approaches, including

those both more and less strongly dependent on integrated modelling.

10.3 Elements of an Integrated Assessment Model

A large number of integrated assessment models, with a wide variety of differing goals and objectives motivating their construction, are now being used to examine the issue of climate change. They vary greatly in their level of detail, but all share the defining trait that they incorporate knowledge from more than one field of study. However, they also vary greatly with regard to their scope. It is therefore important to distinguish between models in terms of this dimension as well as their level of detail. Models that attempt to represent the full range of issues raised by climate change are referred to as "full-scale" IAMs.

Full-scale IAMs must grapple with all the complexity of an IPCC assessment. This is an intimidating array of concerns. But although an IAM for climate change must consider a wide variety of issues, the number of issues is bounded. For the purpose of exposition, we group the principal considerations into four general categories, depicted in Figure 10.1:

- human activities
- atmospheric composition
- climate and sea level
- ecosystems

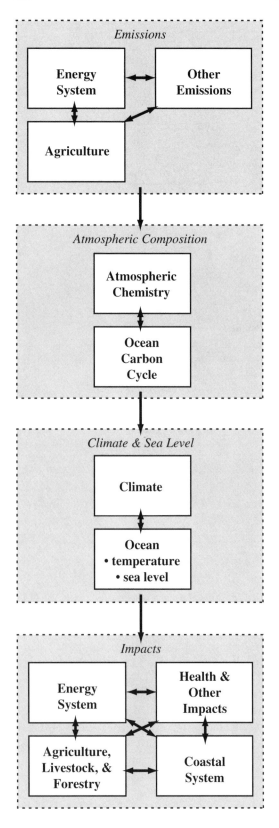

Figure 10.2: End-to-end characterization of IAMs.

is the "end-to-end" characterization depicted in Figure 10.2. In this organizational formulation there are also four categories, this time beginning with emissions and ending with impacts. The principal organizational difference is that human activities and ecosystems are partitioned, with some features of each contained in the emissions and impacts components. This characterization deemphasizes the interactive character of the IAMs, in particular the fact that the same human and natural systems that produce emissions also suffer impacts.

Human systems interact with natural systems in two ways. Human activities are responsible for the emissions of greenhouse-related gases that are the centre of concern in the climate change issue. Human activities are also affected by climate change, either directly, for example, through changes in temperature, which affect demands for space heating and cooling, or indirectly, for example, through changes in sea level, crop productivity, or biodiversity.

Full-scale IAMs must consider the issue of emissions of greenhouse-related gases. The array of gases that matter from the perspective of emissions differs slightly from the array of gases that matter from the perspective of climate. From the perspective of climate change only, the set of gases and particles that have the capacity to change the radiative balance of the planet needs to be considered. At present the set consists principally of the following: water vapour (H_2O), ozone (O_3), carbon dioxide (CO_2), methane (CH_4), nitrous oxide (N_2O), sulphur aerosols, and the chlorofluorocarbons (CFCs) and their substitutes.

The set of gases that must be considered from the perspective of emissions is strongly overlapping, but includes some important differences. Water vapour and O_3 are not emitted in sufficient quantities by human activities to matter. Ozone concentrations are, however, affected by the emissions of other greenhouse-related gases such as carbon monoxide (CO), odd-nitrogen (NO_x), and nonmethane hydrocarbons (NMHC), whereas water vapour concentrations are influenced by the effect of temperature change on the water cycle. Likewise, sulphur aerosols are not emitted but are formed in the atmosphere at a rate that depends on emissions of sulphur oxides and particulate matter as well as other aspects of atmospheric chemistry.

With regard to the emissions of greenhouse-related gases the following human activities figure prominently:

- energy systems
- agriculture, livestock, and forest systems
- industrial systems

The role of energy systems is the single most critical component determining emissions in IAMs. Not only are energy systems associated with the greatest anthropogenic release of carbon to the atmosphere, but they are also associated with the largest anthropogenic release of sulphur compounds as well.

Systems that determine rates of land use change figure importantly, though the relationship between specific human actions and land use change is less well defined than the relationship between energy production and use and the release of

Figure 10.1 is not the only possible depiction of the climate change system. An infinite number of aggregations are possible and a great many "wiring diagrams" already exist. This particular diagram has the virtue of including both human and natural system components. One alternative organization

greenhouse-related gases. Agriculture, livestock, and forestry represent the most extensive anthropogenic uses of land. In addition, agriculture and livestock are important determinants of CH_4 and N_2O releases.

Finally, full-scale IAMs must consider the array of other greenhouse-related emissions to the atmosphere. Most prominent among these are the chlorofluorocarbons and their substitutes, although there are others.

From the perspective of the consequences of climate change, an overlapping but somewhat different list of issues must also be dealt with by IAMs. The problem of climate change impacts is more difficult to deal with in IAMs because impacts are anticipated to affect a wide array of human activities, with no single activity thought to be substantially more vulnerable than others. IAMs thus frequently confront the impacts issue abstractly, using "damage functions," rather than explicitly. Nevertheless, underlying any treatment of impacts within an IAM are, at a minimum, the following human activities:

- agriculture, livestock, and forest systems
- energy systems
- coastal zones
- water systems
- human health
- the value of local air quality
- the values of unmanaged ecosystems[1]

The second information set that a full-scale IAM must generate is the concentrations of greenhouse gases, which the model must translate from both natural emissions and the emission flows generated by human activities. Greenhouse gas concentrations also depend on natural sources and sinks. In general, greenhouse gases can be segregated into CO_2 and other gases. The non-CO_2 greenhouse-related gases are controlled by atmospheric processes. Their sinks are predominantly in the atmosphere. CO_2, on the other hand, is governed by the processes of the carbon cycle. The concentration of CO_2 in the atmosphere is determined predominantly by interactions between atmospheric concentrations and the oceans and terrestrial systems.

Models deal with CO_2 in a variety of ways, ranging from simple airborne fraction models, which use a proportional approximation method to determine atmospheric concentrations, to interactive process models of the atmosphere and biosphere. The present understanding of both the carbon cycle and atmospheric chemistry have been surveyed in Volume 1 of the present report and in previous IPCC scientific reports (see IPCC, 1990, 1992, 1995).

Full-scale IAMs should ultimately also consider the problem of local air quality, as the removal rates for local air pollutants depend on weather conditions, and greenhouse gas abatement influences local air quality. These factors, in turn, interact with the economic value of changes in health conditions. The inclusion of local air quality is not yet possible, however, because of the totally different spatial and temporal scales and aggregation levels of the climate change and local

air pollution problems. At the moment, such analyses can only be done through case studies, such as those being done by RIVM (the Dutch National Institute of Public Health and Environmental Protection) for 25 megacities around the world. Chapter 6 assesses these so-called "secondary benefits" of greenhouse gas abatement.

The third information set that a full-scale IAM must generate is the state of climate and sea level. Climate cannot be derived without dealing in one way or another with oceans. Oceans are an important determinant of the timing of climate change, as they represent an enormous heat sink. Thus, ocean-atmosphere feedbacks also influence the rate of sea level rise. In addition, interactions between the atmosphere and cryosphere affect climate change and sea level. Sea level calculations, for example, must include changes in the volume of meltwater from the major land-based ice sheets. Furthermore, the ocean that interacts with atmospheric processes in determining climate and sea level change also absorbs carbon that has been accounted for in the atmospheric composition model.

In Figure 10.1, the fourth category of IAM information is ecosystems. This category includes information associated with natural emissions of greenhouse-related gases, the terrestrial carbon cycle, and the effect of climate change, sea level rise, and CO_2 on crops, pastures, grazing lands, forests, hydrology, and unmanaged ecosystems.

These systems are strongly interactive. Some models handle them in a holistic manner, explicitly considering the interactions of natural system emissions, the status of unmanaged ecosystems, hydrology, ground cover, crop, and forest productivity. Other models treat them as if they were independent. The managed biosphere interacts strongly with human systems, which determine the selection of crop and managed forest species and the allocation of water resources among competing ends. Interactions between ecosystems and the climate and sea level functions are presently thought to be of second-order importance and are not dealt with in a majority of IAMs.

In addition to the degree of complexity (including disaggregation) considered within and between modules, another major design consideration in an integrated assessment model is the treatment of the considerable uncertainties about virtually every major relationship in the climate change assessment system. Future population and economic growth are uncertain; future greenhouse gas emissions, given population and economic activity, are uncertain; future greenhouse gas concentrations, given emissions, are uncertain; future climate, given atmospheric concentrations of greenhouse gases, is uncertain; future physical impacts of climate change are uncertain; and the future valuation of the physical impacts attributable to climate change is uncertain.

Uncertainty can be handled in a number of ways in integrated assessment modelling. Extensive sensitivity analyses can be performed on key model inputs and parameters, or explicit subjective probabilities can be assessed for these inputs and parameters and fed into a formal risk or decision analysis framework. If a formal risk or decision analysis approach is pursued, it is generally possible to calculate the value of information with respect to wholly or partially resolving the uncertainty associated with each key input or parameter. Such calculations can provide a useful screening of uncertainties to determine where research expenditures may or may not have large net expected benefits. Combined with estimates of research costs and success probabilities, they can help in setting research priorities in a rational way. Of course, these priorities can be expected to change over time as research itself changes perceptions of research costs and benefits.

10.4 Overview of Existing Integrated Assessment Models

Prior to 1992, only two integrated assessment models of climate change had appeared in the literature (Nordhaus, 1989, 1991; Rotmans, 1990). Since 1992 a host of new models has emerged. Table 10.1 lists twenty-two integrated assessment models that are in active use or under active development; in addition, a number of other modelling efforts are underway, so the number of existing integrated assessment models might be expected to at least double in the next few years. Even within the group of models listed in Table 10.1, though, there is a wide variation in level of model maturity. Some models are fully operational and documented. Others are up and running but not yet fully operational or documented. Still others are in module development and testing phases, with some modules not yet fully specified. It is anticipated that all the models shown here will be fully operational, albeit in preliminary versions in some cases, by the end of 1995. The modelling in this area is so active that even models that are fully operational are continually being refined and updated substantially every three to six months. Table 10.2 summarizes the current development status and most recent documentation available for the twenty-two models listed in Table 10.1.

The models included in Table 10.1 can be compared structurally according to the amount of emphasis they place on each of the blocks shown in Figure 10.1. The results of this process are shown in Table 10.3 (adapted from Rotmans *et al.,* 1995). Note that some of the models do not explicitly consider the relationships included in each of the blocks. In particular, several of the key models omit direct modelling of economic activity and rely on exogenous greenhouse gas emission trajectories. In addition, more than half the existing models consider both the physical impacts and their valuation only through aggregate damage functions that relate global mean temperature change directly to economic damage.

10.4.1 State of the art in integrated assessment modelling

It is difficult to characterize simply the state of the art in integrated assessment modelling of climate change – a great deal of model development is underway at present, involving a large number of research teams, with members drawn from a myriad of relevant disciplines, focussing on different dimensions of the problem, and using different types of methodologies. Nonetheless, a focus on the trade-offs between the

complexity of natural systems models, the complexity of economic models, and the effort devoted to the explicit incorporation of uncertainty can help us understand the model development completed so far, as well as that occurring today or planned or anticipated for the future.

There are two broad classes of integrated assessment models: *policy evaluation models* that project the physical, ecological, economic, and social consequences of policies and *policy optimization models* that optimize key policy control variables (e.g., carbon emission control rates or carbon taxes) given formulated policy goals such as maximizing welfare or minimizing the cost of meeting a carbon emission or concentration target. There are two general types of policy evaluation models: deterministic projection models, in which each input and output takes on a single value, and stochastic projection models, in which at least some inputs and outputs are treated stochastically. Policy optimization models can be divided into three general types: models that optimize responses, given targets for emissions or climate change impacts; models that seek to balance the costs and benefits of climate policies; and models of sequential climate decision making under uncertainty. Each approach has strengths and weaknesses, and each produces particular insights regarding climate change and potential policy responses to it. Some of the more advanced models can be used for several of the above purposes.

Policy optimization IAMs focus on balancing the marginal costs of controlling greenhouse gas emissions and adapting to any climate change impacts that may occur with the damages that result after implementation of the mitigation and adaptation policies. These models reflect the strict cost-benefit paradigm discussed in Chapter 5. In this approach any constraint on human activities is explicitly represented and costed out. At present, models of this type include highly aggregated representations of climate damages, generally representing economic losses as a function of mean global surface temperatures but sometimes disaggregating these losses into market and nonmarket damage components.[2] However, as additional research on climate change impacts proceeds, it may be determined that these measurements are inaccurate. Moreover, it may be difficult to get policymakers to implement policies based on aggregate damages, as they are more likely to be able to relate to impacts on particular countries, regions, or sectors (e.g., agriculture or biodiversity in tropical rain forests) which are not explicitly represented in the current cost-benefit type of integrated assessment models. Early models of this type were also so complicated that it was difficult to incorporate explicit representation of uncertainty (and risk aversion) within the model structures. As discussed below, this situation has improved somewhat over the last couple of years.

The policy evaluation IAMs add detail on the physical impacts of climate change on various market and nonmarket sectors in different countries or regions, based in part on the impacts and mitigation areas addressed in Volume 2 of this report. Economic values have generally not yet been put on these impacts, an omission that reflects both the paucity of valuation studies in some sectors and the modellers' percep-

Table 10.1. *Integrated assessment models*

Model	Modellers
AS/ExM (Adaptive Strategies/Exploratory Model)	R. Lempert, S. Popper (Rand); M. Schlesinger (U. of Illinois)
AIM (Asian-Pacific Integrated Model)	T. Morita, M.Kainuma (National Inst. for Environmental Studies, Japan); Y. Matsuoka (Kyoto U.)
CETA (Carbon Emissions Trajectory Assessment)	S. Peck (Electric Power Research Institute) T. Teisberg (Teisberg Assoc.)
Connecticut (also known as the Yohe model)	G. Yohe (Wesleyan University)
CRAPS (Climate Research And Policy Synthesis model)	J. Hammitt (Harvard U.); A. Jain, D. Wuebbles (U. of Illinois)
CSERGE (Centre for Social and Economic Research on the Global Environment)	D. Maddison (University College of London)
DICE (Dynamic Integrated Climate and Economy model)	W. Nordhaus (Yale U.)
FUND (The Climate Framework for Uncertainty, Negotiation, and Distribution)	R.S.J. Tol (Vrije Universiteit Amsterdam)
DIAM (Dynamics of Inertia and Adaptability Model)	M. Grubb (Royal Institute of International Affairs), M.H. Dong, T. Chapuis (Centre Internationale de recherche sur l'environnement et développement)
ICAM-2 (Integrated Climate Assessment Model)	H. Dowlatabadi, G. Morgan (Carnegie-Mellon U.)
IIASA (International Institute for Applied Systems Analysis)	L. Schrattenholzer, Arnulf Grubler (IIASA)
IMAGE 2.0 (Integrated Model to Assess the Greenhouse Effect)	J. Alcamo, M. Krol (Rijksinstitut voor Volksgezondheid Milieuhygiene, Netherlands)
MARIA (Multiregional Approach for Resource and Industry Allocation)	S. Mori (Sci. U. of Tokyo)
MERGE 2.0 (Model for Evaluating Regional and Global Effects of GHG Reductions Policies)	Alan Manne (Stanford U.), Robert Mendelsohn (Yale U.), R. Richels (Electric Power Research Institute)
MiniCAM (Mini Global Change Assessment Model)	J. Edmonds (Pacific Northwest Lab), R. Richels (Electric Power Research Institute), T. Wigley (University Consortium for Atmospheric Research (UCAR))
MIT (Massachusetts Institute of Technology)	H. Jacoby, R. Prinn, Z. Yang (Massachusetts Institute of Technology)
PAGE (Policy Analysis of the Greenhouse Effect)	C. Hope (Cambridge U.); J. Anderson, P. Wenman (Environmental Resources Management)
PEF (Policy Evaluation Framework)	J. Scheraga, S. Herrod (EPA); R. Stafford, N. Chan (Decision Focus Inc.)
ProCAM (Process Oriented Global Change Assessment Model)	J. Edmonds, H. Pitcher, N. Rosenberg (Pacific Northwest Lab); T. Wigley (UCAR)
RICE (Regional DICE)	W. Nordhaus (Yale U.); Z. Yang (MIT)
SLICE (Stochastic Learning Integrated Climate Economy Model)	C. Kolstad (U. of California, Santa Barbara)
TARGETS (Tool to Assess Regional and Global Environmental and Health Targets for Sustainability)	J. Rotmans, M.B.A. van Asselt, A. Beusen, M.G.J. den Elzen, M. Janssen, H.B.M. Hilderink, A.Y. Hoekstra, H.W. Koster, W.J.M. Martens, L.W. Niessen, B. Strengers, H.J.M. de Vries (Rijksinstitut voor Volksgezondheid en Milieuhygiene, Netherlands)

Table 10.2. *Development Status of Integrated Assessment Models (June 1995)*

Model	Status	Reference
AS/ExM	Preliminary version operational	Lempert *et al.* (1994, 1995)
AIM	Operational	Morita *et al.* (1994); Matsuoka *et al.* (1995)
CETA	Operational, with regional and uncertainty variants	Peck and Teisberg (1992, 1993, 1994, 1995)
Connecticut	Operational	Yohe (1995a,b); Yohe and Wallace (1995)
CRAPS	Preliminary version operational	Hammitt (1995a,b); Jain *et al.* (1994)
CSERGE	Preliminary version operational	Maddison (1995)
DICE	Operational, with regional and uncertainty variants under development	Nordhaus (1994)
FUND	Operational	Tol *et al.* (1995)
DIAM	Analytic version operational	Grubb *et al.* (1993, 1995)
	Numeric version operational	Chapuis *et al.* (1995)
ICAM-2	ICAM-1 operational; ICAM-2 operational	Dowlatabadi and Morgan (1993); Dowlatabadi (1995)
IIASA	Energy, economy, and agriculture modules operational	WEC/IIASA (1995)
IMAGE 2.0	Operational	Alcamo (1994)
MARIA	Operational	Mori (1995a,b)
MERGE 2.0	Operational, with uncertainty variant under development	Manne *et al.* (1993)
MiniCAM	Operational	Edmonds *et al.* (1994a,b); Wigley *et al.* (1993)
MIT	Various stages of module testing	MIT (1994)
PAGE	Operational	Commission of the European Communities (1992)
PEF	Prototype operational, enhanced version under development	Cohan *et al.* (1994)
ProCAM	Most modules in testing phase	Edmonds *et al.* (1994a,b)
RICE	Operational	Nordhaus and Yang (1995)
SLICE	Operational	Kolstad (1993, 1994a,b,c)
TARGETS	Targets 1.0 operational	Rotmans *et al.* (1995)

tion that policymakers feel more comfortable trading off natural and physical impacts than dollars. In addition, the targets can be set to avoid certain types of risks, perhaps according to the precautionary principle. On the other hand, there is no guarantee that the marginal cost of implementing the mitigation and adaptation measures resulting from the individual targets will equal the marginal benefit (if that can be assessed) of the impacts avoided. Furthermore, because of the large size of these models, only limited amounts of sensitivity analysis can be performed, and more explicit representations of uncertainty (and risk aversion) have generally not been possible, except for the ICAM-2 model (Dowlatabadi, 1995) and the TARGETS model (Van Asselt and Rotmans, 1995).

Reflecting the high level of uncertainty about the future evolution of socioeconomic and natural systems, some analysts have put the analysis of climate change into explicit frameworks, of the kind discussed in Chapter 2, for decision making under conditions of uncertainty. These models have generally been either the result of a relatively complete uncertainty representation of all key parameters within simplified models or the result of adding a limited number of alternative states to more complex policy evaluation and policy optimization models. In addition, many of these models allow policies to be changed as uncertainties are resolved through time, although the process by which uncertainties will be resolved is usually represented quite simplistically. Stochastic models can generate multiple scenarios that in some cases have probabilities associated with them. Then, the (usually more com-

plex) deterministic models can be run to investigate specific scenarios further. Table 10.4 places the models listed in Table 10.3 into the two primary categories and relevant subcategories discussed above.

10.5 First Results from Integrated Assessment Models

Most integrated assessment models of climate change have been constructed since 1992. By the end of 1994, however, results from a number of these models had already been published. This section gives an overview of these results, highlighting the insights that seem most relevant to the current debate on appropriate global change policies. The variety of different approaches employed to study the climate change issue makes comparison and reconciliation difficult.

In what follows, we group the available model results into two main categories: (1) results from policy evaluation models that include many linkages and interactions between the several key elements of the climate/biosphere system and (2) results from policy optimization models that directly consider the costs and benefits of potential climate change policy responses or minimize costs subject to constraints on emissions, concentrations, climate change, or climate impacts.

There are also large differences in the outputs that individual modellers report from their integrated analyses and the time periods for which those outputs are reported. Some of the more common outputs from the policy optimization models are projections of the cost of controlling greenhouse gas emis-

Table 10.3. *Summary characterization of integrated assessment models*

Model	Forcings 0. CO_2 1. other GHG 2. aerosols 3. land use 4. other	Geographic Specificity 0. global 1. continental 2. countries 3. grids/basins	Socioeconomic Dynamics 0. exogenous 1. economics 2. technology choice 3. land use 4. demographic	Geophysical Simulation[a] 0. Global ΔT 1. 1-D ΔT, ΔP 2. 2-D ΔT, ΔP 3. 2-D Climate	Impact Assessment[b] 0. ΔT 1. Δ sea level 2. agriculture 3. ecosystems 4. health 5. water	Treatment of Uncertainty 0. None 1. Uncertainty 2. Variability 3. Stochasticity 4. Cultural Perspectives	Treatment of Decision Making 0. optimization 1. simulation 2. simulation with adaptive decisions
AS/ExM	0	0	0	0	0	1	2
AIM	0,1,2,3	2,3	1,2,3,4	1,2	0,1,2,3,5	0	1
CETA	0,1	0	1,2	0	0	0 or 1	0
Connecticut	0	0	1	0	0	1	0
CRAPS	0	0	1	0	0	1	2
CSERGE	0	0	1	0	0	1	0
DICE	0	0	1	0	0	0 or 1	0
FUND	0,1	1	1,4	0	0,1,2,3,4	0 or 1	0
DIAM	0	0	1,2	0	0	0 or 1	0
ICAM-2	0,1,2,3	1,2	1,3,4	1,2	0,1,3	1,2,3	1,2
IIASA	0	0	1	1	2	0	0
IMAGE 2.0	0,1,2,3	3	0,2,3	2	1,2,3	1	1
MARIA	0	0,1	1	0	0	0	0
MERGE 2.0	0,1	1	1,2	0	0	0 or 1	0
MiniCAM	0,1,2,3	2,3	1,2,3	2	0	0	1
MIT	0,1,2,3	2,3	1	2,3	0,2,3	1	0,1
PAGE	0,1	1,2	1	0	0,1,2,3,4	2	1
PEF	0,1	1,2	1	0	0	2	1
ProCAM	0,1,2,3	2,3	1,2,3,4	2	0,2,3,5	1	1
RICE	0	1	1	0	0	0	0
SLICE	0	1	1	0	0	1	2
TARGETS	0,1,2,3,4	0	1,2,3,4	2	1,2,3,4	4	1,2

[a]TARGETS includes ozone depletion, soil erosion, acid rain, and toxic and hazardous pollutant releases.
[b]In AIM, FUND, IMAGE, PAGE, and ProCAM, the impacts are calculated separately for each sector.
Source: Adapted from Rotmans et al. (1995).

sions, the damages resulting from climate change, the "control rate," stated in terms of the percentage reduction in greenhouse gas emissions in each year relative to level of emissions projected to occur in the absence of new policy initiatives, and the carbon tax required in each year to limit greenhouse emissions to the levels specified in the scenario under consideration. Policy evaluation models, on the other hand, tend to report physical changes in emissions, concentrations, temperature, and sea level, as well as changes in land use by activity (e.g., agriculture, forestry, etc.), and/or physical impacts like ecosystems at risk, coastal land area lost, fresh water requirements, and mortality rates.

10.5.1 Results from policy evaluation models – contributions to the scientific debate

Policy evaluation models are rich in physical detail and have produced useful insights, for example, into the potential for deforestation as a consequence of interactions between demographics, agricultural productivity, and economic growth and into the relationship between climate change and the extent of potentially malarial regions (see Volume 2, Chapter 25).

10.5.1.1 *Balancing the carbon budget*

To assess the impact of a number of feedback mechanisms within the global carbon cycle, an integrated assessment model has been used to balance the past and present carbon budget. They show that both a historical and a present carbon balance can be obtained in many different ways, resulting in different biospheric fluxes and, thus, in considerably different atmospheric projections. The CO_2-fertilization feedback appears to determine the balance and to dominate the temperature-related feedbacks, whereas the feedback from net biological primary production seems to counterbalance the soil and respiration feedback effect. Future projections based on the IPCC's 1990 "business-as-usual" scenario show that the CO_2 concentrations calculated with the integrated assessment models are lower than the IPCC values, reaching a difference

Table 10.4. *Integrated assessment models by type*

Policy Evaluation Models

Deterministic Projection Models
AIM
IIASA
IMAGE 2.0
MIT
ProCAM
TARGETS

Stochastic Projection Models
PAGE
ICAM-2
TARGETS

Policy Optimization Models

Cost-Benefit and Target-Based Models
CETA
Connecticut
CSERGE
DICE
FUND
DIAM
MARIA
MERGE 2.0
MiniCAM
RICE

Uncertainty-Based Models
AS/ExM
CETA
CRAPS
CSERGE
DICE
FUND
ICAM-2
MERGE 2.0
PEF
SLICE

of about 15% (Rotmans and Den Elzen, 1993; Wigley, 1993). This difference can be explained by the fact that most global carbon cycle models used by the IPCC were unbalanced; the balanced models do not produce terrestrial fluxes that correspond to observations.

10.5.1.2 Integrated land-use analysis
A first attempt to integrate the various aspects of the global land use problem on a geographically explicit base has been made using the IMAGE 2.0 model. The model represents the transformation of land cover as it is affected by climatic, demographic, and economic factors and links these explicitly with the flux of CO_2 and other greenhouse gases between the biosphere and atmosphere. Conversely, it also takes into account the effect of productivity changes in the terrestrial and oceanic biospheres. The integration of agricultural and land cover calculations can provide new insights about shifts in agricultural areas related to climate and the influence that changing land cover has on climate. The first, preliminary, re-

sults show that there may be some validity to the hypothesis that regional demands for land can serve as a surrogate for measuring local land cover changes, and that land use rules can be used to represent the forces driving land conversions. Other results relate to the vulnerability of protected areas under shifting vegetation zones, the consequences for biodiversity and nature conservation, and the determination of risks associated with current productivity levels of specific crops with shifting agricultural patterns. These analyses could in due time assist regional policymakers in assessing the seriousness of climate change impacts (Alcamo, 1994).

10.5.1.3 Global warming potentials
A slightly improved version of the IMAGE 1.0 model has been used to investigate the input and parameter uncertainties as well as methodological uncertainties associated with Global Warming Potentials (GWPs) for greenhouse gases (Den Elzen, 1993; Rotmans and Den Elzen, 1992). In particular, the role of the emission scenario used and the difference between transient and equilibrium GWPs have been discussed. Although integrated assessment models have structural limitations, they can produce estimates for at least the direct impact of greenhouse gases as well as some of the indirect effects.

The advantages of using integrated assessment models of climate change in estimating GWPs are twofold: (a) they can calculate GWPs for each conceivable scenario, so the influence of the emission scenario selected can be stated explicitly; and (b) they also deal with the rates of change of all kinds of targeting processes, so the cumulative effect can be combined with the rate at any time. The results show that the GWPs calculated with integrated assessment models differ from the ones previously published by the IPCC. Considering a time horizon of 100 years, the difference might be as much as 5-10%. This difference demonstrates the crucial role of the chosen scenario in calculating GWPs and cannot be addressed by analytical methods.

10.5.1.4 The sulphate aerosols debate
As discussed at length in Volume 1 of this report, the presence of sulphate aerosols in the atmosphere is now thought to have a strong local cooling effect. This effect is manifested through three pathways: scattering and absorption of shortwave (solar) radiation, cloud reflectivity, and cloud persistence. By incorporating a simplified mathematical expression of the relationship between sulphate aerosols and radiative forcing into integrated assessment models, some of the sulphate aerosol effect can be taken into account. In this way, the sensitivity of the climate system to simultaneous changes in SO_2 and CO_2 emissions can be examined. The first calculations show that over the next decade, it is conceivable that the increased radiative forcing due to SO_2 concentration changes could more than offset reductions in radiative forcing due to reduced CO_2 emissions (Edmonds *et al.*, 1994b), depending on the rate of reduction and a number of other assumptions. Therefore, policies that reduce fossil fuel use may not be so effective in reducing near-term average radiative forcing as a simple calculation based on greenhouse gas emissions alone might imply. The proper treatment of SO_2 is, therefore, an important considera-

tion in the integrated analysis of the consequences of technology development and deployment for climate change.

10.5.1.5 IPCC scenarios

In 1989, a U.S.-Netherlands expert group of the IPCC was asked to develop four different pathways for future global emissions of CO_2, CH_4, N_2O, halocarbons, and the ozone precursors NO_x and CO. The expert group used two alternative integrated assessment models to construct these scenarios: the ASF model from the U.S. Environmental Protection Agency and the IMAGE 1.0 model from the RIVM, the Dutch National Institute of Public Health and Environmental Protection (Rotmans, 1990). Three scenarios were designed in such a way that they would lead to a doubling of the CO_2-equivalent concentration in the atmosphere in the years 2030, 2060, and 2090. These were referred to as the "Business-as-Usual," "2060 Low Emissions," and "Control Policies" scenarios, respectively. The fourth scenario, the "Accelerated Policies" scenario, leads to stabilization of the CO_2-equivalent concentration in the atmosphere at well below doubling of preindustrial concentrations. Each scenario is based on a set of assumptions for key factors, including population growth, economic growth, the costs of technology used to convert energy from one form to another, energy end-use efficiency levels, deforestation rates, CFC emissions, and agricultural emissions.

10.5.1.6 Delayed response analysis

The IMAGE 1.0 model (Rotmans, 1990) was used by the IPCC to analyze delayed policy response options in which the start of the international policy response was delayed to 2000, 2010, 2020, and 2030 respectively. It was calculated that delaying implementation of the "Control Policies" scenarios by 10 years would result in only a minor increase in global mean temperatures, but that it would require a reduction of global CO_2 emissions of 20% with respect to year 2000 levels, whereas starting immediately would require only a 5% reduction with respect to 1990 levels over the same period. This integrated analysis shows that the timing of the climate response policies is crucial for the control of climate change, and that the feasibility of the required transition decreases over time.

10.5.1.7 Risk assessment

The Advisory Group on Greenhouse Gases (AGGG) recommended a maximum rate of global mean temperature increase of 0.1°C per decade, together with a maximum temperature increase of 2°C above the preindustrial global mean temperature level. These temperature targets might be considered as limits beyond which damages to sensitive ecosystems and coral areas might be expected to increase rapidly. One difficulty with these targets is that they are global, whereas large regional variations in temperature change and impacts are likely. Moreover, the targets need to be reviewed periodically in light of potential feedbacks and nonlinearities that may produce surprises and unexpected changes. The "Risk Assessment" calculations showed that (1) all IPCC 1990 and 1992 scenarios except the 1990 "Control Policies" scenario lead to

Table 10.5. *Key results from deterministic cost-benefit analyses*

Model	Control Rate (percentage reduction relative to baseline emissions)	Carbon Tax (1990 U.S. dollars/tonne)
	1990–2000	1990–2000
CETA		
Linear damages	0–1	7–8
Cubic damages	0–2	8–12
DICE	9	5

temperature increases and rates of temperature change greater than the target values, and (2) even the IPCC 1990 "Control Policies" scenario leads to mean global temperature changes that are close to the targets (Den Elzen, 1993).

10.5.2 Results from cost-benefit policy optimization models

In this section we consider results from cost-benefit integrated assessment models run with all inputs and parameters set at their median or best-guess values. Notwithstanding the immense uncertainties inherent in the climate change issue, a number of analysts have suggested that the results from these deterministic analyses provide a useful benchmark for near-term decision making, if not an adequate approximation of the results obtained from more complex approaches that explicitly include consideration of the key uncertainties.

Table 10.5 shows some key results from two models that balance the costs and benefits of greenhouse gas emission reductions. For example, the "optimal run" results from the DICE model (Nordhaus, 1994) show a 1995 control rate (i.e., percentage reduction in emissions relative to baseline greenhouse gas emissions) of about 8.8% with an associated carbon tax[3] of $5.29 per tonne of carbon. This programme leads to an increase in the discounted present value of consumption of 271 billion 1989 dollars or about .04% of discounted baseline consumption. Similar results are obtained from the CETA, MERGE, and SLICE models when run under similar assumptions.

In *The economics of global warming*, Cline (1992) analyzes the time profile of abatement and damage costs under a policy of limiting global carbon emissions to 4 Gt annually and similarly reducing other greenhouse gas emissions. The abatement cost curve is low at first, then peaks at about 3.5% of gross world product, and thereafter declines to a plateau of about 2.5% as a consequence of widening technological alternatives.

The cost-benefit decision for greenhouse policy involves a trade-off between substantial abatement costs early in the horizon and avoidance of potentially large damages later in the horizon. The discounting of future costs and benefits relative to current ones is critical in such a trade-off. On the grounds that policymakers would be risk averse, Cline also weights a high-damage case three times as heavily as a low-damage case. Discounting at a zero rate of time preference, he

Table 10.6. *Key sensitivities from deterministic cost-benefit analyses*

Sensitivity Model and Cases		Control Rate (percentage) 1990–2000	Carbon Tax (1990 U.S. dollars/tonne) 1990–2000
CETA			
Warming per 2XCO$_2$			
Low	1°	0	2
Baseline	3°	0–2	9–12
High	5°	0–7	22–29
Damage Function Power			
Low	1	0–1	8–9
Baseline	2	0–2	9–12
High	3	0–2	10–13
Utility Discount Rate			
Low	2%	8	19–24
Baseline	3%	5	9–12
High	4%	1	5–7
DICE			
Warming per 2XCO$_2$			
Baseline	3°	9	5
High	4.5°	11	8
Damage Function Power			
Baseline	2	9	5
High	4	9	5
Utility Discount Rate			
Low	1%	19	24
Baseline	3%	9	5

finds that the overall benefit-to-cost ratio for aggressive action limiting carbon emissions to 4 Gt is a favourable 1.3. Thus, Cline endorses a much more aggressive control policy than calculated in most of the other pure cost-benefit studies. Much of the difference in results stems from Cline's assumption of risk aversion on the part of national and international policymakers and his use of a zero rate of pure time preference, whereas the other studies generally employ a pure rate of time preference of about 3% and no risk aversion by policymakers. In fact, Cline has shown that the optimal control rate in 2100 in the DICE model would be 50% if a zero rate of pure time preference is employed as opposed to the 15% reported for the 3% rate of time preference in the DICE baseline. Moreover, as discussed below, Nordhaus (1994) reports that the pure rate of time preference is the input to which DICE results are most sensitive. The subject of the appropriate rate of pure time preference is a major focus of Chapter 4 of this report.

Although optimal control rates and carbon taxes vary widely for the year 2100, results from the two models, as shown in Table 10.6, are not all that disparate in the 1990s, though the sensitivity analysis shows a variation in the tax. This reflects the time dynamics of climate change. The costs of control are related to decreases in the rate of emissions as

soon as the controls are applied. The benefits of control, on the other hand, are related to temperature change, which responds to changes in atmospheric concentrations of greenhouse gases with a long lag, whereas atmospheric concentrations respond only slowly to changes in emission rates because of the large stock and long lifetimes of greenhouse gases already in the atmosphere. Thus, the marginal costs of controlling greenhouse gases tend to be highly nonlinear with respect to the control rate, whereas most of the marginal benefits tend to be delayed by several decades.

10.5.3 Cost-effective strategies for stabilizing atmospheric CO$_2$ concentrations

There have been several interesting applications of integrated assessment modelling to the issue of concentration targets. The ultimate goal of the UN Framework Convention on Climate Change is the "stabilization of greenhouse gas concentrations in the atmosphere at a level that would prevent dangerous anthropogenic interference with the climate system." Under the terms of the Convention, mitigation costs are to play a limited role in establishing the concentration target. The permissible concentration level will depend on our understanding of the greenhouse effect and its potential consequences.

Mitigation costs are a more important consideration in determining how the target is to be achieved. The Convention states that "policies and measures to deal with climate change should be cost-effective so as to ensure global benefits at the lowest possible cost." A particular concentration target can be met in a variety of ways. For example, Figure 10.3(a), drawn from the IPCC Synthesis Report, shows trajectories for stabilizing CO$_2$ concentrations at 450, 550, 650, 750, and 1,000 ppmv. Figure 10.3(b) shows two alternative emission paths for reaching each of the four lowest CO$_2$ concentrations. Some ways of meeting concentration targets will be more costly than others. Integrated assessment modelling can help identify emission paths that minimize the costs of meeting a prespecified concentration level (see Chapter 9).

Richels and Edmonds (1995) have examined the question of cost-effectiveness in achieving a particular concentration target. They found that the emission timepath can be as important as the concentration level itself in determining the ultimate price tag. Specifically, they examined alternative emission profiles for limiting CO$_2$ concentrations to 500 ppmv in the year 2100. Employing two widely used energy-economy models (the Edmonds-Reilly model and Global 2100), they found that emission timepaths involving modest reductions in the early years followed by sharper reductions later were less expensive than those involving substantial reductions in the short term. A similar conclusion can be found in Kosobud *et al.* (1994).

There are several reasons why shifting emission reductions into the outer years can reduce mitigation costs. As noted in Wigley *et al.* (1996)

> to a first approximation, a concentration limit defines a "carbon budget" (e.g., an allowable amount of carbon that can be released into the atmosphere between now

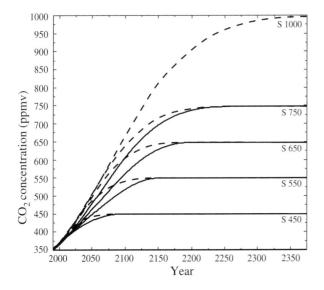
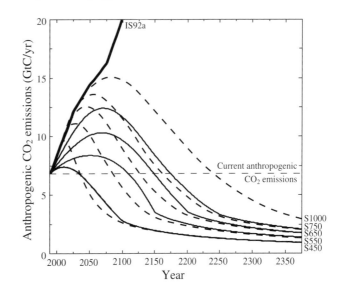

(a) CO_2 concentration profiles leading to stabilization at 450, 550, 650 and 750 ppmv following the pathways defined in IPCC (1995) (solid curves) and for pathways that allow emissions to follow the IS92a scenario (IPCC, 1992) until at least the year 2000 (dashed curves). A single profile that stabilizes CO_2 concentration at 1,000 ppmv and follows IS92a emissions at least until 2000 is also shown.

(b) CO_2 emissions leading to stabilization at concentrations of 450, 550, 650, 750 and 1,000 ppmv following the profiles shown in (a) from a mid-range carbon cycle model. Results from other models could differ from those shown by up to approximately ±15%. For comparison, the CO_2 emissions for IS92a and current emissions (dotted line) are also shown.

Figure 10.3: Emission profiles consistent with stabilization of CO_2 concentrations at levels from 450 to 1,000 ppmv.

and the date at which the target is to be achieved). The issue is how the carbon budget is to be allocated over time. Several factors argue for drawing more heavily on the budget in the early years. With the economy yielding a positive return on investment, emission reductions in the future will be cheaper than emission reductions today. That is, a smaller amount of today's resources needs to be set aside to finance them. As a result, the same level of cumulative emission reductions can be achieved at a lower total cost to society. In addition, slowing the transition away from fossil fuels provides valuable time to develop low-cost, carbon-free alternatives, to allow the capital stock to adapt, and to remove carbon from the atmosphere via the carbon cycle. Cumulative emissions for a 550 ppmv ceiling can differ by more than 60 PgC, with higher cumulative emissions associated with higher near term emissions. (See Volume 1)

Building on the earlier work of Nordhaus (1979), Manne and Richels (1992, 1993) have explored least-cost mitigation paths for achieving concentration targets of 450-750 ppmv. Figure 10.4 shows results from their MERGE model. In each instance, the least-cost path allows for some growth in global emissions in the early years, but this is followed by sharp reductions later on.

These studies should not, however, be seen as supporting a "do nothing" or "wait and see" strategy. First, each concentration path still requires that future capital equipment be less

carbon-intensive than under a scenario with no carbon limits. Given the long-lived nature of energy-producing and -using equipment, this has implications for current investment decisions. Second, new supply options typically take many years to enter the marketplace. To have sufficient quantities of low-cost, low-carbon substitutes in the future would require a sustained commitment to research, development, and demonstration today. Third, any available no-regrets measures for reducing emissions are assumed to be adopted immediately. Finally, it is clear that emissions must ultimately be reduced. One cannot go on deferring emission reductions indefinitely. The lower the concentration target, the more substantial the required emission reductions.

Other authors cite reasons for more mitigation sooner. These include the prospect of inducing further cost reductions through abatement action, the prospect of avoiding being locked in to more carbon-intensive patterns of development, and the extent to which inertia may amplify the costs of having to make more rapid emission reductions later.

Models that emphasize inertia and induced innovation (e.g., Hourcade and Chapuis, 1995) place greatest emphasis on the need to avoid investments that tend to "lock in" a higher carbon future and on the fact that evasive action now reduces both the climate risks and the possibility of having to take more rapid action later. His results show that for an atmospheric limit of 500 ppmv, delaying the response by 20 years could double the subsequent required rate of abatement. A parallel study of CFCs showed that if the phase-out had begun

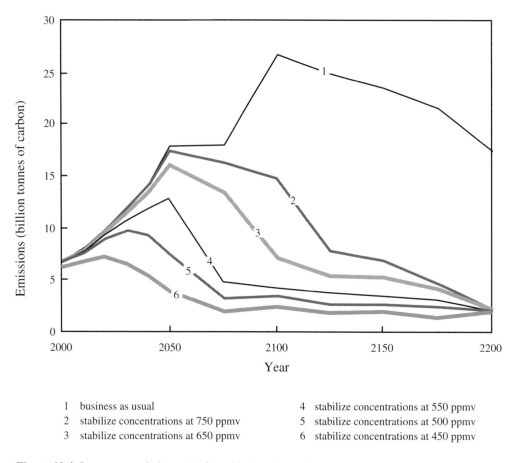

1   business as usual   4   stabilize concentrations at 550 ppmv
2   stabilize concentrations at 750 ppmv   5   stabilize concentrations at 500 ppmv
3   stabilize concentrations at 650 ppmv   6   stabilize concentrations at 450 ppmv

Figure 10.4: Least-cost emission paths for achieving alternative atmospheric concentration targets.

ten years earlier, it would have allowed much slower reductions of CFC use. The level of protection of the ozone layer resulting from the London Amendments to the Montreal Protocol could have been achieved while CFC use continued during the 1990s and with far less need to scrap capital stock.

Another line of analysis is developed by Grubb *et al.* (1994, 1995), drawing on studies of energy systems and the observation that much innovation comes from "learning by doing." Such innovation represents an external benefit that is not captured in market signals. Their model focusses on inertia and induced innovation and they conclude that induced innovation amplifies the benefits of acting sooner rather than later. If induced innovation is sufficient for systems to adapt to emission constraints over a period of a few decades, then the optimal near-term control rate is likely to be considerably larger than projected with models that do not include induced innovation. However, this is a result derived from a cost-benefit analysis in which many of the benefits from stronger early action arise from reduced impacts. The model has not been run to a fixed limit on the concentration of CO_2 in the atmosphere.

Note also that the focus of these analyses is on mitigation costs. Consequently, they provide only partial guidance for policy making. Different emission profiles yield different concentration levels and rates of change in the years leading up to a particular concentration target. The implications for damages need to be considered. Unfortunately, the knowledge base is not yet available for preparing an optimal strategy con-

sidering the full array of costs and benefits. Integrated assessment models that include the full range of factors that bear on the optimal timing of emission reductions have not yet been developed, and the relative importance of the various economic issues that bear on the question is still a matter of debate.

10.5.3.1 *International cooperation*
Integrated assessment models show there is a strong need for international cooperation because developed nations cannot independently reduce atmospheric CO_2 concentrations on their own (OTA, 1994; Bradley *et al.*, 1994; Edmonds *et al.*, 1995; Manne and Richels, 1992; Manne *et al.*, 1993; Nordhaus and Yang, 1995; and Tol *et al.*, 1995; see also Chapter 9).

Regarding the resolution of political conflicts over climate change policy between developed and developing countries, Read (1994b) points to the potential for biofuel production in developing countries. Financed by the developed economies, a biofuel initiative could generate beneficial multiplier effects in underemployed and cash-constrained developing rural economies.

10.5.4 *Results from uncertainty-oriented policy optimization models*

Policy optimization modellers have pursued a number of alternative approaches to incorporating the large uncertainties inherent in the various elements of the climate system into their analyses. The discussion here deals with results obtained from these approaches in the following order:

(1) Sensitivity analyses of key model inputs and parameters
(2) Analyses in which all model inputs and parameters are treated stochastically
(3) Uncertainty analyses that focus on the implications of a small number of uncertainties that seem particularly relevant to the policy issues being addressed

These results also suggest a number of modelling challenges that have been identified as high priority areas for future improvements in integrated assessment modelling.

10.5.4.1 Sensitivity analyses

Given the sizable uncertainties inherent in virtually every major input and model parameter employed in any analysis of global climate change, it is important to assess the implications of the key uncertainties on model results. A common first step in this effort is sensitivity analysis, which involves looking at how key model outputs respond to changes in input or parameter values over plausible ranges.

Table 10.6 shows control rate and carbon tax sensitivities for two models. For example, for the CETA model the initial control rate for the years 1990-2000 moves from 0% below baseline emissions in the base case to 7% when the sensitivity of global mean surface temperature to a CO_2 doubling is increased from 1°C to 5°C. In addition, the initial carbon tax rate for 1990-2000 changes from $2 per tonne to $29/tonne over the same range of global mean surface temperature sensitivities. Similarly, in the DICE model the initial control rate for 1995 changes from 9% below baseline emissions in the base case to 19% when the pure rate of time preference is changed from its base value of 3% to 1%.

Another type of sensitivity analysis involves focussing on a small number of more carefully designed scenarios that are expected to lead to fundamental changes in key model outputs. A recent analysis of carbon-free advanced energy technologies was performed by Edmonds *et al.* (1994b). In this study the implications of advanced energy technologies (including very low-cost biomass fuels) for greenhouse gas emissions and temperature change were investigated. Obviously, the introduction of very low cost noncarbon fuels leads to lower carbon emissions and less temperature rise in the long run (post-2050). A somewhat surprising result of this analysis, though, is an increase in temperature prior to 2050 as the replacement of carbon-based fuels with carbon-free fuels leads to a reduction in sulphur emissions and, therefore, fewer climate-cooling sulphur aerosols in the atmosphere. Lower sulphur emissions, however, would produce benefits in the form of reduced acidic deposition.

10.5.4.2 Baseline projections and uncertainty

Manne and Richels (1993) have argued that any deterministic projection of baseline carbon emissions may be upwardly biased because individual energy consumers should already be reducing their consumption of carbon-based fuels because of the possibility of constraints on carbon emissions in the future. They compute the implicit carbon tax in the year 2000 as a function of the probability that U.S. carbon emissions will be limited to 1990 levels. For example, if consumers feel there is a 50-50 chance that carbon emissions will be constrained in 2010, they will reduce carbon emissions in 2000 as if a carbon tax of $17.50 per tonne of carbon were already in place.

10.5.4.3 Results from stochastic simulation models

Stochastic simulation models generalize the sensitivity analysis idea by including probability distributions for all major inputs and model parameters. Each input distribution is sampled, the value chosen is used in the subsequent calculations of the model, and the process is continued until probability distributions are derived for each output variable of the model. The PAGE (Hope *et al.*, 1993) and ICAM-2 models (Dowlatabadi, 1995) are prominent examples of integrated assessment models that take this approach.

An analysis with PAGE of no control and stringent control options results in a recommendation for adaptation rather than mitigation as a first-best policy initiative. Conditions under which both adaptation and aggressive mitigation options ought to be pursued are also identified.

It is also possible to do a more comprehensive type of sensitivity analysis with the stochastic simulation approach by computing the partial rank correlation coefficient of output measures of interest with respect to variations in each input. For example, Hope *et al.* (1993) report that "cheaper preventative costs of CO_2" and "no action CO_2 emissions (i.e., baseline emissions) of CO_2" have the greatest effect on total cost uncertainty, and "global temperature sensitivity to doubling of CO_2" and "half life of global warming response to change in forcing" have the greatest effect on total impact uncertainty.

The methods described above are unable to make the uncertainties associated with disagreement and subjectivity explicit. Relating the concept of uncertainty to differences in individual perspectives, Van Asselt and Rotmans (1995) arrived at the concept of perspective-based alternative model routes as a methodology to make uncertainties within IAMs visible and tangible. Alternative model routes can be considered as model interpretations in which not only parameters but also relationships are varied according to the bias and preferences of a particular perspective, resulting in alternative model structures.

10.5.4.4 Uncertainty, decision analysis, and the value of information

Climate change may have severe impacts on individuals and societies. On the other hand, the impacts may not be very severe at all. Individuals and societies often attempt to reduce the impact of low probability/high consequence events through various means. As shown in Table 10.7, Nordhaus (1994) groups activities designed to mitigate the effects of uncertainty on economic welfare into three categories: (1) traditional insurance, (2) consumption smoothing, and (3) precautionary investments.

Table 10.7. *Alternative policies to mitigate uncertainty*

Category	Source of Uncertainty	Policy

1. Traditional Insurance	1. Diversifiable (individual) risk	1. (a) Private insurance
		   (b) Social insurance
		       (against terms-of-trade or income losses)

2. Consumption smoothing over time	2. Risk of large or catastrophic loss	2. Investment (for a rainy day)

3. Precautionary investments	3. Uncertain scope of damage or abatement costs	3. (a) Precautionary abatement
		       (e.g., higher carbon tax)
		   (b) Precautionary adaptation
		       (e.g., retreat from coastline)
		   (c) Investment in knowledge (e.g., geophysical and social science research)

Source: Nordhaus (1994).

Traditional insurance involves pooling together large groups of people, each of whom is subject to a small probability of a large loss (such as having a house burn down). Thus, each individual pays the *a priori* expected value of a loss plus a small transaction fee to get compensated for the catastrophe should it occur. The pooling necessary to implement this approach requires that the occurrence of the catastrophe among the members of the population be more or less independent. This is not the case with climate change, however, for which the impacts are likely to be pervasive throughout the globe. On the other hand, since some individuals (e.g., people who live on coastlines) will be affected more severely than others, traditional insurance may help allocate the costs of the total damages resulting from climate change in a way that improves welfare. Even this capability to reallocate the costs of climate change through traditional insurance may also be somewhat limited, though, because those who are most vulnerable may be known in advance or can be easily identified when the impacts of climate change start to be felt.

Consumption smoothing over time amounts to the social equivalent of "saving up for a rainy day." If there are significant thresholds in the impacts of climate change, it is possible that societies will incur large adjustment or mitigation costs. Thus, welfare may be improved by saving capital now to consume when the threshold impacts occur.

Precautionary investments in mitigation, adaptation, or information represent the third type of policy that can be used to mitigate the uncertainty associated with climate change. Such actions enable societies to hedge against the possibility of bad climate outcomes before the major uncertainties determining the severity of the climate change problem have been resolved.

It appears that all three types of policies for coping with climate change uncertainty are valuable. In terms of overall payoff, however, the traditional insurance approach is the most tactical, in that it simply redistributes the costs of any climate change impacts that might occur, and the precautionary societal investment is the most strategic, in that it involves national or international investments now that can significantly reduce the total worldwide costs of climate change impacts in the future. Thus, a number of precautionary investment analyses have started to appear in the literature. One innovative example is the analysis, based on a stochastic optimization model, that is included in Chapter 4 of *Buying greenhouse insurance: The economic costs of CO_2 emission limits* (Manne and Richels, 1992). This analysis deals explicitly only with the cost of CO_2 emission reductions, but it is assumed that U.S. decision makers must act initially without knowing what ultimate limit on carbon emissions will emerge from further scientific research and international negotiation. However, it is assumed that by 2010 it will be revealed whether (1) no limits will be necessary, (2) a 20% emission reduction will be required, or (3) a 50% reduction will be required. Each of these future policy outcomes is assigned a probability of occurrence. This formulation makes the idea of hedging against a range of future outcomes explicit, with the initial control rate and carbon tax for the optimal hedging strategy (the one that maximizes the expected future utility of consumption) lying between the maximum control and no-control strategies, and with the exact level depending on the probabilities assigned to the different control outcomes. Put differently, there is a risk premium associated with emitting carbon, owing to the fact that carbon emissions may be constrained (and possibly severely constrained) in the future.

A study by Hammitt, Lempert, and Schlesinger (1992) traces alternative control strategies for attaining certain temperature constraints. Although not determining an optimal path, this study shows that a "moderate control strategy" is less costly than an "aggressive" approach if either the temperature sensitivity to a doubling of CO_2 is low or the allowable temperature change is above 3° C.

In *Managing the global commons: The economics of climate change,* Nordhaus (1994) performs a decision analysis with his dynamic global cost-benefit model (DICE) using a representation (derived from an extensive stochastic simulation analysis with the model) of the relevant uncertainties associated with climate change. He concludes that "roughly speaking, the optimal carbon tax doubles when uncertainty is taken into account, and the optimal control rate increases by slightly less than half. The increased stringency of controls results from the interaction of different uncertain variables, whereby extreme events may cause significant economic costs."

10.6 Strengths and Limitations of Current Integrated Assessments

The five biggest challenges facing integrated assessment modellers are

(1) Developing a credible way to represent and value the impacts of climate change
(2) Developing a credible way to handle low probability but potentially catastrophic events
(3) Developing realistic representations of the dominant processes and policies in the developing countries
(4) Integrating and managing a large and diverse array of data and models from many researchers and many disciplines
(5) Improving the relevance of the models to policy needs and the presentation of their results to policymakers and the public

10.6.1 Representation and valuation of impacts of climate change

A major problem in attempting to analyze and value climate change impacts is that the projections from most general circulation models, until recently at least, have been based on a hypothetical steady-state situation (a doubled-CO_2 climate). In reality, however, greenhouse gas concentrations are not steady and will not necessarily stabilize at a level equivalent to a doubling of preindustrial CO_2 concentrations. Moreover, there are uncertainties about many elements of these projections, especially at a regional level. The process of projecting transient regionalized changes in the key climate variables – such as temperature and precipitation – that lead to impacts on economies and ecosystems is in its infancy and is thus a source of additional uncertainties. Furthermore, the climate information required to most effectively project the impacts has in many cases not yet been determined, nor have the most appropriate measures of climate impacts, and ecosystems may not currently be in equilibrium. Finally, it may be necessary for this information to be analyzed using valuation methods that are still under development and not tightly linked to the impacts on natural systems in order to provide policymakers with the information they need to decide what to do.

10.6.2 Consideration of low probability/high consequence events

The first results from integrated assessment models, which considered only the expected costs and benefits of controlling greenhouse gases, have generally concluded that only a modest current level of control is warranted. However, it may not be expected conditions that should be our main concern but, rather, relatively low probability catastrophic events that are irreversible or from which it would be very difficult to recover. Unfortunately, lack of data, lack of understanding of the relevant processes, and analytical intractability have prevented such events from receiving adequate attention in the integrated assessments that have been performed to date (see Chapter 6). The implications of these low probability/high consequence events for current decisions have just started to be investigated through the use of integrated assessment models (see Nordhaus, 1994; Peck and Teisberg, 1994; Lempert *et al.*, 1994, 1995).

10.6.3 Critical issues in developing countries

In general, the processes and policy options relevant to climate change are easier to assess in the twenty-four countries of the OECD. This stems from the facts that these countries have been studied more intensively and that their populations and economies are growing relatively slowly. The data and understanding of critical processes and issues in the 140-odd non-OECD countries are more limited. Many of these countries are in a state of rapid development or dynamic change, making projections of key economic drivers and social organizations over even short periods of time extremely difficult. Moreover, the contribution of these countries to climate change and their responses to it are likely to be influenced by other more immediately pressing concerns. Three of the most critical such issues in the developing countries are land use, land tenure, and population.

The way land is used is a key determinant of the net emissions and accumulation of greenhouse gases in the atmosphere and of the impacts of climate change. However, land use and land tenure decisions in the developing countries will be driven by development goals and local pollution concerns rather than climate change concerns over at least the next several decades. Therefore, it is important to track trends in land use and land tenure in order to project the contribution of the developing countries to global climate change and how they will be affected by any changes that might occur. Only a few of the operational integrated assessment models (e.g., Alcamo, 1994; Morita *et al.*, 1993) track land use at all, and even those models are limited by lack of good data regarding current land use patterns in the developing countries, as well as a lack of understanding about who controls land use decisions at present, who is likely to control them in the future, and what criteria will be used in allocating land to alternative uses.

Another fundamental uncertainty that complicates assessments of the magnitude of the global climate change problem and the effectiveness of policy responses to it is future population growth, especially in the developing countries. In general, more population means more economic activity and more greenhouse gas emissions. Again, though, trends and policies regarding future population growth will depend more on other phenomena (the spread of diseases, the level of income, the cultural norms) and policies (e.g., regarding education, health care, and birth control) than on explicit consideration of the implications of population for climate change in the future. Virtually all the existing integrated assessment models take future population growth as given, although the

TARGETS model of Rotmans *et al.* (1994) has recently become the first exception (Van Vienen *et al.*, 1994). Moreover, the projections used generally all come from one or two international agencies.

The extent to which a better understanding and modelling of land use, land tenure, and population growth in the developing countries will alter the insights regarding the climate change problem and potential policy responses to it produced by the current set of aggregate integrated assessment models is an open question. There is no doubt though that there is an urgent need to add detail in these areas that would better reflect the reality of developing countries and thus improve the credibility of the models.

10.6.4 Model integration and management

The complexity and multidisciplinary nature of the climate change issues create another challenge for integrated assessment modellers – that of linking a vast amount of data, analysis, and computer code developed by different researchers from different disciplines into a unified whole. It is particularly important to maintain consistency between the assumptions made in different parts of the analysis and to preserve the integrity of the information passed from one module to another. For example, some of the early integrated assessments made very optimistic assumptions about technical change in some parts of the analysis but not in others.

Another important issue in integrated modelling is the compatibility of the many modules included in the model, each reflecting the modelling approaches and abilities of a distinct set of disciplines. A comparative static model, describing the difference between two equilibrium states (characteristic of many climate and climate impact models run to date), cannot readily be tied to a dynamic model like those used to project economic activity and carbon emissions. But even two dynamic models can work on two entirely different timescales; for instance, larger economic models are at best seasonal whereas general circulation models operate in time steps of tens of minutes. Spatial scales can also differ, not only in resolution, but also conceptually. Economists, for instance, tend to think in terms of nations and geopolitical regions, whereas ecologists think in terms of habitats and life zones. A third difference is the degree to which models approximate the real world. Normative models, which describe how systems should operate (a paradigm reflected in some economic models) cannot be easily integrated with descriptive models of how the world actually operates (common in climate and ecological modelling). The compatibility issue is at present being dealt with through trial and error. Continued feedback with the mother disciplines is required to ensure that modules are not used or changed in an inappropriate manner.

In addition to the specific problems of integrating information across disciplines, modellers have to deal with a number of challenges that need to be addressed in any large-scale modelling enterprise (Karplus, 1992). One issue is separability, or which links to include? This issue was already touched on in the discussion of Figures 10.1 and 10.2. A number of models, for example, neglect the cooling effect of sulphate aerosols (see Section 10.5.1.4), which can have important implications for the temperature profile. In addition, the influence of another link between climate and fossil fuel combustion, hot spells and ozone formation, has only been included parametrically (see Chapter 6), if at all, without having been studied with a full-fledged model. These are just two examples of known links, one with a known effect, one with an unknown effect, that could profitably receive more attention.

Related to the issue of separability is the question of selectivity. Is it appropriate to study the enhanced greenhouse effect in isolation, or should it be studied simultaneously with other major environment and development problems? The TARGETS model (Rotmans *et al.*, 1994) is the first to make such an attempt at integrating these issues. This model tries to address the concept of sustainable development from a world perspective, covering the global issues of human health and demographic dynamics, energy resources, global element cycles, and land- and water-related problems. In addition, the discussion of the secondary benefits of emission control in Chapter 6 and the first results of the FUND model (Tol *et al.*, 1995) indicate that it is worthwhile to tie the analysis of global warming to conventional air pollution issues.

Counteracting the call for more causal links and further integration is the curse of dimensionality. The larger a model, the less transparent it is, and the harder it is for analysts and policymakers to interpret its results. The sheer size of the model renders full sensitivity analyses impossible, and it becomes more difficult for the modellers themselves to oversee what is happening.

A further general problem of modelling, but one that is particularly relevant in the analysis of global change issues, involves the need to consider the consequences of discontinuous climatic or ecological responses. Inputs to IAMs reflect the world as we know it or as we might expect it to evolve, but climate change may bring surprises. Large uncertainties in our knowledge add to the need to consider discontinuous system responses. Atmospheric physics and chemistry seem to react relatively smoothly to external changes. However, ecological and, to some extent, economic responses could potentially be quite discontinuous. In a full uncertainty analysis, low probability events, such as the drying up of the U.S. corn belt, should be considered. The difficulty with such events, however, is that they are unprecedented and therefore hard to model.

The final problem is how to deal with chaotic behaviour of the model itself. A model is chaotic if small changes in its inputs cause large, nonsystematic changes in its output. Because chaos is associated with nonlinear dynamics, integrated assessment models run the risk of being chaotic, yielding advice that arbitrarily depends on how they are calibrated.

10.6.5 Relevance and presentation

The fifth big challenge of integrated assessment modelling is how to improve the capability of modellers to answer the questions that are of greatest concern to politicians and the general public and to present the results in such a manner that they understand the outcome and its limitations. Although this is the eventual aim of integrated assessment, it is not a trivial matter. The majority of the problems obviously arise from the immature state of the current generation of models. Most current models, for example, do not give insight into income distribution or employment issues. Nonmarket impacts can be included only after having been econometrically valued, thus implying substitutability. Economic models of the costs of emission controls often consider only market-based instruments and assess only efficiency. As others have argued in this report, policymakers tend to have a broader outlook that embraces more than economics. On the other hand, integrated assessment models that are more biased towards the natural sciences provide a weaker representation of the societal forces driving emissions and impacts. Evaluation and optimization are often not represented. In addition, some models calculate changes on the basis of potential rather than actual outcomes without considering transitional problems.

With respect to improving the presentation of results, policymakers generally do not welcome voluminous compilations of model results, nor can they usually interpret a set of detailed maps or technical diagrams, nor do they like to have measures prescribed for them, and yet these are broadly the three approaches taken so far. What is needed is an interface where model outcomes can be concisely and understandably represented and perhaps further evaluated and optimized. This implies a further step in integration and the use of information from another discipline: decision support systems.

10.6.6 The state of the art in integrated assessment modelling

A number of approaches to integrated assessment of climate change are being pursued. Each of these has strengths and weaknesses relative to the others. Moreover, individual modelling teams have chosen to focus on different aspects of the climate change issue. At this time, the significant complexities and uncertainties associated with the operation of the climate system, and how it impacts – and is impacted by – human activities, make it impossible to know exactly what to focus on and what methodology to employ. Thus, there is an advantage to the use of multiple research teams pursuing a plethora of alternative approaches. The approaches may provide complementary insights into the causes and effects of climate change or provide identical reenforcing results that increase our confidence in the results from any one approach.

There is also a natural complementarity between the different types of analyses, in that the more aggregate models (particularly if embedded in a probabilistic framework) can be used to focus the development of the more complex models. The more complex models can, in turn, be used as one source of parameter values for the more aggregated models and as a means of testing the effects of the aggregation employed on specific results. Moreover, the simple models can be used to cross-check results from the more complex models for consistency (i.e., they can be used as benchmarks) and to help communicate results from them to the policy development community and to the public. Finally, as each research team continually modifies its work plan and builds on the work of the others, all the approaches may tend to converge. Even if this were to be the case at some point in the future, it is not clear which of the approaches being pursued today would lead most efficiently to that ultimate model. This once again suggests the efficacy of pursuing a multitude of alternative analytic approaches to the study of climate change and the potential responses to it.

Endnotes

1. The following types of values of unmanaged ecosystems are identified in Chapter 6 of this report: (1) direct and indirect use values (e.g., plant inputs into medicine and the role of mangrove forests in coastal protection), (2) option value (preserving a species to retain the possibility that it may be of economic use in the future), and (3) existence value (e.g., the value of knowing that there are still blue whales).

2. An exception is the FUND model (Tol *et al.,* 1995), which has separate damage functions for each of the damage categories discussed in Chapter 6.

3. These "carbon taxes" are actually the marginal costs of efficiently reducing carbon emissions by the optimal amounts. Efficiency in this context means simply that lower cost emission reduction measures are always implemented before higher cost ones.

References

Alcamo, J. (ed.), 1994: *Image 2.0: Integrated modeling of global climate change,* Kluwer, Dordrecht, The Netherlands.

Alcamo, J., R. Shaw, and L. Hordijk (eds.), 1990: *The RAINS model of acidification: Science and strategies in Europe,* Kluwer, Dordrecht, The Netherlands.

Bradley, R.A., J.A. Edmonds, M. Wise, and H. Pitcher, 1994: Controlling carbon: Equity, efficiency, and participation in possible future agreements to control fossil fuel carbon emissions. In *Equity and social considerations related to climate change,* ICIPE Science Press, Nairobi, Kenya.

Brewer, G.D., 1986: Methods for synthesis: Policy exercises. In *Sustainable development of the biosphere,* W.C. Clark and R.E. Munn, eds., pp. 455–473, Cambridge University Press, New York.

Chapuis, T., M. Duong, and M. Grubb, 1995: The greenhouse cost model: An exploration of the implications for climate change policy of inertia and adaptability in energy systems, International Energy Workshop, International Institute for Applied Systems Analysis, Laxenburg, Austria, 20–22 June.

Clark, W.C. (ed.), 1982: *Carbon dioxide review: 1982,* Oxford University Press, New York.

Clark, W.C., and R.E. Munn (eds.), 1986: *Sustainable development of the biosphere,* Cambridge University Press, New York.

Clark, W.C., and J. van Eijndhoven (eds.), 1996: Response assessment. In The Social Learning Group, *Learning to manage global environmental risks,* MIT Press, Cambridge, MA, forthcoming.

Cline, W. R., 1992: *The economics of global warming,* The Institute for International Economics, Washington, DC.

Cohan, D., R.K. Stafford, J.D. Scheraga, and S. Herrod, 1994: The global climate policy evaluation framework, proceedings of the 1994 A&WMA Global Climate Change Conference: Phoenix, 5-8 April, Air & Waste Management Association, Pittsburgh.

Commission of the European Communities, 1992: *PAGE user manual,* Brussels.

Den Elzen, M.G.J., 1993: Global environmental change: An integrated modelling approach, Ph D thesis, University of Maastricht, The Netherlands.

Dowlatabadi, H., 1995: Integrated assessment climate assessment model 2.0, technical documentation, Department of Engineering and Public Policy, Carnegie-Mellon University, Pittsburgh.

Dowlatabadi, H., and M.G. Morgan, 1993: A model framework for integrated assessment of the climate problem, *Energy Policy,* **21,** 209–221.

Edmonds, J., J.M. Reilly, J.R. Trabalka, and D.E. Reichle, 1984: *An analysis of possible future atmospheric retention of fossil fuel CO_2,* TR013, DOE/OR/21400-1, National Technical Information Service, U.S. Department of Commerce, Springfield, VA.

Edmonds, J., and J.M. Reilly, 1985: *Global energy: Assessing the future,* Oxford University Press, New York.

Edmonds, J., H. Pitcher, N. Rosenberg, and T. Wigley, 1994a: Design for the global change assessment model, proceedings of the International Workshop on Integrative Assessment of Mitigation, Impacts, and Adaptation to Climate Change, International Institute for Applied Systems Analysis, Laxenburg, Austria, 13-15 October.

Edmonds, J., M. Wise, C. MacCracken, 1994b: Advanced energy technologies and climate change: An analysis using the global change assessment model (GCAM), presentation to the Air and Waste Management Meeting, 6 April, Tempe AZ, Air and Waste Management Association, Pittsburgh.

Edmonds, J., D. Barns, M. Wise, and M. Ton, 1995: Carbon coalitions: The cost and effectiveness of energy agreements to alter trajectories of atmospheric carbon dioxide emissions, *Energy Policy,* **23**(4/5), 309–336.

Fujii, Y., Y. Kaya, and K. Yamaji (1993): The New Earth 21 model, working paper, Department of Systems Engineering, University of Tokyo.

Grobecker, A.J., S.C. Coroniti, and R.H. Cannon, Jr., 1974: The report of findings: The effects of stratospheric pollution by aircraft, DOT-TST-75-50, U.S. Department of Transportation, Climatic Impact Assessment Program, National Technical Information Service, Springfield, VA.

Grubb, M. (1991): *Energy policies and the greenhouse effect,* Royal Institute of International Affairs, Dartmouth Publishing, Aldershot, Hants., U.K.

Grubb, M., M. Koch, A. Munson, F. Sullivan, and K. Thomson (1993): *The Earth Summit Agreements: A guide and assessment,* Earthscan, London.

Grubb, M., M.H. Duong, and T. Chapuis, 1994: Optimizing climate change abatement responses: On inertia and induced technical change. In *Integrative assessment of mitigation, impacts and adaptation to climate change,* N. Nackicenovic, W.D. Nordhaus, R. Richels, and R.L. Toth, eds., pp. 205–218, IIASA, Laxenburg, Austria.

Grubb, M., M.H. Duong, and T. Chapuis, 1995: The economics of changing course, *Energy Policy,* **23**(4/5), 417–432.

Häfele, W., J. Anderer, A. McDonald, and N. Nakicenovic, 1981: *Energy in a finite world,* Ballinger, Cambridge, MA.

Hammitt, J.K., 1995a: Outcome and value uncertainties in global-change policy, *Climatic Change,* **30**(2), 125–145.

Hammitt, J.K., 1995b: Expected values of information and cooperation for abating global climate change. In *Wise choices: Games, decisions and negotiations,* R. Keeney, J. Sebenius, and R. Zeckhauser, eds., Harvard Business School Press, Boston.

Hammitt, J.K. and J.L. Adams, 1995: The value of international cooperation in abating climate change, American Economics Association/Association of Environmental and Resource Economists Annual Meeting, Washington, DC, 6-8 January.

Hammitt, J.K., R.J. Lempert, and M.E. Schlesinger, 1992: A sequential-decision strategy for abating climate change, *Nature,* **357,** 315–318.

Hope, C., J. Anderson, and P. Wenman, 1993: Policy analysis of the greenhouse effect: An application of the PAGE model, *Energy Policy,* **21,** 327–338.

Hordijk, L., 1991: Use of the RAINS model in acid rain negotiation in Europe, *Environmental Science and Technology,* **25**(4), 596–602.

Hourcade, J.C., and T. Chapuis, 1995: No-regret potentials and technical innovation, *Energy Policy,* **23**(4/5), 433–446.

Hulme, M., S.C.B. Raper, and T.M.L. Wigley, 1995: An integrated framework to address climate change (ESCAPE) and further developments of the global and regional climate modules (MAGICC). *Energy Policy,* **23**(4/5), 347–355.

IPCC (Intergovernmental Panel on Climate Change), 1990: *Climate Change: The IPCC Scientific Assessment,* J.T. Houghton, G.J. Jenkins, and J.J. Ephraums, eds., Cambridge University Press, Cambridge.

IPCC (Intergovernmental Panel on Climate Change), 1992: *Climate change 1992: The supplementary report to the IPCC scientific assessment,* J.T. Houghton, B.A. Callander, and S.K. Varney, eds., Cambridge University Press, Cambridge.

IPCC (Intergovernmental Panel on Climate Change), 1995: *Climate change 1994: Radiative forcing of climate change and an evaluation of the IPCC IS92 emission scenarios,* J.T. Houghton, L.G. Meira Filho, J. Bruce, Hoesung Lee, B.A. Callander, E. Haites, N. Harris, and K. Maskell, eds., Cambridge University Press, Cambridge.

Jain, A., H. Kheshgi, and D. Wuebbles, 1994: *Integrated science model for assessment of climate change,* Lawrence Livermore Laboratory, Livermore, CA, May.

Karplus, W.J., 1992: *The heavens are falling – The scientific prediction of catastrophes in our time,* Plenum Press, New York.

Kolstad, C.D., 1993: Looking vs. leaping: The timing of CO_2 control in the face of uncertainty and learning. In *Costs, impacts, and benefits of CO_2 mitigation,* Y. Kaya, N. Nakicenovic, W.D. Nordhaus, and R.L. Toth, eds., pp. 63–82, IIASA, Laxenburg, Austria.

Kolstad, C.D., 1994a: George Bush vs. Al Gore: Irreversibilities in greenhouse gas accumulation and emissions control investment, *Energy Policy,* **22,** 771–778.

Kolstad, C.D., 1994b: The timing of CO_2 control in the face of uncertainty and learning. In *International Environmental Economics,* E.C. Van Ierland, ed., pp. 75–96, Elsevier, Amsterdam.

Kolstad, C.D., 1994c: Mitigating climate change impacts: The conflicting effects of irreversibilities in CO_2 accumulation and emission control investment. In *Integrative assessment of mitigation, impacts and adaptation to climate change,* N. Nackicenovic, W.D. Nordhaus, R. Richels, and R.L. Toth, eds., pp. 205–218, IIASA, Laxenburg, Austria.

Kosobud, R., T. Daly, D. South, and K. Quinn, 1994: Tradable cumulative CO_2 permits and global warming control, *The Energy Journal,* **15**(2), 213–232.

Lashof, D.A., and D.A. Tirpak, 1989. Policy options for stabilizing global climate, draft report to Congress, U.S. Environmental Protection Agency, Office of Policy, Planning and Evaluation, Washington, DC.

Lempert, R.J., M.E. Schlesinger, and J.K. Hammitt, 1994: The impact of potential abrupt climate changes on near-term policy choices, *Climatic Change,* **26,** 351–376.

Lempert, R.J., M.E. Schlesinger, and S. Bankes, 1995: When we don't know the costs or the benefits: Adaptive strategies for abating climate change, *Climate Change,* forthcoming.

Levy, M.A., 1993: European acid rain: The power of tote-board diplomacy. In *Institutions for the earth: Sources of effective international environmental protection,* P.M. Haas, R.O. Keohane, and M.A. Levy, eds., MIT Press, Cambridge, MA.

MacCracken, J.C., and F.M. Luther, 1985a: *Detecting the climatic effects of increasing carbon dioxide,* DOE/ER-0235, National Technical Information Service, U.S. Department of Commerce, Springfield, VA.

MacCracken, J.C., and F.M. Luther, 1985b: *Projecting the climate effects of increasing carbon dioxide,* DOE/ER-0237, National Technical Information Service, U.S. Department of Commerce, Springfield, VA.

Maddison, D., 1995: A cost-benefit analysis of slowing climate change, *Energy Policy,* **23**(4/5), 337–346.

Manne, A.S., and R.G. Richels, 1992: *Buying greenhouse insurance: The economic costs of CO_2 emission limits,* MIT Press, Cambridge, MA.

Manne, A.S., and R.G. Richels, 1993: CO_2 hedging strategies – The Impact of uncertainty upon emissions, paper presented at the OECD/IEA Conference on the Economics of Climate Change, Paris, 14-16 June.

Manne, A.S., R. Mendelsohn, and R.G. Richels, 1993: MERGE: A model for evaluating regional and global effects of GHG reduction policies, *Energy Policy,* **23**(1), 17–34.

Massachussetts Institute of Technology (MIT), 1994: Center for Global Change Science and Center for Energy and Environmental Policy Research, *Joint program on the science and technology of global climate change,* Cambridge, MA.

Matsuoka, Y., M. Kainuma, and T. Morita, 1995: Scenario analysis of global warming using the Asian-Pacific integrated model, *Energy Policy,* **23**(4/5), 357–372.

Meadows, D.H., D.L. Meadows, J. Randers, and W.W. Behrens, 1972: *The limits to growth,* Universe Books, New York.

Meadows, D.H., J. Richardson, and G. Bruckmann, 1982: *Groping in the dark: The first decade of global modelling,* Wiley, Chichester, UK.

Mesarovic, M.D., and E. Pestel, 1974: *Mankind at the turning point: The second report to the Club of Rome,* Dutton, New York.

Mintzer, I., 1987: *A matter of degrees: The potential for controlling the greenhouse effect,* Research Report 5, World Resources Institute, Washington, DC.

Mori, S., 1995a: A long-term evaluation of nuclear power technology by extended DICE + e model simulations – Multiregional approach for resource and industry allocation, *Progress in Nuclear Energy,* **29**, 135–142.

Mori, S., 1995b: *Long-term interactions among economy, environment, energy, and land-use changes – An extension of MARIA Model,* Technical Report IA-TR-95-04, Science University of Tokyo, Japan.

Morita, T., Y. Matsuoka, M. Kainuma, H. Harasawa, and K. Kai, 1993: *AIM – Asian-Pacific integrated model for evaluating policy options to reduce GHG emissions and global warming impacts,* National Institute for Environmental Studies, Tsukuba, Japan, September.

Morita, T., M. Kaihuma, H. Harasawa, K. Kai, L. Dong-Kumand, and Y. Matsuoka, 1994: *Asian-Pacific integrated model for evaluating policy options to reduce GHG emissions and global warming impacts, interim report,* National Institute for Environmental Studies, Tsukuba, Japan.

Nakicenovic, N., W.D. Nordhaus, R. Richels, and F.L. Toth (eds.), 1994: *Integrative assessment of mitigation, impacts, and adaptation to climate change,* CP-94-9, Proceedings of a workshop held 13-15 October 1993 at IIASA, Laxenburg, International Institute for Applied Systems Analysis, Laxenburg, Austria.

Nordhaus, W.D., 1979: *The efficient use of energy resources,* Yale University Press, New Haven, CT.

Nordhaus, W.D., 1989: The economics of the greenhouse effect, paper presented to the June 1989 meeting of the International Energy Workshop, Laxenburg, Austria, 20-22 June.

Nordhaus, W.D., 1991: To slow or not to slow: The economics of the greenhouse effect, *The Economic Journal,* **101**, 920–937.

Nordhaus, W.D., 1994: *Managing the global commons: The economics of climate change,* MIT Press, Cambridge, MA.

Nordhaus, W.D., and G.W. Yohe, 1983: Future carbon dioxide emissions from fossil fuels. In *Changing climate: Report of the carbon dioxide assessment committee,* National Academy Press, Washington, DC.

Nordhaus, W.D., and Z. Yang, 1995: *RICE: A regional dynamic general equilibrium model of optimal climate change policy,* Yale University Press, New Haven, CT.

OTA (Office of Technology Assessment), 1994: Climate treaties and models: Issues in the international management of climate change, OTA-BP-ENV-128, U.S. Government Printing Office, Washington, DC.

Parson, E.A., 1994: Searching for integrated assessment: A preliminary investigation of methods and projects in the integrated assessment of global climatic change, paper presented to the third meeting of the Harvard-CIESIN Commission on Global Environmental Change Information Policy, NASA Headquarters, Washington DC, 17-18 February.

Parson, E.A., 1995a: Integrated assessment and environmental policy-making: In pursuit of usefulness, *Energy Policy,* **23**(4/5), 463– 475.

Parson, E.A., 1995b: *Why study hard policy problems with simulation gaming?* International Institute for Applied Systems Analysis, Laxenburg, Austria..

Parson, E.A., and Richard Zeckhauser, 1995: Cooperation in the unbalanced commons. In *Barriers to the negotiated resolution of conflict,* K. Arrow, R. Mnookin, L. Ross, A. Tversky, and R. Wilson, eds., Norton, New York.

Peck, S.C., and T.J. Teisberg, 1992: CETA: A model for carbon emissions trajectory assessment, *The Energy Journal,* **13**(1), 55–77.

Peck, S.C., and T.J. Teisberg, 1993: Global warming uncertainties and the value of information: An analysis using CETA, *Resource and Energy Economics,* **15**(1), 71–97.

Peck, S.C., and T.J. Teisberg, 1994: Optimal carbon emissions trajectories when damages depend on the rate or level of warming, *Climatic Change,* **30,** 289–314.

Peck, S.C., and T.J. Teisberg, 1995: Optimal CO_2 control policy with stochastic losses from temperature rise, *Climatic Change,* **31,** 19–34.

Read, P., 1994a: *Responding to global warming: The technology, economics and politics of sustainable energy,* ZED Books, London and Atlantic Highlands, NJ.

Read, P., 1994b: Optimal control analysis of policy regret under adverse climate surprise, International Energy Workshop, East-West Center, Honolulu, 21-22 June.

Richels, R., and J. Edmonds, 1995: The costs of stabilizing atmospheric CO_2 concentrations, *Energy Policy,* **23**(4/5), 373–378.

Rosenberg, N.J. (ed.), 1993: *Towards an integrated impact assessment of climate change: The MINK study,* Kluwer, Boston.

Rosenberg, N.J., and P.R. Crosson, 1991: The MINK project: A new methodology for identifying regional influences of and responses to increasing atmospheric CO_2 and climate change, *Environmental Conservation,* **18**(4), 313–322.

Rotmans, J., 1990: *Image: An integrated model to assess the greenhouse effect,* Kluwer, Dordrecht, The Netherlands.

Rotmans, J., 1995: *TARGETS in transition,* RIVM report, Rijksinstitut Voor Volksgezondheid En Milieuhygiene (National Institute of Public Health and the Environment), Bilthoven, The Netherlands.

Rotmans, J., and M.G.J. den Elzen, 1992: A model-based approach to the calculation of global warming potentials (GWPs), *International Journal of Climatology,* **12,** 865–874.

Rotmans, J., and M.G.J. den Elzen, 1993: Modeling feedback mechanisms in the carbon cycle: Balancing the carbon budget, *Tellus,* **45B**(4), 301–320.

Rotmans, J., M.B.A. van Asselt, A.J. de Bruin, M.G.J. den Elzen, J. de Greef, H. Hilderink, A.Y. Hoekstra, M.A. Janssen, H.W. Koster, W.J.M. Martens, L.W. Niessen, and H.J.M. de Vries, 1994: *Global change and sustainable development: A modelling perspective for the next decade,* Rijksinstitut Voor Volksgezondheid En Milieuhygiene (National Institute of Public Health and Environmental Protection), Bilthoven, The Netherlands.

Rotmans, J., H. Dowlatabadi, J.A. Filar, and E.A. Parson, 1995: Integrated assessment of climate change: Evaluation of methods and strategies. In *Human choices and climate change: A state of the art report,* Batelle Pacific Northwest Laboratory, Washington, DC.

Strain, B.R., and J.D. Cure, 1985: *Direct effects of increasing carbon dioxide on vegetation,* DOE/ER-0238, National Technical Information Service, U.S. Department of Commerce, Springfield, VA.

Tol, R.S.J., 1993: *The climate fund: Modelling costs and benefits,* W-93/17, Institute for Environmental Studies, Vrije Universiteit Amsterdam.

Tol, R.S.J., 1995: *The climate fund sensitivity, uncertainty, and robustness analyses,* W-95/02, Institute for Environmental Studies, Vrije Universiteit Amsterdam.

Tol, R.S.J., T. van der Burg, H.M.A. Jansen, and H. Verbruggen, 1995: *The climate fund – Some notions on the socio-economic impacts of greenhouse gas emissions and emission reduction in an international context,* Institute for Environmental Studies, Report R95/03, Vrije Universiteit Amsterdam.

Trabalka, J. (ed.), 1985: *Atmospheric carbon dioxide and the global carbon cycle,* DOE/ER-0239, National Technical Information Service, U.S. Department of Commerce, Springfield, VA.

U.S. National Research Council, 1983: *Changing Climate,* National Academy Press, Washington, D.C.

Van Asselt, M.B.A., and J. Rotmans, 1995: *Uncertainty in integrated assessment modelling: A cultural perspective–based perspective,* RIVM Report No. 461502009, Rijksinstitut Voor Volksgezondheid En Milieuhygiene (National Institute of Public Health and Environmental Protection), Bilthoven, The Netherlands.

Van Vienen, H.A.W., F.J. Willekens, J. Hutter, M.B.A. van Asselt, H.B.M. Hilderink, L.W. Niessen, and J. Rotmans, 1994: *Fertility change: A global integrated perspective,* RIVM Report No. 461502008, Rijksinstitut Voor Volksgezondheid En Milieuhygiene (National Institute of Public Health and the Environment), Bilthoven, The Netherlands.

WEC (World Energy Council) and IIASA (International Institute for Applied Systems Analysis), 1995: *Global energy perspectives to 2050 and beyond,* World Energy Council, London.

Wigley, T.M.L., 1993: Balancing the carbon budget: Implications for projections of future carbon dioxide concentration changes, *Tellus,* **45B**(5), 409–425.

Wigley, T., 1995: Global mean temperature and sea level consequences of greenhouse gas stabilization, *Geophysical Research Letters,* **22**(1), 45–48.

Wigley, T., M. Hulme, and S. Raper, 1993: *MAGICC: Model for the assessment of greenhouse induced climate change,* University Consortium for Atmospheric Research, Boulder, CO.

Wigley, T., R. Richels, and J. Edmonds, 1996: Economic and environmental choices in the stabilization of atmospheric CO_2 concentrations, *Nature,* **379,** 240–243.

Yohe, G., 1995a: Exercises in hedging against extreme consequences of global change and the expected value of information, Department of Economics, Wesleyan University, Wesleyan, CT.

Yohe, G., 1995b: Strategies for using integrated assessment models to inform near-term abatement policies for greenhouse gas emissions. In *Proceedings of NATO Advanced Research Workshop on Integrated Assessment of Global Environmental Change: Science and Policy, Durham, North Carolina,* Department of Economics, Wesleyan University, Wesleyan, CT.

Yohe, G., and R. Wallace, 1995: Near-term mitigation policy for global change under uncertainty: Minimizing the expected cost of meeting unknown concentrations, Department of Economics, Wesleyan University, Wesleyan, CT.

11

An Economic Assessment of Policy Instruments for Combatting Climate Change

Lead Authors:
B.S. FISHER, S. BARRETT, P. BOHM, M. KURODA, J.K.E. MUBAZI,
A. SHAH, R.N. STAVINS

Contributors:
E. Haites, M. Hinchy, S. Thorpe

The lead authors would like to acknowledge the assistance of L. Goulder, A. Jaffe, R. Newell, and R. Repetto during the preparation of this chapter.

CONTENTS

SUMMARY

A clear distinction needs to be drawn between the costs and benefits of actions taken to reduce the impacts of climate change and the costs and benefits of actions taken to reduce emissions. The first involves a consideration of adaptation policies such as developing new drought-resistant plant cultivars, whereas the second involves a consideration of policies designed to mitigate climate change. The final policy mix adopted by countries is likely to contain both adaptation and mitigation policies.

The world economy and individual national economies suffer from other distortions than those possibly leading to global climate change. Any of these may prevent economies from attaining efficient outcomes. In many cases, correcting for those other distortions would lead to actions that would also serve to reduce the expected damage from climate change. Plainly, such distortions should be corrected, and many governments are already taking steps to do so.

To effect a substantial reduction in net greenhouse gas emissions, such as would be required to stabilize atmospheric concentrations, requires policies expressly designed to mitigate global climate change. The associated policy instruments must be identified at two different levels: those that might be used by a coalition of countries and those that might be used by individual nations unilaterally or to achieve compliance with a multilateral agreement on greenhouse gas emission targets.

Governments may have different sets of criteria for assessing international as well as domestic greenhouse policy instruments. Among these criteria are efficiency or cost-effectiveness, effectiveness in achieving stated environmental targets, distributional (including intergenerational) equity, flexibility in the face of new knowledge, understandability to the general public, and consistency with national institutions and traditions. The choice of instruments may also partly reflect a desire on the part of governments to achieve other objectives such as meeting fiscal targets or influencing pollution levels indirectly related to greenhouse gas emissions. Governments may also be concerned about the effects of policy on competitiveness.

A coalition of nations may choose one or a mix of policy instruments, including tradable quotas, feasible forms of joint implementation, harmonized domestic carbon taxes, international carbon taxes, nontradable quotas, and various international standards. At both the international and national levels, market-based approaches are likely to be more cost-effective than other instruments.

At the international level, all the potentially efficient tax or quota solutions should be available to facilitate future negotiations. Under a harmonized carbon tax, incentives exist for countries to alter related policies to reduce the domestic implications of the tax (for example, by introducing offsetting production subsidies). This possibility could make harmonized carbon taxes less effective than tradable quotas in reducing emissions.

For a global treaty, a tradable quota system is the only potentially cost-effective arrangement where an agreed level of emissions is attained with certainty (subject to enforcement). The initial quota allocation could provide a means of compensation to countries – particularly developing countries – that would bear substantial costs in implementing international response measures. This attribute would provide the opportunity to encourage developing countries to participate actively in global action.

In principle, individual countries can choose from among a large set of available instruments, including carbon taxes, tradable permits, deposit refund systems, and subsidies, as well as technology standards, performance standards, product bans, direct government investment, and voluntary agreements. A choice of tradable quotas at the international level would provide maximum flexibility for instrument choice at the domestic level.

A tradable quota or permit system has the disadvantage of making the marginal cost of emission reductions uncertain, whereas a carbon tax has the disadvantage of leaving the level at which emissions will be controlled uncertain. The weight given to the importance of reducing these different types of uncertainty will be crucial in determining the final choice between competing market-based instruments. Regardless of the final mix of instruments adopted, there will remain a high degree of uncertainty about the physical effects of different levels of emissions.

The consequences of climate change policy will be determined by the choice of the mix of policy instruments, the design and implementation of those policies, and the institutional framework in which the policies must operate. For example, regulatory instruments are likely to have a different impact on innovation than market-based instruments. Furthermore, the welfare effects of a carbon tax or the government sale of tradable permits will depend on whether and how the associated revenues are recycled. In some countries monitoring and enforcement may be more difficult than in others, and such differences could have a direct impact on the effectiveness of some policy instruments.

11.1 Introduction

11.1.1 Guidelines from the FCCC

The aim in this chapter is to provide an economic assessment of possible policy instruments for managing greenhouse gas emissions under the Framework Convention on Climate Change (FCCC). The Framework Convention contains several key guidelines for policy implementation. First is the emphasis given to the need for developed countries to demonstrate that they are taking the leading role in policies to control greenhouse emissions. In essence, developed country signatories (as listed in Annex I of the Convention) have accepted the goal, but not necessarily the requirement, of stabilizing greenhouse gas emissions at 1990 levels by 2000 (Article 4.2). In ratifying the Convention, Annex I countries have effectively accepted a quantitative emission target, although the level of that target remains uncertain. Developing country signatories (non-Annex countries) are under no such obligation. Rather, the economic needs and special circumstances of developing countries (Articles 4.8 and 4.10), and of countries highly dependent on incomes from fossil fuels (Article 4.10), will be taken into account in determining specific commitments to control emissions.

To achieve greenhouse emission reductions, Annex I countries have the option to implement greenhouse policy measures jointly with other parties to the Convention (Article 4.2). This provision is consistent with another guiding principle in the Convention that stipulates that all greenhouse policy measures should be cost-effective – that is, that they should achieve policy goals at least cost (Article 3). Other key principles in Article 3 require the parties to promote sustainable development, to take precautionary measures to minimize the costs of greenhouse uncertainties and risks (noting that where there are risks of serious or irreversible damage, lack of scientific certainty should not be used to justify policy deferral), and to ensure that measures taken to combat climate change do not amount to unfair trade restrictions.

Prior to the development of the FCCC, the most closely related international conventions were the Vienna Convention for Protection of the Ozone Layer (concluded in 1985) and the 1987 Montreal Protocol on Substances that Deplete the Ozone Layer (the text of which was revised in 1990 and again in 1992). The experience of the parties to the Montreal Protocol provides valuable information about the implementation of policy approaches within the FCCC.[1]

In this chapter, the factors affecting the policy mix for the control of greenhouse gas emissions are reviewed in the light of these guiding principles and the general international legal framework in which the Convention must operate.

11.1.2 The greenhouse policy problem

Two characteristics of the greenhouse problem are central to the design of policy responses. The first key feature is that it is a global problem. It is the total accumulation of greenhouse gases in the atmosphere that could cause global warming over the next century (IPCC 1990; 1992), regardless of the geo-graphic source of emissions. In addition, there is a long time lag, up to fifty years, between emission reductions and their impact on atmospheric concentrations. Thus, the greenhouse problem is a pollution problem over space and time, and one in which increased absorption can reduce atmospheric concentrations of greenhouse gases as effectively as reduced emissions. Any benefits from controlling concentrations will depend on long-term global cooperation, and the costs of collective control will be incurred long before any potential benefits are realized.[2]

The second key feature of the greenhouse problem is that both the extent of any climate change and the nature of its effects are uncertain. This means that potential greenhouse policies must be assessed using a decision-making framework that explicitly incorporates risk, uncertainty, and the capacity to learn about evolving climatic and economic conditions around the world.

A basic principle in public policy (as for any financial decision) is to time the introduction of the policy to maximize the expected discounted value of the stream of net benefits from the initiative. In this context, there may be benefits from waiting to reduce uncertainties before implementing greenhouse policies.[3] The value of the information gained from waiting could allow greenhouse policy to be properly tailored to the most likely damage scenario in order to avoid excessive control costs (Peck and Teisberg, 1993; Leary and Scheraga, 1994; Richels and Edmonds, 1994). Conversely, there could be significant costs in waiting, if waiting makes excessive damage costs more likely (Chichilnisky and Heal, 1993) or results in the need for urgent action at some future time, with associated disproportionate adjustment costs.

Greenhouse policy assessment must therefore take into account the existence of opposing risks. Indeed, one important avenue for policy assessment is to examine the extent to which policy can be directed to reduce the costs of uncertainty and the costs of risk from natural damage caused by the enhanced greenhouse effect and from mitigation measures in response to climate change concerns. An immediate response to the greenhouse problem is to invest in research and development to reduce greenhouse uncertainties and subsequently to provide new information to decision makers.

An efficient greenhouse policy would ensure that the costs of greenhouse uncertainties, and of associated risks, and the costs of emission reductions and adaptation strategies are balanced, at the margin, with the benefits from avoiding damage from global warming. One implication of this efficiency criterion is that the optimal policy is the one that achieves a global greenhouse target at least cost in the face of risk, uncertainty, and the need for further knowledge about the causes and effects of climate change. However, regardless of the policy approach adopted, considerable physical uncertainty about both the effects and extent of climate change will remain.

11.2 Greenhouse Policy Instruments and Criteria for Policy Assessment

The variety of instruments available to policymakers to control greenhouse gas emissions is outlined here. Such instruments include conventional regulatory instruments, market-based instruments such as taxes, subsidies, and tradable quotas and permits, and other complementary policies. In this chapter the term "tradable quota" is used to describe internationally traded emission allowances. The term "permit" is more commonly used in the literature to describe domestic trading schemes, and that convention is adopted here. Existing global climate change research that has analyzed a broad range of policy instruments includes Mintzer (1988); US Congress, Office of Technology Assessment (1991); IPCC (1992); National Academy of Sciences (1992); and McCann and Moss (1993).

11.2.1 Domestic policy instruments

11.2.1.1 Conventional regulatory instruments

One way of controlling activities that both directly and indirectly lead to greenhouse gas emissions is to set standards and to regulate the activities of firms or individuals. By mandating standards, governments attempt to ban or alter the use of materials and equipment considered to be damaging. Standards are typically applied in areas such as buildings (energy efficiency, for example), fuel use by motor vehicles, energy efficiency of household appliances, and the content of fuels. Standards may be voluntary or mandatory. They may be fixed or set as targets, or "rolling standards" might be adopted (Grubb, 1991).

11.2.1.2 Market-based instruments

In using market-based instruments, governments attempt to alter price signals to ensure that emitters face direct-cost incentives to control emissions. The primary market-based instruments for greenhouse management are emission taxes and tradable emission permits.

11.2.1.2.1 Taxes and subsidies

Under an emission tax, those who produce emissions face a tax per unit of emissions.[4] All fossil fuels should be taxed at the same rate per unit of their long-term global warming potential. A tax on energy content measured in British thermal units, the so-called BTU tax, would not satisfy this criterion, as it relates to energy use per se rather than to any externalities associated with end products of combustion (see Poterba, 1993). A tax on the carbon content of fossil fuels, on the other hand, would approximate this criterion. Implementing such taxes at a uniform rate per tonne of carbon content of fossil fuels to curtail carbon emissions assumes that existing excises on energy products are levied at the optimal level based on minimizing the excess burden of taxation and internalization of environmental externalities. If such an assumption does not hold in practice, the design of carbon taxes becomes more complicated.

Subsidies might be offered for adopting particular technologies or practices. Such subsidies might be directed at fostering emission abatement or the creation of additional sinks by, for example, subsidizing tree planting.

A subsidy scheme could be linked to a tax scheme by applying the subsidy to reductions in emissions below a baseline and a tax on emissions above the baseline. The rate of subsidy would be applied per unit of emissions at the same rate as the emission tax. A tax/subsidy scheme would mean that firms would not pay a tax on every unit of emissions, but it would involve an additional administrative burden in setting the baseline for every firm.

11.2.1.2.2 Tradable permits

Under an emission-trading scheme, emitters are given permits to emit (the total allocation is the aggregate emission cap for the country) and have the option of buying or selling permits in the marketplace. Although there are important and often subtle differences between taxes and tradable permits, under some restrictive circumstances the outcome can be the same. Both may target full user-cost pricing of the atmosphere to dispose of net greenhouse gas emissions from human activity. When traded on a national market, permit prices are established that show the costs of marginal emissions, just as an emission tax does. The difference is that the tax is exogenous (in this case, set by the government) and its effects on emissions endogenous, whereas emissions are exogenously determined in the case of a permit system and, hence, permit prices are endogenous.

A tradable-permit system could be used in combination with either an international tradable-quota system or an international carbon tax. In the former case, the domestic permit system could either be integrated with an international quota system, where the permit-liable parties (say, a limited set of wholesale fossil fuel dealers) trade directly on the international quota market (see Grubb and Sebenius, 1991; Sandor *et al.,* 1994) or be run as a separate subsystem providing the national government with a net excess demand for (or supply of) emission quotas at the ruling international quota price. In the case where governments paid an international tax on carbon emissions, it would be up to each government to determine beforehand the volume of domestic permits available per period. Ideally, this volume should be such that the resulting permit price would be equal to the tax rate. If not, nationally as well as internationally, too little or too much abatement would take place.

11.2.1.3 Other complementary policies

A range of other complementary instruments exists that might be adopted to moderate greenhouse gas emissions or to promote adaptation to climate change. Education and provision of new information – by promoting research, for example – may be valuable in changing consumer behaviour with respect to energy consumption and the development and adoption of new technology. Family planning may play an indirect role in reducing total energy demand in the future, as might more general education directed specifically at women in developing countries. Modifications of trade policy and reductions in energy production and consumption subsidies (and other market distortions) may also have indirect consequences for greenhouse gas emissions. Changes in migration policies in some countries may allow more flexibility for developing countries to adapt to regional population pressures that may arise as a consequence of changes in the incidence of occurrences such as severe drought.

11.2.2 *International policy instruments*

The use of the available policy instruments will only lead to a cost-effective global outcome if certain conditions are met. First, unless individual countries undertake cost-effective domestic greenhouse policy measures that are compatible with the goal of global efficiency, the policy instruments adopted internationally will not lead to that goal. Each individual country is free to choose its own instrument or combination of instruments to meet its international obligations, but the choice of international instruments will, to some extent, dictate the choice of policy instruments at the domestic level. This is clearly so in the case of the harmonized carbon/energy tax proposed by the European Commission for member states of the European Union. Under this regime, every member of the Union would impose the same tax rate, although states would be free to decide for themselves what to do with the revenues from the tax. If nations themselves were taxed by an international agency, it would not necessarily follow that nations would choose to impose the same tax domestically. For example, they might instead choose to reduce emissions domestically by means of a tradable permit scheme. Similarly, in the case of a system of internationally tradable quotas, individual countries might choose to implement their obligations by means of a domestic tradable permit scheme or through a domestic emission tax.

Second, given that information is not perfect and that distortions already exist in both international and domestic markets, the actual market outcomes from the implementation of particular greenhouse policies will not necessarily be efficient. The importance of the policy environment is discussed in Section 11.3, and implementation issues surrounding the adoption of market-based instruments are outlined in Section 11.6.

11.2.2.1 *Regulatory instruments*

It is conceivable that uniform standards could be established among countries participating in an international emission reduction agreement. But it is likely to be difficult to achieve wide agreement about any large set of specific instruments of this type. For example, individual countries may adopt standards for housing insulation, but it is most unlikely that the same standards would be applicable in both temperate and tropical countries. Moreover, such an approach would limit the domestic policy choices of individual countries and, hence, their flexibility in adjusting their emissions under an international greenhouse gas reduction agreement. Another regulatory approach involves agreements by countries on fixed national emission levels (a "nontradable emission quota" system), much in the tradition of the European Union's Large Combustion Plants Directive, which specifies reductions in the emissions of sulphur dioxide and oxides of nitrogen from plants with a thermal rating of 50 MW or larger. Such an approach would mean that marginal emission abatement costs among participating countries would tend to be different and, hence, total abatement costs, globally speaking, would be unnecessarily high.

11.2.2.2 *International taxes and harmonized domestic taxes*

If countries agreed to apply the same level of domestic greenhouse or carbon taxes (harmonized domestic taxes), marginal abatement costs would tend to be equalized among countries. Such an agreement may have to include side payments from rich to poor countries if the latter are to be encouraged to participate.

An alternative type of international policy to reduce emissions could be an agreement to levy a uniform international tax on greenhouse or carbon emissions in each of the participating countries. The total international tax revenue would be shared among the participating countries according to rules established in the agreement.

If an international tax agreement did not cover all countries, world fossil fuel prices would decrease and fossil fuel use increase in nonparticipating countries (so-called carbon leakage). In addition, since carbon-intensive products would be less expensive in such countries, exports of them to the participating countries would likely rise. A policy instrument might then be introduced by the latter countries to control carbon leakage (see Section 11.6.5 for further discussion). For example, a carbon tariff might be imposed at a rate corresponding to the tax rate on imported products on the basis of their estimated carbon contents.

In the case of a domestic carbon tax imposed by international agreement, the national commitment to impose the tax will also vary because perspectives on global warming vary from one country to another. If a country has signed such an agreement under international pressure, that country could make the carbon tax ineffective by reducing existing energy taxes, by taxing substitutes for fossil fuels (for example, hydroelectricity), by providing subsidies to complements or products that are fossil-fuel-energy intensive, and by lax enforcement of the tax (see Hoel, 1993). Thus, by following a suitable strategy, a free ride becomes possible. A global carbon tax imposed by an international agency, on the other hand, would impinge on national sovereignty and would therefore be difficult to negotiate.

If global carbon taxes were levied as producer taxes instead of consumer taxes, tax revenue would be collected in fossil fuel producer countries instead of consumer countries and, hence, would shift the burden between the two types of countries (Whalley and Wigle, 1991). The distributional effects of a "producer cartel" solution may be unacceptable to a great many countries and, if used, could give rise to retaliatory trade policy measures. (Neither carbon producer taxes nor producer quota systems are further discussed in detail in this chapter.)

11.2.2.3 *Tradable quotas*

Another potentially cost-effective international solution would be one in which countries agree to an allocation of carbon emission quotas, perhaps reflecting an overall emission target. In a practical sense, signatories to the FCCC have implicitly accepted such a quantitative target. International quota trading (Sandor *et al.*, 1994) would establish a quota price – an implicit international tax rate – that would tend to equalize marginal abatement costs among countries. A carbon leakage problem similar to that mentioned above would arise to the extent that such an agreement was not global.

In the case of an international tradable quota scheme, participating countries could use whatever domestic policies they preferred in order to stay within their final quota entitlements once all quota trades were complete. For example, they might employ tradable permits, domestic taxes, or regulations. If a domestic carbon tax were used, the efficient tax rate for the coming period would be the (unknown) quota price level for that period.

11.2.2.4 Other complementary policies

Technology transfer from industrialized to developing countries potentially has a large part to play in reducing future emissions. One mechanism to facilitate such transfers is joint implementation, to the extent that it proves practically feasible. Joint implementation aims at minimizing the joint costs of emission reductions for a group of emitters. In the context of emitters who are committed to targets, it could lead to the development of a tradable quota scheme.

11.2.3 Criteria for policy assessment

In this chapter a range of criteria is used to assess policy instruments to manage the greenhouse problem. Two important criteria are economic efficiency and distributive justice. The efficiency objective or cost-benefit principle is to maximize the global net benefits from the use of resources. Both the global greenhouse emission target and the preferred policy instruments to achieve it are choice variables for satisfying this criterion. However, there is considerable uncertainty regarding the effects of unconstrained greenhouse gas emissions at this time. Hence, there is uncertainty regarding the benefit and cost functions, and as a consequence of this, there is considerable uncertainty regarding the optimal (economically efficient) level of control. One practical response is to employ a cost-effectiveness objective – that is, to minimize the costs of achieving a given global greenhouse emission target.

The cost-effectiveness of achieving a given but potentially time-varying target is a criterion that is employed in much of this chapter. Policies may differ in their ability to achieve an emission target under changing conditions. A policy that consistently "hits the target" (achieves environmental effectiveness) and remains cost-effective is desirable.

11.2.3.1 The choice of policy instruments under uncertainty

In the absence of uncertainty, emission taxes and quantity controls, such as a tradable quota system, are equivalent. Indeed, it would be neither harder nor easier to specify the appropriate tax than the appropriate quantity of quotas. This is because the same information is required to specify both. However, both the science and economics of climate change involve many uncertainties. It is not known precisely how the climate will change given different emission trajectories. Nor

is the cost of following each of these trajectories known. It is therefore important to compare these different instruments under uncertainty.

Perhaps surprisingly, uncertainty with respect to the benefits of abatement on its own does not favour either instrument. If the marginal abatement cost curve is known, then choice of a tax will result in a known quantity of emissions and choice of a quantity of tradable quotas will result in a known quota price (under the usual assumptions). This means that the outcome in terms of both emissions and marginal cost can be determined as easily by one instrument as by the other. Although uncertainty about the benefits of abatement makes choosing the appropriate target difficult, one instrument works as well as the other once the target is chosen.

If the policy goal is to meet a particular emission target, then tradable quotas or an equivalent quantity-based instrument will be preferred, insofar as they can guarantee that the emission target is met. However, from the point of view of efficiency this instrument may not be best.

In an important paper, Weitzman (1974) showed that uncertainty with respect to the costs of abatement does affect the choice between these instruments if the goal of policy is to maximize the net benefits of abatement. A substantial literature in the context of environmental policy followed, including major works by Adar and Griffin (1976), Yohe (1977), and Watson and Ridker (1984). Where there is uncertainty about abatement costs, use of tradable quotas will guarantee that emissions do not exceed the quantity of quotas allocated (assuming full compliance), irrespective of the costs of doing so. Conversely, an emission tax would guarantee that marginal abatement costs did not exceed the magnitude of the tax, no matter how large or small was the resulting level of emissions. What Weitzman shows is that, if the marginal benefit and marginal cost curves are linear, the two instruments will be equivalent only if the slopes of these curves are equal (in absolute value terms). If the marginal cost curve is steeper than the marginal benefit curve, emission taxes will result in a more efficient outcome. Conversely, if the slope of the marginal cost curve is less than the slope of the marginal benefit curve, then tradable quotas would be preferred.

The available evidence indicates that marginal abatement costs will be steep once abatement becomes substantial (see, for example, Nordhaus, 1991b), although this curve may flatten out considerably if a "backstop technology" becomes available. By contrast, little is known about how marginal abatement benefits vary with the level of abatement. There is, however, some concern that a threshold may exist in the damages associated with greenhouse gas concentrations (which depend, in turn, on emissions and the rate of sequestration). Hence, there are arguments that can be made in favour of both instruments.

Although benefit uncertainty on its own has no effect on the identity of the optimal (efficient) control instrument, in the presence of simultaneous uncertainty in both marginal benefits and marginal costs and with some statistical dependence between them, the usual Weitzman result can be reversed, depending on the magnitudes of benefit and cost uncertainty and the degree and sign of the correlation between them (Stavins, 1996). A positive correlation will always tend to favour a quantity instrument and a negative correlation will tend to favour a price instrument.

However, these two instruments are not mutually exclusive, and it turns out that a mixed system can be preferable to either of the pure instrument options (see Roberts and Spence, 1976). Under a mixed system, a certain quantity of quotas may be made available. In addition, both a tax and a subsidy are imposed, with the tax being higher than the subsidy. If costs turn out to be higher than expected, polluters may pay the tax instead of purchasing more costly quotas. The tax thus serves to cap marginal abatement costs. If, on the other hand, costs turn out to be lower than expected, polluters can reduce their emissions even further in order to obtain the subsidy. The mixed system performs better than either pure system under cost uncertainty because the mixed system effectively has two instruments at its disposal.

A nonlinear emission tax can do better still. Under such a tax the marginal tax rate varies with the quantity of emissions. The tax schedule should approximate the marginal benefit curve. If the marginal benefit curve were known with certainty, then obviously a marginal tax schedule identical to the marginal benefit curve would ensure a fully efficient outcome, irrespective of the uncertainties regarding abatement costs. The marginal benefit curve is not known in the case of climate change. However, enough may be known to specify two or three steps in the curve.

For the remainder of this chapter, it will be assumed that a given ceiling on emissions has been identified as a target, and that governments seek to minimize the costs of meeting this target. The main concern here is thus with cost-effectiveness rather than efficiency as such. This assumption is made partly for analytical convenience (only pure tax and tradable quota systems are considered in detail) and partly because so much of the policy debate about global climate change has focussed on the appropriate emission targets. Indeed, the Framework Convention on Climate Change refers explicitly to such targets.

11.2.3.2 Other considerations

As a practical matter it is also important to distinguish static from dynamic cost-effectiveness. Static efficiency refers to a short-term operating environment in which technology options and aggregate primary resource availabilities are fixed; dynamic efficiency pertains to a long-term operating environment in which technology options and primary resource availabilities change. A policy that is cost-effective in the short run may not be cost-effective in the long run.

In addition to the application of the global least-cost principle to policies for emission control, the other main criterion for policy assessment is the objective of distributive justice. This requires that the total net benefits (costs) generated by the policy should be "equitably" shared.

Central to the analysis of the performance of any greenhouse policy is the recognition that the real-world operating environment involves major sources of greenhouse uncertainty and associated risk regarding future economic and ecological conditions. Decision makers will face the costs of

Table 11.1. *A Taxonomy of costs of environmental regulation*

Government Administration of Environmental Statutes and Regulations
  Monitoring enforcement

Private Sector Compliance Expenditures
  Capital
  Operating

Other Direct Costs
  Legal and other transactional
  Shifted management focus
  Disrupted production

Negative Costs
  Natural resource inputs
  Worker health
  Innovation stimulation

General Equilibrium Effects
  Product substitution
  Discouraged investment
  Retarded innovation

Transition Costs
  Unemployment
  Obsolete capital

Social Impacts
  Loss of middle-class jobs
  Economic security impacts

Source: Jaffe *et al.* (1995).

uncertainty and risk from making incorrect decisions. But it is possible to reduce these costs by ensuring that policy is flexible and reversible in response to new information about the most cost-effective future strategy. The ability to modify policy settings and introduce new policies without generating major costs of adjustment are key criteria for greenhouse management.

Potential net benefits from a policy initiative must also take into account the administrative costs of the programme. Whether an efficiency criterion (maximizing net benefits) or a cost-effectiveness criterion (minimizing aggregate costs) is being employed, it is essential that the full measure of costs include both implementation costs (typically borne by governments) and transaction costs (typically borne by the private sector).

Costs need to be measured correctly. A taxonomy of the costs of environmental regulation, beginning with the most obvious and moving towards the least direct, is provided in Table 11.1.[5] First, many policymakers, and much of the general public, would identify the on-budget costs to government of administering (monitoring and enforcing) environmental laws and regulations as *the* cost of environmental regulation. However, most analysts would identify the capital and operating expenditures associated with regulatory compliance as the fundamental part of the overall costs of regulation, although a substantial share of compliance costs for some environmental regulations fall on government rather than private firms – one example being regulations for contaminants in drinking water. Additional direct costs include legal and other transaction costs, the effects of refocussed management attention, and the possibility of disrupted production.

Next, the potential "negative costs" (in other words, non-environmental benefits) of environmental regulation, including the productivity impacts of a cleaner environment and the potential effects of regulation on innovation, should also be considered. General equilibrium effects associated with product substitution, discouraged investment,[6] and retarded innovation constitute another important layer of costs, as do the transition costs of real-world economies responding over time to regulatory changes. Finally, there are potential social impacts, such as those on jobs and economic security, that are given substantial weight in political forums.[7]

This discussion of some of the special difficulties of assessing the cost-effectiveness of alternative policy instruments should not be taken to mean that this single criterion is of exclusive or paramount importance. On the contrary, individual nations will inevitably choose their own criteria to distinguish between competing policy instruments. The specific criteria chosen will always be a function of the individual socioeconomic and cultural context, but in many cases the following set of criteria will be among those considered:

(a) probability that the environmental goal will be achieved
(b) efficiency or cost-effectiveness
(c) dynamic incentives for innovation and the diffusion of improved technologies
(d) flexibility and adaptability to exogenous changes in technology, resource use, and consumer tastes
(e) distributional equity
(f) feasibility in terms of political implementation and administration.

Finally, in assessing policy options in this chapter both the spatial and temporal aspects of emission coverage are considered. The general design features of economic instruments are categorized by the coverage of net greenhouse gas emissions, the scope and level of participation, and the point of application. For example, an emission control objective could involve imposing a target on all or a subset of all human sources and sinks and all or a subset of all greenhouse gases. In addition, the scope of a policy instrument could involve all or a subset of countries. The level of participation in a scheme refers to the economic unit responsible for meeting a target. Options range from the level of the country to that of individuals or companies. The point of application of the policy simply refers to the point in the production or consumption chain of a good or service at which greenhouse gas emissions are to be counted or proxied.

11.2.4 Coverage of greenhouse gases

11.2.4.1 The need for comprehensive targets
Should an initial international programme include all greenhouse gases or focus on CO_2 alone? This question has received considerable debate in the literature (see, for example, Cristofaro and Scheraga, 1990; Victor, 1991; Stewart and Wiener, 1992). The advantage of the more comprehensive approach is the additional flexibility it introduces into the system, and hence the potential it creates for even greater cost-effectiveness. However, the sources and sinks of methane and nitrous oxide emissions are as yet poorly understood. Currently, important anthropogenic sources of emissions of methane appear to include domesticated ruminant animals, rice cultivation, landfills, and mining. For nitrous oxide, they appear to include legume crops and nitrogen fertilizers (Howden and Munro 1994; Pearce and Warford 1993). Clearly, countries with a comparative advantage in agricultural industries could be significantly affected by either the exclusion or use of a multiple gas scheme. Although CO_2 is the main source of past and present greenhouse concerns, methane and nitrous oxide are also significant in radiative forcing.

By including all the major greenhouse gases (sources and sinks) in setting global and any national greenhouse management targets, policymakers would avoid throwing away valuable knowledge (Schmalensee, 1993). Given a set of weights relating the radiative forcing potential of each greenhouse gas to a common base (say CO_2), a multiple gas market policy would only need to involve one quota market or one emission tax scheme and one permit market or domestic tax scheme as well as one control obligation for each country. At this stage, however, these weights are uncertain and may vary with both environmental and economic conditions (Hoel and Isaksen, 1993).

If the administrative burden is deemed to be too great initially for the incorporation of net emissions of greenhouse gases other than CO_2 in an international greenhouse management programme, the programme still needs to be flexible enough for this to be done when implementation costs fall. Indeed, it could provide incentives to generate such an outcome. Care must be taken not to worsen problems for future greenhouse management. The international target must therefore be comprehensive, as must targets for countries within a coalition that adopts any international market-based policy regime.

11.2.4.2 Initial coverage of CO_2 sources and sinks
The coverage issue in the case of CO_2 has been widely discussed (see, for example, UNCTAD, 1992; OECD, 1992a, b). One major question relates to whether consideration should be limited only to changes in emissions of CO_2, or whether it should also include changes in CO_2 sinks, such as expanding forests? Another concerns whether and how an international agreement might help retard deforestation and promote reforestation (Dudek and LeBlanc, 1992). Deforestation could be treated as equivalent to emissions, whereas afforestation activities could be a source of emission abatement credits. A good deal of care is needed in establishing the accounting methodology to ensure that the net effects of land clearing and revegetation with alternative species are measured and that domestic consumption and exports of wood products are separated.

Whether one is considering net CO_2 emissions from fossil fuel burning or net deforestation, it is helpful and, in some policy contexts, even necessary to have a baseline level for net emissions in the absence of any policy change, which can be used to assess the effectiveness of the greenhouse policy in place. This baseline has yet to be determined for the forestry sectors of the world's economies, and cost-effective monitoring techniques have not yet been proven. One of the major difficulties is that many trees exist outside forests, and measuring their contribution to carbon sequestration with remote sensing devices is extremely difficult. A further and more pervasive issue is associated with specifying *ex ante* a "status quo" timepath, against which "improvements" can be measured.

The coverage of the scheme may have a major impact on the incentive of different countries to participate. With an international market programme for CO_2, countries like Brazil and Indonesia might find it economically attractive, as well as environmentally sound, to retard the depletion of their forests or to implement reforestation programmes. Under an international tradable quota regime, CO_2 emission credits would amount to a valuable export commodity from the seller's viewpoint and would be an equally valuable import from the viewpoint of the buyer (a country that would otherwise have a CO_2 emission deficit). Under an international CO_2 emission tax, net emissions would also be treated symmetrically.

Currently, there are significant difficulties in measuring the carbon stored in trees and how it varies over time (Houghton, 1992). Hollinger *et al.* (1994) have made some progress toward a standardized carbon accounting system for a single-species plantation forest established on previously cleared agricultural land in New Zealand. Use of this accounting system within an international system of tradable emission credits has been explored by both MacLaren *et al.* (1993) and the Tasman Institute (1994). Current estimates of the costs of carbon sequestration through tree planting vary widely (ranging from US$1 to US$50 per short ton of carbon abated). These differences reflect the opportunity costs of alternative land uses (Stavins, 1995b) as well as the effects of uncertainty (see Sedjo, 1994, and the references cited there). However, estimates at the low end of the range refer to developing countries and, aside from the limitation of uncertainty, appear promising in terms of shifting any long-term need for high-cost carbon-free backstops further into the future. According to current estimates, such backstops become economic when the marginal cost of emissions is around US$250 per ton of carbon (Manne and Richels, 1991).

Satellite imagery is a critical tool in monitoring forestry systems. Given the potential stimulus of global net CO_2 emission trading, it could be tailored to ensure that coverage is complete and backed by verification (OECD, 1992b). Further, such a price stimulus for reduced deforestation (and increased afforestation) could yield complementary gains in terms of sustainable land management practices and global gains in biodiversity values. Indeed, the global nature of biodiversity values has prompted one suggestion for an international tradable quota system in global forestry management (see Sedjo 1994).

Hence, an important option value would be preserved by including in any international agreement to control CO_2 sources a provision for all parties to review the adoption of

sinks at fixed points in time. However, it would be important to ensure that the integrity of the existing policy regime be preserved as new sinks were included.

11.3 The Domestic Policy Context

From the basic theorems of welfare economics it can be deduced that if an economy is perfectly competitive, if there is a full set of markets, and if information is perfect, then the resulting equilibrium (if it exists) is efficient in the sense that no one could be made better off without making someone else worse off (Pareto efficiency). The real world does not satisfy these conditions. There exist many externalities, of which climate change is only one. Competition is not perfect, nor in many cases is information, and markets are not complete. What is more, even if all the conditions of perfect competition were satisfied, the resulting Pareto-efficient outcome might not accord with society's view of a distribution of resources that is equitable or "fair." If certain other conditions hold, then an alternative, feasible, Pareto-efficient allocation could be sustained as a competitive equilibrium with appropriate lump-sum taxes and transfers, and so the objectives of efficiency and equity need not necessarily clash. But lump-sum taxes and transfers are typically infeasible, and, as a consequence, distorting taxes and transfers are employed virtually everywhere.[8]

The above observations about distortions are important because many analyses of climate change policy assume that the externality of climate change is the only distortion that exists. In fact, climate change policy must be considered in the context of real economies, already rife with distortions. A market economy can function effectively only if governments define property rights and provide for the enforcement of contracts. In some countries, even these basic requirements are not met. The extent to which governments can provide these basic requirements and correct market failures will in part determine GNP. (The importance of GNP, and other economic and social factors, in determining future emissions is highlighted in Chapter 8.)

It is also necessary to take into account any distortions introduced by governments. In some cases, government interventions can undermine net national income and cause environmental damage (see Binswanger, 1989). Another important determinant of future emissions, as discussed in Chapters 3 and 8, is population. Here, too, both market and government failures play a role. For example, high rates of fertility have been linked to the absence of effective capital markets, which makes it difficult or impossible for people to obtain social security (see Dasgupta 1993).

One purpose of this section is to draw attention to the importance of the domestic policy context in evaluating climate change policy proposals. The merits of any given proposal depend on this context. Equally, changes in the context have implications for emissions. Sometimes these two different observations are confused, and another purpose here is to clarify the distinction between them. Obviously, the types of policies that might warrant discussion here are many. However, the discussion is restricted to a few areas that seem particularly important and on which some research has already been done.

11.3.1 Preexisting market distortions

11.3.1.1 Energy subsidies

The emissions abated by a climate change policy will depend not only on the policy itself but also on whether the consumption of energy is subsidized and the magnitude of such subsidies. It is the combination of the climate change policy and these subsidies (and indeed other policies) that will determine relative prices and hence the incentives to adopt substitutes for carbon-intensive fuels.

In some regions such subsidies are significant (Larsen and Shah, 1995). Using border prices as a benchmark, Larsen and Shah (1992) calculated that primary fossil fuel subsidies worldwide are equivalent to a negative carbon tax of US$40 a ton. Larsen and Shah (1995) estimate that global CO_2 emissions would be reduced by between 4 and 5% if all energy subsidies were removed.[9] At the same time, eliminating such subsidies would increase real incomes by improving efficiency. The reason is that the subsidies distort prices; users pay less for fossil fuels than it actually costs to supply them.

An OECD study using its GREEN model arrives at a similar result. The OECD estimates that the removal of energy subsidies would reduce global emissions by 18% compared with the level that would otherwise be attained by 2050 (Burniaux *et al.*, 1992a). Elimination of subsidies could increase global real incomes by 0.7% annually, and real incomes in non-OECD countries could rise by 1.6% annually.

Fossil fuels may also receive indirect subsidies from electricity generation. Electricity is typically subsidized in both developed and developing countries. In the U.S., for example, government regulation frequently restricts electricity prices from privately owned utilities to a level equal to long-run average costs, which are often below marginal costs. Power is sold by the federal government at approximately 25% below even these levels, because of interest and tax subsidies (DCEIA 1992). In developing countries, electricity prices declined in real terms by 25% during the 1980s, and by 1988 were at an average level just over half as large as the average level in OECD countries, even though long-run marginal costs in real terms were higher in developing countries (Schramm, 1992). In 80% of developing countries, electricity prices are, on average, 30% below long-run marginal costs (World Bank 1990, 1992). Such distortions lead to excessive expenditure on new capacity, failure to generate sufficient internal funds to maintain or expand service, excessive energy consumption, and excessive environmental impacts from power generation.

11.3.1.2 The "local" environmental benefits of climate change policy

So far, the effect that the removal of energy subsidies can have on income, as conventionally measured by GNP, has been stressed. But the effects are likely to be felt more widely. One consequence of actions to reduce CO_2 emissions will be a reduction in other pollution. For example, Bye *et al.*, (1989) estimate that a policy that reduced CO_2 emissions in Norway by 20% would have the incidental effect of reducing SO_2 emissions by 21% and NO_x emissions by 14%. Larsen and Shah (1994) calculate that for Pakistan, for example, a carbon tax could be justified on the basis of the benefits of reductions in local pollutants alone, despite the fact that Pakistan already has high energy-related taxes.

11.3.1.3 Information and energy conservation

There has long been concern that apparently cost-effective energy conservation technologies were being adopted and diffused only very gradually and that market penetration rates for such technologies were not as high as engineering-based models predicted. This may be due partly to imperfect capital markets. Another possible reason may be the failure of the market to supply appropriate and credible information about these technologies (Hassett and Metcalf, 1992; Jaffe and Stavins, 1994a).

Empirical work dating back to Hausman (1979) shows that purchases of energy-saving technologies often reflect high rates of discount (see Treadwell *et al.*, 1994).[10] In other words, purchasers of such technologies may insist on earning a higher rate of return on this investment than on alternative investments. In an econometric analysis, Hassett and Metcalf (1992) show that future uncertainty regarding energy prices, due to past volatility in those prices, can attribute a large option value to waiting before investing in energy-conserving capital. On the other hand it could be argued that uncertainty about future energy prices creates an incentive to invest in energy-efficient equipment to minimize the share of energy costs in total costs. This would reduce risk exposure if energy prices were more uncertain than other input prices.

One means of avoiding this dilemma is for the company manufacturing the technology to offer a warranty on the product's performance. But there is a problem in that the performance of the good may depend on how it is used by the consumer, as well as on its intrinsic qualities (the moral hazard problem). A full warranty would therefore create an incentive for the consumer to misuse the good. Another problem is that there may be a tendency for the users most likely to purchase a more expensive good with a full warranty also to be those most likely to misuse the good (the adverse selection problem). For both these reasons, warranties may not be able to convey the information that would benefit both consumers and the firms manufacturing the technology.

Empirical evidence in the United States (Horowitz and Haeri, 1990; Sutherland, 1991) indicates that when information on energy efficiency is widely available, the real estate market functions efficiently – consumers show a willingness to pay more for houses with energy-saving features, all else being equal. However, Jaffe and Stavins (1994a, c) demonstrate that information problems can directly inhibit the diffusion of energy-efficient technologies in new housing. They also show that decisions on such investments depend on expectations about the future. If the price of such technologies is expected to fall, or the availability of information about the performance of such technologies is expected to increase, then consumers may delay making such purchases. Though individually rational, such behaviour can lead to less investment than is socially desirable, depending on whether true market failures are involved (Jaffe and Stavins, 1994b). This creates a potential role for public policy. The provision of home energy ratings is an example of such a policy designed to encourage the purchase of more efficient homes.

A number of projects designed to convey information to rural people exist in developing countries. One such project is the Mount Elgon Conservation and Development Project in Uganda, which is designed to provide new information to local people to enhance the cost-effective use of local fuel resources and minimize damage to forest reserves (Ugandan Ministry of Finance and Economic Planning, 1993).

A more general approach to the provision of information could be to use "eco-labelling." In this way, final consumers could be informed of the total contribution to greenhouse gas emissions as a result of the production of particular consumer durables or other items.

11.3.1.4 *Transport*
A large and growing fraction of CO_2 emissions arises from transport fuel use. Full social (user cost) pricing can promote greater efficiency in transport while reducing these emissions substantially. Most countries tax gasoline (petrol) to finance highways and other public automotive transport services. However, some countries do not collect enough from drivers to pay the full social costs of automotive travel (Repetto *et al.,* 1992).

Appropriately designed road user charges should also reflect the peak-period costs of using congested road capacity (Cameron, 1994). Congestion costs, in the form of time delays, accidents, excess fuel costs, and pollution, are an increasingly serious urban problem in developed and developing countries. In the U.S., the cost of time delays alone has been estimated to be $50 billion a year (Repetto *et al.,* 1992). User charges set at an appropriate level could lead to a change in the allocation of resources to the transport task and, coincidentally, might also lead to a reduction in greenhouse gas emissions, assuming that such charges could be collected cost-effectively.

11.3.1.5 *Agriculture and forestry*
Distortions in agriculture and forestry are common. As already noted, government subsidies and tax policies have encouraged deforestation in the Amazon (see Binswanger, 1989; Mahar, 1988). But the distortions in agriculture go further than this. The external environmental costs of wood harvesting, including loss of soil cover and fertility, are substantial (Newcombe, 1989), and yet user charges for rights to harvest timber on public lands typically do not even cover the replacement costs of the wood. In many countries, land must be cleared to gain land rights (Pearce and Warford, 1993). In sub-Saharan Africa, farmers and nomads carry extra cattle as an insurance against droughts and as an asset. Herd size may also

be taken as a measure of status. Herds are therefore larger than they would be if capital and insurance markets were fully developed (Dasgupta and Göran-Mäler, 1994).

Underpricing of water in agriculture leads to inefficiency in water use, excessive expenditure on irrigation, and a variety of local environmental and social costs, including soil waterlogging and salinization and the degradation of riverine and estuarine environments. Irrigation charges in a sample of six developing countries covered only 1-23% of storage and conveyance costs during the 1980s. Charges for federally supplied irrigation water in the U.S. cover only 5-20% of these costs (Repetto 1986). These charges fail to reflect the marginal opportunity cost of water in alternative urban and industrial uses, which is typically an order of magnitude higher than its value in agriculture. When irrigation water is underpriced, farmers grow more rice than they otherwise would. These practices increase methane emissions (Ranganathan *et al.,* 1994). Proper pricing for water could well generate global benefits as well as significant domestic gains.

11.3.1.6 *Policies affecting adaptation*
The net adverse effects of climate change will depend not only on the extent of climate change, but also on the extent to which economies successfully adapt to any change. Some existing policies may mitigate against adaptation or increase the vulnerability of some sectors of the economy to climate change. For example, subsidized drought or flood insurance may encourage investment in high-risk areas and reduce incentives for self-reliance. Similarly, some agricultural support policies might discourage farmers from shifting to enterprises and production systems better suited to an altered climate.

11.3.2 *Revenue recycling*

The abatement achieved by a carbon tax, and the effect of the tax on an economy, will depend on what is done with the tax revenue. Likewise, a tradable permit scheme can raise the same government revenue as a carbon tax if the government auctions the permits. The impact of such a scheme on the economy will again depend on what is done with the revenue. The direct impact on government revenue of the two policies can be made equivalent across a variety of cases. No direct impact on government revenue would occur if permits were grandfathered (that is, allocated on the basis of some historical record) or tax revenue redistributed to emitters. Intermediate cases would be represented by partial grandfathering or partial redistribution of revenue to emitters. Thus, in principle, the same revenue recycling issues apply regardless of whether a tax or tradable permit scheme is used (Bohm, 1995a), although taxes have been studied in more detail in the literature.

There is widespread agreement that revenue recycling can significantly lower the costs of a carbon tax (Koopmans *et al.,* 1992; Shackleton *et al.,* 1992; Goulder, 1992, 1993, 1995; Bovenberg and de Mooij, 1994; European Commission, 1994; Jorgenson and Wilcoxen, 1994). Some researchers have suggested further that all the abatement costs associated with a carbon tax can be eliminated through revenue recycling in the form of cuts in income taxes or taxes on payrolls. However, at least in the case of cuts in income taxes, research by Goulder (1992, 1993, 1995) and related theoretical work by Bovenberg and de Mooij (1994) reject this stronger claim. Their work indicates that the recycling of revenues through income tax cuts only partly offsets the total general equilibrium abatement costs implied by a carbon tax (also see Chapters 8 and 9 of this report).

None of the above research denies the possibility that raising revenue from a carbon tax or a permit scheme may increase national income when combined with reductions in a burdensome existing tax. However, such a result is an argument for reform of the taxation system rather than for the introduction of a carbon tax or permit scheme (Bohm, 1995a). It may be that some tax other than a carbon tax could result in a greater efficiency gain in raising public revenue. If there are efficiency arguments for a carbon tax or tradable permit scheme and reform of the tax system, then both changes should be introduced.

11.3.3 The broader context for climate change policies

11.3.3.1 "No-regrets" policies
Policies or policy reforms that improve the efficiency of an economy while also reducing greenhouse gas emissions have sometimes been described as "no-regrets" policies because they offer sufficient benefits in other contexts that their adoption could not be regretted even if climate change were later shown not to be detrimental (also see Chapters 8 and 9).

The reduction of virtually any distortion is to be encouraged, provided equity concerns can be safeguarded. Policy reforms that help in this regard are therefore also to be welcomed, whatever the consequences for climate change. If greenhouse gas emissions are reduced as well, then extra potential gains exist. But even where this is not so, the removal of distortions can lead to greater economic welfare. In general, such policies should not be linked directly to climate change policy. However, their significance for climate change policy needs to be recognized, and the prospects of "double dividends" from carbon taxes or tradable permit systems can give the extra political impetus needed to reduce existing distortions. These policies will influence both the "business-as-usual" emission scenario and the effectiveness of any climate change policy.

Although there may be political advantages (particularly in terms of providing impetus for reform) in linking so-called no-regrets policies to climate change policy, such linkage may serve to confuse the policy debate. For example, the observation that a carbon tax (or auctioned tradable permits) may increase national income when combined with a tax reform is really an observation that the structure of taxation could be improved. A carbon tax is not the only device available for improving public finance, and indeed it may not be the best device available; it is possible that a different tax could raise revenue more efficiently than a carbon tax. What this means is that the merits of a carbon tax will depend on whether the tax is evaluated taking all other existing policies as given, or whether it is instead evaluated against the background of a (second-best) efficient tax regime. Put differently, estimates of the full consequences of a carbon tax must take account of how the revenues are to be employed.

It is not obvious, however, how a carbon tax should be evaluated. If the tax is evaluated against the background of a (second-best) efficient tax regime, and yet the actual tax regime is different, then the evaluation could lead to inappropriate public policy. A better approach would be to demonstrate how the performance of a carbon tax (reflecting both the emissions and level of economic activity associated with the tax) depends on the policy context, including the regime for raising public revenue and the presence of distortions in energy pricing.

11.3.3.2 Adaptation policies
It is important to draw a clear distinction between the costs and benefits of actions taken to reduce the impacts of climate change and the costs and benefits of actions taken to reduce emissions. The first reduces the potential damage caused by climate change directly, whereas the second has the effect of reducing emissions of greenhouse gases now, which, in turn, will have an impact on future climate. Examples of adaptation include such actions as increasing irrigation water availability in regions where the climate has become drier, improving refrigeration to offset the effects of a warmer climate, and relocating economic activities away from the coast where sea levels have risen. Measuring the cost of adaptation is somewhat problematic, given that both ecosystems and economic systems are changing all the time and will to some extent adapt autonomously to climate change.

The term "adaptation" may be confusing because actions falling into this category can be counted as a cost of climate change. To see why adaptation can yield a net benefit, consider the problem of estimating the costs of climate change. An estimate might be based on the assumption that there will be no adaptation – that is, if sea level rises by say 50 cm, then all shoreline property less than 50 cm above sea level will be lost. Alternatively, it might be assumed that there will be some adaptation – that dikes might be built, for example. The latter response is indeed a cost of climate change: It would not need to be taken in the absence of climate change. However, the response may reduce the damage associated with climate change. If the reduction in such damage exceeds the costs of adapting, then adaptation should be undertaken (Fankhauser, 1993).

Unlike the case of abatement of greenhouse gases, adaptation typically involves private goods. If the climate were to become drier, a demand would be created for drought-resistant crop varieties, and the firms that developed these would be rewarded by the market. If climate were to become more variable, then individuals would seek to insure themselves against such changes. Such responses belong in the realm of the private sector. However, even when dealing with strictly private goods, there may be a role for the state. Dasgupta (1993), for example, argues that some assistance should be given to the assetless in developing countries who are not able to command sufficient purchasing power to convert their potential labour power into actual labour power. Such assistance, perhaps in the form of agrarian reform, not only re-distributes income but results in an increase in the rate of growth of aggregate incomes. Although such policies are already needed in some countries, the need for them may be increased in the event of climate change.

However, the principal role for government in the context of adaptation is in supplying public goods. Public infrastructure projects such as building dikes or funding resettlement programs are cases in point, as is the funding of public research and development of carbon-free technologies where there would otherwise be strong free-riding incentives. Furthermore, if there are risks of increased environmental hazards (drought, flood, fire, famine, pests) then greater hazard insurance is an appropriate defensive action. Another type of "insurance" could be purchased by increasing public research expenditure (for example, increasing efforts in plant breeding in order to develop plant cultivars better adapted to new climatic conditions).

Any adaptation policies should be designed in concert with mitigation policies. Both types of policies are aimed at minimizing the expected damage from climate change. Adaptation will be undertaken up to the point where the damage avoided by an incremental increase in adaptation equals the associated incremental cost. Abatement will be undertaken up to the point where the reduction in damage effected by an incremental unit of abatement equals the incremental cost. However, the two types of policies are not entirely equivalent. First, the benefits of adaptation are likely to be felt much more quickly than the benefits of mitigation, and, second, some types of adaptation policies will not be subject to the same problems of free riding (see Section 11.6.5) as abatement. For example, if a country defends its shoreline by building seawalls, its own population, in most cases, will receive all the benefits. This is not true of unilateral abatement.

Recent climate research indicates that local changes in climate may depend not only on global phenomena such as the increase in atmospheric concentrations of greenhouse gases but also on local phenomena such as emissions of sulphates. It might be argued that sulphate emissions should be reduced to prevent damage from acid rain, and that policy on sulphate emissions should not be linked to global climate change policy. However, where local climate can be influenced by local policy, it seems that such linkages may nevertheless be made, not least because local climate modifications may be less costly and may not suffer from free-riding problems. To date, research has not considered the economic and policy implications of local climate modification (but see Section 11.4.2).

11.4 Regulations, Voluntary Agreements, and other Nonmarket-Based Instruments

The conventional approach to environmental policy in many countries has employed policy instruments in the form of uniform standards (based on technology or performance) and direct government expenditures on projects that are designed to improve the environment (Baumol and Oates, 1979; OECD, 1989; Hahn and Stavins, 1991). Like market-based incentives, the first of these strategies requires that polluters undertake pollution abatement activities; under the second strategy

the government itself expends resources on environmental quality. Both these strategies figure prominently in current and proposed policy measures to address global climate change.[11] For the reasons already mentioned in Section 11.2.2.1, the discussion of regulations is confined to their application in a domestic policy context.

11.4.1 Uniform technology and performance standards

Uniform regulatory standards (often described as "command-and-control" regulations) can be loosely categorized as either *technology-based* or *performance-based,* although the distinction between these two categories of instruments is often unclear. Technology-based (or design) standards typically require the use of specified equipment, processes, or procedures. In the context of climate change policy, technology-based standards could require that particular types of energy-efficient motors, combustion processes, or landfill gas collection technologies be utilized by firms.

Performance-based standards are more flexible than technology-based standards, specifying allowable levels of pollutant emissions or polluting activities, but leaving the specific methods of achieving those levels to the regulated entities. Examples of performance standards for greenhouse gas abatement include minimum levels of energy efficiency for appliances, maximum allowable levels of carbon dioxide emissions from combustion, and maximum levels of methane emissions from landfills.

Uniform standards can also take the form of outright bans of certain products or processes, such as aerosol sprays containing ozone-depleting substances. Although bans may appear to be the strictest form of regulation, they may actually be a relatively cost-effective policy instrument if low-cost substitutes for targeted products are available. Moreover, bans or other more proactive design standards may make economies of scale in the production of substitutes materialize faster than if market mechanisms are used by themselves (Bohm and Russell, 1985).

Although uniform technology and performance standards may be effective in achieving established environmental goals and standards, they typically lead to economically inefficient outcomes in which firms use unduly expensive means to control pollution (Tietenberg, 1985; Hahn, 1989; Hahn and Stavins, 1991). Because the costs of controlling pollution vary greatly among and even within firms, any given aggregate pollution control level can be met at minimum aggregate control cost only if pollution sources are controlled at the same *marginal cost,* as opposed to the same *emission level.* Indeed, depending on the age and location of emission sources and available technologies, the cost of controlling a unit of a given pollutant may vary by a factor of 100 or more across a range of sources (Crandall, 1984). Nonetheless, because performance standards give economic agents additional flexibility to make choices based on economic criteria, performance-based standards will generally be more cost-effective than technology-based standards. On the other hand, if there is essentially only a single means of achieving a particular performance standard, a technology-based standard may save on information and administration costs.

In theory, the government could achieve a cost-effective allocation of pollution control responsibility among different sources if it assigned source-specific control levels that equated the marginal costs of control across these sources. This approach would, however, require detailed information on the pollution control cost functions of individual firms and sources – data that governments usually lack and could obtain only at great cost, if at all. Although they are not typically designed to address the cost-effectiveness issue, source-specific or firm-specific permit programmes are one approach traditionally taken to adjust regulatory standards to individual circumstances. If pollutants exhibit highly localized effects, such an approach may have distinct advantages over a tax or a more general permit system. Global climate change is not, however, a localized problem; a unit of greenhouse gas emission will have roughly the same impact regardless of where it is emitted.

Even if governments were able to use conventional technology and uniform performance standards to achieve a cost-effective allocation of pollution control at present, such standards would not necessarily provide continuous *dynamic* incentives for the development, adoption, and diffusion of environmentally and economically superior control technologies in the future (Bohm and Russell, 1985; Jaffe and Stavins, 1995).

All forms of intervention have the potential for inducing or forcing some amount of technological change because, by their very nature, they induce or require firms to do things they would not otherwise do. Performance and technology standards can be explicitly designed to be "technology forcing," mandating performance levels that are not currently viewed as technologically feasible or mandating technologies that are not fully developed (Jochem and Gruber, 1990). The problem with this approach, however, is that while regulators can assume that *some* amount of improvement over existing technology will always be feasible, it is impossible to know just how much. Standards must either be made unambitious or else run the risk of being ultimately unachievable, leading to political and economic disruption (Freeman and Haveman, 1971). Another difficulty with a regulatory approach to environmental protection is that the regulatory agency may, over time, develop such a close working relationship with the regulated industry that it relaxes its enforcement standards in the interests of the industry itself. This phenomenon is sometimes referred to as "regulatory capture."

Once a performance standard has been satisfied, there is little benefit to the individual firm from developing and/or adopting even cleaner technology. In addition, regulated firms may fear that if they do develop a cleaner technology, the performance standard will be tightened. Technology standards are even worse than performance standards in inhibiting innovation, since, by their very nature, they constrain the technological choices available, and may thereby remove all incentives to develop new technologies that are environmentally beneficial (Magat, 1979). For example, when vehicle emissions standards requiring the use of catalytic converters were adopted by the European Union,[12] incentives to develop lean-burn engines were reduced. This disincentive occurred because the technologies are presently incompatible, at least in the sense that a lean-burn engine cannot be fitted with a three-way catalytic converter.[13] Lean-burn technology (with two-way converters) capable of meeting Japanese and European standards is now available. However, this technology does not meet present U.S. emission standards or the stricter standards expected to emerge in the near future.[14] However, lean-burn remains an important and developing approach. Not only does it have the potential advantage of reducing CO_2 emissions significantly,[15] but it also offers the prospect of reducing other emissions more effectively over the lifetime of the automobile.[16]

Under some circumstances, however, a performance standard may provide greater incentives for technological adoption than a marketable permit system (Malueg 1990). There are better and worse ways of establishing performance standards. To take an example, the Corporate Average Fuel Economy (CAFE) standards in the U.S. are applied to the fleet *average* of every manufacturer and importer. CAFE may thus be binding on a manufacturer that sells many larger cars as well as some small cars, but not on a manufacturer that sells only small cars. The problem here is that the former manufacturer may sell small cars that are more energy-efficient than the latter manufacturer. In other words, the innovation and manufacture of more efficient automobiles may not be rewarded by these standards. If CAFE differentiated the standards according to the market segment of each vehicle (sub-compact, compact, mid-size, etc.), then firms that sold vehicles that were more efficient for their class than required by the standard would be rewarded by not having to pay the penalty for which the manufacturers of less efficient cars were liable. Better still, if manufacturers were allowed to trade in energy efficiency credits, then even the manufacturer of the most efficient cars would have a continuous incentive to develop even more energy-efficient cars. Finally, a tax on gasoline (petrol) would not only provide incentives for the manufacture of more fuel-efficient cars, but would also provide incentives for vehicle owners to reduce their fuel consumption.

As with virtually all policy instruments, the administration of uniform standards typically includes programmes for compliance monitoring and enforcement. Although technology-based standards may seem to be the least cost-effective of the policy instruments, if monitoring costs are high in some particular circumstances they may have an advantage because they are relatively easy to monitor and enforce. An inspector can simply check whether a particular piece of equipment has been installed, rather than continuously monitor information on emission levels. Performance standards, in general, and pollution charges and marketable permits for non-CO_2 greenhouse gas emissions all require more detailed monitoring systems. These can suffer from the following problems (Beavis and Walker, 1983):

(a) Once emissions leave the source they are usually lost to measurement.

(b) Emissions may be random, rather than fixed values, and may vary depending on equipment breakdowns, shifts in product mix and input quality, or changes in production levels.

(c) Monitoring instruments may be imprecise.

(d) Unless monitoring is continuous, polluters may adjust emissions up or down according to the likelihood of inspection.[17]

11.4.2 Government investment

Direct government expenditures also play a major role in both current and prospective environmental programmes in many countries. As Baumol and Oates (1975) have noted, such government activities include projects to

(a) prevent, mitigate, or adapt to changes in environmental quality

(b) disseminate information

(c) conduct research

(d) educate specialists and the general public

Government purchasing policies may also be used in some instances to attempt to achieve secondary goals such as influencing environmental quality. Many environmental ends that could be achieved through incentive-based policy instruments or uniform standards can also be met directly through government-funded projects or programmes.

The two primary economic rationales for the inclusion of direct government investment in an effective overall government policy are the public good character of many environmental services and the possibility of economies of scale. Public goods arguments for direct government expenditure to disseminate information, conduct research, and sponsor education programmes are common in debates about much public policy. The potential role for government research and information provision regarding renewable energy and energy-efficient technologies is particularly prominent in the climate change context (Jaffe and Stavins, 1995). Proposed methods for addressing climate change through climatic engineering or "geoengineering" options are also likely to require direct government involvement, but at this stage far more research needs to be conducted before such options can be contemplated. For details of some suggested approaches see Nordhaus (1991a), National Academy of Sciences (1992), and Clinton and Gore (1993).

Government or institutional investment at the international level also has a part to play in mitigating climate change. For example, the Global Environment Fund's portfolio on renewable energy includes support for promising backstop technologies such as gasification of wood and crop residue coupled with advanced gas turbines in Brazil and anaerobic digestion of organic residues from agriculture and urban households in India and Pakistan (World Bank, 1993).

11.4.3 Voluntary agreements

Beyond mandatory policy instruments, voluntary agreements can also play an important role in an overall greenhouse gas reduction strategy. The threat of mandatory government intervention may be enough to encourage voluntary agreements. Forward-looking firms may undertake some steps in controlling greenhouse gas emissions if they fear more costly mandatory controls in the absence of voluntary reductions. This could explain why voluntary agreements have arisen in some cases in domestic energy management.[18] The vast majority of greenhouse gas reductions from the actions announced or expanded through, for example, the U.S. *Climate Change Action Plan* (Clinton and Gore, 1993) come from voluntary initiatives aimed at increasing the energy efficiency of the industrial, commercial, residential, and transport sectors.

11.4.4 Demand-side management

Demand-side management may be defined as any activity by an electric utility to influence customer use of electricity in ways that will produce desired changes in the utility's load shape.[19] Demand-side management programmes may affect the quality of service. For example, a lighting retrofit may improve lighting levels as well as reduce electricity consumption, or electricity supply to water heaters may be interrupted during periods of system peak demand. There is an extensive literature on demand-side management (EPRI, 1984, 1991; Katz, 1992; Kahn, 1992; Hirst, 1993; Gellings and Chamberlin, 1993).

Demand-side management programmes may enable a utility to reduce or defer capital expenditures. Regulators evaluate the economic costs and benefits of demand-side management programmes from the perspective of society, all customers, participants, nonparticipants, and utility costs.[20] One criterion applied to these programmes is the Rate Impact Measure (RIM) test, which measures the ability of a programme to reduce costs more than revenues. Programmes that pass the RIM test lead to lower rates immediately and are attractive to nonparticipants. Most regulators, however, focus on the Total Resource Cost (TRC) test, which measures the aggregate benefit to all customers. A programme that passes the TRC test but fails the RIM test (and most fall into this category) raises rates in the short run but reduces them from what they otherwise would have been in the long run.[21]

Literature on the effects of demand-side management programmes on greenhouse gas emissions is sparse. It is too simplistic to assume, however, that demand-side management programmes that reduce aggregate demand for electricity automatically lead to lower emissions of greenhouse gases. For example, peak period demand that would be supplied by hydraulic or natural gas units could be shifted to periods where it is supplied by coal-fired units, resulting in a net increase in emissions (Faruqui and Haites, 1991; Haites, 1993). Nevertheless, demand-side management programmes are generally expected to reduce greenhouse gas emissions. Indeed, some argue that such programmes can be regarded as disguised environmental impact taxes (Sioshansi, 1992).

Evaluations indicate that many demand-side management programmes can be made more cost-effective. There is disagreement in the literature about whether well-designed and delivered demand-side management programmes yield net benefits (Nadel, 1990) or not (Joskow and Marron, 1993). Changes to the electric utility structure in the U.S. have led a number of utilities to scale back their demand-side management efforts. The prospect of a competitive generation market creates pressure to reduce costs and defers plans for capacity additions, thereby reducing the economic justification for demand-side management programmes aimed at reducing aggregate demand for electricity. Demand-side management programmes aimed at smoothing demand for electricity continue to be attractive.

11.4.5 Distributional impacts

Any greenhouse policy will have distributional effects on firms and households. Regulations impose quantity limits on the use of particular inputs or outputs in production and consumption activities. The direct cost of regulation is the reduction in profits and consumer welfare due to the regulatory constraint on choice. The distribution of this cost is often hidden, but that does not make it unimportant. For example, poorer households tend to own appliances that are cheaper to purchase but more costly to run. That is, poorer households tend to make appliance purchasing decisions reflecting higher effective individual discount rates (Hausman, 1979). A regulation that required all households to purchase more efficient appliances would thus disproportionately affect the poor, even though by outward appearances it might seem not to. To the extent that regulation is not a least-cost option to control greenhouse emissions, the excess burden it implies must also be distributed. However, with quantity limits – whether in the form of conventional regulations or tradable permit systems – individual emitters do not face an environmental cost for emissions that are less than the limit. This is in contrast with the distributional consequences of a system that taxes all emissions.

Although the initial incidence of a greenhouse regulation will fall on greenhouse-intensive energy users, the final incidence will depend on the ability of firms to pass the costs of the regulation to others through higher prices for goods and services, and these distributive impacts are likely to differ over time. For example, over the the long run the application of increasingly strict greenhouse pollution controls to newer cars might be regressive (i.e., with costs falling disproportionately on the poor), as low-income earners face substantial increases in used car costs; in the short run, though, low-income earners might gain a relatively larger capital reward on selling their cars (Tietenberg, 1992). Where the costs of regulation are regressive, compensatory transfers may be used, funded from a specific tax or general government finances, or the regulation might be modified, exempting some individuals from the regulatory net.

To date, most applied economics research on potential greenhouse policy responses has concentrated on analyzing the overall costs and incidence of market-based instruments. Multicountry and national studies on the economic impacts of regulatory measures are needed. Researchers may need to give more emphasis to incorporating real world conditions in these models, such as uncertainty and asymmetric information in energy markets. This would allow for any differences in the informational role of price and quantity controls (as well as policy mixes) in determining the size and distribution of greenhouse control costs.

11.5 Market-Based Policy Instruments

Because of the considerable potential costs of meeting greenhouse gas emission targets, one of the central issues for parties to the Framework Convention is the identification of least-cost measures, that is, policies that minimize the costs of achieving a given greenhouse gas emission target. In this section attention is focussed on emission taxes and tradable quotas and permits, the two core market-based instruments that directly target least-cost measures to meet greenhouse gas emission targets. Mention is also made of joint implementation, an instrument that can facilitate technology transfer and is equivalent to bilateral "trading" in emissions.

There is an extensive literature on the principles underlying the use of market-based policy instruments for greenhouse management (for example, Bohm and Russell, 1985; Baumol and Oates, 1988; Stavins, 1988; OECD, 1989, 1993; Tietenberg, 1990; Epstein and Gupta, 1990; Dornbush and Poterba, 1991; Stavins, 1991; Bureau of Industry Economics, 1992; Cropper and Oates, 1992; IIASA, 1992a, b; Pillet *et al.*, 1993; Hahn and Stavins, 1995). For a summary and assessment of the cost and environmental effectiveness of emission taxes and tradable quotas in national applications see Howe (1994).

In a perfectly competitive marketplace, under an emission tax or tradable quota scheme, emitters would reduce emissions up to the point where the marginal cost of control equals the emission tax rate or the equilibrium price of an emission quota. Both instruments would promote dynamic efficiency (cost minimization over the long term, when factors of production are variable and technological change may be stimulated), as each provides a continuous incentive for research and development in emission abatement technologies to avoid the tax or quota purchases. Under competitive markets and certainty, an emission tax is identical to a tradable emission quota scheme in which quota rights are auctioned and revenues are redistributed in the same way (Rajah and Smith, 1993).

11.5.1 Domestic carbon taxes

With market-based policies, there can be incentives to reduce greenhouse emissions through the development and use of new technologies and by making changes to existing production and consumption practices. The aim in using an emission tax (a tax per unit of emissions) is to minimize the total economic costs of achieving a given emission target. In principle, both static and dynamic efficiency gains can be fostered under an emission tax. These gains arise where emitters have different opportunities for emission control (have different marginal abatement cost curves) both in the short run, when factors of production and technological opportunities are largely fixed, and in the long run, when they vary endogenously.

Most research has focussed on the carbon content of primary fossil fuels consumed as the most practicable base for a tax on greenhouse emissions (Pearce, 1991; Boero *et al.*, 1991; Jorgenson and Wilcoxen, 1992; Repetto *et al.*, 1992; Dower and Zimmerman, 1992; OECD, 1992a, 1993; Jones and Tobler, 1993; Pillet *et al.*, 1993; Boyd *et al.*, 1994). A carbon tax is not, however, a perfect proxy for a tax on CO_2 emissions. For example, a carbon tax on fossil fuels provides an incentive to reduce the use of carbon-based fuels, but not to reduce CO_2 emissions by such means as capture (fixation) and disposal of the emissions at source (on carbon removal, see Section 11.5.1.1). There may also be, due to leakage or incomplete combustion, emissions of other carbon compounds that differ from CO_2 in their greenhouse effects (notably methane). In addition, to be consistent, accounting would need to apply to domestic emissions resulting from the processing of fuels from one energy form to another (as in electricity generation).

There is a variety of points in the "product cycle" for fossil fuels, from production to end use, at which a carbon tax could be applied. End use is obviously the point at which emissions occur, but monitoring points covered under the policy would be fewest, and hence implementation costs lowest, if carbon contents were measured and policy applied at the wholesale level.

A carbon tax is a more efficient instrument for reducing energy sector CO_2 emissions than taxes levied on some other basis, such as energy content of fuels or the value of energy products (*ad valorem* energy tax). For example, model simulations of the U.S. economy indicate that an energy tax could be between 20 and 40% more costly, and an *ad valorem* tax two to three times more costly, than a carbon tax for equivalent reductions in emissions (Scheraga and Leary, 1992; Jorgenson and Wilcoxen, 1992). This is because an energy tax raises the price of all forms of energy, whether or not they contribute to CO_2 emissions, and would make it more costly to substitute lower-emitting or nonemitting energy sources for high-emitting energy sources. As a corollary, a combined carbon/energy tax would be less efficient and more costly than a pure carbon tax, due to its energy tax component.

11.5.1.1 *Related instruments: Deposit refund systems for carbon removal*

As already mentioned carbon can be removed from the atmosphere by enhancing natural sinks. It could also be removed by technical means if cost-effective technologies can be developed. One policy instrument that might be considered to provide an incentive to undertake such carbon removal is a deposit refund scheme. A deposit refund system can take one of several forms. One variant combines a tax (deposit) on a commodity with a subsidy for the socially least-cost disposal option (a refund). Another uses mandated deposits, which require private sellers of a commodity to add to the price a deposit that will be refunded under certain conditions. Yet another uses a performance bond, which requires an agent engaging in specified production activities to avoid certain negative consequences of these activities. For surveys of current uses of deposit refund systems and descriptions of potential new areas of application, see Bohm (1981); Hahn (1988); Russell (1988); Stavins (1988); OECD (1989); Anderson *et al.,* (1990); Hahn and Stavins (1991); Sigman (1991); Stavins (1991); and U.S. Environmental Protection Agency (1991).[22]

For a deposit refund system to be a feasible means of encouraging carbon fixation, there must exist alternative actions that decision makers can take to avoid creating the environmental externality in question. This could be a choice between controlling emissions at source and "end-of-pipe" emissions removal.

Currently, potential emitters of CO_2 – purchasers of fossil fuels for combustion, for example – do not have the option of choosing to remove the carbon from emissions.[23] On the other hand, new techniques for carbon removal may eventually become economically feasible. (In 1993, for example, Japan's Ministry of Trade and Industry launched a project on CO_2 fixation – see MITI, 1993.)[24] When and if this option becomes available, incentives will exist to choose carbon removal when this is less expensive than reducing fossil fuel combustion through improvements in energy generation or use efficiency. This would open up the possibility of applying the deposit refund concept to CO_2 emission reduction. It could, for example, provide a mechanism for the inclusion of sinks, such as the development of new forests and other changes in land management, in a market-based permit or tax system.

This suggests that it is important, in the meantime, to maintain appropriate incentives for research and development of technologies that can eventually provide cost-effective options for carbon removal. The appropriate incentive will exist if future carbon removal is known to be subsidized at a level equal to the (tax, tradable quota price, or shadow) cost of carbon emissions.[25]

Thus, it may be possible to introduce clauses into a future climate change protocol that would validate subtractions of carbon removal from emissions. This would, in effect, be equivalent to creating an international deposit refund system – a tax/subsidy scheme – where nations would be credited for certified carbon removal (estimated carbon emissions avoided) by equally large additions to their emission quotas. These quotas would be measured by the carbon content of fossil fuel use, equal to production plus imports minus exports. Moreover, if fossil fuel use is taxed domestically (directly or indirectly through a tradable permit system), and carbon removal is subsidized (credited by a refund of the deposit, equal to the tax), the resulting policy package will be an international deposit refund system.[26]

Other greenhouse gases, such as CFCs, that could be recovered when servicing cooling equipment such as refrigerators or air conditioners, could be made subject to a system with general deposits (taxes) on CFCs and a refund when and if CFCs are recovered, provided that the transaction costs of such a system are not prohibitively high (Stavins, 1988; Miller and Mintzer, 1986). Obviously, if CFCs are successfully phased out in the near future, the role of policy instruments such as deposit refund systems to control CFC emissions will be very limited. Still, the problem of CFCs remaining in discarded cooling equipment, perhaps for some twenty years after their use in new production had been phased out, may be significant enough to justify the use of a CFC deposit refund system (Bohm, 1981, 1990). Note, however, that for products that contain CFCs and that have already been purchased by final users, the deposit refund system will be one with zero deposits and a positive subsidy. In other words, it will collapse into a pure subsidy system. Then, one of the advantages of deposit refund systems, that ordinary taxes are not required for financing the refund/subsidy incentive, would no longer hold.

11.5.2 Tradable permits

A powerful theoretical feature of a perfectly competitive tradable domestic permit scheme is that, no matter what the initial permit allocation, equilibrium permit prices will be the same and the final allocation after domestic trade will be the one that minimizes the cost of reducing emissions. Firms will want to buy permits if abatement costs exceed the permit price and sell permits in the opposite case. In this way, trade will continue until all firms reach a position of indifference between buying and selling permits – that is, between marginal abatement and additional fossil fuel use. When this state is reached, an *ex post* allocation of permits that minimizes the costs of reducing emissions has also been reached.[27]

An international tradable quota scheme could coexist with domestic permit schemes within each country, or particular countries might choose to meet their emission targets by some other means, such as taxes or regulatory systems. In the case of a domestic tradable permit scheme, a national government would issue emission permits (perhaps time-limited) to wholesale dealers in fossil fuels or producers and importers of fossil fuels and allow them to trade on a domestic permit market. The government could also allow permit holders to trade directly on an existing international market. Alternatively, to the extent that both international quota and domestic permit markets existed for a particular country, the government could trade on the international market and set a definite or preliminary domestic limit on the volume of domestic permits for some period ahead.

A government could choose one of two main ways to distribute permits to individual firms. In the first case, firms would be given shares of the total permit volume based on some historical record ("grandfathering") such as their recent fossil fuel sales. The second alternative would be for the government to auction permits. Some combination of these two approaches might also be feasible.

The two approaches differ primarily in two respects. First, grandfathering implies a "transfer" of wealth, equal to the value of the permits, to existing firms, whereas, when permits are auctioned by government, this wealth is transferred to the government. The government would then collect revenue similar to that from a domestic tax producing the same volume of emissions. As with tax receipts, auction revenues could be used to reduce preexisting distorting taxes as outlined in Section 11.3. Second, since grandfathering improves the wealth of incumbent firms and, given uncertainty, may keep them in business longer than otherwise, this allocation approach may reduce the rate of entry of new firms and slow technological change (Bohm, 1994b).

To date most tradable permit systems have made use of eternal permits. However, there are several reasons for preferring a system of time-limited permits in the case of climate change applications. First, to the extent that permits may be initially grandfathered, the negative effects mentioned above would be mitigated – after emitters were given sufficient time to adjust, subsequent allocations of permits could be made by auction. Second, potential future policy changes about emission targets in response to new information, for example, could cause significant problems for permit price formation if eternal permits were used. An alternative approach would be for the government to retain ownership of the permits and lease them to firms for a fixed period.[28]

Allowing permits to be banked, that is, allowing permits for emissions during a given period (e.g., a year) to be used at a later date, is important for both the efficiency and political acceptability of a tradable permit scheme. Without a banking option permit-liable firms would be confronted with greater end-of-period permit price uncertainty.

Stavins (1995a) considers a market for emission permits in which costs are associated with the exchange of permits, and he models several alternative types of transaction cost functions. He finds that transaction costs reduce trading levels and increase abatement costs and, most important, that in some cases, equilibrium permit allocations and hence aggregate control costs are sensitive to initial permit distributions. Thus, in the presence of transaction costs, the initial distribution of permits can matter in terms of efficiency, as well as in terms of equity.

By contrast with international tradable quota systems, which have so far been applied on a small scale only (for example, under the Montreal Protocol for the international CFC production quota trade and for the CFC consumption quota trade within the European Union), there is considerable experience with the use of tradable permit schemes within countries (see OECD, 1992b; UNCTAD, 1992). In most of these applications, permits have been allocated by grandfathering. Many of these applications have been designed to deal with local air pollution problems, and, as a consequence, the permit markets have often been relatively small and far from perfect. This contrasts with the case of a tradable carbon permit system, which would be nationwide. The contrast would be even starker for an international scheme involving many governments and possibly large firms as well.

11.5.3 International carbon taxes

International action would be required to meet a global emission target. One possibility is that a carbon tax could be imposed on nation states themselves by an international agency. In this case, the agreement would specify not only tax rate(s) but also a formula for reallocating the revenues from the tax. Cost-effectiveness would demand that the tax rate be uniform across all countries (assuming full participation), but the reallocation of revenues would not have a direct bearing on cost-effectiveness. As an alternative, the agreement could stipulate that all countries should levy the same domestic carbon tax, so-called harmonized domestic carbon taxes. In both cases, the tax rate that achieved the coalition's emissions target could only be struck through trial and error. The tax rate would also need to be adjusted over time as economic conditions change and as more scientific information becomes available.

Uniform tax rates are required for reasons of cost-effectiveness. But the resulting distribution of costs may not conform to principles of equity and justice. For this reason, transfers of resources may be required. In principle, the two versions of an international tax agreement could involve the same actual financial transfers, although the transfer principles may differ. Under the harmonized tax system, the agreement could involve fixed lump-sum payments from rich to poor countries, whereas under the first-mentioned international tax system, the agreement could specify what shares of the total international tax revenues would go to each participating country (Hoel, 1993). (In a tradable quota system, the financial transfers could again, in principle, be the same, but would then be represented by sales of time-limited quotas by poor countries and quota purchases by rich countries.)

11.5.4 Tradable quotas

Under an international tradable emission quota scheme, all coalition countries would be allocated a quota for emissions (for whatever emission is being controlled). A quota could define either a right to repeated emissions (for example, one tonne of carbon per year over the indefinite future) or a right to emit a given volume once only. Thus a quota system could comprise either "eternal" quotas of the first type or a series of "noneternal" quotas (for example, quotas for five-year periods) or some combination of both. In the case of either type of quota, emission rights could be "banked." In other words, any unused right to emit during a given year could be kept and used at a later time.

In each period, countries would be free to buy and sell quotas on an international exchange (on the spot or forward market) in order that neither buyer nor seller need be identified. Time-limiting the quotas would probably be necessary not only to account for uncertainty about the extent of the enhanced greenhouse problem but also to give credibility to the system. More specifically, it would be necessary to avoid a situation where a government sold quotas, that is, part of the nation's wealth, to an extent that would not be honoured by future governments in the country. Time-limited quotas would also reduce the risk of large countries gaining market power on the quota trade market or the need for measures (such as limits to quota holdings) to ensure that such market imperfections were avoided (Bohm 1995b).

An efficient international tradable quota system presupposes a market organization for quota trade (see Sandor *et al.*, 1994). In the case of a system for the control of emissions of CO_2, quotas would have to be denominated according to the carbon content of the fossil fuel used. If quotas were to be established for the full range of greenhouse gases, it would be necessary to weight gases according to their estimated (and agreed) global warming potential. For any such scheme to be effective in controlling emissions, it is clear that there must be a reasonable probability of detecting and penalizing those responsible for unauthorized emissions. This, however, does not distinguish a tradable quota system from any other international agreement on emission reductions. In what follows, the focus is on quota systems for carbon emissions only.

Negotiations on initial quota allocation are likely to be facilitated by reference to some criteria such as GNP, real GNP, total population, adult population, land area, "basic needs" (defined by industry structure and/or local climate), dependence on fossil fuel production, and others (for an overview, see UNCTAD, 1992; also see Grubb and Sebenius, 1991; Bertram, 1992; Bohm and Larsen, 1993; Hinchy *et al.*, 1993). There are numerous other possibilities (see Chapter 3). Evaluation of proposed rules would need to take account of their international trade repercussions.

Each of the criteria will have adherents, largely those with larger allocations under that criterion.[29] Several criteria may need to be blended to create international consensus on emission allocations.[30] For example, it is clear that the developing countries have relatively little incentive to participate unless they see clear economic benefits from an agreement. At the same time, the wealthy countries will want to make sure their burdens are divided in ways that are perceived as equitable. Whatever the initial allocation, subsequent trading can lead to a cost-effective outcome.[31] This potential for pursuing distributional objectives while assuring cost-effectiveness is an important attribute of the tradable quota approach.[32]

Compared to an international tax agreement, where the effect on emissions is uncertain but related lump sum transfer payments are known, the tradable quota system has a known effect on emissions but quota prices are uncertain and, hence, the distributional effects through quota trade are also uncertain. (This is true for fully global agreements; if only a limited set of countries is involved, carbon leakage must be taken into account in both cases.) This means that the benefits of known effects on emissions in a tradable quota system must be bought at the price of some distributional uncertainty. Thus, if a decision is taken that a poor country should be offered some minimum compensation, then initial quota allocations to that country would have to be increased as compared with the case under certainty. Alternatively, the agreement would have to include some co-insurance system.

Countries allocated quotas surplus to their emission requirements would be able to use the revenue from the sale of these surplus quotas to increase their imports relative to their exports (Chichilnisky *et al.*, 1993). Countries allocated quotas less than their requirements would have to reduce imports relative to exports to pay for additional quotas. In this way a tradable quota scheme would tend to reallocate world production. The allocation of tax revenue from an international carbon tax scheme would have similar effects.

Providing large initial quotas to poor countries for compensatory reasons implies that they would be selling quotas primarily to rich countries. Since quota permit prices represent an implicit or explicit tax on all participating countries, the terms of trade within the coalition for countries with the same carbon intensities in production would remain unaffected. Giving some tariff or other protection from competition from nonparticipating countries, when the agreement does not involve all countries, means that industrialization in poor countries would not have been made more difficult, relatively speaking, aside from the inevitable consequences of reduced global fossil fuel use. In addition, reducing fossil fuel use would emerge as a potentially important "export industry" for the poor countries. From a distributional point of view, the end result would be that poor countries would be perhaps fully compensated, whereas rich countries would have to pay, first, for their own emission reductions as called for by the quota price and, second, for carbon reductions imported through quota purchases from abroad.

11.5.5 Joint implementation

Joint implementation, provided for by Article 4.2 (a) of the FCCC, involves cooperation between two countries, with one funding emission reduction in the other to help the first meet its reduction commitments.[33] Pilot joint implementation projects are now being undertaken by a number of countries. Although many of these involve intergovernmental agreements, the private sector may also be involved directly. The U.S. Initiative on Joint Implementation, for example, involves private sector proposals being approved by an interagency panel.

The potential economic merits and demerits of joint implementation proposals have been widely discussed (Hanisch, 1991; Hanisch *et al.,* 1992; Hanisch *et al.,* 1993; Anderson, 1993; Barrett, 1995; Jones, 1993; Johnson, 1993; Parikh, 1994a, b; Reddy, 1993; Bohm, 1994a; Loske and Oberthür, 1994; Jepma, 1995). In essence, there are three potential roles for joint implementation: (a) as the first step toward establishing an international tradable quota system for greenhouse management among parties that have made a firm commitment to limit their emissions; (b) as a cost-effective option for developed countries to fund emission reduction projects in developing countries that have made no such commitments; and (c) as an activity for exploring when it is cost-effective to bring new emission sources or sinks into an existing international greenhouse management scheme.

A system of joint implementation trades in carbon reduction projects could develop automatically into an international tradable carbon quota system for countries that have carbon targets (Bohm, 1994a). Where aggregate targets exist, the joint level of aggregate carbon emissions reported is a sufficient monitoring statistic for the actual aggregate abatement levels undertaken by each country under a joint carbon reduction policy. Cost incentives are such that emission reductions below the target in one country, when less costly than in another country, would be purchased by, and credited to, the latter. In particular, when several countries are committed to carbon target trading in this fashion, the incremental costs of emission reductions will tend to be equalized across economies. As a result, an international tradable quota scheme could be established once countries commit to binding targets, which would in effect then become their tradable quotas.

The literature on joint implementation has focussed mainly on low-cost emission reduction projects in developing countries. Bohm (1994a) also considers the case of joint implementation between developed and developing countries, where the former commit to binding targets but the latter do not. In the absence of binding targets in developing countries it would be difficult to determine the net emission reduction effects due to a specific joint implementation project, since nationwide indirect and direct effects on emissions must be counted (see Tietenberg and Victor, 1994). The net emission reduction effects of low-cost abatement projects are particularly uncertain, since such projects may be close to being profitable and, hence, may be carried out by the market itself in the near future. In addition to these systematic risks, there are incentives to misrepresent the effectiveness of projects. Parties to a joint implementation project may exaggerate the project's nationwide net emission reduction effects. A clearinghouse version of joint implementation trades between developed and developing countries would eliminate these incentives on the part of the buyer countries. The role of the clearinghouse would be to screen and aggregate all projects from potential sellers before they are offered as anonymous carbon credits to buyers at a market clearing price.

Another potential role for joint implementation is as a complementary exploratory tool for gathering information to expand an existing international regime that initially involves only international carbon trading resulting from fossil fuel emissions. Participating nations could agree to revisit the coverage issue at fixed intervals, modifying strategies to accord with the underlying science and economics of climate change. Successful joint implementation programmes are those that provide the necessary information to incorporate a truly cost-effective new emission source or sink into an international market-based policy. However, this application of joint implementation raises problems of estimating project emission baselines similar to those mentioned in the preceding case, and hence difficulties in ascertaining the nationwide net emission reduction effects of individual noncarbon projects. Current and future joint implementation demonstration projects could provide additional insight into these estimation problems, provided there were sufficient incentives for developed countries to finance them.

The potential driving force behind joint implementation is that both buyer and seller countries, developed as well as developing, would benefit from this particular trade as they do from other forms of voluntary transactions. However, monitoring is a problem in using joint implementation as an instrument for significant cost-effective operations, except for the limited case where parties to the FCCC have committed themselves to emission targets, primarily for carbon emissions. This type of joint implementation amounts to an international tradable quota system, the cost-effectiveness of which is determined by the number of countries that have made such commitments and want to engage in operations under such a system.

11.5.6 Distributional impacts of market-based measures

The literature on industrialized countries typically portrays carbon taxes and other market-based instruments such as

Table 11.2. *A Summary of empirical evidence on the redistributive impact of economic instruments*

Instrument	Author	Country	Model	Results
Carbon tax	Bull, Hassett, & Metcalf (1993)	U.S.	Computable dynamic general equilibrium model; spending behaviour may adjust	Tax burden is nearly proportional with respect to lifetime income
Carbon tax	DeWitt, Dowlatabadi, & Kopp (1991)	U.S.	Partial equilibrium model; spending behaviour may adjust; expenditure data; no recycling of tax revenues	Distributional impact is regressive and varies across regions
Carbon tax	Jorgenson, Slesnick, & Wilcoxen (1992)	U.S.	Computable general equilibrium model; three stages; intertemporal optimization for household consumption	Carbon tax is either mildly progressive or regressive depending on the welfare function used
Carbon tax	Poterba (1991)	U.S.	Partial equilibrium model; expenditure and income data	Carbon tax is regressive, but the impact is smaller if expenditure data are used
Carbon tax	Schillo, Giannarelli, Kelly, Swanson, & Wilcoxen (1992)	U.S.	DECO aggregate macroeconomic model; expenditure data	Depending on the compensation system adopted, the carbon tax is regressive to neutral
Carbon tax	Schillo, Giannarelli, Kelly, Swanson, & Wilcoxen (1992)	U.S.	Urban Institute's TRIM2 microsimulation model; two compensation systems	Carbon tax is regressive with respect to pretax income in both scenarios, but it becomes regressive to neutral relative to posttax income
Carbon tax	Pearson (1992)	Europe	Partial equilibrium model; spending behaviour may/may not adjust (2 models); Eurostat data	Both models indicate that tax on domestic fuels is regressive, while tax on motor fuels is mildly progressive
Carbon tax	Pearson & Smith (1991)	Europe	IFS model of consumer expenditures; compensation system; spending behaviour may adjust	Ireland and UK show a regressive impact. For other countries, the burden is weakly related to income
Carbon tax	Shah & Larsen (1992)	Pakistan	Partial equilibrium model; income and expenditure data; three scenarios	Carbon tax incidence is either proportional or progressive in a developing country context
Carbon tax	Hamilton & Cameron (1994)	Canada	CGE model and simulation model of household expenditure	Tax burden is moderately regressive with the greatest effect on low-income married couples
BTU tax	Bull, Hassett, & Metcalf (1993)	U.S.	Computable dynamic general equilibrium model; spending behaviour may adjust	Tax burden is nearly proportional with respect to lifetime income
Gasoline tax	Greening, Schipper, & Jeng (1993)	U.S.	Partial equilibrium model; expenditure data	Gasoline tax affects negatively mainly older married couples with dependent children. Income distributional results not reported
Gasoline tax	Krupnick, Walls, & Hood (1993)	U.S.	Partial equilibrium econometric model; limited adjusting behaviour	Gasoline taxes are regressive, much more than previous studies since income data are used
Gasoline tax	Poterba (1990)	U.S.	Partial equilibrium model; expenditure data	Gasoline tax is broadly regressive if the lowest income class is ignored; this class devotes a smaller share of its budget to gasoline than the lower-middle income class
Tax on GHG emissions	International Energy Agency (1993)	Europe	Partial equilibrium model; expenditure data	Regressive effect on households if no compensatory measures with respect to domestic heating; less clear result for motor fuels
Tax on direct fuel expenditure	Smith (1992a, b)	Europe	Partial equilibrium model; expenditure data; spending behaviour may/may not adjust; two compensatory systems	Carbon tax is regressive, but if spending behaviour adjusts its impact is smaller; only lump-sum transfers make the impact progressive
Tax on industrial energy use	Smith (1992b)	UK	Input/output tables plus consumer spending simulation program	Modest effect of changes in prices on consumer spending, but negative especially for low-income households

Source: Larsen and Shah (1995).

Table 11.3. *Carbon tax incidence in developing countries*

Institutional considerations	Implications for tax shifting	Tax incidence with respect to		
		Income	Expenditure	Lifetime income
a. Foreign ownership and control	Borne by foreign treasury through foreign tax credits	Nil	Nil	Nil
b. Full market power	Full forward shifting (100% on final consumption)	Regressive (pro-rich)	Less regressive	Less regressive
Perfectly inelastic demand or perfectly elastic supply				
c. Price controls and legal pass-forward of the tax disallowed	Zero forward shifting (100% on capital income)	Progressive (pro-poor)	Progressive	Progressive
Completely inelastic supply	Reduced rents	Progressive	Progressive	Progressive
Import quotas and rationed foreign exchange	No effect on prices (100% on capital income)	Progressive	Progressive	Progressive
d. An intermediate case of (a) and (b) above	Partial forward shifting (31% to capital income, 69% to final consumption)	Proportional	Progressive	Progressive

Source: Shah and Larsen (1992).

emission taxes and gasoline or BTU taxes as regressive because outlays on fossil fuel consumption as a proportion of current annual personal income tend to fall as incomes rise. But recent studies (see Table 11.2) using U.S. and European data show that carbon taxes are considerably less regressive relative to lifetime income or annual consumption expenditures than to annual income (see Poterba, 1991, 1993; Jorgenson *et al.,* 1992; Smith, 1992a, b).

Jorgenson *et al.* (1992) provide the most detailed assessment to date for the U.S. They decompose the simulated equity and efficiency impacts of a carbon tax, using an explicit national social welfare function. In this decomposition, the negative efficiency effect dominates when the carbon tax revenue is rebated as a lump sum, and the much smaller equity effect is either mildly progressive or regressive, depending on assumptions regarding the nation's aversion to income inequality and the measure of progressivity used.

There is evidence that the same holds true for the rest of the world, although for quite a different reason. In developing countries, institutional factors play an important role (see Shah and Larsen, 1992). In the developing world, progressivity, or at least low regressivity, could be fostered by three mechanisms. First, a significant tax burden could be passed on to foreign treasuries, producers, and consumers where there is a significant degree of foreign direct investment from countries where investors are allowed foreign tax credits against domestic liabilities. Second, price controls could be applied to limit the ability of producers to pass the tax on to consumers in terms of higher prices. Finally, combined with binding import quotas or rationed foreign exchange, a tax could reduce the excess profits made by the privileged class.

The existence of factors such as market power are likely to lead to regressivity. In this situation, producers could increase product prices in order to pass on a carbon tax to consumers. As it turns out, in most developing countries, there is some

combination of the above elements, creating a situation where taxes can be only partially shifted to consumers. This means that a carbon tax would be either progressive or much less regressive than often suggested (see Table 11.3). Further, it is likely to be regressive only for the lowest income groups, which could be protected through direct subsidies or alternative measures. In addition, the overall tax structure could be made even less regressive by using a carbon tax to reduce other more regressive taxes.

In principle, if a policy is introduced in order to achieve a particular outcome and it is found to be regressive, then the theoretically appropriate policy instrument to deal with the equity issue is a lump sum transfer to the affected parties. This rests on the assumption that there are no nongreenhouse market distortions. If such distortions exist, then nonlump-sum transfers can improve welfare (Jorgenson *et al.,* 1992). As Schillo *et al.* (1992) explain, if carbon tax revenue is used to reduce labour taxes, or a blend of labour and capital taxes, then simulated household welfare improves as the adverse efficiency effect is reversed by the rebate, but at the cost of reducing the equality of wealth. A capital rebate is shown to neutralize the efficiency effect but has uneven distributional effects.

11.6 Policy Implementation Issues

In assessing any of the wide range of instruments as potential devices for addressing global climate change, it is imperative to give due consideration to the implementation issues that can so severely affect real-world outcomes. Such issues need to be considered in the design of practical policies, whether at the national, multinational, or global level.

In the case of tradable permit systems, as applied to local air pollution problems in the U.S., the claims made for their cost-effectiveness have in some cases exceeded what can reasonably be anticipated. Tietenberg (1980) assimilated the results from ten analyses of the costs of air pollution control, and, in a frequently cited table, indicated the ratio of cost of actual regulatory programmes to least-cost benchmarks. Unfortunately, the resulting ratios (which ranged from 22.0 to 1.1) have sometimes been taken by others to be directly indicative of the potential gains from adopting specific ("cost-effective") mechanisms such as tradable emission permits. A more realistic and appropriate comparison would be one between actual regulatory policies and either actual trading programmes or *reasonably constrained* theoretical permit programmes (Hahn and Stavins, 1992).

A number of factors can adversely affect the performance of tradable permit systems: concentration in the permit market (Hahn, 1984; Misiolek and Elder, 1989); concentration in the output market (Malueg, 1990); transaction costs (Stavins, 1995a); nonprofit-maximizing behaviour, such as sales or staff maximization (Tschirhart, 1984); the preexisting regulatory environment (Bohi and Burtraw, 1992); and the degree of monitoring and enforcement (Keeler, 1991). In the case of taxes, research on implementation issues has focussed on administrative costs (Polinsky and Shavell, 1982), monitoring (Russell, 1990), and enforcement (Harford, 1978; Russell *et al.*, 1986).

In the following sections a review is undertaken of what is known about some prominent issues regarding the implementation of (carbon) taxes and/or tradable permit systems. Where appropriate, reference is also made to regulatory systems. It is important to note that most of the research on tradable permit schemes is based on experience in the U.S., where such schemes were superimposed on regulatory policies. These schemes typically involve a limited number of participants and eternal permits allocated by grandfathering. To control carbon emissions, however, tradable permit schemes might operate where there was no former regulation. They would also involve a larger number of participants, and permits might be time-limited and allocated by auction. Thus, experience in the U.S. may not generalize to such schemes with rather different characteristics.

11.6.1 The "currency" of regulation

Because of the monitoring and enforcement burden associated with regulating actual carbon dioxide emissions, the most practical "currency" for a tax or tradable permit system would presumably be the carbon content of fossil fuels. Given the proportional relationship between carbon content and CO_2 emissions and the present lack of practical means of sequestering these stack gases, this is a highly appropriate approach. Monitoring could rely partly on self-reporting, supplemented by international access to national fossil fuel inventories. Under an international carbon tax or tradable permit scheme, implementation costs may be least where incentives to comply are self-enforcing. An effective enforcement system makes ultimate sanctions credible, so that penalties would rarely need to be imposed. However, in many countries even monitoring fossil fuel consumption is not a trivial problem.[34]

11.6.2 Market power

There are two components to the market power problem for tradable permit systems (Bureau of Industry Economics, 1992). The first is the potential for some economic agents to influence the permit price. The second is the potential for some economic agents to use permits to exercise market power in the output market for the product that "generates" emissions. Market power in the permit market is sufficient but not necessary for market power in the output market (BIE, 1992). Malueg (1990) argues that market power in the output market may reduce economic welfare under a tradable permit scheme even if tradable permits are fully cost-effective.

The degree of competition in the tradable permit market will affect the extent to which potential control cost savings are likely to be realized. Hahn (1984) considers the case of a monopsonist (a price-setting buyer) who forces down the real permit price below the competitive level. This behaviour is not cost-effective, since the monopsonist buys too few permits (spends too much on abatement) whereas the competitive agents will buy too many (thus spending too little on abatement). Misiolek and Elder (1989) consider the converse case of a monopolist (a price-setting seller) who forces up the real permit price above the competitive level. That is, the monopolist spends too little on abatement whereas competitive agents spend too much. To the extent that market power derives from the initial allocation of permits, one solution may be to limit the number of permits held by any player (Tietenberg, 1985). This may be achieved by widening the market to many players, using a variety of means, including limiting the temporal duration of permits.

An emission tax designed to make emitters face the full costs of production will tend to result in a cost-effective allocation of the control burden, provided that emitters are all price takers (that is, small relative to the size of the market). But this outcome does not necessarily occur where emitters have some degree of monopoly power in emission-intensive output markets (Buchanan, 1969). In principle, a monopolist in an output market for an emission-intensive commodity will tend to reduce output below the competitive level in order to raise its profits. Hence, the welfare gains from reduced emissions must more than offset the losses from the monopolist's reduced output for an emission tax to be worthwhile (Cropper and Oates, 1992). Which effect dominates is an empirical issue.[35] For example, Oates and Strassman (1984) find that monopoly power in the output market is unlikely to be a key concern in their study of U.S. industries.

In the case of a scheme to reduce global emissions, there typically will be many more participants than under the schemes operating in the U.S. This would reduce the problem of market power. In an international scheme, firms that are large domestic emitters may be required to hold permits as a means of broadening the market. If market power poses a problem in domestic schemes, government intervention may be required, or a tax scheme may be a preferable option.

The issue of market power warrants further consideration in the context of either tax or permit schemes. Whether or not OPEC is involved in a scheme, the potential oligopolistic response of oil producers could interfere with the effectiveness of attempts to control global emissions (Sinclair, 1992).

11.6.3 Transaction costs

Transaction costs are potentially important in the performance of tradable permit markets (Baumol and Oates, 1988; Tripp and Dudek, 1989; Hahn and Hester, 1989a). Stavins (1995a) identifies three potential sources of transaction costs in tradable permit markets:

(a) search and information
(b) bargaining and decision
(c) monitoring and enforcement

The magnitude of these transaction costs will depend on the structure of the market and the extent to which individual transactions require regulatory approval. If a full market is developed, a market price for permits will be known, and this price will convey all the information parties need in order to decide whether to trade. One party would not need to search for another to trade with, and the terms of trade would not need to be negotiated. Even where such a full market does not develop, innovations may serve to keep transaction costs low. For example, where there are search and information costs, brokers may provide information about pollution control options and potential trading partners in order to exploit potential gains from trade. The third source of transaction costs – monitoring and enforcement – can be significant, but these costs are typically borne by the responsible government authority and not by trading partners and, hence, do not fall within the notion of transaction costs incurred by firms as defined here.

There is abundant anecdotal evidence indicating the prevalence of significant transaction costs in some U.S. trading programmes involving mainly local pollution. Atkinson and Tietenberg (1991) survey six empirical studies that found trading levels – and hence cost savings – in permit markets to be lower than anticipated by theoretical models. Liroff (1989) suggests that this experience with permit systems "demonstrates the need for . . . recognition of the administrative and related transaction costs associated with transfer systems."[36] For example, under the U.S. Environmental Protection Agency's emission trading programme for "criteria air pollutants," there is no ready means for buyers and sellers to identify one another, and as a result buyers frequently pay substantial fees to consultants who assist in the search for available permits (Hahn, 1989). At the other extreme, the high level of trading that took place under the programme of lead rights trading among refineries as part of the U.S. EPA's leaded gasoline phasedown has been attributed to the programme's minimal administrative requirements and the fact that the potential trading partners (refineries) were already experienced at striking deals with one another (Hahn and Hester, 1989a). Hence, transaction costs were kept to a minimum and there was little need for intermediaries.

Another source of indirect evidence of the prevalence of transaction costs in permit markets comes from the well-known bias in actual trading toward "internal trading" within firms, as opposed to "external trading" among firms. It has been hypothesized that the crucial difference favouring the internal trades and discouraging the external trades is the existence of significant transaction costs that arise when the trades are between one firm and another (Hahn and Hester 1989b). The existence of commercial brokers charging significant fees to facilitate transactions lends further credence to this suggestion.

However, although this U.S. experience provides valuable data for the assessment of trading regimes, it must be borne in mind that it is not necessarily relevant to a policy for reducing greenhouse gas emissions. Most of the U.S. programmes have concerned local pollution problems where the market for trading was very thin and where substantial regulatory oversight was required to ensure that the associated environmental objectives were met. Furthermore, it has partly been concerns about future regulatory uncertainty that have discouraged intercompany trading. The lesson from this experience is not that trading involves large transaction costs but that trading regimes should be designed partly with the aim of keeping transaction costs low.

The effects of transaction costs should be ameliorated in markets with relatively large numbers of potential trading sources or where formal international or domestic trading exchanges have been established. As the pool of potential trading partners increases, it should be easier for sources to identify potential trading partners, even in the absence of formal exchanges, thereby lowering transaction costs. A larger number of firms can also mean more frequent transactions, as a result of which more information is generated and uncertainty is reduced.

Economists have tended to give greater emphasis to the symmetry between tradable permits and pollution charges than to their differences, although the two approaches are not symmetrical under conditions of uncertainty (Weitzman, 1974), in the presence of transaction costs (Stavins, 1995a), or under a number of other conditions (Stavins and Whitehead, 1992). Analyses that have compared taxes and permits have assumed zero transaction costs. Systems of pollution taxes can also involve substantial administrative costs, both fixed (per firm) and variable (Polinsky and Shavell, 1982).

11.6.4 Free riding and emission leakage problems

Can a unilateral policy by one country alone or by a group of cooperating countries prove effective in abating global greenhouse gas emissions? This is an important question, for it is total emissions, and not individual country emissions by themselves, that determine global concentrations of greenhouse gases, and yet some countries seem more willing than others to adopt abatement policies.[37] The answer depends on how the other ("noncooperating") countries respond to the unilateral policies adopted by the "cooperating" countries. These responses in turn reflect two phenomena: "leakage" and "free riding."

As Barrett (1994a) explains, free riding and leakage can undermine any international greenhouse management initiatives, whether they be market-based or rely on regulatory measures. Free riding arises when countries that benefit from global abatement do not contribute towards its provision. Leakage arises when abatement by the cooperating countries alters relative world prices (including shadow prices) in a way that leads noncooperating countries to increase their emissions. Leakage thus undermines the competitiveness of cooperating countries as well as the environmental effectiveness of their efforts.

11.6.4.1 *Policies to reduce free riding*

As long as participation in an international greenhouse management policy is voluntary, countries will have incentives to free ride, sharing in the benefits from such a policy without sharing in the costs. Even if there were no leakage, free riding would result in abatement being less than would be globally optimal, in the sense that the benefit of a small increase in global abatement would exceed the associated cost. This issue has been examined in a number of studies, including Barrett (1992a), Hoel (1992), and Parson and Zeckhauser (1992). None of the existing empirical models has been used to estimate the magnitude of potential free riding, although some insights into the gains from full cooperation are provided by Barrett (1992a, b, c), Bohm and Larsen (1993), Hinchy *et al.* (1994a, b), and Hoel (1992).

As Hoel (1992), Carraro and Siniscalco (1993), and Barrett (1994b) have shown, a stable coalition of cooperating countries may exist in spite of free-rider incentives. The size of this coalition will depend on the ability of the cooperating countries to punish countries that might withdraw from the coalition and to reward countries that might accede to it. However, to be effective, such punishments and rewards must be both substantial and credible, and these requirements often clash. As a result, the size of the stable coalition may be quite small. In fact, these analyses do not consider international trade, and where there is trade, free riding will be exacerbated by leakage. On the other hand, trade may also provide a vehicle for deterring free riding.

For example, Barrett (1994a) explains how the threat of a complete ban on trade in carbon-based fuels and products between cooperating and noncooperating countries could work to support full participation in a greenhouse management scheme. The key to this agreement, as with the Montreal Protocol on Substances that Deplete the Ozone Layer, is that the threatened trade ban would come into effect once a threshold level of countries agreed to participate in the scheme. However, the threshold is determined so that all countries would gain from participation once it is reached. In other words, it is not necessary that trade be restricted, but rather that the threat to impose trade restrictions be credible.

An actual ban on trade introduces a distortion in the global economy and is in this sense undesirable. However, free riding is itself a distortion, and if trade restrictions reduce free riding, they may be beneficial overall. The implication for climate change policy is that trade restrictions should not necessarily be prohibited. Whether, and under what circumstances they should be allowed is a different matter and one requiring additional research.

Finally, Heal (1992) argues that the free-riding problem can be exaggerated. He notes that in repeated games it is more likely that players will cooperate than in one-shot games. There may be reinforcing effects that strengthen the chances of forming a stable coalition. An example would be countries sharing in the costs of developing abatement technology. Since developing countries may benefit from abatement technology created in developed countries, this may increase their incentive to abate.

11.6.4.2 *The severity of emission leakage*

There are two main channels through which emission leakage may be transmitted. First, the implementation of a carbon abatement policy by a coalition of cooperating countries would shift comparative advantage in carbon-intensive goods towards noncooperating countries. As a result, production of such goods, and emissions, would rise outside the coalition. Second, the unilateral policy would have the effect of lowering world demand for carbon-intensive fuels, and thereby reduce the world price for such fuels traded in international markets. As a result (and ignoring income effects), demand for such fuels, and emissions, are likely to rise outside the coalition. It should be emphasized that these two responses by noncooperating countries do not result from any deliberate policy to increase emissions, but rather result from the absence of a policy to reduce emissions.

Barrett (1994a) surveys several global simulation studies that provide positive leakage estimates, including GREEN, 12RT, Global 2100, and the Whalley-Wigle model. The leakage rate is defined as the increase in emissions by noncooperating countries divided by the reduction in emissions by cooperating countries. The evidence of positive emission leakage varies widely and is strongly dependent on the model used. For example, positive leakage rates are low in GREEN (Oliveira-Martins *et al.,* 1992) and high in the Whalley-Wigle model (Pezzey, 1992).

In particular, Pezzey estimates that a 20% reduction in carbon emissions within the European Union alone (relative to a baseline trend) would be associated with a leakage rate of 80%. In other words, for every 10 tonnes of carbon abated by the EU, global emissions would fall by only 2 tonnes. Pezzey also calculates that a 20% reduction in OECD emissions would be associated with a leakage rate of 70%. These leakage rates suggest that unilateral policy would be largely ineffective. On the other hand, Oliveira-Martins *et al.* (1992) estimate much lower leakage rates for policies aimed at stabilizing carbon emissions at their 1990 levels. They estimate leakage rates for a unilateral EU policy of 11.9% in 1995 and 2.2% in 2050, and for a unilateral OECD policy of 3.5% in 1995 and 1.4% in 2050. These leakage rates suggest that leakage does not render unilateral policy ineffective. These rates are more in accord with those reported by Hanslow *et al.* (1995), who estimate that a policy-induced 20% reduction in CO_2 in Annex I countries would be associated with a leakage rate of 3.8%.

One reason for the difference in these leakage estimates is that emission leakage is greater where a country's fossil fuel-intensive products and fossil fuels have close substitutes.

Trade flows in GREEN (Oliveira-Martins *et al.*, 1992) are based on the Armington assumption that goods produced from different countries are imperfect substitutes (implying an exporter has some degree of market power). By contrast, in the Whalley-Wigle model goods from different countries are assumed to be perfect substitutes, thus making trade flows more sensitive to relative price changes. In MEGABARE, the model reported in Hanslow *et al.* (1995), imported goods from different sources are imperfect substitutes. Hinchy and Hanslow (1995) report an extension to that model where, in its dynamic version, capital is internationally mobile. As would be expected, the reported emission leakage rates from the dynamic version of MEGABARE are higher than for the comparative static version of the model and lie about halfway between those reported from GREEN and those from the Whalley-Wigle model.

As already noted, other studies estimate leakage rates somewhere between these two sets of estimates. Oliveira-Martins *et al.* (1992) estimate negative leakage rates for some regions in some years, and Horton *et al.*, (1992) argue that leakage may exceed 100% in some cases. Currently, there is no consensus among economists about the magnitude of leakage. More research is needed, and it would be particularly helpful if leakage rates were calculated for identical simulations employing a consistent set of assumptions, as has already been done in estimating the costs of climate change policies (see Dean and Hoeller, 1992). What can be said now is that leakage is a potentially serious problem for unilateral policies.

The above studies ignore a possible third channel for leakage transmission. Under certain assumptions, noncooperating countries will abate their emissions up to the point where their own national marginal benefit of abatement equals their own marginal cost of abatement (see Barrett, 1994b). In the extreme, this optimization rule will mean that noncooperating countries will not abate their emissions at all. However, more generally, this rule implies that noncooperating countries will abate their emissions by less than they would if they cooperated. Where noncooperating countries do undertake positive unilateral abatement, and where the marginal benefit of abatement to noncooperating countries decreases with the level of global abatement, an increase in abatement by cooperating countries will create an incentive for noncooperating countries to reduce their abatement. Hence, leakage may occur even in the absence of trade.

11.6.4.3 *Policies to reduce leakage*

What can be done to reduce emission leakage? What is currently known stems from the basic general equilibrium model of trade so commonly used. As background, take the somewhat simpler case of the international incidence of a global carbon tax. Most of the results from greenhouse modelling studies to date are based on the simplifying assumption that the global cost-effectiveness of a common carbon tax (or tradable quota system) does not depend on the distribution of incomes between or within countries; that is, only relative prices matter for this cost-effectiveness result. (This is not to say that the international distribution of the impacts from these policies does not depend critically on what is done with tax revenues or initial quota allocations.)

Hence, as is consistent with basic trade theory, a uniform global production tax (quota) on greenhouse emissions is equivalent to a uniform global consumption tax (quota) on these emissions. Both minimize the cost to global economic welfare of achieving a given global emission target where there is perfect foresight. However, there are terms-of-trade gains to net fossil fuel exporters under a production tax, and terms-of-trade losses to importers. Under a consumption tax, the converse holds true. That is, fossil fuel exporters are worse off under a consumption tax on emissions, whereas net importers are better off. Precisely such results are illustrated in the Whalley-Wigle model (Dean and Hoeller, 1992).

What happens in this standard framework when coalition membership is less than global? This issue has been examined by Markusen (1975), Krutilla (1991), Bohm (1993), Hoel (1994, 1995), and Barrett (1994a). Treating the coalition as a single entity and the rest of the world as another single entity suggests that, if the coalition is a net importer of carbon-intensive products in the absence of the carbon tax, then a tariff should be imposed on its imports to reduce emission leakage through the terms of trade. If the coalition is a net exporter of these products in the absence of a carbon tax, then it should subsidize its exports. This response minimizes the coalition cost of meeting the greenhouse constraint. Precisely the same argument holds for leakage through trade in carbon-based fossil fuels. In addition, instead of using an import tariff (export subsidy) the equivalent production subsidy (tax) and consumption tax (subsidy) could be applied in the coalition. Hoel (1995) shows that if an optimal tariff (subsidy) or its equivalent can be employed, then the carbon tax should be uniform across all sectors in all coalition countries, but, if not, then differential tax rates and exemptions may be required.

Although border tax adjustments may theoretically be appropriate for reducing leakage, their application poses a number of practical problems. How are the emissions associated with the manufacture of a particular product to be determined? The Montreal Protocol includes a provision for restricting trade in products made using ozone-depleting substances, such as electronics components that are made using CFCs as a solvent. However, this provision has not been implemented, and the Protocol Secretariat was advised in 1993 that to do so would not be feasible (Van Slooten, 1994). To implement the provision would require either sophisticated equipment capable of detecting trace residues of CFCs or certification of the manufacturing facilities of industries in countries that are not parties to the agreement. In the case of global climate change, similar adjustments would be even harder to implement, as virtually all production results in some greenhouse gas emissions.

Furthermore, the appropriate border tax adjustments may not be compatible with current multilateral trading rules. World Trade Organization (WTO) rules allow for border tax adjustments where the taxed or controlled inputs are physically incorporated in the final product. However, in the case of greenhouse gases, the concern is typically with the carbon emitted in the process of manufacturing a good. A GATT dispute panel has ruled (in the Superfund case) that adjustments may be allowed when the use of inputs can be inferred by assuming that the product was manufactured using the "predominant production method." However, a similar approach would not be appropriate in the case of climate change, not least because production methods vary so widely. The recently completed Uruguay Round allows energy taxes to be remitted on exports of manufactured goods, although there is some question about the generality of this provision and whether it could be extended to include imports. Plainly, the rules for applying border tax adjustments need to be clarified.

In summary, all the results from basic trade theory hold in analyzing emission leakage from a carbon tax or quota (see, for example, Woodland, 1982; Vousden, 1990). However, as with a customs union, determining the optimal tariff (subsidy) to reduce positive emission leakage from the carbon tax will be a complicated calculation, given the extensive but differential use of carbon-based fuels in all economies and the differential ability of some countries to exercise market power. Trade compliance with WTO rules will also need to be considered. In addition, further research on the leakage problem is warranted to consider strategic interactions between greenhouse policies in coalition and noncoalition countries.

11.6.5 Compliance

Free-rider deterrence is concerned with securing broad participation in an agreement, and leakage reduction is concerned with making abatement by cooperating countries more effective. A related concern is compliance, or the incentives that countries have to fulfil their pledges under an international agreement. Some international agreements contain explicit compliance measures such as trade sanctions. However, it is more usual for agreements to seek alternative means for securing compliance (see Chayes and Chayes, 1993).

Indeed, it is a fundamental norm of international law that treaties are to be obeyed, and as a rule countries do not negotiate, sign, and ratify agreements with the intention that they will not comply fully with all relevant provisions. Hence, compliance is not so great a problem as it is sometimes taken to be. More difficult are the problems of negotiating an agreement that requires real sacrifices by the parties and of getting countries to sign the agreement in the first place.

Where compliance is a problem, the reasons are usually innocent. For example, four years after the Montreal Protocol was signed, only about half the parties to the agreement had complied fully with the reporting requirements of the treaty. This was not because these countries hoped to get away with noncompliance, but rather because they did not have the resources and technical know-how needed to carry out their obligations. On the other hand, compliance with certain oil pollution treaties once proved more worrying because noncompliance was linked to the difficulty of monitoring and verifying the amount of oil discharged by tankers at sea. However, once an equipment standard was established requiring all new tankers to have separate ballast tanks, monitoring became easy and problems of noncompliance subsided. Indeed, monitoring of international agreements may be the more

important problem (U.S. General Accounting Office, 1992). The lessons seem to be that treaties should be designed to facilitate easy monitoring, and that they should also ensure that all parties have the means to comply with the requirements of the agreements, given the will to do so.

11.6.6 Information issues, the role of brokers, and risk management

Policy instruments should be designed to provide needed information. In the case of tradable permit systems, there are three ways this can be done:

(a) Government can take actions that directly reduce regulatory uncertainty.
(b) Barriers to private brokerage services can be reduced.
(c) Allowance can be made for the development of futures markets.

In the first case, at a minimum, government authorities can avoid creating regulatory barriers (such as requirements for government preapproval of trades) that drive up transaction costs and discourage trading.

Private provision of brokerage services can also play an important role in information provision. Thus, although commercial brokers can certainly be recipients of transaction costs, their activities reduce transaction costs below what they would otherwise be (Stavins, 1995a). Intermediaries, in general, can contribute to social welfare by helping parties economize on transaction costs. Brokers can play the role of consultants, adding value by understanding the regulatory process and by maintaining information about prospective suppliers and demanders of permits. Under the more conventional function of bringing together buyers and sellers ("brokering deals" by matching buy orders and sell orders), these firms both absorb and reduce transaction costs. Finally, brokers may assume risk by buying, holding, and selling permits.

An important merit of an international tradable quota system (compared with an emission tax scheme) is that it can be used as a risk management tool to reduce the costs of greenhouse risks. Some simple examples of the risk management potential of a quota scheme are given below. In these examples, quotas themselves are used as the hedging instrument. If a sophisticated market were to develop, "derivatives" of quotas, such as options and forward and futures contracts, might be used to perform these risk management functions more efficiently. However, in the first instance, the logic of the risk management potential of quotas can be brought out most simply by taking quotas as the instrument.

In the first example, the use of quotas to reduce risks will be considered for investments that have the potential to reduce emissions. Such investments include research and development into new abatement technologies and the transfer of abatement technology across countries. These investments may either succeed or fail and, hence, the return to the investment is uncertain. The key to the risk-reducing role of quotas is that the quota price will be negatively correlated with the success of the investment.

Suppose that quotas (or permits) of some finite duration are widely traded. When a new investment with emission reducing potential is announced, the price of quotas will tend to fall. This is because there is some probability that the investment will succeed and it will be expected that there will be less need for quotas in the future, resulting in reduced demand. If the investment actually fails, this reduced demand will not eventuate and the price of quotas will rise. On the other hand, if the investment succeeds it will not be the probability but the certain success that will influence demand. Quota prices will fall.

Investors can use this negative correlation between the price of quotas and the success of the project to reduce the variability of their returns. If quotas are bought at the start of the project and it fails, investors will be able to sell their quotas above the purchase price. Such a profit will help to offset their losses on the project. If the project succeeds, investors will suffer a loss on their quota sales. However, by narrowing the gap in expected profits between a successful and unsuccessful outcome, quota operations will reduce the risks for investors (see Epstein and Gupta, 1990, for a numerical example). Risk-averse investors will be willing to trade higher average profits for more certain profits. The availability of quotas as a risk management tool increases the probability that the investment will be undertaken.

In the second example, the use of quotas (or permits) to manage the temporal nature of risk is emphasized. Investments in some activities such as coal-fired power stations have a long payback period and generate emissions. If emissions are taxed, future tax levels may have a critical bearing on the return on the investment. There may be sufficient uncertainty about future tax levels to deter some risk-averse investors from the project. However, under a quota scheme, if there are quotas of sufficient duration, the costs of future emissions to investors can be known with certainty. However, in a scheme dominated by long-lived quotas, there would be inflexibility in adapting emissions to changing information on desirable emission levels. A number of ways to reduce this problem have been proposed.[38]

As mentioned above, a forward or futures market could provide a more effective risk management tool than the direct use of a quota market in the simple examples described. A forward market for quotas could develop if there were a sufficient number of hedgers and speculators, and none of them had excessive influence over price signals. The latter requirement could be met by a rule fixing the maximum share of total net emission entitlements held by any one country. By reducing the costs of risk and uncertainty, time-limited quotas would tend to reduce any nation's incentive to pursue strategic behaviour (such as quota hoarding) and could also reduce validity forecasting problems.

Once the parties to the agreement know the basis on which quotas will be allocated over time, a futures market could develop. Provided contracts are standardized, one of the main differences between the forward contracts market described in the previous paragraph and a futures market is that the latter offers greater liquidity if contracts can be settled by the monetary equivalent of quota transactions, as opposed to the delivery of the quota itself. As a consequence, more transactions are likely in the futures market and this should lead to greater information flows, reducing uncertainty and risk and transaction costs. As a consequence, spikes in the quota price would tend to be rapidly smoothed out as market players took up speculative and hedging positions.

A major factor favouring a tradable quota scheme, therefore, is that a forward and/or futures market based on (net) emission quota contracts would provide a way of efficiently reducing the elements of uncertainty and risk in greenhouse costs. This would reduce the costs of control and stimulate investment in the research, development, and use of least-cost mechanisms for net emission control. For example, suppose a country invests in a risky technology transfer project as part of a strategy to meet its national greenhouse target. It can hedge against the risk of project failure by buying futures. Any profit on the futures market transaction can partly compensate for any rise in the spot price if it needs to buy quotas.

In the U.S. an auction system for forward sales of emission quotas is provided for in the 1990 amendment to the Clean Air Act for controlling domestic sulphur dioxide (SO_2) emissions from fossil fuel fired power plants. As Howe (1994) notes, the amended legislation requires total SO_2 emissions from the U.S. electricity power sector to be around 50% below the 1980 level by the year 2000. After January 1, 1995, each of the 111 power plants directly affected in the first phase of measures must hold tradable quotas covering its total annual emissions target (its quota allocation), capped at about 50% of the 1980 level. Currently, allowances may be traded to any party or credited ("banked") for future use. In the second phase, beginning January 1, 2000, most electric power utilities will be brought within the system. In addition to receiving an annual target allowance, quotas for excess emissions may be bought directly from other plants or through auctions held by the Chicago Board of Trade for the United States Environmental Protection Agency.

The temporal component of any pollution problem can be important, but this is particularly so in the case of "stock pollutants," which tend to accumulate in the environment at a rate that significantly exceeds their natural rate of decay. Accumulations of greenhouse gases are of this nature and thus raise a set of time-related issues. If the overall goal of some public policy were to limit the rate or degree of climate change, then significant trade-offs would exist with regard to the timing of any proposed reductions in greenhouse gas emissions. Earlier reductions would have the effect of slowing the potential onset of climate change.

Within the context of a tradable permit system, these temporal considerations can be addressed, to some degree through provisions for (or restrictions on) "banking," a mechanism that enables firms or nations to make early emission reductions in exchange for the right to emit a comparable amount at some later date. This notion could be extended to sinks as well as sources. It could be advantageous to allow nations to engage in banking of greenhouse gas allowances, since this would allow for intertemporally efficient market exchanges and could tend to delay the onset of global climate change.

11.6.7 *Implementation issues for economies in transition*

Nations with economies in transition from centrally planned to market-based systems are likely to exhibit certain characteristics relevant to the choice, design, and implementation of greenhouse policy instruments. A small but rapidly growing literature has begun to investigate issues of particular concern for implementation of environmental policy in transition economies, including matters such as the adaptation of existing environmental tax systems to changing conditions (OECD, 1994; Semeniene and Kundrotas, 1994; Markandya, 1994), historical, institutional, and fiscal factors (Zylicz, 1994a, b), the use of economic instruments to raise revenue for highly constrained government budgets (Zylicz, 1994c, OECD, 1994), the environmental impacts and cost-effectiveness of instruments for air pollution control in specific regions (London Economics, 1993; Csermely, Kaderjak, and Lehoczki, 1994; Dudek, Kulczynski, and Zylicz, 1993), and environmental liability (Bell and Kolaja, 1993).

High rates of economic growth and price inflation could affect the attractiveness of alternative policy instruments over time, due, for example, to the rapid inflation of relative permit prices or the erosion of unit-based taxes (Stavins and Whitehead, 1992). Situations in which a large portion of the economy is state-owned or the private sector is in its infancy suggest the need to validate the usual assumption that emission sources (firms and individuals) are cost-minimizers and that markets are relatively complete (Stavins and Zylicz, 1994). Enterprises may be protected from bankruptcy, facing only "soft budget constraints," or they may have the ability to avoid environmental requirements through negotiation (OECD, 1994).

Concentration of product or emission permit markets due to inherited industry mixes and possible barriers to entry (for example, imperfect capital markets) may also impede the efficient operation of a tradable permit system (Hahn, 1984). Significant structural adjustment, including privatization, shifts in industrial sector shares in the economy, and disruptions in international trading relationships could also affect the stability and predictability of greenhouse gas emissions resulting from alternative policy instruments. Effective taxes may, for example, increase bankruptcies in a period of severe economic problems (OECD, 1994). Even after privatization, many enterprises may be unable to respond efficiently to policy because they lack information on technological options for pollution control and their cost-effectiveness (OECD, 1994).

Other noneconomic characteristics of transition economies may also be relevant to the implementation of policy instruments to manage greenhouse gas emissions. Problems may arise from the legal and administrative constraints inherited from central planning, making monitoring and enforcement difficult. A relatively undeveloped sense of corporate responsibility, a lack of public awareness of environmental issues, and a lack of pressure from nongovernment organizations could further impede effective implementation. Government personnel may lack the necessary administrative skills, due to

a shortage of economic, financial, and accounting skills and an inability on the part of government agencies to offer competitive salaries (OECD, 1994). Finally, high *ex ante* levels of pollution, high levels of desired reduction, and a concentrated pattern of pollution exposure may present environmental conditions that are more extreme than in many advanced industrialized countries, but they may also provide abundant opportunities for low-cost abatement.

11.7 Comparative Assessment of Greenhouse Policy Instruments

In this section an attempt is made to outline the issues that need to be considered in determining any greenhouse policy mix. Countries differ in their institutional structures, their resource endowments, and their levels of industrialization. Differences in economic and technical capacities among countries offer the potential for emission abatement cost saving under a harmonized international greenhouse management scheme but, at the same time, add complexities in terms of reaching final agreement about appropriate policy approaches and burden sharing.

Economic instruments are considered by policy makers in a political environment. This has several important implications for the nature of the instruments finally adopted, as well as for the potential for reaching an international agreement on climate change.

First, to some extent the choice of instrument will be dictated by existing institutional infrastructure and experience. For example, market-based instruments are likely to be seen as less appropriate in an economy with a high level of central planning than in one with a long history of free enterprise.

Second, the ability to enforce the different instruments is likely to vary across nations. In addition, nations are unlikely to grant significant authority to a supranational body that would allow for consistent enforcement across countries.

Third, to the extent that domestic policy is affected, the choice of policy instruments at the international level could affect the likelihood that an agreement will be reached.[39] For example, some countries may be unwilling to accept an agreement involving the use of international taxes or harmonized domestic taxes. On the other hand, as pointed out earlier, a tradable quota scheme leaves open the choice of domestic instruments.

Fourth, any approach that is implemented to control greenhouse gases may vary from the textbook application of these concepts. There are many reasons why both market-based and regulatory approaches deviate from their ideal. Departure of actual instruments from their theoretical ideal, however, is not sufficient cause for rejection of an approach.

Fifth, the adoption of any international instruments will have some impact on the distribution of wealth between countries, as will domestic instruments on the distribution of wealth within them. Negotiations about distributional issues are likely to be crucial in determining the final policy mix that is chosen. In the case of domestic taxes and tradable permits, some of the government revenue may be returned to the affected parties. Thus, for example, many charging systems in

Europe are designed to limit pollution recycle revenues to the participants or earmark the revenue for specific tasks. Similarly, in the U.S., tradable permits for protecting the environment are distributed according to the historical pattern of emissions (grandfathering). Although the precise nature of the distribution will be the subject of vigorous political discussions, countries and special interest groups (including environmental groups) are unlikely to accept an agreement that substantially shifts the distribution of wealth or political power. Since all instruments probably will have to, and also can, be connected with compensatory measures – side payments or specific quota/permit allocations – no difference between them would arise in this regard. For example, an international tax or tradable quota scheme might be designed in such a way as to encourage developing countries to join a coalition in order that they benefit from international transfers of income.

Sixth, governments are likely to attach more stringent monitoring and enforcement requirements to a market-based approach for limiting noncarbon greenhouse gas emissions than to a regulatory system. For example, environmentalists bargained successfully for the installation of continuous emission monitors as a condition for allowing a tradable allowance system for reducing SO_2 emissions in the U.S. A similar strategy is likely to be applied if market-based approaches are implemented for limiting noncarbon greenhouse gases or for controlling carbon sequestration. One notable difference between the two control problems is that technology for accurately monitoring many sources and sinks of greenhouse gases has not yet been developed.

And finally, there are several reasons why politicians have traditionally taken a regulatory approach, rather than an economic incentive-based approach to environmental policy (Bohm and Russell, 1985; Hahn and Stavins, 1991). First, industry tends to favour direct regulation over incentive mechanisms because (a) if a tax instrument is used, the polluter must pay fees in addition to controlling costs, although the acceptance of this approach will be influenced by any revenue recycling, as mentioned above; and (b) firms may have greater influence over the specifics of uniform standards. Second, the effects of quantity regulation are likely to be perceived to be more certain than pollution charges, whose effect will depend on abatement cost functions, which are typically unknown. Third, economic efficiency arguments often rely on a relatively sophisticated understanding of market operation and price effects which seem indirect when compared with regulation of the polluting activity. Finally, in many countries, economists play a minor role in the development of environmental policy, compared with the number of decision makers with backgrounds in law, natural science, or engineering.

11.7.1 Comparing regulatory systems and market-based instruments

Regulatory policies may be defined as those where the authorities determine the level of permissible emissions from an emission source. Market-based policies may be defined as those where firms are free to determine their level of emis-

sions but must pay some penalty (such as a tax or the purchase of an emission permit) determined by the authorities for their level of emissions. To minimize the total costs of abatement, the level of abatement at each source needs to be chosen to equalize the marginal costs of abatement for given output and input prices. If the authorities had complete information about the marginal costs of abatement at each source, regulatory policies could be determined to minimize the total costs of abatement. Given that the authorities will not have such complete information and typically cannot acquire it at a reasonable cost, regulatory approaches tend not to be cost minimizing.

On the other hand, it is necessary to consider the public as well as the private costs of control (Stavins, 1995a). In other words, the total costs to be minimized by a truly cost-effective environmental policy instrument include both the costs of abatement (typically borne by private industry and including transaction costs) and the costs of administration (typically borne by government in the form of monitoring and enforcement costs). When monitoring and enforcement needs are particularly burdensome, performance-based standards in general may not be cost-effective. On the other hand, certain forms of technology standards, which are typically relatively high-cost in terms of abatement, can involve only minimal needs for adequate monitoring and enforcement. Finally, in addition to such concerns about static or allocative cost-effectiveness, it is important to consider the relative effects of alternative policy instruments on the invention, innovation, and diffusion of new technologies. That is, in the long term, it is the dynamic efficiency properties of environmental policy instruments that are likely to be most important.

In the international context, monitoring and enforcement requirements would hardly differ with respect to fossil fuel use, since fossil fuel output plus imports minus exports would have to be reported for each participating country under all systems.

Tradable permits (for emissions during a given time period) and taxes are the two major domestic market-based policies. With tradable permits a national permit exchange would develop among permit-liable fossil fuel producers and importers (or wholesale dealers in fossil fuels) after the initial allocation of permits through recurring government permit auctions or (temporary) grandfathering. In this connection transaction costs would arise. Under a tax scheme administrative costs would be incurred in payment and collection of the tax. The issue of how the costs of violation detection and enforcement would differ between policies has not been studied, and there does not appear to be any empirical evidence that could be applied to the study of these questions.

There have been a number of empirical studies that suggest significant potential cost savings from the adoption of truly cost-effective instruments instead of regulatory approaches (Tietenberg, 1985), although most of these studies have contrasted actual regulatory instrument costs with a theoretical cost-minimizing alternative rather than an actual market-based policy instrument (Hahn and Stavins, 1992). In the final analysis, governments are likely to choose a portfolio of instruments including both some regulatory and some market-based approaches.

In economies without well-developed market systems, there may be net efficiency gains from applying regulatory approaches over a wider range of emission sources. On the other hand, the adoption of a market-based approach may speed the development of the market system. Net efficiency gains may favour the development of market-based systems at an earlier date than otherwise would be the case.

At the international level, there is little scope for using direct regulation of emissions over and above nontradable emission quotas. Such a quota system would clearly entail extra costs to the extent that marginal abatement costs differ among countries. Cost-effective candidates for an international agreement are tradable quotas and international or harmonized domestic carbon taxes.

11.7.2 Comparing domestic tradable permits and domestic tax systems

Both taxes and tradable permits impose costs on industry and consumers.[40] In effect, they force firms to internalize the costs of their pollution. Practically speaking, this means that firms will experience financial outlays, either through expenditures on pollution controls or through cash payments (buying permits or paying taxes). Taxes and permit prices (especially when permits are auctioned by the government) tend to make these costs more visible to industry and the public. This may be problematic for political reasons, although in the long run it may have the advantage of clearly signalling and educating the public about the costs and tradeoffs associated with various levels of environmental control.

In principle, there need be no difference between domestic carbon taxes and tradable carbon permits from a distributional point of view. Moreover, the tax recycling and "double dividend" benefits associated with carbon taxes can exist to the same extent for a permit system. Tradable permits may be grandfathered, in the short run, to (partly) compensate existing firms that may not have been sufficiently forewarned about the new policy. This choice corresponds to a tax scheme where, in a period of transition, all carbon tax revenues are redistributed to the firms that would have received free permits under a permit scheme. Alternatively, or after a period of transition is over, no compensation at all would be paid. This would amount to a tax system where the government kept the tax revenue (and used it for unrelated purposes) or a permit scheme where all permits were auctioned and the government retained the sales revenue. Partial matching versions of each type of scheme might also be imagined.

The difficulty of controlling emission levels through taxes could be a distinct disadvantage in terms of an international agreement. Taxes would have to be varied frequently, given the inadequate information base of the authorities, to determine the appropriate tax level and the need for adjustments in response to changes in the level of economic activity and changes in relative and absolute price levels. The need for frequent changes in tax levels would add to business uncertainty and to the practical difficulty in a political sense of implementing such a policy.

Tradable permit systems may be more susceptible to "strategic" behaviour than tax systems. In order for a tradable permit system to work effectively, relatively competitive conditions must exist in the permit (and product) market. The degree of competition will help determine the amount of trading that occurs and the cost savings that will be realized. Should any one firm control a significant share of the total number of permits, its activities may influence permit prices. Firms might attempt to manipulate permit prices to improve their positions in the permit market (say, by withholding permits and forcing others to cut production or keeping new entrants out). These risks would be reduced by (a) using time-limited permits – that is, permits for emissions for a period of, say, five years, which could be compatible with a corresponding international tradable quota scheme;[41] and (b) government auctioning of permits.[42]

Tradable permits have some advantage over taxes when time and uncertainty are introduced into the analysis. A tradable permit scheme can be designed to reduce uncertainty about the future in a number of ways. One approach would be to issue permits with different durations (Bertram, 1992) or for a set of future (for example, five-year) periods. Firms undertaking emission-intensive investments with long payback periods would be able to reduce uncertainty about future costs by buying permits for the desired number of periods. The development of a forward or futures market for permits (that could be coupled with permits of different duration) would provide an even better mechanism to spread the risks associated with uncertainty about future emission policy. Firms undertaking research and development into technologies to reduce emissions would be able to hedge the risks associated with the payoff from such technologies through operations in the futures market (Epstein and Gupta, 1990). Similarly, firms investing in emission-intensive activities would be able to hedge against the risks of future policy changes through the operations of futures markets.

To summarize, permits are more effective than taxes in achieving given emission targets. The difficulty of controlling emissions through taxes could be a disadvantage. The frequent changes in taxes that may be required would add to business uncertainty. Permits may be more susceptible to strategic manipulation than taxes, but this problem can be reduced, as explained above. Permits appear to have a distinct advantage in creating the basis for a futures market that could enable a more efficient spreading of the risks of future policy uncertainty.

11.7.3 Comparing international tradable quotas and tax systems

As outlined above, economic incentive policies can lead in many situations to lower total pollution control costs and spur greater technological innovation than conventional regulatory approaches. Which incentive-based instrument is most effective, however, will depend on a number of specific factors. The broadest set of possible international applications is considered below.

A system of harmonized domestic carbon taxes would involve an agreement about compensatory international financial transfers as well as the precarbon tax net tax rates on fossil fuels. These taxes would represent (at least) what amounts to an estimate of the domestic environmental effects of fossil fuel combustion. Internationally acceptable estimates of these basic tax levels, which would tend to differ between countries, would be difficult to establish. Moreover, no design seems feasible and generally acceptable where participants are not allowed to undertake policies on their own which indirectly affect fossil fuel use, such as levying a tax on substitutes for carbon and subsidizing complements to carbon (Hoel, 1993). Thus, there are significant risks that a tax harmonization agreement would either never be adopted or fail after implementation.

A system of international taxes, where all participating countries were liable to pay a given carbon tax, could include an agreement on how tax receipts would be shared among the participants. Under such a system countries might retain all or part of the taxes raised domestically, and some participants (low-income countries) might receive a transfer (Hoel, 1993). Each country would have a good knowledge of the amount of tax revenue likely to be raised internally. However, less information would be available about other countries' tax revenues and, hence, there would be uncertainty about the size of the net transfers to and from each country.

A tradable quota scheme leaves each participant to decide what domestic policy to use. Such a scheme does not require any ongoing side payments. Here, the initial allocation of quota entitlements among countries reflects distributional considerations. A disadvantage of a tradable quota scheme is that the (endogenous) future prices in international quota trade are unknown when an agreement on the quota allocation is reached. Hence, the exact distributional implications cannot be known beforehand. This is the price paid for the main advantage of such a scheme, namely that the resulting global emissions will be known with certainty for a global agreement and, net of carbon leakage, for a nonglobal agreement.

Thus, the choice between a tax and a quota regime remains ambiguous. As Yohe (1992) points out, nations facing different circumstances could favour different control strategies. For example, if a case can be made that the marginal social cost of climate change is relatively flat in industrial countries because of their comparative advantage in applying technology to adapting to change, then such countries might favour a tax instrument. On the other hand, developing countries that are likely to face much steeper marginal social cost schedules because of their lower capacity to adapt may favour a system of tradable quotas (perhaps regardless of the initial quota allocation).

Endnotes

1. For a comprehensive legal review of the Convention, see Bodansky (1993).

2. However, it is worth noting that there may be side benefits (in health-related factors, for example) as a consequence of any reduc-

tions in pollution arising from reductions in greenhouse gas emissions.

3. For a general review of the literature on investment under uncertainty, see Pindyck (1991). See also Chapter 10 of this report.

4. Strictly speaking, the term emission "charge" or "fee" would be more appropriate because this is a payment for a right to emit. However, the term emission "tax" is adopted here because the term "carbon tax" is so widely used in the literature.

5. For a very useful breakdown and analysis of the full costs of environmental regulations, see Schmalensee (1994). Conceptually, the cost of an environmental regulation is equal to "the change in consumer and producer surpluses associated with the regulations and with any price and/or income changes that may result" (Cropper and Oates, 1992).

6. For example, if a firm chooses to close a plant because of a new regulation (rather than installing expensive control equipment), this would be counted as zero cost in typical compliance cost estimates.

7. For a fuller explanation of these different categories of environmental protection costs, see Jaffe *et al.* (1995).

8. Although lump sum taxes and transfers are typically infeasible in a single economy, market-based instruments such as a tradable quota scheme or a carbon tax can be designed to achieve transfers of goods and services between countries to implement the equity criteria listed in Section 11.2.3.

9. They estimate that a carbon tax of US$60-70 per tonne of carbon would be required in OECD countries to achieve an equivalent reduction in global emissions.

10. For a comprehensive coverage of discount and social time preference rates, see Chapter 4 of this report.

11. In the case of the United States, for example, see U.S. Environmental Protection Agency (1989); National Academy of Sciences (1992); Clinton and Gore (1993).

12. This is no longer the case even in Europe, which now has performance-based standards as in the U.S. and Japan.

13. These technologies remain incompatible because of the stringent requirements for NO_x control.

14. Currently there are lean-burn Japanese-made vehicles that can meet present Japanese standards and that are also available on the European market (H. Watson, Melbourne University, personal communication). The lower temperatures obtainable in lean-burn combustion reduce NO_x production but result in less complete combustion of hydrocarbons, forcing a continued reliance on at least two-way catalytic converters. These vehicles therefore also incorporate two-way catalysts (which control hydrocarbons and carbon monoxide), with NO_x control being left to careful engine management and exhaust gas recirculation.

15. Lean-burn engines are potentially more fuel-efficient, since by definition they use less fuel in the air/fuel mix. A 10% improvement in fuel efficiency has been reported for Toyota's lean-burn control system with two-way converter (Watson 1994). Such an improvement in fuel efficiency would have to be offset against other effects of lean-burn technology, such as lower temperatures of operation and less smooth running in the absence of more sophisticated engine-control systems. Any disincentive to the technological development of lean-burn engines is of particular concern in the context of reducing carbon dioxide emissions from conventionally fuelled cars, since vehicle-based "engineering" advances in this respect depend essentially on improving fuel efficiencies.

16. Lean-burn technologies also have the potential to reduce toxic emissions more effectively over the lifetime of the automobile; because they are a more durable technology than currently available catalytic converters, which are constructed with an expected lifetime equal to that of the U.S. car (60,000 miles or 100,000 km). However, many cars have lifetimes well in excess of this – for example, the Australian experience is 240,000 km. The resulting number of older cars with malfunctioning converters is a concern, since even a relatively small number of them may be major sources of toxic emissions.

17. For discussions of relevant enforcement issues see, for example Harford (1978), Shibata and Winrich (1983), Polinsky and Shavell (1979).

18. See, for example, the experience of the Netherlands cited in Lenstra and Bonney (1994).

19. Natural gas and water utilities also implement demand-side management programmes, but they have been most common in electric utilities because the cost of meeting peak demand is highest in the case of electricity.

20. These are the co-called California Standard Practice tests (see EPRI, 1991). The tests have been criticized on the grounds that they provide an incomplete cost-benefit analysis (Herman, 1994).

21. Electric transmission and distribution systems are regarded as natural monopolies. Historically, generation technology has exhibited economies of scale, which have produced declining marginal costs. Utilities tended to integrate generation and transmission, and to a lesser extent distribution, to realize the economies of scale with minimal risk. To enable the utility to recover its full costs, rates were based on average costs, which are higher than marginal costs. Marginal costs vary with the demand for electricity and, during peak periods, exceed the average cost. Demand-side management programmes that shift demand from peak periods to lower-cost periods reduce costs with little loss of revenue. Such programmes can reduce current demand and still pass the RIM test. Load growth creates a need to add capacity. This affects future rates regardless of whether the marginal costs of the new capacity are higher or lower than those of existing capacity. Demand-side management programmes that lower load growth defer the need to add capacity and so reduce costs. Since the demand-side management costs are incurred earlier than the expenditures for new capacity, they lead to an initial increase in rates. However, assuming all the estimates to be accurate, the rates should ultimately decrease.

22. A related instrument is the tradable absorption/abatement obligation. See Read (1994).

23. It is precisely for this reason that a carbon tax (with relatively low monitoring costs) is feasible. In contrast, a true CO_2 emission tax would obviously be extremely costly to monitor and enforce. (Compare this with the case of sulphur dioxide (SO_2). Controlling SO_2 emissions by means of a sulphur content tax on coal would be problematic, since it would fail to provide incentives for flue gas desulphurization (scrubbing), even when this would be the cost-effective route to reducing SO_2 emissions.)

24. For a further discussion of decarbonization of fuels and flue gases, see Chapter 19 of the IPCC Working Group II Second Assessment Report.

25. An option of this kind was used to some extent in the Montreal Protocol (Article 1.5). For an analysis of the cost-effective attributes of the Protocol, see Bohm (1990).

26. In addition, uses of fossil fuels for purposes other than combustion, for example, as chemical feedstocks, could be subjected to similar deposit refund treatment to "keep the prices right." Likewise, products such as lubrication oil could be subjected to the carbon tax whereas waste lubrication oil would be entitled to a tax refund if returned for oil recovery or disposal other than by incineration (Bohm, 1981).

27. In a static model that has fixed prices and neglects the public goods aspect of abatement, transfers have purely distributional ef-

fects. However, this is not the case where economies are growing and production technology differs across countries (in the sense that different quantities of capital and emissions are required to produce a given output of consumption or capital goods). In this case, transfers make it possible to raise the growth rate of an economy above the maximum determined by its original productive capacity. This is illustrated by Hinchy and Hanslow (1994), using an *n*-country generalization of a model developed by Tahvonen and Kuuluvainen (1993).

28. When permits are leased from government or when time-limited permits are auctioned by government, the revenue implications of permit schemes approach those of taxes. Theoretical analysis indicates that this is not true in the case where eternal permits are auctioned by government (Bohm, 1994b).

29. For example, under an allocation system related to population levels, the big players in the market would likely be India and China, as permit sellers, and the U.S. and perhaps the former Soviet Union, as buyers. (See Epstein and Gupta, 1990.)

30. For example, the Canadians proposed using population and GNP combined as allocation criteria when CFC reduction obligations were being considered in the development of the Montreal Protocol.

31. This assumption excludes the potential consequences of significant transaction costs.

32. For a general discussion, see Bohm (1992). "Appropriate" initial allocations can serve as an effective device to draw countries – particularly, developing countries – into an international agreement. On this, see Barrett (1992c) and Hinchy *et al.* (1994a, b). Most proposals for allocating control obligations among nations call for proportionately higher rates of reduction in emissions by the industrialized countries (and, among the industrialized countries, by the U.S.) and substantial reductions in the predicted rates of increase in CO_2 emissions by most developing countries. See, for example, Krause (1989), Flavin (1989), and Wirth and Lashof (1990).

33. For a comprehensive review of the legal and practical aspects of joint implementation, see Kuik *et al.* (1994).

34. Monitoring and enforcement are discussed in detail by Tietenberg and Victor (1994).

35. Although there is no cutoff point, it is unlikely that firms or nations could engage in price-setting behaviour if they controlled less than 10% of the market (see Scherer, 1980). Ultimately, the question is whether other firms present credible threats of entry to the market – that is, whether the market is "contestable." If so, it is less likely that anticompetitive behaviour can thrive (see Baumol *et al.*, 1982).

36. Alternative explanations of low observed trading levels have also been advanced. These include lumpy investment in pollution-control technology; concentration in permit or product markets; the sequential and bilateral nature of the trading process (in the context of a nonuniformly mixed pollutant) leading to some initial trades that then preclude better trades from being carried out subsequently (Atkinson and Tietenberg, 1991); and the regulatory environment. Some of these "alternative explanations" of low trading levels can be viewed as special cases of transaction costs, broadly defined.

37. This problem may exist within groups of countries as well as more generally. For a discussion of issues surrounding the attempted introduction of a harmonized carbon tax in the European Union and a possible alternative policy, see Bergesen *et al.* (1994).

38. One proposal is to limit quotas to, say, ten years, with one year overlapping for banking and for practical "end-of-period" reasons (see Bertram, 1992; OECD, 1992b). In particular, one-tenth of quotas could expire each year and could be replaced by a new issue according to a procedure that could be modified, say, every five years. This would reduce the costs of uncertainty by providing flexibility to adapt to new information, including the entry of new sources and sinks into the system. It also establishes a market in forward contracts to reduce the costs of greenhouse risks. Quotas of different duration would coexist in the market to cover both short- and long-term risks, and the quota price would reflect the costs of risk.

39. The likelihood of reaching an international environmental agreement will be affected by several factors. See Sebenius (1991).

40. Compared to conventional regulations, both taxes and permits provide an explicit price signal about the marginal cost of limiting emissions.

41. Granting permits for moving periods of, say, five years, instead of issuing eternal permits, would reduce the possibility of hoarding by a monopolist who could "forever" expose future buyers to leasing permits at monopoly prices.

42. Auctioning could be used to help avoid a situation that may arise under grandfathering in which one large firm is allocated a significantly large share of permits. However, it could be argued that grandfathering would provide an asset that could be sold by a firm wishing to leave an industry and would thus facilitate adjustment.

References

Adar, Z., and J. M. Griffin, 1976: Uncertainty and the choice of pollution control instruments, *Journal of Environmental Economics and Management,* **3,** 178–188.

Anderson, R., 1993: Joint implementation of climate change measures: An examination of some issues, paper presented at a Workshop on the Experience of the Norway/Mexico/Poland Joint Implementation Demonstration Projects, Mexico, 28–29 October, Global Environmental Facility, Washington, DC.

Anderson, R., L. Hofmann, and M. Rusin, 1990: The use of economic incentive mechanisms in environmental management. Research Paper no. 51, American Petroleum Institute, Washington, DC, June.

Atkinson, S., and T. Tietenberg, 1991: Market failure in incentive-based regulation: The case of emissions trading. *Journal of Environmental Economics and Management,* **21,** 17–31.

Barrett, S., 1992a: *Convention on climate change: Economic aspects of negotiations,* OECD, Paris.

Barrett, S., 1992b: Reaching a CO_2-emission limitation agreement for the community: Implications for equity and cost-effectiveness, *European Economy,* Special edition **1,** 3–24

Barrett, S., 1992c: "Acceptable" allocations of tradeable carbon emission entitlements in a global warming treaty. In *Combating global warming: Study on a global system of tradeable carbon emission entitlements,* UNCTAD, ed., pp. 85–113, United Nations, New York.

Barrett, S., 1994a: Climate change policy and international trade. In *Climate change: Policy instruments and their implications,* proceedings of the Tsukuba workshop of IPCC Working Group III, A. Amaro, B. Fisher, M. Kuroda, T. Morita, and S. Nishioka, eds., pp. 15–33, IPCC, Geneva.

Barrett, S., 1994b: Self-enforcing international environmental agreements, *Oxford Economic Papers,* **46,** 878–894.

Barrett, S., 1995: *The strategy of "joint implementation" in the Framework Convention on Climate Change,* UNCTAD/GID/10, United Nations, New York.

Baumol, W., and W. Oates, 1975: *The theory of environmental policy,* Prentice-Hall, Englewood Cliffs, NJ.

Baumol, W., and W. Oates, 1979: *Economics, environmental quality, and the quality of life,* Prentice-Hall, Englewood Cliffs, NJ.

Baumol, W., and W. Oates, 1988: *The theory of environmental policy,* Cambridge University Press, New York.

Baumol, W., J. Panzer, and R. Willig, 1982: *Contestable markets and the theory of industrial structure,* Harcourt Brace Jovanovich, New York.

Beavis, B., and M. Walker, 1983: Random wastes, imperfect monitoring and environmental quality standards, *Journal of Public Economics,* **21,** 377–387.

Bell, R.G., and T.A. Kolaja, 1993: Capital privatization and the management of environmental liability in Poland, *The Business Lawyer,* **48,** (3), 943–961.

Bergesen, H.O., M. Grubb, J.-C. Hourcade, J. Jaeger, A. Lanza, R. Loske, L. Sverdrup, and A. Tudini, 1994: *Implementing the European CO_2 commitment: A joint policy proposal,* The Royal Institute of International Affairs, London.

Bertram, G., 1992: Tradeable emission permits and the control of greenhouse gases, *Journal of Development Studies,* **28**(3), 423–446.

BIE (Bureau of Industry Economics), 1992: *Environmental regulation: The economics of tradeable permits – A survey of theory and practice,* AGPS, Canberra.

Binswanger, H., 1989: *Brazilian policies that encourage deforestation in the Amazon,* World Bank Environment Department Paper no. 16, Washington, DC.

Bodansky, D., 1993: The United Nations Framework Convention on Climate Change: A commentary, *Yale Journal of International Law,* **18,** 451–558.

Boero, G., R. Clarke, and L.A. Winters, 1991: *The macroeconomic consequences of controlling greenhouse gases: A survey,* London, HMSO.

Bohi, D.R., and D. Burtraw, 1992: Utility investment behavior and the emission trading market, *Resource Engineering,* **14,** 129–153.

Bohm, P., 1981: *Deposit–refund systems: Theory and applications to environmental, conservation and consumer policy,* Johns Hopkins University Press, Baltimore, for Resources for the Future, Washington DC.

Bohm, P., 1990: Efficiency issues and the Montreal Protocol on CFCs, Environment Working Paper no. 40, World Bank, Washington, DC.

Bohm, P., 1992: Distributional implications of allowing international trade in CO_2 emissions quotas, *The World Economy,* **15**(1), 107–114.

Bohm, P., 1993: Incomplete international cooperation to reduce CO_2 emissions: alternative policies, *Journal of Environmental Economics and Management,* **24** (3), 258–271.

Bohm, P., 1994a: On the feasibility of joint implementation of carbon emissions reductions. In *Climate change: Policy instruments and their implications,* Proceedings of the Tsukuba workshop of IPCC Working Group III, A. Amaro, B. Fisher, M. Kuroda, T. Morita, and S. Nishioka, eds., pp. 181–198, IPCC, Geneva.

Bohm, P., 1994b: Efficiency aspects of government revenue implications of carbon taxes and tradeable carbon permits, paper presented at International Institute of Public Finance, 50th Congress, Cambridge, MA, 22–25 August.

Bohm, P., 1995a: Environmental taxation and the double dividend: Fact or fallacy? CSERGE Working Paper, University College, London.

Bohm, P., 1995b: *An analytical approach to evaluating the national net costs of a global system of tradeable carbon emission entitlements,* UNCTAD/GID/9, Geneva.

Bohm, P., and B. Larsen, 1993: Fairness in a tradeable-permit treaty for carbon emissions reductions in Europe and the former Soviet Union, *Environmental and Resource Economics,* forthcoming.

Bohm, P., and C. Russell, 1985: Comparative analysis of alternative policy instruments. In *Handbook of natural resource and energy* *economics,* vol. 1, A. Kneese and J. Sweeney, eds., North-Holland, Amsterdam.

Bovenberg, L., and R. de Mooij, 1994: Environmental levies and distortionary taxation, *American Economic Review,* **84** (4), 1085–1089.

Boyd, R., K. Krutilla, and W.K. Viscusi, 1994: Energy taxation as a policy instrument to reduce CO_2 emissions: A net benefit analysis, *Journal of Environmental Economics and Management,* forthcoming.

Buchanan, J., 1969: External diseconomies, corrective taxes, and market structure, *American Economic Review,* **59**(1), 174–177.

Bull, N., K. Hassett, and G. Metcalf, 1993: Who pays broad-based energy taxes? Computing lifetime and regional incidence, Working Paper no. 142, Division of Research and Statistics, Board of Governors of the Federal Reserve System, Washington, DC, September.

Burniaux, J.-M., J.P. Martin, and J. Oliveira-Martins, 1992a: The effect of existing distortions in energy markets on the costs of policies to reduce CO_2 emissions: Evidence from GREEN, *OECD Economic Studies,* **19,** Winter, 141–165.

Burniaux, J.-M., J.P. Martin, G. Nicolleti, and J. Oliveira-Martins, 1992b: The costs of reducing carbon dioxide emissions: Evidence from GREEN, Working Paper no. 115, Economics Department, OECD, Paris.

Bye, B., T. Bye, and L. Lorentsen, 1989: *SIMEN: Studies of industry, environment and energy towards 2000,* Discussion Paper no. 44, Central Bureau of Statistics, Oslo.

Cameron, M., 1994: *Efficiency and fairness on the road,* Environmental Defence Fund, Oakland, CA.

Carraro, C., and D. Siniscalco, 1993: Strategies for the international protection of the environment, *Journal of Public Economics,* **52,** 309–328.

Chayes, A., and A.H. Chayes, 1993: On compliance, *International Organization,* **47,** 175–205.

Chichilnisky, G., and G. Heal, 1993: Global environmental risks, *Journal of Economic Perspectives,* **7**(4), Fall, 65–86.

Chichilnisky, G., G. Heal, and D. Starett, 1993: International emission permits: Equity and efficiency, mimeo, Columbia University, November.

Clinton, W.J. and A. Gore, Jr., 1993: *The climate change action plan,* The White House, Washington, DC.

Crandall, R.W. 1984: The political economy of clean air: Practical constraints on White House review. In *Environmental policy under Reagan's Executive Order: The role of benefit–cost analysis,* V. Smith, ed., Resources for the Future, Washington, DC.

Cristofaro, A., and J. Scheraga, 1990: Policy implications of a comprehensive greenhouse gas budget, mimeo, United States Environmental Protection Agency, Washington, DC, September.

Cropper, M., and W. Oates, 1992: Environmental economics: A survey, *Journal of Economic Literature,* **30**(2), 675–740.

Csermely, A., P. Kaderjak, and Z. Lehoczki, 1994: Direct impacts of industrial restructuring on air pollutant and hazardous waste emissions in Hungary, paper presented at the Fifth Annual Conference of the European Association of Environmental and Resource Economists, Budapest, May.

Dasgupta, P., 1993: *An inquiry into well-being and destitution,* Clarendon Press, Oxford, England.

Dasgupta, P., and K. Göran-Mäler, 1994: *Poverty, institutions and the environmental resource base,* Environment Paper no. 9, World Bank, Washington, DC.

DCEIA (Department of Commerce, Energy Information Administration), 1992: *Federal energy subsidies: Direct and indirect interventions in energy markets,* Energy Information Administration, Washington, DC.

Dean, A., and P. Hoeller, 1992: Costs of reducing CO_2 emissions: Evidence from six global models, *OECD Economic Studies,* Paris, **19,** Winter, 15–47.

DeWitt, D., H. Dowlatabadi, and R. Kopp, 1991: *Who bears the burden of energy taxes?* Discussion Paper QE91–12, Resources for the Future, Washington, DC.

Dornbush, R., and J.M. Poterba, 1991: *Global warming: Economic policy responses,* MIT Press, Cambridge, MA.

Dower, R.C., and Zimmerman, M.B. 1992: *The right climate for carbon taxes: Creating economic incentives to protect the atmosphere,* World Resources Institute, August, Washington, DC.

Dudek, D.J., and A. LeBlanc, 1992: *Preserving tropical forests and climate: The role of trees in greenhouse gas emissions trading,* The Environmental Defence Fund, New York.

Dudek, D., Z. Kulczynski, and T. Zylicz, 1993: Implementing tradeable permits in Poland: A case study of Chorzow. In Conference Proceedings of the Third Annual Meeting of the European Association of Environmental and Resource Economists, L. Presisner, ed., Krakow, Poland, May.

EPRI (Electric Power Research Institute), 1984: *Demand-side management,* Vol. 1, *Overview of key issues,* EPRI, EA/EM-3597, Palo Alto, CA, August.

EPRI (Electric Power Research Institute), 1991: End-use technical assessment guide, vol. 4, Fundamentals and methods, EPRI, CU-7222, Palo Alto, CA, April.

Epstein, J., and R. Gupta, 1990: *Controlling the greenhouse effect: Five global regimes compared,* Brookings Occasional Papers, Brookings Institution, Washington, DC.

European Commission, 1994: Taxation, employment and environment: Fiscal reform for reducing unemployment, European Economy, **56,** 137–177.

Fankhauser, S., 1993: Greenhouse economics and the costs of global warming, PhD thesis, Department of Economics, University College, London.

Faruqui, A., and E. Haites, 1991: Impact of efficient electricity use and DSM programs on United States electricity demand and the environment. In *Proceedings: DSM and the global environment,* Arlington, VA., 22–23 April, The Conference Connection, Bala Cynwyd, PA.

Flavin, C., 1989: *Slowing global warming: A worldwide strategy,* Worldwatch Paper 91, Worldwatch Institute, Washington, DC, October.

Freeman, A., and R. Haveman, 1971: Water pollution control, river basin authorities, and economic incentives: Some current policy issues, *Public Policy,* **19**(1), 53–74.

Gellings, C.W., and J. Chamberlin, 1993: *Demand-side management: Concepts and methods* (2d ed.), The Fairmont Press, Lilburn, GA.

Goulder, L., 1992: Carbon tax design and US industry performance. In *Tax policy and the economy,* vol. 6, J. Poterba, ed., MIT Press, Cambridge, MA.

Goulder, L., 1993: Energy taxes: Traditional efficiency effects and environmental implications. In *Tax policy and the economy,* vol. 8, J. Poterba, ed., MIT Press, Cambridge, MA.

Goulder, L., 1995: Effects of carbon taxes in an economy with prior tax distortions, *Journal of Environmental Economics and Management,* forthcoming.

Greening, L., L. Schipper, and H. Jeng, 1993: Evaluation of the distributional impacts of energy taxes across regions and socioeconomic groups of the United States. Annual Conference on Energy, pp. 432–443.

Grubb, M., 1991: *Energy policies and the greenhouse effect,* vol. 1, *Policy Appraisal,* Royal Institute of International Affairs, Dartmouth, England.

Grubb, M., and J.K. Sebenius, 1991: Participation, allocation and adaptability in international tradeable emission permit systems for greenhouse gas control, *Proceedings of OECD Workshop on Tradeable Emission Permits to Reduce Greenhouse Gases,* Organisation for Economic Co-operation and Development, Paris, June.

Hahn, R.W., 1984: Market power and transferable property rights, *Quarterly Journal of Economics,* **99**(4), 753–765.

Hahn, R.W., 1988: An evaluation of options for reducing hazardous waste, *Harvard Environmental Law Review,* **12,** 201–221.

Hahn, R.W., 1989: Economic prescriptions for environmental problems: How the patient followed the doctor's orders, *Journal of Economic Perspectives,* **3,** 95–114.

Hahn, R.W., and G.L. Hester, 1989a: Marketable permits: Lessons for theory and practice, *Ecology Law Quarterly,* **16,** 361– 406.

Hahn, R.W., and G.L. Hester, 1989b: Where did all the markets go? An analysis of EPA's emissions trading program, *Yale Journal of Regulation,* **6,** 109–153.

Hahn, R.W., and Stavins, R., 1991: Incentive-based environmental regulation: A new era from an old idea? *Ecology Law Quarterly,* **18,** 1–42.

Hahn, R.W., and Stavins, R., 1992: Economic incentives for environmental protection: Integrating theory and practice, *American Economic Review,* **82,** 464–468.

Hahn, R.W., and Stavins, R., 1995: Trading in greenhouse permits: A critical examination of design and implementation issues. In *Shaping national responses to climate change: A post-Rio guide,* H. Lee, ed., Island Press, Washington, DC.

Haites, E., 1993: Emissions reductions due to the DSM programs of Canadian electric utilities, *Proceedings: Demand-side management and the global environment,* Arlington, VA., 24–25 June, The Conference Connection, Bala Cynwyd, PA.

Hamilton, K., and G. Cameron, 1994: Simulating the distributional effects of a Canadian carbon tax, *Canadian Public Policy,* **20**(4), 385–399.

Hanisch, T., 1991: Joint implementation of commitments to curb climate change, Policy Note no. 1991:2, CICERO (Centre for International Climate and Energy Research), Oslo, Norway.

Hanisch, T., R. Pachauri, D. Schmitt, and P. Vellinga, 1992: The climate convention: Criteria and guidelines for joint implementation, Policy Note no. 1992:2, CICERO (Centre for International Climate and Energy Research), Oslo, Norway.

Hanisch, T., R. Selrod, A. Torvanger, and A. Aaheim, 1993: Study to develop practical guidelines for "Joint Implementation" under the UN FCCC, mimeo, CICERO (Centre for International Climate and Energy Research), Oslo, Norway.

Hanslow, K., M. Hinchy, J. Small, B.S. Fisher, and D. Gunasekera, 1995: Trade and welfare effects of policies to address climate change. In *Equity and social considerations related to climate change,* Proceedings of a WG III IPCC Workshop, Nairobi, Kenya, 18–22 July, R. Odingo, A. Alusa, F. Mugo, J. Njihia, A. Heidenreich, and A. Katama, eds., pp. 313–330, WMO/UNEP, Geneva.

Harford, J.D., 1978: Firm behavior under imperfectly enforceable pollution standards and taxes, *Journal of Environmental Economics and Management,* **5,** 26–43.

Hasset, K., and G. Metcalf, 1992: *Energy tax credits and residential investment,* Working Paper no. 4020, National Bureau of Economic Research, Cambridge, MA.

Hausman, J., 1979: Individual discount rates and the purchase and utilization of energy-using durables, *Bell Journal of Economics,* **10,** 33–54.

Heal, G., 1992: International negotiations on emission control, *Structural Change and Economic Dynamics,* **3,** 223–240.

Herman, P., 1994: *The value test: Its context, description, calculation, and implications,* California Energy Commission, Sacramento, CA, May.

Hinchy, M., and K. Hanslow, 1994: Optimal greenhouse policy in a multi-country world of growing economies, ABARE paper presented to the Conference of Economists, Economic Society of Australia, Gold Coast, Queensland, 25–28 September.

Hinchy, M., and K. Hanslow, 1995: Modelling the impact of OECD carbon taxes with the MEGABARE model, ABARE paper presented to the 18th Annual Conference of the International Association of Energy Economics, Washington DC, 5–8 July.

Hinchy, M., K. Hanslow, and B. S. Fisher, 1994a: A dynamic game approach to greenhouse policy: Some numerical results. In *Climate change: Policy instruments and their implications,* Proceedings the Tsukuba workshop of IPCC Working Group III, A. Amaro, B. Fisher, M. Kuroda, T. Morita, and S. Nishioka, eds., pp. 117–131, IPCC, Geneva.

Hinchy, M., K. Hanslow, and B.S. Fisher, 1994b: A dynamic game approach to greenhouse policy: More numerical results, ABARE paper presented to the Tasman Institute Conference on Environmental Health and Economic Wealth: Conflict or concord? Canberra, 15–16 March, ABARE, Canberra.

Hinchy, M., S. Thorpe, and B.S. Fisher, 1993: *A tradable emissions permit scheme,* ABARE Research Report 93.5, Canberra, Australia.

Hirst, E., 1993, Price and cost impacts of utility DSM programs, *The Energy Journal,* **13**(4), 75–94.

Hoel, M., 1992: International environmental conventions: The case of uniform reductions of emissions, *Environmental and Resource Economics,* **2,** 141–159.

Hoel, M., 1993: Harmonization of carbon taxes in international climate agreements, *Environmental and Resource Economics,* **3**(3), June, 221–232.

Hoel, M., 1994: Efficient climate policy in the presence of free riders, *Journal of Environmental Economics and Management,* **27**(3), 259–274.

Hoel, M., 1995: Should a carbon tax be differentiated across sectors? *Journal of Public Economics,* forthcoming.

Hoel, M., and I. Isaksen, 1993: Efficient abatement of different greenhouse gases, Memorandum no. 5, Department of Economics, University of Oslo, April.

Hollinger, D.Y., J.P. MacLaren, P.N. Beets, and J. Turland, 1994: Carbon sequestration by New Zealand's plantation forests, *New Zealand Journal of Forestry Science,* **23**(2), 194–208.

Horowitz, M., and H. Haeri, 1990: Economic efficiency v energy efficiency: Do model conservation standards make good sense? *Energy Economics,* **12**(2), 122–131.

Horton, G., J. Rollo, and A. Ulph, 1992: The implications for trade of greenhouse gas emission control policies, working paper prepared for UK Department of Trade and Industry, London.

Houghton, R., 1992: A blueprint for monitoring the emissions of carbon dioxide and other greenhouse gases from tropical deforestation, mimeo, Woods Hole Research Center, Woods Hole, MA.

Howden, M., and R. Munro, 1994: Methane from Australian livestock: implications for industry and climate change, *Resource Sciences Interface,* **4,** 5–9.

Howe, C., 1994: Taxes versus tradable discharge permits: A review in the light of the US and European experience, *Environmental and Resource Economics,* **4**(2), 151–169.

IIASA (International Institute for Applied Systems Analysis),

1992a: *Proceedings of the International Conference on Economic Instruments for Air Pollution Control, October 18-20, Volume 1, Plenary Sessions,* IIASA, Laxenburg, Austria.

IIASA, 1992b: *Proceedings of the International Conference on Economic Instruments for Air Pollution Control, October 18-20, Volume 2, Parallel Sessions,* IIASA, Laxenburg, Austria.

International Energy Agency, 1993: Taxing energy: Why and how, Economic Analysis Division, OECD, Paris.

IPCC (Intergovernmental Panel on Climate Change), 1990: *Climate change: The IPCC scientific assessment,* Cambridge University Press, Cambridge.

IPCC, 1992: *Climate change: The supplementary report to the IPCC scientific assessment,* Cambridge University Press, Cambridge.

Jaffe, A., and R. Stavins 1994a: The energy paradox and the diffusion of conservation technology, *Resource and Energy Economics,* **16**(2), 91–122.

Jaffe, A., and R. Stavins, 1994b: The energy efficiency gap: What does it mean? *Energy Policy,* **22**(10), 804–810.

Jaffe, A., and R. Stavins, 1994c: Energy-efficiency investments and public policy, *Energy Journal,* **15**(2), 1–23.

Jaffe, A., and R. Stavins, 1995: Dynamic incentives of environmental regulation: The effects of alternative policy instruments on technology diffusion, *Journal of Environmental Economics and Management,* **29,** S43–S63.

Jaffe, A., P. Peterson, P. Portney, and R. Stavins, 1995: Environmental regulation and the competitiveness of U.S. manufacturing, *Journal of Economic Literature,* **33,** 132–163.

Jepma, C.J. (ed.), 1995: *The feasibility of joint implementation,* Kluwer Academic Publications, Dordrecht, the Netherlands.

Jochem, E., and E. Gruber, 1990: Obstacles to rational electricity use and measures to alleviate them, *Energy Policy,* **18,** 340–350.

Johnson, I., 1993: Operational criteria for joint implementation: A response, paper presented to the International Conference on the Economics of Climate Change, OECD/IEA, Paris, 14–16 June.

Jones, B.P., and P. Tobler, 1993: Policy measures to achieve greenhouse targets, ABARE paper presented at the National Agricultural and Resources Outlook Conference, Canberra, 2–4 February, ABARE, Canberra.

Jones, T., 1993: Operational criteria for joint implementation, paper presented at the International Conference on the Economics of Climate Change, OECD/IEA, Paris, 14–16 June.

Jorgenson, D., and P. Wilcoxen, 1992: Reducing US carbon dioxide emissions: An assessment of different instruments, Discussion Paper no. 1590, Harvard Institute of Economic Research, April.

Jorgenson, D., and P. Wilcoxen, 1994: The economic effects of a carbon tax. In *Climate change: Policy instruments and their implications,* proceedings of the Tsukuba workshop of IPCC Working Group III, A. Amaro, B. Fisher, M. Kuroda, T. Morita, and S. Nishioka, eds., pp. 60–72, WMO/UNEP, Geneva.

Jorgenson, D., D. Slesnick, and P. Wilcoxen, 1992: Carbon taxes and economic welfare. In *Brookings papers on economic activity: Microeconomics,* pp. 393–431, Brookings Institution, Washington, DC.

Joskow, P., and D. Marron, 1993: What does a negawatt really cost? Evidence from utility conservation programs, *The Energy Journal,* **13**(4), 41–74.

Kahn, A., 1992: Least cost planning generally and DSM in particular, *Resources and Energy,* **14,** 177–185.

Katz, M., 1992: Demand side management: Reflections of an irreverent regulator, *Resources and Energy,* **14,** 187–203.

Keeler, A.G., 1991: Noncompliant firms in transferable discharge

permit markets: Some extensions, *Journal of Environmental Economics and Management,* **21,** 180–189.

Kolstad, C., 1982: Looking vs. leaping: The timing of CO_2 control in the face of uncertainty and learning, mimeo, Department of Economics, University of Illinois, Urbana, IL.

Koopmans, G., M. Mors, and J. Scherp, 1992: The likely economic impact of the proposed carbon/energy tax in the European Community, paper presented to an International Workshop on Costs, Impacts and Possible Benefits of CO_2 Mitigation, Laxenburg, Austria, 28–30 September.

Krause, F., 1989: *Energy policy in the greenhouse: From warming fate to warming limit – Benchmarks for a global climate convention,* Dutch Ministry of Housing, Physical Planning and Environment and the European Environmental Bureau, The Hague, the Netherlands.

Krupnick, A., M. Walls, and H. Hood, 1993: The distributional and environmental implications of an increase in the federal gasoline tax, Discussion Paper ENR 93–24, Resources for the Future, Washington, DC.

Krutilla, K., 1991: Environmental regulation in an open economy, *Journal of Environmental Economics and Management,* **20**(2), 127–142.

Kuik, O., P. Peters, and N. Schrijver (eds.), 1994: *Joint implementation to curb climate change: Legal and economic aspects,* Kluwer Academic Publishers, Dordrecht/Boston/London.

Larsen, B., and A. Shah, 1992: World fossil fuel subsidies and global carbon emissions, Policy Research Working Paper Series no. 1002, World Bank, Washington, DC.

Larsen, B., and A. Shah, 1994: Global tradeable carbon permits, participation incentives and transfers, *Oxford Economic Papers,* **46,** 841–856.

Larsen, B., and A. Shah, 1995: Global climate change, energy subsidies, and national carbon taxes. In *Public economics and the environment in an imperfect world,* L. Bovenberg and S. Cnossen, eds., pp. 113–132, Kluwer, Dordrecht.

Lashof, D., and D. Ahuja, 1990: Relative contributions of greenhouse gas emissions to global warming, *Nature,* **344,** 529–531.

Leary, N., and J. Scheraga, 1994: Policies for the efficient reduction of CO_2 emissions, *International Journal of Global Energy Issues,* **6** (1/2), 103–107.

Lenstra, W., and M. Bonney, 1994: The role of regulation in Dutch climate change policy. In *Climate change: Policy instruments and their implications,* proceedings of the Tsukuba workshop of IPCC Working Group III, A. Amaro, B. Fisher, M. Kuroda, T. Morita, and S. Nishioka, eds., pp. 241–247, IPCC, Geneva.

Liroff, R.A., 1989: The evolution of transferable emission privileges in the United States, paper presented at the Workshop on Economic Mechanisms for Environmental Protection, Jelenia Gora, Poland.

London Economics, 1993: *Study of economic policy instruments for the control of air pollution in Poland,* London.

Loske, R., and S. Oberthür, 1994: Joint implementation under the climate change convention, *International Environmental Affairs,* **6**(1), Winter.

MacLaren, J.P., S.J. Wakelin, and P.N. Beets, 1993: *Projected carbon sequestation by forestry plantings commencing in 1991,* a contract report for the Ministry of Forestry, New Zealand Forest Research Institute Ltd., Wellington.

Magat, W.A., 1979: The effects of environmental regulation on innovation, *Law and Contemporary Problems,* **43,** 4–25.

Mahar, D., 1988: Government policies and deforestation in Brazil's Amazon Region, Environment Department Working Paper no. 7, World Bank, Washington, DC.

Malueg, D., 1990: Welfare consequences of emission credit trading programs, *Journal of Environmental Economics and Management,* **18**(1), 66–77.

Manne, A.S., and R.G. Richels, 1991: International trade in carbon emission rights: A decomposition procedure, *AEA Papers and Proceedings,* **81**(2), 135–139.

Markandya, A., 1994: Environmental taxation: A review of experience in economies in transition, paper presented at the 50th Congress of the International Institute of Public Finance, Cambridge, MA.

Markusen, J., 1975: International externalities and optimal tax structure, *Journal of International Economics,* **5**(1), 15-29.

McCann, R.J., and S.J. Moss, 1993: *Nuts and bolts: The implications of choosing greenhouse-gas emission reduction strategies,* Policy Study no. 171, Reason Foundation, Los Angeles, CA.

Miller, A., and I. Mintzer, 1986: *The sky is the limit: Strategies for protecting the ozone layer,* Research Report no. 3, World Resources Institute, Washington, DC.

MITI (Ministry of Trade and Industry), 1993: Technology renaissance for the environment and energy, mimeo, Special Committee on Energy and Environment, Government of Japan, Tokyo, September.

Mintzer, I., 1988: Living in a warmer world: Challenges for policy analysis and management, *Journal of Policy Analysis and Management,* **7**(3), 445–459.

Misiolek, W., and E. Elder, 1989: Exclusionary manipulation of markets for pollution rights. *Journal of Environmental Economics and Management,* **16**(2), 156–166.

Nadel, S., 1990: Electric utility programs: A review of the lessons taught by a decade of program experience. In *Proceedings: ACEEE 1990 summer study on energy efficiency in buildings,* American Council for an Energy-Efficient Economy, vol. 8, pp. 179–205, Washington, DC.

National Academy of Sciences, 1992: *Policy implications of greenhouse warming: Mitigation, adaptation, and the science base,* National Academy Press, Washington, DC.

Newcombe, K., 1989: An economic justification for rural afforestation. In *Environmental management and economic development,* G. Schramm and J. Warford, eds., Johns Hopkins Press for the World Bank, Washington, DC.

Nordhaus, W., 1991a: The costs of slowing climate change: A survey. *Energy Journal,* **12**(1), 37–64.

Nordhaus, W., 1991b: To slow or not to slow: The economics of the greenhouse effect, *Economic Journal,* **101**, 920–937.

Oates, W., and D. Strassman, 1984: Effluent fees and market structure, *Journal of Public Economics,* **24**(1), 29–46.

OECD (Organisation for Economic Co-operation and Development), 1989: *Economic instruments for environmental protection,* Paris.

OECD, 1992a: *Climate change: Designing a practical tax system,* Paris.

OECD, 1992b: *Climate change: Designing a tradeable permits system,* Paris.

OECD, 1993: *International economic instruments and climate change,* Paris.

OECD, 1994: *Taxation and the environment in European economies in transition,* Paris.

Oliveira-Martins, J., J.-M. Burniaux, and J.P. Martin, 1992: Trade and the effectiveness of unilateral CO_2 abatement policies: Evidence from GREEN, *OECD Economic Studies,* **19**, Winter, 123–140.

Parikh, J., 1994a: Joint implementation and sharing commitments: A southern perspective, proceedings of a workshop on "Integrated Assessment for Climate Change," International Institute for Applied Systems Analysis, Laxenburg, Austria, 16–18 October.

Parikh, J., 1994b JI: survey II: Supporting North-South cooperation, *Climate Change Bulletin,* Issue 4, 3d quarter, pp. 5–6.

Parson, E., and R. Zeckhauser, 1992: Co-operation in the unbalanced commons, mimeo, John F. Kennedy School of Government, Harvard University, Cambridge, MA. In *Barriers to conflict resolution,* K.J. Arrow *et al.,* eds., Norton, New York.

Pearce, D., 1991: The role of carbon taxes in adjusting to global warming, *Economic Journal,* **101,** July, 938–948.

Pearce, D., and J. Warford, 1993: *World without end,* Oxford University Press, England.

Pearson, M., 1992: Equity issues and carbon taxes, *Climate change: Designing a practical tax system,* OECD Documents, Paris.

Pearson, M., and S. Smith, 1991: The European carbon tax: An assessment of the European Commission's proposals, Institute for Fiscal Studies, London.

Peck, S., and T. Teisberg, 1993: Global warming uncertainties and the value of information: An analysis using CETA, *Resource and Energy Economics,* **15**(1), 87–107.

Pezzey, J., 1992: Analysis of unilateral CO_2 control in the European Community and OECD, *Energy Journal,* **13**(3), 159–171.

Pillet, G., W. Hediger, S. Kypreos, and C. Corbaz, 1993: *The economics of global warming,* Paul Scherrer Institute, Villagen PSI, Switzerland.

Pindyck, R., 1991: Irreversibility, uncertainty, and investment, *Journal of Economic Literature,* **29**(3), 1110–1148.

Polinsky, A.M., and S. Shavell, 1979: The optimal trade-off between the probability and magnitude of fines, *American Economic Review,* **59**(5), 880–891.

Polinsky, A.M., and S. Shavell, 1982: Pigouvian taxation with administrative costs, *Journal of Public Economics,* **19,** 385–394.

Poterba, J.M., 1990: Is the gasoline tax regressive? NBER Working Paper no. 3578, National Bureau of Economic Research, Cambridge, MA.

Poterba, J.M., 1991: Tax policy to combat global warming: On designing a carbon tax, Working Paper no. 3649, National Bureau of Economic Research, Cambridge, MA.

Poterba, J.M., 1993: Global warming policy: A public finance perspective, *Journal of Economic Perspectives,* **7**(4), 47–63.

Rajah, N., and S. Smith, 1993: Taxes, tax expenditures, and environmental regulation, *Oxford Review of Economic Policy,* **9**(4), 41– 64.

Ranganathan, H., H.-U. Neue, and P. Pingali, P. 1994: Global climate change: Role of rice in methane emissions and prospects for mitigation, International Rice Research Institute, Manila, March.

Read, P., 1994: *Responding to global warming: The technology, economics, and politics of sustainable energy,* ZED Books, London.

Reddy, A., 1993: Operational criteria for joint implementation: A response, paper presented to the International Conference on the Economics of Climate Change, OECD/IEA, Paris, 14–16 June.

Repetto, R., 1986: *Skimming and water: Rent-seeking and the performance of public irrigation systems,* World Resources Institute, Washington, DC.

Repetto, R., R. Dower, R. Jenkins, and J. Geoghegan, 1992: *Green fees: How a tax shift can work for the environment and the economy,* World Resources Institute, Washington, DC.

Richels, R., and J. Edmonds, 1994: The economics of stabilizing atmospheric CO_2 concentrations. In *Climate change: Policy instruments and their implications,* Proceedings of the Tsukuba workshop of IPCC Working Group III, A. Amaro, B. Fisher, M. Kuroda, T. Morita, and S. Nishioka, eds., pp. 73–81, IPCC, Geneva.

Roberts, M.J., and M. Spence, 1976: Effluent charges and licenses under uncertainty. *Journal of Public Economics,* vol. 5, pp. 193–208.

Russell, C.S., 1988: Economic incentives in the management of hazardous wastes, *Columbia Journal of Environmental Law,* **13,** 257–274.

Russell, C.S., 1990: Monitoring and enforcement. In *Public policies for environmental protection,* P. Portney, ed., ch. 7, Johns Hopkins University Press, Baltimore.

Russell, C.S., W. Harrington, and W.J. Vaughan, 1986: *Enforcing pollution control laws,* Johns Hopkins University Press, Baltimore, MD.

Sandor, R., J. Cole, and E. Kelly, 1994: *Study on model rules and regulation for a global CO_2 emissions credit market,* United Nations Conference on Trade and Development (UNCTAD), Geneva.

Scheraga, J., and N. Leary, 1992: Improving the efficiency of policies to reduce CO_2 emissions, *Energy Policy,* **20**(5), 394–404.

Scherer, F.M., 1980: *Industrial market structure and economic performance,* Rand-McNally College Publishing, Chicago.

Schillo, B., L. Giannarelli, D. Kelly, S. Swanson, and P. Wilcoxen, 1992: The distributional impacts of a carbon tax, drafting working paper, Energy Policy Branch, US Environmental Protection Agency, February.

Schmalensee, R., 1993: Comparing greenhouse gases for policy purposes, *Energy Journal,* **14,** 245–255.

Schmalensee, R., 1994: The costs of environmental protection. In *Balancing economic growth and environmental goals,* M.B. Kotowski, ed., pp. 55–75, American Council for Capital Formation Center for Policy Research, Washington, DC.

Schramm, G., 1992: Issues and problems in the power sector of developing countries, mimeo, World Bank, Washington, DC.

Sebenius, J.K., 1991: Negotiating a regime to control global warming. In *Greenhouse warming: Negotiating a global regime,* R. Benedick, A. Chayes, D. Lashof, J. Matthews, W. Nitze, E. Richardson, J. Sebenius, P. Thacher, and D. Wirth, eds., pp. 69–98, World Resources Institute, Washington, DC.

Sedjo, R., 1994: Forests to capture carbon: Tradeable permit schemes. In *Climate change: Policy instruments and their implications,* proceedings of the Tsukuba workshop of IPCC Working Group III, A. Amaro, B. Fisher, M. Kuroda, T. Morita, and S. Nishioka, eds., pp. 145–154, IPCC, Geneva.

Semeniene, D., and A. Kundrotas, 1994: From theory to practice: Environmental taxes in a time of economic uncertainty, paper presented at the 50th Congress of the International Institute of Public Finance, Cambridge, MA.

Shackleton, R., M. Shelby, A. Cristofaro, R. Brinner, J. Yanchar, L. Goulder, D. Jorgenson, P. Wilcoxen, P. Pauly, and R. Kaufman, 1992: The efficiency value of carbon tax revenues, mimeo, United States Environmental Protection Agency, Washington, DC.

Shah, A., and B. Larsen, 1992: Carbon taxes, the greenhouse effect and developing countries, Working Paper WPS 957, Policy Research Department, World Bank, Washington, DC.

Shibata, H., and S. Winrich, 1983: Control of pollution when the offended defend themselves, *Economica,* **50,** 425–437.

Sigman, H., 1991: A comparison of public policies for lead recycling, mimeo, Department of Economics, MIT, Cambridge, MA, November.

Sinclair, P., 1992: High does nothing and rising is worse: Carbon taxes should decline to cut harmful emissions, *Manchester School of Economic and Social Studies,* **60**(1), 41–52.

Sioshansi, F., 1992: Demand side management and environmental externalities: Ramifications on utility resource planning, *Utilities Policy,* **2**(4), 321–325.

Smith, S., 1992a: The distributional consequences of taxes on energy and the carbon content of fuels. In *European economy: The economics of limiting CO_2 emissions,* Special Edition no. 1, Commission for the European Communities, Brussels.

Smith, S., 1992b: Distributional effects of a European carbon tax, Nota di Lavoro 22.92, Fondazione ENI Enrico Mattei, Milan, Italy.

Stavins, R.N. (ed.), 1988: *Project 88 – Harnessing market forces to protect our environment: Initiatives for the new president,* Public Policy Study sponsored by Senator Timothy E. Wirth, Colorado, and Senator John Heinz, Pennsylvania, Washington, DC, December.

Stavins, R.N., (ed.), 1991: *Project 88 – Round II, incentives for action: Designing market-based environmental strategies,* Public Policy Study sponsored by Senator Timothy E. Wirth, Colorado, and Senator John Heinz, Pennsylvania, Washington, DC, May.

Stavins, R.N., 1995a: Transaction costs and tradeable permits, *Journal of Environmental Economics and Management,* **29,** 133–148.

Stavins, R.N., 1995b: The costs of carbon sequestration: A revealed-preference approach, working paper, John F. Kennedy School of Government, Harvard University, Cambridge, MA, September.

Stavins, R.N., 1996: Correlated uncertainty and policy instrument choice, *Journal of Environmental Economics and Management,* forthcoming.

Stavins, R., and B. Whitehead, 1992: Pollution charges for environmental protection: A policy link between energy and environment, *Annual Review of Energy and the Environment,* **17,** 187–211.

Stavins, R.N., and T. Zylicz, 1994. Environmental policy instrument design in a transition economy, working paper, John F. Kennedy School of Government, Harvard University, Cambridge, MA.

Stewart, R., and J. Wiener, 1992: The comprehensive approach to global climate policy: Issues of design and practicality, *Arizona Journal of International and Comparative Law,* **9,** 85–113.

Sutherland, R., 1991: Market barriers to energy-efficiency investments, *Energy Journal,* **12**(3), 15–34.

Tahvonen, O., and J. Kuuluvainen, 1993: Economic growth, pollution and renewable resources, *Journal of Environmental Economics and Management,* **24**(2), 101–118.

Tasman Institute, 1994: *A framework for trading carbon credits from New Zealand's forests,* Research Report No. C6, Tasman Institute, Melbourne.

Tietenberg, T. H., 1980: Transferable discharge permits and the control of stationary source air pollution: A survey and synthesis, *Land Economics,* **56,** 391–416.

Tietenberg, T.H., 1985: *Emissions trading: An exercise in reforming pollution policy,* Resources for the Future, Washington, DC.

Tietenberg, T.H., 1990: Economic instruments for environmental regulation, *Oxford Review of Economic Policy,* **6**(1), 17–33.

Tietenberg, T.H., 1992: *Environmental and natural resource economics,* Harper Collins, New York.

Tietenberg, T.H., and D. Victor, 1994: Implementation issues for a tradeable permit approach to controlling global warming. In *Climate change: Policy instruments and their implications,* proceedings of the Tsukuba workshop of IPCC Working Group III, A. Amaro, B. Fisher, M. Kuroda, T. Morita, and S. Nishioka, eds., pp. 155–172, IPCC, Geneva.

Treadwell, R., S. Thorpe, and B. Fisher, 1994: Should we have second thoughts about "no regrets"? *The economics of climate change: Proceedings of an OECD/IEA Conference,* pp. 255–272, OECD/IEA, Paris.

Tripp, J.T.B., and D.J. Dudek, 1989: Institutional guidelines for designing successful transferable rights programs, *Yale Journal of Regulation,* **6,** 369–391.

Tschirhart, J.T., 1984: Transferable discharge permits and profit-maximizing behavior. In *Economic perspectives on acid deposition control,* T.D. Crocker, ed., Butterworth Publishers, Boston, MA.

Ugandan Ministry of Finance and Economic Planning, 1993: *Rehabilitation and development plan 1993/94–1995/96,* vol. 2 – mimeo, Priority Projects, December.

Ulph, A., and D. Ulph, 1994: Global warming: Why irreversibility may not require lower current emissions of greenhouse gases, Discussion Paper No. 9402, Department of Economics, University of Southampton, Southampton, UK.

UNCTAD (United Nations Conference on Trade and Development), 1992: *Combating global warming: Study of a global system of tradeable carbon emission entitlements,* United Nations, New York.

U.S. Congress, Office of Technology Assessment, 1991: *Changing by degrees: Steps to reduce greenhouse gases,* OTA, Washington, DC.

U.S. Environmental Protection Agency, 1989: Policy options for stabilizing global climate, draft report to Congress, Washington, DC.

U.S. Environmental Protection Agency, 1991: *Economic incentives: Options for environmental protection,* Office of Policy, Planning, and Evaluation, Economic Incentives Task Force, 21P-2001, Washington, DC, March.

U.S. General Accounting Office, 1992: *International agreements are not well monitored,* GAO, GAO/RCED-92-43, Washington, DC.

Van Slooten, R., 1994: The case of the Montreal Protocol. In *Trade and environment: Processes and production methods,* pp. 87–91, OECD, Paris.

Victor, D., 1991: Limits of market-based strategies for slowing global warming: The case of tradeable permits, *Policy Sciences,* **24**(2), pp. 199–222.

Vousden, N., 1990: *The economics of trade protection,* Cambridge University Press, Cambridge.

Watson, H., 1994: Engines of the future, paper delivered to the Society of Automotive Engineers, 1994 Automotive Industry Outlook Conference, World Congress Centre, Melbourne, October 6.

Watson, W.D., and R.G. Ridker, 1984: Losses from effluent taxes and quotas under uncertainty, *Journal of Environmental Economics and Management,* **11,** 310–326.

Weitzman, M.L., 1974: Prices vs. quantities, *Review of Economics Studies,* **41,** 477–491.

Whalley, J., and R. Wigle, 1991: The international incidence of carbon taxes. In *Global warming: Economic policy responses,* R. Dornbusch and J. Poterba, eds., pp. 71–97, MIT Press, Cambridge, MA.

Wirth, D., and D. Lashof, 1990: Beyond Vienna and Montreal – Multilateral agreements on greenhouse gases, *Ambio,* **19,** 305–311.

Woodland, A., 1982: *International trade and resource allocation,* North-Holland, Amsterdam.

World Bank, 1990: Review of electricity tariffs in developing countries, mimeo, Industry and Energy Department, Washington, DC.

World Bank, 1992: The bank's role in the electric power sector, mimeo, Washington, DC.

World Bank, 1993: *The World Bank and the environment – Fiscal 1993,* Washington, DC.

Yohe, G., 1977: Comparisons of price and quantity controls: A survey, *Journal of Comparative Economics,* **1,** 213–233.

Yohe, G., 1992: Carbon emissions taxes: Their comparative advantage under uncertainty, *Annual Review of Energy and Environment,* **17,** 301–326.

Zylicz, T., 1994a: Implementing environmental policies in Central and Eastern Europe. In *Investing in natural capital: The ecological economics approach to sustainability,* A. Jansson, M. Hammer, C. Folke, and R. Costanza, eds., Island Press, Washington, DC.

Zylicz, T., 1994b: In Poland, it's time for economics, *Environmental Impact Assessment Review,* **14**(2, 3), 79–94.

Zylicz, T., 1994c: Pollution taxes as a source of budgetary revenues in economies in transition, paper presented at the 50th Congress of the International Institute of Public Finance, Cambridge, MA.

Appendix 1

CONTRIBUTORS TO THE IPCC WG III SECOND ASSESSMENT REPORT

Persons who served as a lead author or contributor for one or more of the chapters of the Working Group III contribution to the IPCC Second Assessment Report or to *Climate Change 1994* are listed below in alphabetical order. The authors and contributors of each chapter are listed at the beginning of the chapter.

A. Aaheim	CICERO, University of Oslo, Norway
A.N. Achanta	University of North Carolina and Tata Energy Research Institute, India
J.M. Alcamo	RIVM – National Institute of Public Health and Environmental Protection, Netherlands
A. Amano	Kwansei Gakuin University, Japan
K.J. Arrow	Stanford University, U.S.
M. Asaduzzaman	Bangladesh Institute of Development Studies, Bangladesh
P.G. Babu	Indira Gandhi Institute of Development Research, India
T. Banuri	Sustainable Development Policy Institute, Pakistan
S. Barrett	London Business School and Centre for Social and Economic Research on the Global Environment (CSERGE), UK
P. Bohm	Stockholm University, Sweden
A. Bouwman	RIVM – National Institute of Public Health and Environmental Protection, Netherlands
J. Byrne	University of Delaware, U.S.
W.U. Chandler	Battelle Institute, U.S.
G. Chichilnisky	Columbia University, U.S.
W.R. Cline	Institute for International Economics, U.S.
Z. Dadi	Beijing Energy Research Institute, China
O. Davidson	University of Sierra Leone, Sierra Leone
H. Dowlatabadi	Carnegie Mellon University, U.S.
J.A. Edmonds	Battelle Institute, U.S.
S. Fankhauser	Global Environment Facility, U.S.
S. Faucheux	Université de Versailles, France
D. Finon	Institut d'Economie et de Politique de l'Energie, CNRS, Grenoble, France
B.S. Fisher	Australian Bureau of Agricultural and Resource Economics (ABARE), Australia
G. Froger	Université de Versailles, France
F. Gassmann	Paul Scherrer Institute, Switzerland
H. Geller	American Council for an Energy Efficient Economy, U.S.
J. Goldemberg	University of São Paulo, Brazil
K. Göran-Mäler	Beijer International Institute of Ecological Economics, Sweden
M. Grubb	The Royal Institute of International Affairs, UK
A. Grübler	International Institute for Applied Systems Analysis, Austria

E.F. Haites	Margaree Consultants Inc., Canada
K. Halsnæs	UNEP Collaborating Centre on Energy and Environment, RISØ National Laboratory, Denmark
W. Hediger	Paul Scherrer Institute, Switzerland
C.A. Hendriks	Institute for Prospective Technologies, Spain
M. Hoel	University of Oslo, Norway
K. Hogan	Environmental Protection Agency, U.S.
S.W. Hong	Construction and Economy Research Institute of Korea, Korea
J.-C. Hourcade	CIRED/CNRS, France
M. Hinchy	Australian Bureau of Agricultural and Resource Economics (ABARE), Australia
M. Jaccard	Simon Fraser University and British Columbia Utilities Commission, Canada
H.K. Jacobson	University of Michigan, U.S.
M. Jefferson	World Energy Council, UK
C.J. Jepma	Open University/University of Groningen, Netherlands
S. Kane	National Oceanographic and Atmospheric Administration, U.S.
S. Kavi Kumar	Indira Gandhi Institute of Development Research, India
A. Kolesov	Tsentr po Effektivnomu Ispolzovanlu Energii, Russia
F. Krause	International Project for Sustainable Energy Paths and Lawrence Berkeley National Laboratory, U.S.
M. Kuroda	Keio University, Japan
E. La Rovere	Federal University of Rio de Janeiro, Brazil
G. Leach	Stockholm Environment Institute, UK
R.S. Maya	Southern Centre for Energy and Environment, Zimbabwe
P. Meier	IDEA, Inc., U.S.
I. Mintzer	University of Maryland at College Park, U.S.
M. Al-Moneef	King Saud University, Saudi Arabia
W.D. Montgomery	Charles River Associates, Inc., U.S.
T. Morita	National Institute for Environmental Studies, Japan
J.K.E. Mubazi	Makerere University, Uganda
M. Munasinghe	The World Bank and University of Colombo, Sri Lanka
P.M. Nastari	Fundação Getulio Vargas and Datagro Ltda, Brazil
R.K. Pachauri	Tata Energy Research Institute, India
J. Parikh	Indira Gandhi Institute of Development Research, India
K. Parikh	Indira Gandhi Institute of Development Research, India
E.A. Parson	Harvard University, U.S.
D.W. Pearce	Centre for Social and Economic Research on the Global Environment (CSERGE), University College of London, United Kingdom
S.C. Peck	Electric Power Research Institute, U.S.
S.A. Pegov	Institute for Systems Studies, Russian Academy of Sciences, Russia
G.J. Pillet	Paul Scherrer Institute, currently ECOSYS SA, Switzerland
A. Qureshi	Climate Institute, U.S.
J. Reilly	Department of Agriculture, U.S.
K.R. Richards	Pacific Northwest National Laboratories, U.S.
R. Richels	Electric Power Research Institute, U.S.
J.B. Robinson	University of British Columbia, Canada
J. Rotmans	RIVM – National Institute of Public Health and Environmental Protection, Netherlands
V.L. Saha	Ministry of Housing, Land, Town, and Country Planning, Mauritius
W. Sassin	Forschungszentrum Jülich, Austria
L. Schrattenholzer	International Institute for Applied Systems Analysis, Austria
R.A. Sedjo	Resources for the Future, U.S.
R.G. Shackleton	Environmental Protection Agency, U.S.
A. Shah	The World Bank, U.S.
X. Shaoxiong	Ministry of Power Industry, China
P.R. Shukla	Indian Institute of Management, Ahmedabad, India
D. Siniscalco	Fondazione ENI Enrico Mattei, Italy
Y. Sokona	ENDA – TM Programme-Energie, Senegal
R. Squitieri	Department of the Treasury, U.S.

R.N. Stavins	Harvard University, U.S.
J. Stiglitz	Council of Economic Advisors, U.S.
P. Sturm	Organization for Economic Cooperation and Development, France
C.E. Suárez	Instituto de Economía Energética/Fundación Bariloche, Argentina
A. Sugandhy	Ministry of Environment, Indonesia
T.J. Teisberg	Teisberg Associates, U.S.
S. Thorpe	Australian Bureau of Agricultural and Resource Economics (ABARE), Australia
R.S.J. Tol	Vrije Universiteit Amsterdam, Netherlands
A. Tudini	ISTAT, Italy
A. van der Veen	University of Twente, Netherlands
P. Vellinga	Vrije Universiteit Amsterdam, Netherlands
J. Weyant	Stanford University, U.S.
F. Yamin	Foundation for International Environmental Law and Development, UK

Appendix 2

REVIEWERS OF THE IPCC WG III REPORT

The persons named below all contributed to the peer review of the IPCC Working Group III Report. Although every attempt was made by the Lead Authors to incorporate their comments, in some cases these formed a minority opinion which could not be reconciled with the larger consensus. Therefore, there may be persons below who still have points of disagreement with areas of the Report.

ALBANIA
E. Demiraj   Hydrometeorological Institute, Academy of Sciences

AUSTRALIA
J. Daley   Business Council of Australia
T. Weir   Department of Primary Industries and Energy

AUSTRIA
T. Balabanov   Institute for Advanced Studies
P.V. Gilli   Institut für Wärmetechnik, TU Graz
I. Ismail   Organization of Petroleum Exporting Countries
B. Okogu   Organization of Petroleum Exporting Countries
Y. Sinyak   International Institute for Applied Systems Analysis
J. van de Vate   International Atomic Energy Agency
D. Victor   International Institute for Applied Systems Analysis

BANGLADESH
M. Asaduzzaman   Bangladesh Institute of Developmental Studies

BELGIUM
J. Delbeke   Commission of the European Communities
J. Dreze   Center for Operations Research and Econometrics
G. Koopman   European Commission
M. Mors   European Commission
C. Pimenta   Globe E.C.

BRAZIL
J. Moreira   Biomass Users Network
L.P. Rosa   COPPE/UFRJ
R. Schaeffer   COPPE/UFRJ

BULGARIA
A. Yotova   Bulgarian Academy of Sciences

CANADA
N. Beaudoin   Environment Canada, Global Air Issues Branch
F.K. Hare   University of Toronto
M. Jaccard   Simon Fraser University
J. Last   University of Ottawa
R. Pocklington   Bedford Institute of Oceanography
J. Robinson   University of British Columbia
G. Wall   University of Waterloo

CHILE
E. Figueroa   CENRE, University of Chile

CHINA
D. Yihui   Chinese Academy of Meteorological Sciences

DENMARK
K. Halsnæs   UNEP Collaborating Centre on Energy and Environment, RISØ National Laboratory
B. Sørensen   Bøskilde University, Physics Department

FRANCE
G. Bernier   Organization for Economic Cooperation and Development
C. Blondin   METEO FRANCE
J.Y. Caneill   Electricité de France, Direction des Etudes et Recherches
S. Faucheux   Université de Versailles
J.-C. Hourcade   CIRED/CNRS
P. Sturm   Organisation for Economic Co-operation and Development

GERMANY
J. Blank   Westfälische Wilhelms-Universität Münster
H. Gottinger   IIEEM
P. Henneike   Wuppertal Institut
E. Jochem   Fraunhofer-Institut für Systemtechnik und Innovationsforschung
M. Leimbach   Potsdam-Institut for Climate Impact Research
R. Pethig   Universität-Gesamthochschule Siegen
W. Strobele   Westfälische Wilhelms-Universität Münster

GREECE
D. Katochianou   Centre of Planning and Economic Research

HONG KONG
W. Barron   University of Hong Kong

HUNGARY
T. Palvolgyi   Ministry for Environment and Regional Policy
T. Farago   Ministry for Environment and Regional Policy

INDONESIA
K. Abdullah   Institut Pertanian Bogor

ITALY
R. Brinkman   Food and Agriculture Organization of the UN
M. Contaldi   ENEA
M. Farrell   Food and Agriculture Organization of the UN
D. Siniscalco   Fondazione ENI Enrico Mattei
A. Tudini   ISTAT
G. Tosato   ENEA

JAPAN
N. Goto   Kanazawa University
Y. Matshoka   Faculty of Engineering, Kyoto University
S. Mori   Science University of Tokyo
Y. Tanaka   Environment Agency
I. Tsuzaka   GISPRI

KENYA
Y. Adebayo   United Nations Environment Programme

LATVIA
I. Shteinbuka   Ministry of Finance

MALAYSIA
H. Chan   Malaysian Institute of Economic Research

MALDIVES
A. Majeed   Department of Meteorology

MALI
M. Diallo Iam   Centre National de la Recherche Scientifique et Technologique

NETHERLANDS
K. Blok   Utrecht University
G. Gelauff   Central Planning Bureau
J. Gupta   Vrije Universiteit Amsterdam
C.W. Lee   University of Groningen, Faculty of Economics
W. Lenstra   Ministry of Housing, Spatial Planning and Environment
H. Merkus   Ministry of Housing, Physical Planning and Environment
J.B. Opschoor   Institute for Social Studies
E. Tellegen   University of Amsterdam
R.S.J. Tol   Vrije Universiteit Amsterdam
T. van der Burg   Erasmus University
E. van Imhoff   NIDI
H. Vollebergh   Erasmus University
D. Wolfson   Scientific Council for Government Policy
E. Worrell   Utrecht University

NEW ZEALAND
P. Maclaren — New Zealand Forest Research Institute
P. Read — Massey University

NORWAY
A. Aaheim — CICERO
O. Benestad — University of Oslo
L. Lorentsen — Royal Ministry of Finance

POLAND
B. Greinert — Technical University of Gdansk
W. Kamrat — Technical University of Gdansk
M. Lissowska — Central School of Economics

PORTUGAL
T. Moreira — University of Evora

RUSSIA
K. Kondratyev — Russian Academy of Sciences

SPAIN
V. Alcantara — Autonomous University of Barcelona
S. Lopez — Autonomous University of Barcelona
J. Pacqual — Autonomous University of Barcelona

SWEDEN
B. Bolin — Chairman, IPCC
P. Soderbaum — Swedish University of Economics

SWITZERLAND
M. Beniston — Department of Geography, ETH-Zurich
G. Fritz — Paul Scherrer Institute
S. Kypreos — Paul Scherrer Institute
W. Seifritz — Chapfstr.4, CH-5200 Windisch
R. Slooff — World Health Organization
N. Sundararaman — IPCC Secretariat

UGANDA
J. Mubazi — Makerere University

UNITED KINGDOM
N. Adger — CSERGE
A. Atkinson — University of Cambridge
S. Boehmer-Christians — University of Sussex
R. Booth — Shell International Petroleum Company Limited
N. Collins — World Conservation Monitoring Centre
T. Cooper — Global Commons Institute
G. Dupont — Shell International Petroleum Company
P. Ekins — Birkbeck College, University of London
K. Gregory — CGS Centre for Business and the Environment
M. Grubb — The Royal Institute of International Affairs
J. Houghton — Working Group I, IPCC
I. Hughes — Coal Research Establishment

M. Jefferson	World Energy Council
A. McCulloch	ICI Chemicals and Polymers Limited
S. Mansoob Murshed	Northern Ireland Economic Research Centre
J. Penman	UK Department of Environment
F. Yamin	Foundation for International Environmental Law and Development

UNITED STATES

J. Ausubel	Rockefeller University
R. Beck	Edison Electric Institute
R. Borgwardt	U.S. Environmental Protection Agency
J. Byrne	University of Delaware
W.T. Christian	ARCO
R. Engelman	Population Action International
B. Flannery	Exxon Research and Engineering Company
R. Ford	Westminster College
D. Hall	California State University
W. Harrison	
D. Hill	Douglas Hill, P.E., P.C.
C. Holmes	National Mining Association
R.O. Jones	American Petroleum Institute
W. Kempton	University of Delaware
L. Kozak	Southern Company Services Inc.
N. Leary	U.S. Environmental Protection Agency
G. Marland	Oak Ridge National Laboratory
R. Mendelsohn	Yale University
A. Miller	University of Maryland
A. Olende	United Nations (DPCSD)
D. Pearlman	The Climate Council
R. Promboin	Amoco Corporation
R. Schmalensee	Massachusetts Institute of Technology
D. Shelor	Department of Energy
J. Shlaes	Global Climate Coalition
T. Siddiqi	East-West Center
D. Spencer	Simteche
J. de Steiguer	U.S. Forest Service
M. Steinberg	Brookhaven National Laboratory
T. Teisberg	Teisberg Associates
S. Vavrik	Yale Law School
G. Yohe	Wesleyan University

VENEZUELA

| L. Perez | Ministerio de Energia y Minas |

VIET NAM

| N. Thi Loc | Hydrometeorological Service |
| N. Van Hai | Hydrometeorological Services |

YUGOSLAVIA

| R. Pesic | University of Belgrade |